Electrochemical Surface Science

Electrochemical Surface Science: Basics and Applications

Special Issue Editors

Gaetano Granozzi
Nicolas Alonso-Vante

MDPI • Basel • Beijing • Wuhan • Barcelona • Belgrade

Special Issue Editors

Gaetano Granozzi
Universita degli Studi di Padova
Italy

Nicolas Alonso-Vante
University of Poitiers
France

Editorial Office
MDPI
St. Alban-Anlage 66
4052 Basel, Switzerland

This is a reprint of articles from the Special Issue published online in the open access journal *Surfaces* (ISSN 2571-9637) from 2018 to 2019 (available at: https://www.mdpi.com/journal/surfaces/special_issues/Electrochemical_Surface_Science)

For citation purposes, cite each article independently as indicated on the article page online and as indicated below:

LastName, A.A.; LastName, B.B.; LastName, C.C. Article Title. *Journal Name* **Year**, *Article Number*, Page Range.

ISBN 978-3-03921-642-0 (Pbk)
ISBN 978-3-03921-643-7 (PDF)

Contents

About the Special Issue Editors

Gaetano Granozzi is Full Professor of Inorganic Chemistry at University of Padova since 1990, where he teaches the courses Inorganic Chemistry of Materials, and Catalysis and Surface Chemistry. He has been Director of the PhD Course in Science and Engineering of Materials and Nanostructures of the University of Padova from 2005 to 2016. He is Leader of the Surface Science and Catalysis Group (SSCG) of the Department of Chemical Science of the University of Padova (http://www.chimica.unipd.it/surfacescience/). Granozzi's research interests are in the areas of materials chemistry, with particular emphasis of heterogeneous catalysis and surface chemistry. Granozzi has authored over 320 scientific publications and his current h-index is 40, representing approximately 7541 citations (Google Scholar).

Nicolas Alonso-Vante has served as Professor at the University of Poitiers since September 1997, where he has been teaching the following graduate courses: Specific Analysis of Solids; Electrocatalysis, Photocatalysis; Conversion and Storage of Chemical Energy. From 1985 to 1997, he was a Senior Scientist at the Hahn-Meitner-Institut-Berlin (now Helmholtz-Zentrum Berlin). His current main research interests at IC2MP UMR CNRS 7285 include (photo)electrochemistry and (photo)electrocatalysis of novel materials using various ex situ and in situ techniques, fuel generation, interfacial characterization, and surface analytical techniques. He is the author of over 250 publications and book chapters, editor of a two-volume e-book on electrochemistry in Spanish, and the author of 2 books and 6 patents. His current h-index 46 corresponding to ca. 6750 citations (RG source).

Editorial

Electrochemical Surface Science: Basics and Applications

Nicolas Alonso-Vante [1],* and Gaetano Granozzi [2],*

[1]   IC2MP-UMR CNRS 7285, University of Poitiers, 86022 Poitiers, France
[2]   Department of Chemical Sciences, University of Padova, 35131 Padova, Italy
*   Correspondence: nicolas.alonso.vante@univ-poitiers.fr (N.A.-V.); gaetano.granozzi@unipd.it (G.G.)

Received: 12 June 2019; Accepted: 1 July 2019; Published: 4 July 2019

The great success of the *Surfaces* Special Issue entitled "Electrochemical Surface Science (EC-SS): Basics and Applications" reflects the great vitality and relevance of the addressed topic. EC-SS stems from the merging of two different disciplines, i.e., surface science (SS) and electrochemistry (EC), which dates back to ca. four decades ago. The two separate disciplines mainly contributed either a methodological approach (SS, aiming at understanding the microscopic processes occurring at surfaces/interfaces at an atomic level) or a selection of the objects to study (EC, aiming at describing the phenomena at electrified interfaces). The combination of the two is nowadays making possible the challenge of unraveling the complex mechanisms of the electron transfer processes occurring at electrodes, i.e., truly interface-driven phenomena which encompass both solid/liquid, solid/solid, and solid/gas interfaces. This hybrid approach has had an enormous impact on fields such as electrocatalysis [1–3], solar energy harvesting [4–6], corrosion, electrochemical energy storage and conversion devices [7], and sensors [8].

Among them, electrocatalysis is omnipresent and plays a key role. Actually, processes at electrodes are often kinetically limited to efficiently run multi-charge transfer reactions. An electrocatalyst is usually needed, i.e., a substance that can reduce the overall activation barrier height of the redox chemical reaction via complex surface-chemistry steps (adsorption/desorption of reactants and products, low kinetic barriers for charge transport) and determine the product selectivity distribution. The figures of merit of an electrocatalyst are nowadays determined following the standard parameters of catalysis, i.e., turnover frequency and number. Determining such parameters and correlating them with the electrocatalyst structure is a task highly facilitated by the adoption of the hybridized EC-SS method.

A historical perspective is useful to better understand the evolution of electrocatalysis. At the beginning of the 20th century (1905), Julius Tafel [9], in Switzerland, reported on the hydrogen evolution reaction (HER) on various electrode materials, thus establishing a quantitative method for HER electrocatalysts benchmarked through the "Tafel equation" [10]. The HER two-electron process, which started to be academically studied in the 1950s, is still under development in many laboratories in the world [11]: the main goal is to provide a sustainable route for the preparation of molecular hydrogen through the electrochemical splitting of water (water splitting; WS). Actually, WS is expected to promote the envisioned hydrogen economy [12], based on molecular hydrogen as an energy vector for the development of a sustainable energy infrastructure established on the efficient interconversion of chemical energy into electricity and vice versa.

One of the current key concepts in electrocatalysis is the replacement of noble-metal-based electrocatalysts with those based on elements that are abundant on Earth [13,14]. The role played by the synergetic EC-SS approach in such a paradigmatic revolution is similar to that it already played in the 1980s, when platinum-based electrocatalysts were optimized [15,16]. In addition, nowadays, two more relevant key concepts have appeared, i.e., in situ and operando techniques. In situ or operando characterization tools aim to monitor the electrochemical reaction while the electron transfer process is occurring [17,18]. The difference between the two terms is subtle. The in situ term relates to experiments where the experimental conditions (e.g., pressure, atmosphere, potential, current, electrolyte, etc.) are controlled during acquisition, but no temporal discrimination is explicitly taken into account. On the other hand, operando tools are related to the study of the system in real life applications. When applied to electrocatalysis, this means obtaining more detailed electrochemical information while monitoring the working electrodes with other techniques, such as X-ray diffraction (XRD), transmission electron microscopy (TEM), atomic force microscopy (AFM), Raman spectroscopy, ultraviolet visible (UV-vis) absorption spectroscopy, X-ray absorption near-edge structure (XANES), nuclear magnetic resonance (NMR), X-ray photoelectron spectroscopy (XPS), etc.

In this Special Issue, a total of 27 scientific papers (one of which is a review) report some of the latest advances in the field of EC-SS. It was a particular target of the Guest Editors to demonstrate the importance and the large scope of EC-SS through examples from a variety of systems and applications. Four papers are related to innovative methods for characterization. As expected, electrocatalysis papers make up the lion's share, but other interesting topics such as sensors, switchable interfaces, potential driven self-assembly on surfaces, and flexible electrodes are also addressed. The Guest Editors hope that the readers will appreciate the different contributions closest to their own field of research.

Conflicts of Interest: The authors declare no conflict of interest.

References

1. Zhu, W.; Zhang, R.; Qu, F.; Asiri, A.M.; Sun, X. Design and Application of Foams for Electrocatalysis. *ChemCatChem* **2017**, *9*, 1721–1743. [CrossRef]
2. Wang, T.; Xie, H.; Chen, M.; D'Aloia, A.; Cho, J.; Wu, G.; Li, Q. Precious metal-free approach to hydrogen electrocatalysis for energy conversion: From mechanism understanding to catalyst design. *Nano Energy* **2017**, *42*, 69–89. [CrossRef]
3. Seh, Z.W.; Kibsgaard, J.; Dickens, C.F.; Chorkendorff, I.; Nørskov, J.K.; Jaramillo, T.F. Combining theory and experiment in electrocatalysis: Insights into materials design. *Science* **2017**, *355*, eaad4998. [CrossRef] [PubMed]
4. Eftekhari, A.; Babu, V.J.; Ramakrishna, S. Photoelectrode nanomaterials for photoelectrochemical water splitting. *Int. J. Hydrog. Energy* **2017**, *42*, 11078–11109. [CrossRef]
5. Grewe, T.; Meggouh, M.; Tüysüz, H. Nanocatalysts for Solar Water Splitting and a Perspective on Hydrogen Economy. *Chemistry* **2016**, *11*, 22–42. [CrossRef] [PubMed]
6. Sivula, K.; van de Krol, R. Semiconducting materials for photoelectrochemical energy conversion. *Nat. Rev. Mat.* **2016**, *1*, 15010. [CrossRef]
7. Lee, J.-S.; Tai Kim, S.; Cao, R.; Choi, N.-S.; Liu, M.; Lee, K.T.; Cho, J. Metal–Air Batteries with High Energy Density: Li–Air versus Zn–Air. *Adv. Energy Mat.* **2011**, *1*, 34–50. [CrossRef]
8. Huang, X.; Yin, Z.; Wu, S.; Qi, X.; He, Q.; Zhang, Q.; Yan, Q.; Boey, F.; Zhang, H. Graphene-Based Materials: Synthesis, Characterization, Properties, and Applications. *Small* **2011**, *7*, 1876–1902. [CrossRef] [PubMed]
9. Tafel, J. Über die Polarisation bei kathodischer Wasserstoffentwicklung. *Zeitschrift für Physikalische Chemie* **1905**, *50*, 641–712. [CrossRef]
10. Burstein, G.T. A hundred years of Tafel's Equation: 1905–2005. *Corros. Sci.* **2005**, *47*, 2858–2870. [CrossRef]
11. Strmcnik, D.; Lopes, P.P.; Genorio, B.; Stamenkovic, V.R.; Markovic, N.M. Design principles for hydrogen evolution reaction catalyst materials. *Nano Energy* **2016**, *29*, 29–36. [CrossRef]
12. Blagojevic, V.A.; Minic, D.G.; Novakovic, J.G.; Minic, D.M. Hydrogen Economy: Modern Concepts, Challenges and Perspectives. In *Hydrogen Energy-Challenges and Perspectives*; IntechOpen: London, UK, 2012. [CrossRef]
13. Tee, S.Y.; Win, K.Y.; Teo, W.S.; Koh, L.-D.; Liu, S.; Teng, C.P.; Han, M.-Y. Recent Progress in Energy-Driven Water Splitting. *Adv. Sci.* **2017**. [CrossRef] [PubMed]
14. Roger, I.; Shipman, M.A.; Symes, M.D. Earth-abundant catalysts for electrochemical and photoelectrochemical water splitting. *Nat. Rev. Chem.* **2007**, *1*, 0003. [CrossRef]
15. Feliu, J.M.; Herrero, E. Surface electrochemistry and reactivity. *Contrib. Sci.* **2011**, *6*, 161–172.

16. Climent, V.; Feliu, J. Thirty years of platinum single crystal electrochemistry. *J. Solid State Electrochem.* **2011**, *15*, 1297–1315. [CrossRef]
17. Weckhuysen, B.M. Preface: recent advances in the in-situ characterization of heterogeneous catalysts. *Chem. Soc. Rev.* **2010**, *39*, 4557–4559. [CrossRef]
18. Tan, J.; Liu, D.; Xu, X.; Mai, L. In situ/operando characterization techniques for rechargeable lithium–sulfur batteries: A review. *Nanoscale* **2017**, *9*, 19001–19016. [CrossRef] [PubMed]

 *surfaces*

Article

Investigation of Photoelectron Properties of Polymer Films with Silicon Nanoparticles

Elizaveta A. Konstantinova [1,2,3,*], Alexander S. Vorontsov [1] and Pavel A. Forsh [1,2,3]

[1] Faculty of Physics, M.V. Lomonosov Moscow State University, Leninskie Gory 1-2, Moscow 119991, Russia; as.vorontsov@physics.msu.ru (A.S.V.); phorsh@mail.ru (P.A.F.)

[2] Department of Nano-, Bio-, Information Technology and Cognitive Science, Moscow Institute of Physics and Technology, Institutskij 9, Dolgoprudny, Moscow 141701, Russia

[3] National Research Center Kurchatov Institute, Akademika Kurchatova 1, Moscow 123182, Russia

* Correspondence: liza35@mail.ru

Received: 21 February 2019; Accepted: 11 May 2019; Published: 13 May 2019

Abstract: Hybrid samples consisting of polymer poly-3(hexylthiophene) (P3HT) and silicon nanoparticles were prepared. It was found that the obtained samples were polymer matrixes with conglomerates of silicon nanoparticles of different sizes (10–10^4 nm). It was found that, under illumination, the process of nonequilibrium charge carrier separation between the silicon nanoparticles and P3HT with subsequent localization of the hole in the polymer can be successfully detected using electron paramagnetic resonance (EPR) spectroscopy. It was established that the main type of paramagnetic centers in P3HT/silicon nanoparticles are positive polarons in P3HT. For comparison, samples consisting only of polymer and silicon nanoparticles were also investigated by the EPR technique. The polarons in the P3HT and P_b centers in the silicon nanoparticles were observed. The possibility of the conversion of solar energy into electric energy is shown using structures consisting of P3HT polymer and silicon nanoparticles prepared by different methods, including the electrochemical etching of a silicon single crystal in hydrofluoric acid solution and the laser ablation of single-crystal silicon in organic solvents. The results can be useful for solar cell development.

Keywords: polymer; silicon nanoparticles; EPR spectroscopy; photoconversion

1. Introduction

Solar cells based on crystalline or polycrystalline silicon are now the dominant technology globally [1–3]. However, the cost of such batteries is quite high, which is leading to the rapid development of technologies using amorphous silicon [4,5]. In addition to the low cost of devices, the use of amorphous silicon can decrease the thickness of solar cells, as well as their weight and material consumption, due to its higher absorption ability. However, the efficiency of solar cells based on amorphous silicon remains quite low, at 14%, compared to crystalline solar cells (approximately 25%) [6]. Increasing the efficiency of amorphous silicon-based solar cells is potentially possible using semiconductor nanocrystals. However, this is a very complex problem because it requires the development of methods for improving the injection and transport of charge carriers in such structures. Therefore, the development of cheap photo-electric converters has become a subject of great interest in the last years [1–4]. One of the perspective directions of the depreciation of phototransformation is the development of solar elements on the basis of polymers [7,8]. However, at the moment, the effectiveness of the phototransformation of such solar elements and their stability are low. The available literary data [7,9] demonstrate the prospects of use of hybrid structures on the basis of silicon nanoparticles (nc-Si) and organic compounds for transformation of solar energy. Such devices have the advantage of solution preparation but at the same time are characterized by a much broader spectral range of absorption because of inorganic semiconductors. It should be borne in mind that the absorption coefficient of nc-Si with a diameter of about 10 nm and more is comparable with that of bulk Si. For small nc-Si with quantum confinement effects, the band gap strongly increases and the absorption coefficient grows. There have only been a few investigations of silicon/organic semiconductor heterojunctions for solar cells [7,9–11]. However, many questions concerning the correlation between the conditions of preparation of nc-Si and the electrophysical and optical properties of such hybrid structures still have no definite answer. Since point defects in solids are the centers of capture of charge carriers and limit the transport of charge carriers, their study is an important problem [12]. A powerful tool for studying the nature and properties of defects is electron paramagnetic resonance (EPR) spectroscopy [13]. The structure and properties of defects in nc-Si and other semiconductor nanoparticles have been successfully studied with EPR spectroscopy, taking advantage of the high specific area of these materials [13,14]. In spite of the different types of nc-Si structures, different preparation and storage conditions, the dominant type of defect (paramagnetic centers) in nc-Si is an Si dangling bond at the Si/SiO_2 interface or P_b center [14,15]. There are two types of P_b centers, P_{b0} (at the (111), (100) Si/SiO_2 interface) and P_{b1} (at the (100) Si/SiO_2 interface), which are characterized by different parameters of EPR spectra [14]. The P_b center concentration is very sensitive to vacuum heating and oxidation [14]. During vacuum heating (at temperatures higher than 300–400 °C), an Si dangling bond similar to that in amorphous silicon is detected in nc-Si [14]. A free electron EPR signal is observed in some samples of nc-Si [14]. EX center defects are detected in high-temperature oxidized nc-Si [14]. In poly-3(hexylthiophene) (P3HT), the main type of defect center is a positive polaron [9]. Therefore, in this work, we investigated the properties of defect states in hybrid samples consisting of polymer P3HT and silicon nanoparticles and in samples consisting only of polymer and silicon nanoparticles for comparison. We also measured the current–voltage characteristics of structures containing silicon nanocrystals in a polymeric matrix.

2. Experimental

Silicon nanoparticles were prepared by two methods [15,16]. The first method was the electrochemical etching of monocrystal plates of silicon of n-type of conduction in a solution of hydrofluoric acid (HF) and ethanol (C_2H_5OH). After etching, the samples were dried, and then silicon particles were removed from the plate mechanically (nc-Si(1)). The second method of manufacturing silicon nanoparticles (nc-Si(2)) was the pulse laser ablation of a substrate of monocrystal silicon in organic solvents (C_6H_5Cl, benzene chloride, and $CHCl_3$, chloroform). For the formation of the structures, the received nanoparticles were immediately added to the P3HT solution. Then, the resulting mixture was applied on the glass substrate containing an ITO layer as one of electrical contacts. The second was the aluminum contact which was sprayed from above onto the P3HT film. The current–voltage characteristics of the samples were measured using Keithley 6487.

The investigation of structure of the prepared samples was performed using scanning electronic microscopy (using a Carl Zeiss Supra 40-30-87 microscope, Oberkochen, Germany). The features of the internal structure of the polymer and the characteristics of the obtained silicon nanoparticles were determined by means of the Raman technique. For the Raman spectra registration, the Horiba Jobin Yvon HR800 spectrometer was used. A helium–neon laser (λ = 632.8 nm) was the source of light excitation. Raman signal recording was made in a configuration of reflection. A digital camera with a charge-coupled device (CCD) matrix was used as the detector. For the direct detection of separation processes of photoinduced charges in structures of nc-Si/P3HT, EPR spectroscopy was used. Measurements were performed using the Bruker EPR-spectrometer ELEXSYS-500, Rheinstetten, Germany (with a frequency of 9.5 GHz and a sensitivity of 5×10^{10} spin/mT). The EPR spectra simulation was carried out using the EasySpin MATLAB toolbox [17]. Illumination was carried out immediately in the spectrometer cavity by means of a mercury lamp with high pressure ($\Delta\lambda$ = 270–900 nanometers, power is 50 W).

3. Results and Discussions

Figure 1 shows an example of the structure of the nc-Si(1)/P3HT. Similar images were observed also for the other samples, nc-Si(2)/P3HT. According to Figure 1, the samples under investigation represent conglomerates of nanoparticles of various sizes (10–10^4 nm) dissolved in a polymeric matrix.

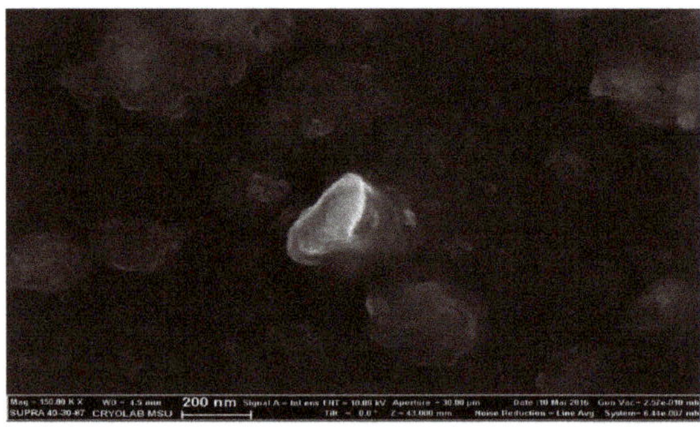

Figure 1. SEM image of nc-Si(1)/poly-3(hexylthiophene) (P3HT) samples.

Raman spectra were similar for both types of samples. The Raman spectrum of the nc-Si(2)/P3HT samples is shown in Figure 2. According to the literature data, the peak near 430 and 520 cm^{-1} in the Raman spectrum corresponds to the C–C deformation mode of the polymer and silicon nanocrystals [11,18], respectively; the peak near 728 cm^{-1} corresponds to an antisymmetric C–S–C ring skeleton deformation in the thiophene ring of the polymer [19]; and the peak near 1390 and 1450 cm^{-1} can be assigned to the C–C skeletal stretch mode and the C=C symmetric stretch mode of the polymer, respectively [20,21]. The other less intense peaks observed in the wavenumber range of 570, 850, 1000, 1100, 1150, 1250 and 1525, 1560 cm^{-1} can be assigned to the C–C rotational mode, C=C deformation mode, C–C stretching mode [21], the C–H bending mode [22], the C–C symmetric stretching mode [23], a combination of the C–C stretching and the C–H stretching mode [21,24,25] and the C = C antisymmetric stretch mode [20,23], respectively. Thus, according to Figure 2, the Raman spectrum consists of lines caused by a polymeric matrix and also a poorly resolved peak, which is characteristic of silicon nanocrystals. The results of a Raman investigation confirmed the data of electronic microscopy, showing that all samples under investigation represent the non-uniform structure of a polymeric matrix with the silicon nanoparticles dissolved in it.

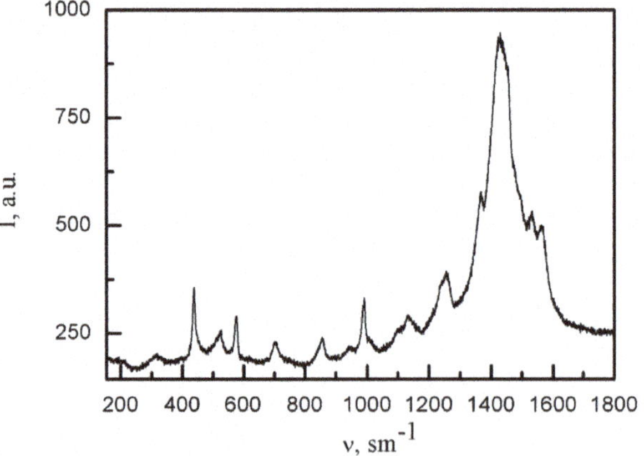

Figure 2. Raman spectrum of the nc-Si(2)/P3HT samples.

For the direct detection of the separation processes of photoinduced charges in the nc-Si/P3HT structures, the EPR-spectroscopy technique was used. In Figure 3a, the EPR spectra of nc-Si(1)/P3HT samples in the dark and under illumination are presented.

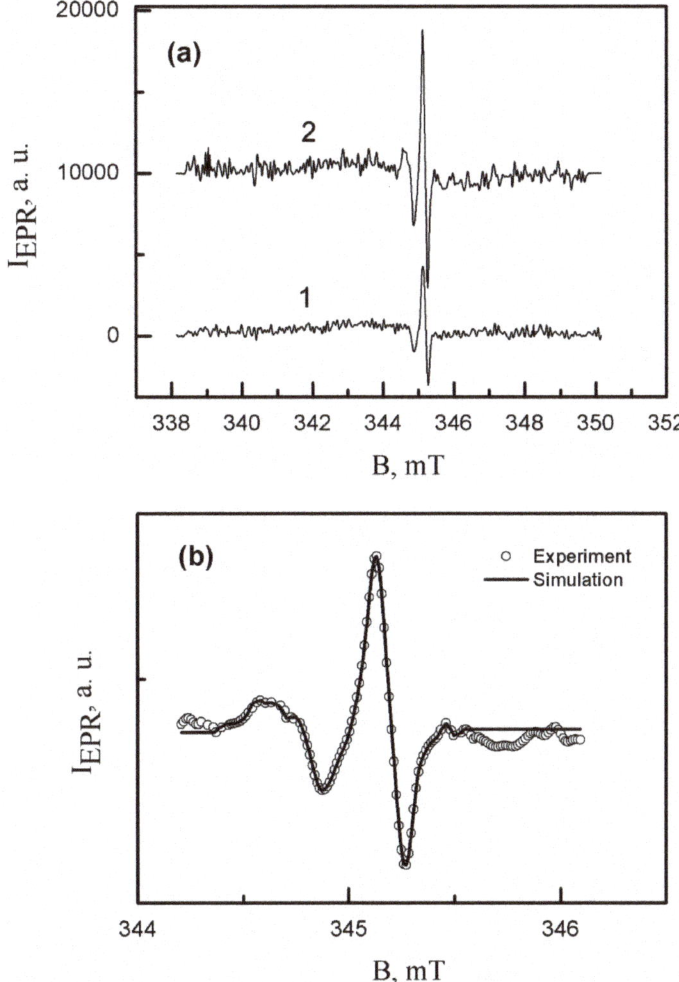

Figure 3. (a) EPR spectra of nc-Si(1)/P3HT samples in the dark (1) and under illumination (2). (b) Experimental and simulated EPR signals of nc-Si(1)/P3HT samples.

The EPR spectra are characterized by two overlapped signals: one is poorly resolved, with a g-factor equal to 2.0050 originating from the silicon dangling bonds [26], while the second one has an anisotropic form and is described by the following parameters: $g_1 = 2.0030$, $g_2 = 2.0022$, $g_3 = 2.0013$, extracted from computer simulation. The result of the computer simulation is shown in Figure 3b. According to the literature data, this EPR signal can be attributed to the positive polarons in P3HT [9,21]. Under illumination, an amplitude of a polaron EPR signal increases, pointing to the separation process of photoexcited charge carriers. To be sure that the positive polarons are localized in P3HT, we have investigated these structures separately. In Figure 4a, the EPR spectra for P3HT samples in similar conditions are shown, both in the dark and under illumination.

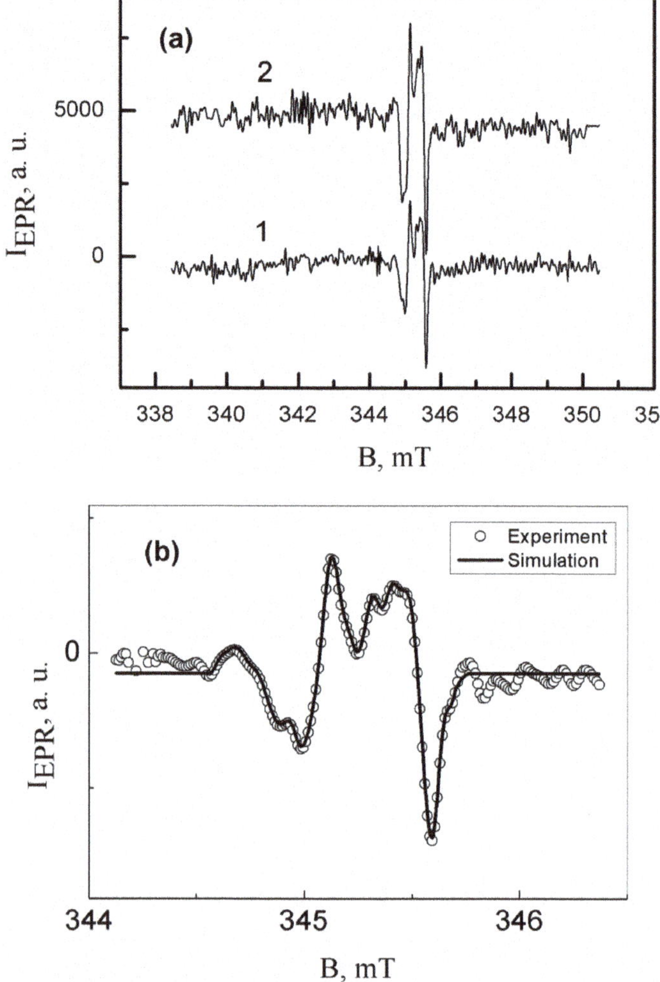

Figure 4. (**a**) EPR spectra of P3HT samples in the dark (1) and under illumination (2). (**b**) Experimental and simulated EPR signals of P3HT samples.

According to the computer simulation, the EPR spectrum consists of two EPR signals with parameters that differ slightly: EPR(I) g_1 = 2.0028, g_2 = 2.0019, g_3 = 2.0009 and EPR(II) g_1 = 2.0031, g_2 = 2.0021, g_3 = 2.0012. The result of the computer simulation is shown in Figure 4b. The values of the g-factors show the polaron nature of the EPR-centers [9,21]. Probably, we observe the polarons in two different orientations. Thus, data from the EPR spectroscopy are direct proof of the formation of polarons in nc-Si/P3HT structures as a result of the separation of charge carriers with the subsequent localization of a hole in P3HT. Note that this process takes place under the conditions of indoor light and are amplified under additional illumination.

The silicon nanoparticles was measured for a comparison of the defects' natures and properties. Figure 5 shows the EPR spectra measured under dark conditions and in the presence of illumination.

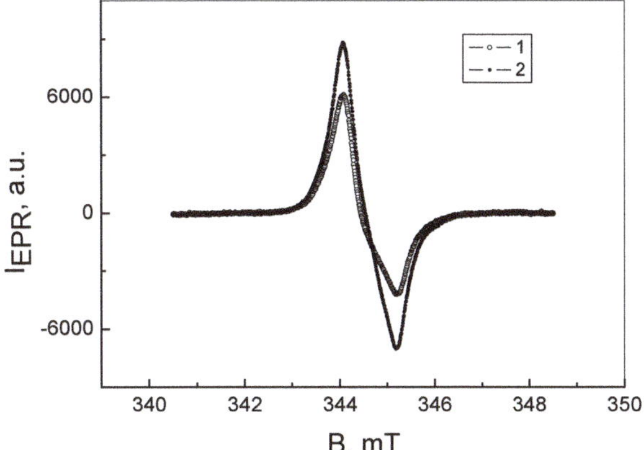

Figure 5. EPR spectra of nc-Si(1) in the dark (1) and under illumination (2).

As follows from the presented data, the EPR spectrum has an anisotropic structure. The following values of the g-tensor were obtained from computer simulation: $g_1 = 2.0021$, $g_2 = 2.0088$. The EPR signal with these anisotropic parameters can be attributed to the paramagnetic centers, which are P_b centers (a silicon dangling bond at the Si/SiO_2 interface) according to the literature data [27,28]. Therefore, an oxidation of silicon nanoparticles takes place in air. Under illumination, the intensity of the EPR spectrum increases. The effect of illumination was reversible, showing recharge processes in the samples.

For the determination of the photovoltaic parameters of the samples and qualitative evaluation of the effectiveness of solar energy conversion, the current density–voltage (abbreviated as current–voltage [8,9]) characteristics were measured (Figure 6). White light illumination was used with an intensity of 50 mW/cm^2. As can be seen from Figure 6, nc-Si(1)/P3HT samples have a higher open-circuit voltage (0,2 V) than nc-Si(2)/P3HT samples (0,16 V). Although these open-circuit voltage values are not a record, it is possible in principle to use nc-Si/P3HT samples for photoconversion.

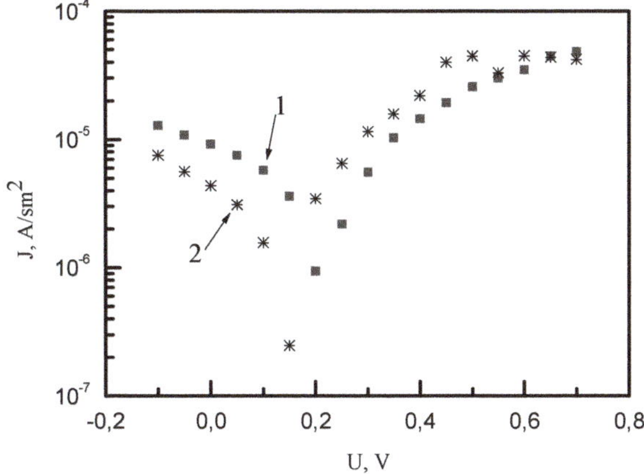

Figure 6. Current–voltage characteristics of the studied samples. Dependence 1—nc-Si(1)/P3HT samples; dependence 2—nc-Si(2)/P3HT samples.

Although the efficiency of the solar energy conversion of the samples in our study was low (not exceeding 5%), we have demonstrated that EPR spectroscopy is an effective tool for the detection of polarons in nc-Si/P3HT prepared by different techniques.

4. Conclusions

The structures consisting of P3HT polymer and silicon nanoparticles, in which silicon nanoparticles are prepared by electrochemical etching and the laser ablation of single-crystal silicon, have the possibility of converting solar energy into electric energy. In nc-Si samples, P_b centers are present according to the EPR data. A lack of P_b centers in the nc-Si/P3HT structures evidences an absence of oxygen in the samples. Positive polarons are detected in P3HT samples. Using EPR spectroscopy, it has been found that it is possible to detect the separation of photoexcited charge carriers in nc-Si/P3HT samples prepared both by the electrochemical etching of a silicon single crystal in hydrofluoric acid solution and by using the laser ablation of single-crystal silicon in organic solvents. The obtained data is useful for solar energy application since they demonstrate the possibility of the preparation of solar cell samples using simple and cheap methods of sample synthesis (for example, the electrochemical etching of silicon monocrystals).

This work was supported by the Russian Foundation for Basic Research (Grant No 18-29-23005).

Author Contributions: Conceptualization, E.A.K. and P.A.F.; Methodology, E.A.K.; Software, A.S.V.; Validation, E.A.K. and A.S.V.; Formal Analysis, P.A.F.; Investigation, E.A.K., A.S.V. and P.A.F.; Resources, A.S.V.; Data Curation, E.A.K.; Writing-Original Draft Preparation, E.A.K.; Writing-Review & Editing, E.A.K., A.S.V. and P.A.F.; Visualization, P.A.F.; Supervision, E.A.K.; Project Administration, A.S.V.; Funding Acquisition, P.A.F.

Funding: This research was funded by the Russian Foundation for Basic Research Grant number 18-29-23005.

Conflicts of Interest: The authors declare no conflict of interest.

References

1. Kazmerski; Lawrence, L. Photovoltaics: A review of cell and module technologies. *Renew. Sustain. Energy Rev.* **1997**, *1*, 71–170. [CrossRef]

2. Meyer, P.V. Technical and economic optimization for CdTe PV at the turn of the millennium. *Prog. Photovolt. Res. Appl.* **2000**, *8*, 161–169. [CrossRef]

3. Rau, U.; Schock, H.W. Cu(In, Ga)Se2 solar Cells. Clean Electricity from Photovoltaics. *Ser. Photoconvers. Solar Energy* **2001**, *1*, 277–345.

4. Gremenok, V.F.; Tivanov, M.S.; Zalesski, V.B. Solar cells based on semiconductors. *Int. Sci. J. Altern. Energy Ecol.* **2009**, *69*, 59–124.

5. Emelyanov, A.V.; Khenkin, M.V.; Kazanskii, A.G.; Forsh, P.A.; Kashkarov, P.K.; Gecevicius, M.; Beresna, M.; Kazansky, P.G. Femtosecond laser induced crystallization of hydrogenated amorphous silicon for photovoltaic applications. *Thin Solid Films* **2014**, *556*, 410–413. [CrossRef]

6. Best Research-Cell Efficiencies. Available online: https://www.nrel.gov/pv/assets/pdfs/best-research-cell-efficiencies-190416.pdf (accessed on 25 April 2019).

7. Hemaprabhaa, E.; Pandeya, U.K.; Chattopadhyaya, K.; Ramamurthya, P.C. Doped silicon nanoparticles for enhanced charge transportation in organic inorganic hybrid solar cells. *Solar Energy* **2018**, *173*, 744–751. [CrossRef]

8. Salikhov, R.B.; Biglova, Y.N.; Yumaguzin, Y.M.; Salikhov, T.R.; Mustafin, A.G. Solar photoconverters based on thin films of organic materials. *Tech. Phys. Lett.* **2013**, *39*, 854–859. [CrossRef]

9. Niesar, S.; Dietmueller, R.; Nesswetter, H.; Wiggers, H.; Stutzmann, M. Silicon/organic semiconductor heterojunction for solar cells. *Phys. Stat. Sol. A* **2009**, *206*, 2775–2781. [CrossRef]

10. Shao, M.; Keum, J.; Chen, J.; He, Y.; Chen, W.; Browning, J.F.; Jakowski, J.; Sumpter, B.G.; Ivanov, I.N.; Ma, Y.-Z.; et al. The Isotopic Effects of Deuteration on Optoelectronic Properties of Conducting Polymers. *Nat. Commun.* **2014**, *5*, 3180. [CrossRef]

11. Liu, C.-Y.; Holman, Z.C.; Kortshagen, U.R. Hybrid solar cells from P3HT and silicon nanocrystals. *Nano Lett.* **2009**, *9*, 449–456. [CrossRef]

12. Kaftelen, H.; Ocakoglu, K.; Thomann, R.; Tu, S.Y.; Weber, S.; Erdem, E. EPR and photoluminescence spectroscopy studies on the defect structure of ZnO nanocrystals. *Phys. Rev. B* **2012**, *86*, 014113-1–014113-9. [CrossRef]

13. Erdem, E. Microwave power, temperature, atmospheric and light dependence of intrinsic defects in ZnO nanoparticles: A study of electron paramagnetic resonance (EPR) spectroscopy. *J. Alloy. Compd.* **2014**, *605*, 34–44. [CrossRef]

14. Von Bardeleben, H.J.; Ortega, C.; Grosman, A.; Morazzani, V.; Siejka, J.; Stievenard, D. Defect and structure analysis of n^+-, p^+- and p-type porous silicon by the electron paramagnetic resonance technique. *J. Lumin.* **1993**, *57*, 301–313. [CrossRef]

15. Riikonen, J.; Salomaki, M.; van Wonderen, J.; Kemell, M.; Xu, W.; Korhonen, O.; Ritala, M.; MacMillan, F.; Salonen, J.; Lehto, V.-P. Surface chemistry, reactivity and pore structure of porous silicon oxidized by various methods. *Langmuir* **2012**, *28*, 10573–10583. [CrossRef]

16. Demin, V.A.; Konstantinova, E.A.; Kashkarov, P.K. Luminescence and photosensitization properties of ensembles of silicon nanocrystals in terms of an exciton migration model. *J. Exp. Theor. Phys.* **2010**, *111*, 830–843. [CrossRef]

17. Stoll, S.; Schweiger, A. EasySpin, a comprehensive software package for spectral simulation and analysis in EPR. *J. Magn. Reson.* **2006**, *178*, 42–55. [CrossRef] [PubMed]

18. Emelyanov, A.; Kazanskii, A.; Kashkarov, P.; Konkov, O.; Terukov, E.; Forsh, P.; Khenkin, M.; Kukin, A.; Beresna, M.; Kazansky, P. Effect of the femtosecond laser treatment of hydrogenated amorphous silicon films on their structural, optical, and photoelectric properties. *Semiconductors* **2012**, *46*, 749–754. [CrossRef]

19. Falke, S.; Eravuchira, P.; Maternyb, A.; Lienaua, C. Raman spectroscopic identification of fullerene inclusions in polymer/fullerene blends. *J. Raman Spectrosc.* **2011**, *42*, 1897–1900. [CrossRef]

20. Motaung, D.E.; Malgas, G.F.; Nkosi, S.S.; Mhlongo, G.H.; Mwakikunga, B.W.; Malwela, T.; Arendse, C.J.; Muller, T.F.G.; Cummings, F.R. Comparative study: The effect of annealing conditions on the properties of P3HT:PCBM blends. *J. Mater. Sci.* **2013**, *48*, 1763–1778. [CrossRef]

21. Konkin, A.; Roth, H.-K.; Scharff, P.; Aganov, A.; Ambacher, O.; Sensfuss, S. K-band ESR studies of structural anisotropy in P3HT and P3HT/PCBM blend polymer solid films: Paramagnetic defects after continuous wave Xe-lamp photolysis. *Solid State Commun.* **2009**, *149*, 893–897. [CrossRef]

22. Saini, V.; Li, Z.R.; Bourdo, S.; Dervishi, E.; Xu, Y.; Ma, X.D.; Kunets, V.P.; Salamo, G.J.; Viswanathan, T.; Biris, A.R.; et al. Electrical, optical, and morphological properties of P3HT-MWNT nanocomposites prepared by in situ polymerization. *J. Phys. Chem. C* **2009**, *113*, 8023–8029. [CrossRef]

23. Baibarac, M.; Lapkowski, M.; Pron, A.; Lefrant, S.; Baltog, I. SERS spectra of poly (3-hexylthiophene) in oxidized and unoxidized states. *J. Raman Spectrosc.* **1998**, *29*, 825–832. [CrossRef]

24. Louarn, G.; Trznadel, M.; Buisson, J.P.; Laska, J.; Pron, A.; Lapkowski, M.; Lefrant, S. Raman spectroscopic studies of regioregular poly (3-alkylthiophenes). *J. Phys. Chem.* **1996**, *100*, 12532–12539. [CrossRef]

25. Klimov, E.; Li, W.; Yang, X.; Hoffmann, J.G.; Loos, G. Scanning Near-Field and Confocal Raman Microscopic Investigation of P3HT-PCBM Systems for Solar Cell Applications. *Macromolecules* **2006**, *39*, 4493–4496. [CrossRef]

26. Emelyanov, A.V.; Konstantinova, E.A.; Forsh, P.A.; Kazanskii, A.G.; Khenkin, M.V.; Petrova, N.N.; Terukov, E.I.; Kirilenko, D.A.; Bert, N.A.; Konnikov, S.G.; et al. Features of the structure and defect states in hydrogenated polymorphous silicon films. *JETP Lett.* **2013**, *97*, 466–469. [CrossRef]

27. Cantin, J.L.; Schoisswohl, M.; Von Bardeleben, H.J. Electron-paramagnetic-resonance study of the microscopic structure of the Si(001)-SiO$_2$ interface. *Phys. Rev. B* **1995**, *52*, R11599–11602. [CrossRef]

28. Minnekhanov, A.A.; Konstantinova, E.A.; Pustovoi, V.I.; Kashkarov, P.K. The Influence of the Formation and Storage Conditions of Silicon Nanoparticles Obtained by Laser-Induced Pyrolysis of Monosilane on the Nature and Properties of Defects. *Tech. Phys. Lett.* **2017**, *43*, 424–427. [CrossRef]

Article

Formic Acid Oxidation on Pd Thin Film Coated Au Nanocrystals

Yongan Tang [1,†] **and Shouzhong Zou** [2,*,†]

1   DOD Technologies Inc., Cary, IL 60013, USA; tangy3@miamioh.edu
2   Department of Chemistry, American University, Washington, DC 20016, USA
*   Correspondence: szou@american.edu
†   Department of Chemistry and Biochemistry, Miami University, Oxford, OH 45056, USA.

Received: 5 April 2019; Accepted: 6 May 2019; Published: 10 May 2019

Abstract: Cubic, octahedral, and rhombic dodecahedral gold nanocrystals enclosed by {100}, {111}, and {110} facets, respectively, were prepared by a seed-mediated growth method at the room temperature. Palladium thin films were coated on these Au nanocrystals by a redox replacement approach to explore their catalytic activities. It is revealed that formic acid and carbon monoxide oxidation in 0.1 M $HClO_4$ on Au nanocrystals coated with one monolayer (ML) of Pd are facet-dependent and resemble those obtained from corresponding Pd single crystals and Pd films deposited on bulk Au single crystals, suggesting epitaxial growth of Pd overlayers on the Au nanocrystal surfaces. As the Pd film thickness increased, formic acid oxidation current density decreased and the CO oxidation potential moved to more negative. The catalytic activity remained largely unchanged after 3–5 MLs of Pd deposition. The specific adsorption of (bi)sulfate was shown to hinder the formic acid oxidation and the effect decreased with the increasing Pd film thickness. These observations were explained in the framework of the d-band theory. This study highlights the feasibility of engineering high-performance catalysts through deposition of catalytically active metal thin films on facet-controlled inert nanocrystals.

Keywords: formic acid oxidation; Au nanocrystals; Pd thin films; electrocatalysis; d-band theory

---

1. Introduction

Direct formic acid fuel cells (DFAFCs) have attracted much attention in the past decade because they have the advantages of high power density, fast oxidation kinetics, high theoretical cell potential, and mild fuel crossover as compared to direct methanol fuel cells [1–5]. As the anode reaction of DFAFCs, electrooxidation of formic acid has been extensively studied [2,5]. Although the nature of the reactive intermediate is under active debate, it is generally accepted that there are two reaction pathways in formic acid oxidation (FAO) [2,6]. The direct pathway (dehydrogenation) involves the adsorption of reactive intermediates and their direct oxidation to carbon dioxide, while the indirect pathway goes through the dehydration of formic acid and forms strongly adsorbed carbon monoxide (CO_{ads}) that poisons catalysts [2,6]. It has been long recognized that the direct pathway is dominant on Pd while the indirect pathway is more prevalent on Pt, which suggests that Pd is a better catalyst than Pt for DFAFCs [6,7]. Indeed, using unsupported and carbon-supported Pd as anode catalysts in DFAFCs, Masel and coworkers obtained better fuel cell performance than Pt-based catalysts at high power densities [8,9]. Despite the higher activity observed on Pd, a significant performance decay was observed after only a few hours of operation [9]. Different adsorbed species, including CO_{ads}, have been proposed to explain the deactivation of the Pd catalysts [2]. In addition, it has been shown that Pd dissolution occurred in an acidic media, resulting in a decreased active surface area [10,11].

Significant efforts have been devoted to develop Pd-based catalysts with high activity as well as long-term durability and stability [2,5,12]. A commonly employed approach is to introduce a second metal either to form an alloy with Pd or as a support for Pd overlayers. Of particular interest in the present work is the Pd–Au combination. Zhou and Lee reported that Au core Pd shell (Au@Pd) nanoparticles showed significantly improved FAO activity and durability, which was attributed to the Au and Pd electronic interactions [13]. More detailed studies by others revealed that the improved FAO activity depends on the Au/Pd ratio [14] and the degree of alloying [15]. Lee et al. found that FAO activity and durability of Pd_3Au nanoparticles can be significantly improved by increasing the surface Pd content through CO-induced surface aggregation [16]. Hong et al. studied FAO on structurally well-controlled PdAu nano-octahedra, and revealed that PdAu alloy octahedra are about 30% more active than similar size Au@Pd octahedra, but both are more active than pure Pd octahedra [17]. By using a seed-mediated growth method, Fermín and coworkers prepared Au@Pd nanoparticles with varying Pd shell thickness from 1 to 10 nm [18]. FAO activity on these core-shell particles was shown to increase with the Pd shell thickness, which was attributed to the lattice strain effect. More recently, by combining differential electrochemical mass spectrometry and in situ Fourier transform infrared spectroscopy with the density functional theory (DFT) calculations, the same group concluded that a faster $HCOO_{ads}$ oxidation step on the thicker Pd shell, induced by the lattice strain effect, might be responsible for the more facile FAO [19].

In addition to the core shell or alloyed nanoparticles, PdAu catalyst can also be prepared by depositing Pd thin films on Au surfaces. Hsu et al. deposited Pd thin films on hollow Au nanospheres through a spontaneous galvanic replacement deposition [20]. Depending on the Pd film's thickness, the Pd-coated Au nanospheres showed enhanced FAO peak current density compared to Pd black, which was ascribed to the electronic coupling between Au and Pd. Obradović and Gojković showed film-thickness-dependent FAO activity on electrochemically deposited Pd films on polycrystalline Au electrodes with the film thickness ranging from 1 to 17 monolayers (MLs) [21]. The maximum peak current density was observed on four ML Pd, but the peak potential is most negative at one ML Pd. All of the Pd films had a higher FAO activity than Pd black, which was attributed to the strain effect from the Au substrate. In an earlier study of FAO on Pd films deposited on various metal surfaces with (111) orientation, Kibler et al. demonstrated nicely the correlation between the d-band center position tuned by the strain effect and the FAO catalytic activity [22]. In their work, Pd monolayer deposited on Au (111) exhibited a higher peak current density than Pd (111), albeit at a higher potential.

One limitation in most of the previous studies of nanoparticles is that the particle surface structure is not controlled and the observed Pd film thickness dependent activity can at least partly arise from the surface structure difference. It has been well-demonstrated that formic acid oxidation is structure-sensitive [7,23–26]. Therefore, controlling the exposed facets of nanoparticles is critical when comparing catalytic activities, and can be used as a tool to improve catalytic performance. In this paper, Pd thin films were deposited on cubic, octahedral, and rhombic dodecahedral (RD) Au nanocrystals that were enclosed by six {100}, eight {111}, and twelve {110} facets, respectively, by a redox replacement approach. The surface-limiting nature of this deposition method ensures the precise control of the film thickness to a monolayer level [27], and pseudomorphic growth of Pd films on Au (111) surface up to several tens of MLs has been demonstrated [28]. The use of nanocrystals with well-defined surface structure minimizes the complication from the surface structural difference of the substrates in film-thickness-dependent studies. Our focus of this study is to explore whether the Pd thin films on Au nanocrystals mimic the electrochemical behaviors of the bulk Pd single crystals and how the FAO activity varies with Pd film thickness. The results reveal that cyclic voltammetric features of hydrogen adsorption/desorption, surface oxidation/reduction, and CO stripping on Au nanocrystals covered with a monolayer of Pd resemble those from the corresponding bulk Pd single crystals in 0.1 M $HClO_4$. FAO on the monolayer Pd-coated Au nanocrystals showed peak shapes in the cyclic voltammograms similar to those of bulk Pd single crystals and Pd film-covered bulk Au single crystals. FAO and CO oxidation were also conducted on Au nanocrystals covered with Pd films with increasing thickness up to 10 MLs. The results show that the FAO peak current density and CO oxidation onset potential decreased with increasing Pd film thickness. These observations were explained with the strain and ligand effects from the d-band theory. This study demonstrates that depositing catalytically active materials as a thin film on facet-controlled nanocrystals is an effective means of obtaining high-performance catalysts because it combines the advantages of optimizing the d-band position by the substrate and the facet-dependent catalytic activity of the overlayer.

2. Experimental Section

2.1. Materials

Hydrogen tetrachloroaurate(III) trihydrate ($HAuCl_4 \cdot 3H_2O$) (99.9%), silver nitrate ($AgNO_3$) (≥99.0%), sodium borohydride ($NaBH_4$) (≥99%), potassium bromide (KBr) (≥99.0%), cetyl trimethylammonium bromide (CTAB) (≥98%), and cetylpyridinium chloride (CPC) (99.0%) were purchased from Sigma-Aldrich. L-Ascorbic acid (AA) (99%) was obtained from Acros Organics. Formic acid (88%) was obtained from Pharmco. Sodium hydroxide (98.5%), and hydrochloric acid (36.5–38%), were received from Fisher Scientific. Double-distilled perchloric acid (70%) and sulfuric acid (95–98%, ultra trace metal grade) were from GFS. Carbon monoxide (99.997%) was from Airgas. All of the chemicals were used without further purification. Ultrapure water from a Milli-Q system (18.2 MΩ·cm) was used in this work.

2.2. Synthesis of Au Seeds

The synthesis of 38 nm Au seed followed previously reported multi-step methods by others with some modifications [29–32]. First, 125 µL of 10 mM $HAuCl_4$ was added to 5 mL of 100 mM CTAB solution at room temperature (22 ± 1 °C), followed by the addition of 300 µL of 10 mM ice-cold $NaBH_4$. The mixture was stirred for 5 min and stored at room temperature for further uses. Second, 24 µL of the above solution was added to a mixture of 20 mL of 100 mM CTAB solution, 1 mL of 10 mM $HAuCl_4$, 120 µL of 10 mM $AgNO_3$, and 160 µL of 100 mM ascorbic acid. The solution was kept at room temperature for 2 h without stirring. The Au nanoparticles were centrifuged at 6500 rpm and redispersed in 20 mL of 100 mM CTAB. Third, 1 mL of 10 mM $HAuCl_4$ and 200 µL of 100 mM ascorbic acid were added in sequence into the solution at 40 °C. The solution was left undisturbed for 1 h then centrifuged at 6500 rpm and redispersed in 20 mL of 10 mM CTAB solution. Lastly, 400 µL of 10 mM $HAuCl_4$ was added into the solution and kept undisturbed at 40 °C overnight. The obtained Au nanoparticles were washed three times with a 100 mM CPC solution by centrifugation at 6500 rpm and then redispersed in 30 mL of 100 mM CPC. The solution was marked as the Au seed solution for further uses.

2.3. Synthesis of Au Nanocrystals

The shape-controlled synthesis of Au nanocrystals mostly followed a reported method [29]. In the synthesis of cubic Au nanocrystals, 5 mL of 100 mM KBr solution, 1 mL of 10 mM $HAuCl_4$ solution, 150 µL of 100 mM ascorbic acid solution, and 2 mL of the Au seed solution were added into 50 mL of 100 mM CPC solution in sequence. The solution was kept at room temperature for 2 h without stirring. The obtained cubic Au nanocrystals were washed five times by centrifugation at 3500 rpm with warm water and then redispersed in 1 mL ethanol. The synthesis of octahedral Au nanocrystals is the same as that of cubes, except that no KBr was used. In the synthesis of RD Au nanocrystals, besides the fact that no KBr was used, the volume of AA with the same concentration was increased to 2 mL and the concentration of CPC was reduced to 10 mM with the same volume.

2.4. Electrochemical Measurement

For a typical measurement, a whole batch of Au nanocrystals were further washed three times with warm water and redispersed in 500 µL ethanol before electrochemical measurement. Ten microliters (10 µL) of Au nanocrystals ethanol suspension was drop-coated on a glassy carbon (GC) disk electrode and used as the working electrode. Cyclic voltammograms (CVs) and chronoamperograms (CAs) were recorded in a conventional two-compartment three-electrode glass cell using a CHI 700C electrochemical analyzer. A Pt wire served as the counter electrode and a KCl-saturated Ag/AgCl electrode was used as the reference electrode. The counter electrode and the reference electrode were in the same compartment that is separated from the working electrode compartment by a fine porous glass frit. The cell resistance was compensated for with the iR compensation function in the analyzer. Before any electrochemical measurements, the Au nanocrystals were subject to an electrochemical cleaning process, which entails electrochemical potential cycling between −0.7 to +0.7 V versus Ag/AgCl for 50 cycles at a scan rate of 0.5 V s^{-1} in 0.1 M NaOH. To prepare Pd-monolayer-coated Au nanocrystals, an atomic layer of Cu was first deposited by underpotential deposition (UPD) in 0.1 M H_2SO_4 + 1 mM $CuSO_4$ solution at 0.05 V for 10 min. Then, the Cu-layer-coated Au nanocrystals were immersed in a deaerated 0.1 M $HClO_4$ + 5 mM $PdCl_2$ for 20 min to replace the atomic layer of Cu. Because of the surface-limiting nature of the Cu UPD, nominally one ML of Pd is deposited after one redox replacement cycle [27,28]. By repeating this procedure, multi-monolayers of Pd were obtained. The electrochemical surface area of Pd films was measured from CO stripping charges by assuming 320, 315, and 315 µC cm^{-2} for complete removal of a saturated CO adlayer on the Pd overlayer deposited on RD, cubic, and octahedral Au nanocrystals [33], respectively. The CO adlayer was formed by purging the gas in the solution for 5 min with the electrode potential being held at -0.10 V versus Ag/AgCl followed by 10 min N_2 purging to remove solution CO. The stripping voltammograms were recorded in 0.1 M $HClO_4$, and the potential was first scanned negatively to -0.17 V and reversed. Typically, two additional cycles between −0.17 and +0.98 V were recorded after CO oxidation, but for clarity, only the CO stripping segment was shown.

2.5. Instrumentation

Scanning electron microscopy (SEM) images were obtained on a Zeiss Supra35 scanning electron microscope operating at 5 kV. High-resolution transmission electron microscopy (HRTEM) and selected-area electron diffraction (SAED) images were taken using a JEOL2100 transmission electron microscope (JEOL USA Inc., Peabody, Massachusetts, USA) operating at 200 kV. X-ray diffraction (XRD) measurements were performed on a Scintag X1 powder diffractometer (Scintag Inc., San Francisco, CA, USA) with Cu Kα radiation (λ = 0.154 nm) operated at 40 kV and 25 mA by step scanning with a step size of 0.02°/step. UV-Vis spectra were recorded by an Agilent 8453 spectrophotometer (Agilent, Santa Clara, CA, USA).

3. Results and Discussion

3.1. Synthesis and Characterization of Au Nanocrystals

Cubic, octahedral, and RD gold nanocrystals enclosed by {100}, {111}, and {110} facets were synthesized following a seed-mediated growth method developed by Niu et al. with some modifications (see experimental section) [29]. The Au seeds used in this approach have a relatively large diameter (38 ± 4 nm), as shown by the SEM image in Figure S1 in the Supplementary Material. Figure 1a–c displays the SEM images of the cubic, octahedral, and RD Au nanocrystals, respectively. The shape difference of Au nanocrystals is clearly observed. The edge length of the Au nanocubes is 61.5 ± 3.4 nm (average length and standard deviation based on measurements from 100 particles). The TEM image and square SAED pattern shown in Figure 1d confirm that these Au nanocubes are enclosed by well-defined {100} facets [34,35]. The edge lengths of Au octahedra and RD are 63.2 ± 4.2 nm and 58.7 ± 2.8 nm, respectively. TEM images and SAED patterns of octahedral and RD Au nanocrystals are presented in Figure 1e,f. The hexagonal dot array in the SAED pattern indicates that the octahedral nanocrystal is enclosed by {111} facets [35]. The RD Au nanocrystals show the elongated hexagonal dot array, which is also consistent with the reported diffraction pattern of Au single crystals along the [011] zone axis [29,36,37]. Compared to those reported in [29], the surfaces of our RDs are smoother, suggesting higher quality of the crystals.

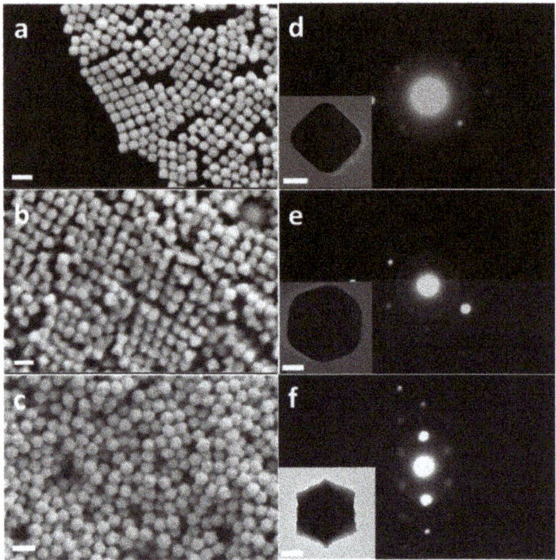

Figure 1. SEM images of (**a**) cubic, (**b**) octahedral, and (**c**) rhombic dodecahedral (RD) Au nanocrystals. Scale bar: 100 nm. SAED patterns of (**d**) cubic, (**e**) octahedral, and (**f**) RD Au nanocrystals. Inset: corresponding TEM images, scale bar: 20 nm.

The structure of the nanocrystals was further examined with powder X-ray diffraction (XRD). The cubic, octahedral, and RD Au nanocrystals show dominant (200), (111), and (220) peaks respectively, confirming the single crystal structure of these Au nanocrystals (Figure 2). The XRD pattern for RD Au nanocrystals showed a much more intense (220) peak compared to the previously reported results by others [29], suggesting that the RD Au nanocrystals synthesized in the present work have a much higher quality, which ensures more accurate activity comparison between different facets. The structural improvement may arise from the lower synthesis temperature (22 °C here versus 30 °C in [29]). Synthesizing the particles at room temperature simplifies the process by eliminating the use of temperature control devices. In addition, we increased the amount of reactants by 10 times, which produces a larger quantity of Au nanocrystals that are needed in the catalytic activity investigations.

The UV-Vis spectra in Figure S2 show shape-dependent surface plasmon resonance peaks. The peak wavelengths of cubic, octahedral, and RD Au nanocrystals are 539 nm, 559 nm, and 598 nm, respectively, which are similar to those reported in ref. 29. The differences in UV-Vis spectra mainly result from the differences in the shape of these Au nanocrystals. It is interesting to note that the peak width of the RD is significantly broader than those of cubes and octahedra. The full width at half maximum (FWHM) for RD is 110 nm, but only about 55 and 60 nm for the cubes and octahedra, respectively.

One of the advantages of the seed-mediated synthesis is that the crystal size can be conveniently tuned through varying the amount of the seed. To demonstrate this point, we decreased the volume of Au seed solution to 100 µL in the growth step, and larger Au nanocrystals were obtained, as shown in Figure S3. The facet edges of these larger nanocrystals are easier to see compared to the smaller ones. The size distributions of these larger cubic, octahedral, and RD Au nanocrystals are 112 ± 13 nm, 117 ± 12 nm, and 135 ± 8 nm, respectively.

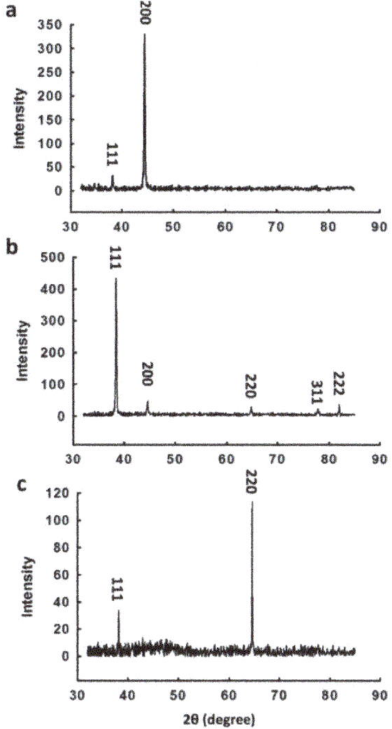

Figure 2. XRD pattern of (**a**) cubic, (**b**) octahedral, and (**c**) RD gold nanocrystals.

Before electrochemical studies, the residual organic adsorbates must be removed from the Au nanocrystal surfaces. This cleaning step is crucial for catalytic applications in order to obtain true shape-dependent activity [38–40]. Several methods have been developed to clean nanocrystal surfaces. However, the ozone removal method has been confirmed to cause severe surface structure distortion due to its high oxidation power [41]. Although carbon monoxide (CO) has been used for effective cleaning of surfactants on Pd nanocrystals [24,42], it does not adsorb strongly on Au surfaces [43,44]. An electrochemical fast potential cycling in a strong basic solution was used here for cleaning these Au nanocrystals, which is similar to the surfactant removal methods reported by others [39,45]. In this cleaning procedure, the as-prepared Au nanocrystals supported on a glassy carbon electrode were scanned in the potential range of −0.70 to +0.70 V versus Ag/AgCl at 0.5 V s^{-1} in 0.1 M NaOH for about 50 cycles until stable CVs, such as those in Figure 3a, were obtained. Before the electrochemical potential cycling treatment, the surface oxidation/reduction charge density was small and the surface oxidation current rise was sluggish as exemplified by the CV from Au nanocubes in Figure 3a (dotted trace), indicating that, even after several times of washing with water, some surfactants remained on the crystal surfaces. After the potential cycling, the surface oxidation current increased sharply between 0.2 and 0.35 V, and the surface oxidation/reduction current peaks showed facet-dependent features. The RD Au nanocrystals showed a surface oxidation peak at around 0.33 V, while on cubes it appeared at 0.4 V and two peaks located at 0.38 and 0.44 V were observed on the octahedra. On the cathodic scan, the surface oxide reduction peak of octahedra appeared at ca. 0.15 V, which is sharper and more positive than those from the cubes and RD. The adsorption peak observed around 0.0 V on the bulk Au (100) single crystal surface, which is a signature of surface reconstruction [46], was not seen on the cube surfaces, suggesting that surface reconstruction does not occur on the nanocube surfaces.

Facet-dependent voltammogram features were also observed in 0.1 M H_2SO_4 (Figure 3b), albeit the differences are more subtle than in the alkaline solutions. These voltammetric features are similar to those observed on bulk Au (111), Au (100), and Au (110) single crystal surfaces [46–48], further confirming that the nanocrystals are enclosed by the three low index facets.

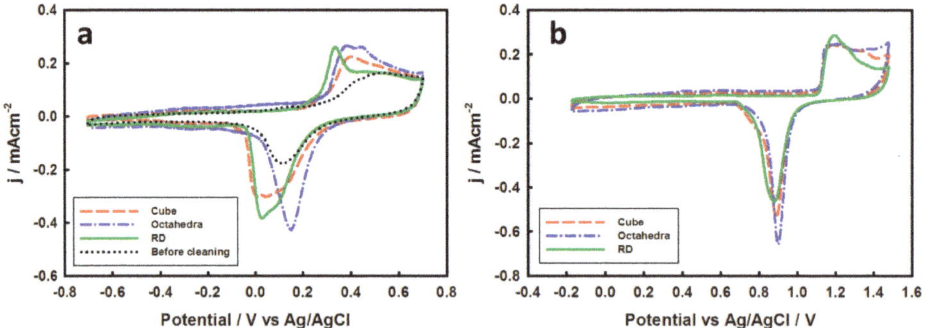

Figure 3. Cyclic voltammograms (CVs) of cubic, octahedral, and RD Au nanocrystals in (**a**) 0.1 M NaOH (**b**) 0.1 M H_2SO_4. The dotted line in (**a**) is the CV of cubic Au nanocrystals before the potential cycling process. Scan rate: $0.100\ \text{V s}^{-1}$.

3.2. Electrochemical Properties of Pd-Monolayer-Coated Au Nanocrystals

As stated in the outset, a central task in this study is to examine whether Pd thin films coated on Au nanocrystals have facet-dependent electrochemical properties, especially formic acid oxidation activity, that mimic those of bulk Pd single crystals or Pd thin film coated bulk Au single crystals [24,33,49]. After electrochemical cleaning in 0.1 M NaOH, the Au nanocrystals were coated with a layer of Pd film by the surface limited redox replacement approach, which nominally deposits one atomic layer of Pd in one deposition-replacement cycle. Figure 4a shows the CVs of different Au nanocrystals coated with a monolayer of Pd recorded in 0.1 M $HClO_4$. Different nanocrystals exhibited different electrochemical features both in the hydrogen adsorption/desorption and Pd oxidation/reduction regions. The Pd monolayer on cubic Au nanocrystals showed hydrogen adsorption/desorption peaks at around 0.0 V, while these peaks appeared at around −0.10 V on RD nanocrystals. A clear shoulder at ca. 0.0 V on the cathodic scan was also observed on the latter nanocrystals. These observations agree with those reported on Pd cubic and RD nanocrystals, respectively [24]. On Pd-monolayer-coated octahedral Au nanocrystals, the hydrogen adsorption/desorption peaks were more reversible than the other two crystals, in accordance with that on bulk Pd (111) single crystal electrodes [33]. At potentials above 0.40 V, a pair of redox peaks appeared at around 0.5 V on Pd-monolayer-coated Au cubes and RDs, which can be assigned to the surface oxidation/reduction of Pd on the basis of similar observations on bulk Pd (100) and (110) surfaces [33]. Interestingly, on Pd-monolayer-coated Au octahedra, the surface oxidation/reduction peaks are much smaller and less reversible, which also agree with the results from bulk Pd (111) single crystals [33,50–52], and can be explained by the relatively inert nature of the Pd {111} facet that has a much higher oxidation potential compared to the Pd {100} and Pd {110} facets [33,49]. The CO stripping voltammograms from monolayer-Pd-coated Au nanocrystals displayed in Figure 4b also show facet-dependent features. The CO oxidation peak potential for the cubic shape nanocrystals is about 50 mV more negative than the other two nanocrystals, which has been observed on Pd (100) bulk single crystal electrodes [33]. In addition to the main CO oxidation peak at 0.75 V, the octahedral nanocrystals show an additional peak at 0.9 V, which has also been observed on a bulk Pd (111) single crystal electrode [51]. The slanted voltammogram observed on the octahedra is from the background current of the supporting GC electrode and only discernable when the nanocrystal coverage is low. The similarities of the observed features in CVs and CO stripping voltammograms between Pd-monolayer-coated Au nanocrystals and the bulk single crystal Pd electrodes/Pd nanocrystals suggest the epitaxial growth of a Pd monolayer on Au nanocrystals.

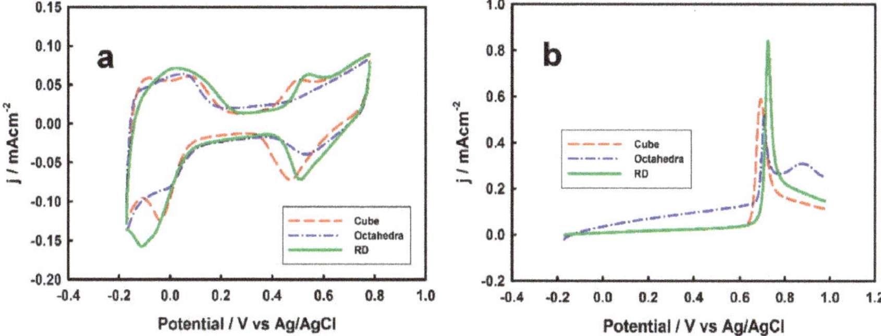

Figure 4. (a) CVs; (b) CO stripping voltammograms of cubic, octahedral, and RD Au nanocrystals coated with one monolayer of Pd recorded in 0.1 M HClO$_4$. Scan rate: 0.100 V s^{-1}.

Figure 5a shows CVs of FAO on Pd-monolayer-coated Au nanocrystals obtained in 0.5 M HCOOH and 0.1 M HClO$_4$. The FAO on these Pd monolayers exhibits shape-dependent features similar to the bulk Pd single crystal electrodes [23] and corresponding Pd nanocrystals [24,26,53]. The oxidation of formic acid on different Pd-coated Au nanocrystals started at around −0.10 V, but the monolayer Pd deposited on cubic Au nanocrystals showed the highest oxidation peak current density at about 0.4 V, which is about double that of the other two nanocrystals. On the bulk electrodes, the formic acid oxidation current density on Pd (100) is about 4 times of the other two low-index single crystal electrodes in H$_2$SO$_4$ [7], but about double that of Pd (111) in HClO$_4$ [23]. In addition, the CVs of Pd-monolayer-coated octahedral Au nanocrystals show similar behavior at the beginning of formic acid oxidation with cubic nanocrystals, but it reaches its maximum current at around 0.3 V, which is much more negative than that of cubic nanocrystals. The RD nanocrystals show a broader maximum current density peak at the potential range of 0.2~0.6 V. These results are consistent with the reported formic acid oxidation on different bulk Pd single crystal electrodes, in which Pd (111) showed a more negative oxidation peak potential than the other two surfaces, Pd (100) had the highest peak current density, and Pd (110) presented a broader peak [7,23]. It is interesting to note that CV features of FAO resembling those from the bulk Pd single crystals were only observed after 2 MLs of Pd deposited on the corresponding bulk Au single crystals [7]. The CVs of formic acid oxidation from these Pd-ML-coated Au nanocrystals also resemble those obtained on the corresponding Pd nanocrystals. Thus, the current density is the highest on cubic Pd nanocrystals and the peak potential is the most negative on Pd octahedra [24,26,53]. The anodic and cathodic potential scans for these Pd-ML-coated Au nanocrystals show nearly identical current densities, which signify the absence of CO poisoning on the time scale of the CV measurements.

To further compare the catalytic activity of Pd monolayers on different Au nanocrystals, formic acid oxidation current densities at 0.38 V (around the peak potential of the cubic particles) and 0.08 V read from the CVs in Figure 5a were plotted in Figure 5c,d. The plots reveal potential dependent catalytic activities of the Pd monolayers. At near peak potential (0.38 V), the Pd monolayer on cubic Au nanocrystals has the highest catalytic activity, which is more than 2 times of the other two nanocrystals. The Pd monolayer on RD nanocrystals is only slightly higher than that of octahedra. At the lower potential (0.08 V), however, the catalytic activity of the Pd monolayer deposited on RD nanocrystals is 1.5 times of the octahedral and cubic nanocrystals (Figure 5d). Moreover, the Pd monolayer on RD nanocrystals shows a better durability with 50% loss of the initial current density at 1000 s, in contrast to 70% in cubic and 90% in octahedral nanocrystals, as shown by the chronoamperometric curves (CAs) in Figure 5b. The higher durability on Pd-coated RD suggests that the surface poisoning is less prominent on this surface at this potential. The surface poisoning species cannot be identified based solely on these results. It has been reported that, at this potential, adsorbed CO can be formed on Pd black and Pd-coated Au electrodes through electroreduction of CO_2 [21,54,55], which is a major product of FAO. It is therefore conceivable that accumulation of adsorbed CO may be responsible for the decay of the activity over time and the CO_2 reduction is structure-sensitive.

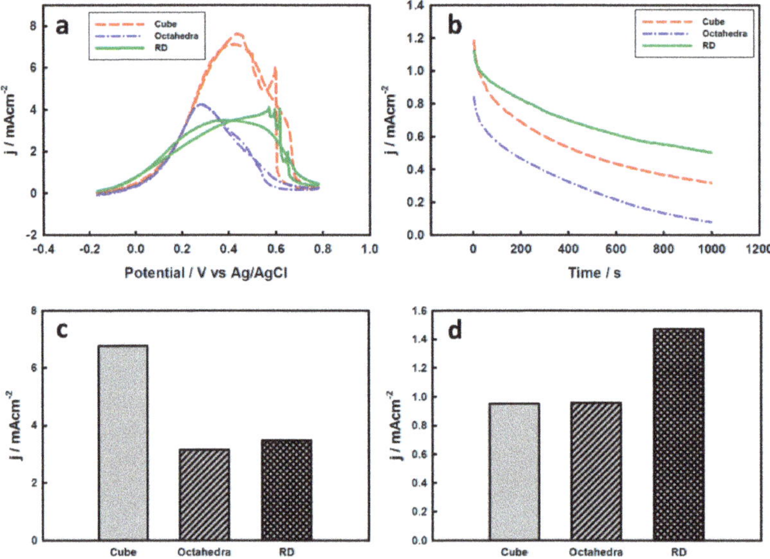

Figure 5. (**a**). CVs of cubic, octahedral, and RD Au nanocrystals coated with one monolayer of Pd recorded in 0.5 M HCOOH + 0.1 M $HClO_4$, scan rate: 0.100 V s^{-1}. (**b**). Corresponding chronoamperometric curves recorded in the same solution at 0.08 V with an initial potential set at -0.10 V. (**c,d**). Comparison of catalytic activities of Pd monolayer at (**c**) 0.38 V and (**d**) 0.08 V.

3.3. Thickness-Dependent Formic Acid Oxidation on Pd Thin Layer Coated Au Nanocrystals

The layer-by-layer growth nature of the redox replacement method for preparing the Pd films allows for the examination of film-thickness-dependent catalytic activity. We examined FAO on Au nanocrystals coated successively with up to 10 MLs of Pd. CO stripping voltammograms were recorded after the formic acid oxidation experiments to assess the electrochemically active surface area and examine Pd film thickness dependence of the CO oxidation. Figure 6 displays thickness-dependent formic acid oxidation activity in terms of current density at 0.38 V in CVs. To facilitate the comparison, the current density was normalized to that of the first layer. The error bars represent the standard deviations of results from three separately prepared Pd films. For each nanocrystal, the highest activity was observed at the first monolayer of Pd. The thickness-dependent activity of all of the nanocrystals examined followed the same trend. As the number of Pd layers increased, the formic acid oxidation activity decreased, and became largely unchanged after five monolayers of Pd. Kibler and Kolb did not observe a clear trend of thickness-dependent FAO on Pd films coated on bulk Au single crystal surfaces, but showed that the FAO current density on the first layer of Pd was in general much lower than the thicker films [7]. Similarly, Obradović and Gojković showed that the first monolayer of Pd film coated on a polycrystalline Au electrode had the lowest FAO activity, but after four monolayers of Pd, the activity remained largely the same [21]. The distinction of the first layer of Pd in the activity series between these previous studies and the present work may come from the different deposition method employed and the use of H_2SO_4 in their studies (vide infra).

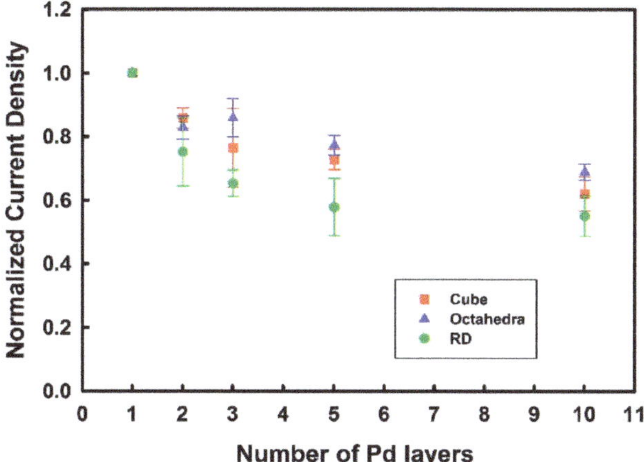

Figure 6. Normalized formic acid oxidation current density at 0.38 V as a function of the Pd film thickness measured in 0.5 M HCOOH + 0.1 M HClO$_4$. The error bars represent the standard deviations of results from three separately prepared Pd films.

Two effects are likely contribute to the thickness-dependent catalytic activity of these Pd overlayers on Au nanocrystals. First, the lattice constant of Au is 5% larger than that of Pd. Pseudomorphic growth of Pd on Au results in an expansion of Pd lattice, which increases the d-band center and therefore the adsorption energy of adsorbates [56]. This is the so-called strain effect. Second, the electronic interaction between two different metals results in electronic effect (the so-called ligand effect), which will cause the energy renormalization of the d-band [57]. A stronger interaction between the overlayer and the substrate atoms will result in an upshift of the d-band center of the overlayer [58]. Decoupling the two effects experimentally is challenging. In density functional theory (DFT) calculations, the two effects could be separated by artificially enlarging the lattice constant of the Pd top layer to account for the tensile strain while changing the underneath substrate from Au to Pd to eliminate the ligand effect [19,59]. All of the calculations show that the tensile strain elevates the d-band center and plays a dominant role [19,56,59,60], but there is no consensus on the ligand effect. Roudgar and Groß predicted that the ligand effect further increases the d-band center position [59], while others showed the opposite [19]. Nevertheless, the net result is an upshifted d-band center for Pd deposited on Au. As the number of Pd layers on Au nanocrystals increases, the lattice constant gradually relaxes back to the value of pure Pd, which has been demonstrated by both scanning tunneling microscopy and surface X-ray scattering [61–64]. The lattice relaxation releases the strain effect. Meanwhile, as the film thickness increases, the electronic interaction between the Pd top layer and the Au substrate becomes weaker and therefore the ligand effect diminishes. Our formic acid oxidation results are in general agreement with these arguments.

The thickness-dependent catalytic activity is also manifested in CO oxidation. Figure 7 displays CO stripping onset potential as a function of Pd overlayer thickness obtained in 0.1 M HClO$_4$. The onset potential here was taken as the potential where CO oxidation current is at 10% of the peak current. In the case where there are multiple peaks, the most negative peak was used. All of the three nanocrystals show a negative potential shift of CO oxidation onset potential when more than one ML of Pd was deposited. After three layers of Pd, the onset potential of CO oxidation remains largely the same. The RD nanocrystals coated with two Pd MLs showed a smaller negative shift compared with the other two Au nanocrystals. The trend of CO oxidation onset potential shift is consistent with catalytic activities shown in Figure 4 and can again be explained by the diminishing strain and ligand effects with increasing film thickness. The thickness-dependent film growth mode may be revealed from the CO stripping voltammograms (Figure S4). As the Pd film thickness increases, one obvious feature of CO oxidation is the appearance of a second oxidation peak at around 0.9 V. This second oxidation peak was observed on octahedral nanocrystals when only one ML of Pd was deposited, but after the deposition of five MLs of Pd it can be clearly seen in all three nanocrystals, as shown in Figure S4. The second CO oxidation peak does not exist on the bulk Pd (100) and Pd (110) electrodes [50,52]. This peak might come from the defect sites formed through three-dimensional (3D) growth on the thicker Pd films, which may play a minor role in the thickness-dependent FAO activity given its smaller magnitude on the cubes and RDs.

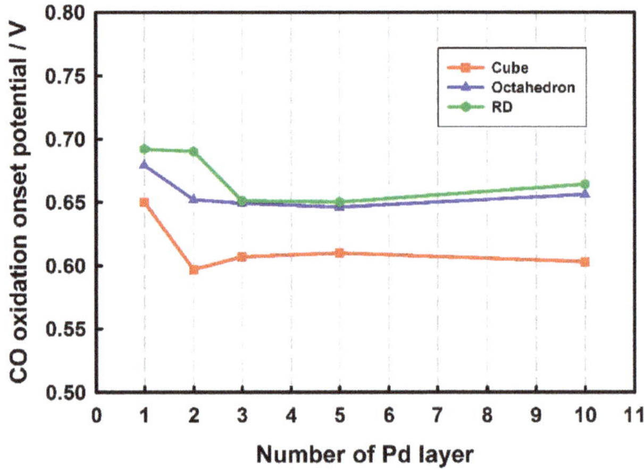

Figure 7. CO oxidation onset potential as a function of the Pd film thickness on different Au nanocrystals in 0.1 M HClO$_4$.

It has been demonstrated that formic acid oxidation rates are higher in HClO$_4$ than in H$_2$SO$_4$ [7,23,26,65], because the strong adsorption of (bi)sulfate anion on Pd surfaces above 0.2 V (versus RHE) [66,67] hinders the formic acid oxidation in H$_2$SO$_4$. It is therefore interesting to examine whether the thickness-dependent adsorption will manifest itself in (bi)sulfate adsorption and hence influences formic acid oxidation. Because of their higher catalytic activity of formic acid oxidation, Pd-film-coated cubic Au nanocrystals were chosen for this study. Figure 8 displays the formic acid oxidation current density at 0.38 V as a function of film thickness in 0.1 M HClO$_4$ or 0.1 M H$_2$SO$_4$ with 0.5 M formic acid. As expected, formic acid oxidation is more facile in 0.1 M HClO$_4$ than in 0.1 M H$_2$SO$_4$ due to the adsorption of (bi)sulfate. Interestingly, the activity disparities in the two electrolytes gradually decreased as the film thickness increased. On the first monolayer of Pd, the current density of FAO in 0.1 M HClO$_4$ is 1.6 times of that in 0.1 M H$_2$SO$_4$, but it decreased to only 1.2 times on 10 monolayers of Pd. This observation can be explained by thickness-dependent (bi)sulfate adsorption on Pd film surfaces. As discussed above, the d-band center upshift is the highest on the first monolayer of Pd deposited on Au nanocubes. As a result, the adsorption energy of (bi)sulfate is the strongest. The strong adsorption of (bi)sulfate anions competes with formate adsorption, and results in a lower catalytic activity compared to the nonspecifically adsorbed supporting electrolytes, such as HClO$_4$. This effect decreases as the thickness of the Pd film increases because the d-band center gradually returns to the value of bulk Pd. While the d-band upshift also increases the adsorption energy of formate, the effect is likely stronger on the (bi)sulfate adsorption. Given that (bi)sulfate adsorption even competes with CO adsorption on Pd [33], this assertion is not unreasonable.

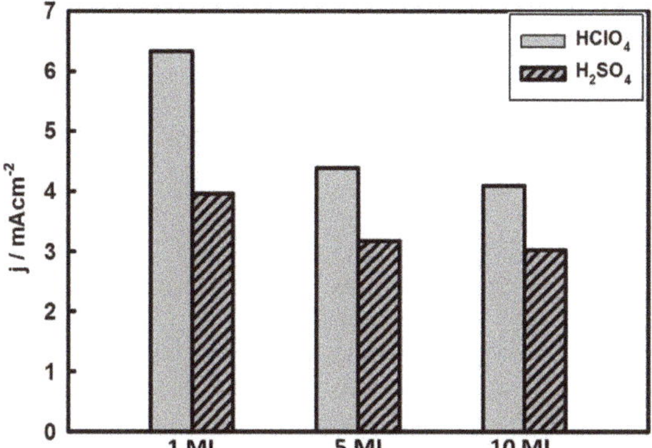

Figure 8. Formic acid oxidation current density at 0.38 V in 0.5 M HCOOH + 0.1 M $HClO_4$ or H_2SO_4 from cubic Au nanocrystals coated with Pd films of different thicknesses.

4. Conclusions

In summary, Pd-monolayer-covered high-quality Au nanocrystals showed facet-dependent formic acid and carbon monoxide oxidation activities that resemble those observed on bulk Pd single crystal surfaces and Pd thin film covered bulk Au single crystal surfaces, suggesting pseudomorphic growth of Pd MLs on these Au nanocrystals. As the Pd film thickness increased, the formic acid oxidation peak current density decreased and the decrement leveled off after about five MLs of Pd. Similarly, the CO oxidation onset potential also gradually decreased with the increase of Pd film thickness and remained largely unchanged after three MLs of Pd. The effects of bi(sulfate) specific adsorption on FAO also decreased with the increasing Pd film thickness. These observations are explained in terms of the d-band center shift with the Pd film thickness as a result of the tensile strain effect and the ligand effect. This study demonstrates that, by combining the facet dependence of catalytic activity with the fine-tuning of d-band position, deposition of catalytically active materials as a thin film on facet-controlled nanocrystals can be an effective approach for engineering high-performance catalysts for a variety of reactions.

Supplementary Materials: The following are available online at http://www.mdpi.com/2571-9637/2/2/27/s1, Figure S1: SEM image of Au seed; Figure S2: UV-Vis spectra of Au nanocrystals; Figure S3: SEM images of larger Au nanocrystals, and Figure S4: CO stripping voltammograms of Au nanocrystals coated with different layers of Pd films.

Author Contributions: Conceptualization, S.Z.; Data curation, Y.T.; Formal analysis, Y.T. and S.Z.; Funding acquisition, S.Z.; Methodology, Y.T.; Project administration, S.Z.; Supervision, S.Z.; Validation, Y.T.; Writing—original draft, Y.T.; Writing—review & editing, S.Z.

Acknowledgments: This work was partially supported by the US National Science Foundation through CHE-1156425 and CHE-1559670.

Conflicts of Interest: The authors declare no conflicts of interest.

References

1. Jeong, K.-J.; Miesse, C.M.; Choi, J.-H.; Lee, J.; Han, J.; Yoon, S.P.; Nam, S.W.; Lim, T.-H.; Lee, T.G. Fuel Crossover in Direct Formic Acid Fuel Cells. *J. Power Sources* **2007**, *168*, 119–125. [CrossRef]
2. Jiang, K.; Zhang, H.-X.; Zou, S.; Cai, W.-B. Electrocatalysis of Formic Acid on Palladium and Platinum Surfaces: From Fundamental Mechanisms to Fuel Cell Applications. *Phys. Chem. Chem. Phys.* **2014**, *16*, 20360–20376. [CrossRef] [PubMed]
3. Uhm, S.; Lee, H.J.; Lee, J. Understanding Underlying Processes in Formic Acid Fuel Cells. *Phys. Chem. Chem. Phys.* **2009**, *11*, 9326–9336. [CrossRef]

4.   Wang, X.; Hu, J.-M.; Hsing, I.M. Electrochemical Investigation of Formic Acid Electro-Oxidation and Its Crossover through a Nafion® Membrane. *J. Electroanal. Chem.* **2004**, *562*, 73–80. [CrossRef]

5.   Yu, X.W.; Pickup, P.G. Recent Advances in Direct Formic Acid Fuel Cells (Dfafc). *J. Power Sources* **2008**, *182*, 124–132. [CrossRef]

6.   Capon, A.; Parsons, R. The Oxidation of Formic Acid at Noble Metal Electrodes Part Iii. Intermediates and Mechanism on Platinum Electrodes. *J. Electroanal. Chem. Interfacial Electrochem.* **1973**, *45*, 205–231. [CrossRef]

7.   Baldauf, M.; Kolb, D.M. Formic Acid Oxidation on Ultrathin Pd Films on Au(hkl) and Pt(hkl) Electrodes. *J. Phys. Chem.* **1996**, *100*, 11375–11381. [CrossRef]

8.   Ha, S.; Larsen, R.; Masel, R.I. Performance Characterization of Pd/C Nanocatalyst for Direct Formic Acid Fuel Cells. *J. Power Sources* **2005**, *144*, 28–34. [CrossRef]

9.   Zhu, Y.; Khan, Z.; Masel, R.I. The Behavior of Palladium Catalysts in Direct Formic Acid Fuel Cells. *J. Power Sources* **2005**, *139*, 15–20. [CrossRef]

10. Solla-Gullon, J.; Montiel, V.; Aldaz, A.; Clavilier, J. Electrochemical and Electrocatalytic Behavior of Platinum–Palladium Nanoparticle Alloys. *Electrochem. Commun.* **2002**, *4*, 716–721. [CrossRef]

11. Rand, D.A.J.; Woods, R. Determination of the Surface Composition of Smooth Noble Metal Alloys by Cyclic Voltammetry. *J. Electroanal. Chem. Interfacial Electrochem.* **1972**, *36*, 57–69. [CrossRef]

12. Antolini, E. Palladium in Fuel Cell Catalysis. *Energy Environ. Sci.* **2009**, *2*, 915–931. [CrossRef]

13. Zhou, W.; Lee, J.Y. Highly Active Core–Shell Au@Pd Catalyst for Formic Acid Electrooxidation. *Electrochem. Commun.* **2007**, *9*, 1725–1729. [CrossRef]

14. Liu, Y.; Wang, L.; Wang, G.; Deng, C.; Wu, B.; Gao, Y. High Active Carbon Supported PdAu Catalyst for Formic Acid Electrooxidation and Study of the Kinetics. *J. Phys. Chem. C* **2010**, *114*, 21417–21422. [CrossRef]

15. Zhang, G.; Wang, Y.; Wang, X.; Chen, Y.; Zhou, Y.; Tang, Y.; Lu, L.; Bao, J.; Lu, T. Preparation of Pd–Au/C Catalysts with Different Alloying Degree and Their Electrocatalytic Performance for Formic Acid Oxidation. *Appl. Catal. B Environ.* **2011**, *102*, 614–619. [CrossRef]

16. Lee, S.Y.; Jung, N.; Cho, J.; Park, H.Y.; Ryu, J.; Jang, I.; Kim, H.J.; Cho, E.; Park, Y.H.; Ham, H.C.; et al. Surface-Rearranged Pd$_3$Au/C Nanocatalysts by Using Co-Induced Segregation for Formic Acid Oxidation Reactions. *ACS Catal.* **2014**, *4*, 2402–2408. [CrossRef]

17. Hong, J.W.; Kim, D.; Lee, Y.W.; Kim, M.; Kang, S.W.; Han, S.W. Atomic-Distribution-Dependent Electrocatalytic Activity of Au–Pd Bimetallic Nanocrystals. *Angew. Chem. Int. Ed.* **2011**, *50*, 8876–8880. [CrossRef] [PubMed]

18. Montes de Oca, M.G.; Plana, D.; Celorrio, V.; Lazaro, M.J.; Fermín, D.J. Electrocatalytic Properties of Strained Pd Nanoshells at Au Nanostructures: CO and HCOOH Oxidation. *J. Phys. Chem. C* **2012**, *116*, 692–699. [CrossRef]

19. Celorrio, V.N.; Quaino, P.M.; Santos, E.; Flórez-Montaño, J.; Humphrey, J.J.L.; Guillén-Villafuerte, O.; Plana, D.; Lázaro, M.J.; Pastor, E.; Fermín, D.J. Strain Effects on the Oxidation of CO and HCOOH on Au–Pd Core–Shell Nanoparticles. *ACS Catal.* **2017**, *7*, 1673–1680. [CrossRef]

20. Hsu, C.; Huang, C.; Hao, Y.; Liu, F. Au/Pd Core–Shell Nanoparticles for Enhanced Electrocatalytic Activity and Durability. *Electrochem. Commun.* **2012**, *23*, 133–136. [CrossRef]

21. Obradovic, M.D.; Gojkovic, S.L. HCOOH Oxidation on Thin Pd Layers on Au: Self-Poisoning by the Subsequent Reaction of the Reaction Product. *Electrochim. Acta* **2013**, *88*, 384–389. [CrossRef]

22. Kibler, L.A.; El-Aziz, A.M.; Hoyer, R.; Kolb, D.M. Tuning Reaction Rates by Lateral Strain in a Palladium Monolayer. *Angew. Chem. Int. Ed.* **2005**, *44*, 2080–2084. [CrossRef] [PubMed]

23. Hoshi, N.; Kida, K.; Nakamura, M.; Nakada, M.; Osada, K. Structural Effects of Electrochemical Oxidation of Formic Acid on Single Crystal Electrodes of Palladium. *J. Phys. Chem. B* **2006**, *110*, 12480–12484. [CrossRef] [PubMed]

24. Zhang, H.X.; Wang, H.; Re, Y.S.; Cai, W.B. Palladium Nanocrystals Bound by {110} or {100} Facets: From One Pot Synthesis to Electrochemistry. *Chem. Commun.* **2012**, *48*, 8362–8364. [CrossRef] [PubMed]

25. Zhang, J.F.; Feng, C.; Deng, Y.D.; Liu, L.; Wu, Y.T.; Shen, B.; Zhong, C.; Hu, W.B. Shape-Controlled Synthesis of Palladium Single-Crystalline Nanoparticles: The Effect of HCl Oxidative Etching and Facet-Dependent Catalytic Properties. *Chem. Mater.* **2014**, *26*, 1213–1218. [CrossRef]

26. Zhang, X.W.; Yin, H.J.; Wang, J.F.; Chang, L.; Gao, Y.; Liu, W.; Tang, Z.Y. Shape-Dependent Electrocatalytic Activity of Monodispersed Palladium Nanocrystals toward Formic Acid Oxidation. *Nanoscale* **2013**, *5*, 8392–8397. [CrossRef]

27. Brankovic, S.R.; Wang, J.X.; Adzic, R.R. Metal Monolayer Deposition by Replacement of Metal Adlayers on Electrode Surfaces. *Surf. Sci.* **2001**, *474*, 173–179. [CrossRef]

28. Sheridan, L.B.; Kim, Y.-G.; Perdue, B.R.; Jagannathan, K.; Stickney, J.L.; Robinson, D.B. Hydrogen Adsorption, Absorption, and Desorption at Palladium Nanofilms Formed on Au(111) by Electrochemical Atomic Layer Deposition (E-ALD): Studies Using Voltammetry and in Situ Scanning Tunneling Microscopy. *J. Phys. Chem. C* **2013**, *117*, 15728–15740. [CrossRef]

29. Niu, W.; Zheng, S.; Wang, D.; Liu, X.; Li, H.; Han, S.; Chen, J.; Tang, Z.; Xu, G. Selective Synthesis of Single-Crystalline Rhombic Dodecahedral, Octahedral, and Cubic Gold Nanocrystals. *J. Am. Chem. Soc.* **2008**, *131*, 697–703. [CrossRef]

30. Nikoobakht, B.; El-Sayed, M.A. Preparation and Growth Mechanism of Gold Nanorods (NRs) Using Seed-Mediated Growth Method. *Chem. Mater.* **2003**, *15*, 1957–1962. [CrossRef]

31. Sau, T.K.; Murphy, C.J. Seeded High Yield Synthesis of Short Au Nanorods in Aqueous Solution. *Langmuir* **2004**, *20*, 6414–6420. [CrossRef] [PubMed]

32. Rodriguez-Fernandez, J.; Perez-Juste, J.; Mulvaney, P.; Liz-Marzan, L.M. Spatially-Directed Oxidation of Gold Nanoparticles by Au(III)-CTAB Complexes. *J. Phys. Chem. B* **2005**, *109*, 14257–14261. [CrossRef] [PubMed]

33. Hara, M.; Linke, U.; Wandlowski, T. Preparation and Electrochemical Characterization of Palladium Single Crystal Electrodes in 0.1 M H_2SO_4 and $HClO_4$ Part I. Low-Index Phases. *Electrochim. Acta* **2007**, *52*, 5733–5748. [CrossRef]

34. Kim, F.; Connor, S.; Song, H.; Kuykendall, T.; Yang, P.D. Platonic Gold Nanocrystals. *Angew. Chem. Int. Ed.* **2004**, *43*, 3673–3677. [CrossRef]

35. Seo, D.; Park, J.C.; Song, H. Polyhedral Gold Nanocrystals with O-H Symmetry: From Octahedra to Cubes. *J. Am. Chem. Soc.* **2006**, *128*, 14863–14870. [CrossRef]

36. Chen, Y.; Milenkovic, S.; Hassel, A.W. Reactivity of Gold Nanobelts with Unique {110} Facets. *ChemPhysChem* **2010**, *11*, 2838–2843. [CrossRef] [PubMed]

37. Lu, F.; Zhang, Y.; Zhang, L.; Zhang, Y.; Wang, J.X.; Adzic, R.R.; Stach, E.A.; Gang, O. Truncated Ditetragonal Gold Prisms as Nanofacet Activators of Catalytic Platinum. *J. Am. Chem. Soc.* **2011**, *133*, 18074–18077. [CrossRef] [PubMed]

38. Tian, N.; Zhou, Z.Y.; Yu, N.F.; Wang, L.Y.; Sun, S.G. Direct Electrodeposition of Tetrahexahedral Pd Nanocrystals with High-Index Facets and High Catalytic Activity for Ethanol Electrooxidation. *J. Am. Chem. Soc.* **2010**, *132*, 7580–7581. [CrossRef] [PubMed]

39. Vidal-Iglesias, F.J.; Aran-Ais, R.M.; Solla-Gullon, J.; Herrero, E.; Feliu, J.M. Electrochemical Characterization of Shape-Controlled Pt Nanoparticles in Different Supporting Electrolytes. *ACS Catal.* **2012**, *2*, 901–910. [CrossRef]

40. Yang, H.; Tang, Y.; Zou, S. Electrochemical Removal of Surfactants from Pt Nanocubes. *Electrochem. Commun.* **2014**, *38*, 134–137. [CrossRef]

41. Vidal-Iglesias, F.J.; Solla-Gullon, J.; Herrero, E.; Montiel, V.; Aldaz, A.; Feliu, J.M. Evaluating the Ozone Cleaning Treatment in Shape-Controlled Pt Nanoparticles: Evidences of Atomic Surface Disordering. *Electrochem. Commun.* **2011**, *13*, 502–505. [CrossRef]

42. Tang, Y.; Edelmann, R.E.; Zou, S. Length Tunable Penta-Twinned Palladium Nanorods: Seedless Synthesis and Electrooxidation of Formic Acid. *Nanoscale* **2014**, *6*, 5630–5633. [CrossRef]

43. Chang, S.C.; Hamelin, A.; Weaver, M.J. Reactive and Inhibiting Adsorbates for the Catalytic Electrooxidation of Carbon-Monoxide on Gold (210) as Characterized by Surface Infrared-Spectroscopy. *Surf. Sci.* **1990**, *239*, L543–L547. [CrossRef]

44. Chang, S.C.; Hamelin, A.; Weaver, M.J. Dependence of the Electrooxidation Rates of Carbon-Monoxide at Gold on the Surface Crystallographic Orientation—A Combined Kinetic-Surface Infrared Spectroscopic Study. *J. Phys. Chem.* **1991**, *95*, 5560–5567. [CrossRef]

45. Lee, Y.; Loew, A.; Sun, S.H. Surface- and Structure-Dependent Catalytic Activity of Au Nanoparticles for Oxygen Reduction Reaction. *Chem. Mater.* **2010**, *22*, 755–761. [CrossRef]

46. Hamelin, A.; Sottomayor, M.J.; Silva, F.; Chang, S.-C.; Weaver, M.J. Cyclic Voltammetric Characterization of Oriented Monocrystalline Gold Surfaces in Aqueous Alkaline Solution. *J. Electroanal. Chem. Interfacial Electrochem.* **1990**, *295*, 291–300. [CrossRef]

47. Hamelin, A. Cyclic Voltammetry at Gold Single-Crystal Surfaces. Part 1. Behaviour at Low-Index Faces. *J. Electroanal. Chem.* **1996**, *407*, 1–11. [CrossRef]

48. Schmidt, T.J.; Stamenkovic, V.; Arenz, M.; Markovic, N.M.; Ross, P.N., Jr. Oxygen Electrocatalysis in Alkaline Electrolyte: Pt(hkl), Au(hkl) and the Effect of Pd-Modification. *Electrochim. Acta* **2002**, *47*, 3765–3776. [CrossRef]

49. Sashikata, K.; Matsui, Y.; Itaya, K.; Soriaga, M.P. Adsorbed-Iodine-Catalyzed Dissolution of Pd Single-Crystal Electrodes: Studies by Electrochemical Scanning Tunneling Microscopy. *J. Phys. Chem.* **1996**, *100*, 20027–20034. [CrossRef]

50. Zou, S.Z.; Gomez, R.; Weaver, M.J. Infrared Spectroscopy of Carbon Monoxide at the Ordered Palladium (110)-Aqueous Interface: Evidence for Adsorbate-Induced Surface Reconstruction. *Surf. Sci.* **1998**, *399*, 270–283. [CrossRef]

51. Zou, S.Z.; Gomez, R.; Weaver, M.J. Infrared Spectroscopy of Carbon Monoxide and Nitric Oxide on Palladium(111) in Aqueous Solution: Unexpected Adlayer Structural Differences between Electrochemical and Ultrahigh-Vacuum Interfaces. *J. Electroanal. Chem.* **1999**, *474*, 155–166. [CrossRef]

52. Zou, S.Z.; Gomez, R.; Weaver, M.J. Coverage-Dependent Infrared Spectroscopy of Carbon Monoxide on Palladium(100) in Aqueous Solution: Adlayer Phase Transitions and Electrooxidation Pathways. *Langmuir* **1999**, *15*, 2931–2939. [CrossRef]

53. Jin, M.S.; Zhang, H.; Xie, Z.X.; Xia, Y.N. Palladium Nanocrystals Enclosed by {100} and {111} Facets in Controlled Proportions and Their Catalytic Activities for Formic Acid Oxidation. *Energy Environ. Sci.* **2012**, *5*, 6352–6357. [CrossRef]

54. Wang, J.-Y.; Zhang, H.-X.; Jiang, K.; Cai, W.-B. From HCOOH to CO at Pd Electrodes: A Surface-Enhanced Infraredspectroscopy Study. *J. Am. Chem. Soc.* **2011**, *133*, 14876–14879. [CrossRef]

55. Zhang, H.-X.; Wang, S.-H.; Jiang, K.; André, T.; Cai, W.-B. In Situ Spectroscopic Investigation of CO Accumulation and Poisoning on Pd Black Surfaces in Concentrated HCOOH. *J. Power Sources* **2012**, *199*, 165–169. [CrossRef]

56. Ruban, A.; Hammer, B.; Stoltze, P.; Skriver, H.L.; Nørskov, J.K. Surface Electronic Structure and Reactivity of Transition and Noble Metals. *J. Mol. Catal. A Chem.* **1997**, *115*, 421–429. [CrossRef]

57. Hammer, B.; Norskov, J.K. Electronic Factors Determining the Reactivity of Metal Surfaces. *Surf. Sci.* **1995**, *343*, 211–220. [CrossRef]

58. Mavrikakis, M.; Hammer, B.; Norskov, J.K. Effect of Strain on the Reactivity of Metal Surfaces. *Phys. Rev. Lett.* **1998**, *81*, 2819–2822. [CrossRef]

59. Roudgar, A.; Gross, A. Local Reactivity of Thin Pd Overlayers on Au Single Crystals. *J. Electroanal. Chem.* **2003**, *548*, 121–130. [CrossRef]

60. Shao, M.H.; Huang, T.; Liu, P.; Zhang, J.; Sasaki, K.; Vukmirovic, M.B.; Adzic, R.R. Palladium Monolayer and Palladium Alloy Electrocatalysts for Oxygen Reduction†. *Langmuir* **2006**, *22*, 10409–10415. [CrossRef]

61. Kibler, L.A.; Kleinert, M.; Kolb, D.M. Initial Stages of Pd Deposition on Au(hkl) Part II: Pd on Au(100). *Surf. Sci.* **2000**, *461*, 155–167. [CrossRef]

62. Kibler, L.A.; Kleinert, M.; Lazarescu, V.; Kolb, D.M. Initial Stages of Palladium Deposition on Au(hkl) Part III: Pd on Au(110). *Surf. Sci.* **2002**, *498*, 175–185. [CrossRef]

63. Kibler, L.A.; Kleinert, M.; Randler, R.; Kolb, D.M. Initial Stages of Pd Deposition on Au(hkl)—Part I: Pd on Au(111). *Surf. Sci.* **1999**, *443*, 19–30. [CrossRef]

64. Takahasi, M.; Tamura, K.; Mizuki, J.; Kondo, T.; Uosaki, K. Orientation Dependence of Pd Growth on Au Electrode Surfaces. *J. Phys. Condens. Matter.* **2010**, *22*, 474002. [CrossRef]

65. Vidal-Iglesias, F.J.; Arán-Ais, R.M.; Solla-Gullón, J.; Garnier, E.; Herrero, E.; Aldaz, A.; Feliu, J.M. Shape-Dependent Electrocatalysis: Formic Acid Electrooxidation on Cubic Pd Nanoparticles. *Phys. Chem. Chem. Phys.* **2012**, *14*, 10258–10265. [CrossRef]

66. Hoshi, N.; Kuroda, M.; Koga, O.; Hori, Y. Infrared Reflection Absorption Spectroscopy of the Sulfuric Acid Anion on Low and High Index Planes of Palladium. *J. Phys. Chem. B* **2002**, *102*, 9107–9113. [CrossRef]

67. Pronkin, S.; Wandlowski, T. ATR-SEIRAS—An Approach to Probe the Reactivity of Pd-Modified Quasi-Single Crystal Gold Film Electrodes. *Surf. Sci.* **2004**, *573*, 109–127. [CrossRef]

Article

Sputtered Platinum Thin-films for Oxygen Reduction in Gas Diffusion Electrodes: A Model System for Studies under Realistic Reaction Conditions

Gustav W. Sievers [1,2,†], Anders W. Jensen [1,†], Volker Brüser [2], Matthias Arenz [3,*] and María Escudero-Escribano [1,*]

[1] Nano-Science Centre, Department of Chemistry, University of Copenhagen, Universitetsparken 5, 2100 Copenhagen Ø, Denmark; sievers@inp-greifswald.de (G.W.S.); awj@chem.ku.dk (A.W.J.)

[2] Leibniz Institute for Plasma Science and Technology, Felix-Hausdorff-Strasse 2, 17489 Greifswald, Germany; brueser@inp-greifswald.de

[3] Department of Chemistry and Biochemistry, University of Bern, 3006 Bern, Switzerland

[*] Correspondence: matthias.arenz@dcb.unibe.ch (M.A.); maria.escudero@chem.ku.dk (M.E.-E.);
Tel.: +41-31631-5384 (M.A.); +45-35-32-83-90 (M.E.-E.)

[†] These authors contributed equally to this work.

Received: 20 March 2019; Accepted: 24 April 2019; Published: 28 April 2019

Abstract: The development of catalysts for the oxygen reduction reaction in low-temperature fuel cells depends on efficient and accurate electrochemical characterization methods. Currently, two primary techniques exist: rotating disk electrode (RDE) measurements in half-cells with liquid electrolyte and single cell tests with membrane electrode assemblies (MEAs). While the RDE technique allows for rapid catalyst benchmarking, it is limited to electrode potentials far from operating fuel cells. On the other hand, MEAs can provide direct performance data at realistic conditions but require specialized equipment and large quantities of catalyst, making them less ideal for early-stage development. Using sputtered platinum thin-film electrodes, we show that gas diffusion electrode (GDE) half-cells can be used as an intermediate platform for rapid benchmarking at fuel-cell relevant current densities (~1 A cm^{-2}). Furthermore, we demonstrate how different parameters (loading, electrolyte concentration, humidification, and Nafion membrane) influence the performance of unsupported platinum catalysts. The specific activity could be measured independent of the applied loading at potentials down to 0.80 V_{RHE} reaching a value of 0.72 mA cm^{-2} at 0.9 V_{RHE} in the GDE. By comparison with RDE measurements and Pt/C measurements, we establish the importance of catalyst characterization under realistic reaction conditions.

Keywords: electrocatalysis; oxygen reduction; ORR; gas diffusion electrode; platinum; fuel cells; thin-films; benchmarking; mass transport

---

1. Introduction

Fuel cell technologies, which convert chemical energy directly into clean electricity, are expected to play a key role in environmentally friendly energy conversion schemes [1]. Polymer exchange membrane fuel cells (PEMFCs) are very promising both for transportation and stationary applications [2,3]. Platinum-based catalysts typically catalyze both the oxidation of hydrogen at the anode and the reduction of oxygen at the cathode of PEMFCs; as a result, the overall cost of the PEMFC technology is closely linked with the platinum loading [4]. The slow kinetics of the oxygen reduction reaction (ORR) causes significant losses in cell potential. Consequently, the loading at the cathode accounts for the majority of platinum usage in PEMFC. In order to reduce the Pt loading, most scientific and industrial research focuses on developing new ORR catalysts that present both high activity and stability for the ORR [5–7].

The ORR activity can be enhanced by modifying the geometric [8] or electronic structure [9] of the Pt-based electrocatalytic surfaces. To improve the catalyst mass activity, we can enhance: i) the electrochemically active surface area (ECSA) [10] and/or ii) the specific catalytic activity by modifying the electronic properties of platinum by alloying [11–13]. As a result, numerous catalyst concepts have been developed with activities far exceeding commercial Pt catalysts using thin-film rotating disk electrode (RDE) measurements in liquid half cells [10]. However, the impressive activity enhancements obtained with RDE have not been translated to real devices [14]. To the best of our knowledge, only a single concept based on dealloyed bimetallic Pt catalysts have exceeded the Department of Energy (DOE) target of activity normalized by mass of precious group metal (PGM) of 0.44 A/mg$_{PGM}$ at 0.9 V$_{RHE}$ in a membrane electrode assembly (MEA) [15].

Benchmarking Pt-based catalysts using the RDE method allows a fast investigation of the trends in electrocatalytic activity of nanoparticles supported on high surface area carbon materials and sputtered thin films [16–18]. However, there are many differences between RDE and MEA measurements. These include different catalyst loadings; in RDE, typical 5–20 µg$_{Pt}$ cm^{-2} are used; on the other hand, MEAs typical employs loadings in the range of 0.1-0.5 mg$_{Pt}$ cm^{-2} [19]. Secondly, different testing parameters are utilized. In RDE measurements, the potential is typically swept with a rate between 5 and 100 mV s^{-1} at room temperature [17]. In contrast, MEA testing is generally carried-out at elevated temperatures with constant currents or potentials [20]. A third difference is the reaction environment at the point of the active site. In MEAs, the oxygen reduction occurs at the triple phase boundary where O$_2$ gas, proton transport mechanisms and electrical conductive particles coexist [21]. Conversely, the catalytic layer in the RDE configuration is immersed in an oxygen-saturated aqueous electrolyte, enabling superior access of oxygen, protons and electrons [22].

Another drawback of RDE benchmarking is the low diffusion and solubility of molecular oxygen in aqueous electrolytes. Consequently, the rate of the reaction is under full mass transport limitation at current densities orders of magnitude lower than actual fuel cells. Using the RDE technique, it is therefore, only possible to measure activities at low overpotentials (typically 0.9 V *versus* the reversible hydrogen electrode, V$_{RHE}$), far from the working potential in real fuel cells (0.6–0.8 V$_{RHE}$). While trends in ORR rate can be investigated at wider potential ranges with use of the Koutecky-Levich analysis [23], it is associated with great uncertainties in weakly adsorbing electrolytes. In MEAs, the catalyst layers are applied on porous gas diffusion electrodes in order to enable diffusion of O$_2$ from the gas phase. However, MEA benchmarking requires large quantities of catalyst and specialized equipment; as a result, it is less accessible for early stage catalysts development.

Since Zatilis et al. showed that superior, near mass transport free current could be achieved with a floating electrode configuration using an artificial gas diffusion interface [24], several gas diffusion electrode (GDE) half-cells designs have been reported for ORR benchmarking [25–28]. These works include measurements in concentrated phosphoric acid at elevated temperature [25], methanol and ethanol oxidation [26] and Pt/C in HClO$_4$ showing comparable activity to MEA, both at room temperature [27] and at elevated temperatures using a Nafion separating membrane [28]. Furthermore, different GDE setups have recently been used for both CO [29] and CO$_2$ reduction [30–32].

In this study, we utilize a GDE setup that enables testing commercial gas diffusion layers with an interchangeable Nafion membrane resembling actual PEMFCs configuration [28]. Using sputtered platinum as a model catalyst, we demonstrate the influence of different testing parameters (electrolyte, scan rate, membrane and humidification) on the performance of the unsupported catalysts. By investigating the effect of different catalyst loadings in both GDE and RDE, we identify key design principles and establish the importance of benchmarking under realistic reaction conditions.

2. Materials and Methods

2.1. Preparation of Pt Thin-Film Electrodes

For the Pt thin film deposition, the magnetron electrode was equipped with a planar target of 99.95 % Pt (Junker Edelmetalle, Waldbüttelbrunn, Germany). It was located at the superior part of the recipient. The RF generator (Advanced Energy, Fort Collins, CO, USA) had a driving frequency of 13.56 MHz. The reactor chamber was evacuated to a base pressure of 5×10^{-3} Pa. An argon plasma was ignited in the chamber at a working pressure of 5 Pa. The platinum loadings of the sputtered Pt TFs were estimated by inductively coupled plasma mass spectrometry (ICP-MS) measurements (Aurora Elite, Bruker, Billerica, MA, USA). For this purpose, the catalysts were digested in aqua regia freshly mixed with 30% HCl (Suprapur, Merck, Darmstadt, Germany) and 65% HNO_3 (Suprapur, Merck, Darmstadt, Germany) in a volumetric ratio of 3:1, respectively. A Dektak 3ST profilometer (Veeco, Plainview, NY, USA) was used to measure the film thickness. Corresponding to the Pt loadings of 11.7 μg_{Pt} cm^{-2}, 16.2 μg_{Pt} cm^{-2}, 29.4 μg_{Pt} cm^{-2} and 66.2 μg_{Pt} cm^{-2}, the films had a thickness of 15 nm, 20 nm, 37 nm and 83 nm, respectively.

2.2. Physical Characterization

The morphology of the sputtered Pt thin-film was investigated with scanning electron microscopy (JSM 7500F, JEOL, Tokyo, Japan) with a field-emission gun, a semi-in-lens conical objective lens and a secondary electron in-lens detector for high-resolution and high-quality image observation of structural features of the deposited films at a maximum specified resolution of 1.0 nm at 15 keV. The technique enables imaging the surface without any preparative coatings. The GDE was directly placed onto the SEM unit holder. The phase identity and crystallite size of the Pt catalyst was investigated by means of x-ray diffraction (XRD) using a Bruker D8 Advance Diffractometer (Bruker, Billerica, MA, USA), with measurements performed over a 2Theta range from 20° to 80°, step width 0.05° and 5 s per step. Cu was used as the anode target.

2.3. Chemicals, Materials and Gases

Deionized ultrapure water (resistivity >18.2 MΩ·cm, total organic carbon (TOC) < 4 ppb) from a Milli-Q system (Millipore, Burlington, MA, USA) was used for acid dilutions, catalyst ink formulation, and the GDE cell cleaning. Perchloric acid (70% HClO4, Suprapur, Merck) were used for electrolyte preparation. The Nafion membrane (Nafion 117, 183 μm thick, Chemours, Wilmington, DE, USA) and the gas diffusion layer (GDL) with a microporous layer (MPL) (H23 C2, Freudenberg, Weinheim, Germany) were employed in the GDE cell measurements. The following gases (Air Liquide, Taastrup, Denmark) were used in electrochemical measurements: Ar (99.999%), O2 (99.999%), and CO (99.97%).

2.4. Gas Diffusion Electrode (GDE)

Electrochemical measurements were carried out using a GDE setup [28]. In contrast to RDE, benchmarking is limited to low overpotentials, the GDE setup allow testing of ORR activity at high current densities under realistic mass transport conditions. The GDE half-cell consists of two cell components (see Figure 1): (i) a lower component made of body of stainless steel with a flow field and gas supply (ii) an upper cell body of polytetrafluoroethylene (PTFE) with electrolyte and counter electrode (platinum mesh) and reference electrode (reversible hydrogen electrode, RHE). For cleaning, the Teflon upper part was soaked in mixed acid (H_2SO_4:HNO_3 = 1:1, v:v) overnight. Subsequently, it was rinsed thoroughly by ultrapure water, and boiled three times.

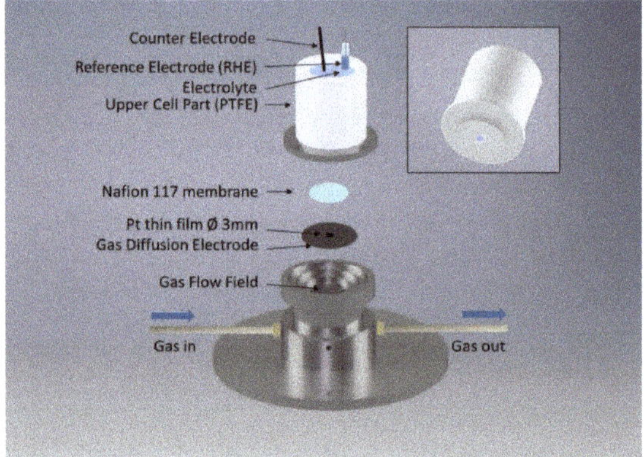

Figure 1. Schematic illustration of the gas diffusion electrode (GDE) half-cell setup, showing the catalyst layer and membrane sandwiched between the upper PTFE with electrolyte and the lower stainless steel body with the gas flow field.

The geometric surface area of the working electrode was 3 mm. The GDE was used with and without membrane between electrolyte and working electrode. When the membrane was used, it was pressed during assembling of the cell to the catalyst layer. For measurements using humidification, the gas was bubbled through a gas humidifier.

2.5. Electrochemical Measurements

All electrochemical measurements were performed using a computer-controlled potentiostat (ECi 200, NordicElectrochemistry, Copenhagen, Denmark). The measurements were performed using 4 M $HClO_4$ and 1 M $HClO_4$ aqueous solutions at room temperature. The high electrolyte concentrations reduce the solution resistance between working electrode, counter electrode and reference electrode. Prior to the measurements, the working electrode was purged from the backside (through the gas diffusion layer) with O_2 gas and the catalyst was conditioned by potential cycles between 0.1 and 1.0 V_{RHE} at a scan rate of 100 mV s^{-1} until a stable cyclic voltammogram could be observed (*ca.* 30 cycles). The ECSA of the catalyst was determined by conducting CO stripping voltammetry. The working electrode was held at 0.05 V during a CO purge through the GDL for 2 min followed by an Ar purge for 10 min. The ECSA was determined from the CO (Q_{CO}) oxidation charge recorded at a scan rate of 50 mV s^{-1} and the respective Pt loading (L_{Pt}) using a fixed conversion coefficient of 390 μC cm$^{-2}_{Pt}$ [16]; ECSA = $Q_{CO}/(L_{Pt} \times 390$ μC cm$^{-2}_{Pt}$). To determine the ORR activity, linear sweep voltammetry (LSV, anodic scan) was conducted by purging the electrode with O_2 from below and scanning the potential at a scan rate of 50 mV s^{-1} or 100 mV s^{-1}. The polarization curves were corrected for the non-faradaic background by subtracting the cyclic voltammograms (CVs) recorded in Ar-purged electrolyte at the identical scan rate. Furthermore, the resistance between the working and reference electrode (~10 Ω) was determined using an AC signal (5 kHz, 5 mV) and thereafter compensated for using analogue positive feedback scheme of the potentiostat. The resulting effective resistance was 1 Ω or less for each experiment.

3. Results and Discussion

3.1. Structural Characterization of Pt Thin-Films

Representative SEM measurements of Pt thin films (TF) are displayed in Figure 2a–c. It can be seen that small Pt domains with a size of ca. 5 nm agglomerate to form cauliflower-like structures with a size between 50 nm and 150 nm. The cauliflower morphology is related to a fractal structure. The resulting surface activity and surface area is determined by this structure. The cross-section in Figure 2b shows the structure of the top-surface layer to be porous and the attaching part to the microporous layer of the gas diffusion electrode to be denser. XRD of Pt TF shows broad diffraction peaks of the indices (111), (200) and (220) crystal structure (see Figure 2d). The average crystallite size estimated by the Scherrer equation is 7.2 nm, which fits to the Pt domain size identified with the electron microscope.

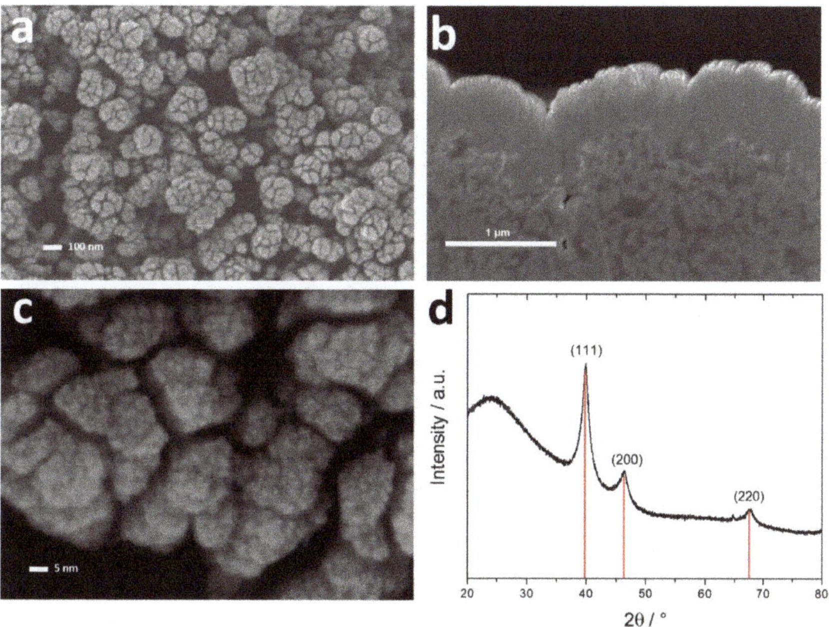

Figure 2. (**a–c**) SEM micrographs of ~500 nm Pt thin films (TF) on a gas diffusion electrode (GDE) with microporous layer (MPL) H23 C2 from Freudenberg prepared by RF unbalanced magnetron sputtering at 15 W and a working pressure of 5 Pa: top view (**a**), cross-section view (**b**) high-magnification top-view (**c**). (**d**) XRD measurements of Pt TF deposited onto a gas diffusion electrode (Freudenberg, H23 C2) with Pt (111), (200) and (220) diffraction pattern (PDF 00-001-1194).

3.2. Electrochemical Characterization

Figure 3 shows the base cyclic voltammograms in argon-saturated electrolyte as well as CO-stripping measurements in order to determine the electrochemically active surface area. Figure 3a displays in detail the cyclic voltammograms of Pt TF with increasing Pt characteristic features for higher loadings, as observed in the hydrogen under potential deposition (0.1–0.4 V_{RHE}) region as well as the region where O-containing species are adsorbed (0.7–1.0V_{RHE}). The linear sweep voltammogram of the CO oxidation charge transfer is shown in Figure 3b for the different loadings. The sweep shows two CO oxidation peaks, one peak between 0.75 and 0.77 V_{RHE} and a second around 0.82–0.85 V_{RHE}, in contrast to the CO oxidation sweep measured in the RDE with one peak at 0.73 V_{RHE}. The integrated CO oxidation charge is plotted against the Pt loading with the ideal linear slope crossing 0 and the second lowest loading; see Figure 3c. The oxidation charge increases with loading. However, it can be seen that the CO oxidation charge as function of the loading deviates from an ideal straight line for the highest loading. This could indicate that there is a mass transport problem due to blocking of CO as a reactant or trapped product; however, no CO oxidation peaks were visible in sub sequential sweeps. Thus it can be assumed that the ECSA measurement with CO adsorption is valid.

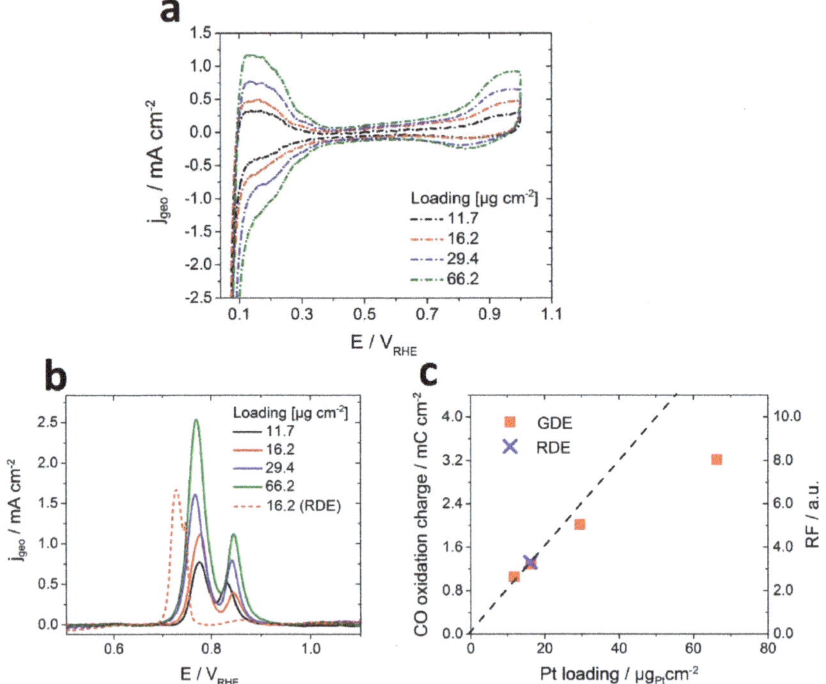

Figure 3. (**a**) Loading dependent electrochemical behaviour of Pt TF in GDE setup (298 K, 4 M HClO$_4$, 0% RH), cyclic voltammograms recorded with 50 mV s^{-1} and Ar saturated flow field. (**b**) CO stripping curves recorded with 50 mV s^{-1}. (**c**) Plot of CO oxidation charge as function of the Pt loading on the GDE and RDE. The linear slope indicates the ideal correlation between oxidation charge and Pt loading.

Our results from CO-stripping experiments therefore, show that the decrease in ECSA for higher Pt loadings is a result of decreasing Pt utilization. This is supported by the cross-section SEM micrograph displayed in Figure 2b with a higher loading. The graph shows a higher porosity at the near surface region, compared to the region near the Microporous Layer (MPL). Increasing the loading/film thickness leads to lower Pt utilization due to lack of long-range porosity for higher loadings of sputtered Pt thin-films. Consequently, the ECSA measurements have to be carefully done for each loading, especially with loadings above 30 μg$_{Pt}$ cm^{-2}. The ECSA of the Pt TF GDE is determined to be 20 m^2 g^{-1}$_{Pt}$ for the low-loaded 16.2 μg$_{Pt}$ cm^{-2} gas diffusion electrodes.

3.3. Performance of Pt Thin-Films in the GDE Setup

Figure 4 shows a comparison between the linear sweep voltammograms during the oxygen reduction reaction for both RDE and GDE. As stated in our previous work [28], the GDE enables the measurement of realistic mass transport conditions and high current densities. In contrast, the RDE is mass transport limited due to low solubility of oxygen in the condense phase/liquid electrolyte, but it's well defined hydrodynamics allow extraction of intrinsic kinetic currents [16,17]. Therefore, it is only possible to measure the kinetic current densities at very low overpotentials.

The ORR activity from RDE measurements in acidic electrolyte is typically reported at 0.9 V$_{RHE}$ for less active platinum catalysts and for highly active catalysts it is often only possible to measure at even higher potentials up to 0.95 V$_{RHE}$ [33]. However, fuel cells typically operate at a working potential of in the range of 0.6–0.8 V$_{RHE}$, as indicated in Figure 4. At these potentials, the current in the

RDE is under full mass transport limitation. While in the GDE the current density is not free of mass transport limitations, GDE measurements reveal more realistic current densities [28].

Our GDE setup allows measuring a maximum oxygen reduction current density of ~1 A cm^{-2} using only 16.2 µg$_{Pt}$ cm^{-2}; this is over two orders of magnitude higher than what is observed in the RDE measurements. The maximum current density is reached at ~0.3 V$_{RHE}$. At potentials below 0.3 V$_{RHE}$ in the GDE it can be seen that the current decreases. This correlates to the region where hydrogen under potential adsorption occurs. Therefore, it is presumed that the decrease in the current density observed between 0.1 V$_{RHE}$ and 0.3 V$_{RHE}$ is due to blocking by protons on the Pt surface which would limit the full oxygen reduction to water. In this region, the catalyst reduces oxygen to hydrogen peroxide, which corresponds to a two-electron transfer compared to a four-electron transfer for the reduction of oxygen to water [34]. It is notable that the maximum current density is reached in the same potential region as the maximum in the extended Koutecky-Levich analysis [23] as well as the potential of zero total charge (PZTC) [35]. The maximum oxygen reduction rate, therefore, correlates with the minimum ion coverage; most likely a competition between protons and oxygen species at the surface and, to a lesser extent, perchlorate anions.

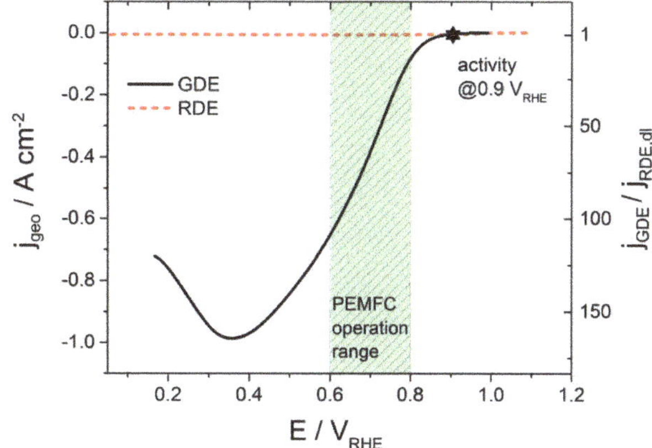

Figure 4. Comparison of the current density-potential curves as measured in a GDE setup and a conventional RDE setup. The activity measurement in the GDE setup with oxygen saturated flow field was conducted at 298 K, 4 M HClO$_4$, Nafion 117 membrane, 0% RH and 50 mV s^{-1}. The measurement of the activity in the RDE setup was conducted at 1600 RPM, 298 K, 0.1 M HClO$_4$ and 50 mV s^{-1}. The catalyst was a Pt thin film prepared by magnetron sputtering with a Pt loading of 16.2 µg$_{Pt}$ cm^{-2} for both RDE and GDE. The right axis shows the current density enhancement as compared to the diffusion limited current observed in RDE.

3.4. Benchmarking Different Pt Loadings in the GDE Setup

The Pt TF catalyst was tested with different Pt loadings on the GDE, as shown in Figure 5. The obtained linear sweep voltammograms are plotted as kinetic current densities normalized by the catalyst geometric surface area, the specific activities (normalized by the roughness factor) and mass activities. The activities normalized by geometric surface area show the expected increase of current density over the full potential range. At 0.3 V$_{RHE}$, the maximum activity is achieved and reaches 0.75 A cm^{-2} for 11 µg$_{Pt}$ cm^{-2} and nearly 1 A cm^{-2} for the highest loading 66 µg$_{Pt}$ cm^{-2}. The onset potential for the ORR is shifting to lower overpotentials. The specific activities were calculated using the roughness factor determined by CO oxidation; it can be seen that the onset potential stays constant with a value of 0.71 mA cm^{-2}$_{Pt}$ at 0.90 V$_{RHE}$. The specific activity starts to differ in the region from 0.75 V$_{RHE}$–0.8 V$_{RHE}$ for the different loadings. Furthermore, the lowest Pt loading leads to the highest specific activity measured. The Pt TF catalyst with 11 µg$_{Pt}$ cm^{-2} achieves a specific activity of 94.0 mA cm^{-2}$_{Pt}$ at 0.65

V_{RHE}. While the GDE with the highest loading of 66 μg_{Pt} cm^{-2} reaches only a value of 38.6 mA cm$_{Pt}^{-2}$ at 0.65 V_{RHE}, corresponding to a "loss" of 59% at 0.65 V_{RHE}. The reason for the potential-dependent specific activity is most likely related to mass transport, which can be distinguished due to oxygen transport, proton transport, H_2O transport or electronic conductivity. In order to distinguish between the different influences, several operation parameters were changed and discussed later in this paper.

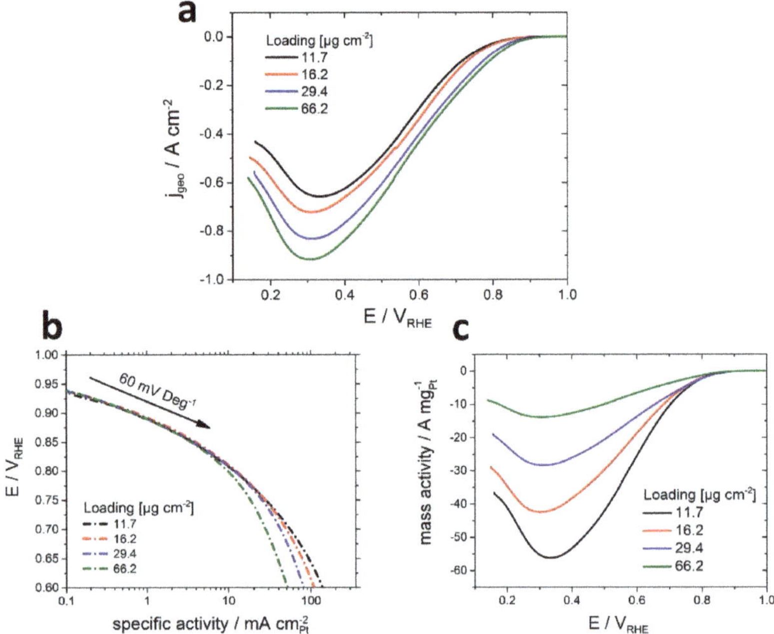

Figure 5. Influence of different Pt loadings on the catalyst layer activity (11.7 μg_{Pt} cm^{-2}, 16.2 μg_{Pt} cm^{-2}, 29.4 μg_{Pt} cm^{-2}, 66.4 μg_{Pt} cm^{-2}). (**a**) Activity normalized by the geometric surface area, (**b**) surface-area specific activity and (**c**) mass specific activity. The measurement of the was conducted in the GDE setup with oxygen saturated flow field (298 K, 4 M HClO$_4$, 0% RH, 100 mV s^{-1}).

Comparing the mass activity for the different loadings, the highest activity can again be achieved with the GDE with the lowest loading of 11 μg_{Pt} cm^{-2}. The activity of the GDE with 11 μg_{Pt} cm^{-2} reaches 17 A mg$^{-1}_{Pt}$ at 0.65 V_{RHE} and 55 A mg$^{-1}_{Pt}$ at 0.3 V_{RHE}. The highest loaded GDE achieves only a mass activity of 5 A mg$^{-1}_{Pt}$ at 0.65 V_{RHE} and 19 A mg$^{-1}_{Pt}$ at 0.3 V_{RHE}. At 0.65 V_{RHE} this is a loss of 75% and at 0.3 V_{RHE} the "loss" is 70% of the mass activity compared to the lowest loading.

3.5. Systematic Change of Operation Parameters

The Pt TF GDE was evaluated with different operational parameters to understand the activity determining factors of the unsupported catalyst layers (see Figure 6). First, Figure 6a shows the ORR polarization curves of the GDE with a separating membrane, as compared with a non-separated GDE setup with 4 M perchloric acid. The separating membrane increases the geometric activity over the full potential window and at 0.65 V_{RHE} by 104%. To clarify the cause of this drastic increase in activity, we performed a set of measurements with 1 M HClO$_4$ instead of 4 M HClO$_4$ (see Figure 6a). Reducing the molarity of the electrolyte greatly enhanced the activity. Interestingly, the polarization curve for Pt in 1 M HClO$_4$ resembles the one obtained in 4 M HClO$_4$ using a Nafion membrane. This suggests that specific anion adsorption is inhibiting the rate of ORR at high electrolyte concentrations. While perchloric acid is generally considered a non-adsorbing electrolyte, it is known to inhibited ORR on Pt at high concentrations [36]. Thus, in the absence of a separating membrane, an electrolyte concentration of 1 M $HClO_4$ would be preferable. While no differences were observed in the pure kinetic region (until ~0.8 V_{RHE}), the membrane slightly enhanced the activity as compared to 1 M $HClO_4$ in the mass-transport-dominated region (0.3–0.8 V_{RHE}), possibly due to enhanced proton transport or reduced flooding of the hydrophilic catalyst layer. At high current densities, oxygen transport is critical. Recently, Kongkanand el. al showed that local O_2 resistance in the ionomer film is essential for the performance supported catalysts [4]. While unsupported catalysts are generally free of ionomer, the degree of wetting would influence the oxygen transport in the catalyst layer. It is, therefore, possible that the membrane reduces flooding as a result of less penetrating electrolyte, as compared to measurements without the Nafion membrane. However, from the presented data, it cannot be concluded if adsorbed water on the surface hinders the reaction.

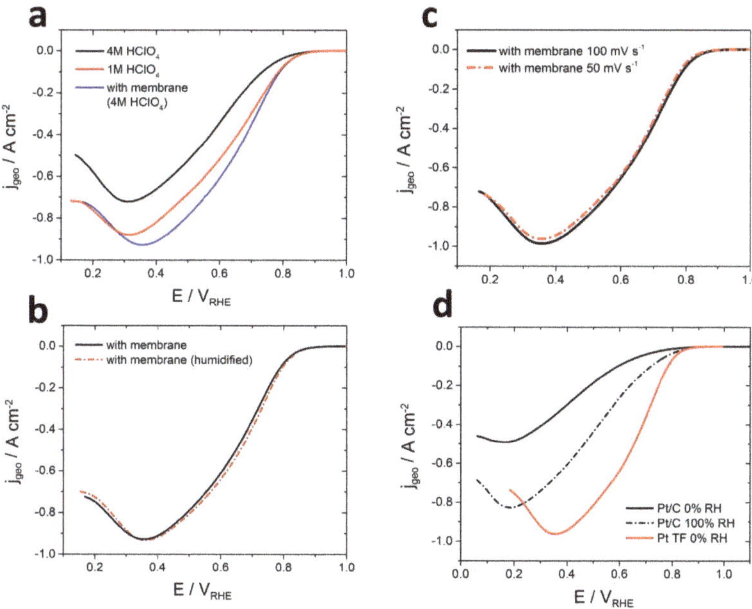

Figure 6. Influence of: (**a**) the separating Nafion 117 membrane, (**b**) the humidification on the activity of Pt TF 0% vs 100% relative humidity (RH), (**c**) the scan rate change between 50 mV s^{-1} and 100 mV s^{-1}. The measurement of the activity was conducted in the GDE setup with oxygen saturated flow field (16.2 μg$_{Pt}$ cm^{-2}, 298 K, 4 M HClO$_4$, 0% RH, 50 mV s^{-1}). (**d**) Influence of catalyst system on the catalyst layer performance measured in the GDE with Nafion 117 as membrane in 4 M HClO$_4$ at RT at 50 mV s^{-1} with Pt TF (16.2 μg$_{Pt}$ cm^{-2}, in red) and Pt/C (10 μg$_{Pt}$ cm^{-2}, in black) [28].

Secondly, the scan rate was varied to 50 mV s^{-1} compared to the 100 mV s^{-1}, the results showed negligible decrease in activity (see Figure 6b). This indicates that the current densities are defined by faradaic currents instead of capacitive currents. Capacitive currents increase as the current is proportional to the applied scan rate. Furthermore, it is a sign for a reaction in equilibrium, where mass transport should not be the significant driver limiting the reaction. If mass transport would limit the reaction rate, as the reactant supply drops, the geometric current density should decrease with lower scan rate.

The third operation parameter studied was the humidification of oxygen. It was previously shown that a change of oxygen humidification can have a significant impact on the gained activity of platinum nanoparticles supported on carbon (Pt/C) [28]. However, there is nearly no change in the activity of Pt TF after humidifying the oxygen with 100% relative humidity (RH); see Figure 6c.

This indicates that proton transport is not a limiting step with unsupported Pt catalysts. This is in contrast to Pt/C in GDE, where we observed a doubling in activity with humidification at room temperature, indicating a limiting behaviour; see Figure 6d. In our previous study, Pt/C catalyst was mixed with Nafion [28]. The Nafion ionomer has to be humidified to work as a sufficient proton conductor. In the unsupported Pt TF catalyst, protons seem to be transported on the metallic surface by either Pt-OH or Pt-H. The transport of protons on platinum was first investigated by McBreen et al. [37] and recently, Zenyuk et al. suggested that the transport mechanism might be potential dependent above 0.7 V, where Pt-O and Pt-OH are present on the surface. At potentials below the point of zero charge, the proton is believed to be transported by surface migration [38]. This speculation was already introduced earlier for oxygen reduction catalysts by Debe et al. for Pt on non-conductive organic whiskers [39], the so-called nano-structured thin films (NSTF). This is further supported by the fact that the catalyst layer is also more hydrophobic compared to Pt/C. Thus, at room temperature, sufficient wetting of catalyst layer can be obtained only from water generated by the ORR reaction. Thus, our results support the claim that proton transport on bare Pt is not a limiting step.

The specific activities and ECSAs for the RDE and GDE benchmarking measurements on Pt TFs are summarised in Table 1, as well as compared to commercial Pt/C in the RDE, GDE and MEA. The specific activity of Pt/C in the GDE cell at 60° C and 0.9 V_{RHE} was found to be comparable to the specific activity of the MEA with commercial Pt/C and decreased at room temperature. In general, the specific activity at 0.9 V_{RHE} of GDE and MEA of Pt/C was found to be smaller compared to the RDE measurements. This is also the case for Pt TF where the specific activity in the RDE was benchmarked with 1.82 mA cm^{-2}_{Pt} at 0.9 V_{RHE} and in the GDE with 0.72 mA cm^{-2}_{Pt} with membrane or 0.71 mA cm^{-2}_{Pt} without membrane. The decrease in specific activity for commercial Pt/C is nearly 50% compared to a loss approximately 60% for Pt TF. In the higher current density region, at 0.65 V_{RHE}, the specific activity with Pt TF is significantly higher than with Pt/C. In contrast, the current density at 0.65 V_{RHE} by mass is only slightly increased for Pt TF - with 24.1 A/mg$_{Pt}$ for Pt/C and 28.6 A/mg$_{Pt}$ for Pt TF. This indicates that unsupported Pt catalysts have an advantage in activity compared to commercial Pt/C catalysts, especially without humidification at low temperatures. In contrast, at higher loadings, where most devices operate, the low Pt utilization inhibits the performance of the unsupported Pt catalysts. The investigations show that with the help of simple benchmarking of different catalyst systems in the GDE, new ways of developing possible applications can be explored. It turns out that different catalyst systems also require different environmental conditions and can play to their strengths under different operating conditions.

Table 1. Comparison of the characterization under different conditions for sputtered Pt thin-film catalyst in GDE and RDE setup and as well as a commercial 46.5 wt % Pt/C catalyst (TEC10E50E, Tanaka) in RDE, GDE and MEA measurements. SA, ECSA and RH stands for ORR specific activity, electrochemical active surface area and relative humidity, respectively.

Catalyst Layer (Reference)	Electrolyte	Loading [$\mu g_{Pt}\ cm^{-2}$]	Temperature/Humidity [°C]	SA @0.9V_{RHE} [mA cm^{-2}_{Pt}]	SA @0.65V_{RHE} [mA cm^{-2}_{Pt}]	ECSA [$m^2\ g^{-1}_{Pt}$]
Pt TF RDE (this work)	0.1 M HClO$_4$	16	rt	1.82	-	21
Pt TF GDE (this work)	Nafion 117/4 M HClO$_4$	16	rt/0% RH	0.72	144	20
Pt TF GDE (this work)	4 M HClO$_4$	16	rt/0% RH	0.71	67.5	20
Pt/C RDE [40]	0.1 M HClO$_4$	14	rt	0.49	-	76
Pt/C GDE [28]	Nafion 117/4 M HClO$_4$	5	rt/100% RH	0.14	25.9	93
Pt/C GDE [28]	Nafion 117/4 M HClO$_4$	5	60/100% RH	0.18	47.9	81
Pt/C MEA [41]	Nafion 117	90	80/100% RH	0.21	-	80

4. Conclusions

In this paper, we present the electrochemical characterization of platinum thin-films in a GDE setup. Unlike the commonly used RDE method, this half-cell method facilitates investigations under triple phase boundary conditions. Moreover, the GDE setup allows achieving current density values comparable to operating current densities in actual PEMFCs.

Herein, we investigate sputtered Pt with very low loadings, achieving high current densities (close to 1 A cm^{-2}) in GDE measurements. Furthermore, we show that different parameters such as loading, electrolyte concentration, humidification and Nafion membrane influence the performance of the Pt catalyst layer. By carefully measuring the activity with different loadings in the GDE, we observe that specific activities at potentials down to 0.75–0.80 V_{RHE} can be determined independently of the applied loading. The values obtained in the GDE were two-and-half times lower than those measured with the RDE technique using the same catalyst. Our results confirm that activity values obtained in liquid half-cells are not directly transferable to real devices. This is mainly due to transport resistance at the triple phase interface and not a result of higher loadings or different testing parameters.

Interestingly, our results show that at fuel-cell relevant overpotentials, the activity is highly dependent on the loading and use of the membrane. Furthermore, we observed that humidification had no influence on the activity of catalyst layer for the unsupported catalyst. This is in contrast to Pt/C, where the use of humidification nearly doubled the catalytic activity. These findings highlight the potential impact of GDE half-cell benchmarking. While RDE remains a valuable tool for observing trends in intrinsic catalytic performance, the properties of the catalyst layer have to be investigated under more realistic conditions. Using sputtered platinum as a model catalyst, we show that the catalyst layer can be efficiently evaluated in the fuel cell operation region using a GDE half-cell setup. It is our hope that this enables accelerated development of fuel cell catalysts by shortening expensive and time-consuming MEA optimization.

Author Contributions: Conceptualization, A.W.J. and G.S.; formal analysis, A.W.J. and G.S.; investigation, A.W.J. and G.S.; data curation, A.W.J. and G.S.; project administration, M.A. and M.E.-E.; resources, V.B.; supervision, M.E.-E. and M.A.; writing—original draft preparation, A.W.J. and G.S.; writing—review and editing, M.A. and M.E.-E.

Funding: This research was funded by the Danish Innovation Fund (4 M center) and the Danish 350 DFF through (grant number 4184-00332). M.E.-E. and A.W.J. gratefully acknowledge the Villum Foundation V-SUSTAIN grant 9455 to the Villum Center for the Science of Sustainable Fuels and Chemicals.

Conflicts of Interest: The authors declare no conflict of interest.

References

1. Debe, M.K. Electrocatalyst approaches and challenges for automotive fuel cells. *Nat. Cell Biol.* **2012**, *486*, 43–51. [CrossRef] [PubMed]
2. Rabis, A.; Rodriguez, P.; Schmidt, T.J. Electrocatalysis for Polymer Electrolyte Fuel Cells: Recent Achievements and Future Challenges. *ACS Catal.* **2012**, *2*, 864–890. [CrossRef]
3. Wagner, F.T.; Lakshmanan, B.; Mathias, M.F. Electrochemistry and the Future of the Automobile. *J. Phys. Chem. Lett.* **2010**, *1*, 2204–2219. [CrossRef]
4. Mathias, M.F.; Kongkanand, A. The Priority and Challenge of High-Power Performance of Low-Platinum Proton-Exchange Membrane Fuel Cells. *J. Phys. Chem. Lett.* **2016**, *7*, 1127–1137.
5. Stamenkovic, V.R.; Fowler, B.; Mun, B.S.; Wang, G.; Ross, P.N.; Lucas, C.A.; Marković, N.M. Improved Oxygen Reduction Activity on Pt3Ni(111) via Increased Surface Site Availability. *Science* **2007**, *315*, 493–497. [CrossRef]
6. Strasser, P.; Koh, S.; Anniyev, T.; Greeley, J.; More, K.; Yu, C.; Liu, Z.; Kaya, S.; Nordlund, D.; Ogasawara, H.; et al. Lattice-strain control of the activity in dealloyed core–shell fuel cell catalysts. *Nat. Chem.* **2010**, *2*, 454–460. [CrossRef] [PubMed]

7. Banham, D.; Ye, S. Current Status and Future Development of Catalyst Materials and Catalyst Layers for Proton Exchange Membrane Fuel Cells: An Industrial Perspective. *ACS Energy Lett.* **2017**, *2*, 629–638. [CrossRef]

8. Strmcnik, D.; Escudero-Escribano, M.; Kodama, K.; Stamenkovic, V.R.; Cuesta, A.; Marković, N.M. Enhanced electrocatalysis of the oxygen reduction reaction based on patterning of platinum surfaces with cyanide. *Nat. Chem.* **2010**, *2*, 880–885. [CrossRef] [PubMed]

9. Stamenković, V.; Mun, B.S.; Mayrhofer, K.J.J.; Ross, P.N.; Marković, N.M.; Rossmeisl, J.; Greeley, J.; Nørskov, J.K. Changing the Activity of Electrocatalysts for Oxygen Reduction by Tuning the Surface Electronic Structure. *Angew. Chem. Int. Ed.* **2006**, *45*, 2897–2901. [CrossRef]

10. Escudero-Escribano, M.; Jensen, K.D.; Jensen, A.W. Recent advances in bimetallic electrocatalysts for oxygen reduction: Design principles, structure-function relations and active phase elucidation. *Curr. Opin. Electrochem.* **2018**, *8*, 135–146. [CrossRef]

11. Stamenkovic, V.R.; Mun, B.S.; Arenz, M.; Mayrhofer, K.J.J.; Lucas, C.A.; Wang, G.; Ross, P.N.; Marković, N.M. Trends in electrocatalysis on extended and nanoscale Pt-bimetallic alloy surfaces. *Nat. Mater.* **2007**, *6*, 241–247. [CrossRef]

12. Greeley, J.; Stephens, I.E.L.; Bondarenko, A.S.; Johansson, T.P.; Hansen, H.A.; Jaramillo, T.F.; Rossmeisl, J.; Chorkendorff, I.; Nørskov, J.K.; Jaramillo, T. Alloys of platinum and early transition metals as oxygen reduction electrocatalysts. *Nat. Chem.* **2009**, *1*, 552–556. [CrossRef]

13. Escudero-Escribano, M.; Malacrida, P.; Hansen, M.H.; Vej-Hansen, U.G.; Velázquez-Palenzuela, A.; Tripkovic, V.; Schiøtz, J.; Rossmeisl, J.; Stephens, I.E.L.; Chorkendorff, I. Tuning the activity of Pt alloy electrocatalysts by means of the lanthanide contraction. *Science* **2016**, *352*, 73–76. [CrossRef] [PubMed]

14. Stephens, I.E.L.; Rossmeisl, J.; Chorkendorff, I. Toward sustainable fuel cells. *Science* **2016**, *354*, 1378–1379. [CrossRef]

15. Han, B.; Kukreja, R.S.; Theobald, B.R.; O'Malley, R.; Wagner, F.T.; Carlton, C.E.; Kongkanand, A.; Gan, L.; Strasser, P.; Shao-Horn, Y. Record activity and stability of dealloyed bimetallic catalysts for proton exchange membrane fuel cells. *Energy Environ. Sci.* **2015**, *8*, 258–266. [CrossRef]

16. Mayrhofer, K.; Strmcnik, D.; Blizanac, B.; Stamenković, V.; Arenz, M.; Marković, N. Measurement of oxygen reduction activities via the rotating disc electrode method: From Pt model surfaces to carbon-supported high surface area catalysts. *Electrochim. Acta* **2008**, *53*, 3181–3188. [CrossRef]

17. Pedersen, C.M.; Escudero-Escribano, M.; Velázquez-Palenzuela, A.; Christensen, L.H.; Chorkendorff, I.; Stephens, I.E. Benchmarking Pt-based electrocatalysts for low temperature fuel cell reactions with the rotating disk electrode: Oxygen reduction and hydrogen oxidation in the presence of CO (review article). *Electrochim. Acta* **2015**, *179*, 647–657. [CrossRef]

18. Zamburlini, E.; Jensen, K.D.; Stephens, I.E.; Chorkendorff, I.; Escudero-Escribano, M. Benchmarking Pt and Pt-lanthanide sputtered thin films for oxygen electroreduction: Fabrication and rotating disk electrode measurements. *Electrochim. Acta* **2017**, *247*, 708–721. [CrossRef]

19. Gasteiger, H.A.; Kocha, S.S.; Sompalli, B.; Wagner, F.T. Activity benchmarks and requirements for Pt, Pt-alloy, and non-Pt oxygen reduction catalysts for PEMFCs. *Appl. Catal. B: Environ.* **2005**, *56*, 9–35. [CrossRef]

20. Gasteiger, H.; Panels, J.; Yan, S. Dependence of PEM fuel cell performance on catalyst loading. *J. Power Sources* **2004**, *127*, 162–171. [CrossRef]

21. O'Hayre, R.; Prinz, F.B. The Air/Platinum/Nafion Triple-Phase Boundary: Characteristics, Scaling, and Implications for Fuel Cells. *J. Electrochem. Soc.* **2004**, *151*, A756. [CrossRef]

22. Schmidt, T.J.; Schmutz, P.; Frankel, G.S. Characterization of High-Surface-Area Electrocatalysts Using a Rotating Disk Electrode Configuration. *J. Electrochem. Soc.* **1998**, *145*, 2354. [CrossRef]

23. Zana, A.; Wiberg, G.K.H.; Deng, Y.-J.; Østergaard, T.; Rossmeisl, J.; Arenz, M. Accessing the Inaccessible: Analyzing the Oxygen Reduction Reaction in the Diffusion Limit. *ACS Appl. Mater. Interfaces* **2017**, *9*, 38176–38180. [CrossRef]

24. Zalitis, C.M.; Kramer, D.; Kucernak, A.R. Electrocatalytic performance of fuel cell reactions at low catalyst loading and high mass transport. *Phys. Chem. Chem. Phys.* **2013**, *15*, 4329. [CrossRef] [PubMed]

25. Hu, Y.; Jiang, Y.; Jensen, J.O.; Cleemann, L.N.; Li, Q. Catalyst evaluation for oxygen reduction reaction in concentrated phosphoric acid at elevated temperatures. *J. Power Sour.* **2018**, *375*, 77–81. [CrossRef]

26. Wiberg, G.K.H.; Fleige, M.; Arenz, M. Gas diffusion electrode setup for catalyst testing in concentrated phosphoric acid at elevated temperatures. *Sci. Instrum.* **2015**, *86*, 024102. [CrossRef]

27. Pinaud, B.A.; Bonakdarpour, A.; Daniel, L.; Sharman, J.; Wilkinson, D.P. Key Considerations for High Current Fuel Cell Catalyst Testing in an Electrochemical Half-Cell. *J. Electrochem. Soc.* **2017**, *164*, F321–F327. [CrossRef]

28. Inaba, M.; Jensen, A.W.; Sievers, G.W.; Escudero-Escribano, M.; Zana, A.; Arenz, M. Benchmarking high surface area electrocatalysts in a gas diffusion electrode: measurement of oxygen reduction activities under realistic conditions. *Energy Environ. Sci.* **2018**, *11*, 988–994. [CrossRef]

29. Jouny, M.; Luc, W.; Jiao, F. High-rate electroreduction of carbon monoxide to multi-carbon products. *Nat. Catal.* **2018**, *1*, 748–755. [CrossRef]

30. Dinh, C.-T.; Burdyny, T.; Kibria, M.G.; Seifitokaldani, A.; Gabardo, C.M.; De Arquer, F.P.G.; Kiani, A.; Edwards, J.P.; De Luna, P.; Bushuyev, O.S.; et al. CO_2 electroreduction to ethylene via hydroxide-mediated copper catalysis at an abrupt interface. *Science* **2018**, *360*, 783–787. [CrossRef] [PubMed]

31. Möller, T.; Ju, W.; Bagger, A.; Wang, X.; Luo, F.; Thanh, T.N.; Varela, A.S.; Rossmeisl, J.; Strasser, P.; Moeller, T. Efficient CO_2 to CO electrolysis on solid Ni–N–C catalysts at industrial current densities. *Energy Environ. Sci.* **2019**, *12*, 640–647. [CrossRef]

32. Higgins, D.; Hahn, C.; Xiang, C.; Jaramillo, T.F.; Weber, A.Z. Gas-Diffusion Electrodes for Carbon Dioxide Reduction: A New Paradigm. *ACS Energy Lett.* **2019**, *4*, 317–324. [CrossRef]

33. Chen, C.; Kang, Y.; Huo, Z.; Zhu, Z.; Huang, W.; Xin, H.L.; Snyder, J.D.; Li, D.; Herron, J.A.; Mavrikakis, M.; et al. Highly Crystalline Multimetallic Nanoframes with Three-Dimensional Electrocatalytic Surfaces. *Science* **2014**, *343*, 1339–1343. [CrossRef]

34. Sepa, D.; Vojnovic, M.; Damjanovic, A. Reaction intermediates as a controlling factor in the kinetics and mechanism of oxygen reduction at platinum electrodes. *Electrochim. Acta* **1981**, *26*, 781–793. [CrossRef]

35. Garcia-Araez, N.; Climent, V.; Feliu, J. Potential-Dependent Water Orientation on Pt(111), Pt(100), and Pt(110), As Inferred from Laser-Pulsed Experiments. Electrostatic and Chemical Effects. *J. Phys. Chem. C* **2009**, *113*, 9290–9304. [CrossRef]

36. Láng, G.; Horanyi, G. Some interesting aspects of the catalytic and electrocatalytic reduction of perchlorate ions. *J. Electroanal. Chem.* **2003**, *552*, 197–211. [CrossRef]

37. McBreen, J. Voltammetric Studies of Electrodes in Contact with Ionomeric Membranes. *J. Electrochem. Soc.* **1985**, *132*, 1112. [CrossRef]

38. Liu, J.; Zenyuk, I.V. Proton transport in ionomer-free regions of polymer electrolyte fuel cells and implications for oxygen reduction reaction. *Curr. Opin. Electrochem.* **2018**, *12*, 202–208. [CrossRef]

39. Debe, M.K. Tutorial on the Fundamental Characteristics and Practical Properties of Nanostructured Thin Film (NSTF) Catalysts. *J. Electrochem. Soc.* **2013**, *160*, F522–F534. [CrossRef]

40. Nesselberger, M.; Ashton, S.; Meier, J.C.; Katsounaros, I.; Mayrhofer, K.J.J.; Arenz, M. The Particle Size Effect on the Oxygen Reduction Reaction Activity of Pt Catalysts: Influence of Electrolyte and Relation to Single Crystal Models. *J. Am. Chem. Soc.* **2011**, *133*, 17428–17433. [CrossRef] [PubMed]

41. Inaba, M.; Suzuki, T.; Hatanaka, T.; Morimoto, Y. Fabrication and Cell Analysis of a Pt/SiO2 Platinum Thin Film Electrode. *J. Electrochem. Soc.* **2015**, *162*, F634–F638. [CrossRef]

 *surfaces*

Article

Electro-Oxidation of CO Saturated in 0.1 M HClO$_4$ on Basal and Stepped Pt Single-Crystal Electrodes at Room Temperature Accompanied by Surface Reconstruction

Kiyotaka Abe [1,2], **Hiroyuki Uchida** [1,*] **and Junji Inukai** [1,*]

[1]   Clean Energy Research Center, University of Yamanashi, 4 Takeda, Kofu, Yamanashi 400-8510, Japan

[2]   Present address: GS Yuasa Corporation, 555 Sakaijuku, Kosai-shi, Shizuoka 431-0452, Japan; kiyotaka.abe@jp.gs-yuasa.com

*   Correspondence: h-uchida@yamanashi.ac.jp (H.U.); jinukai@yamanashi.ac.jp (J.I.); Tel.: +81-55-220-8185 (J.I.)

Received: 28 December 2018; Accepted: 11 April 2019; Published: 16 April 2019

Abstract: The electro-oxidation of CO on Pt surface is not only fundamentally important in electrochemistry, but also practically important in residential fuel cells for avoiding the poisoning of Pt catalysts by CO. We carried out cyclic voltammetry on Pt(111), (110), (100), (10 10 9), (10 9 8), (10 2 1), (432), and (431) single-crystal surfaces using a three compartment cell to understand the activity and durability towards the electro-oxidation of CO saturated in 0.1 M HClO$_4$. During the potential cycles between 0.07 and 0.95 V vs. the reversible hydrogen electrode, the current for the electro-oxidation of CO at potentials lower than 0.5 V disappeared, accompanied by surface reconstruction. Among the electrodes, the Pt(100) electrode showed the lowest onset potential of 0.29 V, but the activity abruptly disappeared after one potential cycle; the active sites were extremely unstable. In order to investigate the processes of the deactivation, potential-step measurements were also conducted on Pt(111) in a CO-saturated solution. Repeated cycles of the formations of Pt oxides at a high potential and Pt carbonyl species at a low potential on the surface were proposed as the deactivation process.

Keywords: CO electro-oxidation; Pt single-crystal electrodes; potential cycling; potential stepping; surface reconstruction

---

1. Introduction

The electrochemical oxidation (electro-oxidation) of CO, particularly on the Pt surface, is one of the most fundamental and most studied electrochemical reactions [1–3]. Many papers have been devoted to the oxidation of a pre-adsorbed monolayer of CO, often referred to as CO stripping, which usually occurs at a relatively high potential, in the 0.7 to 1.0 V range, vs. the reversible hydrogen electrode (RHE), within a narrow potential range, far from the equilibrium potential for the reaction, −0.104 V vs. RHE [4,5]. On the other hand, a current can often be observed in the lower potential region, from 0.3 to 0.5 V vs. RHE, depending both on the characteristics of the Pt surface and also to some extent on the potential at which CO was initially adsorbed. This current has been referred to as "pre-peak," "pre-wave," "pre-oxidation", or "pre-ignition" current. The origins of this current are still controversial [6–15]. With the use of an electrolyte solution saturated with CO (bulk CO), the current can flow steadily, since CO is being supplied continuously from the solution, in contrast to the case of pre-adsorbed CO, which is only a monolayer of CO adsorbed on the surface.

Strmcnik et al., using ex situ scanning tunneling microscopy (STM), presented results that have provided insight into the electro-oxidation of bulk CO in the lower potential region and found that the presence of Pt islands and other low coordination structures on the Pt(111) surface is associated with high activity for CO oxidation at low potentials [16]. They found that, during the potential cycling process, these islands were removed. Thus, the authors proposed that the steps associated with islands, particularly at the nanometer level, are important catalytic sites, due to their low coordination number (CN). We reported the electro-oxidation of bulk CO saturated in 0.1 M HClO$_4$ on stepped Pt(111) surfaces: Pt(10 10 9) with (111) steps and Pt(20 19 19) with (100) steps (ball models in Figure 1), combined with imaging by operando STM during the electro-oxidation of bulk CO [15]. Our results, in agreement with those of Strmcnik et al. [16], suggest that the activity increases in the order: Terraces < steps < kinks < step-adatoms (single adatoms at steps), i.e., increasing activity with decreasing CN [17]. One of the most striking findings of ours was that potential cycling in the presence of CO tends to strongly favor the transformation of steps to maximize the occurrence of (111) steps, so that, for example, initially disordered (111) steps become perfectly straight, and initially straight (100) steps become zigzagged with (111) microsteps [15]. Both steps become highly inert, again due to the superb ordering of the CO adlayer across the step. The atomic-level insight, with a direct link between surface structure and electrocatalytic activity, provided a step forward in guiding the design of new, more active catalysts and electrocatalysts for CO oxidation and CO-tolerant hydrogen oxidation.

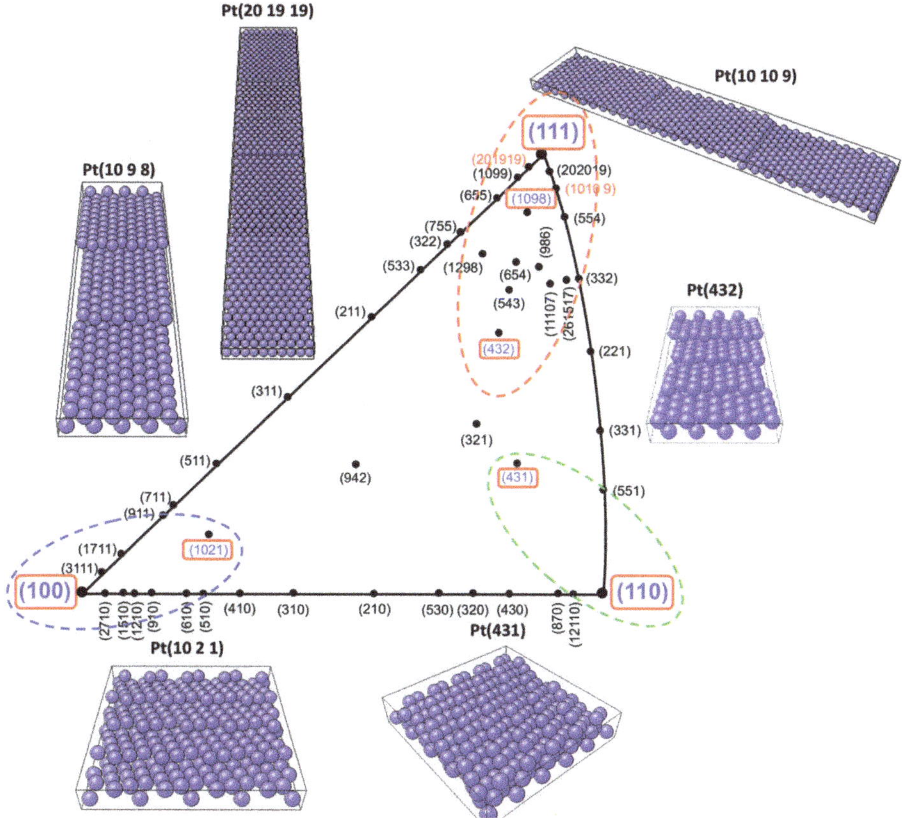

Figure 1. Unit triangle of stereographic projection, showing the location of various Pt surfaces. Ball models of (20 19 19), (10 10 9), (10 9 8), (10 2 1), (432), and (431) planes are depicted in blue.

In spite of the fundamental and practical importance of the electro-oxidation of bulk CO in electrolyte solutions in the lower potential region, the reaction on single-crystal electrodes was studied only on some Pt(111)-related surfaces [12,15–18], Pt(110) [16], and Pt(100) [16,19]. However, in order to increase the reaction rate of the bulk CO electro-oxidation in the lower potential region, the reaction on surfaces with "atomically-controlled" defects also must be elucidated. The information thus obtained could be applied to the development of catalysts for residential fuel cells by avoiding the poisoning of Pt catalysts with hydrogen-rich fuel gas (reformate) produced from hydrocarbon fuels. In this work, we carried out potential cycling in a CO-saturated $HClO_4$ solution not only on basal Pt(111), (110), and (100), but also on surfaces with atomically zigzag steps of Pt(10 10 9), (10 9 8), (10 2 1), (432), and (431). On all the surfaces, the electro-catalytic activity in the pre-peak region decreased and disappeared after the potential cycling between 0.07 and 0.95 V in CO-saturated $HClO_4$. By examining cyclic voltammograms (CVs) in N_2-saturated $HClO_4$ before and after the potential cycling in the CO-saturated solution, we propose the surface reconstruction on all the Pt electrodes. Potential-step experiments were also carried out on Pt(111) to elucidate the possible mechanism of the Pt deactivation during the potential cycles.

2. Materials and Methods

Single-crystal beads of Pt, approximately 3 mm in diameter, were made by crystallization of a molten ball formed at the ends of Pt wires in a hydrogen/oxygen flame as reported previously [15,20–22]. The reflection of a laser beam from the crystal facets were used to determine the orientation of the single-crystal bead for exposing either Pt(111) or the (100) plane (Figure S1, Supporting Information). Each plane was then mechanically polished with successively finer-grade diamond pastes down to 0.25 μm with an accuracy of <0.2°. For Pt(110), (10 10 9), (10 9 8), (10 2 1), (432), and (431) planes, single-crystal beads were installed in a 4-axis X-ray diffractometer (Rigaku, Tokyo, Japan) for determining those planes prior to the mechanical polishing [15,23]. All of the samples were treated in a hydrogen/oxygen flame for two hours to eliminate surface damage caused by the mechanical polishing. Each sample was placed in an infrared image furnace filled with hydrogen [15,22], heated up to 1450 K, and gradually cooled down to 473 K, at which the gas was exchanged to Ar. The sample was taken out and placed in ultrapure water (18.2 MΩ, Milli-Q, Merck Millipore, Billerica, MA, USA). The Pt electrode covered with ultrapure water was carefully and immediately transferred into an electrochemical cell filled with a 0.1 M $HClO_4$ (ultrapure grade, Kanto Chemical, Tokyo, Japan) solution saturated either with N_2 or CO. The continuous potential cycling in a CO-saturated solution was carried out between 0.07 and 0.95 V at 50 mV s^{-1} (35.2 s for each cycle), whereas the CVs in a N_2-saturated solution were recorded between 0.05 and 0.95 V at 50 mV s^{-1} before and after the 30 potential cycles in a CO-saturated solution. On Pt(111), the potential-step measurements were carried out. The sample was first exposed to a CO-saturated solution at 0.05 V for 2 min. Then, the potential was stepped and kept at 0.60 V for 6 min, then at 0.95 V for 2 min, and back at 0.60 V for 6 min. This 16-min procedure was repeated 26 times, and the current density was simultaneously measured. All electrochemical measurements were carried out at 293 K using a potentiostat (PGSTAT128N with an analog scan module, Metrohm Autolab, Herisau, Switzerland) with the hanging meniscus method in a three-compartment electrochemical cell (Figure S2, Supporting Information). All electrode potentials in this work are referred to the RHE.

3. Results and Discussion

3.1. CVs on Pt Single-Crystal Surfaces in CO-Saturated $HClO_4$

The continuous potential cycling was carried out on Pt samples in a 0.1 M $HClO_4$ solution saturated with CO. After 30 cycles, the steady-state CV curves showed no change on all samples. Figure 2(a-1)–(c-1) show CVs obtained in N_2-saturated $HClO_4$ before (black line) and after (red line) the 30 potential cycles in CO-saturated $HClO_4$ on Pt(111), (10 9 8), and (432), respectively. Those three planes are related to Pt(111) (Figure 1). On the Pt(111) surface used in this study, a small amount of (111) steps existed as reported in our previous study [15]. On (10 9 8) and (432), ideally, steps are composed of alternating (111) and (100) steps with one-atomic width (Figure 1). The average step density was estimated to be 1/9 on Pt(10 9 8) and 1/4 on Pt(432). The hydrogen and hydroxide adsorption/desorption regions are seen at potentials lower and higher than 0.5 V, respectively. A subtle growth of the spikes at around 0.8 V was seen on Pt(111), which is indicative of a small increase in the flatness of the terraces [15,23]. On (10 9 8) and (432), very small changes are observed at the peaks observed in both hydrogen and hydroxide adsorption/desorption regions. In H_2SO_4 solutions, the peaks around 0.13 and 0.28 V have been reported to be associated with (110) and (100) planes [23]. If this is also the case in $HClO_4$, both surfaces had characteristics of (110) and (100) planes, which is evident from the ball models shown in Figure 1, although detailed structural analysis should be carried out based on techniques, such as in situ STM [15]. Figure 2(a-2)–(c-2) show CVs on Pt(111), (10 9 8), and (432), respectively, obtained in a 0.1 M $HClO_4$ solution saturated with CO. The potential was swept repetitively from 0.07 to 0.95 V with 30 cycles at a scan rate of 50 mV s^{-1}. Only the CVs for positive-going scans are shown in the figures. Figure 2(a-3)–(c-3) are enlargements of the lower potential range in 2(a-2), 2(b-2), and 2(c-2), respectively. The current due to CO oxidation initially commenced at 0.37 V on Pt(111), at 0.43 V on (10 9 8), and at 0.42 V on (432). A broad anodic oxidation feature, often termed a "pre-peak", but referred to more properly as "low-potential CO oxidation current", at about 0.5 to 0.7 V was seen, followed by a sharp spike at around 0.8 V, known as the main peak. After 30 cycles, practically no current was seen at the lower potential region than the main peak on all surfaces. Because the CVs in the N_2-saturated $HClO_4$ solution (Figure 2(a-1)–(c-1)) were similar before and after the potential cycles, the surface structures might not be drastically changed; only atomic structures of steps might have been changed as reported on Pt(111) and stepped Pt(111) with either (111) or (100) steps [15]. In the case of (111) steps on Pt(111), the increase of the step density increase the "low-potential CO oxidation current" [15]. However, for Pt(10 9 8) and (432), the increase in the step density by a factor of 2.3 rather decreased the current (Figure 2). This could be explained by the difference in the actual configurations at the steps at Pt(10 9 8) and (432), although in the ideal models shown in Figure 1, the steps are identical on both surfaces. The ratio of the peak currents at 0.13 and 0.28 V are different on Pt(10 9 8) and (432) in an N_2-saturated solution, which also indicates that the step structures on Pt(10 9 8) and (432) are different. Interestingly, the decrease of the "low-potential CO oxidation current" was similar on Pt(111) and (10 9 8). Therefore, the step structure on Pt(10 9 8) might be similar to that on Pt(111).

Figure 3 shows CVs on Pt(110) and (431). Pt(431) is oriented with (110), ideally with only a 2-atomic width of the (111) terraces (see Figure 1). On Pt(110), the CV in N_2-saturated $HClO_4$ before the potential cycles (black line in Figure 3(a-1)) was identical to those reported previously [24–26]. By in situ STM, the surface prepared in the same manner as in this study was reported to consist of parallel atomic rows aligned along the [110] direction, with an atomically-resolved (1 × 1) structure [22]. After the potential cycles in CO-saturated $HClO_4$, many characteristic peaks in the CV were lost (red line in Figure 3(a-1)). Therefore, the surface structure must be largely reconstructed. The lost peaks might indicate a reconstruction into "dim" surface structures with no distinct steps and terraces after the potential cycles. The CVs on Pt(431) in N_2-purged 0.1 M $HClO_4$ obtained before and after the potential cycles in a CO-saturated solution showed a large peak at 0.1 V in the hydrogen adsorption/desorption region. Little change was observed before and after the potential cycles (Figure 3(b-1)) in contrast to the results on Pt(110) (Figure 3(a-1)). This is indicative of little change in the surface structure after the potential cycles on Pt(431), because of the unique surface morphology of Pt(431); the defect sites were densely and uniformly arranged on the surface, which cause the surface energy to be rather small. Figure 3(a-2),(b-2) show CVs on Pt(110) and (431) obtained in a 0.1 M $HClO_4$ solution saturated with CO, respectively. Figure 3(a-3),(b-3) are enlargements of the lower potential ranges. On Pt(110), a small "low-potential CO oxidation current" with an onset of 0.35 V was seen to decrease as the potential was cycled. On Pt(431), the "low-potential CO oxidation current" commencing at 0.40 V during the first scan was comparable to those on Pt(111)-related surfaces (Figure 2), which abruptly decreased even at the second cycle. Because we expected a large "low-potential CO oxidation current" on this highly-defected surface, the results were surprising. In our previous paper, we concluded that bulk CO electro-oxidation at potentials lower than 0.5 V proceed at surface sites not fully covered by CO molecules [15]. If this reaction model is correct, a small amount of sites not covered by CO existed before the potential cycles on Pt(431), but the sites quickly disappeared and the surface was efficiently covered by CO during the first cycle. To conclude, bulk CO electro-oxidation at low potentials is not necessarily high at a surface with many defects; the absorption structure of CO molecules must be simultaneously taken into account, too.

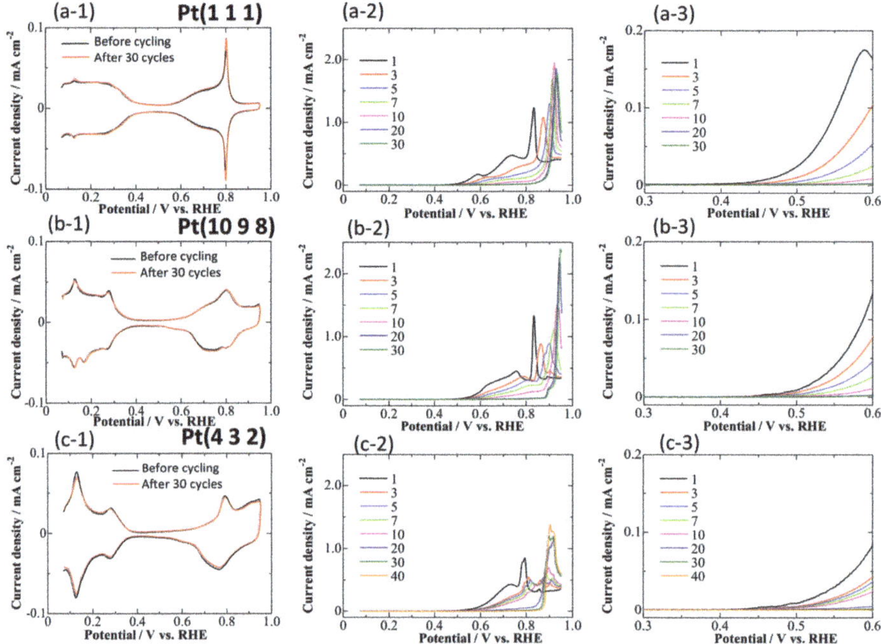

Figure 2. (**a-1**), (**b-1**), and (**c-1**) show CVs obtained in N_2-saturated $HClO_4$ before (black line) and after (red line) the 30 potential cycles in CO-saturated $HClO_4$ on Pt(111), (10 9 8), and (432), respectively. Scan rate = 50 mV s^{-1}. (**a-2**), (**b-2**), and (**c-2**) show the results of the CVs on Pt(111), (10 9 8), and (432), respectively, obtained in a 0.1 M $HClO_4$ solution saturated with CO. Scan rate = 50 mV s^{-1}. (**a-3**), (**b-3**), and (**c-3**) are enlargements of the lower potential range in (**a-2**), (**b-2**), and (**c-2**), respectively.

Figure 4 shows CVs on Pt(100) and (10 2 1). On (10 2 1), the steps are ideally all (111), but kinks are located on the four-fold sites of Pt(100) (CN = 6), which are unique to this surface. On Pt(100), the CV in N_2-saturated $HClO_4$ before the potential cycles (black line in Figure 3(a-1)) was identical to those reported previously [26]. In situ STM images with Pt atoms forming (100)–(1 × 1) structure were already reported on the surfaces after the similar sample preparation procedures [27,28]. On Pt(100), the electro-oxidation of bulk CO commenced at 0.29 V, the lowest potential among the surfaces investigated in this study. On Pt(10 2 1), the onset potential for CO oxidation was 0.35 V, higher by 0.06 V than on Pt(100). After the potential cycles in CO-saturated solution, the shape of the CVs changed on both surfaces [15], but not in an extinctive manner as on Pt(110). Although the onset potential was very low on Pt(100), the "low-potential CO oxidation current" became non-observable even at the second cycle. Therefore, the sites for the electro-oxidation of bulk CO were extremely active, but unstable on Pt(100). Since the onset potential of Pt(100) was lower than on Pt(10 2 1) with kinks, the active sites on Pt(100) might be those with lower CN than kinks, or isolated Pt adatoms (CN = 4) or clusters composed of a few Pt atoms (CN = 4–6) as proposed previously [15,16,27,29]. However, as already mentioned, the reaction rate of the electro-oxidation of bulk CO is not determined only by the numbers of sites with smaller CNs, but also by the structure of the CO adlayer [15,28].

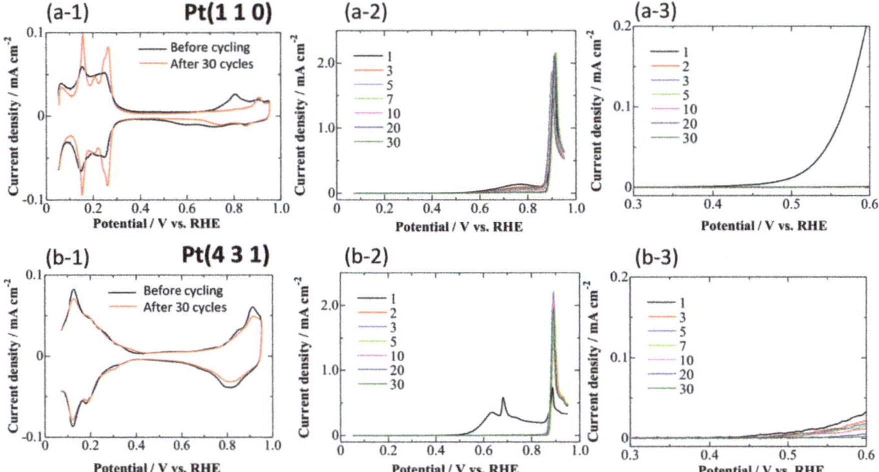

Figure 3. (**a-1**) and (**b-1**) show CVs obtained in N_2-saturated $HClO_4$ before (black line) and after (red line) the 30 potential cycles in CO-saturated $HClO_4$ on Pt(110) and (431), respectively. Scan rate = 50 mV s^{-1}. (**a-2**) and (**b-2**) show the results of the CVs on Pt(110) and (431), respectively, obtained in a 0.1 M $HClO_4$ solution saturated with CO. Scan rate = 50 mV s^{-1}. (**a-3**) and (**b-3**) are enlargements of the lower potential range in (**a-2**) and (**b-2**), respectively.

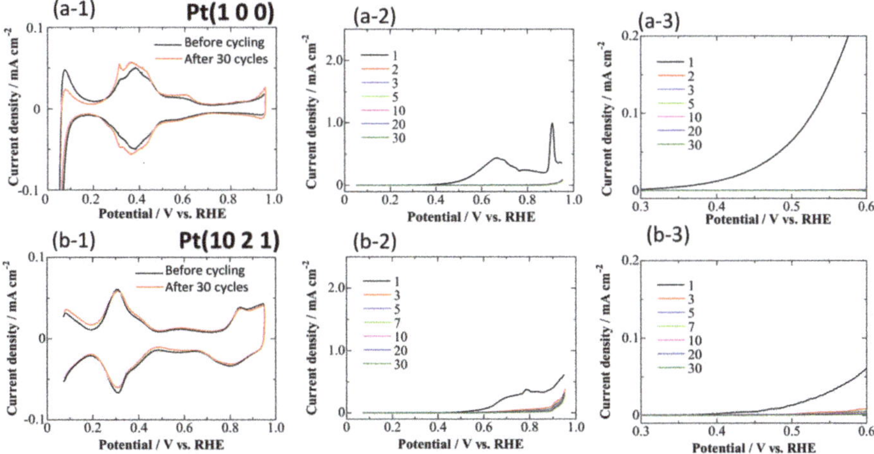

Figure 4. (**a-1**) and (**b-1**) show CVs obtained in N_2-saturated $HClO_4$ before (black line) and after (red line) the 30 potential cycles in CO-saturated $HClO_4$ on Pt(100) and (10 2 1), respectively. Scan rate = 50 mV s^{-1}. (**a-2**) and (**b-2**) show the results of the CVs on Pt(100) and (10 2 1), respectively, obtained in a 0.1 M $HClO_4$ solution saturated with CO. Scan rate = 50 mV s^{-1}. (**a-3**) and (**b-3**) are enlargements of the lower potential range in (**a-2**) and (**b-2**), respectively.

In Table 1, the onset potentials during the first potential scans are listed on all the Pt electrodes. It is clear that the numbers of defects shown in the ideal structures (Figure 1) are not related to the order of the onset potentials. This might be indicating that the structures of actual active sites are mostly not as those shown in Figure 1. In our previous paper, it was reported that the active sites on Pt(111) existed preferentially at steps and were very small in number [15]. Even by using single-crystal surfaces with zigzag steps in this study, the active sites for bulk CO electro-oxidation were not increased. Interestingly, the onset potentials were always lower on the basal planes rather than on stepped surfaces on Pt(111)-, Pt(110)-, and Pt(100)-oriented surfaces. This might imply that those highly-active sites were adatoms or clusters, having low CNs, on the basal planes or at steps, as discussed on Pt(100). The instability of those sites is discussed in the following section.

Table 1. Onset potentials for bulk CO electro-oxidation in 0.1 M $HClO_4$ saturated with CO.

| | Pt(111)-Oriented | | | Pt(110)-Oriented | | Pt(100)-Oriented | |
	Pt(111)	Pt(10 9 8)	Pt(432)	Pt(110)	Pt(431)	Pt(100)	Pt(10 2 1)
Onset potential/V vs. RHE	0.37	0.43	0.42	0.35	0.40	0.29	0.35

3.2. Potential Stepping on Pt(111) in CO-Saturated HClO$_4$

On all Pt surfaces with different types of defects, the current for the CO electro-chemical oxidation in the lower potential region disappeared after potential cycles. By ex situ [16] and operando [15] STM, it has been reported that the loss in the activity for the CO electro-oxidation was related to the transformation of Pt surface structures by a reduction of the surface defects [16]. In our previous report, the deactivation of the electrodes was connected with the formation of the single atomic Pt species [15]:

Anodic scan (up to 0.95 V):

$$Pt(OH)_2 \rightarrow PtO_2 + 2H^+ + 2e^- \tag{1}$$

$$PtO + H_2O \rightarrow PtO_2 + 2H^+ + 2e \tag{2}$$

Cathodic scan (lower than 0.80 V):

$$PtO_2 + 2H_2O + 4H^+ + 2e^- \rightarrow Pt(H_2O)_4{}^{2+} \tag{3}$$

$$Pt(H_2O)_4{}^{2+} + 4CO \rightarrow Pt(CO)_4{}^{2+} + 4H_2O \tag{4}$$

$$Pt(CO)_4{}^{2+} + 2e^- \rightarrow Pt(CO)_2 + 2CO \tag{5}$$

At 0.95 V, bulk CO electro-oxidation and the formation of Pt oxides must competitively proceed on surface, where Pt atoms with low CNs are expected to be easily oxidized. When the potential was scanned in the cathodic direction, those (single or several atomic) oxides were reduced to $Pt(CO)_4{}^{2+}$ or $Pt(CO)_2$ on the surface. The single atomic Pt species, either as $Pt(CO)_4{}^{2+}$ or $Pt(CO)_2$, would then have the opportunity to redeposit at the nearby site in an energetically favorable configuration. In the previous section, the active sites on Pt(100) were proposed as adatoms or atomically small clusters. $Pt(CO)_4{}^{2+}$ or $Pt(CO)_2$ could be preferentially formed from those isolated Pt adatoms with small CNs, therefore, those active sites should be very unstable as shown in Figure 4.

In addition to CV measurements (Figures 2–4), we carried out potential stepping [18,30]. The Pt(111) surface was used because the atomically-resolved structural information was obtained only on Pt(111)-related single-crystal electrodes during bulk CO electro-oxidation [15]. The sample was exposed to a CO-saturated solution at 0.05 V for 2 min, stepped to 0.60 V and left for 6 min, to 0.95 V for 2 min, then back to 0.60 V for 6 min. The current density obtained at 0.60 V was chosen as a measure for the CO electro-oxidation in the lower potential region. This 16 min procedure was cycled 26 times. The current density was simultaneously recorded. Figure 5 shows the current density during the potential-step cycles. The current density was initially 0 mA cm^{-2} at 0.05 V, then abruptly increased to 0.17 mA cm^{-2} as the potential was stepped to 0.60 V at 2 min (black line in Figure 5). The current density slowly decreased to 0.06 mA cm^{-2} during the 6-min interval. Subsequently, the current density increased sharply to 0.5 mA cm^{-2} upon the potential step to 0.95 V at 8 min and decreased to 0.25 mA cm^{-2} at 10 min. When the potential was switched down to 0.60 V at 10 min, after a cathodic spike, the current density very slowly decreased down to 0.02 mA cm^{-2} in 6 min. Interestingly, after the sample was again kept at 0.05 V for 2 min, the current density jumped to 0.14 mA cm^{-2} accompanied by the potential step to 0.60 V, slightly lower than the first jump (0.17 mA cm^{-2}). The current decreased to 0.04 mA cm^{-2} at 0.60 V after 6 min (or at 8 min) in the second cycle. Indeed, the current density at 0.60 V between 2 and 8 min in the (N + 1)th (N = 1 to 25) cycle was always larger than that at 0.60 V between 10 and 16 min in the Nth cycle, showing the partial recovery of the activity in the lower potential region. Between 2 and 8 min and between 10 and 16 min at 0.6 V, the current density at the corresponding period in the (N + 1)th cycle was always smaller than that in the Nth cycle (Figure 5), showing a continuous deactivation. Between 8 and 10 min at 0.95 V, the current density was not in an order. This discrepancy at 0.95 V might be due to the diffusion of the solution not controlled in the experimental configuration shown in Figure S2. Rotation disk electrode measurements are needed for detailed analyses.

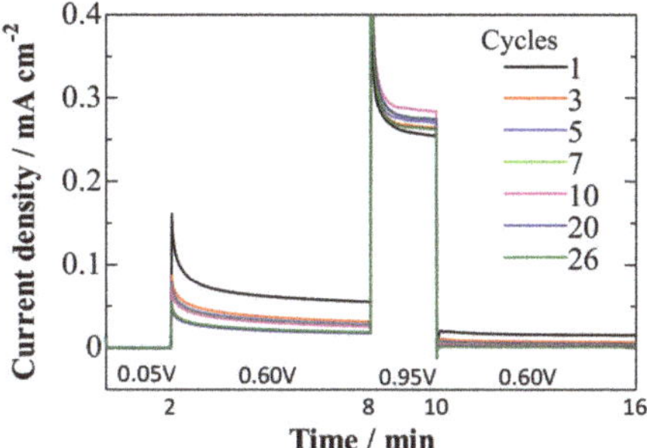

Figure 5. Current density obtained on Pt(111) during the potential steps in CO-saturated HClO$_4$.

Operando STM measurements were carried out on Pt(111) in CO-saturated HClO$_4$ at 0.05, 0.60, and 0.95 V. At 0.95 V, the atomic arrangements were not able to be obtained because of the continuous electro-oxidation of CO, although the terraces were observed to be flat. At 0.05 and 0.60 V, the (2 × 2)-3 CO structure was steadily observed as shown in Figure 6. Therefore, the CO adlayer was rigid and stable between 0.05 and 0.60 V. It is also understood that the bulk CO electro-oxidation at 0.6 V proceeded at steps. The reconstruction of steps without potential cycling either at 0.05 or 0.60 V was not observed by STM, indicating that the mobile sites were very limited in number.

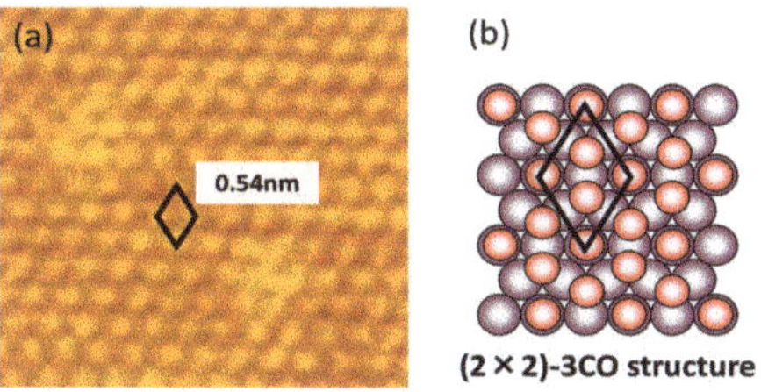

Figure 6. (**a**): Operando STM image of a CO adlayer on a Pt(111) terrace forming (2 × 2)-3CO obtained in 0.1 M HClO$_4$ saturated with CO at 0.60 V vs. RHE during the CO electro-oxidation. A unit cell is depicted as a rhombus. The length of the sides was 0.54 nm, and the angle was 120°. (**b**): Structural model of a (2 × 2)-3CO adlayer on Pt(111)–(1 × 1). The red and purple circles represent CO molecules and Pt atoms, respectively.

Based on the results obtained above, the deactivation processes during the potential steps were modeled.

1. At 0.95 V, the surface was continuously deactivated by the simultaneous formation of surface oxides (Figure 5, Equations (1) and (2)). Pt adatoms or Pt clusters with low CNs were preferentially oxidized. Those oxidized species were immobile on the surface.

2.   The current density at 0.60 V, after being treated at 0.95 V, became much smaller (Figure 5) along with the formation of a stable CO adlayer on the terraces (Figure 6). $Pt(CO)_4^{2+}$ (Equation (4)) or $Pt(CO)_2$ (Equation (5)) species were also formed after the potential step by the reduction of the (single or several atomic) Pt oxides. We believe that, at least partially, $Pt(CO)_4^{2+}$ remained on the surface at 0.60 V after the potential step from 0.95 V, because Pt oxides were not completely reduced to metal Pt at 0.60 V after going through 0.95 V, as seen in the CVs obtained in pure $HClO_4$ (Figures 2–4). Because the reactivity did not change very much at 0.60 V, the surface morphology did not change either. Therefore, the planar and electronically charged $Pt(CO)_4^{2+}$ species, inactive towards the bulk CO electro-oxidation, might be immobile on the Pt surface.

3.   At 0.05 V, all $Pt(CO)_4^{2+}$ species became $Pt(CO)_2$. The current density at 0.60 V after treatment at 0.05 V became larger than that after treatment at 0.95 V (Figure 5), because of the formation of $Pt(0)(CO)_2$ from $Pt(II)(CO)_4^{2+}$; metallic Pt species can be active towards bulk CO electro-oxidation. The reaction rate, on the other hand, was continuously lowered at 0.60 V. This is because of the deposition of mobile $Pt(CO)_2$ at the energetically favorable adjacent site on the surface, thus lowering its CN and eventually its reaction activity.

4.   Repeating the treatments at 0.05 and 0.95 V gradually decreased the reaction rate of the bulk CO electro-oxidation in the lower potential region (Figure 5). The potential cycling between 0.07 and 0.95 V much enhanced the deactivation process on Pt electrodes (Figures 2–4) compared with the potential-step treatment (Figure 5), possibly because of the continuous morphological change during the potential cycling.

4. Conclusions

The electro-oxidation of CO was investigated on Pt single-crystal electrodes with different surface orientations in CO-saturated 0.1 M $HClO_4$. On all Pt(111)-, (110)-, and (100)-oriented surfaces, the "low-potential CO oxidation current" disappeared during the potential cycles between 0.07 and 0.95 V in a CO-saturated solution. During the potential cycles, the surface structures changed. On Pt(111)-oriented surfaces, the reconstruction proceeded only at steps [15]. On Pt(110), the reconstruction was very large, probably changing the terrace structures too. On Pt(110)-oriented Pt(431), the change in surface structure was larger than Pt(111)-oriented surfaces, but much smaller than Pt(110), in spite of the largest numbers of steps and kinks, probably because of the less structural irregularity and smaller surface energy compared with Pt(110)–(1 × 1). On Pt(100)-related surfaces, the electro-oxidation of bulk CO proceeded at lower potentials than on other surfaces (Table 1). Especially on Pt(100), the onset potential was as low as 0.29 V. The unstable Pt adatoms or clusters were proposed to exist on the surface.

The potential-step measurements were carried out on Pt(111) in CO-saturated $HClO_4$. At 0.95 V, the Pt atoms with low CNs were preferentially oxidized, which was then reduced to $Pt(CO)_4^{2+}$ or $Pt(CO)_2$ upon the potential step to 0.60 V. $Pt(CO)_4^{2+}$ was immobile at 0.60 V and inactive towards the bulk CO electro-oxidation. At 0.05 V, only $Pt(CO)_2$ existed as a single Pt species on the surface, which were mobile and decreased its CN by the self-deposition at a stable site; the surface reactivity thus decreased. On pure Pt electrodes, unfortunately, it seems improbable to maintain the "low-potential CO oxidation current" after potential cycles in CO-saturated $HClO_4$. The active sites, allowing the adsorption of water molecules through the CO adlayer [15], must be structurally maintained eventually by a surface modification, such as by alloying.

The systematic investigation of the atomically-arranged defects on electrodes has started only recently [15,16]. The combination of defects and adsorbates may create highly-active sites, but are very small in number. As reported in this paper, those created sites could be unstable and changeable too. New analytical technologies and theoretical backbones must be established and applied to this new field.

Supplementary Materials: The following are available online at http://www.mdpi.com/2571-9637/2/2/23/s1, Figure S1. Schematic illustration of the adjustment of (111) and (100) facets by a laser beam reflection method.

Figure S2. Schematic illustration of a tree-compartment cell. The electrochemical measurements were carried out with the hanging meniscus method.

Author Contributions: Conceptualization, J.I.; Methodology, H.U. and J.I.; Validation, H.U. and J.I.; Investigation, K.A.; Resources, H.U. and J.I.; Writing—original draft preparation, J.I.; Writing—review and editing, H.U. and J.I.; Visualization, K.A. and J.I.; Supervision, H.U. and J.I.; Project administration, J.I.; Funding acquisition, J.I.

Funding: This study was supported by Grant-in-Aid for Scientific Research (KAKENHI) and the New Energy and Industrial Technology Development Organization (NEDO), Japan.

Acknowledgments: The authors thank Prof. Nagakazu Furuya of the University of Yamanashi for the stepped-sample preparation

Conflicts of Interest: The authors declare no conflict of interest.

References

1. Markovic, N.M.; Ross, P.N. Surface science studies of model fuel cell electrocatalysts. *Surf. Sci. Rep.* **2002**, *45*, 117–229. [CrossRef]
2. Koper, M.T.M.; Lai, S.C.S.; Herrero, E. Mechanisms of the Oxidation of Carbon Monoxide and Small Organic Molecules at Metal Electrodes. In *Fuel Cell Catalysis*; John Wiley & Sons, Inc.: Hoboken, NJ, USA, 2008; pp. 159–207.
3. Korzeniewski, C.; Climent, V.; Feliu, J.M. Electrochemistry at Platinum Single Crystal Electrodes. In *Electrochemistry: A Series of Advances*; Bard, A.J., Zoski, C., Eds.; CRC Press: Boca Raton, FL, USA, 2012; Volume 24, pp. 75–169.
4. Li, Q.-X.; Huo, S.-J.; Ma, M.; Cai, W.-B.; Osawa, M. Ubiquitous Strategy for Probing ATR Surface-Enhanced Infrared Absorption at Platinum Group Metal–Electrolyte Interfaces. *J. Phys. Chem. B* **2005**, *109*, 7900–7906. [CrossRef]
5. Atkins, P.W. *Physical Chemistry*, 6th ed.; Oxford University Press: Oxford, UK, 1998.
6. Lopez-Cudero, A.; Cuesta, A.; Gutierrez, C. Potential dependence of the saturation CO coverage of Pt electrodes: The origin of the pre-peak in CO-stripping voltammograms. Part 2: Pt(100). *J. Electroanal. Chem.* **2006**, *586*, 204–216. [CrossRef]
7. Lopez-Cudero, A.; Cuesta, A.; Gutierrez, C. Potential dependence of the saturation CO coverage of Pt electrodes: The origin of the pre-peak in CO-stripping voltammograms. Part 1: Pt(111). *J. Electroanal. Chem.* **2005**, *579*, 1–12. [CrossRef]
8. Lebedeva, N.P.; Koper, M.T.M.; Herrero, E.; Feliu, J.M.; van Santen, R.A. Cooxidation on stepped Pt[n(111)×(111)] electrodes. *J. Electroanal. Chem.* **2000**, *487*, 37–44. [CrossRef]
9. Lebedeva, N.P.; Koper, M.T.M.; Feliu, J.M.; van Santen, R.A. Role of Crystalline Defects in Electrocatalysis: Mechanism and Kinetics of CO Adlayer Oxidation on Stepped Platinum Electrodes. *J. Phys. Chem. B* **2002**, *106*, 12938–12947. [CrossRef]
10. Lebedeva, N.P.; Rodes, A.; Feliu, J.M.; Koper, M.T.M.; van Santen, R.A. Role of Crystalline Defects in Electrocatalysis: CO Adsorption and Oxidation on Stepped Platinum Electrodes as Studied by in situ Infrared Spectroscopy. *J. Phys. Chem. B* **2002**, *106*, 9863–9872. [CrossRef]
11. Chen, Q.-S.; Berna, A.; Climent, V.; Sun, S.-G.; Feliu, J.M. Specific reactivity of step sites towards CO adsorption and oxidation on platinum single crystals vicinal to Pt(111). *Phys. Chem. Chem. Phys.* **2010**, *12*, 11407. [CrossRef] [PubMed]
12. Farias, M.J.S.; Camera, G.A.; Feliu, J.M. Understanding the CO Preoxidation and the Intrinsic Catalytic Activity of Step Sites in Stepped Pt Surfaces in Acidic Medium. *J. Phys. Chem. C* **2015**, *119*, 20272–20282. [CrossRef]
13. Mayrhofer, K.J.J.; Blizanac, B.B.; Arenz, M.; Stamenkovic, V.R.; Ross, P.N.; Markovic, N.M. The Impact of Geometric and Surface Electronic Properties of Pt-Catalysts on the Particle Size Effect in Electrocatalysis. *J. Phys. Chem. B* **2005**, *109*, 14433–14440. [CrossRef]
14. Marković, N.M.; Grgur, B.N.; Lucas, C.A.; Ross, P.N. Electrooxidation of CO and H₂/CO Mixtures on Pt(111) in Acid Solutions. *J. Phys. Chem. B* **1999**, *103*, 487–495. [CrossRef]
15. Inukai, J.; Tryk, D.A.; Abe, T.; Wakisaka, M.; Uchida, H.; Watanabe, M. Direct STM Elucidation of the Effects of Atomic-Level Structure on Pt(111) Electrodes for Dissolved CO Oxidation. *J. Am. Chem. Soc.* **2013**, *135*, 1476–1490. [CrossRef]

16. Strmcnik, D.S.; Tripkovic, D.V.; van der Vliet, D.; Chang, K.C.; Komanicky, V.; You, H.; Karapetrov, G.; Greeley, J.; Stamenkovic, V.R.; Markovic, N.M. Unique Activity of Platinum Adislands in the CO Electrooxidation Reaction. *J. Am. Chem. Soc.* **2008**, *130*, 15332–15339. [CrossRef] [PubMed]

17. Somorjai, G.; Li, Y. *Introduction to Surface Chemistry and Catalysis*, 2nd ed.; John Wiley & Sons: Hoboken, NJ, USA, 2010.

18. Batista, E.A.; Iwasita, T.; Viestich, W. Mechanism of Stationary Bulk CO Oxidation on Pt(111) Electrodes. *J. Phys. Chem. B* **2004**, *108*, 14216–14222. [CrossRef]

19. Rednev, A.V.; Wandlowski, T. An influence of pretreatment conditions on surface structure and reactivity of Pt(100) towards CO oxidation reaction. *Russ. J. Electrochem.* **2012**, *72*, 259–270. [CrossRef]

20. Itaya, K. In situ scanning tunneling microscopy in electrolyte solutions. *Prog. Surf. Sci.* **1998**, *58*, 121–247. [CrossRef]

21. Wakisaka, M.; Ohkanda, T.; Yoneyama, T.; Uchida, H.; Watanabe, M. Structures of a CO adlayer on a Pt(100) electrode in HClO4 solution studied by in situ STM. *Chem. Commun.* **2005**, 2710–2712. [CrossRef]

22. Wakisaka, M.; Asizawa, S.; Yoneyama, T.; Uchida, H.; Watanabe, M. In Situ STM Observation of the CO Adlayer on a Pt(110) Electrode in 0.1 M HClO4 Solution. *Langmuir* **2010**, *26*, 9191–9194. [CrossRef]

23. Furuya, N.; Koide, S. Hydrogen adsorption on platinum single-crystal surfaces. *Surf. Sci.* **1989**, *220*, 18–28. [CrossRef]

24. Huerta, F.; Mollaron, E.; Quijada, C.; Vezquez, J.L.; Perez, J.M.; Aldaz, A. The adsorption of methylamine on Pt single crystal surfaces. *J. Electroanal. Chem.* **1999**, *467*, 105–111. [CrossRef]

25. Dederichs, F.; Friedrich, K.A.; Daum, W. Sum-Frequency Vibrational Spectroscopy of CO Adsorption on Pt(111) and Pt(110) Electrode Surfaces in Perchloric Acid Solution: Effects of Thin-Layer Electrolytes in Spectroelectrochemistry. *J. Phys. Chem. B* **2000**, *104*, 6626–6632. [CrossRef]

26. Wakisaka, M.; Morishima, S.; Hyuga, Y.; Uchida, H.; Watanabe, M. Electrochemical behavior of Pt–Co(111), (100) and (110) alloy single-crystal electrodes in 0.1 M HClO4 and 0.05 M H2SO4 solution as a function of Co content. *Electrochem. Commun.* **2012**, *18*, 55–57. [CrossRef]

27. Sashikata, K.; Sugata, T.; Sugimasa, M.; Itaya, K. In Situ Scanning Tunneling Microscopy Observation of a Porphyrin Adlayer on an Iodine-Modified Pt(100) Electrode. *Langmuir* **1998**, *14*, 2896–2902. [CrossRef]

28. Kibler, L.A.; Cuesta, A.; Kleinert, M.; Kolb, D.M. In-situ STM characterisation of the surface morphology of platinum single crystal electrodes as a function of their preparation. *J. Electroanal. Chem.* **2000**, *484*, 73–82. [CrossRef]

29. Arenz, M.; Mayrhofer, K.J.J.; Stamenkovic, V.; Blizanac, B.B.; Tomoyuki, T.; Ross, P.N.; Markovic, N.M. The Effect of the Particle Size on the Kinetics of CO Electrooxidation on High Surface Area Pt Catalysts. *J. Am. Chem. Soc.* **2005**, *127*, 6819–6829.

30. Giz, M.J.; Batista, E.A.; Vielstich, W.; Iwasita, T. A topographic view of the Pt(111) surface at the electrochemical interface in the presence of carbon monoxide. *Electrochem. Commun.* **2007**, *9*, 1083–1085. [CrossRef]

 *surfaces*

Article

Spatially Resolved XPS Characterization of Electrochemical Surfaces

Benedetto Bozzini [1], **Danjela Kuscer** [2], **Matteo Amati** [3], **Luca Gregoratti** [3,*], **Patrick Zeller** [3], **Tsvetina Dobrovolska** [4] **and Ivan Krastev** [4]

[1] Dipartimento di Ingegneria dell'Innovazione, Università del Salento, v. Monteroni, 73100 Lecce, Italy; benedetto.bozzini@unisalento.it

[2] Jožef Stefan Institute, Jamova cesta 39, 1000 Ljubljana, Slovenia; Danjela.Kuscer@ijs.si

[3] Elettra - Sinctrotrone Trieste S.C.p.A. S.S. 14, km 163.5 in Area Science Park, 34149 Trieste-Basovizza, Italy; matteo.amati@elettra.eu (M.A.); patrick.zeller@elettra.eu (P.Z.)

[4] Institute of Physical Chemistry, Bulgarian Academy of Sciences, 1113 Sofia, Bulgaria; tsvetina@ipc.bas.bg (T.D.); krastev@ipc.bas.bg (I.K.)

* Correspondence: luca.gregoratti@elettra.eu

Received: 20 February 2019; Accepted: 9 April 2019; Published: 15 April 2019

Abstract: Synchrotron-based scanning photoelectron microscopy (SPEM) has opened unique opportunities for exploiting processes occurring at surfaces and interfaces, which control the properties of materials for electrochemical devices, where issues of chemical and morphological complexity at microscopic length scales should be faced and understood. The present article aims to demonstrate the present capabilities of SPEM to explore the surface composition of micro- and nano-structured materials, focusing on cases relevant to electrochemical technologies. We report and discuss a selection of recent results about three different systems, targeting hot topics in the fields of electrochemical energy storage and electrochemical fabrication: (i) an in-depth analysis of Ag-In electrodeposited alloys exhibiting dynamic pattern formation, (ii) the analysis of electrochemical processes at the electrodes of a self-driven solid oxide fuel cell and (iii) an *operando* characterization of a single-chamber solid oxide fuel cell. The last example has been performed at near-ambient pressure conditions using a unique specially designed setup which extends the traditional capabilities of scanning photoemission microscopes in the ultra-high and high-vacuum regimes to operating conditions that are closer to realistic ones, contributing to overcome the so-called "pressure gap".

Keywords: *operando*; near ambient pressure XPS; scanning photoelectron microscopy; solid oxide fuel cells; surface science; electrodeposited alloys

---

1. Introduction

The ubiquitous and enabling role of electrochemical technologies in many different fields of technology [1], but especially in energy storage [2], cannot be overemphasized. Electrochemical processes and the devices in which they are implemented are highly complex systems in which successful functioning is the result of the cooperation of different types of physics interacting on a range of diverse space- and time-scales. The complex nature of electrochemical systems generally makes the task of understanding and optimizing them an extremely challenging one. One of the main difficulties underlying electrochemistry is the necessity of transferring charge between electrically and ionically conducting phases in processes that are coupled with chemical reactions and phase-transformations. Moreover, these complex processes occur mainly at the electrode–electrolyte interface and involve thin layers of material. In these regions structural, chemical and electronic properties of the materials may significantly deviate from the intrinsic bulk ones. An additional obstacle in the characterization of the reactive interfaces is their position: they are often buried and not easily accessible without disturbing the functionally relevant chemical and structural ambient. In the last decades, traditional bulk-sensitive analysis tools, in particular, X-ray diffraction or fluorescence spectromicroscopy, largely employed in the characterization of electrochemical systems, have been more extensively supported by X-ray based surface sensitive techniques such as X-ray photoelectron spectroscopy (XPS) which can maximize the information confined in thin layers [3]. XPS was initially conceived as a spectroscopy technique necessarily requiring operation in ultra-high vacuum (UHV) or at least high-vacuum (HV) ambient and it was developed and implemented with this approach for many decades. But, as for other fields of science and technology, a state-of-the-art method for the investigation of electrochemical systems should tackle the so-called "material gap" and "pressure gap". The first one is linked to the development of nanomaterials and nanotechnology and the corresponding need of understanding processes at the nanoscale, while the latter aims to analyze an electrochemical reaction in the real pressure working regime. The latter is a challenge for many vacuum-based characterization techniques. In recent years a lot of efforts have been put in trying to overcome the two limitations for XPS, with the successful outcome of developing novel approaches.

The fast progress in the development and implementation of photoelectron spectromicroscopy techniques, i.e., adding spatial resolution and imaging capabilities to XPS, started in the last decade of the 20th century with the construction of low emittance X-ray synchrotron machines, providing tunable energy photon beams with very high brightness and variable polarization [4,5]. Thanks to the high photon intensity provided by insertion devices it has become possible to add sub-micrometer lateral resolution to XPS, realizing new instruments called photoelectron microscopes [6]. In X-ray photoelectron microscopes high spatial resolution is achieved using either full-field or scanning methods. In scanning instruments, called scanning photoelectron microscopes (SPEM), the photon beam is demagnified to sub-micrometer dimensions using suitable photon optics. Modern SPEMs can achieve spatial resolutions as low as 70–100 nm, allowing full material characterization at the nanoscale [7,8]; despite the fact that full-field photoemission microscopes can achieve even better lateral resolutions (20–30 nm), they impose severe constraints in the sample geometry and morphology which, instead, do not affect the performance of scanning ones. This flexibility with respect to sample requirements is particularly crucial for the *operando* characterization of electrochemical systems. Since its construction in the mid of 90's, the soft X-ray SPEM at the Elettra synchrotron research center has investigated many electrochemical systems based on solid state electrolytes and electrodes in the standard UHV or HV environment, often in *operando* conditions [9–11] demonstrating how the material gap can be successfully overcome in the characterization of electrochemical systems. With respect to the pressure gap problem in XPS, a fundamental contribution to overcome it was achieved with the development of the differentially-pumped electron analyzers at the beginning of this century; after almost 20 years of advances, XPS analysis at pressures as high as several tens of millibar can be performed in gas and liquid environments not only at synchrotron beamlines but also with conventional instruments [12,13].

In addition to standard near-ambient pressure XPS approaches (generally denoted by the acronym APXPS) to the characterization of electrochemical systems, now largely diffused [14–16], it is worthwhile to mention a couple of developments particularly adaptable to electrochemistry: (i) the implementation of hard X-rays XPS, which allows spectroscopy investigation of deeper layers [17] such as buried interfaces and (ii) the "dip and tip" method to study solid-liquid interfaces [18,19], available at few synchrotron beamlines. In spite of their innovation all these approaches fail in combining APXPS with submicron spatial resolution; this technological problem started to be solved as recently as in the last decade thanks to the efforts of the team of the Escamicroscopy beamline at Elettra which allowed the development and implementation of innovative technical solutions for SPEM, capable of performing near-ambient pressure photoemission spectromicroscopy while keeping the typical SPEM performance [20,21].

This manuscript is presenting three examples of electrochemical experiments performed with SPEM in three different environmental conditions. In the first experiment, the characterization of an ex situ electrodeposited Ag-In alloy, exhibiting a dynamic pattern formation mechanism, has been performed. In the second experiment the electrochemical processes at the electrodes of a self-driven solid oxide fuel cell (SOFC) have been analyzed; in this case the *operando* condition was achieved in HV pressure regimes, while the last case reports on the *operando* characterization of a single-chamber solid oxide fuel cell (SC-SOFC) at near-ambient pressure conditions, using a newly developed cell.

2. Materials and Methods: the Soft X-rays Scanning Photoemission Microscope of Elettra

2.1. The Standard UHV—HV Setup

The standard configuration of the SPEM was originally designed to work in UHV conditions; a schematic view of the focusing optics and photoelectron detection systems of the SPEM of the Escamicroscopy beamline at Elettra is shown in Figure 1. The X-ray beam is focused on the sample using a diffractive lens called zone plate (ZP), which allows to achieve a spot diameter of ~130 nm as best performance. In the typical experiment the X-ray spot impinges normally to the sample even if a polar rotation is possible. While samples are raster-scanned with respect to the X-ray microprobe, the photoelectrons generated at the sample surface are collected and energy-selected by means of a hemispherical electron analyzer (HEA), equipped with a 48-channels electron detector [22]. The takeoff angle of photoelectrons is fixed to 30°. This configuration strongly enhances the surface sensitivity of the SPEM because due to inelastic collisions only photoelectrons generated in the topmost layers can reach the HEA. A SPEM can be operated in two modes: (i) imaging spectromicroscopy and (ii) microspot spectroscopy. The imaging mode maps the lateral distribution of elements by collecting photoelectrons within a selected kinetic energy window while scanning the specimen with respect to the microprobe. By properly setting the energy range, the distribution of the chemical states of a selected element can be measured, in addition to that of the bare elemental concentration. The microspot mode is similar to conventional XPS, i.e., energy distribution curves are measured from the illuminated local micro-spot area selected within an image.

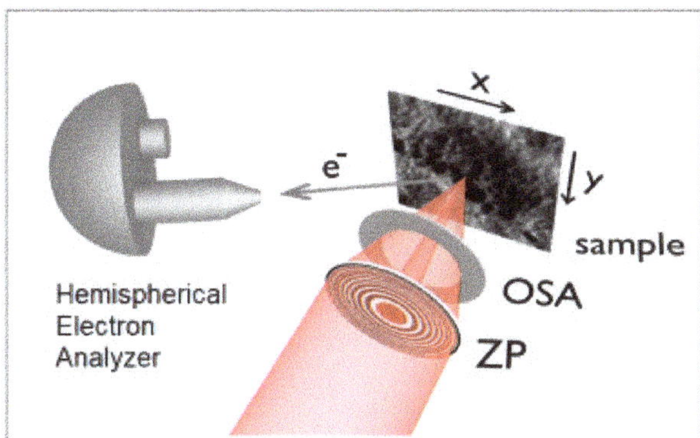

Figure 1. An illustrative sketch of the SPEM instrument. The X-ray beam is focused with a zone-plate (ZP) and an order-sorting aperture (OSA), down to a spot of 130 nm and the sample is raster scanned in front of it.

In this configuration, the samples are exposed to the common environment of the analysis chamber, which hosts the focusing optics, the HEA and other ancillary equipment; the highest allowable pressure in the case of experiments where samples need to be exposed to gases is about 1×10^{-5} mbar.

2.2. Novel Solutions for Operando Near-ambient Pressure Approaches

The typical photon energies and ZP dimensions used in the SPEM imply distances between ZP and samples of a few mm. Moreover, an additional pinhole, the order-sorting aperture (OSA), located between the ZP and the sample, is necessary to select a single diffracted spot. These geometrical constraints are the main reason why the technical solutions found to overcome the "pressure gap" for XPS setups could not be transferred straightforwardly to SPEMs. At Elettra two special setups have been recently developed, allowing experiments in near-ambient pressure conditions, preserving the best achievable spatial resolution: (i) the Dynamic High Pressure (DHP) setup and (ii) the Near-Ambient Pressure Cell (NAP-Cell).

The DHP system uses a highly collimated gas jet as a supersonic jet expansion into the SPEM vacuum chamber [23]; gases are confined into a narrow beam and point to the sample surface by using a pulsed jet instead of a continuous one. Similar setups have been already used in other research fields to control the amount of injected gases [24,25]. The DHP set-up has been designed to fit into the SPEM chamber: the gas jet is generated in the region around the sample by a thin needle connected to a pulsed valve which produces a series of gas jet shots at a fixed repetition rate. Each shot gives rise to a short burst of pressure in a small and confined volume near the sample and to an increase of the background pressure in the SPEM chamber. By finely tuning (i) the shot duration, (ii) the sample-to-needle distance and (iii) the DHP gas line pressure, the pressure at the sample can be controlled, while the injected gas is efficiently diluted into the large chamber volume, so that the background pressure does not exceed the HV limit in the analyzer. Figure 2 shows a typical single-shot pressure time-profile at the sample surface for a pulsed valve aperture time of 3.2 ms and the corresponding SPEM background pressure. The distance between the sample and the needle tip was fixed to 2 mm and the pressure of the gas in the DHP gas line was 3.5 bar. As shown in Panel (a), in these conditions the sample surface is experiencing a gas environment of ~10 mbar for a few ms. The corresponding change in the SPEM pressure is reported in Panel (b). A valve aperture frequency of 0.35 Hz is needed to keep the SPEM background pressure within the limit of ~1×10^{-5} mbar. More details can be found in ref [21].

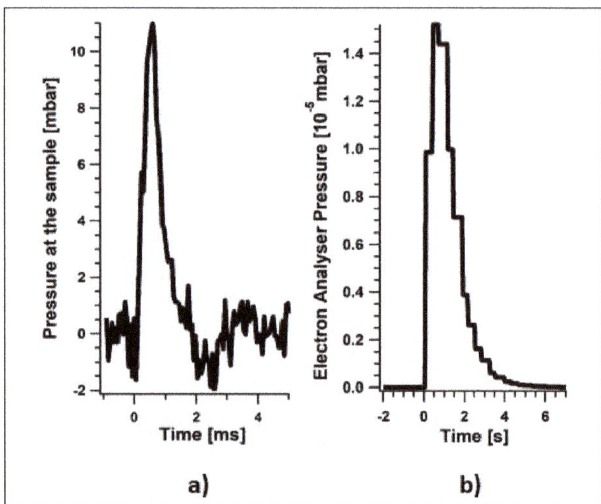

Figure 2. (a) Time-profile of a single pressure shot at the sample surface for a pulsed valve aperture time of 3.2 ms. (b) Corresponding pressure change in the SPEM chamber.

To address experiments where a constant pressure value is needed, the NAP-Cell has been developed. This device consists of a small vacuum-sealed cell-in which the sample is encapsulated-equipped with small pinholes as photon-in/photoelectron-out apertures [20]. A sectional view of this setup is shown in Figure 3. The gases needed in the experiment are delivered into the cell through a flexible metal bellow which terminates the dosing line located inside the SPEM chamber. A gas-dosing system allows regulation of the gas flow into the line at the desired flux/pressure. An encapsulated heater allows to heat up the sample, at the back of which it is placed, to temperatures up to 900–950 K. Additional electrical contacts are available for biasing samples in electrochemical setups. A pressure gap between the SPEM vacuum chamber and the inner volume of the NAP-Cell is established by the impedance induced by the small pinholes, which makes it possible to reach NAP conditions (~0.1 mbar) inside the cell while maintaining HV condition outside it (10^{-5} mbar). It is worth noting that, at variance with a standard NAP setup for spectroscopy, a differentially-pumped electron analyzer is not needed.

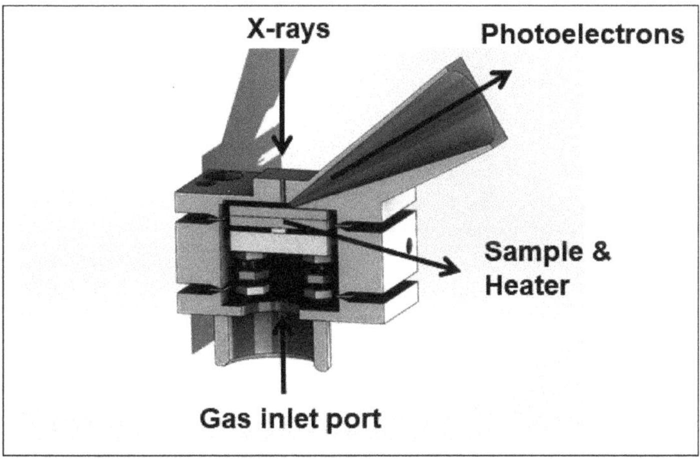

Figure 3. Working principle of the Near-Ambient Pressure Cell (NAP-Cell) technology for SPEM. The reactive gases needed in the experiment are delivered into the cell from the bottom aperture. Two separate holes are designed for the incoming X-rays and the outgoing photoelectrons. An encapsulated heater and the electrical contacts are located inside the cell.

Regarding the microscopy capability, the entire cell is scanned with respect to the X-rays beam, as for conventional UHV experiments. The "visible" area of samples hosted in the NAP-Cell is defined by the size and orientation of the two holes and it is approximately a circle of 0.4 mm diameter. A special design, characterized simply by one smaller pinhole, allows to reach up to 1 mbar, but at the expense of a limitation in the field of view, down to 200 μm.

It is important to highlight that real devices such as catalytic or electrochemical systems operate at pressure equal or higher than one atmosphere; therefore, it is the dream of each technique of analysis to reach *operando* conditions similar to the realistic ones. The NAP-Cell developed at Elettra—in line with most of the currently available NAP photoelectron systems around the world—may operate at pressure levels which are still a few orders of magnitude lower than realistic ones; nevertheless these conditions may activate chemical and physical processes which are pressure-dependent and which do not occur at the typical UHV or HV pressure levels of standard XPS. An example can be found in the last example of this manuscript: the partial conversion of the Sr signal from "lattice" to "surface" found at NAP conditions could not be detected while operating at HV pressures. We are aware of the fact that modern APXPS electron analyzers for spectroscopy can achieve pressures of a few tens of mbar—which nevertheless, as hinted at above, does not essentially shrink the gap with respect to practical operating conditions—but with a spatial resolution limited to hundreds of μm. The SPEM at Elettra is the only available XPS microscope where a pressure value of 0.1 mbar can be achieved for *operando* measurements.

2.3. Sample Preparation

2.3.1. Ag-In Alloys

A spontaneously patterned Ag-In electrodeposit was obtained from an aqueous electrolyte of composition: $KAg(CN)_2$ (Umicore, Schwäbisch Gmünd, Germany) 30 mM, $InCl_3$ (Alfa Aesar, Kandel, Germany) 0.1 M, D(+)-glucose (Fluka) 0.1 M, KCN (Merck, Darmstadt, Germany) 0.5 M. The alloy was grown onto a Cu foil cathode (2×1 cm^2) that was electrochemically degreased and pickled in H_2SO_4 20 vol % prior to electrodeposition. To avoid Ag displacement-plating, the Cu cathode was brought in contact with the electrolyte undercurrent. Alloy plating was carried out at room temperature in a 2-electrode cell using a large-area (20 cm^2) platinized-titanium anode. Pattern formation can be controlled by appropriately adjusting: (i) the In^{3+}/Ag^+ ratio in the plating bath, (ii) the current density and (iii) the fluid dynamic conditions of the electrolyte. Figure 4 reports the experimental conditions ensuring pattern formation in stagnant (diamonds) and stirred (crosses, magnetic stirring at 50 rpm, 100 mL cell) electrolytes. The alloy was grown by electrodeposition at 0.5 A/dm^2 for 2.5 h, yielding a coating thickness of ca. 40 μm. More experimental and theoretical details, as well as comprehensive reference to relevant literature, can be found in ref. [26]. The sample analyzed in this research (Figure 5) shows a double-spiral pattern with a spiral-core distance of ca. 0.5 mm and branch spacing of 0.15 mm. The sample exhibits a 3-phase crystallographic structure, featuring $\alpha-$Ag-rich terminal solid solution (Figure 6; marked with red dots) and the Ag$_3$In and AgIn$_2$ intermetallics, as shown by the Powder Diffraction Files (PDF) Cards No: 29-0677 and No: 65-1552, respectively, in keeping with the literature [27,28]. The same phase structure was found in the single-spiral sample studied in [26]: the quantitative differences in phase composition and preferential orientations are compatible with differences in electrodeposition current density.

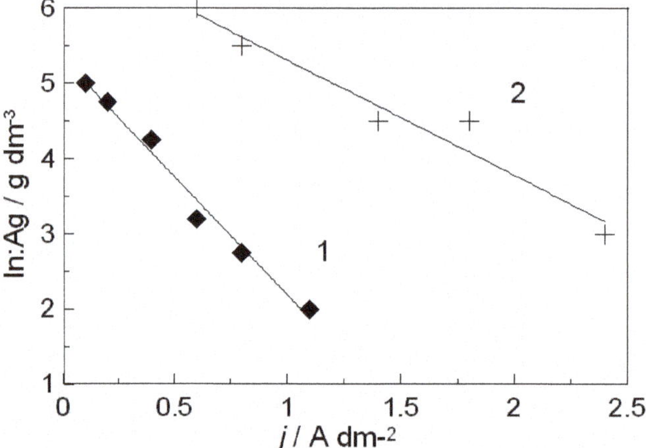

Figure 4. Electrodeposition conditions yielding pattern formation in Ag-In alloys: metal concentration ratio in the electrolyte (Y-axis); current density (X-axis) and fluid-dynamic conditions of the bath (1: stagnant, 2: stirred).

The photon energy employed was 652 eV, yielding an energy resolution of about 0.4 eV for imaging. The intrinsic surface sensitivity of soft-X ray photoelectron spectroscopy, enhanced by the grazing acceptance detection of SPEM (see Section 2.1 above), provides an effective probing depth limited to 1 nm, with dominating contribution from the top two atomic layers. Details on the image processing procedure are reported in the Supporting Information of ref. [26]. Ar$^+$ bombardment (1.5×10^{-5} mbar Ar, 1–1.5 kV, sample current of 0.8 μA) erosion was employed for depth-profiling purposes; the ion beam was delivered at 45° with respect to the sample. Sequential sputtering periods were applied, corresponding to depths of 2, 4, 10, 20, 50, 65 and 90 nm, estimated according to standard models for sputtering depth ([29] and references therein): for full methodological and quantitative details, refer to ref. [26]. These sputtering steps effectively allow for monitoring of progressive depth variations in the alloy pattern. Possible sputter-induced roughening of the original surface topography [30] was monitored by AFM and found to be negligible and therefore to have no impact on the measured variations in the composition. Moreover, the evident chemical changes taking place in-depth demonstrate that intermixing and re-deposition of sputtered material are negligible.

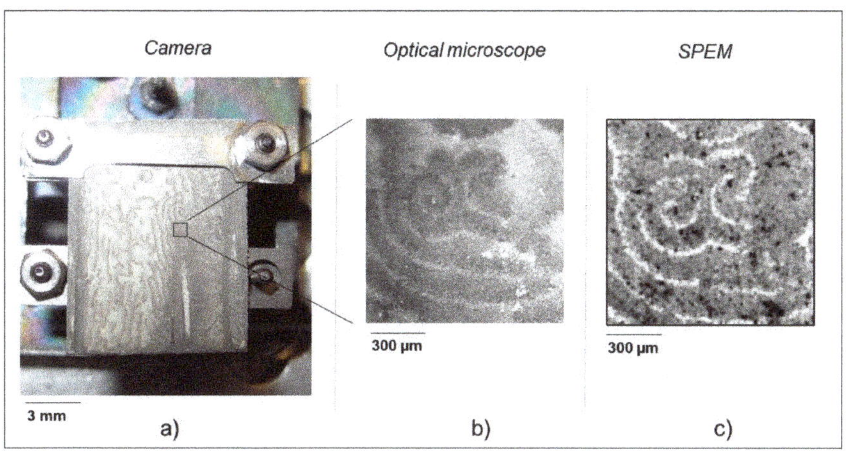

Figure 5. The electrodeposited Ag-In alloy exhibiting a double-spiral pattern, subjected to depth-dependent SPEM analysis. (**a**) The analyzed sample mounted on the SPEM sample-holder. (**b**) Optical-microscopy image. (**c**) SPEM chemical-state map at the Ag 3d energy, showing the Ag intermetallic/Ag solid solution ratio at a depth of ca. 50 nm.

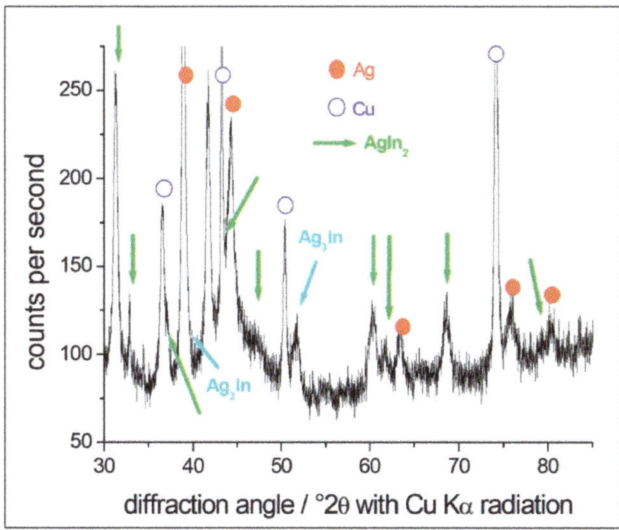

Figure 6. X-ray diffractogram of the Ag-In electrodeposited sample analyzed by SPEM, exhibiting the double-spiral pattern. The phase-structure features an α-Ag terminal solid solution and two Ag-In intermetallics; the remaining reflexes correspond to Cu reflections from the substrate.

2.3.2. Mn-Ni SOFC at High Vacuum

For this experiment, we have used planar, YSZ(100)-supported cells with Ni and Au/Mn layers: the electrodes were fabricated by evaporating 100 nm thick Ni layers, and Au (70 nm)–Mn (20 nm) bi-layers grown onto a Cr adhesion buffer (10 nm): the cell scheme is depicted in Figure 7 (for details on fabrication, see, e.g., ref. [31]). Prior to measurements, the samples were subjected to Ar$^+$ sputtering to remove contaminant layers introduced during fabrication, followed by annealing at 650 °C in 1×10^{-5} mbar O_2 to restore the initial oxidation state of the materials. Electrochemical measurements were run with a PAR potentiostat. The cell was short-circuited through a potentiostat working in a ZRA (Zero-Resistance Ammetry) mode for simultaneous monitoring of the current response.

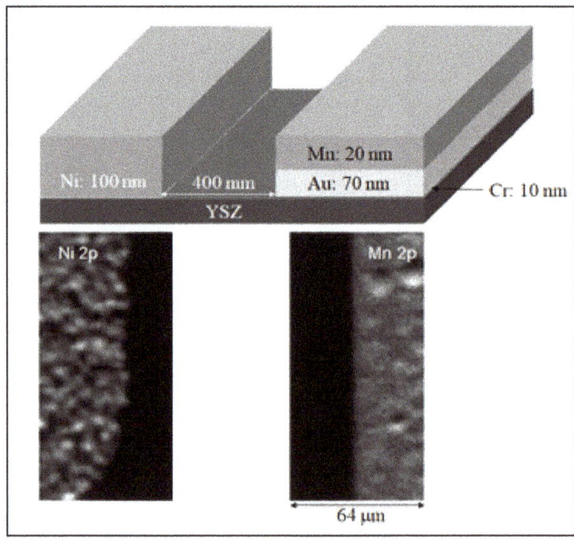

Figure 7. Scheme of the electrolyte (YSZ)–supported the planar cell with Ni and Mn electrodes for in situ electrochemical SPEM measurements. The images at the bottom are SPEM maps at the Ni 2p and Mn 2p energies of the electrode-electrolyte interfaces.

To optimize the photoemission yield of the main core levels, a photon energy of 1000 eV was set. As for all the experiments reported in this paper, a 5.0 purity grade of the gases was used.

2.3.3. LSM-NiO Single Chamber SOFC

The experiments on the single chamber SOFC were performed on (100)-oriented single crystal YSZ (ZrO_2 stabilized with 9.5 % Y_2O_3) electrolyte on top of which a mixture of 50 wt % of LSM-YSZ ($(La_{0.8}Sr_{0.2})_{0.95}MnO_3$ and 50 wt % of $(Y_2O_3)_{0.08}(ZrO_2)_{0.92}$ was used as cathode and a mixture of 60 wt % NiO and 40 wt % of $(Y_2O_3)_{0.08}(ZrO_2)_{0.92}$ (Ni–YSZ) as anode. The LSM-YSZ and Ni-YSZ powder mixtures were blended with an organic vehicle and screen-printed on the electrolyte, forming two symmetrical rectangles separated by a gap of ∼ 1.5 mm and sintered at 1200 °C for 2 h. An Pt grid was also screen printed onto the electrodes and subsequently fired at 1100 °C for 30 min to provide a more uniform distribution of potentials and more robust electrical contact. The SC-SOFC experiments were carried out at 923 K with a CH_4/O_2 gas mixture with a volume ratio of 2/1, in conformity with the largest proportion of the available literature. A more detailed description of the sample preparation can be found in ref [32].

To optimize the photoemission yield of the main core levels a photon energy of 750 eV was set. In the results section of this paper, a collection of photoemission spectra and chemical maps will be shown; in most of the measurements, an overall energy resolution of 0.35 eV was set. The deconvolution of the spectra was performed using Doniach–Sunijch functions convoluted with Gaussians, following well-established procedures.

Electrochemical measurements have been carried out in a two-electrode configuration with a Versastat potentiostat.

The planar configuration, with anode and cathode on the same plane, is a classical SC-SOFC configuration also for practical applications (see, among others, the following reviews [33–35]). The relative arrangement of the electrodes in practical systems is often of the interdigitated type, by the main reason for this is the extent of the electrodes in larger devices and this has no impact on microkinetics. One could argue about the optimization of the inter-electrode distance, but this aspect is exclusively relevant if energetic optimization of the device is at stake. In electroanalytic terms, however, the contribution of the ohmic drop of an unoptimized electrolyte gap can be very easily separated from charge-transfer and mass-transport contribution by standard electrochemical measurements, such as electrochemical impedance spectrometry. Therefore, with a straightforward ohmic correction, all the electrokinetic information can be accurately retrieved. In particular, the link between electronic conditions and surface chemistry can be established with utmost accuracy with our device.

3. Results

3.1. In-Depth Analysis of Ag-In Ex Situ Electrodeposited Alloys Exhibiting a Dynamic Pattern Formation Mechanism

A characteristic aspect of electrodeposited alloy patterns is that these structures can be imaged very precisely by visible-light microscopy, but exhibit a strikingly poor contrast in SEM, owing to their extreme surface-confinement. As proved in ref. [26,36], SPEM, owing to its high chemical-state sensitivity combined with surface sensitivity and lateral resolution, is the method of choice for unravelling the morphochemical details of electrodeposited alloy patterns. In view of completing and further validating the scenario set up in the two publications mentioned just above, in this study we have explored the 3D space elemental and chemical-state distribution of an Ag-In sample, fabricated under different conditions with respect to the one investigated in ref. [26]. Insight into the chemical speciation of Ag and In had been gained in the detailed study reported in ref. [26], were we disclosed that Ag is present in two chemical states: elemental (solid solution) and intermetallic, while In exhibits an oxidized and an intermetallic form. This chemical-state scenario has been confirmed in the analysis of the new sample and this information has been directly used for image processing of the Ag 3d and In 3d SPEM maps, with the method expounded in the Supporting Information of ref [26]. Briefly, the compositional inhomogeneity of the sample has been quantified by selecting the parts of the spectrum recorded in correspondence of each pixel of the image, representing one of the two components of the two key elements of the alloy. Figure 8 shows a representative selection of Ag $3d_{5/2}$ and In $3d_{5/2}$ SPEM images of the electrodeposited double-spiral, obtained by eroding the sample to the indicated depths. The maps depicted in Panels (a) and (c) were derived from the total intensity of the Ag and In $3d_{5/2}$ spectra and represent the spatial distribution of the Ag and In surface concentrations. Panels (b) and (d) show the Ag and In chemical maps, obtained by selecting the spectral regions representative of different chemical states. For Ag (Panel (b)) the contrast is due to the combination of intermetallic- and solid solution-type metallic states. Regarding In (Panel (d)), instead, contrast is determined by the distribution of In between Ag-In intermetallics and In^{3+} oxides/hydroxides. Coherently with our findings of ref. [26], it is clear that elemental contrast (Panels (a), (c)) tends to fade out rapidly in depth, while chemical-state contrast (Panels (b), (d)) is markedly more persisting. In addition, it is worth noting that the Ag and In elemental (Panels (a), (c)) and chemical-state distributions (Panels (b), (d)) are anticorrelated, in keeping with the surface concentration ratio distribution, imaged in Panel (e). Quantitative details, highlighting in-depth changes, can be more evidently pinpointed by averaging over space, as shown in Figure 9. Our results confirm the alloy electrodeposition mechanism proposed in [26] and further validates the mathematical DIB model of electrochemical pattern formation, recently developed in our group ([37,38] and references therein). More specifically, the present results provide additional support of the model capability of predicting complex spiral-based morphologies. Briefly, co-electrodeposition of In with Ag in amounts exceeding the solubility of the former element in the latter, gives rise to the formation of Ag-In intermetallics following a two-step process. Ag^+ and In^{3+} are reduced to the elemental state, forming an α -Ag terminal solid solution together with Ag-In intermetallics, but simultaneously basic In^{3+} salts precipitate at the electrode-electrolyte interface. The basic-salt film, in turn, can be electro-reduced forming fresh elemental In that can react with α-Ag to form more intermetallic. This two-step process effectively explains why In^{3+} oxi-hydroxides can be found at the alloy surface, but they are absent in the bulk of the layer. As far as pattern formation is concerned, the DIB model provides a simple framework for the rationalization of this phenomenon, in that it predicts that the coupling of electrodeposit morphology with adsorption of an electroactive intermediate can give rise to the formation of dynamic patterns, among which double-spirals [39]. SPEM maps provide factual spectroscopic evidence of the presence of adsorbed In^{3+} electroactive intermediates and of their organization in a typical pattern. In addition, a set of DIB-model parameters can be identified, yielding a computed pattern with a remarkably good matching between the theoretical map and the experimental pattern (Figure 10). The difference in the numerical values of the parameters that correspond to the double-spiral pattern with respect to those found for the single spiral studied in [26], is coherent with the different current-density values employed for the growth of the two alloys (for details on the correspondence between model parameters and experimental variables, see [40]).

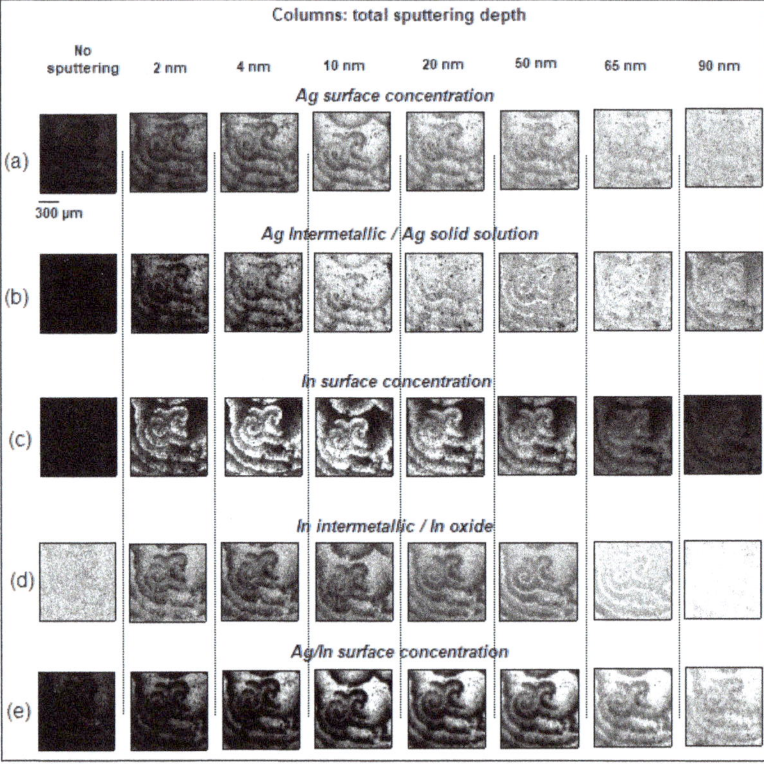

Figure 8. Ag $3d_{5/2}$ (**a,b**) and In $3d_{5/2}$ (**c,d**) SPEM surface concentration (**a**), (**c**) and chemical-state (**b**), (**d**) maps, recorded at different depths. Topographical contributions have been removed from all maps. The Ag/In ratio maps (**e**) outlines the compositional variations in depth.

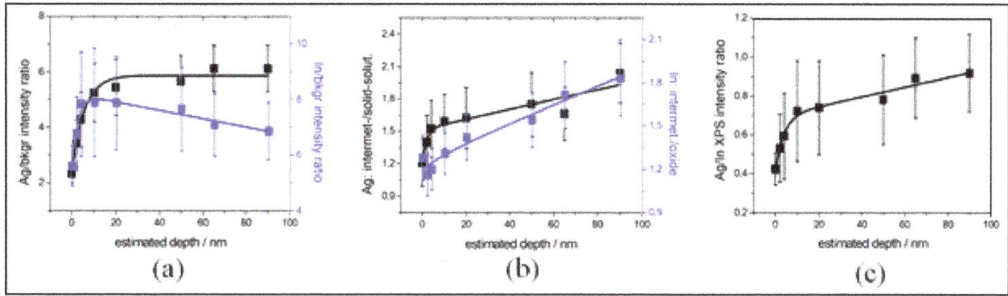

Figure 9. Spatial averages, with respective standard deviations of compositional and chemical-state observables from the maps of Figure 8. (**a**) Elemental amounts rationed to the background, from Panels (a) and (c) of Figure 8; (**b**) intermetallic fractions, from Panels (b) and (d) of Figure 8; (**c**) Ag/In ratios from Panel (e) of Figure 8.

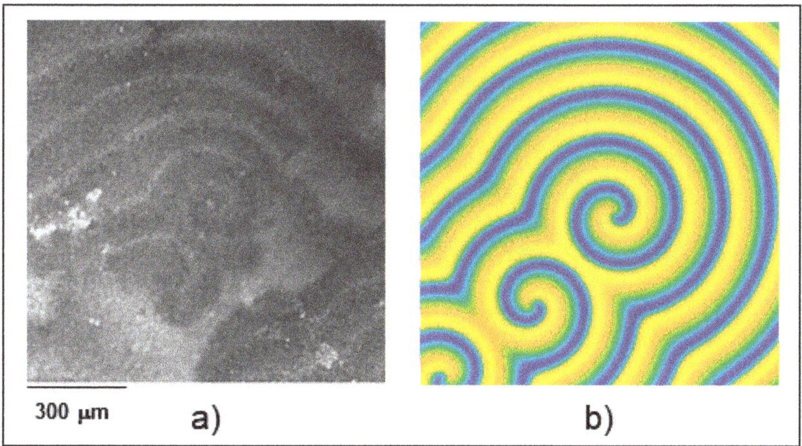

Figure 10. Comparison between optical-microscopy image (left, **a**) and computed (right, **b**) pattern from the DIB model, obtained with the following parameter values: $\alpha = 0.5$; $\gamma = 0.2$; $k_2 = 2.5$; $k_3 = 1.5$; $A_1 = 10$; $A_2 = 30$; $B = 22$; $C = 2.2$.

3.2. Spectromicroscopy Analysis of Electrochemical Processes at the Electrodes of a Self-driven Solid Oxide Fuel Cell

The present section contributes to the scantily covered field of in situ SPEM studies of self-driven electrochemical systems, in which chemical energy is spontaneously converted into electrical energy. In situ studies of self-driven fuel cells and batteries, though more difficult than those of externally-driven systems such as symmetric cells or cells with a single electrode running the real reaction of the energetic device, is notably important because the couplings between the anodic and cathodic phenomena are known to play a crucial role for the efficiency and durability of devices for electrochemical energetics. In the present study, we investigate a cell with a solid cathodic reagent, MnO_2, and an anodic gas, H_2, reacting at a Ni/NiO electrocatalyst. In particular, we compare operating conditions with active gas pressures in the high vacuum and NAP ranges.

Sample annealing in 10^{-5} mbar of O_2 at 923 K in the SPEM chamber, resulted in the oxidation of the starting materials to NiO and MnO_2 that, by exposure to different pressures of H_2, can transform into Ni and MnO, generating different redox couples at the two electrodes, yielding in turn electrical power, according to the scheme of Figure 11. The electrokinetics of the system was followed with transient experiments consisting in switching from O_2 10^{-5} mbar to H_2 10^{-5} and 1 mbar, and recording time-dependent spectra and current time-series (Figure 12). In order to emphasize the current response, the cell was short-circuited and the current was measured with a ZRA. Curve (A) in Figure 12 shows a representative transient corresponding to switching from 10^{-5} mbar O_2 to 10^{-5} mbar H_2. In keeping with the thermodynamic analysis of ref. [41], H_2 is oxidized to H_2O and MnO_2 is reduced to MnO, giving rise to the electrochemical reaction indicated in Figure 12, whereby O^{2-} is transported through the electrolyte. In correspondence, also Ni forms, enhancing the H_2 oxidation rate, as witnessed by the progressive rise in current, in keeping with the literature [42]. Curve (B) is obtained by switching from 10^{-5} mbar O_2 to 1 mbar H_2. In these conditions, on the one hand, the electrochemical reaction is faster, but, on the other hand, also the chemical reduction of MnO_2 is anticipated, resulting in a progressive decrease of the cell current.

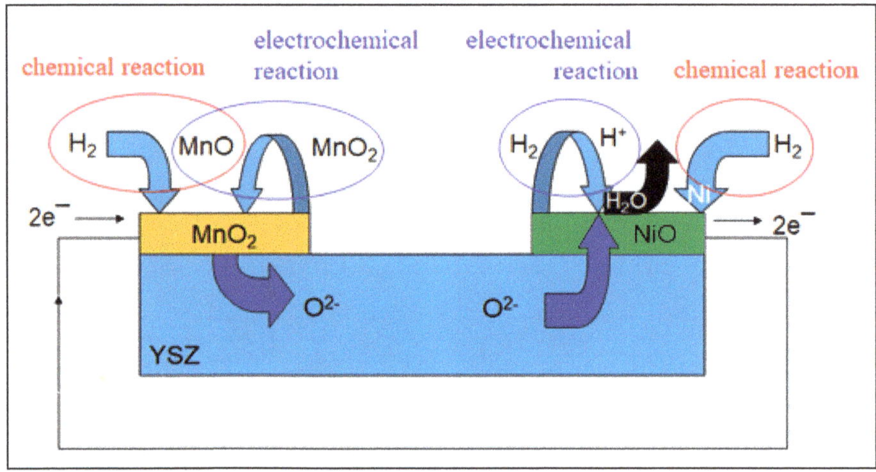

Figure 11. Reaction scheme for the MnO_2/YSZ/NiO cell in H_2 ambient.

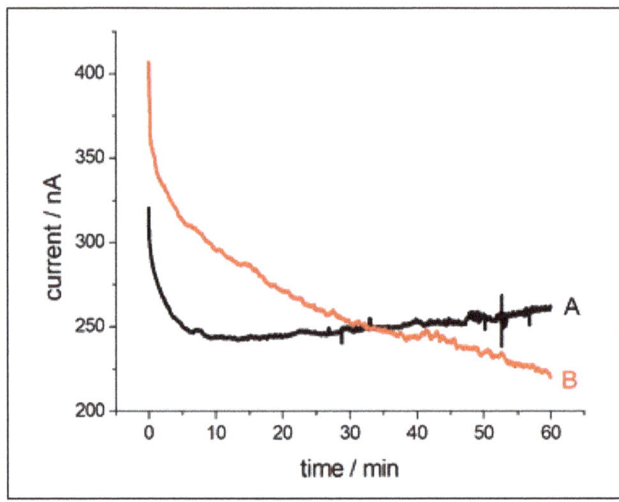

Figure 12. Current transients corresponding to O_2/H_2 gas-switching experiments with a MnO_2|YSZ|NiO cell at 923 K: (A) from 10^{-5} mbar O_2 to 10^{-5} mbar H_2; (B) from 10^{-5} mbar O_2 to 1 mbar H_2. The image is elaborated from Sci. Rep. 3 (2013) 2848, with permission.

Panels (A) and (B) of Figure 13 report representative sequences of Ni 2p µXPS spectra acquired after having switched the gas ambient from 10^{-5} mbar O_2 to 10^{-5} and 1 mbar H_2, respectively. Replicated experiments confirm in detail the observed kinetics. Ni in the initial state is fully oxidized and a progressive reduction can be noticed, that is notably faster in the case of the lower-pressure experiment. The reduction rates are summarized in Panel (C). Panel (D) reports the Mn 2p µXPS spectra acquired for the initial and final states of the cathode at the two H_2 pressures investigated: formation on MnO can be clearly assessed from the characteristic satellite structure [43]. Coherently with the recorded cell-currents and compatibly with the thermodynamic framework of ref. [41], at high pressure, the electrochemical reaction with H_2 oxidation at NiO and MnO_2 reduction to MnO is faster than the chemical Ni reduction and for this reason NiO is stabilized at short times. At low pressure instead, the chemical reduction of Ni is favoured, resulting in a progressive enhancement of the electrocatalytic performance of the anode.

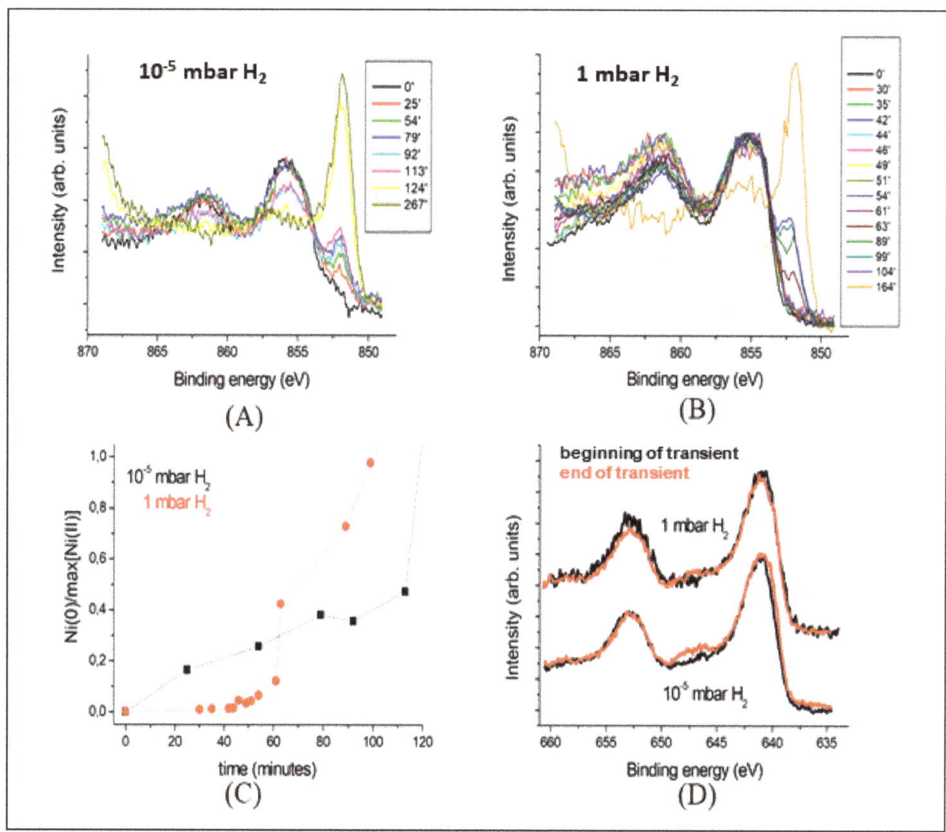

Figure 13. (**A,B**) Time-dependent Ni 2p µXPS spectra recorded at the indicated times after gas-switching from 10^{-5} mbar O_2 to: (**A**) 10^{-5} mbar H_2, (**B**) 1 mbar H_2. (**C**) Time dependence of the intensity of the elemental Ni peak normalized on the maximum of the oxidized Ni peak. (**D**) Mn 2p µXPS spectra measured at the beginning and at the end of the gas-switching transients.

3.3. Operando Characterization of a Single-chamber Solid Oxide Fuel Cell at Near-ambient Pressure

3.3.1. Electrochemical Measurements

Since limited literature information is available on the performance of SC-SOFC at the pressures of interest for APXPS, preliminarily to *operando* spectroscopy, we carried out calibration work in a range of vacuum conditions spanning the interval including the actual operating conditions and slightly higher ones, that allow the achievement of electrochemical responses that match the ones obtained at conventional working pressures. In Figure 14 we report a representative selection of current transients obtained by short-circuiting the cell in the reactive gas mixture: the actual operating condition during SPEM. More information including pressure-dependent open-circuit potentials and electrochemical impedance spectrometry can be found in ref. [26]. We have chosen to operate the SC-SOFC under short-circuited conditions because, on the one hand, even though the current that can be drawn from the device is rather low, it can be measured accurately and demonstrates the correct SOFC operation and, on the other hand, since short-circuit represents the most aggressive condition, successful operation in this condition is an *a fortiori* proof of the feasibility of the approach. These experiments consisted in measuring the cell current with the ZRA method in $CH_4/O_2 = 2/1$ mixtures of defined pressure. One can observe that a sudden current surge is followed by a relaxation, the rate of which negatively correlates with the pressure. After the transient, the current tends to a well-defined steady-state: this allows to run NAP-SPEM experiments under perfectly controlled electrochemical conditions.

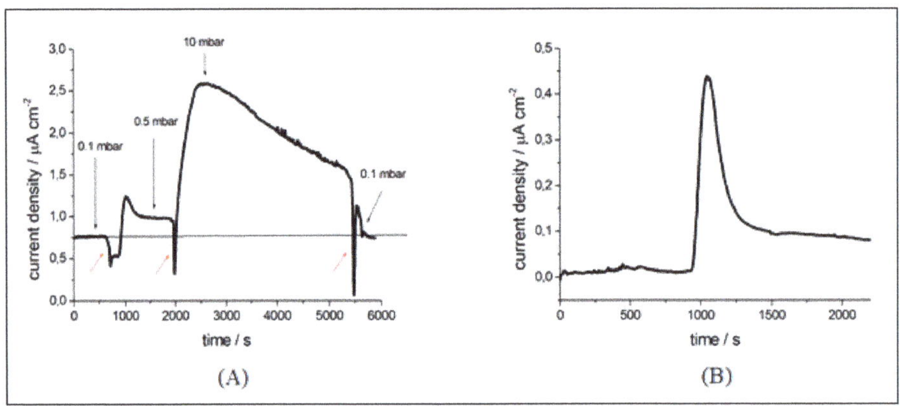

Figure 14. Current-density transients for the short-circuited SC-SOFC, recorded with the Zero-Resistance Ammetry (ZRA) technique in $CH_4/O_2 = 2/1$ ambient at the indicated pressures at 924 K. (**A**) Experiments carried out in a mock-up of the analysis chamber. Red arrows indicate pump-down transients. (**B**) Data recorded during *operando* NAP-SPEM measurements.

3.3.2. Spatially Resolved Photoemission Measurements at Near Ambient Pressure

In order to prove the validity of our approach as robustly as possible, we have decided to carry out *operando* XPS concentrating on the cathode of the NiO/YSZ/LSM SC-SOFC, since the chemical state of the elements of this perovskite is far less sensitive to electrochemically-induced changes in comparison with Ni [41]. In the SPEM, the cell temperature was set to 923 K for the whole measurement time. To avoid any reduction of the electrode/electrolyte surface due to the combination of vacuum/high temperature, in the stand-by periods between electrochemical measurements, a partial pressure of 5×10^{-2} mbar of O_2 was introduced in the NAP-Cell. In these conditions, we have measured several µXPS survey and high-resolution spectra as well as SPEM maps in different locations to probe the chemical composition of the cathode and of the interface region and its morphology of pristine cells. In addition to the main constituents of the cathodic/electrolyte material (La, Sr, Mn, Zr, Y, O) we also assessed the presence of two impurities: C and Si, the latter being typical for YSZ [42]. Figure 15 reports a representative selection of maps at the Sr 3d and La 4d energies (Panel (A)) and μXPS spectra (Panel (B)) in the region across the interface between the LSM-YSZ patch (left) and the YSZ electrolyte (right). These elemental maps show a clear topographic contrast in the cathode region, highlighting micrometric and submicrometric granular morphology, combined with evident chemical contrast. In addition, signal from Sr and La can be detected inside the electrolyte, proving diffusion of these elements into the YSZ region. Differently from Sr and La, Mn remains strictly confined to the cathode region. At the right of Panel (A), we show intensity profiles for Sr 3d and La 4d signal extracted from the SPEM maps along the dotted lines and in Panel (B) we report μXPS spectra measured at different positions along the same line, up to 240 μm from the electrode/electrolyte interface. The La 4d spectrum is partly overlapping the Si 2p one, that provides a convenient internal reference for relative intensity variations of this perovskite component. The spatial dependence of the Sr and La signal shows that the surface coverage of YSZ with Sr is constant in the investigated zone, while a La gradient can be appreciated, highlighting a difference in the surface diffusion coefficients of these two components of the cathodic material on the electrolyte. In this work, we concentrated on the evolution of the cathode elements in SC-SOFC ambient and under electrochemical operating conditions. In particular, we carried out a fine analysis of the La 4d, Zr 3d, O 1s and Mn 2p core levels, acquired on different grains of the cathode, as well as in representative locations of the electrolyte, in the following ambients: (i) 0.05 mbar O_2; (ii) 0.1 mbar H_2 and (iii) 0.1 mbar CH_4/O_2 = 2/1. The electrochemical conditions tested were: (a) OCP in all three ambients, (b) short-circuited cell in ambient (iii). Ambient (iii) is the reactive gas mixture employed for self-driven SC-SOFC operation (for details, see [26,41]); as hinted-at above, ambient (i) was used to preserve the original chemistry of the cell components in stand-by conditions between electrochemical measurements and ambient (ii) was set for 1 h before starting the SC-SOFC operation in order to activate the NiO anode [38]. Briefly, detailed analysis of the La 4d, Zr 3d and Mn 2p core level spectra acquired on different grains of the cathode and on the electrolyte under all investigated conditions, did not exhibit any effect either of the ambient or of the electrochemical polarization. These spectra showed the typical binding energies and spectral shape reported in the literature for similar materials [43–46] and the only differences were in the intensities, as a result of the heterogeneous stoichiometry of the LSM-YSZ grains. Instead, the Sr 3d line exhibited chemical-state sensitivity to the gas ambient: specifically, spectral changes have been found in H_2 and CH_4/O_2 ambients, as detailed below. It is worth noting that the chemical state of Sr in SOFC materials has been investigated by ex situ XPS in some detail in the literature [44,46–48]. Briefly, on the one hand, Sr has been described in La-Sr perovskites with transition metals and, on the other hand, as a species diffused on YSZ. The former state—characterized by an XPS spectrum with peaks at BEs between 131.4 and 132.5 eV—, corresponds to the perovskite lattice. The latter state—exhibiting peaks at BEs between 132.8 and 133.8 eV—denotes the separation of a secondary oxidic phase. In the cases reported in the literature, these two prototypical chemical states are typically found in combination within the probed volume.

In our experiments carried out in H_2 atmosphere, a slight, but well-defined spectral change in the low BE tail of Sr spectra located on the cathode patch was found during the change from O_2 to H_2 atmosphere, while the same element, in the form expressed when diffused into the electrolyte region, did not respond to the change (Figure 16, spectra a), b), d) and e)). This chemical change of Sr in LSM cannot be reliably accounted for on the basis of the available literature, but can be justified in terms of partial reduction.

As anticipated in ref. [26], the introduction of SC-SOFC CH_4/O_2 2/1 gas feed in the NAP-Cell brings about a notable, irreversible change in the Sr 3d spectrum measured at the cathode (Figure 16, spectrum (c)), while no changes are found in Sr diffused onto the electrolyte patch (Figure 16, spectrum (f)). The spectral change found in the cathode region occurs at OCP and it is not further modified by applying electrochemical polarization within the range of self-driven SC-SOFC operation. The μXPS spectrum recorded in the cathode region exhibits a strong surface contribution, while that measured on the electrolyte patch can be entirely accounted for by lattice Sr. The observed irreversible surface segregation of Sr by XPS is in keeping with literature results reporting, on the one hand, the formation of a SrO surface phase by crystallographic means [49] and, on the other hand, electrocatalytic degradation due to blocking of the ORR active sites [50].

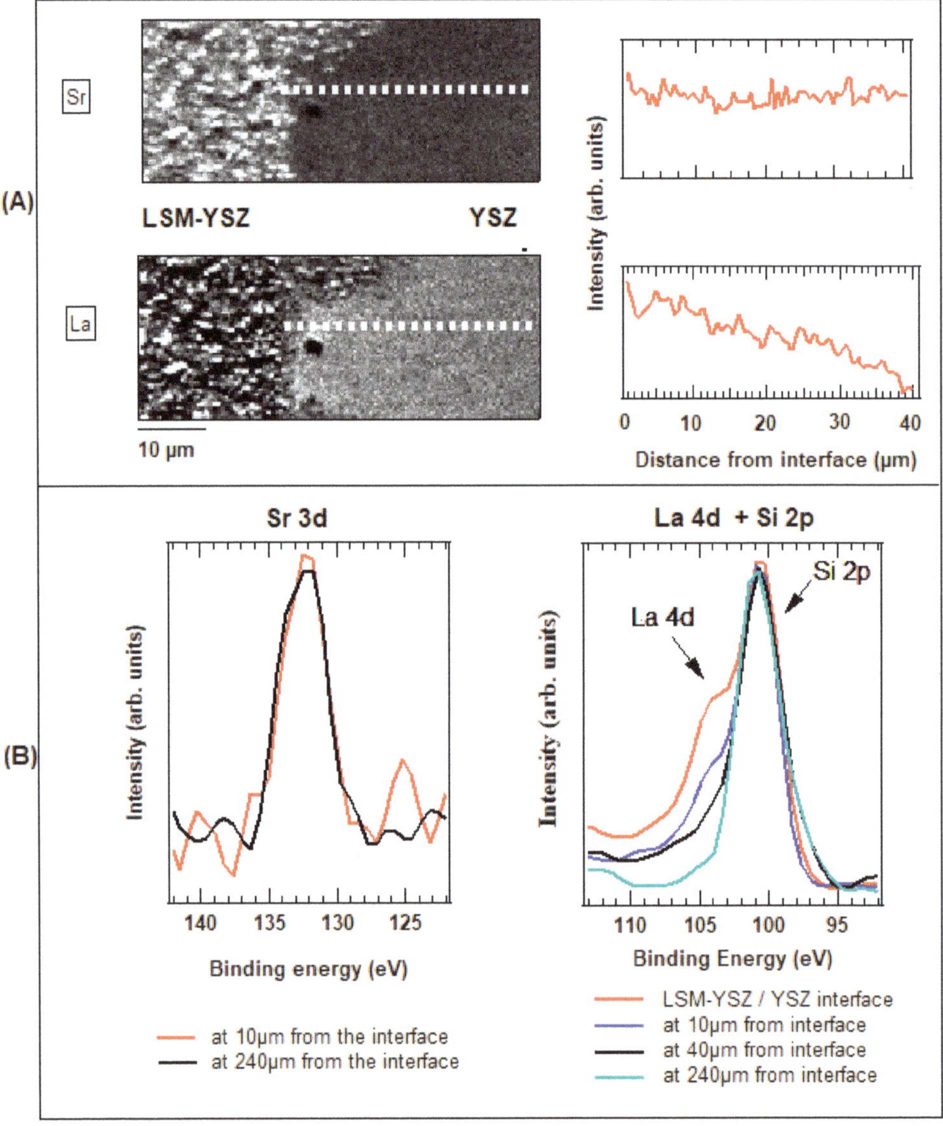

Figure 15. (**A**) SPEM maps at the Sr 3d and La 4d energies and elemental concentration profiles, recorded with a pristine cell in 0.05 mbar of O_2 at 924 K. The imaged region is the interface between the LSM-YSZ cathode (left) and the YSZ electrolyte (right). (**B**) Sr 3d and La 4d µXPS spectra measured at the indicated distances from the cathode/electrolyte interface along the dotted line shown in Panel (A).

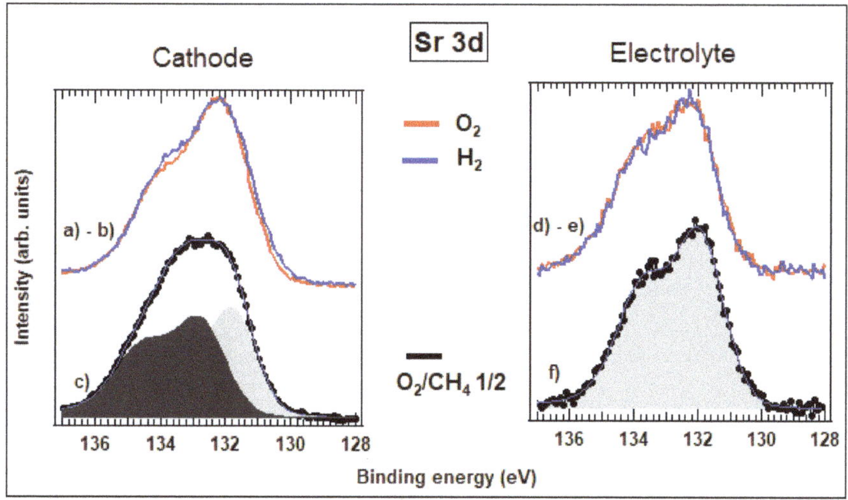

Figure 16. Sr 3d µXPS spectra recorded on the cathode ((a), (b) and (c)) and electrolyte ((d), (e) and (f)) patches at 923 K. Spectra (a), (b), (d) and (e): comparison of OCP conditions in 0.05 mbar of O$_2$ and 0.1 mbar H$_2$. Spectra (c) and (f): comparison between electrochemically inert 0.05 mbar O$_2$ ambient and reactive SC-SOFC mixture CH$_4$/O$_2$ 2/1 0.1 mbar.

4. Conclusions

Scanning photoelectron spectromicroscopy has undergone a fast development at the third-generation synchrotron light sources, becoming a true microscopic tool for probing material properties, mainly at their surface. In the field of electrochemistry this approach has been proved capable of following morphological and chemical effects occurring at surfaces and interfaces. The present paper reviews three different experiments showing how SPEM can overcome the "material" and "pressure" gaps in the characterization of electrochemical surfaces. In the first experiment the fascinating spatiotemporal patterns formed during an ex situ electrodeposition of an Ag-In alloy has been investigated. A depth profile mapping of the distribution of the chemical states of Ag and In with high spatial and spectral resolution has been performed providing a 3D frame of the patterns. The experimental data perfectly match the mathematical predictions of the DIB model of electrochemical pattern formation.

The other two cases reported in the manuscript deal with the analysis of electrochemical processes at the electrodes of solid oxide fuel cells. In the first case, an in situ study of a self-driven cell was performed. The cell was formed by a solid cathodic material, MnO$_2$, and a mixture of Ni/NiO as anode. The system was analyzed at high vacuum and near ambient environmental pressure conditions. The main achievement is that, coherently with the recorded cell currents at high H$_2$ pressure, its oxidation at NiO and the corresponding MnO$_2$ reduction to MnO is faster than the chemical Ni reduction and for this reason, NiO is stabilized at short times. At low pressure instead, the chemical reduction of Ni is favoured, resulting in a progressive enhancement of the electrocatalytic performance of the anode.

In the last experiment, we have shown an *operando* analysis of a single-chamber solid oxide fuel cell at near ambient pressure regime by using a unique specially designed setup which extends the traditional capabilities of scanning photoemission microscopes in the ultra-high and high-vacuum regimes to more realistic conditions of environmental pressure. The response of the cathodic site of a NiO/YSZ/LSM SC-SOFC exposed to O$_2$, H$_2$ and a mixture of CH$_4$ and O$_2$ showed a chemical-state sensitivity of Sr: a well-defined spectral change in the low BE tail of Sr spectra located on the cathode patch was found during the change from O$_2$ to H$_2$ atmosphere, while the same element, when diffused into the electrolyte region, did not respond to the same change. This chemical change of Sr in LSM cannot be reliably accounted for on the basis of the available literature, but can be justified in terms of partial reduction. Moreover, the CH_4/O_2 2/1 gas feed brings about a notable, irreversible change in the Sr 3d spectrum measured at the cathode, while no changes are found in Sr diffused onto the electrolyte patch.

We believe the reported examples demonstrate the great potential of SPEM to gain insight into electrochemical and chemical processes taking place in energy conversion and other systems, which can guide the design of next-generation devices with improved performance.

Author Contributions: D.K. prepared the single-chamber SOFC test samples. All authors characterized the samples, revised and edited the manuscript.

Funding: The research leading to these results has received funding from the European Community's Seventh Framework Programme (FP7/2007-2013) under grant agreement n° 312284 and the European Community's Horizon 2020 Framework Programme under grant agreement n° 730872. The author D.K. acknowledges the financial support from the Slovenian Research Agency (research core funding No. P2-0105).

Conflicts of Interest: The authors declare no conflict of interest.

References

1. Nasirpouri, F. An Overview to Electrochemistry. In *Electrodeposition of Nanostructured Materials*; Springer Series in Surface Science 62; Springer: Berlin, Germany, 2017; Volume 62, pp. 43–73.

2. Buckley, D.N.; O'Dwyer, C.; Quill, N.; Lynch, R.P. Electrochemical energy storage. In *Energy Storage Options and Their Environmental Impact*; Hester, R.E., Harrison, R.M., Eds.; The Royal Society of Chemistry: London, UK, 2019; pp. 115–149.

3. Lin, F.; Liu, Y.; Yu, X.; Cheng, L.; Singer, A.; Shpyrko, O.G.; Xin, H.L.; Tamura, N.; Tian, C.; Weng, T.-C.; et al. Synchrotron X-ray Analytical Techniques for Studying Materials Electrochemistry in Rechargeable Batteries. *Chem. Rev.* **2017**, *117*, 13123–13186.

4. Margaritondo, G.; Cerrina, F. Overview of soft-X-ray photoemission spectromicroscopy. *Nuclear Instrum. Methods A* **1990**, *191*, 26–35. [CrossRef]

5. Ade, H.; Kirz, J.; Hulbert, S.; Johnson, E.; Anderson, E.; Kern, D. Scanning photoemission microscopy with synchrotron radiation. *Vac. Sci. Technol.* **1991**, *A9*, 1902.

6. Günther, S.; Kaulich, B.; Gregoratti, L.; Kiskinova, M. Photoelectron microscopy and applications in surface and materials science. *Prog. Surf. Sci.* **2002**, *70*, 187–260. [CrossRef]

7. Amati, M.; Barinov, A.; Feyer, V.; Gregoratti, L.; Al-Hada, M.; Locatelli, A.; Mentes, T.O.; Sezen, H.; Schneider, C.M.; Kiskinova, M. Photoelectron microscopy at Elettra: Recent advances and perspectives. *J. Electron. Spectrosc. Relat. Phenom.* **2018**, *224*, 59–67. [CrossRef]

8. Horiba, K.; Nakamura, Y.; Nagamura, N.; Toyoda, S.; Kumigashira, H.; Oshima, M.; Amemiya, K.; Senba, Y.; Ohashi, H. Scanning photoelectron microscope for nanoscale three-dimensional spatial-resolved electron spectroscopy for chemical analysis. *Rev. Sci. Instrum.* **2011**, *82*, 113701-1. [CrossRef]

9. Bozzini, B.; Amati, M.; Gregoratti, L.; Kazemian, M.; Prasciolu, M.; Tondo, E.; Trygub, A.L.; Kiskinova, M. In Situ Electrochemical X-ray Spectromicroscopy Investigation of the Reduction/Reoxidation Dynamics of N-Cu Solid Oxide Fuel Cell Anodic Material in Contact with a Cr Interconnect in 2×10^{-6} mbar O_2. *J. Phys. Chem. C* **2012**, *116*, 7243–7248. [CrossRef]

10. Huber, A.; Falk, M.; Rohnke, M.; Luerssen, B.; Gregoratti, L.; Amati, M.; Janek, J. In situ study of electrochemical activation and surface segregation of the SOFC electrode material $La_{0.75}Sr_{0.25}Cr_{0.5}Mn_{0.5}O_{3+/-\delta}$. *Phys. Chem. Chem. Phys.* **2012**, *14*, 751–758. [CrossRef]

11. Backhaus-Ricoult, M.; Adib, K.; Work, K.; Badding, M.; Ketcham, T.; Amati, M.; Gregoratti, L. In-situ scanning photoelectron microscopy study of operating (La,Sr)FeO₃-based NOx-sensing surfaces. *Solid State Ion.* **2012**, *225*, 716–726. [CrossRef]

12. Jirgensen, A.; Esser, N.; Hergenrçder, R. Near ambient pressure XPS with a conventional X-ray source. *Surf. Interface Anal.* **2012**, *44*, 1100–1103. [CrossRef]

13. Bluhm, H.; Havecker, M.; Knop-Gericke, A.; Kiskinova, M.; Schlögl, R.; Salmeron, M. In Situ X-ray Photoelectron Spectroscopy Studies of Gas-Solid Interfaces at Near-Ambient Conditions. *MRS Bull.* **2007**, *32*, 1022–1030. [CrossRef]

14. Crumlin, E.J.; Bluhm, H.; Liu, Z. In situ investigation of electrochemical devices using ambient pressure photoelectron spectroscopy. *J. Electron. Spectrosc. Relat. Phenom.* **2013**, *190*, 84–92. [CrossRef]

15. Starr, D.E.; Liu, Z.; Hävecker, M.; Knop-Gericke, A.; Bluhm, H. Investigation of solid/vapor interfaces using ambient pressure X-ray photoelectron spectroscopy. *Chem. Soc. Rev.* **2013**, *42*, 5833–5857. [CrossRef]

16. Crumlin, E.J.; Liu, Z.; Bluhm, H.; Yang, W.L.; Guo, J.H.; Hussain, Z. X-Ray Spectroscopy of Energy Materials under in Situ/Operando Conditions. *J. Electron. Spectrosc. Relat. Phenom.* **2015**, *200*, 264–273. [CrossRef]

17. Takagi, Y.; Uruga, T.; Tada, M.; Iwasawa, Y.; Yokoyama, T. Pressure Hard X-ray Photoelectron Spectroscopy for Functional Material Systems as Fuel Cells under Working Conditions. *Acc. Chem. Res.* **2018**, *51*, 719–727. [CrossRef]

18. Favaro, M.; Valero-Vidal, C.; Eichhorn, J.; Toma, F.M.; Ross, P.N.; Yano, J.; Liu, Z.; Crumlin, E.J. Elucidating the alkaline oxygen evolution reaction mechanism on platinum. *J. Mater. Chem. A* **2017**, *5*, 11634–11643. [CrossRef]

19. Lichterman, M.F.; Richter, M.H.; Hu, S.; Crumlin, E.J.; Axnanda, S.; Favaro, M.; Drisdell, W.; Hussain, Z.; Brunschwig, B.S.; Lewis, N.S.; et al. An electrochemical, microtopographical and ambient pressure X-ray photoelectron spectroscopic investigation of Si/TiO$_2$/Ni/Electrolyte Interfaces. *J. Electrochem. Soc.* **2016**, *163*, H139–H146. [CrossRef]

20. Sezen, H.; Alemán, B.; Amati, M.; Dalmiglio, M.; Gregoratti, L. Spatially Resolved Chemical Characterization with Scanning Photoemission Spectromicroscopy: Towards Near-Ambient-Pressure Experiments. *ChemCatChem* **2015**, *7*, 3665–3673. [CrossRef]

21. Amati, M.; Abyaneh, K.M.; Gregoratti, L. Dynamic High Pressure: A novel approach toward near ambient pressure photoemission spectroscopy and spectromicroscopy. *J. Instrum.* **2013**, *8*, T05001. [CrossRef]

22. Gregoratti, L.; Barinov, A.; Benfatto, E.; Cautero, G.; Fava, C.; Lacovig, P.; Lonza, D.; Kiskinova, M.; Tommasini, R.; Mahl, S. 48-Channel electron detector for photoemission spectroscopy and microscopy. *Rev. Sci. Instrum.* **2004**, *75*, 64–68. [CrossRef]

23. Miller, D.R. *Atomic and Molecular Beam Methods*, 1st ed.; Oxford University Press: Oxford, UK, 1988.

24. Soukhanovskii, V.A.; Kugel, H.W.; Kaita, R.; Majeski, R.; Roquemore, A.L. Supersonic gas injector for fueling and diagnostic applications on the national spherical torus experiment. *Rev. Sci. Instrum.* **2004**, *75*, 4320–4323. [CrossRef]

25. Piseri, P.; Podestà, A.; Barborini, E.; Milani, P. Production and characterization of highly intense and collimated cluster beams by inertial focusing in supersonic expansions. *Rev. Sci. Instrum.* **2001**, *72*, 2261–2267. [CrossRef]

26. Bozzini, B.; Amati, M.; Dobrovolska, T.; Gregoratti, L.; Krastev, I.; Sgura, I.; Taurino, A.; Kiskinova, M. Depth-dependent scanning photoelectron microspectroscopy unravels the origin of dynamic pattern formation in alloy electrodeposition. *J. Phys. Chem. C* **2018**, *112*, 15996–16007. [CrossRef]

27. Dobrovolska, T.; Jovic, V.D.; Jovic, B.M.; Krastev, I. Phase Identification in Electrodeposited Ag-In Alloys by ALSV technique. *J. Electroanal. Chem.* **2007**, *611*, 232–240. [CrossRef]

28. Dobrovolska, T.; Beck, G.; Krastev, I.; Zielonka, A. Phase Composition of Electrodeposited Silver-Indium Alloys. *J. Solid State Electrochem.* **2008**, *12*, 1461–1467. [CrossRef]

29. Briggs, D.; Seah, M.P. *Practical Surface Analysis, Auger and X-ray Photoelectron Spectroscopy*, 2nd ed.; Wiley: Chichester, UK, 1990; Volume 1.

30. Bozzini, B. Effect of Sputter-Induced and other Roughness on Auger Electron Intensities. *Il Vuoto* **1997**, *26*, 18–29.

31. Bozzini, B.; Amati, M.; Gregoratti, L.; Kiskinova, M. In-situ Photoelectron Microspectroscopy and Imaging of Spontaneous Electrochemical Processes at the Electrodes of a Self-Driven Cell. *Sci. Rep.* **2013**, *3*, 2848–2852. [CrossRef]

32. Bozzini, B.; Kuscer, D.; Drnovšek, S.; Al-Hada, M.; Amati, M.; Sezen, H.; Gregoratti, L. Spatially resolved photoemission and electrochemical characterization of a single-chamber solid oxide fuel cell. *Top. Catal.* **2018**, *61*, 2185–2194. [CrossRef]

33. Yano, M.; Tomita, A.; Sano, M.; Hibino, T. Recent advances in single-chamber solid oxide fuel cells: A review. *Solid State Ion.* **2007**, *77*, 3351–3359. [CrossRef]

34. Kuhn, M.; Napporn, T.W. Single-Chamber Solid Oxide Fuel Cell Technology—From Its Origins to Today's State of the Art. *Energies* **2010**, *3*, 57–134. [CrossRef]

35. Hibino, T. Single Chamber Solid Oxide Fuel Cell. In *Encyclopedia of Applied Electrochemistry*; Kreysa, G., Ota, K., Savinell, R.F., Eds.; Springer: New York, NY, USA, 2014.

36. Bozzini, B.; Amati, M.; Gregoratti, L.; Lacitignola, D.; Sgura, I.; Krastev, I.; Dobrovolska, T. Intermetallics as key to Spiral Formation in In-Co Electrodeposition. A Study based on Photoelectron Microspectroscopy, Mathematical Modelling and Numerical Approximations. *J. Phys. D* **2015**, *48*, 395502. [CrossRef]

37. Lacitignola, D.; Bozzini, B.; Peipmann, R.; Sgura, I. Cross-diffusion effects on a morphochemical model for electrodeposition. *Appl. Math. Model.* **2018**, *57*, 492–513. [CrossRef]

38. Zhang, C.; Zheng, Y.; Ran, R.; Shao, Z.; Jin, W.; Xu, N.; Ahn, J. Initialization of a methane-fueled single-chamber solid-oxide fuel cell with NiO+ SDC anode and BSCF+ SDC cathode. *J. Power Sources* **2008**, *179*, 640–648. [CrossRef]

39. Lacitignola, D.; Bozzini, B.; Sgura, I. Spatio-temporal organization in a morphochemical electrodeposition model: Analysis and numerical simulation of spiral waves. *Acta Appl. Math.* **2014**, *132*, 377–389. [CrossRef]

40. Lacitignola, D.; Bozzini, B.; Frittelli, M.; Sgura, I. Turing Pattern Formation on the Sphere for a Morphochemical Reaction-Diffusion Model for Electrodeposition. *Commun. Nonlinear Sci. Numer. Simul.* **2017**, *48*, 484–508. [CrossRef]

41. Zhang, C.; Lin, Y.; Ran, R.; Shao, Z. Activation of a single-chamber solid oxide fuel cell by a simple catalyst-assisted in-situ process. *Electrochem. Commun.* **2009**, *11*, 1563–1566. [CrossRef]

42. Idris, M.A.; Bak, T.; Li, S.; Nowotny, J. Effect of Segregation on Surface and Near-Surface Chemistry of Yttria-Stabilized Zirconia. *J. Phys. Chem. C* **2012**, *116*, 10950–10958. [CrossRef]

43. Han, S.W.; Lee, J.D.; Kim, K.H.; Song, H.; Kim, W.J.; Kwon, J.; Lee, H.G.; Hwang, C.; Jeong, J.I.; Kang, J.S. Electronic structures of the CMR perovskites $R_{1-x}A_xMnO_3$ (R = La, Pr; A = Ca, Sr, Ce) using photoelectron spectroscopy. *J. Korean Phys. Soc.* **2002**, *40*, 501–510.

44. Huber, A.K.; Falk, M.; Rohnke, M.; Luerssen, B.; Amati, M.; Gregoratti, L.; Hesse, D.; Janek, J. In situ study of activation and de-activation of LSM fuel cell cathodes—Electrochemistry and surface analysis of thin-film electrodes. *J. Catal.* **2012**, *294*, 79–88. [CrossRef]

45. Konysheva, E.Y.; Kuznetsov, M.V. Fluctuation of surface composition and chemical states at the hetero-interface in composites comprised of a phase with perovskite structure and a phase related to the Ruddlesden-Popper family of compounds. *RSC Adv.* **2013**, *3*, 14114–14122. [CrossRef]

46. Crumlin, E.J.; Mutoro, E.; Liu, Z.; Grass, M.E.; Biegalski, M.D.; Lee, Y.L.; Morgan, D.; Christen, H.M.; Bluhm, E.; Shao-Horn, Y. Surface strontium enrichment on highly active perovskites for oxygen electrocatalysis in solid oxide fuel cells. *Energy Environ. Sci.* **2012**, *5*, 6081–6086. [CrossRef]

47. Aphale, A.; Liang, C.; Hu, B.; Singh, P. Cathode degradation from airborne contaminants in solid oxide fuel cells: A Review. In *Solid Oxide Fuel Cells Lifetime and Reliability*, 1st ed.; Brandon, N.P., Ruiz-Trejo, E., Boldrin, P., Eds.; Academic Press: London, UK, 2017; pp. 106–108.

48. Chen, K.; Hyodo, J.; Al, N.; Ishihara, T.; Jiang, S.P. Boron deposition and poisoning of $La_{0.8}Sr_{0.2}MnO_3$ oxygen electrodes of solid oxide electrolysis cells under accelerated operation conditions. *Int. J. Hydrogen Energy* **2016**, *41*, 1419–1431. [CrossRef]

49. Hu, B.; Keane, M.; Mahapatra, M.K.; Singh, P. Stability of strontium-doped lanthanum manganite cathode in humidified air. *J. Power Sources* **2014**, *248*, 196–204. [CrossRef]

50. Wang, W.; Jiang, S.P. A mechanistic study on the activation process of (La, Sr)MnO_3 electrodes of solid oxide fuel cells. *Solid State Ion.* **2006**, *177*, 1361–1369. [CrossRef]

Article

Probing the Surface of Noble Metals Electrochemically by Underpotential Deposition of Transition Metals

Nolwenn Mayet, Karine Servat, K. Boniface Kokoh and Teko W. Napporn *

IC2MP, UMR 7285 CNRS, Université de Poitiers, 4 rue Michel Brunet B27 TSA 51106, 86073 Poitiers CEDEX 09, France; nolwenn.mayet@univ-poitiers.fr (N.M.); karine.servat@univ-poitiers.fr (K.S.); boniface.kokoh@univ-poitiers.fr (K.B.K.)
* Correspondence: teko.napporn@univ-poitiers.fr

Received: 15 February 2019; Accepted: 2 April 2019; Published: 9 April 2019

Abstract: The advances in material science have led to the development of novel and various materials as nanoparticles or thin films. Underpotential deposition (*upd*) of transition metals appears to be a very sensitive method for probing the surfaces of noble metals, which is a parameter that has an important effect on the activity in heterogeneous catalysis. Underpotential deposition as a surface characterization tool permits researchers to precisely determine the crystallographic orientations of nanoparticles or the real surface area of various surfaces. Among all the work dealing with *upd*, this review focuses specifically on the main *upd* systems used to probe surfaces of noble metals in electrocatalysis, from poly- and single-crystalline surfaces to nanoparticles. Cu_{upd} is reported as a tool to determine the active surface area of gold- and platinum-based bimetallic electrode materials. Pb_{upd} is the most used system to assess the crystallographic orientations on nanoparticles' surface. In the case of platinum, Bi and Ge adsorptions are singled out for probing (1 1 1) and (1 0 0) facets, respectively.

Keywords: underpotential deposition (*upd*); Au; Pt; Pd; nanoparticles; cyclic voltammetry; electrocatalysis

---

1. Introduction

Surface of materials is a key parameter in heterogeneous catalysis. Moreover, in electrocatalysis, it is the reaction site where the exchange of electrons occurs, thereby the structure of the materials or any roughness affect their electrochemical response. Referring the current density to the geometric surface area does not consider the surface structure of the material, however, the active sites at the surface of a material depend on its structure, therefore, the characterization of material surface in electrocatalysis becomes an important step in understanding the reactions that occur and evaluating their activity. Several electrochemical methods are used for characterizing the material surface. The underpotential deposition (*upd*) appears to be a powerful tool for probing the surface of electrocatalysts. The *upd* is sensitive to local order, contrary to electron diffraction methods such as Low Energy Electron Diffraction (LEED) or Reflexion High Energy Electron Diffraction (RHEED) which are more sensitive to long-range order [1]. The *upd* can be carried out on polycrystalline surfaces, single-crystals, and recently nanoparticles. The latter have different shapes and sizes that involve different surface structures leading to a variety of activities observed in heterogeneous catalysis and in electrocatalysis. The development of nanoscaled materials in electrochemistry implies the increase of the utilization of *upd* for various applications, from surface probing to the synthesis of nanomaterials.

2. Generalities about *upd*

2.1. Principles and Thermodynamics of the upd Process

The *upd* of a metal is defined as the deposition of a metal onto a foreign metallic substrate in a potential region where a new and pure phase does not form on the substrate.

Figure 1 shows a schematic representation of the voltammetric profile for *upd* and the overpotential deposition (*opd*) of a metal on a foreign substrate. The *upd* occurs at higher potentials than the potential of reduction ($E_{upd} > E$) because of the strong interactions between the deposited metal and the substrate surface structure. In the *upd* region, the metal is deposited as a monolayer on the substrate through self-limiting deposition. In the *opd* region, the metal is deposited as multilayers (bulk electrodeposition) in the absence of self-limiting behavior, creating rough structures [2].

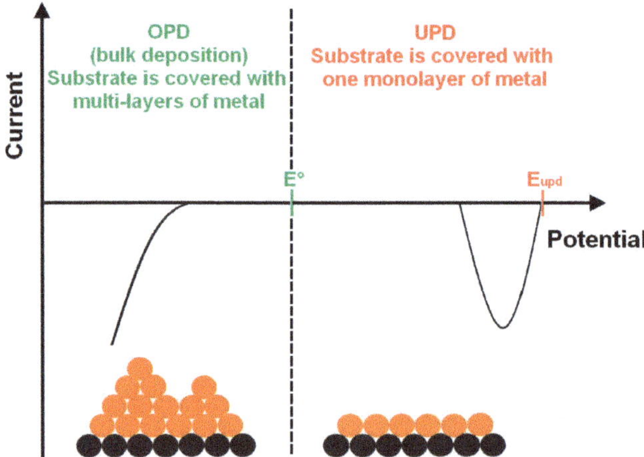

Figure 1. Schematic profile of a voltammetric curve showing the underpotential deposition (*upd*) and overpotential deposition (*opd*) processes of a metal (in orange) on a foreign substrate (in black).

The reduction potential E_{eq} of the metal (M) associated with the reaction $M^{n+} + ne^- \rightarrow M$ is determined by the Nernst equation:

$$E_{eq} = E° + \frac{RT}{nF} ln \frac{a_{Mn+}}{a_M}, \tag{1}$$

where E_{eq} is the Nernst potential of the reaction, $E°$ is the standard potential of the reaction, R is the universal gas constant, T is the temperature, n is the number of exchanged electrons in the reaction, F is the Faraday constant, a_{Mn+} is the activity of M^{n+} ions, and a_M is the activity of the condensed phase M ($a_M = 1$).

The difference between the *upd* onset potential (E_{upd}) and the bulk electrodeposition potential (E) is correlated to the work function of the deposited metal and the substrate [3]. The underpotential shift ($\Delta U = E_{upd} - E$) can be plotted against the difference in work function between bulk substrate and bulk deposited metal $\Delta \Phi$ [3]:

$$\Delta U = \alpha \Delta \Phi, \alpha = 0.5 \text{ V eV}^{-1}. \tag{2}$$

This linear relation between underpotential shift and work function difference suggests that the covalent part of the ad-atom-substrate bond does not differ appreciably to the bond strength between the ad-atom and the surface of the same metal.

The thermodynamic aspects of *upd* has been extensively reported on both polycrystalline and single-crystalline surfaces [4–6]. It shows clearly that the interaction between particle and substrate has a strong effect on the changes in work function and also the interaction between particles, which leads to several energy states. The effect of partially charged *upd* species that leads to the absence of discreteness of charge has been also reported [6]. The theoretical aspect of *upd* has been reported to understand thermodynamics behind this phenomenon [7,8]. It is well known that crystallographic orientation, defects (steps, kinks, grain boundaries), surface reconstruction [9], and alloying play a crucial role in the initial steps of metal deposition. Figure 2 shows the three low Miller indices for a faces-centered cubic (*fcc*) system. The plane (1 1 1) is the most compact arrangement by comparison to the more opened (1 1 0) plane.

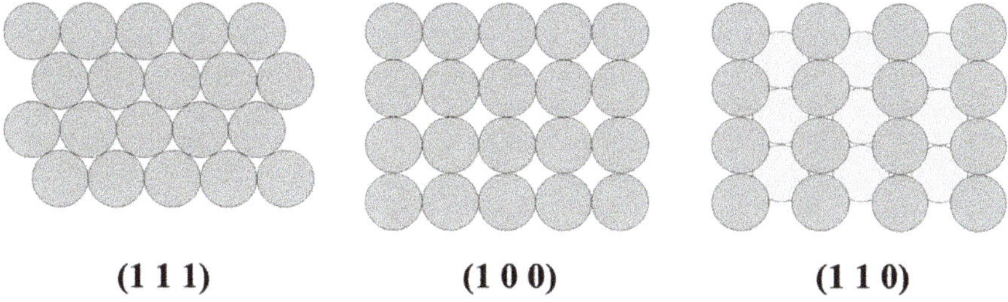

Figure 2. Representation of the three low Miller indices for a faces-centered cubic system.

2.2. A Method of Characterization: Cyclic Voltammetry

Different techniques can be employed to characterize *upd* layers [10]. Cyclic voltammetry is the method most used for carrying out the *upd* of a metal over a substrate. It consists in linearly applying a potential with a constant scan rate *(dE/dt)* on the electrode, between two fixed potential limits. The current is monitored during the scans, the current-potential curves (voltammograms) show the oxidation-reduction processes that occur at the surface. The *upd* curves display cathodic and anodic peaks, for the deposition and dissolution of the deposited metal, respectively. The presence of several distinct adsorption peaks on the voltammograms indicates that the formation of the monolayer takes place at different energetic adsorption steps. The structure of the peak strongly depends on the crystallographic orientation of the substrate and the defects. The peaks are not well defined for polycrystalline electrodes, which present different crystallographic orientations and a high density of defects (kinks, grain boundaries, steps, etc.)—by comparison with single-crystal electrodes.

2.3. The upd Metals

Various substrates were used to perform the *upd* of a metal. The studies began with polycrystalline substrates [11,12], which have a mixed surface structure. To understand and explain the phenomenon, well defined surfaces are proposed. Therefore, *upd* of a metal monolayer onto polycrystalline or single-crystal substrates of a foreign metal has been studied intensively. These investigations included various substrates such as Au, Pt, Ag, Cu, Rh, Ru, Pd, and many deposited ions like Ag^+ [13,14], Cu^{2+} [15–24], Cd^{2+} [25], Pb^{2+} [26–30], Tl^+ [31–34], Cd^{2+} [35,36], Hg^{2+} [37], Sn^{2+} [38,39], Bi^{3+} [30,40–45], Ge^{4+} [46–48], Sb^{2+} [49–51], and Ni^{2+} [52]. The charge corresponding to the *upd* permits researchers to estimate the amount of deposited metal. It is assumed that the metal ions in the sub-monolayers are completely discharged, even if partially charged deposits have been reported. In a large number of studies, the effect of a foreign metal monolayer deposited by *upd* on a metallic substrate was investigated in relation to electrocatalytic reactions [53,54]. Indeed, the *upd* of a monolayer of a foreign metal on a substrate modifies its surface structure. This modification has the advantage of affecting the adsorbed species on the surface and thus the electrochemical reaction. The metal deposited is called ad-atom. It was used in electrosynthesis to modulate the fabrication of different reaction products from the same compound [55,56]. The *upd* process also plays a crucial role in the synthesis of nanoparticles. The layer obtained by *upd* can be the precursor for fabricating nanoparticles of noble metals by galvanic replacement [57]. In this short review, *upd* of transition elements on noble metals' surfaces will be addressed. It includes how surface crystallographic orientations are probing and describes the determination of electrochemically active surface area (ECSA).

3. The M_{upd} on Noble Metals: Au, Pt, and Pd

3.1. The Case of Au

3.1.1. Cu on Au

The underpotential deposition of Cu (Cu_{upd}) on low-index facets of Au appears to be the most extensively studied system. A large number of studies concerns the Cu_{upd} on gold single-crystal Au(1 1 1) [16,18,20,23,58–65]. Investigations on the electrode surface permitted to provide evidences of the Cu_{upd} species at the strained gold surface [24,66]. It is difficult to calculate the *upd* shift of copper ions using electrolytes containing different anions because of their effect. Indeed, anions provide a great influence on the Cu_{upd} on Au, leading to different surface structures and onset potentials of the *upd* process [21,22,67,68]. Cu_{upd} has been investigated on different gold surface orientations (low and high index). Therefore, Cu_{upd} on Au(1 0 0) has been scrutinized by scanning tunneling microscopy (STM) in hydrochloric and sulfuric acids. A (1 × 1) copper ad-layer structure was reported in sulfuric acid, while an incommensurate (2 × 1) one-dimensional structure was observed when hydrochloric acid was added [59]. These anions are specifically adsorbate species because chloride anions are more strongly bound on the gold surface than sulfate anions. Moreover, in the presence of bromide anions on Au(1 1 1) and Pt(1 1 1), a study realized has demonstrated once again that the anions have an influence on the *upd* process and the stability of the deposited ad-layer [17]. It also strongly depends on the substrate because the different behaviors on Cu_{upd} originate from geometric constraint imposed by the lattice structure relative to the deposited ad-layer rather than from energetic considerations. The studies of Cu_{upd} on Au(1 1 1) in sulfuric acid media suggest the co-adsorption of 2/3 monolayer of copper and 1/3 of sulfate (SO_4^{2-}) [15]. This gives a pseudomorphic ($\sqrt{3} \times \sqrt{3}$) structure, followed by a further deposition of 1/3 monolayer of copper and completion of a full epitaxial (1 × 1) monolayer of copper which is still covered by a ($\sqrt{3} \times \sqrt{3}$) layer of sulfate anions [18]. A recent study goes further into the mechanism of Cu_{upd} in the presence of sulfates [21,22]. Indeed, cyclic voltammetry investigations were performed on Au(1 1 1) in both *upd* and *opd* regions. In the potential region where the Cu_{upd} takes place, the first peaks which correspond to the phase transition of sulfate anions on Au(1 1 1) substrate were observed. During the cathodic potential sweep, the adsorption of 2/3 monolayer and then 1/3 monolayer of copper, both with the co-adsorption of sulfate anions, were respectively obtained. During the anodic potential sweep, desorption of the formed monolayer occurs. When the CV is recorded in both *upd* and *opd* regions at lower potentials, the process of adsorption/desorption of multilayer copper is observed. According to these authors, the EC-STM images reveal the nucleation of the ($\sqrt{3} \times \sqrt{7}$) sulfate structure. At −500 mV vs. Pt/PtO, the ($\sqrt{3} \times \sqrt{3}$) structure of SO_4^{2-} anions on 2/3 ML of copper can be seen on HR-STM image. This structure is also observed on two-dimensional gold islands. These islands are provided from the lifting of the (1 1 1) reconstruction. This study shows that different structures are observed on terraces resulting from different local thickness of copper. A ($\sqrt{3} \times \sqrt{3}$)-like structure of sulfate is typical of an incomplete second layer of copper. Figure 3 shows a model (top view) of the sulfate structure ($\sqrt{3} \times \sqrt{3}$), and 2/3 of monolayer of Cu, and the sulfate structure ($\sqrt{3} \times \sqrt{7}$) on the first pseudomorphic monolayer of Cu deposited on Au (1 1 1).

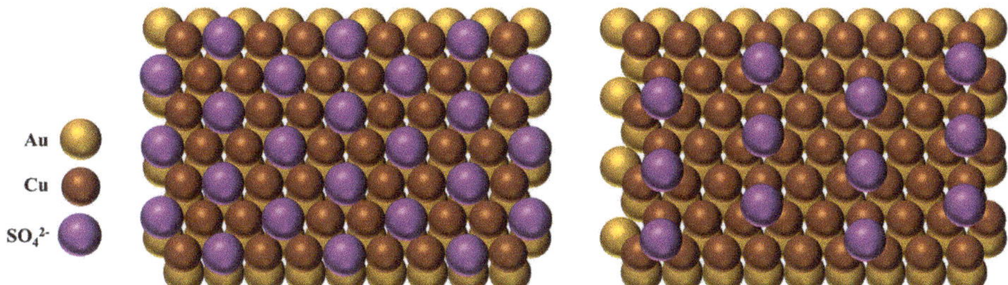

Figure 3. Model of the ($\sqrt{3} \times \sqrt{3}$) sulfate structure, with 2/3 of Cu monolayer (left) and ($\sqrt{3} \times \sqrt{7}$) sulfate structure on the first pseudomorphic monolayer of Cu underpotentially deposited (right) on Au (1 1 1).

Cu$_{upd}$ studies were carried on stepped surfaces of gold with [(n−1)(1 1 1) × (1 1 0)] (n = 6, 7, 10) structures [19]. Figure 4 shows the voltammograms for Cu$_{upd}$ on (3 3 2), (7 7 5), and (5 5 4) surfaces of Au. These structures are made of (1 1 1) terraces and (1 1 0) steps. The voltammogram of Au(3 3 2) displays two broad and overlapping peaks in the first adsorption/desorption region (at 0.520 and 0.560 V vs. RHE). When the step density decreases (the width n of the terrace increases), the peak at 0.560 V vs. RHE is decreasing and leads to assign it to (1 1 0) steps. This potential is close to the main *upd* peak on the Au(1 1 0) electrode (0.570 V vs. RHE). In the case of the peak at 0.520 V vs. RHE, the comparison with copper *upd* on Au(1 1 1) indicates the formation of ($\sqrt{3} \times \sqrt{3}$) R30° structure on the (1 1 1) terraces. It can be seen that this peak becomes better defined when the width of (1 1 1) terraces increases. By deconvoluting these peaks with Gaussian functions, it is found that the charge calculated for (1 1 0) steps on Au(3 3 2) is 17%, with respect to the total *upd* charge. By comparison, the relative geometric area of the (1 1 0) steps domain covers 16.7 % of the surface, which is in good agreement with the relative calculated charge. However, since Cu$_{upd}$ is a complex function of sulfate coverage, it should be noticed that there is no significant change in the sulfate coverage in this potential range. It is necessary to measure the charge of the co-adsorbed sulfate anions in order to perform a full quantitative analysis of the copper *upd* charge on stepped surfaces. On the other hand, the voltammograms in Figure 4 display peaks between 0.350 V and 0.400 V vs. RHE. These peaks are attributed to the ($\sqrt{3} \times \sqrt{3}$) R30° → (1 × 1) phase transition on the (1 1 1) terraces. The splitting of this peak is also influenced by the step density but can be a consequence of the reconstruction of stepped surfaces. This study clearly shows that the steps have an influence on the energetic/kinetics of phase transition. It has been suggested that this influence could be either electronic or structural (Smoluchowski effect).

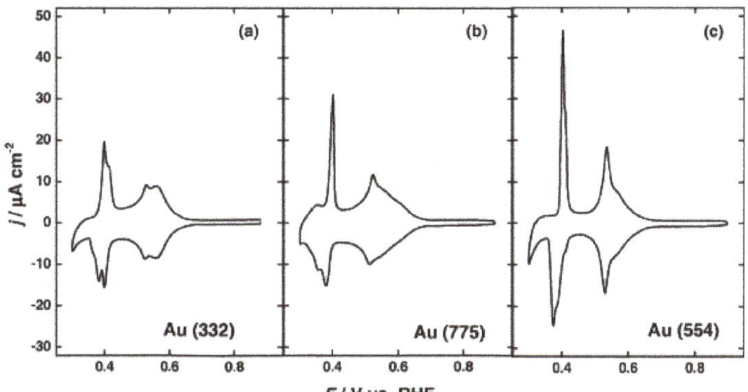

Figure 4. Cyclic voltammograms of (**a**) Au(3 3 2); (**b**) Au(7 7 5); and (**c**) Au(5 5 4) in 0.05 mol L^{-1} H$_2$SO$_4$ + 1 mmol L^{-1} CuSO$_4$ at 5 mVs^{-1}. Reprinted from reference [19]. Copyright (2004) with permission from Elsevier.

The Cu_{upd} can be employed to quantify the surface area of nanoporous gold films [69]. The values obtained by this method are in agreement with those obtained by integrating the reduction peak of gold oxides. However, it should be noted that nanoporous gold films contain residual silver since their preparation consists of the dealloying of bimetallic films. The presence of silver can influence the Cu_{upd}, since this process does not occur on silver.

Many papers reported the effect of metal *upd* on the electrocatalytic activity of a material. In some cases, the underpotential deposited metal can inhibit or even totally deactivate the surface towards electrocatalytic reactions. Cu_{upd} has been used in order to elucidate the mechanism of hydrogen evolution reaction on palladium-modified stepped gold surfaces [70]. The Cu_{upd} allows us to block (deactivate) the palladium step sites towards hydrogen evolution reaction, which is not favored in the presence of copper. Figure 5 shows ball models for Au(h k l)/Pd surface, with steps blocked by copper ad-atoms.

Figure 5. Ball models of Au(h k, l)/Pd (h = k, l =k−1) surface where the Pd steps sites are blocked by Cu modified with 1 ML of Pd.

A technique named Dynamic Electro-Chemo-Mechanical Analysis (DECMA) has been developed, which permits assessment of the strain effect on the adsorption process during Cu_{upd} [24].

3.1.2. Pb on Au

Many studies of the surface structure of gold electrodes have been carried out through the *upd* of lead (Pb_{upd}). These studies allow us to assign the different voltammetric peaks of different surfaces to the deposition processes that occur on low index planes or terraces and steps. The influence of the crystallographic orientation on the process has been largely discussed [28]. Indeed, Pb_{upd} has been reported on polycrystalline gold [71], low-index facets, on (1 0 0), (1 1 1), and vicinal facets [43,72], as well as a large variety of stepped surfaces [73].

Lead and gold atoms have very different sizes (lead is about 20% larger than gold), which favor the formation of incommensurate ad-layers. Indeed, the analysis of the structure of a lead monolayer on Au(1 1 1) by X-ray diffraction shows a hexagonal structure [29]. Figure 6 shows a schematic representation of the structure of a lead monolayer on Au(1 1 1).

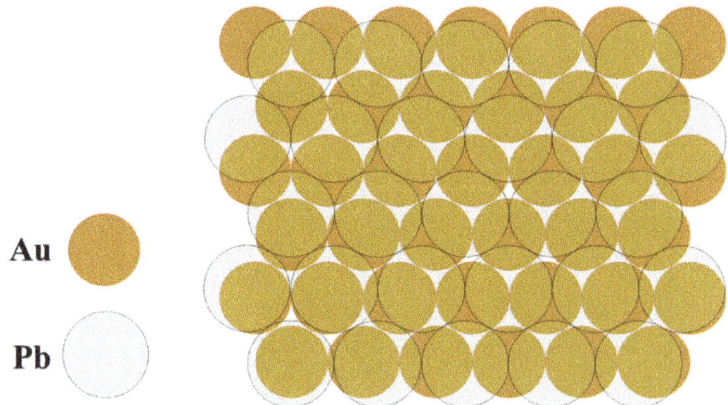

Figure 6. Representation of the hexagonal monolayer of lead on Au(1 1 1).

One of the interesting features of Pb_{upd} on gold is the specific adsorption at different potentials depending on the surface orientation, as illustrated on a polycrystalline gold electrode (Figure 7). During the negative potential sweep (from high to low potentials), lead is deposited on (1 1 0) facets in the potential range from 0.55 to 0.47 V vs. RHE. The deposition on (1 0 0) facets occurs at around 0.43 V vs. RHE and at 0.38 V vs. RHE on (1 1 1). The peak profile (width, shift) depends on the domain width or the presence of terraces. This feature permits assessing the coverage of lead on various facets of gold and also the percentage of each facet on this surface [74].

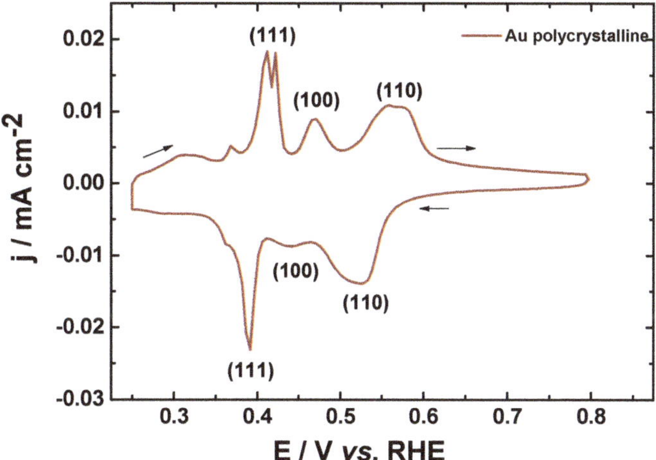

Figure 7. Cyclic voltammogram of Au polycrystalline in 0.1 mol L^{-1} NaOH + 1 mmol L^{-1} Pb(NO$_3$)$_2$, recorded at 20 mV s^{-1} and at 20 °C.

The lead stripping from other gold surfaces with high-index facets (Figure 8) is investigated by Hamelin et al. [27]. The evolution of the curves through a zone of stereographic triangle (in Figure 8) is shown. Each stepped surface is also described as TLK (terrace, ledge, kink) notation. In Figure 8a (zone (1 1 1)-(1 1 0)), the characteristic peak of lead stripping from (1 1 1) terraces decreases gradually, while a broader peak increases due to the stripping from the steps. An inversion of the nature of the step and terrace takes place at (3 3 1) facets. It should be taken into account that (1 1 0) is a stepped surface, as is 2(1 1 1)-(1 1 1). In Figure 8b (zone (1 1 0)-(1 0 0)), the stripping peak of lead from (1 0 0) terraces decreases progressively, as the facets are closer to (1 1 0). The (2 1 0) facet appears as a turning point of the zone (1 1 0)-(1 0 0). In Figure 8c (zone (1 0 0)-(1 1 1)), the evolution of the stripping curves are the same as the previous one, again with a turning point for the (3 1 1) facet.

Pb_{upd} on single-crystals has also been investigated in alkaline media for surface characterization applications. It was observed that the pH affects the position of the deposition/stripping peaks because of the different chemical states of lead: Pb^{2+} in acid media and Pb(OH)$_3^-$ in alkaline media. The charge values associated with the deposition of a monolayer of lead in 0.1 mol L^{-1} NaOH + 1 mmol L^{-1} Pb(NO$_3$)$_2$ media are 444 μC cm^{-2} for (1 1 1), 340 μC cm^{-2} for (1 0 0) and 330 μC cm^{-2} for (1 1 0) [75], respectively.

Figure 9 shows the cyclic voltammograms of three low-index Au single-crystals in alkaline media [75]. Different behaviors were observed for each orientation. On Au(1 1 1), a single sharp deposition peak (0.380 V vs. RHE) was observed, while its associated stripping peak occurs at 0.435 V vs. RHE. In the case of Au(1 0 0) electrode, two peaks were associated with the Pb_{upd}. The main peak occurs at 0.430 V vs. RHE and the smaller one at 0.380 V vs. RHE. The two peaks associated with the dissolution of lead are recorded at 0.385 and 0.475 V vs. RHE. For the Au(1 1 0) electrode, the peaks associated with the deposition and stripping of lead are much broader and less defined than those of the previous orientations. The process is more sluggish on an Au(1 1 0) surface.

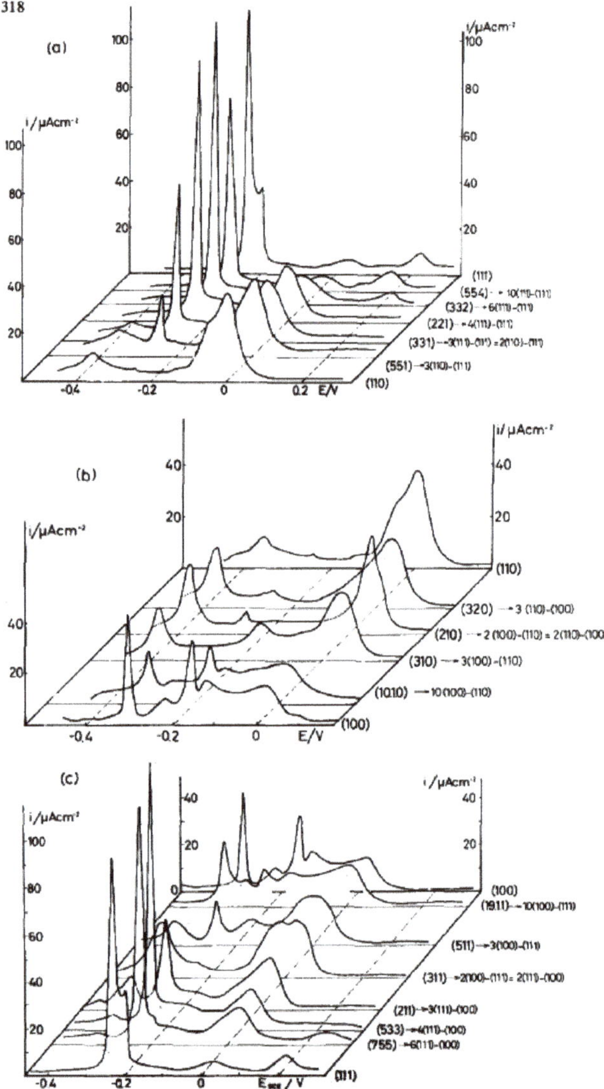

Figure 8. Cyclic voltammograms for the stripping of Pb_{upd} on Au(h k l) electrodes in 10 mmol L^{-1} $HClO_4$ + 1 mmol L^{-1} PbF_2. Au(h k l) surfaces are from the three main zones of the projected stereographic triangle: (**a**) (1 1 1)-(1 1 0) zone, (**b**) (1 1 0)-(1 0 0) zone, and (**c**) (1 0 0)-(1 1 1) zone. All the surfaces are identified with Miller index notation and with the step notation. Reprinted from reference [27]. Copyright (1984) with permission from Elsevier.

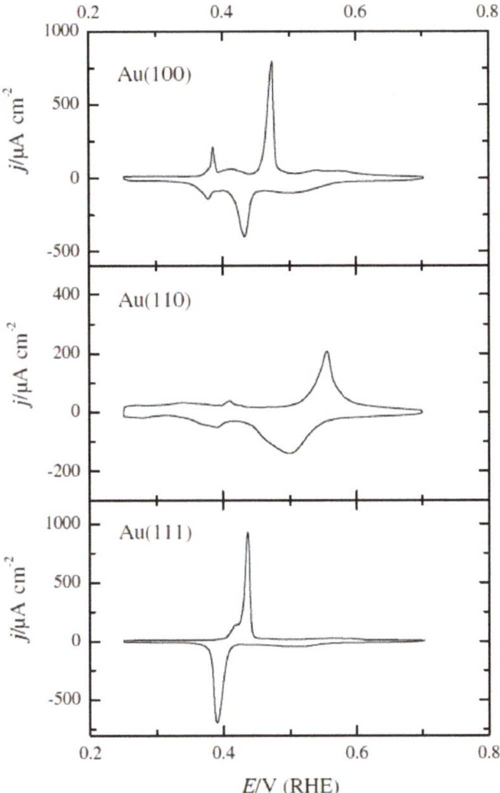

Figure 9. Cyclic voltammograms of Au(1 0 0), Au(1 1 0), and Au(1 1 1) in 0.1 mol L^{-1} NaOH + 1 mmol L^{-1} Pb(NO$_3$)$_2$ at 50 mV s^{-1}. Reprinted from reference [75]. Copyright (2004) with permission from Elsevier.

Pb$_{upd}$ has been reported to characterize bimetallic surface like for Au-Ag, as the Pb$_{upd}$ is also favorable on silver surfaces [35,76]. This technique can be used to quantify the surface coverage of gold and silver in an Au-Ag catalyst, based on the dependence of the deposited lead charge density on the nature of the underlying substrate. Indeed, the coverage of silver on the gold electrode affects the Pb$_{upd}$. Two peaks are observed during the stripping of adsorbed lead (Figure 10), the first one at −0.02 V vs. SHE, which is clearly influenced by the silver coverage (θ_{Ag}). The calculations of the charge densities corresponding to the Pb$_{upd}$ can be made from the cyclic voltammograms. A charge density of 290 µC cm^{-2} was obtained for a bare gold electrode, while a value of 385 µC cm^{-2} was calculated on Ag$_{upd}$-modified gold. The difference of charge densities for the two surfaces may be due to the difference in the availability of sites of lead deposition or to the different adsorption energies [4]. The coverage of silver can be estimated with the following equation:

$$\theta_{Ag}Q_{Pb\text{-}Ag} + \theta_{Au}Q_{Pb\text{-}Au} = Q_{net}, \tag{3}$$

where Q$_{Pb\text{-}Ag}$ represents the charge density of Pb$_{upd}$ on silver (385 µC cm^{-2}), Q$_{Pb\text{-}Au}$ is the charge density of lead *upd* on gold (290 µC cm^{-2}), Q$_{net}$ is the total charge density (subtracted with background current), and θ_{Au} and θ_{Ag} are the gold and silver surface coverages, respectively (with $\theta_{Au} + \theta_{Ag} = 1$).

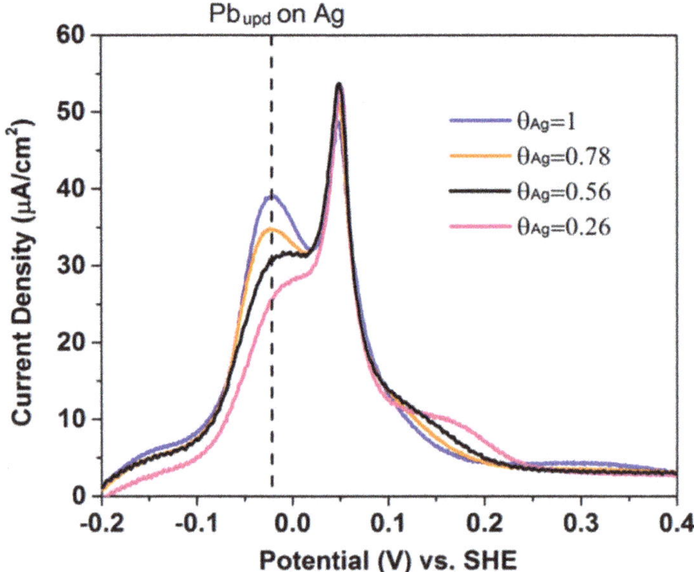

Figure 10. Lead stripping voltammetric curves of Ag$_{upd}$-modified Au with different coverages (θ_{Ag}) in 10 mmol L^{-1} HClO$_4$ + 0.2 mmol L^{-1} Pb(ClO$_4$)$_2$ at 20 mV s^{-1}. Reprinted from reference [76]. Copyright (2017) with permission from Elsevier.

3.1.3. Probing Gold Nanoparticles Surface by upd

Pb$_{upd}$ appears to be an efficient tool to characterize the surface of gold nanoparticles (AuNPs). However, contrary to single-crystal surfaces, AuNPs exhibit generally different planes on their surface. Moreover, synthesis of nanoparticles requires the use of shape-directing agents as surfactants. Pb$_{upd}$ in alkaline media can be employed to verify the cleanliness of NPs' surface and oxidize adsorbed organics on their surface [75]. The surfactant species are oxidized at potential higher than 1.20 V vs. RHE. The advantage of working in alkaline media is that PbO$_2$ deposition takes place at potentials lower than 1.10 V vs. RHE, which is lower than the onset potential for the formation of gold oxides. Gold surface is protected by these species, avoiding its oxidation which can modify its surface, while organic species are oxidized.

The Pb$_{upd}$ has been performed on AuNPs with different shapes and crystallographic orientations [74,77–79]. Hebie et al. [74] have clearly shown the shape dependence of Pb deposition/stripping on AuNPs. Indeed, Figure 11 shows cyclic voltammograms of Pb$_{upd}$ on different AuNPs—nanorods (AuNRs), nanocubes (AuNCs), and nanospheres (AuNSs). The peak associated with the lead stripping from (1 0 0) facets is clearly seen for AuNCs. It is well known that cubic NPs are enclosed by (1 0 0) facets [80]. As the AuNCs studied in this case possess truncated edges, lead stripping is also observed for (1 1 1) and (1 1 0) planes.

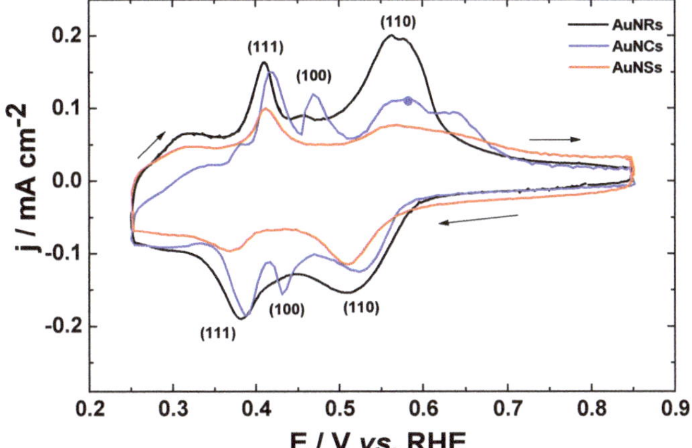

Figure 11. Cyclic voltammograms of different AuNPs (nanospheres AuNSs, nanorods AuNRs, and nanocubes AuNCs) in 0.1 mol L^{-1} NaOH + 1 mmol L^{-1} Pb(NO$_3$)$_2$ recorded at 20 mV s^{-1} and at 20 °C.

A recent study dealt with the Pb$_{upd}$ on quasi-spherical and faceted AuNPs [81]. These investigations have revealed a typical split stripping peak associated with Au(1 1 1) facets in the case of faceted AuNPs, and a single peak in the case of spherical ones. The split peak results from a sluggish kinetics of lead dissolution [82]. It appears that in the case of small terraces with more steps and defects, this doublet stripping peak is not observed, contrary to the peak obtained with faceted NPs. The study suggests that introducing a degree of roughness on (1 1 1) terraces by potential cycles can suppress the splitting profile. The splitting of the peak associated with the desorption from (1 1 1) terraces can be used as an indication of the surface crystallinity of polyfaceted AuNPs.

Thallium is also used for *upd* process on gold nanoparticles. It is known to be used for characterizing AuNPs with defined shapes (nanocrystals) [83,84]. Tl$_{upd}$ has been employed to check the surface cleanliness and the crystallinity of these nanocrystals before performing an oxygen reduction reaction.

3.2. Case of Pt

3.2.1. H$_{upd}$ on Pt

The famous *upd* process widely studied on platinum is the *upd* of hydrogen (H$_{upd}$) [85,86]. H$_{upd}$ is not strictly an *upd* process but the adsorption of hydrogen on the surface. The H$_{upd}$ region acts as a fingerprint for platinum surfaces. Indeed, the profile of the hydrogen adsorption/desorption region obtained by cyclic voltammogram permits an indication of the facets, to assess the active surface area and also to ensure the cleanliness of electrode and cell. Combined with the adsorption of organic molecule, it permits determining the surface coverage. The underpotential deposition of metals can give complementary information to quantify the crystallographic orientations [47]. Traditionally, the Electrochemical Active Surface Area (ECSA) of platinum is determined by the charge corresponding to the H$_{upd}$. Indeed, one platinum atom has the ability to adsorb one hydrogen atom. The charge associated with H$_{upd}$ indicates the number of platinum atoms at the surface. The density of charge associated with the formation of a monolayer of hydrogen depends on the crystallographic orientations, and the value of 210 μC cm^{-2} is the average value for polycrystalline platinum [87,88]. CO-stripping is also used as a method to calculate the ECSA of platinum in acid media [89]. It is important to notice that in acid media, the number of electrons per site during the adsorption of CO on platinum is equal to 2.

The development of bimetallic surfaces for catalysis requests to determine ECSA values. In the case of ruthenium-containing catalysts, H$_{upd}$ is not suitable because of the overlap of the hydrogen and ruthenium oxidation currents [90]. Moreover, hydrogen can be absorbed in the Ru oxide lattice, leading to more than one monolayer of hydrogen. CO-stripping on bimetallic Pt-Ru surface appears to be suitable, however, the adsorption modes on both platinum and ruthenium needs to be deeply elucidated. The mode of adsorption is an important parameter to calculate the charge.

3.2.2. Bi on Pt: A Probe for (1 1 1) Planes

It has been shown that bismuth is spontaneously adsorbed on a platinum surface when the electrode is simply put in contact with a solution containing a salt of Bi(III) species [44]. When a linear potential sweep is applied to the electrode, the oxidized and reduced forms of this compound stay adsorbed at the platinum surface. For the planes (1 1 1), (1 0 0), and (1 1 0), the oxidation peaks are located at 0.610 V vs. RHE, 0.825 V vs. RHE, and 0.915 V vs. RHE, respectively. Bi_{upd} appears as a tool to highlight the presence of (1 1 1) facets, especially when the surface exhibits a combination of several planes. It has been proposed that one atom of bismuth covers three atoms of platinum [47]:

$$Pt_3(1\ 1\ 1)\text{-}Bi + 2H_2O = Pt_3(1\ 1\ 1)\text{-}Bi(OH)_2 + 2H^+ + 2e^-. \tag{4}$$

In the case of stepped surfaces, the redox peak associated with the adsorption of bismuth on (1 1 1) terraces sites gives a charge which is proportional to the number of (1 1 1) terrace sites, without interference from adsorbed bismuth to another site [47]. The adsorption of bismuth suppresses the H_{upd} region, which means that H adsorption sites on platinum are fully blocked.

To characterize the (1 1 1) sites belonging to terrace domains, the charge density values of each surface are obtained by integrating the voltammetric peaks. Figure 12 shows the plots of the charge underneath the bismuth redox peak (q_{Bi}) *versus* the calculated terrace charge [47]. The calculation of the terrace charge depends on how bismuth is adsorbed on the step and terrace sites. Bismuth is preferentially adsorbed on step sites, so the adsorption takes place on the terraces when all the step sites have been covered [91]. A linear equation is obtained and can be used to calibrate the (1 1 1) terrace sites:

$$q_{Bi} = (0.64 \pm 0.02)q_t(1\ 1\ 1) \tag{5}$$

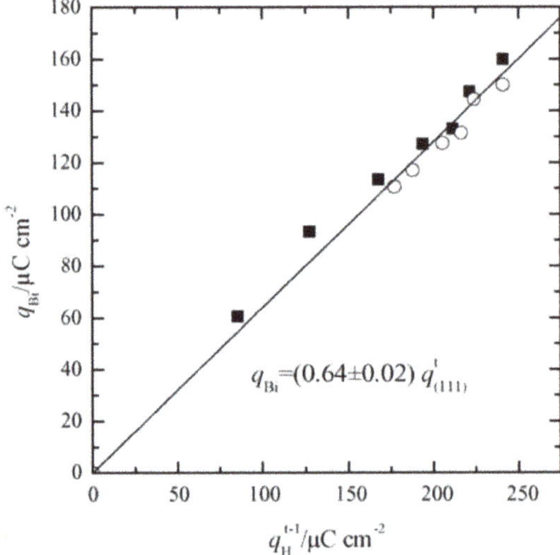

Figure 12. Charge density values of the bismuth redox peak vs. the charge associated with the (1 1 1) sites on the terraces, assuming that one electron is exchanged per site (■: Pt(n,n,n−2), ○: Pt(n+1,n−1,n−1) electrodes). Reprinted from reference [47]. Copyright (2005) with permission from Elsevier.

3.2.3. Ge on Pt: A Probe for (1 0 0) Planes

The *spontaneous deposition* of germanium (Ge_{upd}) acts as a complementary tool to Bi_{upd} to characterize platinum surfaces. It has been proposed that one atom of germanium covers four atoms of platinum and the equation of the oxidation of adsorbed germanium on platinum can be written as follows:

$$Pt_4(1\ 0\ 0)\text{-}Ge + H_2O = Pt_4(1\ 0\ 0)\text{-}GeO + 2H^+ + 2e^-. \tag{6}$$

As for Bi_{upd}, the oxidation of a spontaneously deposited layer of germanium on a platinum surface will give information about the crystallographic orientations of the surface [46]. The H_{upd} is completely inhibited when a full layer of germanium is deposited (Figure 13). Conversely, the H_{upd} is partially inhibited when the surface is not fully covered by germanium. The sharp reversible peaks at 0.540 V vs. RHE are associated with germanium ad-atoms on the (1 0 0) plane. During continuous cycling between 0.060 and 0.600 V vs. RHE, the voltammetric profile remains stable, meaning that the germanium species are still adsorbed on the surface. Moreover, the pH value of the electrolyte does not affect the peaks, suggesting that a germanium oxide is formed.

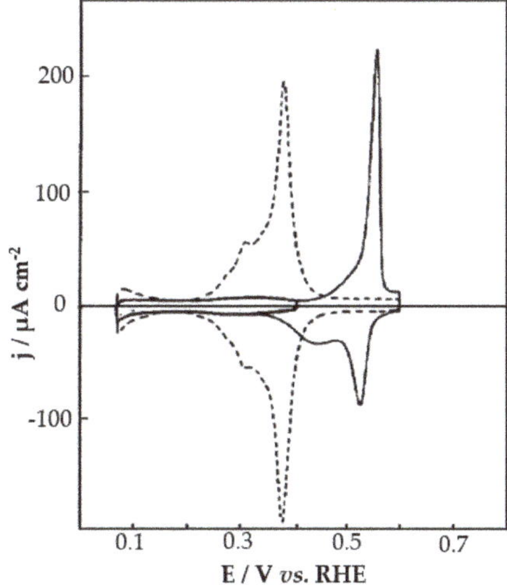

Figure 13. Voltammograms of a Pt(1 0 0) surface fully covered with Ge (continuous line) and Pt(1 0 0) blank. Reprinted from Reference [46]. Copyright (1992) with permission from Elsevier.

The (1 1 0) plane does not show a reversible oxidation/reduction process but a quick desorption is observed. For (1 1 1) planes, the oxidation peak is centered at 0.730 V vs. RHE. However, the oxidation makes a partial dissolution of the germanium layer.

Germanium appears to be sensitive to (1 0 0) planes. Conversely to bismuth, germanium is not irreversibly adsorbed on Pt and it is possible to recover the hydrogen region following the experiment.

As for the study related to (1 1 1) terraces toward bismuth adsorption, adsorbed germanium can be used to characterize (1 0 0) terraces sites of stepped platinum surfaces [47,48].

A linear equation can also be obtained to calibrate terrace sites, as for bismuth on (1 1 1) terraces:

$$q_{Ge} = (0.56 \pm 0.03)qt(1\ 0\ 0). \tag{7}$$

3.2.4. Cu on Pt: A Tool to Characterize Bimetallic Surfaces

As it was reported in this mini review, Cu_{upd} is so far, one of the most studied systems on low-index single-crystals as well as on stepped surfaces [92]. As for gold surfaces, Cu_{upd} on platinum is strongly dependent of the anions present in the electrolyte [17,93]. However, there are only few investigations about its application as a tool for surface characterization. Cu_{upd} is a suitable technique to calculate electrochemical active surface area (ECSA) of Pt and alloys containing Pt. Indeed, Cu_{upd} has been proposed to characterize the bimetallic Pt-Ru surfaces. The integration of the copper stripping peak area permits the estimation of ECSA. It has been shown that Cu_{upd} is an accurate tool to estimate ECSA of each platinum and ruthenium in Pt–Ru electrodes. The coverage of ruthenium can also be determined by this method. Cu_{upd} has also been reported for the calculation of ECSA of Pt-Pd alloys [94]. Indeed, copper is a suitable ad-atom to deposit because its adsorption occurs at a potential region where no other Faradaic reaction takes place on both platinum and palladium.

3.2.5. Probing the Surface of Pt Nanoparticles

Bi_{upd} and Ge_{upd} were also investigated on platinum nanoparticles to probe their surface [47,95–97]. In the case of nanoparticles containing (1 1 1) and (1 0 0) planes, the fraction of each domain is obtained by the ratio of the Faradaic charge of germanium ($Q_{f,Ge}$) or bismuth ($Q_{f,Bi}$) with the charge associated with the hydrogen desorption process ($Q_{f,H}$). The characterization of PtNPs prepared in different conditions has been reported [98]. The use of Ge_{upd} and Bi_{upd} permitted determining the percentage of (1 0 0) and (1 1 1) facets, respectively. The fraction of (1 1 0) sites has been estimated. Indeed, the peak of H_{upd} associated with (1 1 0) facets decreases, while the difference from 100% of the sum of (1 1 1) and (1 0 0) facets decreases.

3.3. Case of Pd

Cu on Pd

There are only a few studies about *upd* of metals on palladium surfaces [11,40]. This is due to the difficulty for metal monolayers to be adsorbed on the double-layer region, which is narrow [99]. Palladium also has the ability to absorb hydrogen in its bulk, this process may mask the *upd* one. It should be noted that there is no study about *upd* of metals to characterize palladium surfaces. For example, Cu grows epitaxially with the Pd substrate.

Cu_{upd} has been studied on polycrystalline and single-crystalline palladium electrodes [100]. This study has shown that, under adequate conditions, Cu adsorption can be restricted to a potential range where no other Faradaic reaction takes place. Cu_{upd} on palladium surfaces occurs at potentials lower than the onset potential of oxides formation, and hydrogen absorption is inhibited by changing the pH of the electrolyte. Figure 14 shows the cyclic voltammograms of *upd*/stripping on palladium (1 0 0) and (1 1 1) single-crystal and polycrystalline palladium. It is clearly seen that the single-crystal electrode present less complex voltammograms than polycrystalline palladium. On Pd(1 1 1), the *upd* and desorption processes give a single sharp peak (A_3/D_3) preceded by a small shoulder. On Pd(1 0 0), two peaks are observed (A_2/D_2 and A_6/D_6) and a small shoulder noted A_1/D_1. The complexity of the voltammogram obtained on polycrystalline palladium is due to the variety of crystal planes.

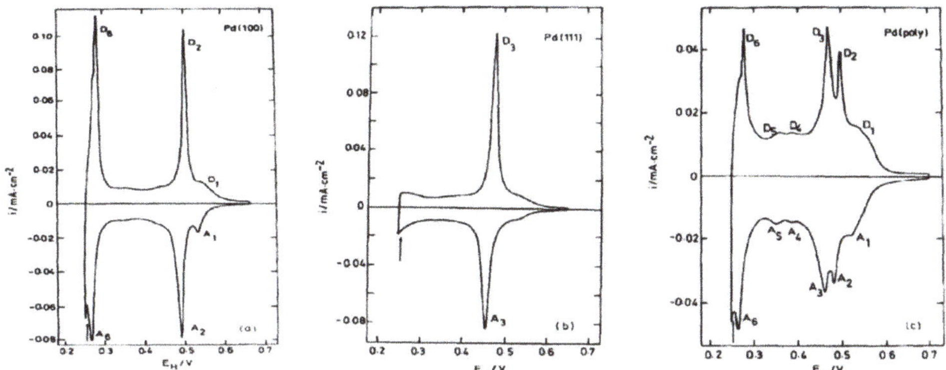

Figure 14. Cyclic voltammograms of Pd single-crystal electrodes in 0.5 mol L^{-1} NaClO$_4$ + 0.01 mol L^{-1} HClO$_4$ + 1 mmol L^{-1} Cu^{2+}, at 10 mV s^{-1}: (**a**) Pd(1 0 0), (**b**) Pd(1 1 1), and (**c**) polycrystalline Pd. Reprinted from reference [100]. Copyright (1988) with permission from Elsevier.

4. Conclusions

Investigations on *upd* of transition metals at noble metal surfaces have been increasing over these last two decades. It appears to be a powerful electrochemical tool to probe and characterize the working electrode surface. Indeed, *upd* techniques have been reported to calculate electrochemical active surface area, especially in the case of bimetallic surface, and to probe the crystallographic orientations among the surface of single-crystals and nanoparticles. Pb$_{upd}$ is mostly used to characterize the orientation of gold nanoparticles and surfaces. In the case of platinum surfaces, the probe of crystallographic orientation is made with bismuth and germanium adsorption. Cu$_{upd}$ is one of the most studied *upd* systems that appear to be useful for characterizing the surface of bimetallic materials.

Author Contributions: All authors contributed equally to this paper.

Funding: The authors acknowledge the Region Nouvelle Aquitaine for the financial support.

Conflicts of Interest: The authors declare no conflict of interest.

References

1. Scortichnini, C.L.; Reilley, C.N. Surface characterization of Pt electrodes using underpotential deposition of H and Cu: Part I. Pt(100). *J. Electroanal. Chem.* **1982**, *139*, 233–245. [CrossRef]
2. Bockris, J.M.; Nagy, Z.; Damjanovic, A. On the deposition and dissolution of zinc in alkaline solutions. *J. Electrochem. Soc.* **1972**, *119*, 285–295. [CrossRef]
3. Kolb, D.; Przasnyski, M.; Gerischer, H. Underpotential deposition of metals and work function differences. *J. Electroanal. Chem.* **1974**, *54*, 25–38. [CrossRef]
4. Kirowa-Eisner, E.; Bonfil, Y.; Tzur, D.; Gileadi, E. Thermodynamics and kinetics of UPD of lead on polycrystalline silver and gold. *J. Electroanal. Chem.* **2003**, *552*, 171–183. [CrossRef]
5. Szabó, S. Underpotential deposition of metals on foreign metal substrates. *Int. Rev. Phys. Chem.* **1991**, *10*, 207–248. [CrossRef]
6. Swathirajan, S.; Bruckenstein, S. Thermodynamics and kinetics of underpotential deposition of metal monolayers on polycrystalline substrates. *Electrochim. Acta* **1983**, *28*, 865–877. [CrossRef]
7. Leiva, E. Recent developments in the theory of metal upd. *Electrochim. Acta* **1996**, *41*, 2185–2206. [CrossRef]
8. Rojas, M.I.; Dassie, S.A.; Leiva, E.P.M. Theoretical-study about the adsorption of lead on (111), (100), (110) monocrystalline surfaces of gold. *Z. Phys. Chem.* **1994**, *185*, 33–50. [CrossRef]
9. Kolb, D.M.; Schneider, J. Surface reconstruction in electrochemistry: Au(100-(5 × 20), Au(111)-(1 × 23) and Au(110)-(1 × 2). *Electrochim. Acta* **1986**, *31*, 929–936. [CrossRef]

10. Oviedo, O.A.; Reinaudi, L.; García, S.G.; Leiva, E.P.M. Experimental techniques and structure of the underpotential deposition phase. In *Underpotential Deposition*; Scholz, F., Ed.; Springer: Cham, Switzerland, 2016; pp. 17–89.

11. Szabó, S.; Bakos, I.; Nagy, F.; Mallát, T. Study of the underpotential deposition of copper onto polycrystalline palladium surfaces. *J. Electroanal. Chem.* **1989**, *263*, 137–146. [CrossRef]

12. Engelsmann, K.; Lorenz, W.J.; Schmidt, E. Underpotential deposition of lead on polycrystalline and single-crystal gold surfaces: Part I. Thermodynamics. *J. Electroanal. Chem.* **1980**, *114*, 1–10. [CrossRef]

13. Garcia, S.; Salinas, D.; Mayer, C.; Schmidt, E.; Staikov, G.; Lorenz, W.J. Ag UPD on Au(100) and Au(111). *Electrochim. Acta* **1998**, *43*, 3007–3019. [CrossRef]

14. Rooryck, V.; Reniers, F.; Buess-Herman, C.; Attard, G.A.; Yang, X. The silver UPD on gold(111) revisited. *J. Electroanal. Chem.* **2000**, *482*, 93–101. [CrossRef]

15. Hachiya, T.; Honbo, H.; Itaya, K. Detailed underpotential deposition of copper on gold(111) in aqueous-solutions. *J. Electroanal. Chem.* **1991**, *315*, 275–291. [CrossRef]

16. Uchida, H.; Hiei, M.; Watanabe, M. Electrochemical quartz crystal microbalance study of copper adatoms on Au(111) electrodes in solutions of perchloric and sulfuric acid. *J. Electroanal. Chem.* **1998**, *452*, 97–106. [CrossRef]

17. Herrero, E.; Glazier, S.; Buller, L.J.; Abruña, H.D. X-ray and electrochemical studies of Cu upd on single crystal electrodes in the presence of bromide: Comparison between Au(111) and Pt(111) electrodes. *J. Electroanal. Chem.* **1999**, *461*, 121–130. [CrossRef]

18. Nakamura, M.; Endo, O.; Ohta, T.; Ito, M.; Yoda, Y. Surface X-ray diffraction study of Cu UPD on Au (1 1 1) electrode in 0.5 M H_2SO_4 solution: The coadsorption structure of UPD copper, hydration water molecule and bisulfate anion on Au (1 1 1). *Surf. Sci.* **2002**, *514*, 227–233. [CrossRef]

19. Kuzume, A.; Herrero, E.; Feliu, J.M.; Nichols, R.J.; Schiffrin, D.J. Copper underpotential deposition at high index single crystal surfaces of Au. *J. Electroanal. Chem.* **2004**, *570*, 157–161. [CrossRef]

20. Danilov, A.I.; Molodkina, E.B.; Rudnev, A.V.; Polukarov, Y.M.; Feliu, J.M. Kinetics of copper deposition on Pt(111) and Au(111) electrodes in solutions of different acidities. *Electrochim. Acta* **2005**, *50*, 5032–5043. [CrossRef]

21. Madry, B.; Wandelt, K.; Nowicki, M. Deposition of copper and sulfate on Au(111): New insights. *Appl. Surf. Sci.* **2016**, *388*, 678–683. [CrossRef]

22. Madry, B.; Wandelt, K.; Nowicki, M. Sulfate structures on copper deposits on Au(111): In situ STM investigations. *Electrochim. Acta* **2016**, *217*, 249–261. [CrossRef]

23. Sebastián, P.; Gómez, E.; Climent, V.; Feliu, J.M. Copper underpotential deposition at gold surfaces in contact with a deep eutectic solvent: New insights. *Electrochem. Commun.* **2017**, *78*, 51–55. [CrossRef]

24. Yang, M.Z.; Zhang, H.X.; Deng, Q.B. Understanding the copper underpotential deposition process at strained gold surface. *Electrochem. Commun.* **2017**, *82*, 125–128. [CrossRef]

25. Dursun, Z.; Ben Aoun, S.; Taniguchi, I. Electrocatalytic oxidation of D-glucose using a Cd ad-atom-modified Au(111) electrode in alkaline solution. *Turk. J. Chem.* **2008**, *32*, 423–430.

26. Adžić, R.; Yeager, E.; Cahan, B.D. Optical and electrochemical studies of underpotential deposition of lead on gold evaporated and single-crystal electrodes. *J. Electrochem. Soc.* **1974**, *121*, 474–484. [CrossRef]

27. Hamelin, A.; Lipkowski, J. Underpotential deposition of lead on gold single-crystal faces: Part 2. General discussion. *J. Electroanal. Chem.* **1984**, *171*, 317–330. [CrossRef]

28. Hamelin, A. Underpotential deposition of lead on single-crystal faces of gold: Part 1. The influence of crystallographic orientation of the substrate. *J. Electroanal. Chem.* **1984**, *165*, 167–180. [CrossRef]

29. Samant, M.G.; Toney, M.F.; Borges, G.L.; Blum, L.; Melroy, O.R. Grazing incidence x-ray diffraction of lead monolayers at a silver (111) and gold (111) electrode/electrolyte interface. *J. Phys. Chem.* **1988**, *92*, 220–225. [CrossRef]

30. Feliu, J.; Fernandez-Vega, A.; Orts, J.; Aldaz, A. The behaviour of lead and bismuth adatoms on well-defined platinum surfaces. *J. Chim. Phys.* **1991**, *88*, 1493–1518. [CrossRef]

31. Wang, J.X.; Adzic, R.R.; Ocko, B.M. X-ray scattering study of Tl adlayers on the Au(111) electrode in alkaline solutions: Metal monolayer, OH⁻ coadsorption, and oxide formation. *J. Phys. Chem.* **1994**, *98*, 7182–7190. [CrossRef]

32. Wheeler, D.R.; Wang, J.X.; Adžić, R.R. The effects of anions on the underpotential deposition of Tl on Pt(111): A voltammetric study. *J. Electroanal. Chem.* **1995**, *387*, 115–119. [CrossRef]

33. Batchelor-McAuley, C.; Wildgoose, G.G.; Compton, R.G. The contrasting behaviour of polycrystalline bulk gold and gold nanoparticle modified electrodes towards the underpotential deposition of thallium. *New J. Chem.* **2008**, *32*, 941–946. [CrossRef]

34. Rodriguez, P.; García-Aráez, N.; Herrero, E.; Feliu, J.M. New insight on the behavior of the irreversible adsorption and underpotential deposition of thallium on platinum (111) and vicinal surfaces in acid electrolytes. *Electrochim. Acta* **2015**, *151*, 319–325. [CrossRef]

35. Bonfil, Y.; Brand, M.; Kirowa-Eisner, E. Characteristics of subtractive anodic stripping voltammetry of lead, cadmium and thallium at silver-gold alloy electrodes. *Electroanalysis* **2003**, *15*, 1369–1376. [CrossRef]

36. Brust, M.; Ramirez, S.A.; Gordillo, G.J. Site-specific modification of gold nanoparticles by underpotential deposition of cadmium atoms. *ChemElectroChem* **2018**, *5*, 1586–1590. [CrossRef]

37. Abaci, S.; Zhang, L.; Shannon, C. The influence of counter anions on the underpotential deposition of mercury(II) on Au(111): Temperature-dependent studies. *J. Electroanal. Chem.* **2004**, *571*, 169–176. [CrossRef]

38. Rodes, A.; Herrero, E.; Feliu, J.M.; Aldaz, A. Structure sensitivity of irreversibly adsorbed tin on gold single-crystal electrodes in acid media. *J. Chem. Soc. Faraday Trans.* **1996**, *92*, 3769–3776. [CrossRef]

39. Meier, L.A.; Salinas, D.R.; Feliu, J.M.; García, S.G. Spontaneous deposition of Sn on Au(1 1 1). An in situ STM study. *Electrochem. Commun.* **2008**, *10*, 1583–1586. [CrossRef]

40. Szabo, S. Investigations of copper, silver and bismuth deposition on palladium in perchloric acid media. *J. Electroanal. Chem.* **1977**, *77*, 193–203. [CrossRef]

41. Adžić, R.; Jovančićević, V.; Podlavicky, M. Optical and electrochemical study of underpotential deposition of bismuth on gold electrode. *Electrochim. Acta* **1980**, *25*, 1143–1146. [CrossRef]

42. Ganon, J.P.; Clavilier, J. Electrical resistance measurement in electrochemical adsorption experiments of lead and bismuth on thin films of gold. II. *Surf. Sci.* **1984**, *147*, 583–598. [CrossRef]

43. Ganon, J.P.; Clavilier, J. Electrochemical adsorption of lead and bismuth at gold single crystal surfaces with vicinal (111) orientations. I. *Surf. Sci.* **1984**, *145*, 487–518. [CrossRef]

44. Clavilier, J.; Feliu, J.; Aldaz, A. An irreversible structure sensitive adsorption step in bismuth underpotential deposition at platinum electrodes. *J. Electroanal. Chem.* **1988**, *243*, 419–433. [CrossRef]

45. Thiel, K.-O.; Hintze, M.; Vollmer, A.; Donner, C. Bismuth UPD on the modified Au (1 1 1) electrode. *J. Electroanal. Chem.* **2010**, *638*, 143–150. [CrossRef]

46. Gómez, R.; Llorca, M.; Feliu, J.; Aldaz, A. The behaviour of germanium adatoms irreversibly adsorbed on platinum single crystals. *J. Electroanal. Chem.* **1992**, *340*, 349–355. [CrossRef]

47. Rodriguez, P.; Herrero, E.; Solla-Gullon, J.; Vidal-Iglesias, F.J.; Aldaz, A.; Feliu, J.M. Specific surface reactions for identification of platinum surface domains - Surface characterization and electrocatalytic tests. *Electrochim. Acta* **2005**, *50*, 4308–4317. [CrossRef]

48. Rodríguez, P.; Herrero, E.; Solla-Gullón, J.; Vidal-Iglesias, F.; Aldaz, A.; Feliu, J. Electrochemical characterization of irreversibly adsorbed germanium on platinum stepped surfaces vicinal to Pt (1 0 0). *Electrochim. Acta* **2005**, *50*, 3111–3121. [CrossRef]

49. Gooyoung, J.; Choong Kyun, R. Two electrochemical processes for the deposition of Sb on Au(100) and Au(111): Irreversible adsorption and underpotential deposition. *J. Electroanal. Chem.* **1997**, *436*, 277–280. [CrossRef]

50. Chen, Y.; Wang, L.S.; Pradel, A.; Ribes, M.; Record, M.C. A voltammetric study of the underpotential deposition of cobalt and antimony on gold. *J. Electroanal. Chem.* **2014**, *724*, 55–61. [CrossRef]

51. Chen, Y.; Wang, L.S.; Pradel, A.; Merlen, A.; Ribes, M.; Record, M.C. Underpotential deposition of selenium and antimony on gold. *J. Solid State Electrochem.* **2015**, *19*, 2399–2411. [CrossRef]

52. Vaskevich, A.; Sinapi, F.; Mekhalif, Z.; Delhalle, J.; Rubinstein, I. Underpotential deposition of nickel on {111}-textured gold electrodes in dimethyl sulfoxide. *J. Electrochem. Soc.* **2005**, *152*, C744–C750. [CrossRef]

53. Hebie, S.; Napporn, T.W.; Kokoh, K.B. Beneficial promotion of underpotentially deposited lead adatoms on gold nanorods toward glucose electrooxidation. *Electrocatalysis* **2017**, *8*, 67–73. [CrossRef]

54. Alvarez-Rizatti, M.; Jüttner, K. Electrocatalysis of oxygen reduction by UPD of lead on gold single-crystal surfaces. *J. Electroanal. Chem.* **1983**, *144*, 351–363. [CrossRef]

55. Parpot, P.; Kokoh, K.B.; Beden, B.; Belgsir, E.M.; Leger, J.M.; Lamy, C. Selective electrocatalytic oxidation of sucrose on smooth and upd-lead modified platinum-electrodes in alkaline-medium. *Stud. Surf. Sci. Catal.* **1993**, *78*, 439–445.

56. Kokoh, K.; Léger, J.-M.; Beden, B.; Lamy, C. "On line" chromatographic analysis of the products resulting from the electrocatalytic oxidation of d-glucose on Pt, Au and adatoms modified Pt electrodes—Part I. Acid and neutral media. *Electrochim. Acta* **1992**, *37*, 1333–1342. [CrossRef]

57. Chen, H.M.; Xing, Z.L.; Zhu, S.Q.; Zhang, L.L.; Chang, Q.W.; Huang, J.L.; Cai, W.B.; Kang, N.; Zhong, C.J.; Shao, M.H. Palladium modified gold nanoparticles as electrocatalysts for ethanol electrooxidation. *J. Power Sources* **2016**, *321*, 264–269. [CrossRef]

58. Vélez, P.; Cuesta, A.; Leiva, E.P.M.; Macagno, V.A. The underpotential deposition that should not be: Cu(1 × 1) on Au(111). *Electrochem. Commun.* **2012**, *25*, 54–57. [CrossRef]

59. Möller, F.; Magnussen, O.M.; Behm, R.J. CuCl adlayer formation and Cl induced surface alloying: An in situ STM study on Cu underpotential deposition on Au(110) electrode surfaces. *Electrochim. Acta* **1995**, *40*, 1259–1265. [CrossRef]

60. Holzle, M.H.; Zwing, V.; Kolb, D.M. The influence of steps on the deposition of Cu onto Au(111). *Electrochim. Acta* **1995**, *40*, 1237–1247. [CrossRef]

61. Hasegawa, Y.; Avouris, P. Manipulation of the reconstruction of the Au(111) surface with the STM. *Science* **1992**, *258*, 1763–1765. [CrossRef] [PubMed]

62. Nichols, R.J.; Kolb, D.M.; Behm, R.J. STM observations of the initial stages of copper deposition on gold single-crystal electrodes. *J. Electroanal. Chem.* **1991**, *313*, 109–119. [CrossRef]

63. Magnussen, O.M.; Hotlos, J.; Nichols, R.J.; Kolb, D.M.; Behm, R.J. Atomic structure of Cu adlayers on Au(100) and Au(111) electrodes observed by in situ scanning tunneling microscopy. *Phys. Rev. Lett.* **1990**, *64*, 2929–2932. [CrossRef]

64. Kongstein, O.E.; Bertocci, U.; Stafford, G.R. In situ stress measurements during copper electrodeposition on (111)-textured Au. *J. Electrochem. Soc.* **2005**, *152*, C116–C123. [CrossRef]

65. Hagenström, H.; Schneeweiss, M.A.; Kolb, D.M. Copper underpotential deposition on ethanethiol-modified Au(111) electrodes: Kinetic effects. *Electrochim. Acta* **1999**, *45*, 1141–1145. [CrossRef]

66. Deng, Q.; Smetanin, M.; Weissmüller, J. Mechanical modulation of reaction rates in electrocatalysis. *J. Catal.* **2014**, *309*, 351–361. [CrossRef]

67. Akhade, S.A.; McCrum, I.T.; Janik, M.J. The impact of specifically adsorbed ions on the copper-catalyzed electroreduction of CO_2. *J. Electrochem. Soc.* **2016**, *163*, F477–F484. [CrossRef]

68. Mrozek, P.; Sung, Y.-E.; Han, M.; Gamboa-Aldeco, M.; Wieckowski, A.; Chen, C.-H.; Gewirth, A.A. Coadsorption of sulfate anions and silver adatoms on the Au (111) single crystal electrode. Ex situ and in situ comparison. *Electrochim. Acta* **1995**, *40*, 17–28. [CrossRef]

69. Rouya, E.; Cattarin, S.; Reed, M.L.; Kelly, R.G.; Zangari, G. Electrochemical characterization of the surface area of nanoporous gold films. *J. Electrochem. Soc.* **2012**, *159*, K97–K102. [CrossRef]

70. Hernandez, F.; Baltruschat, H. Hydrogen evolution and CuUPD at stepped gold single crystals modified with Pd. *J. Solid State Electrochem.* **2007**, *11*, 877–885. [CrossRef]

71. Hamelin, A.; Katayama, A.; Picq, G.; Vennereau, P. Surface characterization by underpotential deposition—Lead on gold surfaces. *J. Electroanal. Chem.* **1980**, *113*, 293–300. [CrossRef]

72. Hamelin, A.; Katayama, A. Lead underpotential deposition on gold single-crystal surfaces—The (100) face and its vicinal faces. *J. Electroanal. Chem.* **1981**, *117*, 221–232. [CrossRef]

73. Hamelin, A. Lead adsorption on gold single-crystal stepped surfaces. *J. Electroanal. Chem.* **1979**, *101*, 285–290. [CrossRef]

74. Hebie, S.; Cornu, L.; Napporn, T.W.; Rousseau, J.; Kokoh, B.K. Insight on the surface structure effect of free gold nanorods on glucose electrooxidation. *J. Phys. Chem. C* **2013**, *117*, 9872–9880. [CrossRef]

75. Hernandez, J.; Solla-Gullon, J.; Herrero, E. Gold nanoparticles synthesized in a water-in-oil microemulsion: Electrochemical characterization and effect of the surface structure on the oxygen reduction reaction. *J. Electroanal. Chem.* **2004**, *574*, 185–196. [CrossRef]

76. Yu, L.; Akolkar, R. Lead underpotential deposition for the surface characterization of silver ad-atom modified gold electrocatalysts for glucose oxidation. *J. Electroanal. Chem.* **2017**, *792*, 61–65. [CrossRef]

77. Hernandez, J.; Solla-Gullon, J.; Herrero, E.; Feliu, J.M.; Aldaz, A. In situ surface characterization and oxygen reduction reaction on shape-controlled gold nanoparticles. *J. Nanosci. Nanotechnol.* **2009**, *9*, 2256–2273. [CrossRef]

78. Hebie, S.; Kokoh, K.B.; Servat, K.; Napporn, T.W. Shape-dependent electrocatalytic activity of free gold nanoparticles toward glucose oxidation. *Gold Bull.* **2013**, *46*, 311–318. [CrossRef]

79. Hebie, S.; Napporn, T.W.; Morais, C.; Kokoh, K.B. Size-dependent electrocatalytic activity of free gold nanoparticles for the glucose oxidation reaction. *ChemPhysChem* **2016**, *17*, 1454–1462. [CrossRef] [PubMed]

80. Niu, W.X.; Zheng, S.L.; Wang, D.W.; Liu, X.Q.; Li, H.J.; Han, S.A.; Chen, J.; Tang, Z.Y.; Xu, G.B. Selective synthesis of single-crystalline rhombic dodecahedral, octahedral, and cubic gold nanocrystals. *J. Am. Chem. Soc.* **2009**, *131*, 697–703. [CrossRef] [PubMed]

81. Jeyabharathi, C.; Zander, M.; Scholz, F. Underpotential deposition of lead on quasi-spherical and faceted gold nanoparticles. *J. Electroanal. Chem.* **2018**, *819*, 159–162. [CrossRef]

82. Schultze, J.; Dickertmann, D. Potentiodynamic desorption spectra of metallic monolayers of Cu, Bi, Pb, Tl, and Sb adsorbed at (111), (100), and (110) planes of gold electrodes. *Surf. Sci.* **1976**, *54*, 489–505. [CrossRef]

83. Lu, F.; Zhang, Y.; Liu, S.Z.; Lu, D.Y.; Su, D.; Liu, M.Z.; Zhang, Y.G.; Liu, P.; Wang, J.X.; Adzic, R.R.; et al. Surface proton transfer promotes four-electron oxygen reduction on gold nanocrystal surfaces in alkaline solution. *J. Am. Chem. Soc.* **2017**, *139*, 7310–7317. [CrossRef]

84. Zhang, Y.; Lu, F.; Liu, S.; Lu, D.; Su, D.; Liu, M.; Zhang, Y.; Liu, P.; Wang, J.X.; Adzic, R.R. Oxygen reduction on gold nanocrystal surfaces in alkaline electrolyte: Evidence for surface proton transfer effects. *ECS Trans.* **2018**, *85*, 93–110. [CrossRef]

85. Conway, B.; Angerstein-Kozlowska, H.; Sharp, W.; Criddle, E. Ultrapurification of water for electrochemical and surface chemical work by catalytic pyrodistillation. *Anal. Chem.* **1973**, *45*, 1331–1336. [CrossRef]

86. Zolfaghari, A.; Chayer, M.; Jerkiewicz, G. Energetics of the underpotential deposition of hydrogen on platinum electrodes I. Absence of coadsorbed species. *J. Electrochem. Soc.* **1997**, *144*, 3034–3041. [CrossRef]

87. Clavilier, J.; Orts, J.M.; Gómez, R.; Feliu, J.M.; Aldaz, A. Comparison of electrosorption at activated polycrystalline and Pt(531) kinked platinum electrodes: Surface voltammetry and charge displacement on potentiostatic CO adsorption. *J. Electroanal. Chem.* **1996**, *404*, 281–289. [CrossRef]

88. Chen, Q.-S.; Solla-Gullón, J.; Sun, S.-G.; Feliu, J.M. The potential of zero total charge of Pt nanoparticles and polycrystalline electrodes with different surface structure: The role of anion adsorption in fundamental electrocatalysis. *Electrochim. Acta* **2010**, *55*, 7982–7994. [CrossRef]

89. Maillard, F.; Savinova, E.R.; Stimming, U. CO monolayer oxidation on Pt nanoparticles: Further insights into the particle size effects. *J. Electroanal. Chem.* **2007**, *599*, 221–232. [CrossRef]

90. Green, C.L.; Kucernak, A. Determination of the platinum and ruthenium surface areas in platinum−ruthenium alloy electrocatalysts by underpotential deposition of copper. I. Unsupported catalysts. *J. Phys. Chem. B* **2002**, *106*, 1036–1047. [CrossRef]

91. Herrero, E.; Climent, V.C.; Feliu, J.M. On the different adsorption behavior of bismuth, sulfur, selenium and tellurium on a Pt (775) stepped surface. *Electrochem. Commun.* **2000**, *2*, 636–640. [CrossRef]

92. Molodkina, E.; Danilov, A.; Feliu, J.M. Cu UPD at Pt (100) and stepped faces Pt (610), Pt (410) of platinum single crystal electrodes. *Russ. J. Electrochem.* **2016**, *52*, 890–900. [CrossRef]

93. Markovic, N.; Ross, P. Effect of anions on the underpotential deposition of copper on platinum (111) and platinum (100) surfaces. *Langmuir* **1993**, *9*, 580–590. [CrossRef]

94. Fiçicioğlu, F.; Kadirgan, F. Characterization of a Pt+ Pd alloy electrode by underpotential deposition of copper. *J. Electroanal. Chem.* **1993**, *346*, 187–196. [CrossRef]

95. Solla-Gullón, J.; Vidal-Iglesias, F.; Herrero, E.; Feliu, J.; Aldaz, A. CO monolayer oxidation on semi-spherical and preferentially oriented (1 0 0) and (1 1 1) platinum nanoparticles. *Electrochem. Commun.* **2006**, *8*, 189–194. [CrossRef]

96. Solla-Gullón, J.; Rodríguez, P.; Herrero, E.; Aldaz, A.; Feliu, J.M. Surface characterization of platinum electrodes. *Phys. Chem. Chem. Phys.* **2008**, *10*, 1359–1373. [CrossRef]

97. Coutanceau, C.; Urchaga, P.; Baranton, S. Diffusion of adsorbed CO on platinum (100) and (111) oriented nanosurfaces. *Electrochem. Commun.* **2012**, *22*, 109–112. [CrossRef]

98. Solla-Gullón, J.; Vidal-Iglesias, F.; Rodriguez, P.; Herrero, E.; Feliu, J.; Clavilier, J.; Aldaz, A. In situ surface characterization of preferentially oriented platinum nanoparticles by using electrochemical structure sensitive adsorption reactions. *J. Phys. Chem. B* **2004**, *108*, 13573–13575. [CrossRef]

99. Chierchie, T.; Mayer, C.; Lorenz, W. Structural changes of surface oxide layers on palladium. *J. Electroanal. Chem.* **1982**, *135*, 211–220. [CrossRef]
100. Chierchie, T.; Mayer, C. Voltammetric study of the underpotential deposition of copper on polycrystalline and single crystal palladium surfaces. *Electrochim. Acta* **1988**, *33*, 341–345. [CrossRef]

Article

Nitrogen-Doped Ordered Mesoporous Carbons Supported Co₃O₄ Composite as a Bifunctional Oxygen Electrode Catalyst

Jing Wang [1], Shuwei Zhang [1], Haihong Zhong [1,2], Nicolas Alonso-Vante [2], Dianqing Li [1], Pinggui Tang [1] and Yongjun Feng [1,*]

[1] State Key Laboratory of Chemical Resource Engineering, Beijing Engineering Center for Hierarchical Catalysts, Beijing University of Chemical Technology, No. 15 Beisanhuan East Road, Beijing 100029, China; 2017200993@buct.edu.cn (J.W.); 2016201069@buct.edu.cn (S.Z.); 2014200922@grad.buct.edu.cn (H.Z.); lidq@mail.buct.edu.cn (D.L.); tangpg@mail.buct.edu.cn (P.T.)

[2] IC2MP, UMR-CNRS 7285, University of Poitiers, F-86022 Poitiers CEDEX, France; nicolas.alonso.vante@univ-poitiers.fr

* Correspondence: yjfeng@mail.buct.edu.cn; Tel.: +86-10-6442-5385

Received: 20 February 2019; Accepted: 26 March 2019; Published: 29 March 2019

Abstract: It is increasingly useful to develop bifunctional catalysts for oxygen reduction and oxygen evolution reaction (ORR and OER) for fuel cells, metal-air rechargeable batteries, and unitized regenerative cells. Here, based on the excellent conductivity and stability of ordered mesoporous carbons, and the best ORR and OER activity of Co₃O₄, the composite Co₃O₄/N-HNMK-3 was designed and manufactured by means of a solvothermal method, using ordered N-doped mesoporous carbon (N-HNMK-3) as substrate, and then the bifunctional electrocatalytic performance corresponding to ORR, OER in alkaline media was carefully investigated. The results showed that Co₃O₄/N-HNMK-3 composite, a non-precious metal centered electrocatalyst, displayed excellent ORR performance (activity, selectivity, and stability) close to that of commercial 20 wt.% Pt/C and a promising OER activity near 20 wt.% RuO₂/C. The outstanding bifunctional activities of Co₃O₄/N-HNMK-3 was assessed with the lowest $\triangle$E value of 0.86 V ($E_{OER,10\,mA\,cm^{-2}}$-$E_{ORR,-3\,mA\,cm^{-2}}$) with respect to the two commercial precious metal-based electrocatalysts.

Keywords: Ordered mesoporous carbon; nitrogen doping; cobalt-based electrocatalyst; bifunctional oxygen electrode; solvothermal method

---

1. Introduction

Hydrogen is deemed as the best way for renewable sources energy-storage to replace petroleum-based energy resources because of its high calorific value, abundant reserve, and near zero release. Yet, both the oxygen reduction reaction (ORR) and oxygen evolution reaction (OER) individually play crucial roles in fuel cells for the practical application of hydrogen, and water electrolysis for the high-efficiency production of high-purity hydrogen [1–4]. Recently, unitized regenerative cells (URCs), integrating water electrolysis and fuel cell in one setup, have attracted increasing attention due to potential applications in aviation, submarines and transportation as a mobile power source. This issue remains key to exploring high performance and low cost. In terms of electrocatalytic activity, Pt/C and Ru-/Ir oxides are generally considered to be the most advanced electrochemical oxygen catalysts for the ORR and OER processes, respectively. However, high cost, low methanol tolerance, and poor stability continue to hamper practical marketing development [5]. Therefore, it remains a big challenge to develop promising bifunctional oxygen electrodes for the development of URCs.

Among non-precious metal electrocatalysts being developed such as cobalt-based chalcogenides, $CoSe_2$, CoS, and Co_3O_4, show outstanding performance toward oxygen electrode reactions including activity, selectivity, and durability in alkaline media [6–9]. However, cobalt-based chalcogenide electrocatalysts have a low electrical conductivity and *per se* a negative effect on the performance of the electrocatalyst. On the other hand, ordered mesoporous carbon as a conductive substrate has attracted more attention because of large surface, high conductivity and long durability. [10–13]. In addition, the doping heteroatom, such as nitrogen [14,15], phosphor [16,17], sulfur [18] and boron [19], can further improve electrocatalytic performance towards oxygen electrode reactions due to new active species. In a previous report, we developed N-doped ordered mesoporous carbons (CMK-3) with high ORR activity, long stability, and outstanding methanol tolerance in 0.1 M KOH [20]. Therefore, it is possible to explore high-performance, low-cost electrocatalysts using N-doped ordered mesoporous carbon as a substrate to support cobalt-based chalcogenides.

Here, we manufactured the N-doped ordered mesoporous carbon substrate (N-HNMK-3) for Co_3O_4 composite (Co_3O_4/N-HNMK-3) by means of a solvothermal route using the home-made ordered mesoporous carbon as carbon source, melamine as a nitrogen source, cobalt acetate as a cobalt source, and urea as the precipitating agent. Figure 1 describes the corresponding synthesis process. In addition, we carefully examined the bifunctional electrocatalytic performance of ORR and OER processes in alkaline media using the rotating disc electrode technique. The compound, based on non-precious metals, shows a promising electrocatalytic performance.

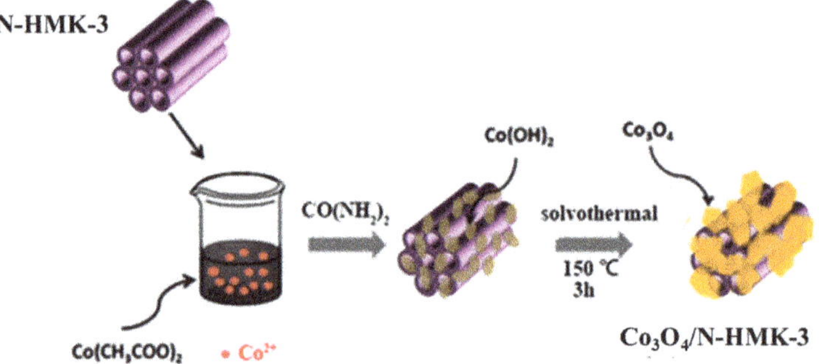

Figure 1. Schematic on preparation process of Co_3O_4/N-HNMK-3.

2. Materials and Methods

2.1. Chemicals

Cobalt acetate ($Co(CH_3COO)_2·6H_2O$), urea ($CO(NH_2)_2$), HNO_3 (6M), melamine ($C_3H_6N_6$), and ethanol (C_2H_5OH) were A.R. grade and were directly used as received from Beijing Chemical Co. Ltd (Beijing, China). CMK-3 was synthesized as reported previously [20].

2.2. Synthesis of Nitrogen-Doped Ordered Mesoporous Carbon (N-HNMK-3)

Nitrogen-doped ordered mesoporous carbon was fabricated by post-treated nitrogen doping method using ordered mesoporous carbon (CMK-3, surface area, 1247.5 m^2/g; pore volume, 1.5 mL/g) as carbon source, and melamine as nitrogen source. To obtain HNMK-3, CMK-3 (0.5002 g) was dispersed in 6.0 M HNO_3 (50.0 mL) under vigorous stirring at 70 °C for 3 h, and the resulting suspension was cleaned to neutral pH using deionized water. Thereafter, 0.5000 g HNMK-3 and 2.5000 g melamine were then mixed during grinding to form the homogeneous mixture. Finally, the obtained mixture was calcined in a nitrogen atmosphere at 900 °C for 4 h to produce nitrogen-doped ordered mesoporous carbon (N-HNMK-3).

2.3. Synthesis of Co$_3$O$_4$/N-HNMK-3 Electrocatalyst

Co$_3$O$_4$/N-HNMK-3 electrocatalyst was carefully generated via the solvothermal route using N-HNMK-3 as conductive support; Co(CH$_3$COO)$_2$ as cobalt source, and CO(NH$_2$)$_2$ as the precipitating agent. Typically, 0.0175 g of the as-prepared N-HNMK-3 was dispersed in ethanol (40.0 mL) by ultrasound for 2h to form the suspension A. Simultaneously, 1.50 g Co(Ac)$_2$ (6.0 mmol) was added in deionized water (40.0 mL) and ultra-sonicated for 20 min to form solution B. Then suspension A and solution B were mixed and ultrasonicated again to produce another suspension. At this point 0.72 g CO(NH)$_2$ was added under vigorous stirring at 80 °C for 24 h to produce a suspension, which was heated at 150 °C in a Teflon-lined stainless-steel autoclave for 3 h. The cooled product was centrifuged and washed with deionized water and ethanol three cycles, and dried at 60 °C overnight. The collected final product was Co$_3$O$_4$/N-HNMK-3. In addition, Co$_3$O$_4$, as a reference, was prepared in the same procedure without N-HNMK-3.

2.4. Characterization

The materials phase was characterized by Rigaku UItima III XRD (Rigaku Corporation, Tokyo, Japan) with Cu K$_\alpha$ radiation (λ = 0.154 nm) at 5 min^{-1} per 2θ. The morphologies were analyzed by Zeiss Supra 55 SEM(Carl Zeiss Jena, Oberkochen, German), a JEOL JEM-2012 TEM (JEOL, Tokyo, Japan), and HRTEM (JEOL, Tokyo, Japan) with a line resolution of 0.19 nm was used. Raman spectra were examined via a Nanofinder 3.0 Raman spectrometer with a He-Ne laser beam of 532 nm. Binding energies of chemical species were collected by XPS using VG ESCALAB 2201 XL spectrometer (Thermo VG Scientific, West Sussex, England).

2.5. Electrochemical Measurements

All electrochemical tests were performed in a standard three-electrode system on the Chen Hua electrochemical workstation (CH Instruments Ins., Shanghai, China) at 25 °C. Here, a saturated calomel electrode and a Pt-wire were employed as the reference and counter electrodes, respectively. In this work, all potentials were referred to a reversible hydrogen electrode (RHE): E_{RHE} = E_{SCE} + 0.99 V (0.1 M KOH); and E_{RHE} = E_{SCE} + 1.06 V (1.0 M KOH).

Prior to catalyst deposition, the GC disk was polished with γ-alumina (5A), and successively ultra-sonically treated in water, and then ethanol for one min in each solvent. The ink was prepared by mixing the catalyst powder (6.7 mg), Nafion (5 wt.%) (40 μL), ethanol (300 μL), deionized water (1160 μL) and ultra-sonicated for 40 min. 4.2 μL of the ink was deposited onto the GC disk (4.0 mm, diameter) with a catalyst mass loading of 150 μg cm^{-2}. The electrolyte was degassed with argon and oxygen for 30 min before cyclic voltammetry (CV) tests. Also, two benchmarks: 20 wt.% Pt/C (Johnson Matthey) and RuO$_2$/C were prepared as references with a corresponding loading of 60 μg cm^{-2} for Pt, and RuO$_2$.

In the ORR test, the LSV curves were recorded in O$_2$-saturated 0.1 M KOH by scanning the disc potential vs. RHE from 0.95 V to 0.15 V at 5 mVs^{-1} with the electrode rotated at 2500, 2025, 1600,1225, 900, 625, and 400 rpm, and analysis performed using the Koutecky-Levich (K-L) equation:

$$\frac{1}{j} = \frac{1}{j_k} + \frac{1}{j_d} = \frac{1}{j_k} + \frac{1}{B\omega^{\frac{1}{2}}} \tag{1}$$

where:

$$B = 0.62nFC_0 D_0^{2/3} v^{-1/6} \tag{2}$$

$$j_k = nFkC_0 \tag{3}$$

j, j_k, j_d are the corresponding measured, kinetic and diffusion limiting current densities, respectively [5]; C_0 and D_0 were the saturated concentration, and the diffusion coefficient of O$_2$ (1.14 $\times$ 10^{-6} mol cm^{-3} and 1.73 $\times$ 10^{-5} cm s^{-1}) in 0.1 M KOH, v was the kinetic viscosity coefficient (0.01 cm^2 s^{-1}).

In the RRDE test, the HO_2^-% and the number of electrons transferred(n) was calculated by the following two equations:

$$HO_2^-\% = 200 * \frac{I_R/N}{I_D + I_R/N} \tag{4}$$

$$n = 4 * \frac{I_D}{I_D + I_R/N} \tag{5}$$

where I_D, I_R and N is disk current, ring current and collection efficiency (0.424), respectively.

For the OER tests, measurements were carried out in O_2-saturated 1.0 M KOH with a rotating speed of 1600 rpm at a scan rate of 5 mV s^{-1} from 1.0 V to 1.65V *vs.* RHE.

3. Results and Discussion

3.1. Structure, Morphology, and Chemical Composition

Figure 2 shows the powder PXRD patterns of N-HNMK-3, Co_3O_4, and Co_3O_4/N-HNMK-3. Compared to CMK-3, the peak of N-HNMK-3 samples at 23.8°/2θ shifts to a lower angle with a higher degree of graphitization, which could be the result of a change in electron distribution following the incorporation of the heteroatom on the surface of carbon materials [21]. Based on the Bragg equation, a series of Bragg reflection peaks in Co_3O_4 and Co_3O_4/N-HNMK-3 samples correspond to (111), (220), (311), (400), (511), and (440) crystal faces, that match well with the cubic spinel phase Co_3O_4 (ICDD PDF card No. 42-1467). This means that the urea-assisted solvothermal method is an available way to prepare the Co_3O_4/N-HNMK-3 composite.

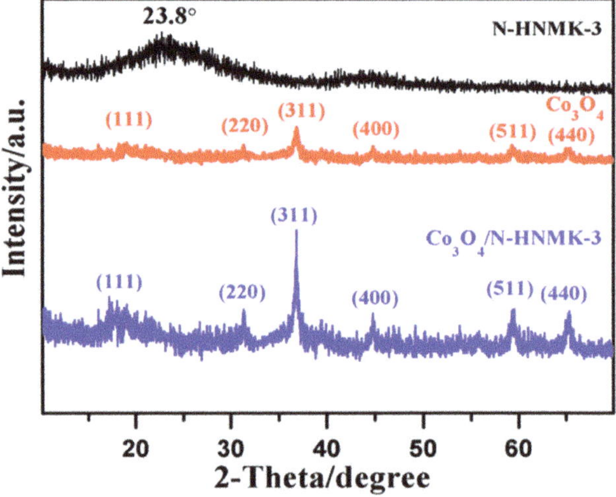

Figure 2. Powder XRD patterns of N-HNMK-3, Co_3O_4, and Co_3O_4/N-HNMK-3 composite.

Figure 3 displays the morphologies and microstructures of Co_3O_4 and Co_3O_4/N-HNMK-3 samples. The Co_3O_4 sample shows different rod-shape in size, Figure 3a. In comparison, the Co_3O_4 particles, in Co_3O_4/N-HNMK-3 composite, are well-dispersed and supported on N-HNMK-3 substrate, Figure 3b,c. In addition, the average particle size for Co_3O_4 was 20–30 nm, Figure 3c. The HRTEM image of Co_3O_4 reveals three lattice distances such as 0.286, 0.244 and 0.202 nm, which individually correspond to (220), (311) and (400) Co_3O_4 crystal planes, Figure 3d.

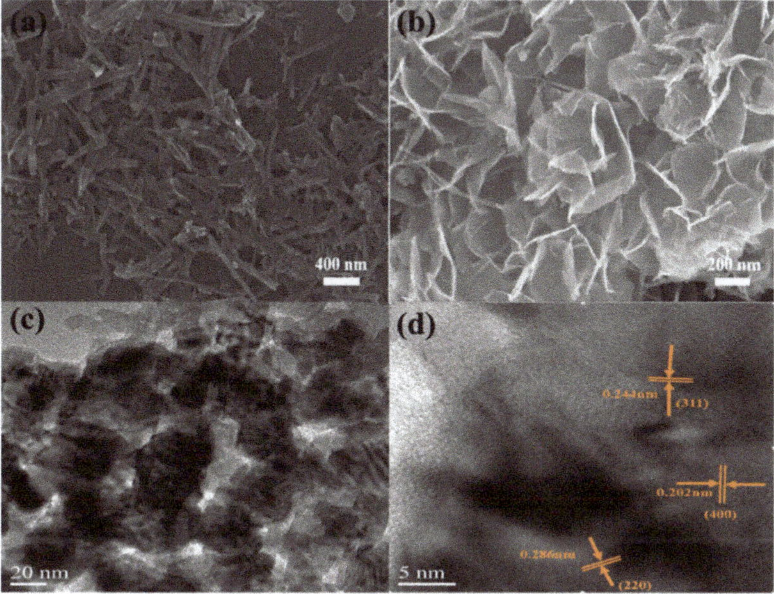

Figure 3. (**a**) The SEM images of Co_3O_4, (**b**) Co_3O_4/N-HNMK-3. (**c**) TEM and (**d**) HRTEM images of Co_3O_4/N-HNMK-3.

In addition, Figure 4 shows the Raman spectra of N-HNMK-3, Co_3O_4, and Co_3O_4/N-HNMK-3 samples as well as the corresponding fitting profiles of N-HNMK-3 and Co_3O_4/N-HNMK-3. One observes two characteristic bands for N-HNMK-3, and Co_3O_4/N-HNMK-3 centered at 1358 (D peak) and 1590 (G peak) cm^{-1}. The former peak (D peak) refers to sp^3-like C atoms defect sites, whereas the latter (G peak) refers to sp^2 C=C stretching mode [22]. Remarkably, other peaks were detected at 192, 468, 516 and 677 cm^{-1}, assessing the presence of Co_3O_4 [23]. The intensity ratio of two bands (I_D/I_G) illustrates the disordered degree of carbonization. A ratio of 2.96 for N-HNMK-3, and 2.78 for Co_3O_4/N-HNMK-3 suggest that the incorporation of Co_3O_4 decreases the disordered degree, increasing the conductivity of the substrate.

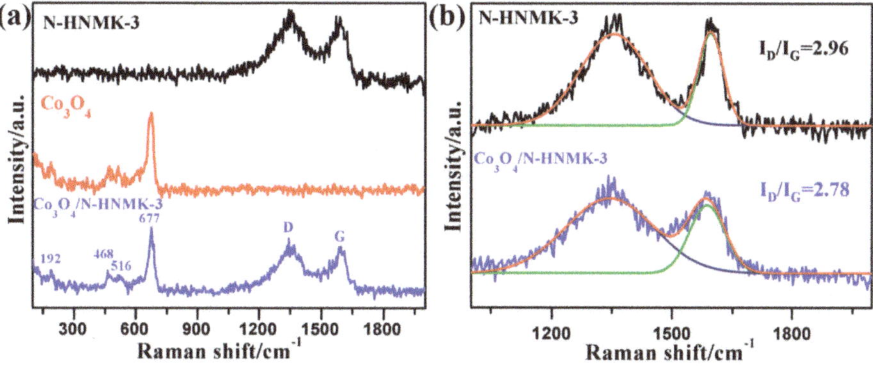

Figure 4. (**a**) Raman spectra of N-HNMK-3, Co_3O_4 and Co_3O_4/N-HNMK-3. (**b**) The corresponding Raman fitting profiles of N-HNMK-3 and Co_3O_4/N-HNMK-3.

Surface elemental properties and electronic valence states were further characterized by XPS. A survey spectrum of Co_3O_4/HNMK-3 displays the existence of C 1s, N 1s, O 1s, Co 2p signals at 284.5, 399.9, 530.5, 780.1 and 796.2 eV, respectively, as shown in Figure 5a. The corresponding atomic percentage of every element of different samples, N-HNMK-3 has a similar nitrogen content with Co_3O_4/HNMK-3, as summarized in Table 1. In the C1s high-resolution XPS spectrum, shown in Figure 5b, the prominent peak centered at 284.6 eV corresponds to sp^2-hybridised carbon atoms in graphitic structure. The series of fitted peaks at 285.7, 288.3, and 288.6 eV are attributed to C=N, O-C=O and C-N, respectively. The XPS spectra of C=N and C-N further assess the presence of N-dopant atoms. As to Co2p spectrum, Figure 5c there are two peaks centered at 779.7 and 795.1 eV attributed to Co $2p_{3/2}$ and $2p_{1/2}$ signals, with an energy separation of 15.4 eV, evidencing the presence of Co_3O_4 in the composites [24]. Two nitrogen species detected in the N 1s high resolution, shown in Figure 5d, at ~398.6 eV and 400.2 eV are assigned to pyridine and Pyrrole-N. Previous results already demonstrated that pyridine N is an active site for the ORR [25]. The relative content of Pyrrole-N in Co_3O_4/HNMK-3 sample is 70.2%, which is higher than that of Pyridine-N (29.8%). It's worth noting that N-HNMK-3 and Co_3O_4/HNMK-3 samples have similar N content.

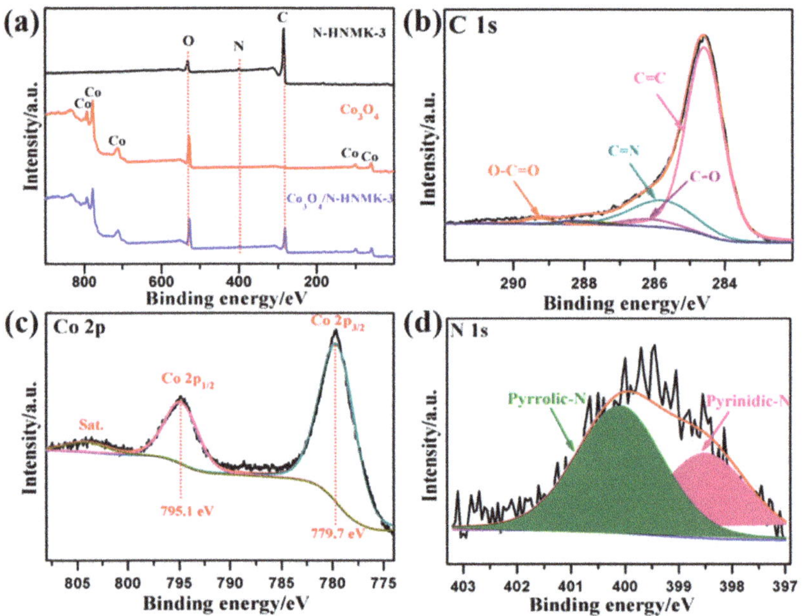

Figure 5. (a) XPS survey scan of samples: N-HNMK-3, Co_3O_4 and Co_3O_4/N-HNMK-3; XPS spectra of (b) C 1s; (c) Co 2p; (d) N 1s of Co_3O_4/N-HNMK-3.

Table 1. The element amount (at. %) in N-HNMK-3, Co_3O_4 and Co_3O_4/N-HNMK-3 samples.

Samples	C	N	Co	O	Pyridinic-N	Pyrrolic-N
N-HNMK-3	88.8	2.6		8.6	31.2	68.8
Co_3O_4			42.8	57.2		
Co_3O_4/N-HNMK-3	60.6	2.2	9.4	27.8	29.8	70.2

3.2. Electrocatalytic ORR Performance

The ORR performance of Co_3O_4/N-HNMK-3 composite was determined by RDE technique, and compared to two commercial reference materials: Pt/C and RuO_2/C. Figure 6 displays the activity towards ORR in 0.1 M KOH electrolyte. Specifically, Figure 6a,b show the ORR activity difference among five samples: N-HMNK-3, physical mixture of N-HMNK-3 and Co_3O_4, Co_3O_4/N-HMNK-3 composite, Pt/C, and RuO_2/C. Herein, it is observed that the Co_3O_4/N-HMNK-3 composite has a much higher ORR activity compared to N-HMNK-3, the physical mixture, and RuO_2/C, and close to Pt/C. Interestingly, the Co_3O_4/N-HNMK-3 composite shows much better ORR activity than the mechanical mixture (Co_3O_4 +N-HNMK-3). This difference in the ORR activity results from the possible synergistic effect between two compositions, and from the high dispersion of Co_3O_4 on the N-HNMK-3 substrate. In addition, cyclic voltammetry measurement, shown in Figure 6c, was conducted to gain deeper insight into the ORR process. In the O_2-saturated electrolyte there is a cathodic peak centered at 0.75 V, which is absent in the Ar-saturated electrolyte. The onset potential of Co_3O_4/N-HMNK was detected at 0.90 V *vs.* RHE. The ORR hydrodynamics on Co_3O_4/N-HMNK-3 between 400 rpm to 2500 rpm is shown in Figure 6d with well-defined sigmoidal curves. The electron transfers number (n) was estimated through the slopes of fitted K-L plots, see the inset in Figure 6d. Within the applied electrode potential from 0.2 to 0.5 V, the estimated transfer number (n) per O_2 molecule was ca. 3.94, suggesting a 4-electron transfer pathway. Meanwhile, we used the RRDE to calculate the hydrogen peroxide and electron transfer number in Figure S1 (See Supporting Information). The low hydrogen peroxide below 15% and transfer number beyond ca. 3.75 in the range of 0.1–0.75 V suggested a close 4-electron transfer pathway.

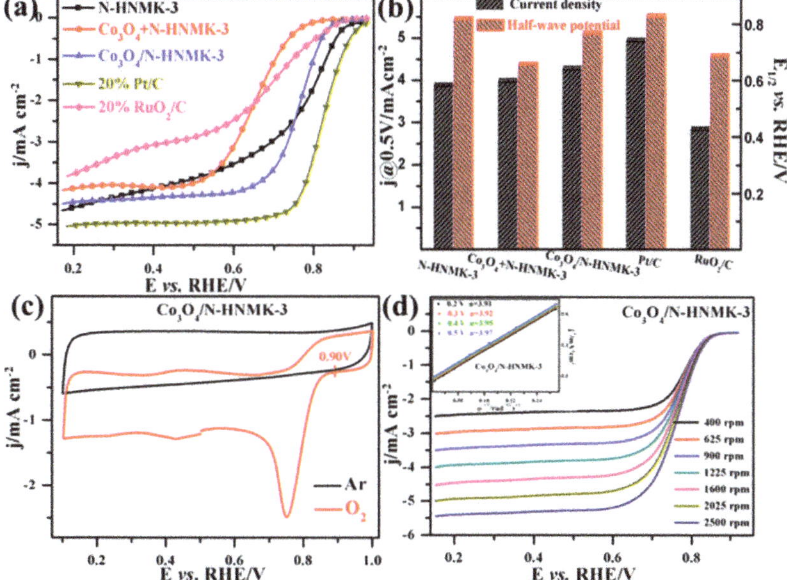

Figure 6. (a) ORR polarization curves of N-HNMK-3, Co_3O_4+N-HNMK-3 (physical mixture), Co_3O_4/N-HNMK-3, Pt/C, and RuO_2/C at 1600 rpm in O_2-saturated 0.1M KOH; (b) the limited current density at 0.50 V/RHE and half-wave potential from (a); (c) CV curves of Co_3O_4/N-HNMK-3 in Ar- and O_2-saturated 0.1M KOH; (d) ORR polarization curves of Co_3O_4/N-HNMK-3 at various rotation rates, the inset shows the corresponding K-L plot of Co_3O_4/N-HNMK-3.

In addition to activity, the selectivity (tolerance to small organics) and the stability are two other key parameters for a high performance electrocatalyst. Figure 7a depicts the electrocatalytic selectivity in the presence and the absence of 3.0 M methanol in 0.1 M KOH electrolyte. After the addition of 3 M methanol, no significant change was observed in the polarization curves in Ar- and/or O_2-saturated electrolyte, see the inset in Figure 7a. In O_2-saturated electrolyte Co_3O_4/N-HNMK-3 showed a complete tolerance to methanol. In contrast, as it is well established, Pt/C has no selectivity towards the ORR in presence of methanol, since this material develops a mixed potential due to methanol oxidation and oxygen reduction. Furthermore, the stability test, as demonstrated by chronoamperometry, Figure 7b, was conducted at 0.70 V/RHE, and revealed that, after 18000 s, the ORR activity of Co_3O_4/N-HNMK-3 dropped 18.2% against 23.9% on Pt/C. Meanwhile, CVs with multiple cycles in oxygen- saturated solution for ORR were shown in Figure S2 (See Supporting Information), after the accelerated aging test, the reduction potential only shifts 36 mV to a negative region for 1500 cycles.

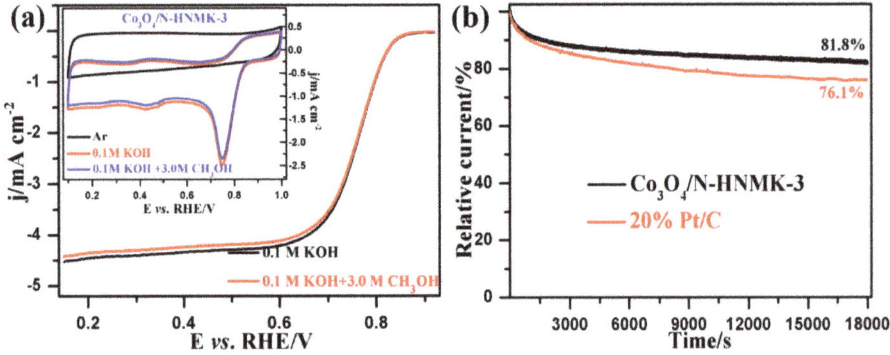

Figure 7. (**a**) CV and LSV curves of Co_3O_4/N-HNMK-3 in O_2- saturated with or without 3.0M methanol in 0.1 M KOH; (**b**) The durability test of Co_3O_4/N-HNMK-3 and 20 wt.% Pt/C at 0.70 V in O_2-saturated 0.1 M KOH at 1600 rpm.

3.3. Electrocatalytic OER Performance

The OER plays a crucial role in water electrolysis and metal-air batteries' systems [26]. In particular, high-performance OER electrocatalysts aim at reducing energy costs to produce high purity hydrogen and promote the development of URCs. Figure 8 shows the activity of Co_3O_4/N-HNMK-3 compared to N-HNMK-3, physical mixture of Co_3O_4 + N-HNMK-3, Pt/C, and RuO_2/C in 1.0 M KOH. Apart from RuO_2/C, the Co_3O_4/N-HNMK-3 has the highest OER activity with a higher current density of 19.56 mA cm^{-2} at 1.65 V, lower overpotential (365 mV) at the current density of 10 mA cm^{-2}, a value close to the best OER electrocatalyst RuO_2/C. The Tafel plots for the OER kinetics are shown in Figure 8b. The slope of 93 mV dec^{-1} for Co_3O_4/N-HNMK-3 is higher than that of 74 mV dec^{-1} for RuO_2/C, and lower than that of the physical mixture. In addition, the OER stability test, via CV, performed on Co_3O_4/N-HNMK-3 is shown in Figure 8c,d. No significant variation was observed after 1500 cycles, and to some extent, Co_3O_4/N-HNMK-3 showed a ca. 4.1% increase in the overpotential at 10 mA cm^{-2} after 1500 cycles. The results confirm that Co_3O_4/N-HNMK-3 is quite durable for the OER in 1.0 M KOH electrolyte.

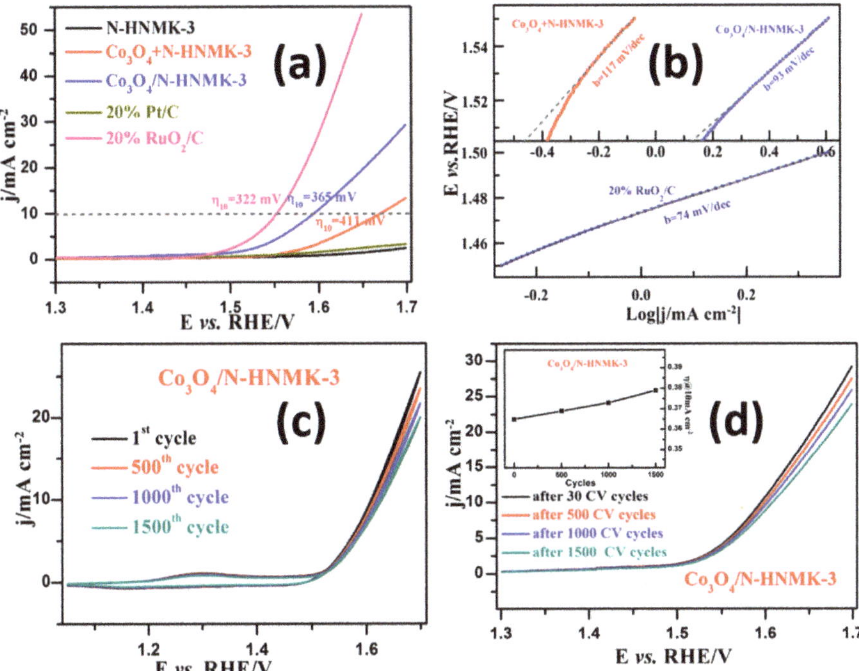

Figure 8. (**a**) LSV curves of N-HNMK-3, Co_3O_4 + N-HNMK-3 (physical mixture), Co_3O_4/N-HNMK-3, 20 wt.% Pt/C, and 20 wt.% RuO_2/C towards the OER in O_2-saturated 1.0 M KOH at 5 mV s^{-1} and 1600 rpm; (**b**) Tafel plots of each samples derived from curves in (**a**); (**c**) CVs of Co_3O_4/N-HNMK-3 from 1.05 V to 1.70 V at 200 mV s^{-1} in O_2-saturated 1.0 M KOH from 1st cycle to 1500th cycle; (**d**) LSV curves of Co_3O_4/N-HNMK-3 from 1.30 V to 1.70 V at 5 mV s^{-1}, the inset presents the variation of η_{10} at different CV cycles.

3.4. The Bifunctional ORR/OER Performance

The overvoltage between ORR and OER is an important parameter to evaluate the bifunctional electrocatalytic activity of the material. The corresponding data are summarized in Table 2. The ΔE value is defined as the difference in potential between the OER current density of 10.0 mA cm^{-2} and the ORR current density of -3.0 mA cm^{-2}. Among the listed catalysts [27–32] reported in the literature, Co_3O_4/N-HNMK-3 shows the best and promising bifunctional activity with a minimum ΔE value, which is attributed to the interaction developed between Co_3O_4 species onto N-doped ordered mesoporous carbons.

Table 2. The oxygen bifunctional electrocatalytic performance of catalysts prepared in this work compared with other bifunctional catalysts from the literature.

Catalysts	E *vs.* RHE at j = −3 mA cm^{-2}	E *vs.* RHE at j = 10 mA cm^{-2}	ΔE (E$_{OER@10\,mA\,cm^{-2}}$)-(E$_{ORR@-3\,mA\,cm^{-2}}$)	Refs.
N-HNMK-3	0.69	2.13	1.44	This work
Co$_3$O$_4$+N-HNMK-3	0.56	1.69	1.13	This work
Co$_3$O$_4$/N-HNMK-3	0.73	1.59	**0.86**	This work
20% Pt/C	0.81	2.07	1.26	This work
20% RuO$_2$/C	0.44	1.55	1.11	This work
Co$_3$O$_4$/Co$_2$MnO$_4$	0.68	1.77	1.09	27
MnCo$_2$O$_4$/Nanocarbon	0.79	1.75	0.96	28
NiCo$_2$O$_4$/Graphene	0.60	1.68	1.08	29
CoFe$_2$O$_4$/Biocarbon	0.69	1.67	0.98	30
O-NiCoFe-LDH	0.62	1.67	1.05	31
Co(OH)$_2$CO$_3$/C	0.81	1.73	0.92	32

4. Conclusions

In this paper, a nitrogen-doped ordered mesoporous carbons supported Co$_3$O$_4$ composite (Co$_3$O$_4$/N-HNMK-3) was prepared through a solvothermal method. The Co$_3$O$_4$/N-HNMK-3 composite showed a much higher ORR performance, which was close to that of the commercial Pt/C in 0.1 M KOH. An excellent OER performance near commercial RuO$_2$/C in 1.0 M KOH was also obtained. Interestingly, Co$_3$O$_4$/N-HNMK-3 has outstanding bifunctional activity with a much lower $\triangle$E value between E$_{OER,10\,mA\,cm^{-2}}$-E$_{ORR,-3\,mA\,cm^{-2}}$ as compared to Pt/C and RuO$_2$/C in alkaline medium. Undoubtedly, the synthesis method developed here to obtain the oxygen electrode based on Co$_3$O$_4$/N-HNMK-3, with non-precious metals will promote the practical development of fuel cells, water electrolysis and unitized regenerative cells.

Supplementary Materials: The following are available online at http://www.mdpi.com/2571-9637/2/2/18/s1, Figure S1. (a) Ring (top) and disk (down) current density from RRDE measurements of Co$_3$O$_4$/N-HNMK-3 samples after annealing at different temperature in O$_2$-saturated 0.1 M KOH at 25 °C with a sweep rate of 5 mV s^{-1} at a rotating speed of 1600 rpm; (b) Molar fraction of HO$_2$$^-$ formation and electron transfer number n from rotating ring-disk electrode (RRDE) curves in (a). Figure S2. (a) CVs of Co$_3$O$_4$/N-HNMK-3 from 0.1 V to 1.0 V at 100 mV s^{-1} in O$_2$-saturated 0.1 M KOH from 1st cycle to 1500th cycle; (b) LSVs of Co$_3$O$_4$/N-HNMK-3 from 0.1 V to 1.0 V at 5 mV s^{-1} in O$_2$-saturated 0.1 M KOH from 1st cycle to 1500th cycle.

Author Contributions: Methodology, N.A.-V. and Y.F.; formal analysis, S.Z., G.Y. and H.Z.; investigation, J.W.; writing—original draft preparation, J.W. and S.Z.; writing—review and editing, N.A.-V., D.L., P.T., and Y.F.; supervision, Y.F.; project administration, Y.F.

Acknowledgments: This work is supported by the National Natural Science Foundation of China, and the Fundamental Research Funds for the Central Universities (12060093063). H.H. Zhong specially thanks the financial support from China Scholarship Council (201806880040).

Conflicts of Interest: The authors declare no conflict of interest.

References

1. Carmo, M.; Fritz, D.L.; Mergel, J.; Stolten, D. A comprehensive review on PEM water electrolysis. *Int. J. Hydrogen* **2013**, *38*, 4901–4934. [CrossRef]
2. Postole, G.; Auroux, A. The poisoning level of Pt/C catalysts used in PEM fuel cells by the hydrogen feed gas impurities: The bonding strength. *Int. J. Hydrogen* **2011**, *36*, 6817–6825. [CrossRef]
3. Zhang, X.; Chan, S.H.; Ho, H.K.; Tan, S.-C.; Li, M.; Li, G.; Li, J.; Feng, Z. Towards a smart energy network: The roles of fuel/electrolysis cells and technological perspectives. *Int. J. Hydrogen* **2015**, *40*, 6866–6919. [CrossRef]
4. Qiao, S.Z.; Jiao, Y.; Zheng, Y.; Jaroniec, M. Design of electrocatalysts for oxygen- and hydrogen-involving energy conversion reactions. *Chem. Soc. Rev.* **2015**, *44*, 2060–2086.
5. Sui, S.; Wang, X.; Zhou, X.; Su, Y.; Riffat, S.; Liu, C.-J. A comprehensive review of Pt electrocatalysts for the oxygen reduction reaction: Nanostructure, activity, mechanism and carbon support in PEM fuel cells. *J. Mater. Chem. A* **2017**, *5*, 1808–1825. [CrossRef]

6.   Feng, Y.; He, T.; Alonso-Vante, N. In situ free-surfactant synthesis and ORR-electrochemistry of carbon-supported Co_3S_4 and $CoSe_2$ nanoparticles. *Chem. Mater.* **2007**, *20*, 26–28. [CrossRef]

7.   Xu, X.; Liang, H.; Ming, F.; Qi, Z.; Xie, Y.; Wang, Z. Prussian Blue Analogues Derived Penroseite (Ni,Co)Se_2 Nanocages Anchored on 3D Graphene Aerogel for Efficient Water Splitting. *ACS Catal.* **2017**, *7*, 6394–6399. [CrossRef]

8.   Zhong, H.; Campos-Roldán, C.A.; Zhao, Y.; Zhang, S.; Feng, Y.; Alonso-Vante, N. Recent Advances of Cobalt-Based Electrocatalysts for Oxygen Electrode Reactions and Hydrogen Evolution Reaction. *Catalysts* **2018**, *8*, 559. [CrossRef]

9.   Jia, X.; Gao, S.; Liu, T.; Li, D.; Tang, P.; Feng, Y. Controllable Synthesis and Bi-functional Electrocatalytic Performance towards Oxygen Electrode Reactions of Co_3O_4/N-RGO Composites. *Electrochim. Acta* **2017**, *226*, 104–112. [CrossRef]

10.   Tan, H.; Li, Y.; Jiang, X.; Tang, J.; Wang, Z.; Qian, H.; Mei, P.; Malgras, V.; Bando, Y.; Yamauchi, Y. Perfectly ordered mesoporous iron-nitrogen doped carbon as highly efficient catalyst for oxygen reduction reaction in both alkaline and acidic electrolytes. *Nano Energy* **2017**, *36*, 286–294. [CrossRef]

11.   Shen, H.; Gracia-Espino, E.; Ma, J.; Tang, H.; Mamat, X.; Wagberg, T.; Hu, G.; Guo, S. Atomically FeN_2 moieties dispersed on mesoporous carbon: A new atomic catalyst for efficient oxygen reduction catalysis. *Nano Energy* **2017**, *35*, 9–16. [CrossRef]

12.   Jiang, T.; Wang, Y.; Wang, K.; Liang, Y.; Wu, D.; Tsiakaras, P.; Song, S. A novel sulfur-nitrogen dual doped ordered mesoporous carbon electrocatalyst for efficient oxygen reduction reaction. *Appl. Catal. B: Environ.* **2016**, *189*, 1–11. [CrossRef]

13.   Ferrero, G.A.; Fuertes, A.B.; Sevilla, M.; Titirici, M.-M.; Solis, M.S. Efficient metal-free N-doped mesoporous carbon catalysts for ORR by a template-free approach. *Carbon* **2016**, *106*, 179–187. [CrossRef]

14.   Zhang, J.; Zhao, Z.; Xia, Z.; Dai, L. A metal-free bifunctional electrocatalyst for oxygen reduction and oxygen evolution reactions. *Nat. Nanotechnol.* **2015**, *10*, 444–452. [CrossRef]

15.   Mengxia, S.; Changping, R.; Yan, C. Covalent entrapment of cobalt-iron sulfides in N-doped mesoporous carbon: Extraordinary bifunctional electrocatalysts for oxygen reduction and evolution reactions. *ACS Appl. Mater. Interfaces* **2015**, *7*, 1207.

16.   Lei, W.; Deng, Y.-P.; Li, G.; Cano, Z.P.; Wang, X.; Luo, D.; Liu, Y.; Wang, D.; Chen, Z. Two-Dimensional Phosphorus-Doped Carbon Nanosheets with Tunable Porosity for Oxygen Reactions in Zinc-Air Batteries. *ACS Catal.* **2018**, *8*, 2464–2472. [CrossRef]

17.   Chai, G.-L.; Qiu, K.; Qiao, M.; Titirici, M.-M.; Shang, C.; Guo, Z.X. Active sites engineering leads to exceptional ORR and OER bifunctionality in P,N Co-doped graphene frameworks. *Environ. Sci.* **2017**, *10*, 1186–1195. [CrossRef]

18.   Patra, S.; Choudhary, R.; Roy, E.; Madhuri, R.; Sharma, P.K.; Sharma, D.P.K. Heteroatom-doped graphene 'Idli': A green and foody approach towards development of metal free bifunctional catalyst for rechargeable zinc-air battery. *Nano Energy* **2016**, *30*, 118–129. [CrossRef]

19.   Han, J.-S.; Chung, D.Y.; Ha, D.-G.; Kim, J.-H.; Choi, K.; Sung, Y.-E.; Kang, S.-H. Nitrogen and boron co-doped hollow carbon catalyst for the oxygen reduction reaction. *Carbon* **2016**, *105*, 1–7. [CrossRef]

20.   Zhong, H.; Zhang, S.; Jiang, J.; Feng, Y. Improved Electrocatalytic Performance of Tailored Metal-Free Nitrogen-Doped Ordered Mesoporous Carbons for the Oxygen Reduction Reaction. *ChemElectroChem* **2018**, *5*, 1899–1904. [CrossRef]

21.   Zhao, X.; Wang, A.; Yan, J.; Sun, G.; Sun, L.; Zhang, T. Synthesis and Electrochemical Performance of Heteroatom-Incorporated Ordered Mesoporous Carbons. *Chem. Mater.* **2010**, *22*, 5463–5473. [CrossRef]

22.   Granozzi, G.; Balbuena, P.B.; Ma, J.; Habrioux, A.; Luo, Y.; Ramos-Sanchez, G.; Calvillo, L.; Alonso-Vante, N. Electronic interaction between platinum nanoparticles and nitrogen-doped reduced graphene oxide: Effect on the oxygen reduction reaction. *J. Mater. Chem. A* **2015**, *3*, 11891–11904.

23.   Wang, C.; Shi, P.; Yao, W. Synergistic effect of Co_3O_4 nanoparticles and graphene as catalysts for peroxymonosulfate-based orange II degradation with high oxidant utilization efficiency. *J. Phys. Chem. C* **2016**, *120*, 336–344. [CrossRef]

24.   Wu, Z.-S.; Ren, W.; Wen, L.; Gao, L.; Zhao, J.; Chen, Z.; Zhou, G.; Li, F.; Cheng, H.-M. Graphene Anchored with Co_3O_4 Nanoparticles as Anode of Lithium Ion Batteries with Enhanced Reversible Capacity and Cyclic Performance. *ACS Nano* **2010**, *4*, 3187–3194. [CrossRef]

25. Liu, Y.; Ai, K.; Lu, L. Polydopamine and Its Derivative Materials: Synthesis and Promising Applications in Energy, Environmental, and Biomedical Fields. *Chem. Rev.* **2014**, *114*, 5057–5115. [CrossRef]

26. Gorlin, Y.; Jaramillo, T.F. A Bifunctional Nonprecious Metal Catalyst for Oxygen Reduction and Water Oxidation. *J. Am. Chem. Soc.* **2010**, *132*, 13612–13614. [CrossRef] [PubMed]

27. Wang, D.; Chen, X.; Evans, D.G.; Yang, W. Well-dispersed Co_3O_4/Co_2MnO_4 nanocomposites as a synergistic bifunctional catalyst for oxygen reduction and oxygen evolution reactions. *Nanoscale* **2013**, *5*, 5312. [CrossRef]

28. Ge, X.; Liu, Y.; Goh, F.W.T.; Hor, T.S.A.; Zong, Y.; Xiao, P.; Zhang, Z.; Lim, S.H.; Li, B.; Wang, X.; et al. Dual-Phase Spinel $MnCo_2O_4$ and Spinel $MnCo_2O_4$/Nanocarbon Hybrids for Electrocatalytic Oxygen Reduction and Evolution. *ACS Appl. Mater. Interfaces* **2014**, *6*, 12684–12691. [CrossRef] [PubMed]

29. Lee, D.U.; Kim, B.J.; Chen, Z. One-pot synthesis of a mesoporous $NiCo_2O_4$ nanoplatelet and graphene hybrid and its oxygen reduction and evolution activities as an efficient bi-functional electrocatalyst. *J. Mater. Chem. A* **2013**, *1*, 4754. [CrossRef]

30. Liu, S.; Bian, W.; Yang, Z.; Tian, J.; Jin, C.; Shen, M.; Zhou, Z.; Yang, R. A facile synthesis of $CoFe_2O_4$/biocarbon nanocomposites as efficient bi-functional electrocatalysts for the oxygen reduction and oxygen evolution reaction. *J. Mater. Chem. A* **2014**, *2*, 18012–18017. [CrossRef]

31. Qian, L.; Lu, Z.; Xu, T.; Wu, X.; Tian, Y.; Li, Y.; Huo, Z.; Sun, X.; Duan, X. Trinary Layered Double Hydroxides as High-Performance Bifunctional Materials for Oxygen Electrocatalysis. *Adv. Mater.* **2015**, *5*, 1500245. [CrossRef]

32. Wang, Y.; Ding, W.; Chen, S.; Nie, Y.; Xiong, K.; Wei, Z. Cobalt carbonate hydroxide/C: An efficient dual electrocatalyst for oxygen reduction/evolution reactions. *Chem. Commun.* **2014**, *50*, 15529–15532. [CrossRef] [PubMed]

MDPI

Article

An Electrochemical Route for Hot Alkaline Blackening of Steel: A Nitrite Free Approach

Maximilian Eckl, Steve Zaubitzer, Carsten Köntje, Attila Farkas, Ludwig A. Kibler and Timo Jacob *

Institute of Electrochemistry, Ulm University, Albert-Einstein-Allee 47, 89081 Ulm, Germany; maximilian.eckl@uni-ulm.de (M.E.); steve.zaubitzer@uni-ulm.de (S.Z.); koentje@gmx.net (C.K.); attila.farkas@uni-ulm.de (A.F.); ludwig.kibler@uni-ulm.de (L.A.K.)
* Correspondence: timo.jacob@uni-ulm.de; Tel.: +49-731-50-25401

Received: 23 February 2019; Accepted: 26 March 2019; Published: 29 March 2019

Abstract: Blackening belongs to the predominant technological processes in preserving steel surfaces from corrosion by generating a protective magnetite overlayer. In place of the commonly used dipping-procedure into nitrite-containing blackening baths at boiling temperatures that are far above 100 °C, here we describe a more environmentally friendly electrochemical route that operates at temperatures, even below 100 °C. After an investigation of the electrochemical behavior of steel samples in alkaline solutions at various temperatures, the customarily required bath temperature of more than 130 °C could be significantly lowered to about 80 °C by applying a DC voltage that leads to an electrode potential of 0.5−0.6 V vs. Pt. Thus, it was possible to eliminate the use of hazardous sodium nitrite economically and in an optimum way. Electrochemical quantification of the corrosion behavior of steel surfaces that were in contact with 0.1 M KCl solution was carried out by linear sweep voltammetry and by Tafel slope analysis. When comparing these data, even the corrosion rates of conventional blackened surfaces are of the same magnitude as a blank steel surface. This proves that magnetite overlayers represent rather poor protective layers in the absence of additional sealing. Moreover, cyclic voltammetry (CV), atomic force microscopy (AFM), scanning electron microscopy (SEM) and auger electron spectroscopy (AES) characterized the electrochemically blackened steel surfaces.

Keywords: Surface Modification; Blackening of Steel; Magnetite; Corrosion Protection; Auger-Electron Spectroscopy

---

1. Introduction

The technology of blackening is of utmost importance for surface finishing and the refinement of various types of steel. Chemical oxidation of an iron surface forms a tenebrous decorative passivation layer, which, in particular, can inhibit corrosion processes [1]. The resulting black-bluish coloration represents a convenient quality feature for steel components in a wide area of applications. The traditional hot alkaline blackening process for steel parts, in the following simply called blackening, involves five consecutive steps [2–4]: (1) degreasing and the removal of oils, (2) first water flushing, (3) actual blackening, where the specimens are immersed into a boiling sodium hydroxide solution that contains sodium nitrite as oxidizing agent and some other additives, (4) second water flushing for the removal of blackening bath components, and (5) surface treatment of the blackened steel by a water-resistant sealant. As a result, the blackening process leads to the oxidation of the surface iron atoms by forming black iron oxide (magnetite Fe_3O_4) [5–7]. In a similar manner to the well-known iron phosphating process [8], corrosion protection by blackening is not just achieved by simple deposition of an additional layer, but by a chemical transformation of the substrate's surface itself, resulting in a so-called conversion coating [2,9]. The blackened surfaces are usually sealed by a waterproof layer (e.g., oil, etc.) to retard the initial stages of corrosion at defects and pores that are formed during the chemical oxidation process [5,6].

The formation of the magnetite surface layer is generally thought to involve the following two steps [6,10,11]: First, iron surface atoms are oxidized to iron(II) hydroxide ($Fe(OH)_2$, see Equation (1)) by nitrite (NO^{2-}), which acts as oxidizing reagent. $Fe(OH)_2$ is then further oxidized to magnetite (Equation (2)) via the Schikorr reaction [12,13], causing the black surface burning [3,5,6,14].

$$Fe_{surf} + 2\,OH^-{}_{(aq)} \rightarrow Fe(OH)_{2,surf} + 2\,e^- \tag{1}$$

$$3Fe(OH)_{2,surf} \rightarrow Fe_3O_{4,surf} + H_{2(g)} + 2\,H_2O_{(l)} \tag{2}$$

Although the Schikorr reaction is spontaneous, i.e., thermodynamically favored, even at room temperature, the formation of magnetite is usually kinetically hindered at temperatures below $100\,°C$ [4].

The simplicity and robustness of the blackening procedure is highly esteemed in many industrial sectors. However, hazardous and toxic sodium nitrite and the high energy costs represent major drawbacks. Hence, the development of a more environmentally friendly and efficient process would be highly desirable and promising for several branches. Attempts to form magnetite layers through an electrochemical reaction have been reported earlier for steel in hot ($70-98\,°C$) ammonium nitrate solution [14,15]. However, this solution is, also considered to be unsafe. Alternatively, a hydrothermal synthesis of magnetite has been proposed [3,4,16], requiring an autoclave and likewise relatively high temperatures ($100-250\,°C$), in comparison to the conventional process.

It is obvious to alternatively study the electrochemical route for the Schikorr-reaction (Equation (3)), which means that electrochemical blackening might be feasible without the use of harmful oxidants [5,6,17–20]:

$$3\,Fe(OH)_{2,surf} \rightarrow Fe_3O_{4,surf} + 2\,H_2O_{(l)} + 2\,H^+{}_{(aq)} + 2\,e^- \tag{3}$$

The anodic oxidation of steel has been known since the 1960s, where solutions of boiling 14.3 M sodium hydroxide and high current densities ($50-100\,mA\cdot cm^{-2}$) are used to fabricate a loose and porous magnetite layer [18,19,21].

In this article, we demonstrate that blackening can easily be electrochemically achieved for steel samples that were immersed into concentrated sodium hydroxide solution under potentiostatic control at temperatures that are much lower than those used in the hitherto existing commercial process. The surface passivation layer that is generated in the absence of a chemical oxidizing agent under mild conditions by this new electrochemical reaction shows decorative properties comparable to the ones that are achieved with the standard technology of the costly and harmful chemical process. The quality and selected properties of the electrochemically blackened surfaces were further investigated while using atomic force microscopy (AFM), scanning electron microscopy (SEM), auger electron spectroscopy (AES), and electrochemical methods.

2. Materials and Methods

2.1. Materials

The steel discs of the standard S235JRG2C+C (geometric area: $2.83\ cm^2$) were grinded before each measurement with 320 SiC abrasive paper (Buehler, ITW Test & Measurement GmbH, Esslingen, Germany), sequentially polished with 9 and 3 µm diamond suspension, 0.05 µm Al_2O_3 suspension, and finally successively rinsed with isopropanol (VWR, analytical grade) and water. Ultrapure water that was obtained from a water purification system (Arium 611 UV; TOC $\leq$ 3 ppb, Sartorius AG, Göttingen, Germany) was used as solvent in all experiments. Sodium hydroxide (NaOH, VWR, 99%), blackening salt (DWE Brünofix GmbH, Rednitzhembach, Germany and CSP—Chemische Spezialprodukte Olaf

Günther, Markkleeberg, Germany), and potassium chloride (KCl, suprapur®, Merck KGaA, Darmstadt, Germany) were used without further purification.

2.2. Blackening

Both the conventional chemical and the electrochemical blackening were performed in the same measurement cell, which could be used in a three-electrode setup. 30 g of blackening salt or sodium hydroxide was dissolved in 30 mL water and then heated in an iron crucible to the desired temperature between room temperature and 120 °C. A platinum counter electrode, a platinum quasi-reference electrode, and the steel sample were immersed into the solution. Following this process, the samples were rinsed with ultrapure water and then dried in a nitrogen stream. A potentiostat (Fritz-Haber-Institut, Berlin, Germany) was used for electrochemical blackening and for recording cyclic voltammograms.

Immersion into a boiling solution (130 °C) of blackening salt dissolved in water (concentration $1\ kg \cdot L^{-1}$) for 12 min chemically blackened steel samples. For electrochemical blackening, the steel samples were immersed into the electrolyte (blackening salt or sodium hydroxide solution) for 12 min, using the following experimental conditions:

- Blackening solution at 80 °C: $E = 0.5$ V vs. Pt
- Blackening solution at 100 °C: $E = 0$ V vs. Pt
- 23 M NaOH at 80 °C: $E = 0.6$ V vs. Pt
- 23 M NaOH at 100 °C: $E = 0.1$ V vs. Pt
- 23 M NaOH at 120 °C: $E = 0.1$ V vs. Pt

At temperatures that were higher than 120 °C, the electrolyte started boiling, in which electrochemical measurements were no longer practicable. Therefore, the maximum temperature for electrochemical measurements was set to 120 °C. It should be noted that the formation of magnetite at temperatures below 80 °C by using more positive potentials ($E > 0.6$ V vs. Pt) was not successful.

2.3. Characterization of the Blackened Samples

Atomic force microscopy (Nanosurf FlexAFM, Nanosurf AG, Liestal, Switzerland) in tapping mode using Tap190 GD-G cantilevers (Budgetsensors, Innovative Solutions Bulgaria Ltd., Sofia, Bulgaria), scanning electron microscopy, and Auger-electron spectroscopy (Phi 660, Physical Electronics, Chanhassen, MN, USA) characterized the blackened surfaces.

The electrochemical characterization and quantification of the corrosion behavior was studied with a conventional three electrode setup, consisting of a working, a reference, and a platinum counter electrode. All of the potentials were recorded against an Ag/AgCl reference electrode (Schott, $c_{KCl} = 3\ mol \cdot L^{-1}$). The working electrode was contacted with the electrolyte by a hanging meniscus configuration, resulting in a geometrical surface area of 0.785 cm^2. The electrolyte was a 0.1 M KCl solution, which was purged with N_2 for at least 1 h before each measurement. Electrochemical characterization of blackened samples was done by cyclic voltammetry between -1.5 V to -0.2 V vs. Ag/AgCl at a scan rate of 20 mV·s^{-1}. The corrosion protection was quantified by linear voltammetry in a potential range between -0.1 and 0.1 V vs. E_{OC} at a scan rate of 1 mV·s^{-1}. E_{OC} is the open circuit potential after 1 h immersion of the sample into the electrolyte. The corrosion potential E_{corr}, Tafel slope coefficient β_c, β_a, and corrosion current density j_{corr} (as indicators of the quality of the corrosion protection) were obtained from the measured potential–current curve by Tafel slope analysis and fitting to the Wagner–Traud equation. The polarization resistance (R_P) was obtained from the slope of the polarization curve at E_{OC} or by calculation while using the Stern–Geary equation.

3. Results and Discussion

3.1. Chemical Blackening

The steel samples were prepared according to the recipe that is described in DIN 50938 in order to compare the electrochemical and the chemical hot alkaline blackening procedures [22]. The polished steel slices were immersed into a boiling solution of blackening salt dissolved in water (concentration $1\,kg\cdot L^{-1}$) at 130 °C for 12 min, yielding a uniform black-blackened surface. Auger electron spectroscopy identified the iron oxide species (see Figure 1a). The resulting oxide species were determined by the intensity ratio of the fine structure of the $Fe(M_{2/3}VV)$ signals at 46 and 54 eV [23]. The relevant intensity ratio of approx. 8 to 10 matches that of Fe_3O_4 (Figure 1b). The magnetite surface appears relatively smooth (profile roughness parameter $R_a = 12 \pm 1$ nm) (see the SEM and AFM images in Figure 1b,c). Besides some cracks, surface defects that have been formed during the blackening process are visible in these images.

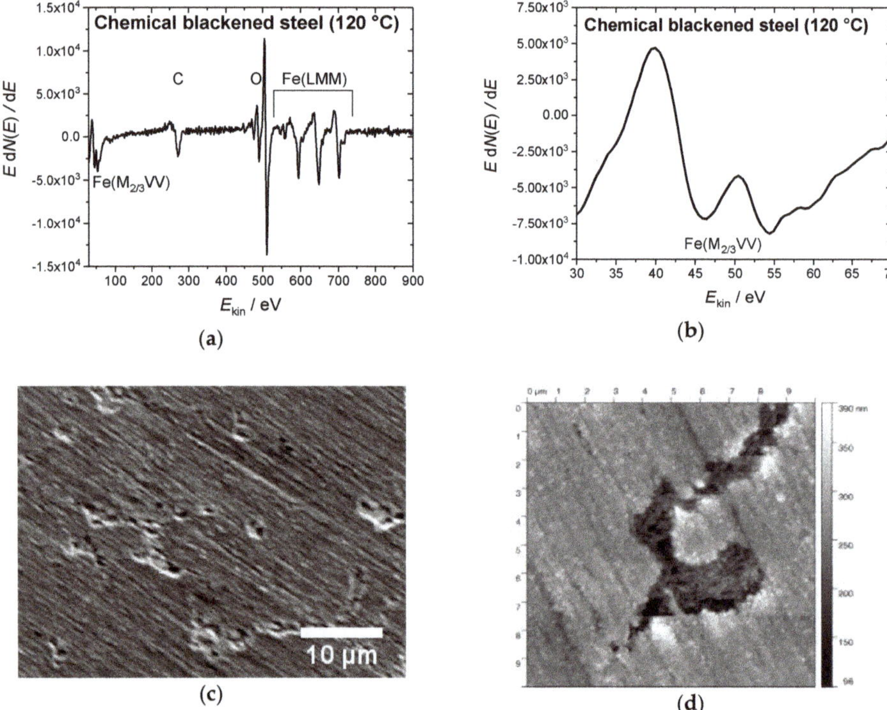

Figure 1. (**a**) Auger electron spectrum of a steel sample blackened at 120 °C in hot alkaline nitrite solution, showing (**b**) the $Fe(M_{2/3}VV)$ signals at 46 and 54 eV with the intensity ratio of 7:10 for magnetite on the surface. (**c**) Scanning electron microscopy (SEM) image of the same sample (10 kV, 2000 magnification) with some defects in the blackened surface; (**d**) The atomic force microscopy (AFM) topography image shows a 10×10 μm^2 area of the steel sample, scaling-up a surface defect similar seen in the SEM image.

3.2. Electrochemical Blackening

Besides conventional chemical blackening at variable temperature, the steel samples were electrochemically blackened both in a nitrite containing blackening solution and in sodium hydroxide solution that is free of nitrite. To determine the optimum temperature and the DC voltage for electrochemical blackening, the electrochemical behavior of steel samples was studied at different temperatures between room temperature and 120 °C in blackening salt and 23 M sodium hydroxide solution by cyclic voltammetry (Figure 2). The study of steel electrodes that are in contact with sodium hydroxide solution was also performed until the boiling temperature of 120 °C.

According to the Pourbaix diagram of iron [17,24] and the well-known electrochemical behavior of iron in alkaline solution [11,25–30], the anodic peak A1 at −0.08 V vs. Pt for steel in 23 M NaOH (Figure 2c,d) was identified as the oxidation of Fe to Fe(II) [28–30]. This is in agreement with the reaction that is given in Equation (1). The peak A2 at around 0.1 V vs. Pt in blackening salt solution and in the range of 0.06 to 0.16 V vs. Pt in sodium hydroxide corresponds to the oxidation of iron hydroxide to magnetite that is in correspondence to the Schikorr reaction (Equation (3)) [11,27–29]. Magnetite is finally oxidized to iron(III) oxide at peak A3 [11,27]. Furthermore, the two cathodic peaks of C2 and C3 are observed for nitrite-free solutions (Figure 2c,d). They correspond to the related oxidation peaks A2 and A3 [29]. Measurements that were obtained with blackening salt solution (Figure 2a,b) only exhibit the anodic peaks A2, A3, and the hydrogen evolution regime. It is noteworthy that the current and charge densities for sodium hydroxide solution are approximately twice as large as those that were measured with blackening salt.

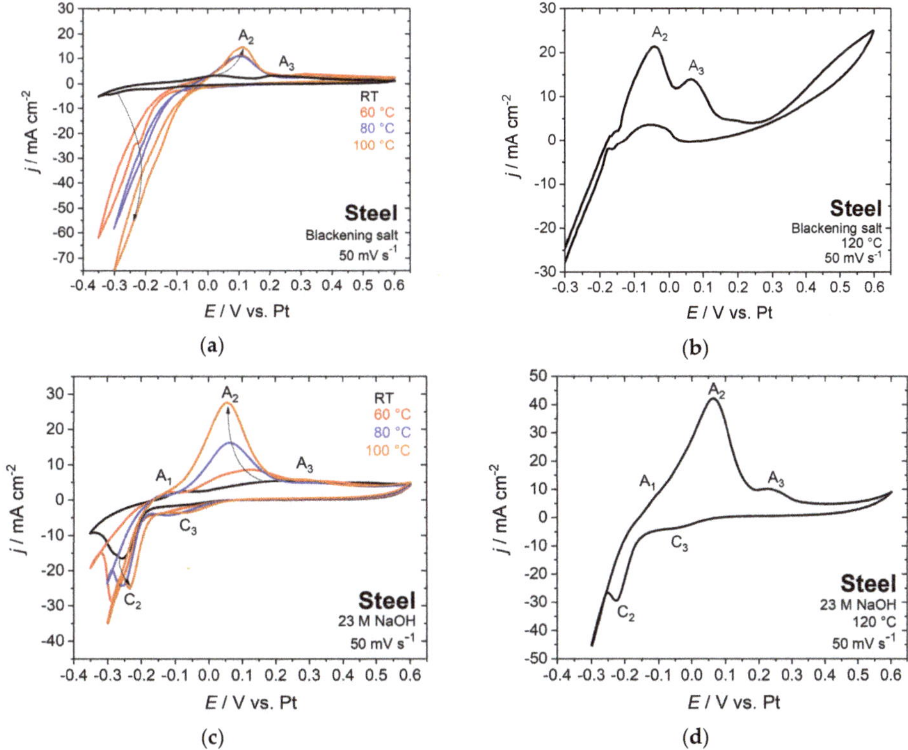

Figure 2. Cyclic voltammograms of steel in blackening salt solution (**a**) at various temperatures between room temperature and 100 °C, (**b**) at 120°C and (**c**) in 23 M sodium hydroxide solution at various temperatures between room temperature and 100 °C, and (**d**) at 120 °C; scan rate 50 mV s^{-1}.

A black coloring of the surface already appeared after the first voltammetric cycle in boiling blackening solution, while at room temperature, even after one hour, no visible changes in color have been observed. Only a few signs of corrosion at grain boundaries were monitored in the AFM topography image (Figure 3). This can be explained by the slow kinetics of the Schikorr reaction, as already mentioned above. The blackening of the surface becomes evident after increasing the temperature to 80 °C. For this reason, the temperature range between 80 and 120 °C was examined in detail, both with sodium hydroxide and blackening salt solution as electrolytes.

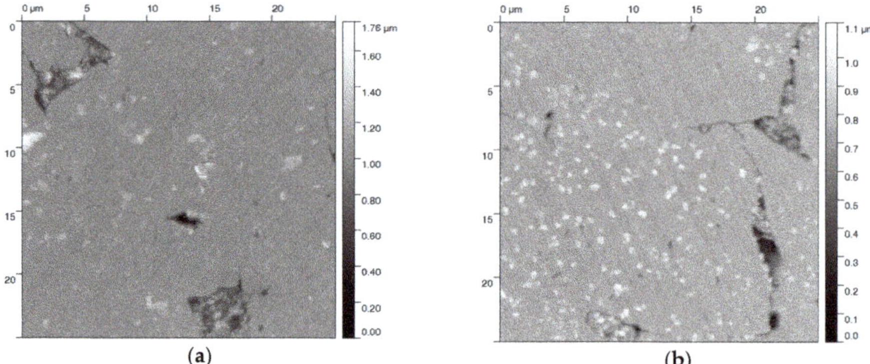

Figure 3. AFM topography images (25 × 25 μm^2) of a steel slice cycled in (**a**) 23 M NaOH and (**b**) blackening salt solution. No significant differences in the surface topography appear. Some corrosion at grain boundaries and the presence of corrosion products resemble each other.

The open circuit potential (E_{OC}) that was established for steel in blackening salt solution after 12 min is −0.05 V vs. Pt, which is located around the onset of peak A2 (Figure 2b). For this reason, the potential was set to 0 V vs. Pt for the electrochemical treatment with blackening solution at 100 °C. Applying a potential of 0.1 V vs. Pt in 23 M NaOH solution at 100–120 °C (Figure 2c), which almost corresponds to peak A2, yields a uniform black magnetite layer after 12 min. With the same potential at 80 °C in blackening salt and 23 M NaOH solution, insufficient blackening is obtained due to the slow kinetics of the Schikorr reaction. It is obvious that the electrode potential and temperature are the main parameters that control the quality of electrochemical blackening. Within the same process time of 12 min, a more positive electrode potential of 0.5 V vs. Pt with blackening salt and 0.6 V vs. Pt with sodium hydroxide solution leads to the formation of a compact Fe_3O_4 overlayer, which Auger electron spectroscopy subsequently characterized (Figure 5).

The potentials and the resulting current densities that were measured after 12 min (j_{final}) for the two electrolytes at temperatures of 80, 100, and 130 °C (boiling temperature of blackening salt solution) are summarized in Table 1.

Table 1. Reaction conditions for electrochemical blackening of steel in blackening salt solution and in 23 M NaOH between 80 and 130 °C, temperature T, electrode potential E, and final current density j_{final}. The profile roughness as obtained from AFM imaging is given in the last column. The use of 23 M NaOH solution at temperatures over 100 °C increases the surface roughness by a factor of 5, in comparison to the chemical hot alkaline blackening.

Electrolyte	T/°C	E/V vs. Pt	j_{final}/mA cm^{-2}	R_a/nm
	80	0.5	1.0	15 ± 4
Blackening salt solution 1 kg·L^{-1}	100	0	0.6	38 ± 6
	130	−0.05 (E_{OC})	(0)	12 ± 1
	80	0.6	0.4	30 ± 11
NaOH (23 M)	100	0.1	1.0	59 ± 7
	120	0.1	19.5	61 ± 11

Electrochemical blackening at temperatures below 80 °C by using more positive potentials was not successful. Apparently, ferrate(VI) ions (FeO_4^{2-}) are formed at higher potentials, resulting in a purple colored electrolyte [28,31,32]. UV-VIS spectroscopy (Figure 4) could confirm the electrochemical formation of the ferrate(VI) ion, showing a band with an absorption maximum at 505 nm (green color, complementary of purple) [32–34].

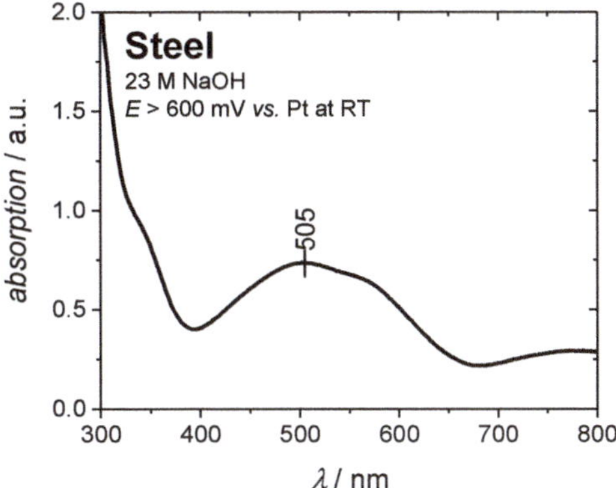

Figure 4. UV-VIS spectrum of 23 M NaOH after electrochemical blackening at potentials E > 0.6 V vs. Pt. The absorption maximum at approx. 500 nm indicates the formation and presence of ferrate(VI) ions.

In order to confirm the chemical composition of the iron oxide overlayers that were formed by electrochemical blackening, the steel samples were examined by Auger electron spectroscopy. Figure 5 shows (exemplarily for all samples under study) the Auger electron spectrum of steel blackened in NaOH at 80 °C and 0.6 V vs. Pt. The intensity ratios for the Fe($M_{2/3}$VV) signals at 46 and 54 eV (Figure 5b) are similar to those that were obtained for the samples with classical chemical treatment (Figure 1a,b). We conclude that the formed black oxide layer is indeed magnetite (Fe_3O_4), which is in agreement with the findings in literature [23].

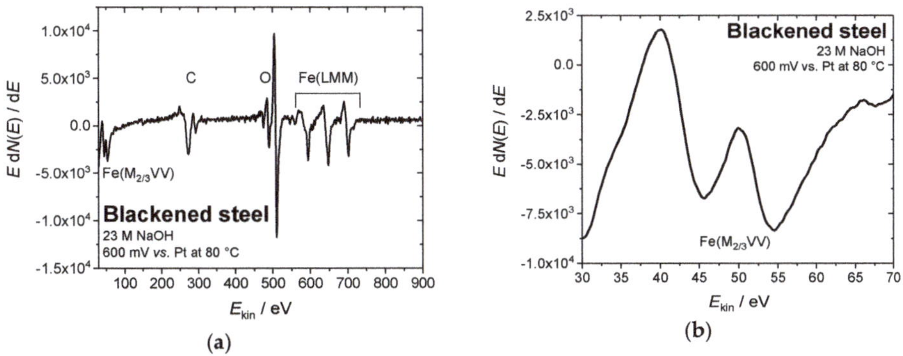

Figure 5. *Cont.*

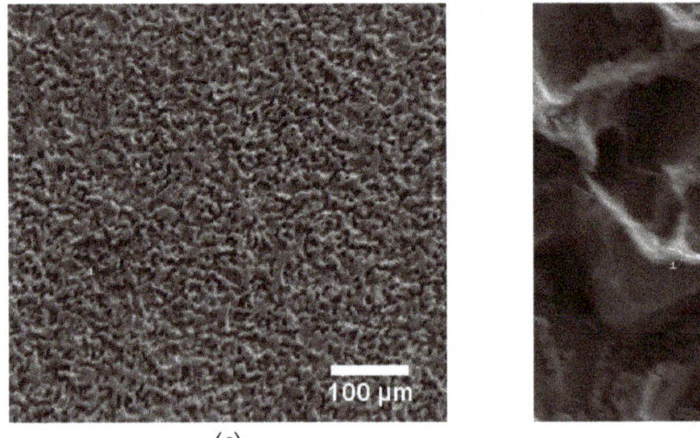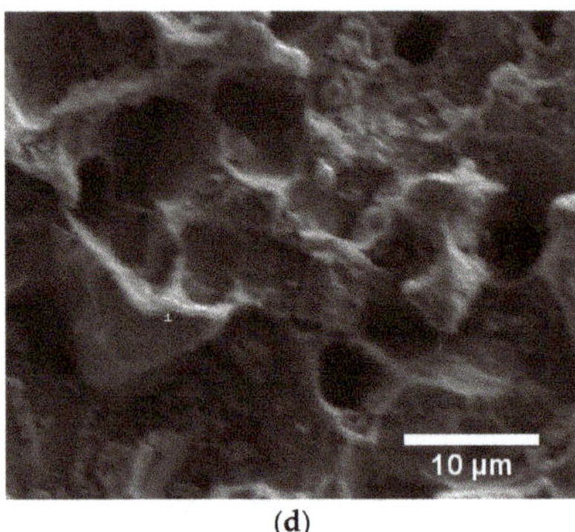

(c)   (d)

Figure 5. (**a**) Auger electron spectrum of steel blackened in NaOH at 80 °C and 0.6 V vs. Pt. (**b**) The intensity ratio of the peaks at 46 and 54 eV indicates the formation of a magnetite overlayer (Fe$_3$O$_4$), (**c**,**d**) SEM images of a steel sample blackened in NaOH at 80 °C and 0.6 V vs. Pt, indicating a corrosive attack during the blackening process (10 kV, (**c**) 100x, and (**d**) 2000x magnification).

In addition to the Auger electron spectroscopy measurements, we also took SEM images, showing sponge-like structures on the surfaces (Figure 5c,d). This indicates corrosive attack and surface roughening by NaOH during the electrochemical blackening process. The electrochemical origin of this corrosion process is a parallel reaction that is competitive with the one sketched in Equation (2). At high pH values, iron(II)hydroxide reacts with excess hydroxide to soluble HFeO$_2^-$ (Equation (4) [17,26].

$$Fe(OH)_{2,surf} + OH^-_{(aq)} \rightarrow HFeO_2^-{}_{(aq)} + H_2O \tag{4}$$

AFM topography images of the blackened samples after different treatments were recorded in order to quantify the surface roughness (Figure 6). Only minor changes could be observed for blackening salt solution at 80 and 130 °C (Figure 6a,b). This is in agreement with a relatively small roughness factor of 12–15 nm (Table 1). In contrast, steel samples that were blackened in a solution of sodium hydroxide at 80 °C, and even more at 120 °C, yield a rougher (R_a = 30–60 nm) and more damaged surface (Figure 6c,d). These surface defects are triggered by the corrosive effect of the sodium hydroxide solution and the high anodic current densities [31]. As expected, the higher the anodic current density, the larger the roughness of the blackened surface. We would like to point out that, by applying a potential in both blackening solutions at 100 °C and 23 M NaOH, the roughness is at least increased by a factor of two. This increase of roughening at higher temperature is possibly caused by the enhanced formation of Fe(OH)$_2$ (Figure 2) and the corrosive attack of NaOH (Equation (4)), in competition with the Schikorr reaction.

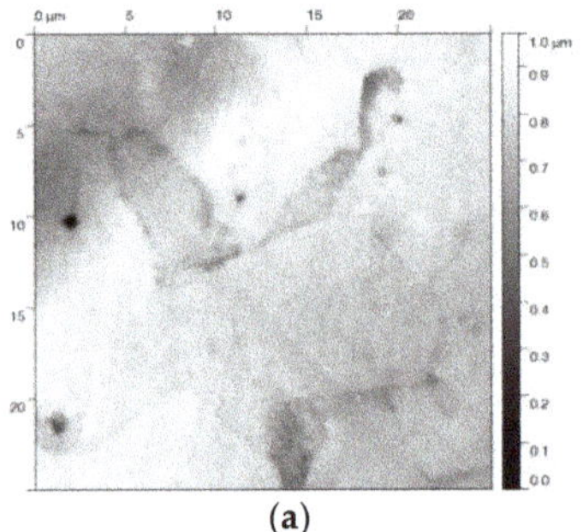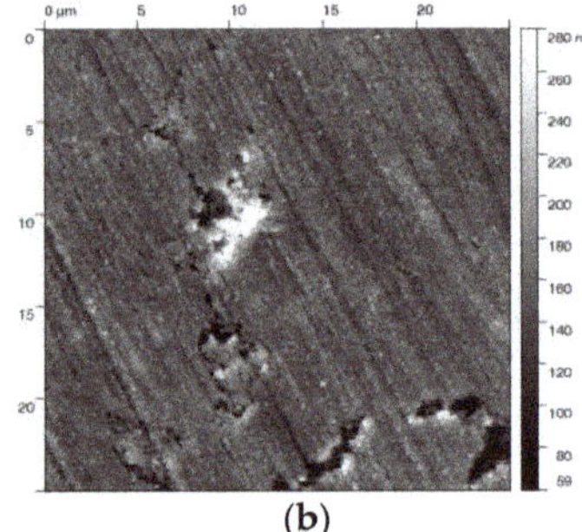

(a)   (b)

Figure 6. *Cont.*

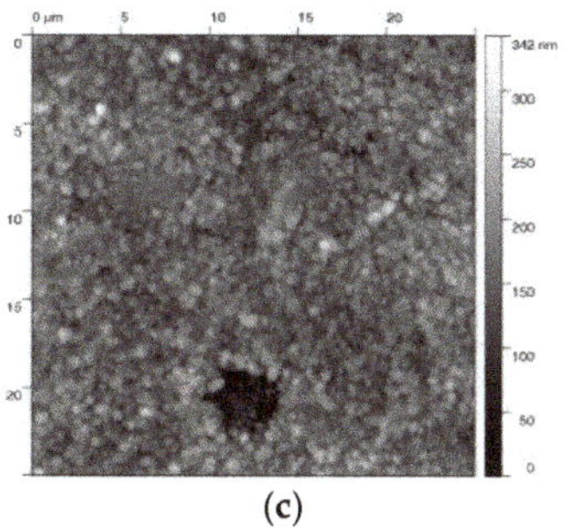
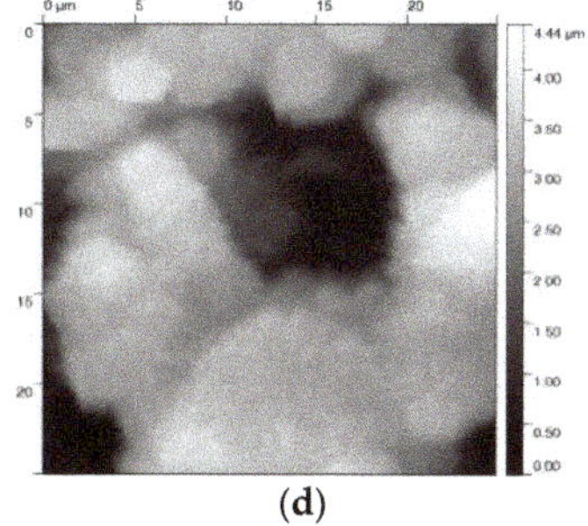

(c) (d)

Figure 6. AFM images of steel samples electrochemically blackened in (**a**) blackening salt solution at 80 °C and (**b**) 130 °C, and (**c**) in NaOH at 80 °C, and (**d**) 120 °C. The images show a 25 × 25 μm² area of the individual samples.

3.3. Electrochemical Characterization

The electrochemical behavior of the blackened samples was studied by cyclic voltammetry in the potential range between −1.5 and −0.2 V vs. Ag/AgCl, with 0.1 M potassium chloride solution at a sweep rate of 20 mV·s⁻¹ (see Figure 7). In the case of the chemically blackened sample, the positive potential limit was expanded to 0 V vs. Ag/AgCl to study an additional anodic current peak at −0.27 V vs. Ag/AgCl. Subsequently, a Tafel slope analysis was performed for the voltammetric data.

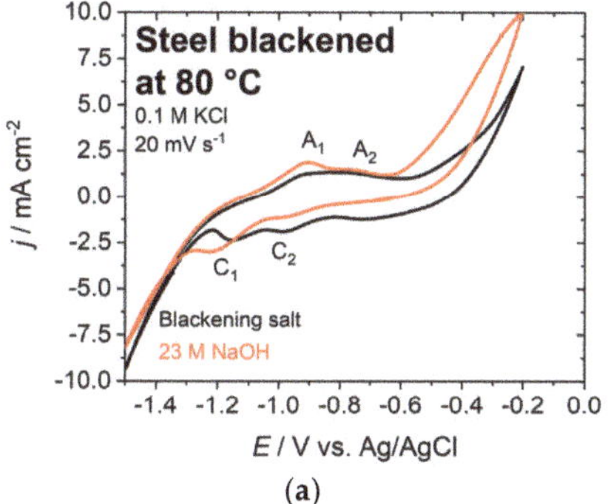
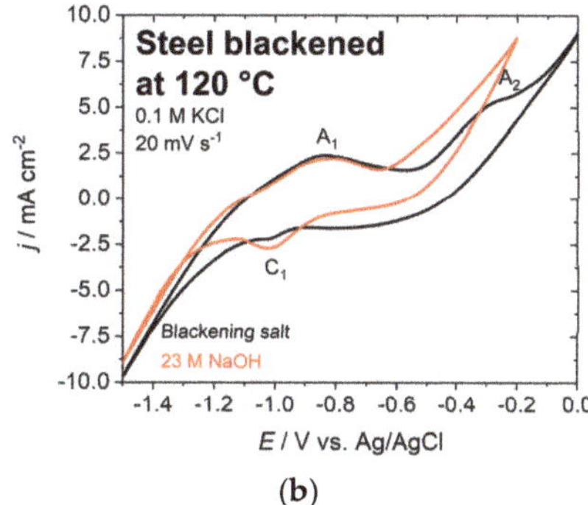

(a) (b)

Figure 7. Cyclic voltammogram of blackened steel discs in 0.1 M KCl solution at (**a**) 80 °C and (**b**) 120 °C; black curve: blackening salt; red curve: NaOH; scan rate 20 mV s⁻¹.

The cyclic voltammograms for the various steel samples that were blackened at different temperatures are all very similar in the potential range between −1.5 and −0.6 V vs. Ag/AgCl (Figure 7). The surfaces that were blackened in NaOH start oxidizing to Fe(III) at −0.6 V vs. Ag/AgCl, while this reaction requires higher potentials for samples that were treated in the blackening salt solution. This can be explained by a more defect-free and smoother steel surface, which is apparent from the images that were obtained by AFM.

The corrosion potentials (E_{corr}), corrosion current densities (j_{corr}), and Tafel slope coefficients (β_c and β_a) for the various blackened samples (listed in Table 2) were obtained by Tafel slope analysis from the polarization curves that are shown in Figure 8. In order to determine the quality of the blackened surfaces and their corrosion behavior (qualitative salt spray test may give similar trends), the corrosion rates (k_{corr}) were calculated while using the Faraday laws. The polarization resistances (R_P) were obtained by calculation using the Stern–Geary equation [35].

$$R_p = \frac{\beta_a |\beta_c|}{2.303(\beta_a + |\beta_c|) \; j_{corr}} \tag{5}$$

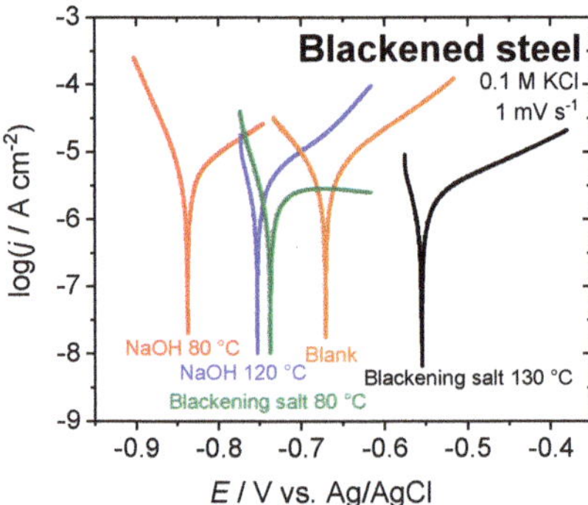

Figure 8. Potentiodynamic polarization curves for pristine and blackened steel surfaces in in 0.1 M KCl; scan rate 1 mV s^{-1}. Electrochemical blackening was performed both in 23 M NaOH and blackening salt solution.

A comparison of the corrosion potentials shows that the electrochemically-blackened samples appear to be less noble than the classically blackened or untreated steel samples. Nevertheless, the corrosion current densities and rates are similar to the classically blackened samples and about 70% lower than for an untreated steel sample. The samples that were treated with sodium hydroxide solution at 80 °C have an approximately 40% lower corrosion current when compared to the blank sample. The polarization resistance supports this trend, illustrating an improvement of the corrosion resistance of the samples that were oxidized in blackening salt solution by a factor of approx. two as compared to a blank surface, while the improvement of electrochemical oxidized samples in 23 M NaOH is about 10–40%.

The rough and porous structure of the magnetite layer possibly caused the less noble behavior and lower corrosion protection of the electrochemical-blackened samples (Figure 6c,d). In this case, chloride ions could pass the oxide layer and attack the metal, forming a local cell beneath the oxide, which reduces the corrosion potential.

Supporting this, the form of the Tafel plots in Figure 8 and the values of the anodic Tafel slope coefficients (β_a) of steel samples chemically and electrochemically blackened in blackening salt solution indicate protection of the surface by passivation due to the formed magnetite layer. In contrast, the anodic Tafel slope coefficients (β_a) of steel samples that were electrochemically blackened in 23 M NaOH seem to be similar to an untreated sample. This behavior suggests that, although the corrosion rate decreased by a reduction of the active surface area, the porous structure of the magnetite layer does not result in a surface protection by passivation. The data in Table 2 indicates that the chemically and electrochemically formed magnetite overlayers represent rather poor protective layers in the absence of additional sealing. Since the blackened surfaces are usually sealed, the rougher surface may increase the adhesion strength of the sealant, providing even better corrosion resistance.

Table 2. Corrosion potential (E_{corr}), corrosion current (j_{corr}), Tafel slope coefficient ($|\beta_c|$ and β_a), polarization resistance (R_P) and corrosion rate (k_{corr}) of steel samples blackened under different conditions. Blank steel samples were used as reference.

Sample	E_{corr}/V vs. Ag/AgCl	j_{corr}/μA cm^{-2}	$|\beta_c|$/V dec^{-1}	β_a/V dec^{-1}	R_P/kΩ cm^{-2}	k_{corr}/mm a^{-1}
Blank	−0.67	7.1	0.07	0.12	2.7	0.082
Blackening salt solution 80 °C	−0.74	1.2	0.04	0.21	6.6	0.014
Blackening salt solution 130 °C	−0.57	2.0	0.03	0.18	5.6	0.023
NaOH 80 °C	−0.84	4.4	0.04	0.12	3.0	0.051
NaOH 120 °C	−0.75	2.8	0.03	0.12	3.7	0.033

4. Conclusions

The electrochemical behavior of steel in blackening salt solution and the concentrated sodium hydroxide solution was investigated between room temperature and 120 °C by cyclic voltammetry. It is shown that the blackening of steel can be electrochemically achieved by applying a potential in the range of 0 to 0.6 V vs. Pt at temperatures between 80 and 120 °C. A further decrease of the bath temperature to room temperature and a simultaneous increase of potential did not yield the desired result due to the formation of Fe(VI) ions and the slow kinetics of the Schikorr reaction. Auger electron spectroscopy confirmed the formation of magnetite on the surface during blackening. AFM studies revealed surface roughening for the given reaction conditions. Electrochemical studies for blackened steel in 0.1 M KCl that is based on cyclic voltammetry and corrosion measurements revealed a similar electrochemical behavior and an equivalent corrosion protection for the following experimental conditions:

- 23 M NaOH, 0.6 V vs. Pt and 80 °C,
- 23 M NaOH, 0.1 V vs. Pt and 120 °C, and
- blackening solution, 0.5 V vs. Pt and 80 °C.

Very similar surface properties are achieved in an efficient electrochemical blackening process at lower temperature when compared to classical chemical blackening at 130 °C with hazardous sodium nitrite. The combination of electrochemical blackening and surface sealing may help to develop an inexpensive, environmentally-friendly, and adequate method to obtain the aspired corrosion protection of steel surfaces, together with a unique appearance.

Author Contributions: Conceptualization, M.E., S.Z. and L.A.K.; Data curation, C.K.; Formal analysis, M.E., S.Z. and C.K.; Funding acquisition, T.J.; Investigation, C.K.; Methodology, A.F.; Resources, T.J.; Supervision, A.F., L.A.K. and T.J.; Validation, A.F.; Writing—original draft, M.E. and S.Z.; Writing—review & editing, L.A.K. and T.J.

Acknowledgments: The authors thankfully acknowledge Kneissler Brüniertechnik GmbH for supporting this research study.

Conflicts of Interest: The authors declare no conflict of interest.

References

1. Kuśmierczak, S.; Makovský, M. Problems of Blackening of Steel Parts in Technical Practice. *MATEC Web Conf.* **2018**, *244*, 01012. [CrossRef]
2. Farrell, R.W. Blackening of Ferrous Metals. *Met. Finish.* **2007**, *105*, 390–396. [CrossRef]
3. Machmudah, S.; Zulhijah, R.; Wahyudiono; Setyawan, H.; Kanda, H.; Goto, M. Magnetite Thin Film on Mild Steel Formed by Hydrothermal Electrolysis for Corrosion Prevention. *Chem. Eng. J.* **2015**, *268*, 76–85. [CrossRef]
4. Fattah-alhosseini, A.; Yazdani Khan, H.; Heidarpour, A. Comparison of Anti-Corrosive Properties between Hot Alkaline Nitrate Blackening and Hydrothermal Blackening Routes. *J. Alloy. Compd.* **2016**, *676*, 474–480. [CrossRef]

5. Hurd, R.M.; Hackerman, N. Kinetic Studies on Formation of Black-Oxide Coatings on Mild Steel in Alkaline Nitrite Solutions. *J. Electrochem. Soc.* **1957**, *104*, 482. [CrossRef]

6. Robertson, J. The Mechanism of High Temperature Aqueous Corrosion of Stainless Steels. *Corros. Sci.* **1991**, *32*, 443–465. [CrossRef]

7. Reghuraj, A.R.; Saju, K.K. Black Oxide Conversion Coating on Metals: A Review of Coating Techniques and Adaptation for SAE 420A Surgical Grade Stainless Steel. In *Materials Today: Proceedings*; Elsevier: Amsterdam, The Netherlands, 2017; Volume 4, pp. 9534–9541.

8. Sankara Narayanan, T.S.N. Phosphate Conversion Coatings—A Metal Pretreatment Process. *Corros. Rev.* **1994**, *12*, 201–238. [CrossRef]

9. Onofre-Bustamente, E.; Domínguez-Crespo, M.A.; Genescá-Llongueras, J.; Rodríguez-Gómez, F.J. Characteristics of Blueing as an Alternative Chemical Conversion Treatment on Carbon Steel. *Surf. Coat. Technol.* **2007**, *201*, 4666–4676. [CrossRef]

10. Chao, C.Y. A Point Defect Model for Anodic Passive Films. *J. Electrochem. Soc.* **1981**, *128*, 1187. [CrossRef]

11. Daub, K.; Zhang, X.; Wang, L.; Qin, Z.; Noël, J.J.; Wren, J.C. Oxide Growth and Conversion on Carbon Steel as a Function of Temperature over 25 and 80°C under Ambient Pressure. *Electrochim. Acta* **2011**, *56*, 6661–6672. [CrossRef]

12. Deiss, E.; Schikorr, G. Über Das Ferrohydroxyd (Eisen-2-Hydroxyd). *Zeitschrift für Anorganische und Allgemeine Chemie* **1928**, *172*, 32–42. [CrossRef]

13. Schikorr, G. Über Eisen(II)-Hydroxyd Und Ein Ferromagnetisches Eisen(III)-Hydroxyd. *Zeitschrift für Anorganische und Allgemeine Chemie* **1933**, *212*, 33–39. [CrossRef]

14. Vershok, D.B.; Misurkin, P.I.; Timofeeva, V.A.; Solov'eva, A.B.; Kuznetsov, Y.I.; Timashev, S.F. On the Formation of Magnetite Coatings on Low-Carbon Steel in Ammonium Nitrate Solution. *Prot. Met.* **2008**, *44*, 721–728. [CrossRef]

15. Kuznetsov, Y.I.; Vershok, D.B.; Timashev, S.F.; Solov'eva, A.B.; Misurkin, P.I.; Timofeeva, V.A.; Lakeev, S.G. Features of Formation of Magnetite Coatings on Low-Carbon Steel in Hot Nitrate Solutions. *Russ. J. Electrochem.* **2010**, *46*, 1155–1166. [CrossRef]

16. Zhu, H.; Cao, F.; Zuo, D.; Zhu, L.; Jin, D.; Yao, K. A New Hydrothermal Blackening Technology for Fe3O4 Coatings of Carbon Steel. *Appl. Surf. Sci.* **2008**, *254*, 5905–5909. [CrossRef]

17. Beverskog, B.; Puigdomenech, I. Revised Pourbaix Diagrams for Iron at 25–300 °C. *Corros. Sci.* **1996**, *38*, 2121–2135. [CrossRef]

18. Fattah-Alhosseini, A.; Khan, H.Y. Anodic Oxidation of Carbon Steel at High Current Densities and Investigation of Its Corrosion Behavior. *Metall. Mater. Trans. B Process Metall. Mater. Process. Sci.* **2017**, *48*, 1659–1666. [CrossRef]

19. Burleigh, T.D.; Schmuki, P.; Virtanen, S. Properties of the Nanoporous Anodic Oxide Electrochemically Grown on Steel in Hot 50% NaOH. *J. Electrochem. Soc.* **2009**, *156*, C45. [CrossRef]

20. Burleigh, T.D.; Dotson, T.C.; Dotson, K.T.; Gabay, S.J.; Sloan, T.B.; Ferrell, S.G. Anodizing Steel in KOH and NaOH Solutions. *J. Electrochem. Soc.* **2007**, *154*, C579. [CrossRef]

21. Fedot'ev, N.P.; Grilikhes, S.Y. Anodizing Cooper, Magnesium, Zinc, Cadmium, Steel and Silver. *Electroplat. Met. Finish.* **1960**, *13*, 413–417.

22. Deutsches Institut für Normung e.V. *Brünieren von Bauteilen Aus Eisenwerkstoffen—Anforderungen Und Prüfverfahren*; Beuth Verlag GmbH: Berlin, Germany, 2016.

23. Müsig, H.-J.; Arabczyk, W. The Interaction of Oxygen with the Iron (111) Surface, Mainly Studied by AES. *Krist. Tech.* **1980**, *15*, 1091–1099. [CrossRef]

24. Misawa, T. The Thermodynamic Consideration for Fe-H$_2$O System at 25 °C. *Corros. Sci.* **1973**, *13*, 659–676. [CrossRef]

25. Dünnwald, J.; Otto, A. Ramanspektroskopie an Oxidschichten Auf Reineisen Im Elektrolyten. *Fresenius' Zeitschrift für Analytische Chemie* **1984**, *319*, 738–742. [CrossRef]

26. Schrebler Guzmán, R.S.; Vilche, J.R.; Arvía, A.J. The Potentiodynamic Behaviour of Iron in Alkaline Solutions. *Electrochim. Acta* **1979**, *24*, 395–403. [CrossRef]

27. Xu, W.; Daub, K.; Zhang, X.; Noel, J.J.; Shoesmith, D.W.; Wren, J.C. Oxide Formation and Conversion on Carbon Steel in Mildly Basic Solutions. *Electrochim. Acta* **2009**, *54*, 5727–5738. [CrossRef]

28. Cekerevac, M.; Nikolic-Bujanovic, L.; Simicic, M. Investigation of Electrochemical Synthesis of Ferrate, Part I: Electrochemical Behavior of Iron and Its Several Alloys in Concentrated Alkaline Solutions. *Hem. Ind.* **2009**, *63*, 387–395. [CrossRef]
29. Mácová, Z.; Bouzek, K. The Influence of Electrolyte Composition on Electrochemical Ferrate(VI) Synthesis. Part III: Anodic Dissolution Kinetics of a White Cast Iron Anode Rich in Iron Carbide. *J. Appl. Electrochem.* **2012**, *42*, 615–626. [CrossRef]
30. Nieuwoudt, M.K.; Comins, J.D.; Cukrowski, I. The Growth of the Passive Film on Iron in 0.05 M NaOH Studied in Situ by Raman Microspectroscopy and Electrochemical Polarization. Part II: In Situ Raman Spectra of the Passive Film Surface during Growth by Electrochemical Polarization. *J. Raman Spectrosc.* **2011**, *42*, 1353–1365. [CrossRef]
31. Beck, F.; Kaus, R.; Oberst, M. Transpassive Dissolution of Iron to Ferrate(VI) in Concentrated Alkali Hydroxide Solutions. *Electrochim. Acta* **1985**, *30*, 173–183. [CrossRef]
32. Schmidbaur, H. The History and the Current Revival of the Oxo Chemistry of Iron in Its Highest Oxidation States: Fe VI-Fe VIII. *Zeitschrift für Anorganische und Allgemeine Chemie* **2018**, *644*, 536–559. [CrossRef]
33. Perfiliev, Y.D.; Benko, E.M.; Pankratov, D.A.; Sharma, V.K.; Dedushenko, S.K. Formation of Iron(VI) in Ozonalysis of Iron(III) in Alkaline Solution. *Inorg. Chim. Acta* **2007**, *360*, 2789–2791. [CrossRef]
34. Rush, J.D.; Bielski, B.H.J. Pulse Radiolysis Studies of Alkaline Iron(III) and Iron(VI) Solutions. Observation of Transient Iron Complexes with Intermediate Oxidation States. *J. Am. Chem. Soc.* **1986**, *108*, 523–525. [CrossRef] [PubMed]
35. Perez, N. *Electrochemistry and Corrosion Science*; Springer International Publishing: Cham, Switzerland, 2016.

Article

Cyclic Voltammetry Characterization of Au, Pd, and AuPd Nanoparticles Supported on Different Carbon Nanofibers

Anna Testolin [1], Stefano Cattaneo [1], Wu Wang [2], Di Wang [2,3], Valentina Pifferi [1,*], Laura Prati [1], Luigi Falciola [1] and Alberto Villa [1,*]

[1] Dipartimento di Chimica, Università degli Studi di Milano, Via Golgi 19, 20133 Milano, Italy; Anna.Testolin@unimi.it (A.T.); Stefano.Cattaneo2@unimi.it (S.C.); Laura.Prati@unimi.it (L.P.); luigi.falciola@unimi.it (L.F.)

[2] Institute of Nanotechnology, Karlsruhe Institute of Technology, Hermann-von-Helmholtz-Platz 1, D-76344 Eggenstein-Leopoldshafen, Germany; wu.wang@partner.kit.edu (W.W.); di.wang@kit.edu (D.W.)

[3] Karlsruhe Nano Micro Facility, Karlsruhe Institute of Technology, Hermann-von-Helmholtz-Platz 1, D-76344 Eggenstein-Leopoldshafen, Germany

* Correspondance: Valentina.Pifferi@unimi.it (V.P.); alberto.villa@unimi.it (A.V.); Tel.: +39-025-031-4222 (V.P.); +39-025-031-4361 (A.V.)

Received: 18 January 2019; Accepted: 13 March 2019; Published: 19 March 2019

Abstract: Three types of carbon nanofibers (pyrolytically stripped carbon nanofibers (PS), low-temperature heat treated carbon nanofibers (LHT), and high-temperature heat treated carbon nanofibers (HHT)) with different graphitization degrees and surface chemistry have been used as support for Au, Pd, or bimetallic AuPd alloy nanoparticles (NPs). The carbon supports have been characterized using Raman, X-ray photoelectron spectroscopy (XPS), and cyclic voltammetry (CV). Moreover, the morphology of the metal nanoparticles was investigated using transmission electron microscopy (TEM) and CV. The different properties of the carbon-based supports (particularly the graphitization degree) yield different electrochemical behaviors, in terms of potential window widths and electrocatalytic effects. Comparing the electrochemical behavior of monometallic Au and Pd and bimetallic AuPd, it is possible to observe the interaction of the two metals when alloyed. Moreover, we demonstrate that carbon surface has a strong effect on the electrochemical stability of AuPd nanoparticles. By tuning the Au-Pd nanoparticles' morphology and modulating the surface chemistry of the carbon support, it is possible to obtain materials characterized by novel electrochemical properties. This aspect makes them good candidates to be conveniently applied in different fields.

Keywords: gold; palladium; bimetallic alloy; carbon nanofibers (CNFs); cyclic voltammetry (CV)

---

1. Introduction

Over the last decades, many studies have focused on metal nanoparticles, thanks to their excellent catalytic, electrocatalytic, optical, and magnetic properties, that often differ from the ones of their correspondent bulk metals [1–3]. There has also been interest in bimetallic nanosystems, because of the possibility to create new materials characterized by specific and innovative properties due to synergistic effects among precursors [4–7]. These effects reflect a general enhancement of the performances of the final products with respect to their monometallic components. The possibility to have different kinds of bimetallic samples (in terms of composition, structure, metal loading, morphology, etc.) has led to widespread applications in a lot of fields such as engineering, electronics, catalysis, electrocatalysis, and sensors [8–10]. Particularly in heterogeneous catalysis and electrocatalysis, metal nanoparticles are often supported on specific substrates [11–17], which in turn possess different characteristics and properties related to the amount of catalyst loaded on the solid material, or to the different nature and functionalization of the substrate itself. However, in many cases, the effect of the support on the properties of the metal nanoparticles and the possible synergistic effects between bimetallic composites and supports are yet to be investigated in depth.

In the present work, we propose an integrated study of three types of carbon nanofibers (PS, LHT, and HHT) modified with Au and/or Pd nanoparticles (AuNPs, PdNPs, and alloyed AuPdNPs). AuPd systems are widely used in electrocatalysis due to their peculiar behavior compared to the corresponding monometallic Au and Pd [18–23]. Besides surface characterization techniques, such as Raman, TEM, and XPS, commonly involved in the study of these kind of materials [24,25], cyclic voltammetry is here used as a supplementary and complementary characterization method [26–29]. The different properties of the supports, particularly in terms of graphitization degree, are reflected in different electrochemical behaviors, in terms of potential window widths, electrocatalytic effects, and preferential sites for anchoring the nanoparticles. Moreover, the results highlight how the bimetallic samples behave differently with respect to the relative Au and Pd monometallic systems, underlying the presence of an intimate contact between the two metals, which provides new materials, completely different from the original ones. This aspect makes them good candidates to be conveniently applied in many different applications.

2. Materials and Methods

2.1. Metal Nanoparticles/CNFs Preparation

Commercial carbon nanofibers (CNFs), PR24-PS (hereafter shortened as PS) from Applied Sciences, inc. (average diameter of 100 ± 30 nm and a specific surface area of 45 m^2/g); CNFs PR24-LHT (LHT) from Applied Sciences, inc. (average diameter of 100 ± 30 nm and a specific surface area of 43 m^2/g); and CNFs PR24-HHT (HHT) from Applied Sciences, inc. (average diameter of 100 ± 30 nm and a specific surface area of 41 m^2/g) were used as pristine carbon materials. $NaAuCl_4 * 2H_2O$ (99%, Aldrich, St. Louis, MO, USA) and Na_2PdCl_4 (98 % Aldrich) were used as metal precursors, while $NaBH_4$ (> 98%, Ventron) and polyvinyl alcohol (PVA, MW = 9000–10,000, 80 % hydrolyzed, Aldrich) were used as reducing and protective agents, respectively.

AuNPs: 0.051 mol of solid $NaAuCl_4 * 2H_2O$ and PVA solution (1% w/w, metal/PVA 1:1 w/w) were added to 100 mL of Milli-Q H_2O. The solution was stirred for few minutes, after which $NaBH_4$ (metal/$NaBH_4$ = 1/4 mol/mol) solution was added under vigorous magnetic stirring. Within a few minutes of sol generation, the sol was immobilized by adding the support (ca. 1 g) under vigorous stirring. The slurry was filtered, and the catalyst washed thoroughly with distilled water and dried at 80 °C for 4 h. The total nominal metal loading was 1 wt %.

PdNPs: 0.094 mol of Na_2PdCl_4 and PVA solution (1% w/w, metal/PVA 1:1 w/w) were added to 100 mL of Milli-Q H_2O. The solution was stirred for few minutes, after which $NaBH_4$ (metal/$NaBH_4$ = 1/8 mol/mol) solution was added under vigorous magnetic stirring. Within a few minutes of sol generation, the sol was immobilized by adding the support (ca. 1 g) under vigorous stirring. The slurry was filtered, and the catalyst washed thoroughly with distilled water and dried at 80 °C for 4 h. The total nominal metal loading was 1 wt %.

AuPdNPs: 0.037 mol of solid $NaAuCl_4 * 2H_2O$, 0.025 mol of Na_2PdCl_4 and PVA solution (1% w/w, metal/PVA 1:1 w/w) were added to 100 mL of Milli-Q H_2O. The yellow solution was stirred for few minutes, after which $NaBH_4$ (metal/$NaBH_4$ = 1/8 mol/mol) solution was added under vigorous magnetic stirring. Within a few minutes of sol generation, the sol was immobilized by adding the support (ca. 1 g) under vigorous stirring. The slurry was filtered, and the catalyst washed thoroughly with distilled water and dried at 80 °C for 4 h. The total nominal metal loading was 1 wt %.

2.2. Electrochemical Characterization

Cyclic voltammetric experiments were performed using an AutoLab PGStat128 (Metrohm AutoLab, Utrecht, The Netherlands) equipped with the NOVA 2.1 Software (Metrohm AutoLab,

Utrecht, The Netherlands). The experimental setting was composed of a conventional three electrodes cell in which a saturated calomel electrode and a platinum wire were used as reference (RE) and counter (CE) electrodes, respectively. A glassy carbon (GC) modified with carbon nanofibers (CNFs) or with mono or bimetallic Au and Pd NPs supported on carbon nanofibers was used as working electrode (WE). Before the modification, GC was polished with diamond powder (1 μm, Sigma Aldrich, Milan, Italy) on a Struers DP-Nap cloth and washed with milli-Q water. 5 mg of the different CNF-NPs powders were suspended in 1 mL EtOH (96 %, Sigma Aldrich) and 4.5 μL Nafion (5% solution in low aliphatic alcohols, Aldrich, Milan, Italy). Then, 20 μL of the suspension was deposited on GC using an automatic micropipette (Kartell, Noviglio, Milan, Italy) and allowed to dry for approximately 30 min. Cyclic voltammetries were recorded in aqueous solution with 0.1 M H_2SO_4 as supporting electrolyte. The potential was scanned from -1 V (SCE) to $+2$ V (SCE), at a scan rate of 100 mV s^{-1} and a step potential of 5 mV.

2.3. Morphological Characterization

Raman spectroscopy was performed with a Renishaw inVia Raman microscope for analysing the graphitisation degree of the carbon nanofibers. Bare supports and fresh and used catalysts were analysed. Typically, a sample of approximately 0.01 g was placed on a metal slide inside the spectrometer. The powder was analysed under an IR class laser (514 nm) with a laser intensity of 50 %. The sample was scanned at an attenuation time of 22 seconds, and 10 scans were carried out to give a spectrum.

X-ray photoelectron spectroscopy (XPS) was performed on a Thermo Scientific K-alpha+ spectrometer (Thermo Fisher Scientific, Waltham, MA, USA). Samples were analysed using a monochromatic Al x-ray source operating at 72 W (6 mA × 12 kV), with the signal averaged over an oval-shaped area of approximately 600 × 400 microns. Data were recorded at pass energies of 150 eV for survey scans and 40 eV for high resolution scan with a 1eV and 0.1 eV step size, respectively. Charge neutralization of the sample was achieved using a combination of both low energy electrons and argon ions (less than 1 eV), which gave a C(1s) binding energy of 284.8 eV.

All data were analysed using CasaXPS (v2.3.17 PR1.1) using Scofield sensitivity factors and an energy exponent of -0.6.

The Au, Pd specimens were characterized by transmission electron microscopy (TEM, FEI Titan 80−300, Thermo Fisher Scientific, Waltham, MA, USA). The samples were directly dispersed on copper grids covered with holey carbon film. Morphology of the catalysts was characterized by high angle annular dark-field (HAADF) scanning transmission electron microscopy (STEM), and its composition information was acquired by EDAX S-UTW EDX detector in an FEI Titan 80−300 microscope operating (Thermo Fisher Scientific, Waltham, MA, USA) at 300 kV. Analysis of STEM-EDX spectrum imaging was carried out by using TEM image and analysis (TIA 4.7 SP3 version) software. Particle size of the specimens was measured on HAADF-STEM images by using the ImageJ software (National Institutes of Health, Bethesda, MD, USA) fitting the particles with ellipsoid shapes.

3. Results and Discussion

3.1. Morphological Characterization

Commercial carbon nanofibers (CNFs) thermally processed at different severity degrees [30,31], were selected as supports. Raman spectroscopy has been used to investigate the graphitization degree of CNFs treated at different temperatures. In particular, the ratio of the integral intensities of D and G bands (I_D/I_G), which is an index of the defectiveness of the graphite layers [32], was measured (Table 1). G band is related to structurally-ordered graphite domains, whereas D band can be associated with the turbostratic and/or disordered carbonaceous structures [32]. As expected, increasing the temperature of the heat treatment results in a higher graphitization degree (I_D/I_G = 0.75, 0.60, 0.11 for PS, LHT, HHT, respectively) (Table 1). X-ray photoelectron spectroscopy (XPS) has been used to confirm the different graphitization degrees. C1s XPS spectra show two main components: the first one at 284 eV, which can be attributed to sp^2-hybridised carbon species, and the second component at 285 eV ascribable to the presence of sp^3-hybridised carbon species [33] (Figure 1 and Tables 1 and 2). Sp^2/sp^3 increased with increasing annealing temperature (4.1, 7.0 and 16.8 for PS, LHT, HHT, respectively) (Table 1). Accordingly, a higher amount of oxygen has been observed decreasing the graphitization degree (C:O ratio of 88.3:11.7, 92.2:7.8, 97.3:2.7 for PS, LHT, HHT, respectively) (Table 2). Table 2 summarizes the oxygen species observed on the surface, their concentration, and the overall elemental composition. Fitting the O1s peak, five contributions were observed (Figure 2). Binding Energy (BE) of 530.2–530.5 eV corresponds to highly conjugated forms of carbonyl oxygen such as quinone [34]. The second contribution (531.6-531.9 eV) can be assigned to a carbon–oxygen double bond, whereas the one at 533.0-533.5 eV corresponds to an ether-like single bond C-O-C and to carbon oxygen single-bonds in hydroxyl groups C-O-H [35]. Oxygen species at 534.6–535.0 eV refers to the presence of carboxylic groups COOH [35]. The last signal at 536.7–537.1 eV can be assigned to adsorbed water [36]. C=O, C-O-H and C-O-C were the most abundant species present in all the CNFs (Table 2 and Figure 2).

Table 1. Atomic content of sp^2 and sp^3 carbon and ratio sp^2/sp^3 from X-ray photoelectron spectroscopy (XPS) and I_D/I_G ratio from Raman for the bare supports.

Catalyst	C sp^2 (%)	C sp^3 (%)	sp^2/sp^3	I_D/I_G
PS	75.8	18.6	4.1	0.75
LHT	83.8	12.0	7.0	0.60
HHT	90.8	3.4	16.8	0.11

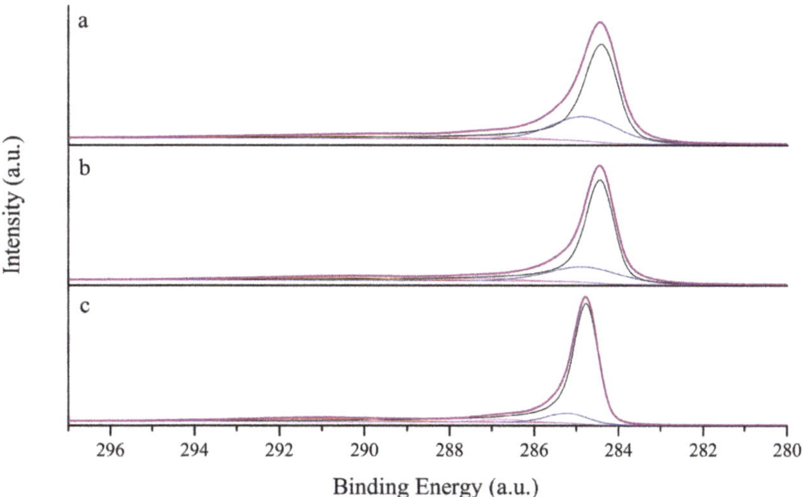

Figure 1. C1s XPS spectra of (**a**) pyrolytically stripped carbon nanofibers (PS), (**b**) low-temperature heat treated carbon nanofibers (LHT), and (**c**) high-temperature heat treated carbon nanofibers (HHT).

Table 2. XPS analysis of the carbon nanofibers (CNFs).

Sample		O1S					C1s				Atomic
		C = O	C-O-H	C-O-C	COOH	H_2O	sp^2	sp^3	C = O	C = C	Ratio % C:O
PS	BE	530.2	531.8	533.0	534.6	536.7	284.4	284.9	287.2	290.0	88.3:11.7
	Atom %	(10.2)	(40.7)	(33.1)	(9.8)	(6.0)	(75.8)	(18.6)	(3.7)	(1.8)	
LHT	BE	530.5	531.6	533.1	535.0	537.1	284.5	284.9	287.4	290.1	92.2:7.8
	Atom %	(11.6)	(34.2)	(40.3)	(8.8)	(5.1)	(83.8)	(12.0)	(2.6)	(1.6)	
HHT	BE	530.2	531.9	533.5	535.0	537.0	284.7	285.2	287.9	290.6	97.3:2.7
	Atom %	(13.9)	(46.2)	(24.0)	(10.1)	(5.8)	(90.8)	(5.4)	(1.6)	(2.2)	

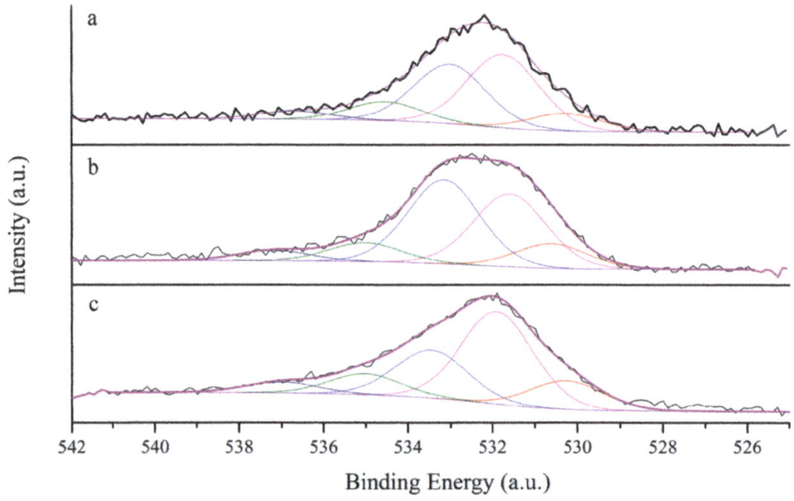

Figure 2. O1s XPS spectra of (**a**) PS, (**b**) LHT, and (**c**) HHT.

Monometallic Au and Pd, together with their corresponding bimetallic system (AuPd), were prepared by sol immobilization method using polyvinyl alcohol (PVA) as protective agent and $NaBH_4$ as reducing agent and immobilized on the three different CNFs. The monometallic and bimetallic systems consist of nanoparticles with an average size of 3–4 nm [37,38]. To investigate the morphology of the bimetallic systems, STEM-EDX spectrum imaging was performed on single nanoparticles to map the interior metal compositions. Element maps of one AuPd particle are reported in Figure 3, as example. Both Pd and Au signals were found in nanoparticles with a similar distribution, indicating the formation of AuPd alloyed nanoparticles. Quantification of the integrated EDX spectra of this nanoparticle indicates that the atomic ratio of Au and Pd is 64/36 close to the nominal 60/40 (Figure 3).

3.2. Electrochemical Characterization

As demonstrated by Raman spectra and XPS analysis (see Tables 1 and 2, Figures 1 and 2 in Section 3.1), the three types of carbon nanofibers differ in the ratio between the sp^2 and sp^3 conformation of carbon atoms, i.e., in the "graphitization" degree, an index of the order degree in carbon-based materials [31,39,40]. These differences are reflected in different voltammetric behaviours, especially in terms of potential window widths (the useful voltage range, between the upper and lower limits where oxidation and reduction of the solvent/supporting electrolyte occur) and electro-catalytic effects (Figure 4).

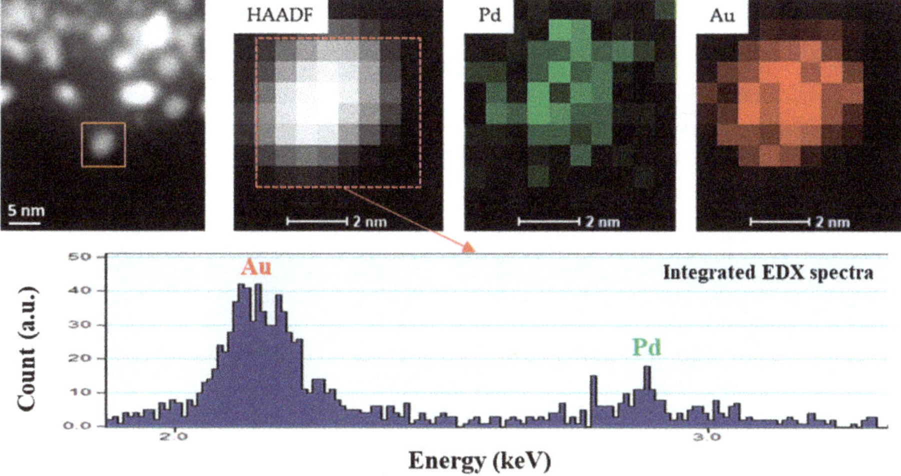

Figure 3. The element maps obtained from Scanning Electron Microscopy- Energy Dispersive X-ray Analysis (STEM)-EDX spectrum imaging in the area marked box of AuPd nanoparticles and the integrated EDX spectra of this nanoparticle.

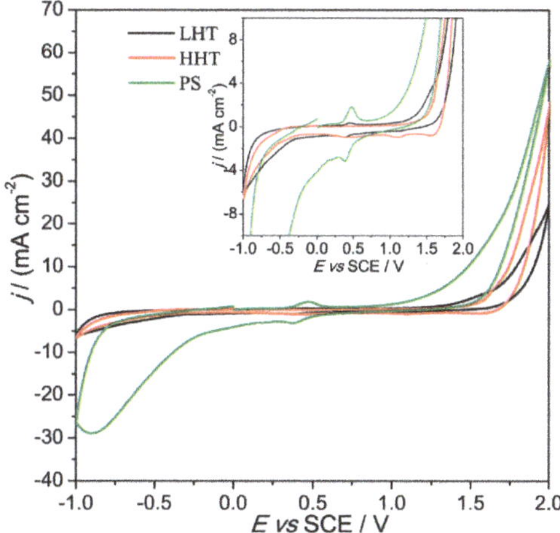

Figure 4. Cyclic voltammetry (CV) patterns of PS, LHT, and HHT nanofibers recorded in 0.1 M H_2SO_4 at 100 mV s^{-1}. Zoom of the current density values in the inset.

The potential window width is in the order PS < LHT < HHT (Figure 4), both in the cathodic and anodic side, pointing out the higher electrochemical stability and inertness of HHT (> LHT > PS) towards oxygen evolution reaction (OER, in the anodic side) and hydrogen evolution reaction (HER, in the cathodic side).

Moreover, PS nanofibers present oxidation and reduction peaks around +0.4 V (SCE), which are less marked for LHT and basically absent in the case of HHT. Considering the surface morphology of the investigated materials (Section 3.1), these peaks have been attributed to the presence of oxygen functionalizations on the nanofibers [41,42], which are in higher concentrations in the case of PS, lower for LHT, and almost absent in HHT (Table 2). It is known, in fact, that the lower the graphitization degree, the higher the presence of surface impurities and defects [43]. The presence of these functionalities can be responsible also for the different potential windows of the materials.

In order to investigate if the various properties of the nanofibers reflect also in a different behavior when they are used as support for the nanoparticles samples, we studied CNFs modified with monometallic Au or Pd NPs and with bimetallic AuPd alloyed NPs.

Figure 5a reports the voltammograms of the Au-modified nanofibers. As previously observed in Figure 4, the smallest potential window is related to PS system, followed by LHT and then HHT. In this case, however, additional oxidation and reduction peaks related to the presence of gold nanoparticles on the carbonaceous fibers are present (inset of Figure 5A). In particular, apart from the same peaks of the bare nanofibers at +0.4 V (SCE), a well-defined signal (for LHT and HHT) at around +1.3 V (SCE) is evident, whose peak-shape confirms the nanosized structure of the Au particles [44]. In the case of PS, the signal is partially covered by the reaction of the background, resulting in a shoulder plateau. Considering the oxidation peak potential position, a small catalytic effect for HHT (peak at +1.25 V vs SCE, with respect to the value of +1.31 for LHT-supported sample) is observed, probably related to a higher interaction between HHT nanofiber and AuNPs, due to the high graphitization degree of the material which reflects in a favored electron transfer.

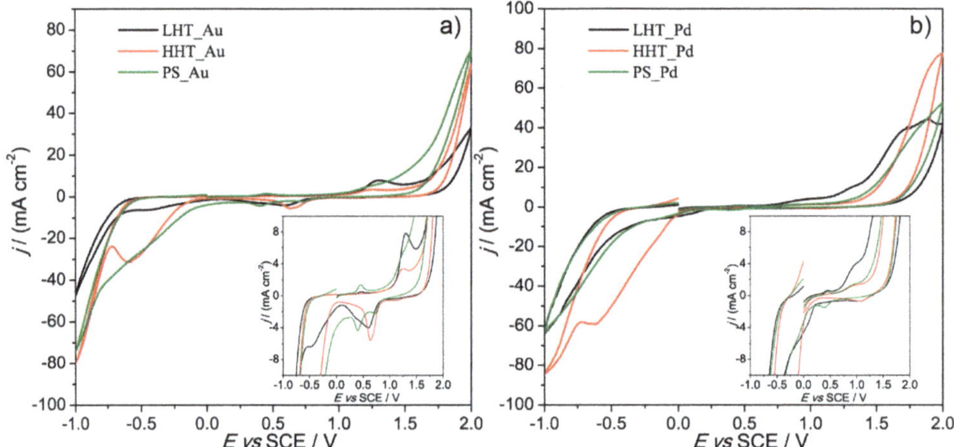

Figure 5. Cyclic voltammetry (CV) of PS, LHT, and HHT modified with AuNPs (**a**) or PdNPs (**b**) recorded in H_2SO_4 0.1 M at 100 mV s^{-1}. In the inset, zoom on the density current values.

The interesting behavior of the supported materials can be outlined considering the main oxide-reduction peaks in the 0.5–0.8 V (SCE) region, to be compared with the peak position (at about +0.8 V (SCE)) of analogous AuNPs system in the absence of the carbon nanofibers support [40]. In the present case, all the three different supported samples show the gold oxide reduction peaks at lower potentials (and in the order PS < LHT < HHT). This behavior can be ascribed to a different electrochemical stabilizing effect exhibited by the different carbonaceous fibers on the nanomaterials that can be further and conveniently exploited in sensing and catalytic applications.

In the case of PdNPs (Figure 5b), the presence of a small oxidation peak around +1 V (SCE) is evident, which is in line with the values of analogous colloidal PdNPs. In the case of LHT_Pd, more intense signals at about 1.73 and 1.89 V (SCE) are present, probably due to the concomitant Pd over-oxidation together with O_2 evolution. In both HHT systems (HHT_Au and HHT_Pd), an evident reduction peak at about −0.6 V (SCE) is present, whose nature is yet to be clarified (Figure S1).

The CV patterns of the bimetallic alloyed AuPd nanoparticles supported on the carbonaceous materials (Figure 6) are more interesting. Also in this case, the broad oxidation peak related to the reaction of the background is present in the PS fibers case, which covers the ones related to the metallic nanoparticles. In the case of LHT, it can be noticed the presence of two oxidation peaks at about +1.25 and +1.74 V (SCE) related to the Pd and Au oxide formation, respectively, both shifted at higher anodic potentials with respect to their monometallic components [45]. This behavior can be attributed to a stabilizing effect that the two metals exhibit when they are alloyed, as already evidenced in the literature [7]. The single reduction signal at about -0.29 V (SCE) is probably related to the reduction of the alloyed AuPd oxide, which occurs at a more cathodically shifted potential with respect to its monometallic references, resulting in a great stabilization of the oxide, probably due to the synergistic effects of the two metals in the alloy, and also due to a contribution by the carbonaceous support. This unexpected potential shift of the main oxide-reduction peak (usually, on bare Pd, in this region it is more likely to have proton absorption), has been observed in other homogeneous noble metals alloys [46] and in other bimetallic Pd-based catalysts supported on carbon nanofibers [47].

For HHT, a small oxidation peak around +1.2 V (SCE) is present (see the inset of Figure 6), which could be related to the formation of the palladium oxide, but no signal related to the oxidation of AuNPs is evident. It seems that the AuNPs oxidation process occurs at higher potential with respect to +2 V (SCE), where the signal of the OER takes place. Also, the reduction peak in this case is not evident (probably extremely cathodically shifted and under the signal of the HER background), once more evidencing the greater stabilizing effect of HHT on the alloyed NPs.

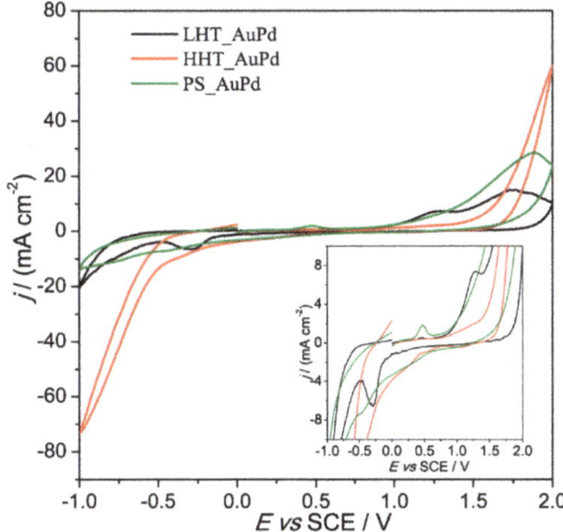

Figure 6. Cyclic voltammetric (CV) features of PS, LHT, and HHT modified with AuPdNPs recorded in H_2SO_4 0.1 M at 100 mV s^{-1}. Zoom of the density current values in the insets.

4. Conclusions

Three types of carbon nanofibers (PS, LHT, and HHT) characterized by a different degree of graphitization (HHT > LHT > PS) have been studied by CV. The different properties of the carbon-based supports yield different electrochemical behaviors, in terms of potential window widths and electrocatalytic effects. Au, Pd, and bimetallic AuPd alloy nanoparticles have been prepared by sol immobilization using PVA as a stabilizer and $NaBH_4$ as a reducer, and deposited on the three carbon nanofibers. The subsequent support of the metal NPs yields the different behavior of the latter in comparison to the non-supported systems. A cathodic and anodic shift in the oxidation and reduction peaks, related to the formation and disruption of the oxide, is evident, particularly in the case of the use of the more graphitized HHT support. This effect can be attributed to a different interaction between the metal systems and the carbon supports, which provides devices characterized by the presence of the metal catalyst, but with unexpected large potential windows, to be conveniently used in electrocatalysis and sensors applications.

Supplementary Materials: The following are available online at http://www.mdpi.com/2571-9637/2/1/16/s1, Figure S1: Cyclic Voltammetric (CV) features of HHT nanofibers modified with Au, Pd and AuPdNPs recorded in H_2SO_4 0.1 M at 100 mV s^{-1}.

Author Contributions: V.P. and A.V. designed the experiments; A.T. performed the CV experiments and wrote the article; A.V. carried out the XPS experiments and helped in the interpretation; W.W. and D.W. carried out the TEM experiments and helped in the interpretation; V.P. and L.F. helped in the interpretation of the CV results; S.C. prepared the materials; and L.P. and L.F. were involved in the writing and editing of the manuscript.

Acknowledgments: We acknowledge Karlsruhe Nano Micro Facility for using the TEM.

Conflicts of Interest: The authors declare no conflict of interest.

References

1. Campbell, F.W.; Compton, R.G. The use of nanoparticles in electroanalysis: An updated review. *Anal. Bioanal. Chem.* **2010**, *396*, 241–259. [CrossRef]

2. Sardar, R.; Funston, A.M.; Mulvaney, P.; Murray, R.W. Gold Nanoparticles: Past, Present, and Future. *Langmuir* **2009**, *25*, 13840–13851. [CrossRef] [PubMed]

3. Silvestri, A.; Mondini, S.; Marelli, M.; Pifferi, V.; Falciola, L.; Ponti, A.; Ferretti, A.M.; Polito, L. Synthesis of Water Dispersible and Catalytically Active Gold-Decorated Cobalt Ferrite Nanoparticles. *Langmuir* **2016**, *32*, 7117–7126. [CrossRef]

4. Rodriguez, J.A.; Goodman, D.W. The nature of the metal-metal bond in bimetallic surfaces. *Science* **1992**, *257*, 897–903. [CrossRef] [PubMed]

5. Toshima, N.; Yonezawa, T. Bimetallic nanoparticles—Novel materials for chemical and physical applications. *New J. Chem.* **1998**, *22*, 1179–1201. [CrossRef]

6. Sankar, M.; Dimitratos, N.; Miedziak, P.J.; Wells, P.P.; Kiely, C.J.; Hutchings, G.J. Designing bimetallic catalysts for a green and sustainable future. *Chem. Soc. Rev.* **2012**, *41*, 8099–8139. [CrossRef] [PubMed]

7. Villa, A.; Wang, D.; Su, D.S.; Prati, L. New challenges in gold catalysis: Bimetallic systems. *Catal. Sci. Technol.* **2015**, *5*, 55–68. [CrossRef]

8. Schrinner, M.; Proch, S.; Mei, Y.; Kempe, R.; Miyajima, N.; Ballauff, M. Stable Bimetallic Gold–Platinum Nanoparticles Immobilized on Spherical Polyelectrolyte Brushes: Synthesis, Characterization, and Application for the Oxidation of Alcohols. *Adv. Mater.* **2008**, *20*, 1928–1933. [CrossRef]

9. Kim, C.; Dionigi, F.; Beermann, V.; Wang, X.; Möller, T.; Strasser, P. Alloy Nanocatalysts for the Electrochemical Oxygen Reduction (ORR) and the Direct Electrochemical Carbon Dioxide Reduction Reaction (CO_2 RR). *Adv. Mater.* **2018**, 1805617. [CrossRef]

10. Hu, Y.; Zhang, H.; Wu, P.; Zhang, H.; Zhou, B.; Cai, C. Bimetallic Pt-Au nanocatalysts electrochemically deposited on graphene and their electrocatalytic characteristics towards oxygen reduction and methanol oxidation. *Phys. Chem. Chem. Phys.* **2011**, *13*, 4083–4094. [CrossRef]

11. Liu, X.Y.; Wang, A.; Zhang, T.; Mou, C.Y. Catalysis by gold: New insights into the support effect. *Nano Today* **2013**, *8*, 403–416. [CrossRef]

12. Bracey, C.L.; Ellis, P.R.; Hutchings, G.J. Application of copper-gold alloys in catalysis: Current status and future perspectives. *Chem. Soc. Rev.* **2009**, *38*, 2231–2243. [CrossRef]

13. Gliech, M.; Klingenhof, M.; Görlin, M.; Strasser, P. Supported metal oxide nanoparticle electrocatalysts: How immobilization affects catalytic performance. *Appl. Catal. A Gen.* **2018**, *568*, 11–15. [CrossRef]

14. Zhang, J.W.; Sun, K.K.; Li, D.D.; Deng, T.; Lu, G.P.; Cai, C. Pd-Ni bimetallic nanoparticles supported on active carbon as an efficient catalyst for hydrodeoxygenation of aldehydes. *Appl. Catal. A Gen.* **2019**, *569*, 190–195. [CrossRef]

15. Peneau, V.; He, Q.; Shaw, G.; Kondrat, S.A.; Davies, T.E.; Miedziak, P.; Forde, M.; Dimitratos, N.; Kiely, C.J.; Hutchings, G.J. Selective catalytic oxidation using supported gold-platinum and palladium-platinum nanoalloys prepared by sol-immobilisation. *Phys. Chem. Chem. Phys.* **2013**, *15*, 10636–10644. [CrossRef] [PubMed]

16. Rodriguez, R.C.; Moncada, A.B.; Acevedo, D.F.; Planes, G.A.; Miras, M.C.; Barbero, C.A. Electroanalysis using modified hierarchical nanoporous carbon materials. *Faraday Discuss.* **2013**, *164*, 147. [CrossRef] [PubMed]

17. Villa, A.; Schiavoni, M.; Prati, L. Material science for the support design: A powerful challenge for catalysis. *Catal. Sci. Technol.* **2012**, *2*. [CrossRef]

18. Romero Hernández, A.; Manríquez, M.E.; Ezeta Mejia, A.; Arce Estrada, E.M. Shape Effect of AuPd Core-Shell Nanostructures on the Electrocatalytical Activity for Oxygen Reduction Reaction in Acid Medium. *Electrocatalysis* **2018**, *9*, 752–761. [CrossRef]

19. Yang, J.; Deng, S.; Lei, J.; Ju, H.; Gunasekaran, S. Electrochemical synthesis of reduced graphene sheet-AuPd alloy nanoparticle composites for enzymatic biosensing. *Biosens. Bioelectron.* **2011**, *29*, 159–166. [CrossRef]

20. Jin, C.; Wan, C.; Dong, R. Electrocatalytic Activity Enhancement of Pd Nanoparticles Supported on Reduced Graphene Oxide by Surface Modification with Au. *J. Electrochem. Soc.* **2017**, *164*, H696–H700. [CrossRef]

21. Shang, L.; Zhao, F.; Zeng, B. Sensitive voltammetric determination of vanillin with an AuPd nanoparticles-graphene composite modified electrode. *Food Chem.* **2014**, *151*, 53–57. [CrossRef] [PubMed]

22. Mueller, J.E.; Krtil, P.; Kibler, L.A.; Jacob, T. Bimetallic alloys in action: Dynamic atomistic motifs for electrochemistry and catalysis. *Phys. Chem. Chem. Phys.* **2014**, *16*, 15029–15042. [CrossRef] [PubMed]

23. Pluntke, Y.; Kibler, L.A.; Kolb, D.M. Unique activity of Pd monomers: Hydrogen evolution at AuPd(111) surface alloys. *Phys. Chem. Chem. Phys.* **2008**, *10*, 3684–3688. [CrossRef] [PubMed]

24. Baer, D.R.; Amonette, J.E.; Engelhard, M.H.; Gaspar, D.J.; Karakoti, A.S.; Kuchibhatla, S.; Nachimuthu, P.; Nurmi, J.T.; Qiang, Y.; Sarathy, V.; et al. Characterization challenges for nanomaterials. In *Surface and Interface Analysis*; John Wiley & Sons, Ltd.: Hoboken, NJ, USA, 2008; Volume 40, pp. 529–537.

25. Villa, A.; Dimitratos, N.; Chan-Thaw, C.E.; Hammond, C.; Veith, G.M.; Wang, D.; Manzoli, M.; Prati, L.; Hutchings, G.J. Characterisation of gold catalysts. *Chem. Soc. Rev.* **2016**, *45*. [CrossRef] [PubMed]

26. Holt, L.R.; Plowman, B.J.; Young, N.P.; Tschulik, K.; Compton, R.G. The Electrochemical Characterization of Single Core-Shell Nanoparticles. *Angew. Chem. Int. Ed.* **2016**, *55*, 397–400. [CrossRef] [PubMed]

27. Plowman, B.J.; Sidhureddy, B.; Sokolov, S.V.; Young, N.P.; Chen, A.; Compton, R.G. Electrochemical Behavior of Gold–Silver Alloy Nanoparticles. *ChemElectroChem* **2016**, *3*, 1039–1043. [CrossRef]

28. Tschulik, K.; Ngamchuea, K.; Ziegler, C.; Beier, M.G.; Damm, C.; Eychmueller, A.; Compton, R.G. Core-Shell Nanoparticles: Characterizing Multifunctional Materials beyond Imaging—Distinguishing and Quantifying Perfect and Broken Shells. *Adv. Funct. Mater.* **2015**, *25*, 5149–5158. [CrossRef]

29. Ranđelović, M.; Momčilović, M.; Matović, B.; Babić, B.; Barek, J. Cyclic voltammetry as a tool for model testing of catalytic Pt- and Ag-doped carbon microspheres. *J. Electroanal. Chem.* **2015**, *757*, 176–182. [CrossRef]

30. Tessonnier, J.P.; Rosenthal, D.; Hansen, T.W.; Hess, C.; Schuster, M.E.; Blume, R.; Girgsdies, F.; Pfänder, N.; Timpe, O.; Su, D.S.; et al. Analysis of the structure and chemical properties of some commercial carbon nanostructures. *Carbon N. Y.* **2009**, *47*, 1779–1798. [CrossRef]

31. Campisi, S.; Sanchez Trujillo, F.; Motta, D.; Davies, T.; Dimitratos, N.; Villa, A. Controlling the Incorporation of Phosphorus Functionalities on Carbon Nanofibers: Effects on the Catalytic Performance of Fructose Dehydration. *C* **2018**, *4*, 9. [CrossRef]

32. Tuinstra, F.; Koenig, J.L. Raman Spectrum of Graphite. *J. Chem. Phys.* **1970**, *53*, 1126–1130. [CrossRef]

33. Xie, F.Y.; Xie, W.G.; Gong, L.; Zhang, W.H.; Chen, S.H.; Zhang, Q.Z.; Chen, J. Surface characterization on graphitization of nanodiamond powder annealed in nitrogen ambient. *Surf. Interface Anal.* **2010**, *42*, 1514–1518. [CrossRef]

34. Martínez, M.T.; Callejas, M.A.; Benito, A.M.; Cochet, M.; Seeger, T.; Ansón, A.; Schreiber, J.; Gordon, C.; Marhic, C.; Chauvet, O.; et al. Sensitivity of single wall carbon nanotubes to oxidative processing: Structural modification, intercalation and functionalisation. *Carbon N. Y.* **2003**, *41*, 2247–2256. [CrossRef]

35. Figueiredo, J.L.; Pereira, M.F.R.; Freitas, M.M.A.; Órfão, J.J.M. Modification of the surface chemistry of activated carbons. *Carbon N. Y.* **1999**, *37*, 1379–1389. [CrossRef]

36. Ketteler, G.; Ashby, P.; Mun, B.S.; Ratera, I.; Bluhm, H.; Kasemo, B.; Salmeron, M. In situ photoelectron spectroscopy study of water adsorption on model biomaterial surfaces. *J. Phys. Condens. Matter* **2008**, *20*, 184024. [CrossRef]

37. Wang, D.; Villa, A.; Su, D.; Prati, L.; Schlögl, R. Carbon-supported gold nanocatalysts: Shape effect in the selective glycerol oxidation. *ChemCatChem* **2013**, *5*, 2717–2723. [CrossRef]

38. Sanchez, F.; Alotaibi, M.H.; Motta, D.; Chan-Thaw, C.E.; Rakotomahevitra, A.; Tabanelli, T.; Roldan, A.; Hammond, C.; He, Q.; Davies, T.; et al. Hydrogen production from formic acid decomposition in the liquid phase using Pd nanoparticles supported on CNFs with different surface properties. *Sustain. Energy Fuels* **2018**, *2*, 2705–2716. [CrossRef]

39. Franklin, R.E. Homogeneous and heterogeneous graphitization of carbon. *Nature* **1956**, *177*, 239. [CrossRef]

40. Chen, X.; Deng, X.; Kim, N.Y.; Wang, Y.; Huang, Y.; Peng, L.; Huang, M.; Zhang, X.; Chen, X.; Luo, D.; et al. Graphitization of graphene oxide films under pressure. *Carbon N. Y.* **2018**, *132*, 294–303. [CrossRef]

41. Wilgosz, K.; Chen, X.; Kierzek, K.; Machnikowski, J.; Kalenczuk, R.J.; Mijowska, E. Template method synthesis of mesoporous carbon spheres and its applications as supercapacitors. *Nanoscale Res. Lett.* **2012**, *7*, 269. [CrossRef]

42. Andreas, H.A.; Conway, B.E. Examination of the double-layer capacitance of an high specific-area C-cloth electrode as titrated from acidic to alkaline pHs. *Electrochim. Acta* **2006**, *51*, 6510–6520. [CrossRef]

43. Tasis, D.; Tagmatarchis, N.; Bianco, A.; Prato, M. Chemistry of Carbon Nanotubes Chemistry of Carbon Nanotubes. *Chem. Rev.* **2006**, *106*, 1105–1136. [CrossRef]

44. Bonanni, A.; Pumera, M.; Miyahara, Y. Influence of gold nanoparticle size (2-50 nm) upon its electrochemical behavior: An electrochemical impedance spectroscopic and voltammetric study. *Phys. Chem. Chem. Phys.* **2011**, *13*, 4980–4986. [CrossRef] [PubMed]

45. Pifferi, V.; Chan-Thaw, C.; Campisi, S.; Testolin, A.; Villa, A.; Falciola, L.; Prati, L.; Pifferi, V.; Chan-Thaw, C.E.; Campisi, S.; et al. Au-Based Catalysts: Electrochemical Characterization for Structural Insights. *Molecules* **2016**, *21*, 261. [CrossRef] [PubMed]

46. Grdeń, M.; Łukaszewski, M.; Jerkiewicz, G.; Czerwiński, A. Electrochemical behaviour of palladium electrode: Oxidation, electrodissolution and ionic adsorption. *Electrochim. Acta* **2008**, *53*, 7583–7598. [CrossRef]

47. Maiyalagan, T.; Scott, K. Performance of carbon nanofiber supported Pd–Ni catalysts for electro-oxidation of ethanol in alkaline medium. *J. Power Sources* **2010**, *195*, 5246–5251. [CrossRef]

Article

"Click" Chemistry on Gold Electrodes Modified with Reduced Graphene Oxide by Electrophoretic Deposition

Vladyslav Mishyn [1], Patrik Aspermair [1,2], Yann Leroux [3], Henri Happy [1], Wolfgang Knoll [2,4], Rabah Boukherroub [1] and Sabine Szunerits [1,*]

[1] Univ. Lille, CNRS, Centrale Lille, ISEN, Univ. Valenciennes, UMR 8520—IEMN, F-59000 Lille, France; vladyslav.mishyn@univ-lille.fr (V.M.); patrik.aspermair@univ-lille.fr or patrik.aspermair@cest.at (P.A.); henri.happy@iemn.univ-lille.fr (H.H.); rabah.boukherrroub@univ-lille.fr (R.B.)

[2] CEST Competence Center for Electrochemical Surface Technology, 3430 Tulln, Austria; wolfgang.knoll@cest.at

[3] Univ. Rennes, CNRS, ISCR—UMR 6226, Campus de Beaulieu, F-35000 Rennes, France; yann.leroux@univ-rennes1.fr

[4] AIT Austrian Institute of Technology, 3430 Tulln, Austria

* Correspondence: sabine.szunerits@unvi-lille.fr; Tel.: +33-3-62-53-17-25

Received: 22 January 2019; Accepted: 8 March 2019; Published: 18 March 2019

Abstract: The coating of electrical interfaces with reduced graphene oxide (rGO) films and their subsequent chemical modification are essential steps in the fabrication of graphene-based sensing platforms. In this work, electrophoretic deposition (EPD) of graphene oxide at 2.5 V for 300 s followed by vapor treatment were employed to coat gold electrodes uniformly with rGO. These interfaces showed excellent electron transfer characteristics for redox mediators such as ferrocene methanol and potassium ferrocyanide. Functional groups were integrated onto the Au/rGO electrodes by the electro-reduction of an aryldiazonium salt, 4-((triisopropylsilyl)ethylenyl)benzenediazonium tetrafluoroborate (TIPS-Eth-ArN) in our case. Chemical deprotection of the triisopropylsilyl function resulted in propargyl-terminated Au/rGO electrodes to which azidomethylferrocene was chemically linked using the Cu(I) catalyzed "click" chemistry.

Keywords: reduced graphene oxide; electrophoretic deposition; surface chemistry; click chemistry

---

1. Introduction

Accurate analysis of the presence of disease-specific biomarkers in biological fluids remains of great importance in clinical settings [1] and electrochemical sensors can reach that goal by converting a chemical or a biological response into a processable and quantifiable electrochemical signal [2]. Graphene and its related derivatives have generated great expectations as a transducing platform in biosensing, due to their good mechanical properties accompanied by biocompatibility, electrical conductivity and fast charge transfer kinetics [3–6]. A mandatory step in the production of biosensors is the modification of graphene-based materials with recognition elements. Covalent and non-covalent strategies have been employed, including amide bond formation and π–π interactions, among others [7–9], to integrate surface functionalities and ligands onto graphene-based transducers. The development of these approaches depends on having robust graphene-coated interfaces at hand. Next to drop-casting and spin-coating of reduced graphene oxide (rGO) suspensions onto electrical interfaces, electrophoretic deposition (EPD) has been shown to be an effective technique for manipulating graphene oxide (GO) suspensions with the aim of producing graphene-related films [9–13]. The ability of EPD to be applied to different materials and to control the thickness of the deposits has been well known for a decade [14,15]. EPD has gained increased interest as an alternative processing technique for the deposition of various nanomaterials ranging from metal oxide particles [16] to carbon nanotubes [17]. EPD is also relevant to the development of graphene-based coatings in a cost-effective manner [13,18]. Its capacity to be used in more complex, integrated electrode systems is an advantage over drop-casting and other less defined deposition techniques [10,11,19–24]. The group of Boccaccini added intensively to this field by deepening our understanding of EPD through the investigation of GO-EPD kinetics as a function of deposition time and potential [12].

To design a powerful electrochemical sensor, working with highly reduced graphene oxide nanosheets formed by EPD is required. Cathodic EPD would be the preferential approach, as it allows the simultaneous deposition and reduction of GO to rGO [9]. The presence of carboxyl and hydroxyl functions on GO results in an overall negatively charged material of about -41.3 ± 0.8 mV for aqueous GO suspensions. Migration to the anode rather to the cathode occurs upon applying a DC voltage [13]. The anodic EPD of GO results in GO with a low degree of reduction during the deposition process. Thermal or chemical reductions [25] are necessary to restore the aromatic network in order to obtain a material with good electron transfer properties. This is one of the reasons why GO is often charged with a cationic polymer (e.g., polyethyleneimine) [8] or metallic cations (Ni^{2+}, Cu^{2+}, etc.) [21–23] to achieve a positively charged GO nanomaterial, which can be deposited by cathodic EPD. The presence of polyethyleneimine (PEI) has been shown to be advantageous for the integration of surface ligands and the formation of an immunosensor for the selective and sensitive electrochemical detection of uropathogenic *Escherichia coli* [8], the detection of dopamine in meat [20] and Ni^{2+} for the construction of non-enzymatic glucose sensors operating in a basic medium. However, the formation of well-reduced rGO by EPD free of metal ions and other surface ligands remains a challenge.

In this study, we evaluate the effect of applied electrical current and applied voltage on the electrochemical behavior of electrophoretically deposited rGO on gold thin film electrodes. It is shown that the use of a voltage bias of 2.5 V for 5 min results in rGO thin films of good electrochemical behavior. These films can also be submitted to further surface modification using diazonium electrochemistry without altering their adhesion characteristics.

2. Materials and Methods

2.1. Materials

Potassium hexacyanoferrate(II) ($[K_4Fe(CN)_6]$), hydrazine hydrate, phosphate buffer tablets (PBS, 0.1 M), tetrabutylammonium fluoride (TBAF), ferrocenemethanol, copper(II) sulfate ($CuSO_4$), L-ascorbic acid, EDTA and N-butylhexafluorophosphate (NBu_4PF_6) were purchased from Sigma-Aldrich and used as received. Graphene oxide (GO) powder was purchased from Graphenea, Spain. 4-((triisopropylsilyl)ethylenyl)benzenediazonium tetrafluoroborate (TIPS-Eth-ArN$_2$$^+$) was synthesized as reported previously [26].

Azidomethylferrocene was synthesized according to Reference [27].

Au thin film electrodes were prepared by thermal evaporation of 5 nm of titanium and 40 nm of gold onto cleaned glass slides.

2.2. Electrophoretic Deposition

Before electrophoretic deposition, the gold electrode was cleaned by UV/ozone for 5 min, rinsed with acetone and water and dried under a nitrogen flow. The deposition took place in a two-electrode system with a platinum foil (1 cm^2) as the cathode and the cleaned gold surface as the anode (0.5 cm^2). The electrodes were placed in parallel to each other at a fixed distance of 1.5 cm. An aqueous GO solution of 1 mg mL^{-1} was used for the EPD. Voltage biases of 1.25 V, 2.5 V, 5 V or 10 V were applied using a potentiostat/galvanostat (Metrohm Autolab, Utrecht, The Netherlands) for 5 min. The modified gold electrodes were slowly withdrawn manually from the solution and dried in a horizontal position under ambient conditions for 1 h. After the deposition was complete, the modified electrodes were placed into a Teflon autoclave (45 mL) and sealed with 1 mL of hydrazine hydrochloride. The autoclave was heated to 80 °C and kept under a constant temperature for 4 h. The interfaces were then removed and gently washed with water.

2.3. Surface Modification

2.3.1. Diazonium Chemistry

The electrografting of 4-((triisopropylsilyl)ethylenyl)benzenediazonium tetrafluoroborate (TIPS-Eth-ArN$_2$$^+$) (1 mM) in 0.1 M NBu$_4PF_6$ in acetonitrile was performed using cyclic voltammetry with a scan rate of 50 mV s^{-1} for five cycles between +0.60 V and −0.75 V vs. Ag/AgCl. The electrodes were rinsed with copious amounts of acetonitrile and acetone and dried under a stream of argon.

2.3.2. "Click" Chemistry

Before "click" chemistry, the TIPS protection group was removed by the immersion of the Au/rGO-TIPS surface into tetrabutylammonium fluoride (TBAF, 0.05 M in THF) for 20 min. The Au/rGO interface was then immersed into an aqueous solution of CuSO$_4$ (10 mM) and L-ascorbic acid (20 mM) in the presence of azidomethylferrocene (0.83 mM in THF) and left for 1 h under an argon atmosphere. The interface was then treated with an aqueous solution of EDTA for 10 min to chelate any remaining Cu^{2+} residues and finally washed copiously with acetone and water and left to dry.

2.4. Surface Characterization Techniques

2.4.1. Scanning Electron Microscopy (SEM)

SEM images were obtained using an electron microscope ULTRA 55 (Zeiss, Paris, France) equipped with a thermal field emission emitter and three different detectors (EsB detector with filter grid, high efficiency in-lens SE detector and Everhart–Thornley Secondary Electron Detector).

2.4.2. X-Ray Photoelectron Spectroscopy (XPS)

X-ray photoelectron spectroscopy (XPS) was performed in a PHI 5000 VersaProbe-Scanning ESCA Microprobe (ULVAC-PHI, Osaka, Japan) instrument at a base pressure below 5 × 10^{-9} mbar. Core-level spectra were acquired at a pass energy of 23.5 eV with a 0.1 eV energy step. All spectra were acquired with 90° between the X-ray source the and analyzer. After the subtraction of the linear background, the core-level spectra were decomposed into their components with mixed Gaussian−Lorentzian (30:70) shape lines using CasaXPS software. Quantification calculations were conducted using sensitivity factors supplied by PHI.

2.4.3. Electrochemical Measurements

Electrochemical measurements were performed with a potentiostat/galvanostat (Metrohm Autolab, Utrecht, The Netherlands). A conventional three-electrode configuration was employed using a silver wire and a platinum mesh as the reference and auxiliary electrodes, respectively.

2.4.4. Micro-Raman Analysis

Micro-Raman spectroscopy measurements were performed on a LabRam HR Micro-Raman system (Horiba Jobin Yvon, Palaiseau, France) combined with a 473 nm laser diode as the excitation source. Visible light was focused by a 100× objective. The scattered light was collected by the same objective in backscattering configuration, dispersed by a 1800 mm focal length monochromator and detected by a CCD.

2.4.5. Atomic Force Microcopy (AFM)

Tapping mode AFM images in air and ambient temperature were recorded using a Bruker Dimension 3100 AFM (Bruker, Champs-sur-Marne, France). The surfaces were imaged with a silicon cantilever (AppNano TM300, typical spring constant: 50 N/m) working at a frequency of 369 kHz. Image treatment and root mean square (RMS) roughness R_a were obtained with WSXM software (Bruker, Champs-sur-Marne, France). Surface roughness of the samples was measured by scanning over a 5×5 μm area.

2.4.6. Profilometry

The thickness of the deposited films was determined by using an optical profilometer (Zygo NewView 6000 Optical Profilometer with MetroPro software). This equipment uses non-contact, three-dimensional scanning white light and optical phase-shifting interferometry, has vertical z-scan measurements ranging from 0.1 nm to 15,000 μm and has capabilities of 1 nm height resolution with step accuracies better than 0.75%. Images were taken with a $10\times$ lens with a 14 mm field of view.

3. Results

3.1. Electrophoretic Deposition

The process for the formation of stable electrochemically active reduced graphene oxide (rGO) thin films on gold thin film electrodes is schematically depicted in Figure 1A. It is based on a two-step process in which, after EPD at 2.5 V for 5 min, the full reduction of GO to rGO is obtained by immersion into hydrazine vapor for 4 h. The use of a potential of 2.5 V proved to be of high importance in the process. Initial investigation revealed that the electrical current signature changes significantly at voltages greater than 2.5 V (Figure 1B), where the electrical current passing through the interface increases during the first 25 s before stabilizing. In contrast, at potential biases of 2.5 V and lower, a decrease of the electrical current is first observed before its stabilization. Indeed, due to the insulating character of GO, the electrical current should decrease during the deposition process as observed for a low voltage bias. The current drop at 2.5 V is slower than that at 1.25 V. The reason for this is not well understood, but could be due to a partial electrochemical reduction of GO under these conditions. A similar behavior was observed by Diba et al. following the deposition of GO at 3 and 5 V [12].

The increase in current at elevated voltages indicates that next to the material deposition an electrochemical reaction occurs, delaying surface passivation [12]. Visual inspection of the interfaces (Figure 1C) shows clearly that the deposition occurred. However, a closer visual inspection of the electrical interfaces (Figure 1D) reveals that the gold thin films were partially destroyed, most likely by the gas evolution during water hydrolysis occurring in parallel at these voltage biases. This results in the swelling of the GO deposit and eventual gold film rupture with poor film attachment.

Figure 2A shows SEM images of the different interfaces. While the films formed at 1.25 V display granular like structures, the films deposited at 2.5 V exhibit the typical rGO-like wrinkle structures with no evident local surface inhomogeneities and a complete and even coating of the gold thin film interface. From the cross-section image (Figure 1A), no protrusions and hollow internal structures are visible, indicating that at a lower potential local inhomogeneities associated with rGO formation are avoided. The average surface roughness (R_a), as determined by tapping-mode atomic force microscopy (AFM) measurements, changed from $R_a = 4.1$ nm (Au) to $R_a = 3.5$ nm for Au/rGO (2.5 V deposition potential).

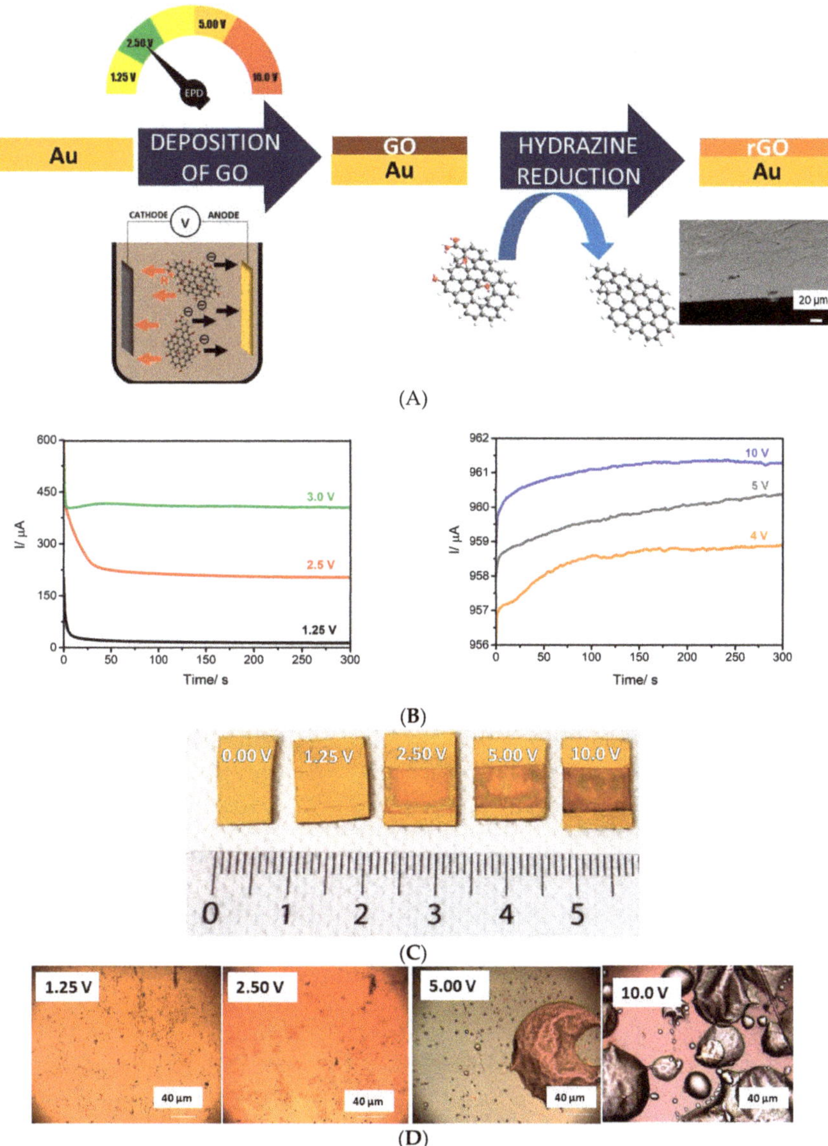

Figure 1. (**A**) Schematic illustration of the formation of an electrochemically active reduced graphene oxide (rGO) thin film by anodic electrophoretic deposition (EPD) from an aqueous graphene oxide suspensions (GO, 1 mg/mL) at a potential bias of 2.5 V for 5 min, followed by chemical treatment in hydrazine vapor; cross-sectional SEM image of a gold interface coated with rGO using EPD at 2.5 V for 5 min. (**B**) Change of current as a function of deposition time using different applied potentials: concentration of GO (1 mg/mL) in water (pH 7.5), deposition interface: gold thin film electrode. (**C**) Photographic images of gold thin film electrodes before (0.00 V) and after anodic EPD of an aqueous solution of GO (1 mg/mL, pH 7.5) for 5 min at 1.25, 2.50, 5.00 and 10.00 V. (**D**) Photographs of the different interfaces obtained (objective: 10×, numerical aperture: 0.9).

To estimate the anodic efficiency of the EPD process, the gold interface was weighed before and after rGO deposition and reduction. To convert the deposited weight (m_{rGO}, 10 µg for 2.5 V for 5 min) into a film thickness (d_{rGO}), Equation (1) was used:

$$d_{rGO}\,(nm) = \frac{m_{rGO}\times}{A \times \rho} \tag{1}$$

where A is the surface area (0.8 cm^2 in our case) and ρ is the density of rGO (2.09 g cm^{-3}) [28]. A film thickness of 59 nm was determined. This is in agreement with profilometry measurements where a film thickness of about 55 nm was determined (Figure 2B). Samples prepared at 1.25 V for 5 min had a deposited mass of 3 µg and an estimated rGO film thickness of 20 nm, below the acceptable accuracy limit of profilometry to be validated. Increasing the potential to 5 V resulted in 120 nm thick films. The Hamakar model [29] correlates the time-dependent (t = 300 s) amount of deposited materials (m_{rGO} in g) with the electrical field strength (E, 1.6 V m^{-1} in our case), the surface area of the electrode (0.8 cm^2), the electrophoretic mobility of GO (1.97 × 10^{-4} cm^2/(Vs)$^{-1}$), the concentration of the particle suspension (1 mg mL^{-1}, 0.001 g cm^{-3}) and the anodic efficiency factor f (Equation (2)) [29].

$$m_{rGO} = c_{rGO} \times A \times \mu \times E \times t \times f \tag{2}$$

In the case of f = 1, the amount of deposited rGO should be equal to m_{rGO} = 47 µg. The determined rGO amount, however, was only 10 µg, which implies an efficiency factor of f = 0.2.

As a voltage bias of 2.5 V seems to be the best condition for the EDP of rGO films onto the gold electrodes, these interfaces were investigated in greater detail. The electrochemical behavior of this interface using ferrocene-methanol redox couple in an aqueous solution is depicted in Figure 2C. Whereas on a bare gold surface a fully reversible voltammogram was observed, an irreversible voltammogram with a very small oxidation peak was observed on GO-coated gold surfaces. To improve the electrochemical behavior of such coated gold surfaces, the interfaces were further treated with hydrazine vapor, known for its strong reducing power. The cyclic voltammograms of hydrazine treated surfaces were largely improved, showing a well-defined redox couple with increased capacitance behavior as expected for rGO materials (Figure 2B, blue curve).

The chemical composition of the deposited graphene matrix was further evaluated using XPS (Figure 2D). The high resolution C1s core level spectrum of the initial GO suspensions showed contributions at 284.2 (C=C sp^2), 285.0 (C–H/C–Csp^3), 286.7 (C–O) 288.7 (C=O) and a small contribution at 291.0 (O–C=O). The C1s XPS core spectra of GO-coated gold surfaces revealed similar contributions with different intensities. In particular, the band at 284.2 eV due to C=C sp^2 increased compared to the GO solution, indicating the partial restoration of the sp^2 network of the deposited graphene material. The XPS of GO-coated gold showed additional bands at 285.0 and 286.7 eV with a large band at 288.0 eV due to C=O. The hydrazine reduction of the GO–gold interfaces resulted in a decrease of the epoxy/ether functions at 286.7 eV, in accordance with a partial reduction of GO mostly likely due to the elimination of CO by a Kolb-like mechanism [12,13]. The ketone groups remained preserved.

The Raman spectrum of GO-coated gold surfaces before and after hydrazine reduction (4 h) is presented in Figure 2E. The increase of the D/G ratio from 0.86 to 0.97 after hydrazine chemical reduction was observed.

Figure 3A summarizes the electrochemical behavior of a neutral redox species, ferrocene methanol, on bare and GO-coated gold surfaces after the hydrazine chemical reduction. The electrochemically active surface area of the different electrodes was determined by plotting the peak current as a function of the square root of the scan rate (Figure 3B), according to Equation (3) [30]:

$$A = slope/(268600 \times n^{3/2} \times D^{1/2} \times c) \tag{3}$$

where n is the number of electrons transferred in the redox event (n = 1), D is the diffusion coefficient of ferrocene methanol (7.5×10^{-6} cm^2 s^{-1}) and c is the concentration of ferrocene methanol ($1\ 10^{-6}$ mol cm^{-3}). Taking into account the experimentally determined slopes (9.265×10^{-5} AV$^{1/2}$ s$^{-1/2}$ for Au and 9.980×10^{-5} AV$^{1/2}$s$^{-1/2}$ for Au/rGO), the active surfaces of 0.126 cm^2 (naked gold) and 0.136 cm^2 (Au/rGO) were determined. By plotting the current density vs. potential (Figure 3B), a larger current density was detected on the EPD coated gold interface due to the excellent electrochemical behavior of the interface. This is in agreement with the deposition of an electrochemically active 3D rGO material.

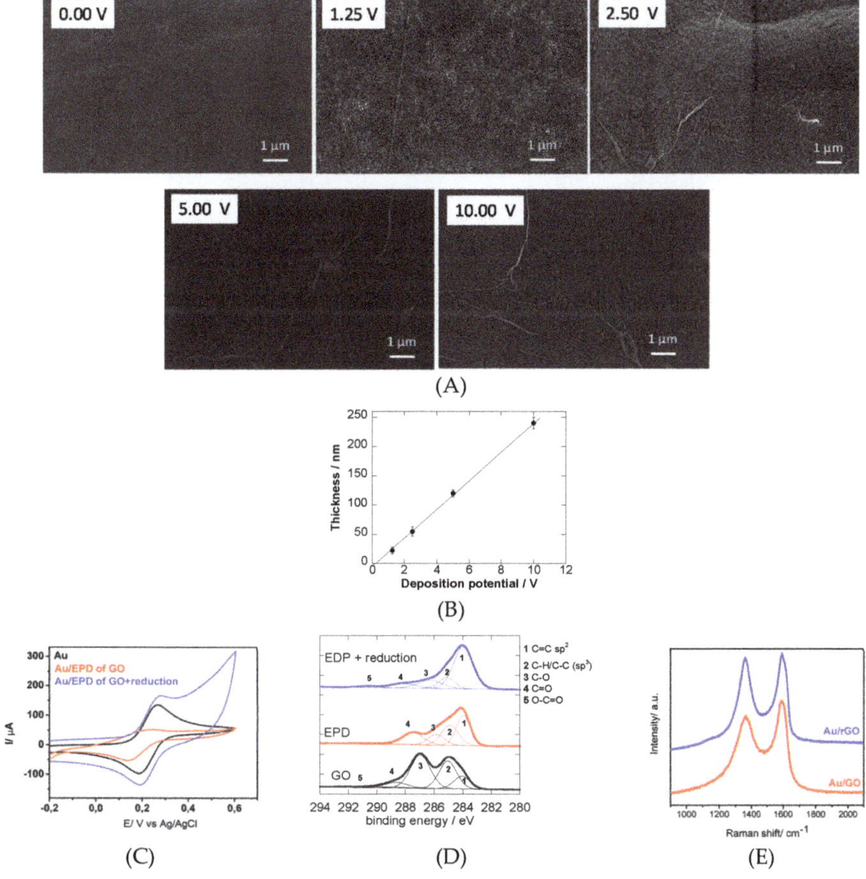

Figure 2. (**A**) SEM image of gold thin films before and after EPD from an aqueous graphene oxide suspension (GO, 1 mg/mL) at different potential biases for 5 min. (**B**) Thickness of reduced graphene oxide as a function of applied potential (t = 5 min). (**C**) Cyclic voltammograms recorded on gold thin film electrodes (black), after EPD from an aqueous graphene oxide suspension (GO, 1 mg/mL) at a potential bias of 2.5 V for 5 min (red); after further reduction with hydrazine (blue) using ferrocenemethanol (1 mM)/PBS (0.1 M), scan rate = 100 mV s^{-1}. (**D**) C1s high-resolution spectra of GO (black), EPD GO (red) and further reduced GO (blue). (**E**) Raman spectra of the EPD film formed at 2.5 V before (red) and after hydrazine reduction (blue).

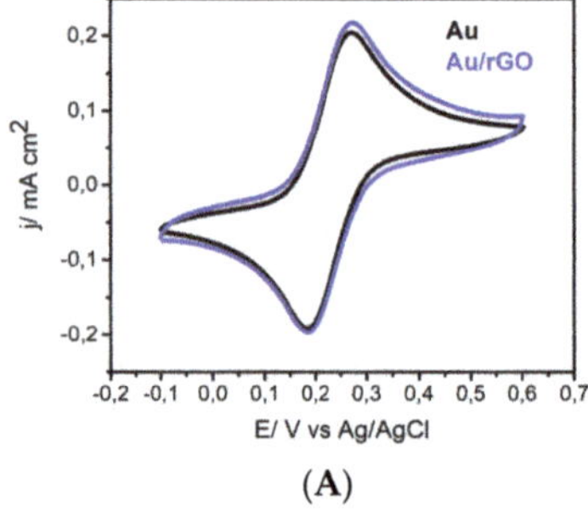
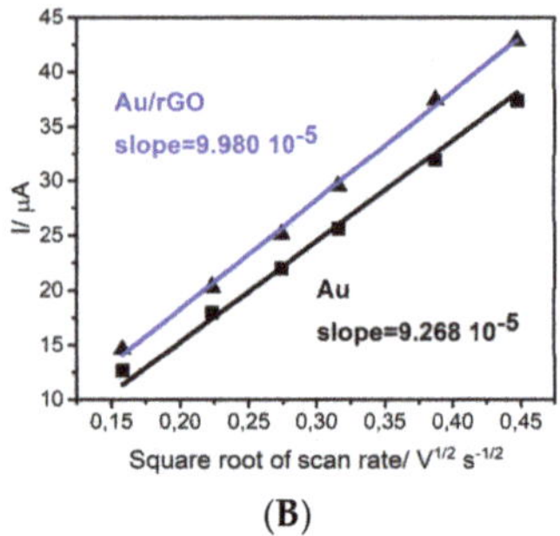

Figure 3. (**A**) Cyclic voltammograms recorded on gold thin film electrodes (black) and after EPD and further reduction with hydrazine using ferrocene methanol (1 mM)/PBS (0.1 M), scan rate = 100 mV s^{-1}. (**B**) Cyclic voltammograms recorded on gold thin film electrodes (black) and after EPD and further reduction with hydrazine using ferrocene methanol (1 mM)/PBS (0.1 M), scan rate = 100 mV s^{-1}.

3.2. Surface Modification Using Diazonium Electrografting

To validate the stability of the formed interface, efficient covalent modification based on the electroreduction of a triisopropylsilyl-protected ethynyl diazonium salt was performed (Figure 4A). Covalent surface modification of graphene-based materials using the electroreduction of aryldiazonium salts is a popular approach as it allows the introduction of different chemical groups. The major drawback of this approach is the difficulty to control the extent of the reaction—notably, to limit the reaction to the formation of a functional monolayer. The highly reactive nature of the formed aryl radical results in the formation of disordered polyaryl multilayers, which in the case of a sensor might limit the dynamic range of sensing. The introduction of bulky substitutions on the ArN_2^+ moieties limits radical addition reactions and allows the formation of ultrathin functional layers [26,31,32]. On the basis of this concept, the precursor 4-((triisopropylsilyl)ethylenyl)benzenediazonium tetrafluoroborate (TIPS-Eth-ArN$_2^+$) was used for the formation of an organic thin film on Au/rGO. The layer was obtained by potential cycling in a solution containing TIPS-Eth-ArN$_2^+$ (Figure 4B). As observed by cyclic voltammetry, the reduction peak of TIPS-Eth-ArN$_2^+$ at -0.2 V decreased rapidly after five cycles, indicating the blocking of the Au/rGO electrode. Compared to glassy carbon electrodes, the reduction peak shifted to more negative potentials [26]. The blocking properties of the layer was further investigated by recording the cyclic voltammogram of the oxidation of ferrocene methanol in water before and after the electrografting process (Figure 4C) and was found to be typical of a totally blocked electrode.

To have access to the acetylene function, the modified Au/rGO-TIP surface was immersed into tetrabutylammonium fluoride (TBAF, 0.05 M in THF) at room temperature. The success of the deprotection step was evidenced by the recovery of the ferrocene oxidation signal, being almost identical to that obtained on Au/rGO (not shown). The deprotected substrate was further treated by "click" chemistry (Huisgen 1,3-dipolar cyclization) using azidomethylferrocene, as first reported by Gooding et al. [33]. The success of the integration of ferrocene units onto the Au/rGO electrodes was seen from the presence of the Fe2p component in the XPS survey spectrum (Figure 5A). In the absence of copper catalyst, no click reaction took place and no Fe2p component could be recorded. Figure 5B shows the cyclic voltammogram of the ferrocene-modified electrode examined in electrolytic ethanol solution. A surface concentration of $\Gamma = 2.5 \times 10^{-10}$ mol cm^{-2} of bound ferrocene groups was derived from these measurements using Equation (4):

$$\Gamma = Q/nFA \tag{4}$$

where Q is the passed charge, n is the number of exchanged electrons (n = 1), F is the Faraday constant and A is the electroactive surface of the electrode determined as 0.136 cm^2. Considering the ferrocene molecules as spheres with a diameter of 6.6 Å, the theoretical maximum coverage for an idealized ferrocene monolayer can be estimated as $\Gamma = 4.8 \times 10^{-10}$ mol cm^{-2} [34]. In our case, about half of the full coverage was achieved.

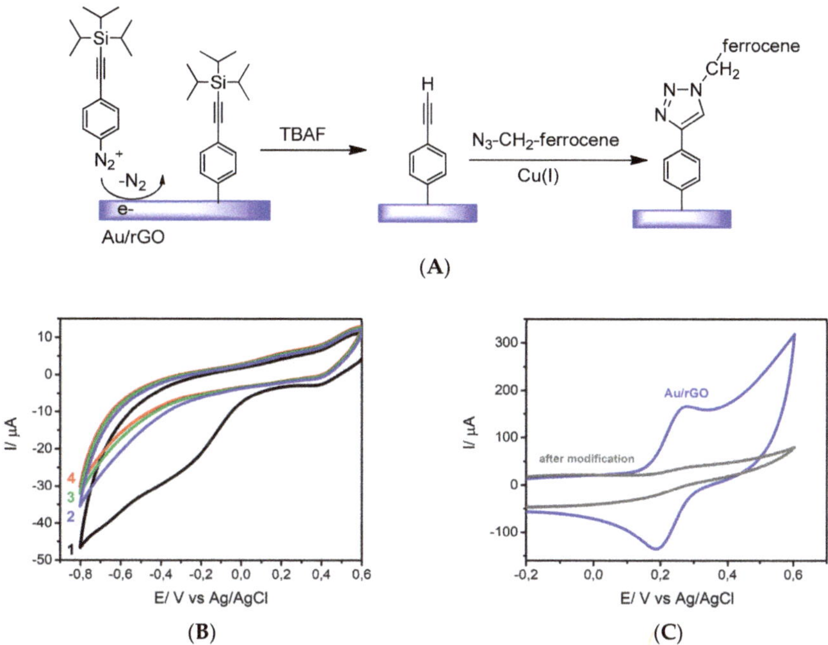

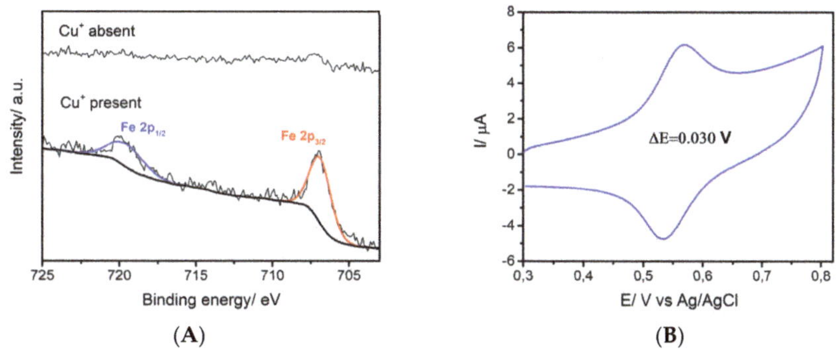

Figure 4. (**A**) Covalent modification of Au/rGO electrodes using a silyl-protected diazonium salt followed by tetrabutylammonium fluoride (TBAF)-based deprotection and "click" reaction using azidomethylferrocene as a model compound. (**B**) Electroreduction of 4-((triisopropylsilyl) ethylenyl)benzenediazonium tetrafluoroborate (TIPS-Eth-ArN$_2^+$) on Au/rGO, scan rate 50 mV s^{-1}. (**C**) Cyclic voltammograms of ferrocene methanol (1 mM)/0.1 M PBS of rGO/Au before (blue) and after modification with TIPS-Eth-ArN$_2^+$.

Figure 5. (**A**) High-resolution Fe2p X-ray photoelectron spectroscopy (XPS) core-level spectrum of ferrocene modified Au/rGO interfaces. (**B**) Cyclic voltammograms in acetonitrile/ NBu$_4$PF$_6$ 0.1 M), scan rate 100 mV s^{-1}.

4. Conclusions

In conclusion, by applying a low EPD voltage of 2.5 V, the deposition of partially reduced graphene oxide film can be formed on gold thin film electrodes in a homogenous, and mechanically and chemically stable manner. Further reduction of the films in hydrazine vapor results in reduced graphene oxide-coated substrates showing excellent electron transfer characteristics. The films proved to be of high robustness and could be further modified via the electroreduction of aryldiazonium salts. Using a protection–deprotection approach allows the immobilization of a functional monolayer which can be further functionalized by Cu(I)-catalyzed "click" chemistry. We have demonstrated that a densely packed ferrocenenyl monolayer can be efficiently integrated on these rGO-modified surfaces. Due to the versatility and mild conditions of the "click" chemistry reaction, a wide range of functional groups can be immobilized on such surfaces. These results pave the way for the use of this technology for the modification of more complex electrode configuration such as screen-printed or flexible integrated electrode arrays and the integration of surface ligands for sensing-related applications.

Author Contributions: V.M.: electrophoretic deposition and electrochemistry; P.A.: click chemistry; Y.L.: surface chemistry; H.H.: characterization; W.K.: editing of the draft; R.B.: characterization, editing of the draft; S.S.; original draft preparation, supervision of work.

Acknowledgments: General financial support from the Centre National de la Recherche Scientifique (CNRS), the University of Lille, the Hauts-de-France region and the EU through FLAG-ERA JTC2015 "Graphtivity" is acknowledged.

Conflicts of Interest: The authors declare no conflict of interest.

References

1. Reina, G.; González-Domínguez, J.M.; Criado, A.; Vázquez, E.; Bianco, A.; Prato, M. Promises, facts and challenges for graphene in biomedical applications. *Chem. Soc. Rev.* **2017**, *46*, 4400–4416. [CrossRef]
2. Szunerits, S.; Boukherroub, R. Graphene-based bioelectrochemistry and bioelectronics: A concept for the future? *Curr. Opin. Electrochem.* **2018**, *12*, 141–147. [CrossRef]
3. Ambrosi, A.; Chua, C.K.; Latiff, N.M.; Loo, A.H.; Wong, C.H.; Eng, A.Y.; Bonanni, A.; Pumera, M. Graphene and its electrochemistry—An update. *Chem. Soc. Rev.* **2016**, *45*, 2458–2493. [CrossRef] [PubMed]
4. Peña-Bahamonde, J.; Nguyen, H.N.; Sofia, K.; Rodrigues, D.F. Recent advances in graphene-based biosensor technology with applications in life sciences. *J. Nanobiotechnol.* **2018**, *16*, 75. [CrossRef] [PubMed]
5. Szunerits, S.; Boukherroub, R. Electrochemistry of graphene: The current state of the art. *Electrochemistry* **2013**, *12*, 211–242.
6. Szunerits, S.; Boukherroub, R. Graphene-based nanomaterials in innovative electrochemistry. *Curr. Opin. Electrochem.* **2018**, *10*, 24–30. [CrossRef]
7. Chekin, F.; Vasilescu, A.; Jijie, R.; Singh, S.K.; Kurungot, S.; Iancu, M.; Badea, G.; Boukherroub, R.; Szunerits, S. Sensitive electrochemical detection of cardiac troponin I in serum and saliva by nitrogen-doped porous reduced graphene oxide electrode. *Sens. Actuators B Chem.* **2018**, *262*, 180–187. [CrossRef]
8. Jijie, R.; Kahlouche, K.; Barras, A.; Yamakawa, N.; Bouckaert, J.; Gharbi, T.; Szunerits, S.; Boukherroub, R. Reduced graphene oxide/polyethylenimine based immunosensor for the selective and sensitive electrochemical detection of uropathogenic Escherichia coli. *Sens. Actuators B.* **2018**, *260*, 255–263. [CrossRef]
9. Wang, Q.; Vasilescu, A.; Wang, Q.; Coffinier, Y.; Li, M.; Boukherroub, R.; Szunerits, S. Electrophoretic Approach for the Simultaneous Deposition and Functionalization of Reduced Graphene Oxide Nanosheets with Diazonium Compounds: Application for Lysozyme Sensing in Serum. *ACS Appl. Mater. Interfaces* **2017**, *9*, 12823–12831. [CrossRef] [PubMed]
10. Subramanian, P.; Lesniewski, A.; Kaminska, I.; Vlandas, A.; Vasilescu, A.; Niedziolka-Jonsson, J.; Pichonat, E.; Happy, H.; Boukherroub, R.; Szunerits, S. Lysozyme detection on aptamers functionalized graphene-coated SPR interfaces. *Biosens. Bioelectron.* **2013**, *50*, 239–243. [CrossRef]
11. Subramanian, P.; Barka-Bouaifel, F.; Bouckaert, J.; Yamakawa, N.; Boukherroub, R.; Szunerits, S. Graphene-coated surface plasmon resonance interfaces for studing the interactions between cells and surfaces. *ACS Appl. Mater. Interfaces* **2014**, *6*, 5422–5431. [CrossRef] [PubMed]

12. Diba, M.; Garcia-Callastegui, A.; Klupp Taylor, R.N.; Pishbin, F.; Ryan, M.P.; Shaffer, M.S.P.; Boccaccini, A.R. Quantitative evaluation of electrophoretic deposition kinetics of graphene oxide. *Carbon* **2014**, *67*, 656–661. [CrossRef]

13. An, S.J.; Zhu, Y.; Lee, S.H.; Stoller, M.D.; Emilsson, T.; Park, S.; Velamakanni, A.; An, J.; Ruoff, R.S. Thin Film Fabrication and Simultaneous Anodic Reduction of Deposited Graphene Oxide Platelets by Electrophoretic Deposition. *J. Phys. Chem. Lett.* **2010**, *1*, 1259–1263. [CrossRef]

14. Besra, L.; Liu, M. A review on fundamentals and applications of electophoretic depositon (EDP). *Prog. Mater. Sci.* **2007**, *52*, 1–61. [CrossRef]

15. Ammam, M. Electrophoretic deposition under modulated electric fields: A review. *RSC Adv.* **2012**, *2*, 7633–7646. [CrossRef]

16. Ata, M.S.; Liu, Y.; Zhitomirsky, I. A review of new methods of surface chemical modification, dispersion and electrophoretic deposition of metal oxide particles. *RSC Adv.* **2014**, *4*, 22716–22732. [CrossRef]

17. Boccaccini, A.R.; Cho, H.; Roether, J.A.; Thoimasz, B.J.C.; Minay, E.J.; Shaffer, M.S.P. Electrophoretic deposition of carbon nanotubes. *Carbon* **2006**, *44*, 3149–3160. [CrossRef]

18. Chavez-Valdez, A.; Shaffer, M.S.P.; Boccaccini, A.R. Applications of Graphene Electrophoretic Deposition. A Review. *J. Phys. Chem. B* **2013**, *117*, 1502–1515. [CrossRef]

19. He, L.; Sarkar, S.; Barras, A.; Boukherroub, R.; Szunerits, S.; Mandler, D. Electrochemically stimulated drug release from flexible electrodes coated electrophoretically with doxorubicin loaded reduced graphene oxide. *Chem. Commun.* **2017**, *53*, 4022–4025. [CrossRef]

20. Kahlouche, K.; Jijie, R.; Hosu, I.; Barras, A.; Gharbi, T.; Yahiaoui, R.; Herlem, G.; Ferhat, M.; Szunerits, S.; Boukherroub, R. Controlled modification of electrochemical microsystems with polyethylenimine/reduced graphene oxide using electrophoretic deposition: Sensing of dopamine levels in meat samples. *Talanta* **2018**, *178*, 32–440. [CrossRef]

21. Maaoui, H.; Singh, S.K.; Teodorescu, F.; Coffinier, Y.; Barras, A.; Chtourou, R.; Kurungot, S.; Szunerits, S.; Boukherroub, R. Copper oxide supported on three-dimensional ammonia-doped porous reduced graphene oxide prepared through electrophoretic deposition for non-enzymatic glucose sensing. *Electrochim. Acta* **2017**, *224*, 346–354. [CrossRef]

22. Wang, Q.; Li, M.; Szunerits, S.; Boukherroub, R. Preparation of reduced graphene oxide-Cu composites through electrophoretic deposition: Application for nonenzyamtic glucose sensing. *RSC Adv.* **2015**, *5*, 15861–15869. [CrossRef]

23. Subramanian, P.; Niedziolka-Jonsson, J.; Lesniewski, A.; Wang, Q.; Li, M.; Boukherroub, R.; Szunerits, S. Preparation of reduced graphene oxide-Ni(OH)$_2$ composites by electrophoretic deposition: Application for non-enzymatic glucose sensing. *J. Mater. Chem. A* **2014**, *2*, 5525–5533. [CrossRef]

24. Wang, Q.; Vasilescu, A.; Wang, Q.; Li, M.; Boukherroub, R.; Szunerits, S. Electrophoretic approach for the modification of reduced graphene oxide nanosheets with diazonium compounds: Application for electrochemical lysozyme sensing. *ACS Appl. Mater. Interfaces* **2017**, *9*, 12823–12831. [CrossRef]

25. Pei, S.; Cheng, H.-M. The reduction of graphene oxide. *Carbon* **2012**, *50*, 3210–3228. [CrossRef]

26. Leroux, Y.R.; Fei, H.; Noel, J.-M.; Roux, C.; Hapiot, P. Efficient Covalent Modification of a carbon surface: Use of a silyl protecting group to form an active monolayer. *J. Am. Chem. Soc.* **2010**, *132*, 14039–14041. [CrossRef] [PubMed]

27. Casas-Solva, J.M.; Vargas-Berenguel, A.; Captian-Vallvey, L.F.; Santoyo-Gonzalez, F. Convenient Methods for the Synthesis of Ferrocene−Carbohydrate Conjugates. *Org. Lett.* **2004**, *6*, 3687–3690. [CrossRef]

28. Park, S.; An, J.; Jung, I.; Piner, R.D.; An, S.J.; Li, X.; Velamakanni, A.; Ruoff, R.S. Colloidal suspensions of highly reduced graphene oxide in a wide variety of organic solvents. *Nano Lett.* **2009**, *9*, 1593–1597. [CrossRef]

29. Hamaker, H.C. Formation of a deposite by electrophoresis. *Trans. Farady Soc.* **1940**, *35*, 279–287. [CrossRef]

30. Ngamchuea, K.; Eloul, S.; Tschulik, K.; Compton, R.G. Planar diffusion to macro disc electrodes—What electrode size is required for the Cottrell and Randles-Sevcik equations to apply quantitatively? *J. Sol. State Electrochem.* **2014**, *18*, 3251–3257. [CrossRef]

31. Nielsen, L.T.; Vase, K.H.; Dong, M.; Besenbacher, F.; Pedersen, S.U.; Daasbjerg, K. Electrochemical Approach for Constructing a Monolayer of Thiophenolates from Grafted Multilayers of Diaryl Disulfides. *J. Am. Chem. Soc.* **2007**, *129*, 1888–1889. [CrossRef]

32. Combellas, C.; Jiang, D.-E.; Kanoufi, F.; Pinson, J.; Podvorica, F.I. Steric Effects in the Reaction of Aryl Radicals on Surfaces. *Langmuir* **2009**, *25*, 286–293. [CrossRef] [PubMed]
33. Ciampi, S.; Le Saux, G.; Harper, J.B.; Gooding, J.J. Optimization of Click Chemistry of Ferrocene Derivatives on Acetylene-Functionalized Silicon(100) Surfaces. *Elecroanalysis* **2008**. [CrossRef]
34. Chidsey, C.E.D.; Bertozzi, C.R.; Putvinski, T.M.; Mujsce, A.M. Coadsorption of ferrocene-terminated and unsubstituted alkanethiols on gold: Electroactive self-assembled monolayers. *J. Am. Chem. Soc.* **1990**, *112*, 4301–4306. [CrossRef]

 *surfaces*

Article

Effects of the Interfacial Structure on the Methanol Oxidation on Platinum Single Crystal Electrodes

Mohammad Ali Kamyabi [1,2], **Ricardo Martínez-Hincapié** [2], **Juan M. Feliu** [2,*] and **Enrique Herrero** [2,*]

[1] Department of Chemistry, Faculty of Science, University of Zanjan, Zanjan 45371-38791, Iran; makamyabi@gmail.com

[2] Instituto de Electroquímica, Universidad de Alicante, Apdo. 99, E-03080 Alicante, Spain; ricuardomartin@gmail.com

Received: 11 February 2019; Accepted: 1 March 2019; Published: 12 March 2019

Abstract: Methanol oxidation has been studied on low index platinum single crystal electrodes using methanol solutions with different pH (1–5) in the absence of specific adsorption. The goal is to determine the role of the interfacial structure in the reaction. The comparison between the voltammetric profiles obtained in the presence and absence of methanol indicates that methanol oxidation is only taking place when the surface is partially covered by adsorbed OH. Thus, on the Pt(111) electrode, the onset for the direct oxidation of methanol and the adsorption of OH coincide. In this case, the adsorbed OH species are not a mere spectator, because the obtained results for the reaction order for methanol and the proton concentrations indicate that OH adsorbed species are involved in the reaction mechanism. On the other hand, the dehydrogenation step to yield adsorbed CO on the Pt(100) surface coincides with the onset of OH adsorption on this electrode. It is proposed that adsorbed OH collaborates in the dehydrogenation step during methanol oxidation, facilitating either the adsorption of the methanol in the right configuration or the cleavage of the C—H bond.

Keywords: methanol oxidation; platinum single crystals; pH and concentration effects; adsorbed OH

---

1. Introduction

The methanol (MeOH) oxidation reaction has attracted significant attention in the last decades, both from a fundamental and applied point of view [1–5]. From the fundamental approach, methanol is a small organic molecule and the knowledge acquired by studying its oxidation mechanism can be later transferred to more complex organic molecules. In fact, its mechanism can serve as a model for the oxidation of more complex alcohols, such as ethanol or ethylene glycol. On the applied side, methanol can be used as a fuel in fuel cells. When compared with hydrogen as fuel, methanol presents several practical advantages: It is liquid at room temperature with a high volumetric energy density and does not require the deployment of new distribution channels for its use. However, the anodic over-potentials are significantly higher than those measured for the hydrogen oxidation reaction. For this reason, the development of more effective electrocatalysis for the oxidation reaction is needed if methanol is going to be employed as fuel. The intelligent design of more efficient electrocatalysts requires a deep knowledge of the reaction mechanism.

Among all pure metals, platinum is the best electrocatalyst for the oxidation reaction. In the complete oxidation, six electrons are exchanged according to:

$$CH_3OH + H_2O \rightarrow CO_2 + 6H^+ + 6e \tag{1}$$

It is now generally accepted that the oxidation mechanism takes place following a dual path mechanism [1]. In one of the paths, adsorbed CO is formed, which is considered a poisoning intermediate, because it is strongly adsorbed on the surface and its oxidation requires high potentials. On the other hand, the second path avoids the formation of CO and goes through an active intermediate [1]. In a reaction in which six electrons are exchanged, several intermediates are involved, and as has been proposed, the two paths may share some intermediates [6,7]. As happens in any reaction in which adsorbed intermediates are involved, the reactivity is very dependent on the surface structure [8–14]. Moreover, not only the surface structure, that is, the local arrangement of the surface atoms, affects reactivity, but also the interfacial structure and composition. First, the comparison between ultra-high vacuum and electrochemical environments indicates that the bonding mode of the methanol molecule is clearly affected by the presence of the interfacial water [10] because the proposed intermediates and their binding modes are different. Second, the presence of specifically adsorbed anions on the electrode surface inhibits the reactivity [11]. Thus, currents in the presence of sulfate or phosphate are significantly smaller than those measured in perchloric acid solutions. Additionally, other factors that affect the interfacial structure should be considered, such as, solution pH, and electrode charge. In all the cases, all these parameters modify the interaction between the methanol molecule and the surface, which in turn controls the reaction mechanism.

In order to have a deeper knowledge of the effects of interfacial on the methanol oxidation reaction, experiments in solutions with pH values ranging between one and five have been carried out. Due to the large impact of the specific adsorption of anions on the methanol oxidation currents [11], electrolyte solutions should have an adequate buffering capacity without containing anions that can adsorb strongly on the surface. For that purpose, mixtures of perchloric acid and fluoride have been used because perchlorate and fluoride anions do not adsorb strongly on the electrode surface and the pKa of HF/F- is 3.15 [15,16]. Furthermore, to better explore the relationship between the interfacial structure and the reactivity, methanol concentrations between 10^{-4} and 10^{-1} M have been used. The low concentration regime will allow detecting the modification of the adsorption behavior of the electrode due to the presence of methanol and relating them with the oxidation mechanism and its selectivity.

2. Materials and Methods

Low index platinum single crystal electrodes, namely Pt(111), Pt(100), and Pt(110), were fabricated from small single crystal beads (ca. 2.5 mm diameter). These beads were oriented, cut, and polished until mirror finish using the procedure described by Clavilier [17,18]. Prior to any experiment, the electrodes were cleaned by flame annealing, cooled down in a H_2/Ar atmosphere and protected with water in equilibrium with this gas mixture [19]. The protected electrodes were immersed in the electrolyte solution at a controlled potential (0.1 V) in meniscus configuration so that only the surface with the desired orientation is in contact with the solution. For the Pt(111) and Pt(100), this procedure leads to the formation of a well-ordered surface with the nominal orientation [20,21]. Due to the reconstruction phenomena undergone by the Pt(110) surfaces, which is strongly affected by the treatment, the surface presented here corresponds to the (1 × 2) or (1 × 1)/(1 × 2) structure with major (1 × 2) contribution [22,23]. Thus, the observed behavior should correspond with that expected for the (1 × 2) surface, although contributions from the (1 × 1) cannot be discarded. The cleanliness of the electrode surface and the solutions were checked by voltammetry in the absence of methanol. The characteristic voltammetric profiles of the different electrodes, which are stable upon cycling when clean solutions are used, were obtained.

Experiments were carried out at room temperature, 22 °C, in classical two-compartment electrochemical cells de-aerated by using Ar (N50, Air Liquide, Paris, France, in all gases used), including a large platinum counter electrode and a Reversible Hydrogen (N50) Electrode (RHE) as reference. The solutions were prepared using concentrated $HClO_4$ (for analysis, Merck KGaA, Darmstadt, Germany), NaF (99.99%, Suprapur, Merck KGaA, Darmstadt, Germany), $KClO_4$ (Suprapur, Merck KGaA, Darmstadt, Germany), and ultrapure water (Purelab Ultra, Elga, High Wycombe, United Kingdom). Solutions with a pH between two and five were prepared using mixtures of NaF and $HClO_4$ so that the buffering capacity is high enough to maintain the interfacial pH constant. Voltammograms were recorded using a potentiostat (EA161, eDAQ, Sydney, Australia) and a digital recorder (ED401, eDAQ, Sydney, Australia) in a hanging meniscus configuration.

3. Results

3.1. Pt(111) Electrode

Figure 1 shows the voltammetric profiles of the Pt(111) electrode in 0.1 M $HClO_4$ with two different methanol concentrations: 10^{-2} and 10^{-4} M and the comparison with that recorded in the absence of methanol for the same electrolyte solution. The comparison between the blank voltammogram of the Pt(111) electrode and those obtained in the presence of methanol allows correlating the interfacial structure with the reactivity. Moreover, relevant information regarding the oxidation mechanism can be obtained from this comparison, depending on the examined region. As aforementioned, it is generally accepted that methanol oxidation reaction occurs through a dual path mechanism, in which a path involves the formation of adsorbed CO, a poisoning intermediate, and a second path going through an active intermediate. Due to the high number of electrons exchanged in the process, both paths should be interconnected and can share some intermediates leading to a complex mechanism [6,7]. The voltammetric profile in the supporting electrolyte can be divided into three regions, hydrogen adsorption region below 0.4 V, the OH adsorption region above 0.6 V, and the double lager region between 0.4 and 0.6 V. The analysis of the modifications of the profiles in these regions will allow obtaining relevant data for the oxidation mechanism.

The profile measured for 10^{-2} M has the typical shape of those obtained at higher methanol concentrations, with an effective onset potential of ca. 0.6 V [14,24], which coincides with the onset of the OH adsorption process. The large currents measured for this concentration prevents detailed analysis of the effects of the surface composition in the methanol oxidation in this region. To diminish oxidation currents and to establish whether the OH adsorption process is affected by the reaction, currents were recorded for 10^{-4} M methanol. For this concentration (Figure 1B), modifications of the blank voltammetry recorded in the absence of methanol are only observed in the region where OH is adsorbed. Thus, in the positive scan direction, a new peak appears at ca. 0.72 V, apparently superimposed with the original OH adsorption signal, because some of the characteristic features of this region, such as the spike at 0.80 V, are still visible. On the other hand, the profile in the negative scan direction shows complex behavior, which is clearly a consequence of the addition of two different signals: the negative currents associated with the reductive desorption of adsorbed OH and the positive current of the methanol oxidation reaction. To determine whether OH adsorption process is affected by methanol, the blank voltammetric profile obtained in the supporting electrolyte has been subtracted from that measured for the 10^{-4} M methanol concentration (Figure 1C). As can be seen, the resulting voltammetric profile has the same shape as that recorded with larger methanol concentrations, and the only difference is the absolute values of the currents, which are proportional to the methanol concentration. It should be stressed that a spike at 0.80 V in the blank voltammogram, which has been associated to changes in the OH adsorption mode depending on the water structure, is very sensitive to the actual composition of the solution [25]. The presence of traces of species in concentrations well below 10^{-6} M alters the currents in this region. In this case, the subtraction of the two profiles shows a bipolar peak in this region that is only due to the change in the composition of the interface, but not related to the oxidative process of methanol. Then, the subtracted voltammetric profile demonstrates that the OH adsorption process between 0.6 and 0.9 V is also occurring in the presence of methanol and that the interfacial structure of the electrode is not significantly affected by the presence of methanol.

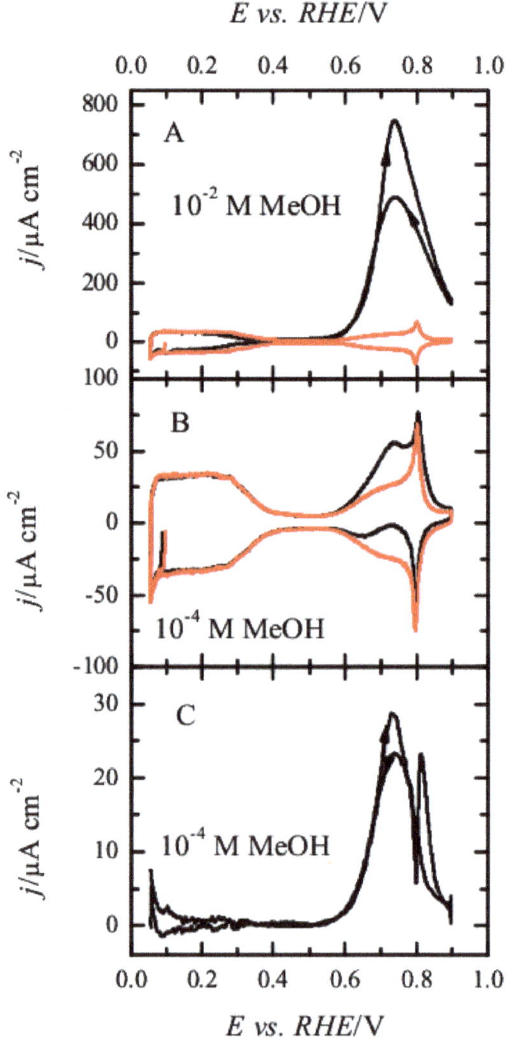

Figure 1. Voltammetric profiles of the Pt(111) electrode in (**A**) 0.1 M HClO$_4$ + 10^{-2} M MeOH and (**B**) 0.1 M HClO$_4$ + 10^{-4} M MeOH. The red trace shows the profile in absence of MeOH. (**C**) Subtraction of the profiles in panel B. Scan rate: 50 mV s^{-1}.

These results clearly show that significant oxidation currents are only recorded in the region where OH is adsorbed, implying that the oxidation of methanol is taking place on a surface partially covered by OH. Moreover, during the spike at 0.80 V, there is a change in the reactivity and the completion of the OH layer at 0.85 V leads to the inhibition of the methanol oxidation reaction. Thus, according to these results, adsorbed OH seems to be required for the oxidation of methanol, but also some free sites where the methanol molecule could interact with the electrode.

The inspection of the hydrogen adsorption and double layer regions can be used to provide information on the type of intermediates that are formed during the oxidation reaction. Probably, CO is the most important intermediate, because its oxidation, which occurs through a well-characterized Langmuir-Hishenlwood mechanism, requires the presence of adsorbed OH on the surface [26]. Thus, rates for the CO oxidation reaction on Pt are only significant above 0.7 V, and if CO is formed below this potential value, adsorbed CO molecules are accumulated on the surface [12,14,27]. The accumulation of CO on the surface blocks hydrogen adsorption and the hydrogen adsorption charge recorded below 0.4 V diminishes [28]. As can be seen, in 10^{-4} M methanol, the accumulation of CO is negligible because the profile in the hydrogen region for this concentration overlaps with that recorded in the absence of methanol. This fact implies that if CO is formed during the oxidation, it is oxidized at the same rate that is produced and the accumulation during the time of the experiment is negligible. For higher concentrations, there is a small diminution in the hydrogen adsorption charge, implying that some CO has been accumulated on the electrode surface. However, the amount of CO accumulated is very small. In fact, chronoamperometric measurements indicate that CO formation takes place exclusively above 0.5 V and that the rate for this surface is the lowest of all the single crystal electrodes studied [14].

Additionally, other possible intermediates may be formed and adsorb on the electrode surface giving rise to characteristic signals. One of the possible intermediates in the oxidation of methanol is formic acid. However, the observed behavior is not compatible with the formation of a measurable amount of formic acid. In the acidic media, formic acid adsorbs on the platinum electrode as formate in the bidentate configuration. The adsorption of formate modifies the voltammogram so that the typical shape corresponding to OH adsorption above 0.6 V disappears and a new signal between 0.4 and 0.6 V can be observed [29]. Moreover, formic acid is readily oxidized on the Pt(111) electrode above 0.35 V [30,31]. None of these new signals appear on the recorded voltammograms, even for the highest recorded concentration (0.1 M). Thus, the measured profile between 0.4 and 0.6 V in the presence of methanol matches that recorded in its absence, implying that formic acid is not being formed at significant concentrations during the oxidation of methanol on this electrode surface in the present conditions. It should be mentioned that a small fraction of formic acid has been detected by HPLC analysis after long electrolysis time (ca. 15 min) and with high methanol concentrations [32]. Clearly, the electrolysis time and the solution concentration affect the product distribution. On the other hand, the presence of methyl formate detected by mass spectroscopy [13,27] appears to be a consequence of the attack of a methanol molecule in solution to some intermediate species adsorbed on the surface and not as a result of the esterification reaction between solution methanol and formic acid produced in the partial oxidation of methanol [33,34].

From all these results, it can be concluded that the absence of significant modifications in the interfacial structure, in the presence of methanol (aside that relate to the small amount of CO accumulated on the surface), indicates that the different adsorbed intermediates of the reaction are not strongly adsorbed and their interaction with the surface must be weak. In fact, when anions are strongly adsorbed, such as sulfate or phosphate added to the supporting electrolyte, the oxidation currents diminish significantly, because methanol adsorbates cannot compete for the adsorption sites with the strongly adsorbed sulfate or phosphate [14,24]. However, the presence of adsorbed OH seems to be required for the oxidation. To further prove this relationship between adsorbed OH and methanol oxidation, the reaction order for methanol was calculated. For this, the current densities measured at 0.65 and 0.75 V after subtraction of the current density at these potentials in the blank voltammogram are plotted vs. the electrode potential in a double logarithmic graph (Figure 2). As can be seen, a linear relationship is observed for 0.65 V with a slope of 0.53. For 0.70 V, there is a clear deviation from linearity, especially for 0.1 M methanol, where currents are smaller than those predicted from the behavior observed for low concentrations, and the measured slope is 0.62. The fractional reaction orders and the dependence with the electrode potential and OH coverage clearly indicate that the reaction mechanism is complex, and OH should be involved in the process.

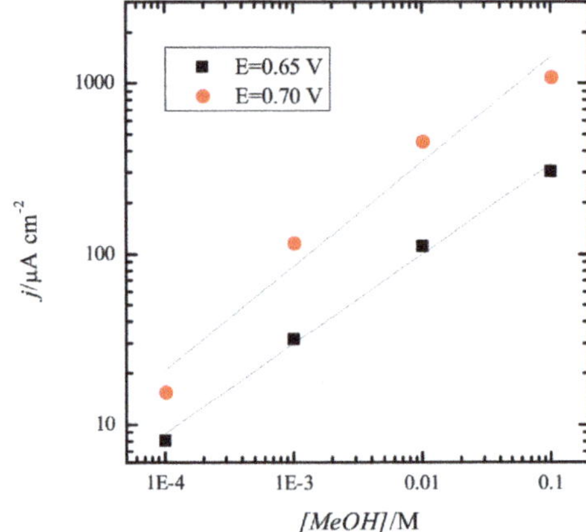

Figure 2. Double logarithm plot for the current density measured at 0.65 and 0.70 V in 0.1 M $HClO_4$ vs. MeOH concentration.

Additionally, the comparison with the observed behavior for formic acid will also provide some additional insight. The onset of formic acid oxidation (in the absence of previously formed CO) matches the onset of formate adsorption because the adsorption of formate in the favorable configuration is required for the oxidation [35,36]. This adsorption process starts at a potential where hydrogen is desorbed and the adsorption strength of formate is larger than that of OH because OH adsorption is blocked in its presence. On the other hand, the results presented here indicate that the interaction of methanol (and that of the intermediates except for CO) is weak. For this reason, it would be expected that the interaction of methanol with the surface was more favorable in the double layer region because there are no other adsorbed species on the surface. However, the currents in this region are negligible and only where OH is adsorbed, oxidation currents are recorded, reinforcing the relationship between methanol oxidation and adsorbed OH. Additionally, it should be mentioned that the detection of soluble intermediate products increases when sulfuric acid is used as supporting electrolyte [13,27,32]. The presence of adsorbed sulfate prevents the adsorption of OH and the possible intermediates, and thus, the oxidation of methanol becomes more difficult and increasing amounts of partial oxidation products are formed.

Regarding the role of adsorbed OH in this process, previous results with isotopically labeled methanol and water indicate that the cleavage of the C-H bond is involved in the rate determining step [24]. Then, adsorbed OH should be involved in positioning the methanol molecule in the right configuration for the cleavage and/or in interacting with one of the hydrogen atoms of the methanol molecule in such a way that the activation energy of this (these) step(s) is (are) diminished. This interpretation agrees with the higher amount of partial oxidation products obtained when the adsorption of OH is hindered by the presence of specifically adsorbed anions. This is not the only role of OH in the oxidation mechanism because it is also required for the oxidation of CO in the Langmuir-Hinshelwood mechanism [26,37,38].

Additional effects of the interfacial structure in the oxidation reaction can be obtained by changing the solution pH. Two interfacial parameters are affected, the electrode surface charge and the water structure. The equilibrium potential for the methanol oxidation reaction shifts 59 mV per pH unit, as the RHE scale does. For this reason, it is expected that the methanol oxidation reaction occurs in the same potential region in the RHE scale. However, as the solution pH increases, the electrode potentials for the oxidation in a pH-independent potential scale, such as the Standard Hydrogen Scale (SHE) scale, shift to more negative potential values. Since the potential of zero free charge of the Pt(111) surface is constant and pH-independent (0.28 V vs. SHE [39]), as the pH increases, the electrode charge becomes more negative, and thus, affecting the interaction of the species with the surface. Additionally, water structure also changes, altering the interaction of the molecules with water, which in turn can influence the reactivity.

To study those effects, the methanol oxidation reaction was investigated in different pH solutions up to a pH value of 5 (Figure 3). As can be seen, the qualitative behavior of the electrode is independent of the solution pH, and the oxidation of methanol is only taking place in the region where OH is adsorbed. Thus, at 10^{-4} M methanol, an oxidation wave is observed as superimposed with the OH adsorption in the positive scan direction, whereas in the negative direction, the profile is more complex, as happens for pH = 1. Additionally, the profile in the hydrogen region overlaps with the measured in the absence of methanol, which implies that CO is not accumulated on this surface. For higher concentrations, the general shape of the voltammogram in methanol containing solutions do not change with the pH, showing a broad wave, whose onset potential coincides with that of OH adsorption and currents in the negative scan direction, which are slightly smaller than those recorded in the positive direction. However, a detailed analysis shows some modifications with pH, as discussed below.

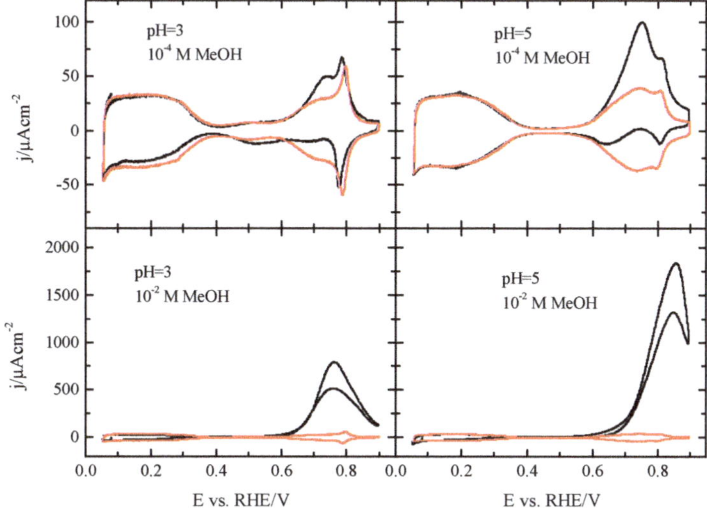

Figure 3. Voltammetric profiles of the Pt(111) electrode in pH = 3 (left column) and pH = 5 (right column) with different concentrations of MeOH. The red trace shows the profile in absence of MeOH. Scan rate: 50 mV s^{-1}.

The detailed analysis of the different effects require that the profiles at different pH values are compared. Figure 4 shows the voltammetric profiles of the Pt(111) electrode in the presence and absence of methanol for the different pH values. As can be seen, the changes in the blank voltammetric profile are small, especially for pH values lower than four. The onset for the OH adsorption is the same and the only major modifications are observed for the spike, which tends to disappear and as the solution pH increases, in a clear evolution to the profile recorded in alkaline media [40]. Despite these minimal changes, the oxidation of methanol is clearly affected by the pH. As shown in Figure 4B, the onset of the oxidation displaces to more positive potentials in the RHE scale and the current increases. Clearly, the changes in the interfacial properties are affecting this reaction. Additional information on that can be obtained calculating the reaction order for protons. For an irreversible reaction, reaction orders should be calculated at a constant potential value, in this case, in the SHE scale. Thus, when plotting the measured current at 0.5 V vs. SHE as a function of the pH in a double logarithm scale, a linear relationship is obtained in which the slope is ca. 0.45 (Figure 5). This value implies that the reaction order with respect to the proton concentration is –0.45. While the fractional orders are difficult to explain with simple mechanisms, the straightforward interpretation would be that protons are released before the rate-determining step. This observed behavior is in good agreement with the isotopic effect observed [11], which also indicate that the release of protons is involved in the rate determining step. An additional effect that should be taken into account when analyzing this behavior is the effect of the surface electrode charge. Since the potential of zero free charge is 0.28 V vs. SHE, as the pH increases, the region in which methanol oxidation takes place moves from positive charge to neutral values. Thus, the interaction between the species involved in the reaction (OH and methanol) and the surface is modified and affects the reaction rates. In this sense, for the oxygen reduction reaction, it has been proposed that higher rates obtained in neutral values are related to the surface charge values [41]. For neutral pH values, the charge of the surface is neutral in the onset of the ORR. The observed behavior for methanol shows a very similar dependence.

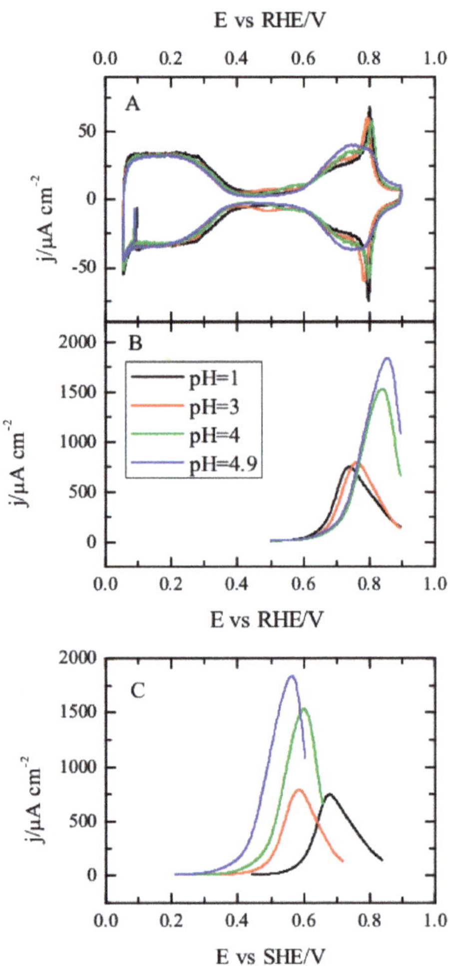

Figure 4. Dependence of the voltammetric profile of the Pt(111) with pH (**A**) in absence of methanol vs. RHE, (**B**) in 10^{-2} M MeOH vs. RHE and (**C**) in 10^{-2} M MeOH vs. SHE. For methanol oxidation, only the positive scan direction is shown. Scan rate 50 mV s^{-1}.

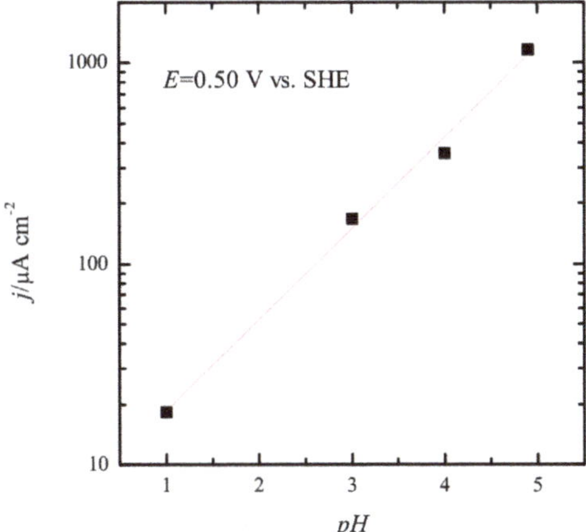

Figure 5. Current density for the oxidation of methanol at 0.5 V vs. SHE as a function of the solution pH.

3.2. Pt(100) and Pt(110) Electrodes

The major difference between the behavior of the Pt(111) electrode and the other two basal planes is the presence/absence of a pure double layer region where no specific adsorption occurs. For the Pt(110) and Pt(100) electrodes, OH is readily adsorbed at low potentials when hydrogen is being desorbed [23,42–45], and thus there is no pure double layer region. Then, any molecule that interacts with the surface should either compete with hydrogen or OH for the adsorption sites. Additionally, the overlapping of the two regions prevents the determination of the potential of zero free charge.

For the Pt(100) electrode (Figure 6), hydrogen is adsorbed on the surface between 0.05 and 0.45 V and the OH adsorption occurs in a broad wave between 0.45 and 0.7 V, just after the desorption of hydrogen from the surface [43,45]. In the presence of 10^{-4} M methanol, the only significant modification is the appearance of a peak at ca. 0.65 V in the positive scan direction, and a small diminution in the hydrogen adsorption charge. Unlike the Pt(111) electrode, this peak is only visible in the positive scan direction, and thus should be related to the oxidation of some CO molecules that have been accumulated on the electrode surface at low potentials. However, the amount of CO accumulated on the surface is very small and negligible currents are recorded in the negative scan direction. As the methanol concentration increases, this CO oxidation peak increases in charge and shifts to more positive potentials. In parallel, hydrogen adsorption charge below 0.45 V diminishes due to the blockage of hydrogen adsorption sites by CO. Additionally, a second oxidation wave after this peak CO emerges both in the positive and negative scan directions. This wave shows a hysteresis between both scan directions, which is clearly visible for the solution containing 10^{-2} M methanol, and should be related to the accumulation of CO in the positive scan direction at potentials below 0.6 V. Thus, the presence of adsorbed CO in the positive scan direction hinders the oxidation of additional molecules of methanol on the surface. After the oxidation of CO, methanol molecules can interact with the surface and give additional oxidation current contributions. In the negative scan direction, any CO formed during the reaction can be readily oxidized at potentials higher than 0.6 V, giving rise to the hysteresis between both scan directions.

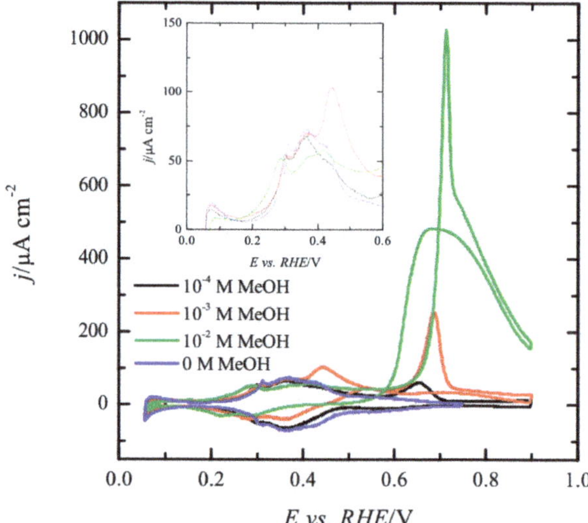

Figure 6. Voltammetric profiles of the Pt(100) electrode in 0.1 M HClO$_4$ with different MeOH concentrations. Inset: magnification of the positive scan direction in the region between o and 0.6 V

The process of CO formation on the electrode surface can be followed for the 10^{-3} M methanol concentration (Figure 7). As can be seen, the voltammogram for the first scan in the presence of 10^{-3} M methanol overlaps with that recorded in the absence of methanol until 0.45 V, implying that no CO has been accumulated on the surface up to this potential value. Coinciding with the onset of OH adsorption, some additional currents are observed. These currents should correspond to the formation of CO from methanol according to the reaction

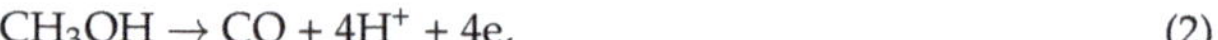

$$CH_3OH \rightarrow CO + 4H^+ + 4e, \tag{2}$$

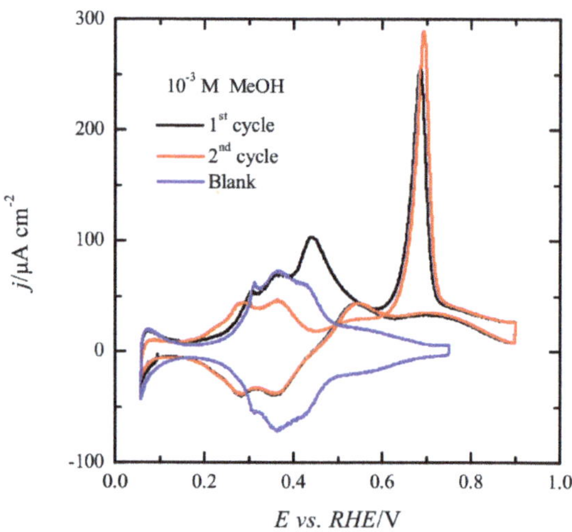

Figure 7. First and second voltammetric scan for the Pt(100) electrode in 0.1 M HClO$_4$ + 10^{-3} M MeOH. The blue trace shows the profile in absence of MeOH. Scan rate: 50 mV s^{-1}.

In fact, the analysis of the transient currents shows that CO is formed between 0.45 and 0.6 V [14], implying that the formation of CO is only taking place in the region where OH is adsorbed. In the negative scan direction, some additional currents are observed in this region, which are also related to the formation of CO, as revealed by the significant diminution of the hydrogen adsorption charge observed below 0.4 V. After this first scan, the second scan does not show the currents at 0.45 V, because CO is already adsorbed on the surface. It should be noted that the following scans are identical to the second scan. For the 10^{-2} M methanol concentration, the process of CO formation is not as visible as in 10^{-3} M (inset of Figure 6). While currents are recorded between 0.4 and 0.6 V, the peak is not well developed. It should be highlighted that, for this concentration, the results are very dependent on how the meniscus is formed. This fact implies that CO is being formed during the immersion of the electrode and the meniscus formation. While the electrode is immersed at a controlled potential (0.1 V), a few ms are required to establish the imposed potential. Due to the higher concentrations and the concomitant higher reaction rates, CO is being formed during this process. As can be seen in Figure 6 for this concentration, hydrogen desorption states have been displaced to lower potential values and hydrogen desorption charge is smaller, which is the characteristic behavior of a Pt(100) surface partially covered by CO [46]. Thus, a lower amount of CO can be formed in the positive scan, resulting in a less defined peak for this process.

Regarding the behavior of the methanol oxidation reaction on the Pt(100) electrode in different pH solutions (Figure 8), the general behavior is the same as that recorded in 0.1 M HClO4. In the positive scan direction, the peak associated with the oxidation of the accumulated CO can be observed, followed by the presence of a second oxidation wave, which is also visible in the negative scan direction. In general, no significant variation of the currents is observed, and the measured currents in the broad wave in the negative scan direction are almost independent of the pH. The observed changes, especially in the shape of the peak associated with the oxidation of CO and the broad wave are probably related to the changes in the OH adsorption mode. The voltammograms in absence of methanol show a progressive evolution from the profile measured at pH = 1 (Figure 7) to that obtained in alkaline solutions [47]. While the OH coverage is not significantly affected by the pH, the differences in the voltammograms reflect that the energetics of the OH adsorption process is altered by pH [47]. If, as proposed, OH involved in the reaction, the changes in the energetics should imply changes in the kinetics, that in this cases are translated in changes of the voltammetric shape. On the other hand, a reaction order cannot be obtained using this wave due to experimental problems. In this case, the raising part of the curve in the negative scan direction (that appearing between 0.6 and 0.7 V) should be used because it is not affected by the presence of adsorbed CO or any other inhibition process related to the formation of a compact OH layer. However, this region is very is very narrow and when curves are transformed in the SHE scale, the regions for the different pH do not overlap, preventing the accurate measurement of the current for the different pH values at a constant potential vs. SHE. Nevertheless, the observed behavior is similar to that observed for oxygen reduction, where currents are almost independent of the solution pH [41].

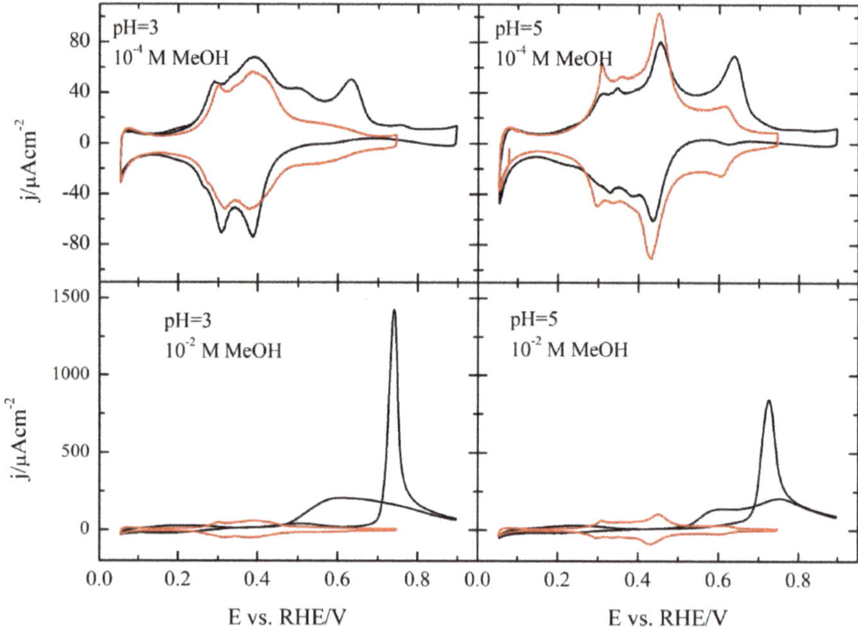

Figure 8. Voltammetric profiles of the Pt(100) electrode in pH = 3 (left column) and pH = 5 (right column) with different concentrations of MeOH. The red trace shows the profile in absence of MeOH. Scan rate: 50 mV s^{-1}.

For the Pt(110) electrode, the voltammogram for the oxidation of methanol shows a wave with a complex shape in the positive scan, whose onset is located at ca. 0.5 V in 10^{-4} M methanol, whereas, in the negative scan direction, the currents are almost negligible (Figure 9). These small currents cannot be assigned to the formation of adsorbed CO since the voltammetric profile in the low potential region overlaps with that measured in the absence of methanol. It should be noted that the signals appearing in the blank voltammogram between 0.06 and 0.30 V contain contributions both from hydrogen and OH adsorption. Then, the low currents recorded in the negative scan direction should be assigned to the oxidation of the surface, which is an irreversible process that occurs for this electrode above 0.8 V. As the concentration increases, the current also increases. However, the peak current difference between the positive and negative scan directions remains, due to the oxidation of the surface. It should be stressed that the oxidation of methanol is taking place, also on this electrode, on a surface partially covered by OH, reinforcing the connection between adsorbed OH and methanol oxidation. On the other hand, the solution pH effect on this surface is very similar to that described for the Pt(100) electrode. It should be noted that the voltammograms for this surface are very sensitive to the actual structure of the electrode, likely related to the fraction of the surface having the (1 × 2) reconstruction. This ratio depends, in turn, on the annealing conditions [23], so that the measured currents for the methanol oxidation reaction for this electrode show large variability.

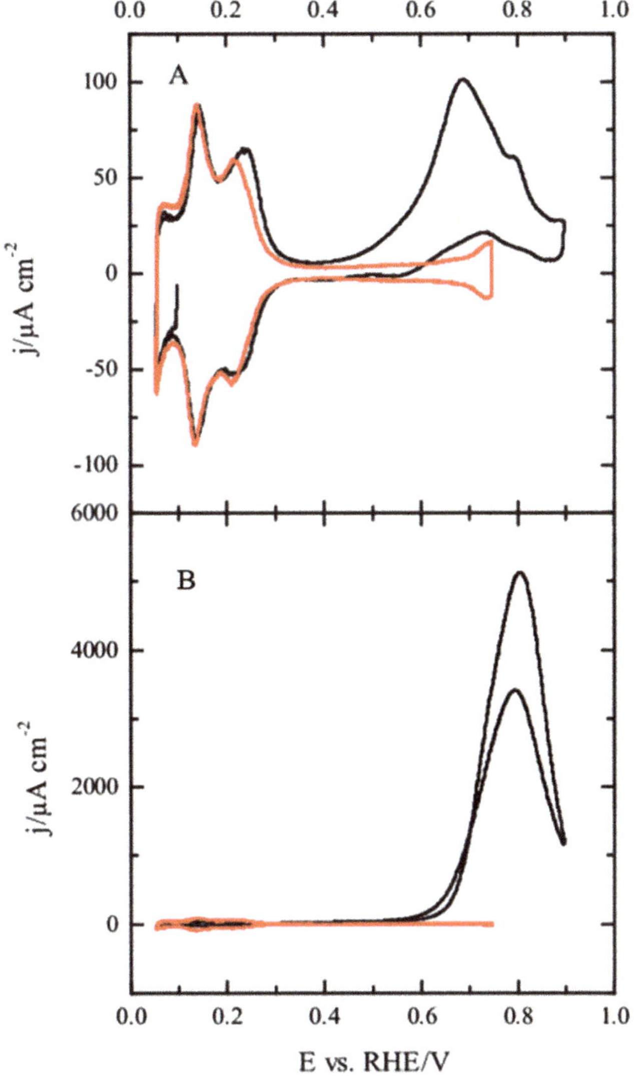

Figure 9. Voltammetric profiles of the Pt(110) electrode in (**A**) 0.1 M HClO$_4$ + 10^{-4} M MeOH and (**B**) 0.1 M HClO$_4$ + 10^{-4} M MeOH. The red trace shows the profile in absence of MeOH. Scan rate: 50 mV s^{-1}.

4. Conclusions

The use of diluted methanol solutions with different pH has allowed establishing relationships between the interfacial structure and the reactivity. First, the results indicate that methanol oxidation is only taking place when the surface is partially covered by adsorbed OH. Thus, the dehydrogenation step to yield adsorbed CO on the Pt(100) surface starts on the same electrode potential for the onset of OH adsorption on this electrode. On the other hand, on the Pt(111) electrode, where CO is not accumulated, the onset for the direct oxidation of methanol and the adsorption of OH coincide. In this case, the adsorbed OH species are not a mere spectator, because the obtained results for the reaction order for methanol and the proton indicate that OH adsorbed species are involved in the reaction mechanism. Since previous results show that the rate determining step in the methanol oxidation reaction is a dehydrogenation step, probably adsorbed OH collaborates in this step, facilitating either the adsorption of the methanol in the right configuration or the cleavage of the C—H bond so that the activation energy is minimized. These results are important for the computational modeling of the reaction because adsorbed OH should be incorporated in the model to reflect the real interfacial structure.

Author Contributions: Conceptualization, J.M.F. and E.H.; methodology, E.H.; validation, M.A.K. and R.M.-H.; formal analysis, E.H. and R.M.-H.; investigation, M.A.K. and R.M.-H.; writing—original draft preparation, E.H.; writing—review and editing, M.A.K., R.M.-H., J.M.F. and E.H.; visualization, R.M.-H. and E.H.; supervision, J.M.F. and E.H.; project administration, J.M.F. and E.H.; funding acquisition, J.M.F. and E.H.

Funding: This research was funded by the MINECO (Spain), grant number CTQ2016-76221-P.

Conflicts of Interest: The authors declare no conflict of interest.

References

1. Parsons, R.; Vandernoot, T. The oxidation of small organic molecules: A survey of recent fuel cell related research. *J. Electroanal. Chem.* **1988**, *257*, 9–45. [CrossRef]
2. Sriramulu, S.; Javi, T.D.; Stuve, E.M. Kinetic Modeling of Electrocatalytic Reactions: Methanol Oxidation on Platinum Electrodes. In *Interfacial Electrochemistry, Theory, Experiments and Applications*; Wieckowski, A., Ed.; Marcel Dekker: New York, NY, USA, 1998; p. 793.
3. Iwasita, T. Electrocatalysis of methanol oxidation (vol 47, pg 3663, 2001). *Electrochim. Acta* **2002**, *48*, 289. [CrossRef]
4. Markovic, N.M.; Ross, P.N. Surface science studies of model fuel cell electrocatalysts. *Surf. Sci. Rep.* **2002**, *45*, 117–229. [CrossRef]
5. Koper, M.T.M.T.M.; Lai, S.C.C.S.; Herrero, E. Mechanisms of the Oxidation of Carbon Monoxide and Small Organic Molecules at Metal Electrodes. In *Fuel Cell Catalysis: A Surface Science Approach*; Koper, M.T.M., Ed.; John Wiley & Sons, Inc.: Hoboken, NJ, USA, 2009; pp. 166–171. ISBN 978-0-470-13116-9.
6. Cohen, J.L.; Volpe, D.J.; Abruna, H.D. Electrochemical determination of activation energies for methanol oxidation on polycrystalline platinum in acidic and alkaline electrolytes. *Phys. Chem. Chem. Phys.* **2007**, *9*, 49–77. [CrossRef] [PubMed]
7. Neurock, M.; Janik, M.; Wieckowski, A. A first principles comparison of the mechanism and site requirements for the electrocatalytic oxidation of methanol and formic acid over Pt. *Faraday Discuss.* **2009**, *140*, 363–378. [CrossRef]
8. Clavilier, J.; Lamy, C.; Leger, J.M. Electrocatalytic Oxidation of Methanol on Single-Crystal Platinum-Electrodes—Comparison with Polycrystalline Platinum. *J. Electroanal. Chem.* **1981**, *125*, 249–254. [CrossRef]
9. Lamy, C.; Leger, J.M.; Clavilier, J.; Parsons, R. Structural effects in electrocatalysis: A comparative study of the oxidation of CO, HCOOH and CH$_3$OH on single crystal Pt electrodes. *J. Electroanal. Chem.* **1983**, *150*, 71–77. [CrossRef]
10. Franaszczuk, K.; Herrero, E.; Zelenay, P.; Wieckowski, A.; Wang, J.; Masel, R.I.I. A comparison of electrochemical and gas-phase decomposition of methanol on platinum surfaces. *J. Phys. Chem.* **1992**, *96*, 8509–8516. [CrossRef]
11. Herrero, E.; Franaszczuk, K.; Wieckowski, A. Electrochemistry of methanol at low index crystal planes of platinum. An integrated voltammetric and chronoamperometric study. *J. Phys. Chem.* **1994**, *98*, 5074–5083. [CrossRef]
12. Housmans, T.H.M.; Koper, M.T.M. Methanol oxidation on stepped Pt n(111) x (110) electrodes: A chronoamperometric study. *J. Phys. Chem. B* **2003**, *107*, 8557–8567. [CrossRef]
13. Lai, S.C.S.; Lebedeva, N.P.; Housmans, T.H.M.; Koper, M.T.M. Mechanisms of carbon monoxide and methanol oxidation at single-crystal electrodes. *Top. Catal.* **2007**, *46*, 320–333. [CrossRef]
14. Grozovski, V.; Climent, V.; Herrero, E.; Feliu, J.M. The role of the surface structure in the oxidation mechanism of methanol. *J. Electroanal. Chem.* **2011**, *662*, 43–51. [CrossRef]
15. Martínez-Hincapié, R.; Sebastián-Pascual, P.; Climent, V.; Feliu, J.M. Exploring the interfacial neutral pH region of Pt(111) electrodes. *Electrochem. Commun.* **2015**, *58*, 62–64. [CrossRef]

16. Sebastián, P.; Martínez-Hincapié, R.; Climent, V.; Feliu, J.M. Study of the Pt(111) | electrolyte interface in the region close to neutral pH solutions by the laser induced temperature jump technique. *Electrochim. Acta* **2017**, *228*, 667–676. [CrossRef]

17. Clavilier, J.; Faure, R.; Guinet, G.; Durand, R. Preparation of monocrystalline Pt microelectrodes and electrochemical study of the plane surfaces cut in the direction of the {111} and {110} planes. *J. Electroanal. Chem.* **1980**, *107*, 205–209. [CrossRef]

18. Korzeniewski, C.; Climent, V.; Feliu, J. Electrochemistry at Platinum Single Crystal Electrodes. In *Electroanalytical Chemistry A Series of Advances: Volume 24*; CRC Press: Boca Raton, FL, USA, 2011; pp. 75–170. ISBN 978-1-4398-3750-4.

19. Clavilier, J.; Elachi, K.; Petit, M.; Rodes, A.; Zamakhchari, M.A. Electrochemical Monitoring of the Thermal Reordering of Platinum Single-Crystal Surfaces after Metallographic Polishing from the Early Stage to the Equilibrium Surfaces. *J. Electroanal. Chem.* **1990**, *295*, 333–356. [CrossRef]

20. Herrero, E.; Orts, J.M.; Aldaz, A.; Feliu, J.M. Scanning tunneling microscopy and electrochemical study of the surface structure of Pt(10,10,9) and Pt(11,10,10) electrodes prepared under different cooling conditions. *Surf. Sci.* **1999**, *440*, 259–270. [CrossRef]

21. García-Aráez, N.; Climent, V.; Herrero, E.; Feliu, J.M. On the electrochemical behavior of the Pt(1 0 0) vicinal surfaces in bromide solutions. *Surf. Sci.* **2004**, *560*, 269–284. [CrossRef]

22. Markovic, N.M.; Grgur, B.N.; Lucas, C.A.; Ross, P.N. Surface electrochemistry of CO on Pt(110)-(1x2) and Pt(110)-(1x1) surfaces. *Surf. Sci.* **1997**, *384*, L805–L814. [CrossRef]

23. Attard, G.A.; Hunter, K.; Wright, E.; Sharman, J.; Martínez-Hincapié, R.; Feliu, J.M. The voltammetry of surfaces vicinal to Pt{110}: Structural complexity simplified by CO cooling. *J. Electroanal. Chem.* **2017**, *793*, 137–146. [CrossRef]

24. Herrero, E.; Wieckowski, A. Electrochemistry of Methanol at Low Index Crystal Planes. *J. Phys. Chem.* **1994**, *98*, 5074–5083. [CrossRef]

25. Berna, A.; Climent, V.; Feliu, J.M. New understanding of the nature of OH adsorption on Pt(111) electrodes. *Electrochem. Commun.* **2007**, *9*, 2789–2794. [CrossRef]

26. Angelucci, C.A.C.A.; Herrero, E.; Feliu, J.M.J.M. Modeling CO oxidation on Pt(111) electrodes. *J. Phys. Chem. C* **2010**, *114*, 14154–14163. [CrossRef]

27. Housmans, T.H.M.; Wonders, A.H.; Koper, M.T.M. Structure Sensitivity of Methanol Electrooxidation Pathways on Platinum: An On-Line Electrochemical Mass Spectrometry Study. *J. Phys. Chem. B* **2006**, *110*, 10021–10031. [CrossRef] [PubMed]

28. Herrero, E.; Fernández-Vega, A.; Feliu, J.M.; Aldaz, A. Poison formation reaction from formic acid and methanol on Pt(111) electrodes modified by irreversibly adsorbed Bi and As. *J. Electroanal. Chem.* **1993**, *350*, 73–88. [CrossRef]

29. Grozovski, V.; Vidal-Iglesias, F.J.; Herrero, E.; Feliu, J.M. Adsorption of formate and its role as intermediate in formic acid oxidation on platinum electrodes. *ChemPhysChem* **2011**, *12*, 1641–1644. [CrossRef] [PubMed]

30. Grozovski, V.; Climent, V.; Herrero, E.; Feliu, J.M. Intrinsic activity and poisoning rate for HCOOH oxidation on platinum stepped surfaces. *Phys. Chem. Chem. Phys.* **2010**, *12*, 8822. [CrossRef] [PubMed]

31. Perales-Rondón, J.V.; Herrero, E.; Feliu, J.M. Effects of the anion adsorption and pH on the formic acid oxidation reaction on Pt(111) electrodes. *Electrochim. Acta* **2014**, *140*, 511–517. [CrossRef]

32. Batista, E.A.; Malpass, G.R.P.; Motheo, A.J.; Iwasita, T. New mechanistic aspects of methanol oxidation. *J. Electroanal. Chem.* **2004**, *571*, 273–282. [CrossRef]

33. Abd-El-Latif, A.A.; Baltruschat, H. Formation of methylformate during methanol oxidation revisited: The mechanism. *J. Electroanal. Chem.* **2011**, *662*, 204–212. [CrossRef]

34. Mostafa, E.; Abd-El-Latif, A.E.A.A.; Baltruschat, H. Electrocatalytic oxidation and adsorption rate of methanol at Pt stepped single-crystal electrodes and effect of Ru step decoration: A DEMS study. *ChemPhysChem* **2014**, *15*, 2029–2043. [CrossRef] [PubMed]

35. Ferre-Vilaplana, A.; Perales-Rondón, J.V.V.; Buso-Rogero, C.; Feliu, J.M.M.; Herrero, E. Formic acid oxidation on platinum electrodes: A detailed mechanism supported by experiments and calculations on well-defined surfaces. *J. Mater. Chem. A* **2017**, *5*, 21773–21784. [CrossRef]

36. Ferre-Vilaplana, A.; Perales-Rondón, J.V.; Feliu, J.M.J.M.; Herrero, E.; Perales-Rondón, J.V.; Feliu, J.M.J.M.; Herrero, E. Understanding the effect of the adatoms in the formic acid oxidation mechanism on Pt(111) electrodes. *ACS Catal.* **2015**, *5*, 645–654. [CrossRef]

37. Bergelin, M.; Herrero, E.; Feliu, J.M.M.; Wasberg, M. Oxidation of CO adlayers on Pt(111) at low potentials: An impinging jet study in H2SO4 electrolyte with mathematical modeling of the current transients. *J. Electroanal. Chem.* **1999**, *467*, 74–84. [CrossRef]

38. Lebedeva, N.P.; Rodes, A.; Feliu, J.M.; Koper, M.T.M.; van Santen, R.A. Role of crystalline defects in electrocatalysis: CO adsorption and oxidation on stepped platinum electrodes as studied by in situ infrared spectroscopy. *J. Phys. Chem. B* **2002**, *106*, 9863–9872. [CrossRef]

39. Rizo, R.; Sitta, E.; Herrero, E.; Climent, V.; Feliu, J.M.J.M. Towards the understanding of the interfacial pH scale at Pt(111) electrodes. *Electrochim. Acta* **2015**, *162*, 138–145. [CrossRef]

40. Vidal-Iglesias, F.J.; Arán-Ais, R.M.; Solla-Gullón, J.; Herrero, E.; Feliu, J.M. Electrochemical characterization of shape-controlled Pt nanoparticles in different supporting electrolytes. *ACS Catal.* **2012**, *2*, 901–910. [CrossRef]

41. Briega-Martos, V.; Herrero, E.; Feliu, J.M.J.M. Effect of pH and Water Structure on the Oxygen Reduction Reaction on platinum electrodes. *Electrochim. Acta* **2017**, *241*, 497–509. [CrossRef]

42. Gamboa-Aldeco, M.E.; Herrero, E.; Zelenay, P.S.; Wieckowski, A. Adsorption of bisulfate anion on a Pt(100) electrode: A comparison with Pt(111) and Pt(poly). *J. Electroanal. Chem.* **1993**, *348*, 451–457. [CrossRef]

43. Climent, V.; Gómez, R.; Orts, J.M.; Feliu, J.M. Thermodynamic analysis of the temperature dependence of OH adsorption on Pt(111) and Pt(100) electrodes in acidic media in the absence of specific anion adsorption. *J. Phys. Chem. B* **2006**, *110*, 11344–11351. [CrossRef]

44. Souza-Garcia, J.; Climent, V.; Feliu, J.M. Voltammetric characterization of stepped platinum single crystal surfaces vicinal to the (110) pole. *Electrochem. Commun.* **2009**, *11*, 1515–1518. [CrossRef]

45. Gómez, R.; Orts, J.M.; Alvarez-Ruiz, B.; Feliu, J.M. Effect of temperature on hydrogen adsorption on Pt(111), Pt(110), and Pt(100) electrodes in 0.1 M HClO4. *J. Phys. Chem. B* **2004**, *108*, 228–238. [CrossRef]

46. Herrero, E.; Feliu, J.M.; Aldaz, A. Poison formation reaction from formic acid on Pt(100) electrodes modified by irreversibly adsorbed bismuth and antimony. *J. Electroanal. Chem.* **1994**, *368*, 101–108. [CrossRef]

47. Arán-Ais, R.M.; Figueiredo, M.C.; Vidal-Iglesias, F.J.; Climent, V.; Herrero, E.; Feliu, J.M. On the behavior of the Pt(1 0 0) and vicinal surfaces in alkaline media. *Electrochim. Acta* **2011**, *58*, 184–192. [CrossRef]

Article

Strategies to Hierarchical Porosity in Carbon Nanofiber Webs for Electrochemical Applications

Svitlana Yarova, Deborah Jones, Frédéric Jaouen and Sara Cavaliere *

Institut Charles Gerhardt Montpellier, UMR CNRS 5253, Agrégats Interfaces et Matériaux pour l'Energie, Université de Montpellier - ENSCM - CNRS, CEDEX 5, 34095 Montpellier, France; svitlana.yarova@etu.umontpellier.fr (S.Y.); deborah.jones@umontpellier.fr (D.J.); frederic.jaouen@umontpellier.fr (F.J.)
* Correspondence: sara.cavaliere@umontpellier.fr; Tel.: +33-46714-9098

Received: 4 February 2019; Accepted: 25 February 2019; Published: 5 March 2019

Abstract: Morphology and porosity are crucial aspects for designing electrodes with facile transport of electrons, ions and matter, which is a key parameter for electrochemical energy storage and conversion. Carbon nanofibers (CNFs) prepared by electrospinning are attractive for their high aspect ratio, inter-fiber macroporosity and their use as self-standing electrodes. The present work compares several strategies to induce intra-fiber micro-mesoporosity in self-standing CNF webs prepared by electrospinning polyacrylonitrile (PAN). Two main strategies were investigated, namely i) a templating method based on the addition of a porogen (polymethyl methacrylate, polyvinylpyrrolidone, Nafion® or $ZnCl_2$) in the electrospinning solution of PAN, or ii) the activation in ammonia of previously formed CNF webs. The key result of this study is that open intra-fiber porosity could be achieved only when the strategies i) and ii) were combined. When each approach was applied separately, only closed intra-fiber porosity or no intra-fiber porosity was observed. In contrast, when both strategies were used in combination all CNF webs showed high mass-specific areas in the range of 325 to 1083 $m^2 \cdot g^{-1}$. Selected webs were also characterized for their carbon structure and electrical conductivity. The best compromise between high porosity and high electrical conductivity was identified as the fibrous web electrospun from PAN and polyvinylpyrrolidone.

Keywords: carbon nanofiber; porous fiber; electrospinning; mesopore; micropore; porogen; ammonia activation; surface area

1. Introduction

The performance of electrodes in electrochemical energy storage and conversion devices strongly depends on two key factors: a) the density of active sites or active area where faradaic or capacitive phenomena occur; and b) the mass- and charge-transport of species (gases, liquids, ions and electrons) to and from the active sites across the porous electrode [1,2]. The design of porous and hollow nanostructures is crucial to obtain high surface areas, nanoscale porous structures and a high density of electrochemically accessible active sites. In particular, carbon is widely used as a support (e.g., for fuel cells) or as an active material (e.g., in supercapacitors and in some batteries) due to its sufficiently high electrical conductivity, low cost and reasonable stability in specific electrochemical potential windows.

Among carbon materials, carbon nanofibers (CNFs) are very promising for the fabrication of electrochemical electrodes due to their high aspect ratio and the possibility to assemble CNFs in unique three-dimensional web structures with very high macropore volume between the fibers. High porosity within electrodes is critical to avoid a slow Knudsen diffusion regime in gas-diffusion electrodes (as in fuel cells), or to avoid reactant/ion depletion (in supercapacitors, redox flow batteries, etc.). A very efficient and upscalable approach to prepare a CNF network is via the electrospinning and carbonization of a polymer solution [3,4]. Furthermore, CNF webs obtained via electrospinning can often be designed to possess sufficient mechanical stability and through-plane electric conductivity to be used as free-standing electrodes, as shown in recent studies on batteries [5,6], supercapacitors [7–9], fuel cells [10–12], vanadium redox-flow [13] and Li–O$_2$ batteries [14]. This technique may not only simplify the electrode fabrication process, but also maximizes the macropore volume in the electrode and avoids binders, otherwise necessary with dispersed CNFs or other carbon powder materials to form the electrode. Such binders or additives often reduce the accessibility of active sites by ions in the electrolyte, by electrons from the current collector and/or by reactants in the gas-phase or in the electrolyte.

Electrospun CNFs are generally prepared by electrospinning a polymer precursor (e.g., polyacrylonitrile (PAN), polyimide, polybenzimidazole or polyvinyl alcohol), followed by at least one thermal treatment to convert the polymer fibrous network into conductive carbon [15]. The nature and molecular weight of the polymer precursor, its concentration in the electrospun solution as well as the thermal treatment conditions all can influence the morphological and physical properties of the CNF web obtained. Typical CNF diameters obtained are in the range 100–500 nm, resulting in specific surface area in the range of 4–40 m$^2 \cdot$g^{-1} if the CNF does not possess internal porosity. In order to improve the specific surface area of CNF-based electrodes obtained by electrospinning, and therefore their electrochemical properties, it is crucial to develop strategies that result in CNF webs with internal porosity inside the fibers. The other possible approach, namely reducing the CNF diameter to <10 nm (to result in an expected specific area of *ca* 200 m$^2 \cdot$g^{-1}), is limited both in terms of the possible increase of specific area and also by the limitations of the mechanical stability of a CNF web with such an ultralow fiber diameter.

Two main strategies have been investigated previously to prepare porous carbon fibers by electrospinning, one in which a porogen is introduced in the electrospinning solution and which results in biphasic polymer-porogen fibers, and another in which the plain CNFs are subjected after their synthesis to a reactive chemical (gas or solid). In the former case (labelled "pre-synthesis"), the carbonization of the biphasic polymer fibers leads to the selective removal of the porogen during the temperature ramp-up, resulting in CNFs having either a closed or open porosity. In the second case (labelled "post-synthesis"), the selective etching of less-organized carbon domains in the CNFs by the reactive chemical may result in the formation of open pores from the outer surface of CNFs and the inward. However if the carbon structure of the CNF is homogeneous the reactive etching may simply result in a thinning of the CNF, without the creation of internal porosity.

In the pre-synthetic approach, the electrospinning technique allows for the introduction of various porogens in the polymer precursor solution, whereby the porogen acts as a hard or soft template during the subsequent carbonization step. The most studied hard templates for preparing CNFs are preformed nanoparticles or nanostructures (silica [8,16], nano-CaCO$_3$ [17]) that generate porosity upon their removal with chemical and/or thermal treatments. The hard template approach allows control of the size and morphology of the pores and minimizes undesired chemical reactions between the template and the carbon structure [18]. Hard templates can also form during the carbonization treatment itself, as is the case when introducing metal salts (e.g., Co(NO$_3$)$_2$ 6H$_2$O, Ni(CH$_3$COO)$_2$, ZnCl$_2$) in the electrospun polymer solution. The transition metal salts form metallic or metal oxide particles (e.g., Co, Ni, ZnO), which then serve as porogen, and in some cases, as a catalyst for enhanced graphitization of the CNF formed from the polymer precursor carbonization [19–23]. Other templates including Prussian blue analogues [24] and metal organic frameworks [25–27] have been used in conjunction with polymer precursors, conferring both porosity and heteroatom doping to CNFs obtained by electrospinning and carbonization. Sacrificial polymers or organic molecules such as polymethyl methacrylate (PMMA) [6,14,28–30], polyvinylpyrrolidone (PVP) [31,32], poly(ethylene oxide) [33], Nafion® [34], polysulfone [35], polystyrene [36], poly-L-lactic acid [37] and beta-cyclodextrin [38] can also be spun together with the main carbon precursor (any polymer forming conductive carbon with high yield during pyrolysis), allowing the formation of porous CNFs upon their removal by solvent or thermal treatment. Different kinds of templates can further be combined to produce hierarchical porosity, which is crucial for the transport of different species (ions, gases, liquids) in the electrodes of various electrochemical devices. For instance, the combination of PVP with Mg salts [22,39] gave rise to hierarchical meso- and microporosity inside CNFs prepared by electrospinning. For the post-synthetic approach, various chemical activations have been investigated to create porosity in preformed CNFs. Immersion of CNFs in concentrated KOH solutions followed by a treatment at high temperature is a recognized approach [40,41]. Other more direct approaches consist of re-pyrolyzing the CNFs in a reactive atmosphere, such as steam [42,43] or ammonia [44].

While a number of studies on CNFs prepared by resorting to either a pre-synthetic approach involving a given porogen and given carbon precursor, or a post-synthetic approach, few studies have compared the merits of using different porogens in otherwise identical preparation conditions for CNFs, and even less studies have reported on the combination of i) porogen-PAN electrospun solution and ii) a post-synthetic treatment to further increase intra-fiber porosity. In this paper, we therefore investigated the formation of intra-fiber porosity in self-standing CNF webs that were derived from electrospun PAN, resorting either to pre-synthetic and/or post-synthetic approaches to create porosity. PAN was selected for its relatively high melting point, high carbon yield and facility to be electrospun [15]. For the pre-synthetic approach, various hard ($ZnCl_2$) and soft templating agents (PMMA, PVP, Nafion®) were co-electrospun with PAN. For the post-synthetic approach, NH_3 pyrolysis was applied to CNF webs derived from electrospun porogen-PAN solutions. The resulting CNF webs were characterized for their properties relevant to electrochemical applications, including not only mass-specific surface area and porosity, but also morphology, structure, composition and electrical conductivity.

2. Materials and Methods

To prepare reference CNFs, PAN (Mw = 150,000 $g \cdot mol^{-1}$, Sigma-Aldrich, Saint Louis, MO, USA) was first dissolved in N,N-dimethylformamide (DMF, pure, Carlo Erba, Val de Reuil, France) for 12 h at 50 °C (10 wt % PAN concentration), after which the solution was cooled to room temperature. The polymer fibers were then electrospun at 20 °C and collected on a drum rotating at 100 rpm (Spraybase ®, Dublin, Ireland). The distance between the tip of the needle (22 gauge) and the collector was 10 cm, and a voltage of 13 kV (Auto-Reversing High Voltage Power Supply Spellman CZE1000R, West Sussex, United Kingdom) was applied to obtain a stable Taylor cone. The flow rate was kept as a constant 1 $mL \cdot h^{-1}$ (syringe pump KDS 100 Legacy Syringe Pump, KD Scientific, Holliston, MA, USA).

The obtained PAN-based fibers were submitted to stabilization and carbonization to give rise to carbon materials [45,46]. In particular, the electrospun PAN fibers were treated in air at 150 °C for 2 h with a heating rate of 2.5 $°C \cdot min^{-1}$, and then at 250 °C for 3 h with a heating rate of 2.5 $°C \cdot min^{-1}$. The stabilized nanofibers were finally carbonized at 1000 °C (ramp rate 5 $°C \cdot min^{-1}$) for 2 h under flowing argon atmosphere. After 2 h at 1000 °C, the heating was stopped and the sample cooled down naturally to room temperature under flowing Ar. The obtained CNFs are labelled PAN_{10}-CNF, the scalar 10 standing for the wt % PAN in DMF solution.

To prepare biphasic polymer fibers comprising PAN, the porogen was solubilized or dissolved in a PAN/DMF solution. The investigated porogens are PMMA (Sigma-Aldrich, Mw = 15,000 or 120,000 $g \cdot mol^{-1}$), PVP (Sigma-Aldrich, Mw ~1,300,000 $g \cdot mol^{-1}$, Steinheim, Germany), Nafion® (NR50, Sigma-Aldrich, Saint Louis, MO, USA) and zinc chloride ($ZnCl_2$ anhydrous, >98%, purchased from Alfa Aesar, Kander, Germany). For PMMA and $ZnCl_2$, the PAN and/or porogen concentrations in DMF were varied, since the introduction of the porogen had an obvious influence on rheological properties of the solution and morphology of the electrospun polymer fiber web. Table 1 details the investigated compositions of the electrospun solutions. Such biphasic polymer fibers were stabilized in air and carbonized in flowing Ar in the exact same conditions as described above for pure-PAN based fibers. The final materials are labelled as $Porogen_x$-PAN_y-CNF, with x standing for the wt % of porogen in the electrospun solution, y the wt % of PAN in the electrospun solution, and CNF indicating the pyrolysis was performed in Ar (see labels in Table 1).

Table 1. Composition of the polyacrylonitrile (PAN)/Porogen electrospun solutions and labels of the resulting carbon nanofibers (CNFs) and NH_3-activated CNF webs (ACNFs). The DMF amount in the electrospun solutions was (100-porogen wt %-PAN wt %).

Porogen (M_w)	Porogen wt %	PAN wt %	CNF Label (after Carbonization in Ar)	CNF Label (after Carbonization in Ar and NH_3 Activation)
None	0	10	PAN_{10}-CNF	PAN_{10}-ACNF
PMMA (15,000)	2	8	15kPMMA$_2$-PAN$_8$-CNF	15kPMMA$_2$-PAN$_8$-ACNF
PMMA (120,000)	4	6	120kPMMA$_4$-PAN$_6$-CNF	-
PMMA (15,000)	2	8	15kPMMA$_2$-PAN$_8$-CNF	-
PMMA (120,000)	4	6	120kPMMA$_4$-PAN$_6$-CNF	-
PVP	5	10	PVP$_5$-PAN$_{10}$-CNF	PVP$_5$-PAN$_{10}$-ACNF
Nafion®	2	8	Nafion$_2$-PAN$_8$-CNF	Nafion$_2$-PAN$_8$-ACNF
$ZnCl_2$	1	10	Zn$_1$-PAN$_{10}$-CNF	-
$ZnCl_2$	3	10	Zn$_3$-PAN$_{10}$-CNF	-
$ZnCl_2$	5	10	Zn$_5$-PAN$_{10}$-CNF	-
$ZnCl_2$	7	10	Zn$_7$-PAN$_{10}$-CNF	Zn$_7$-PAN$_{10}$-ACNF

Chemical activation of selected CNFs was performed by applying a flash pyrolysis in ammonia at 900 °C. Unless otherwise indicated, the pyrolysis duration in ammonia was 15 min. The resulting materials are identified by their label ending with ACNF, instead of CNF for non-activated material.

The morphology of the polymer fibers and/or CNFs was investigated by field emission-scanning electron microscopy (FE-SEM) using a Hitachi S-4800 microscope (Hitachi Europe SAS, Velizy, France). Data analysis and fiber diameter distribution were performed using an image processing software Image J 1.48 v (U. S. National Institutes of Health, Bethesda, MD, USA). CNFs were analyzed by transmission electron microscopy (TEM) using a JEOL 2200FS (Source: FEG) microscope operating at 200 kV equipped with a CCD camera Gatan USC (16 MP) (Tokyo, Japan). For TEM cross-sectional analysis, a microtome was used on resin-encapsulated sample and slices were deposited on carbon-coated copper grids (Agar Scientific, Stansted, Essex, United Kingdom).

Surface area and porosimetry of the samples were analyzed with N_2 physisorption with a Tristar II Micromeritics instrument (Norcross, GA, USA) at 77 K. Prior to analysis, all samples were outgassed overnight at 120 °C under vacuum. The resulting isotherms being of type I according to the IUPAC classification, BET plots were drawn below the relative pressure of 0.1 from the adsorption branches and employed to evaluate the BET specific surface (S_{BET}). The alpha-plot method was utilized to determine the mesoporous (V_{meso}) and microporous volumes (V_{micro}) as well as average pore diameter (d_{pore}).

C, H, N, O elemental analysis was performed with a Vario MICRO Element Analyzer (Elementar Analysensysteme, Hanau, Germany).

Raman spectra were recorded on a LabRAM Aramis IR2 Horiba Jobin Yvon spectrometer (Villeneuve d'Ascq, France) equipped with a He/Ne laser (λ = 633 nm) and a long work distance objective ×50. The spectra were fitted with five bands using Origin software (OriginLab Corporation, Northampton, UK).

The in-plane electrical conductivity of self-standing CNF webs was measured using a 2400 Keithley in a four-electrode configuration on a 5 mm × 40 mm × 0.05 mm carbon electrode strip in a Fumatech MK3-L cell operated in the current range of 0–100 mA.

3. Results and Discussion

3.1. Characterization of Porogen-PAN Fibrous Webs After Carbonization in Ar

As a reference, the CNF web prepared only from PAN was characterized (labelled PAN_{10}-CNF). The resulting CNF web was self-standing and flexible, as already reported by the authors [10].

The morphology of PAN_{10}-CNF was investigated by SEM and TEM. The material was observed to be randomly oriented cylindrical fibers (Figure 1a) with an average diameter of 200 nm (Figure 2a). These fibers were smooth and dense, as demonstrated by the TEM cross-section in Figure 3a. According to the latter image and to previous reports [47], the porosity and specific surface area developed by such fibers is very low (*ca* 20 $m^2 \cdot g^{-1}$), mainly attributed to the outer surface area of the fibers, with no internal porosity. The nearly flat N_2 adsorption isotherm demonstrates the lack of porosity inside the fibers (Figure 4).

In order to create internal porosity within fibers, a range of porogens were added to the precursor PAN solution prior to electrospinning, as described in the experimental methods. The investigated polymer templates (PMMA, PVP, Nafion®) were selected for their relative stability and rigidity. They are known to be stable up to 250 °C in air, the conditions used during the stabilization step. When heated to much higher temperatures in inert gas, they decompose and form mainly volatile products, thereby acting as porogens, while PAN decomposes as well but is converted to carbon in high yield.

The results obtained with PMMA are discussed first. PMMA of two different molecular weights (15,000 and 120,000 $g \cdot mol^{-1}$) and in different ratios to PAN (1:4 and 2:3) was added in the electrospun DMF-based solution. Those ratios were selected on the basis of a previous study on PMMA-PAN composite fibers [36]. In all the cases, the addition of PMMA resulted in CNFs with slightly smaller average diameter (compare Figure 1a,b) than for the reference PAN_{10}-CNF web (average diameter of 175 nm, see Figure 2b), but with otherwise practically identical features and properties (Figure S1). Figure 1b depicts the sample $15kPMMA_2$-PAN_8-CNF as an example. The decrease in fiber diameter is most probably due to the lower concentration of PAN in the electrospun solution relative to the reference PAN_{10}-CNF (8% vs. 10%), and was also expected due to the change in precursor viscosity already observed upon addition of this polymer [6]. The developed porosity is clearly visible on the cross-sectional TEM micrograph (Figure 3b), where pores with an average diameter of 10 nm appeared within the fibers. The modification of the preparation parameters (higher molecular weight for PMMA, or different PMMA:PAN ratio) did not have significant effects on the obtained morphology and porosity (images not shown). The question as to whether the formed porosity is open or closed is discussed later.

The results obtained with PVP porogen are now discussed. A PVP:PAN ratio of 1:2 was employed (sample labelled PVP_5-PAN_{10}-CNF). In these conditions, carbon fibers with considerably greater diameter compared to the reference PAN_{10}-CNF (average diameter of 750 nm) (Figures 1c and 2c) were obtained. This can be attributed to a modification of the viscosity of the electrospun solution, affected by the presence of PVP aggregates and consequent decrease in chain entanglement. The pores generated inside the CNFs presented an average size around 3 nm (Figure 3c).

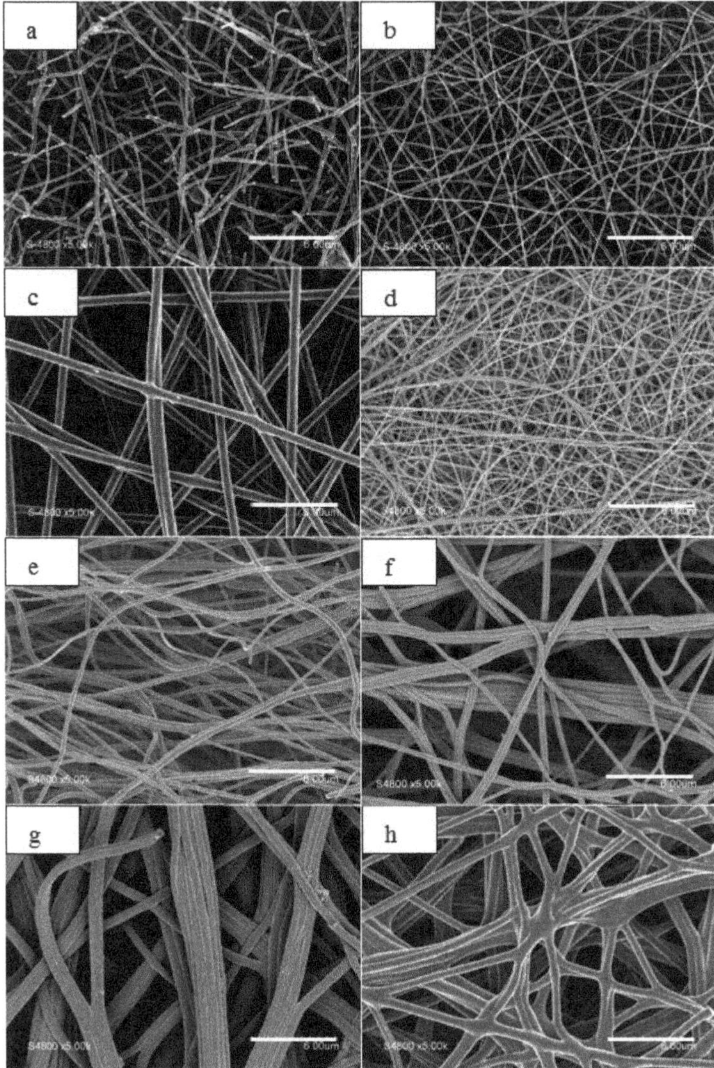

Figure 1. FE-SEM micrographs of: (**a**) PAN_{10}-CNF; (**b**) $15kPMMA_2$-PAN_8-CNF; (**c**) PVP_5-PAN_{10}-CNF; (**d**) $Nafion_2$-PAN_8-CNF; (**e**) Zn_1-PAN_{10}-CNF; (**f**) Zn_3-PAN_{10}-CNF; (**g**) Zn_5-PAN_{10}-CNF; (**h**) Zn_7-PAN_{10}-CNF. The scale bar corresponds to 6 μm.

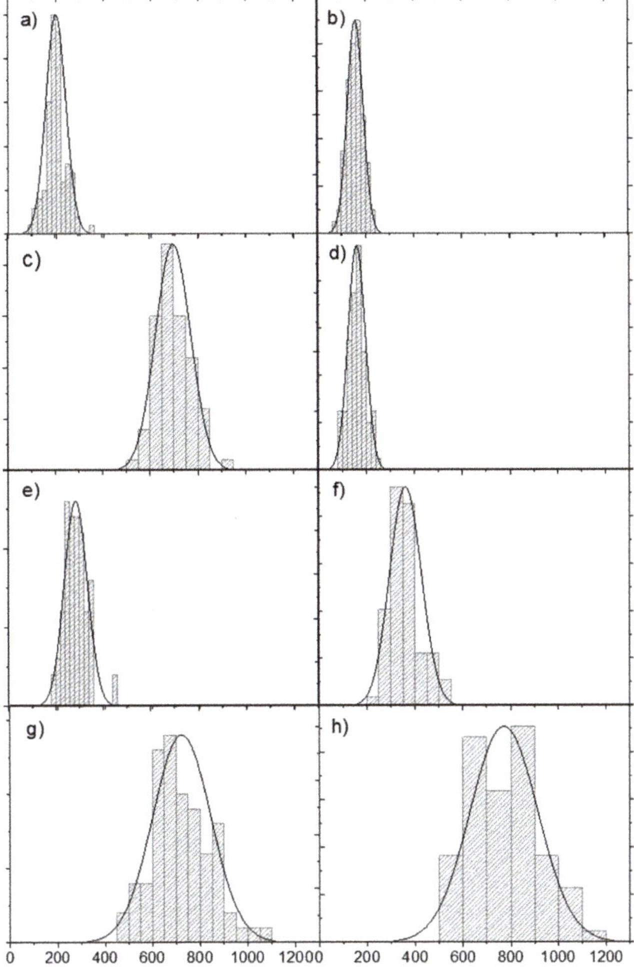

Figure 2. Histograms of the fiber diameter distribution for: (**a**) PAN_{10}-CNF; (**b**) $15kPMMA_2$-PAN_8-CNF; (**c**) PVP_5-PAN_{10}-CNF; (**d**) $Nafion_2$-PAN_8-CNF; (**e**) Zn_1-PAN_{10}-CNF; (**f**) Zn_3-PAN_{10}-CNF; (**g**) Zn_5-PAN_{10}-CNF; (**h**) Zn_7-PAN_{10}-CNF.

The third polymer template investigated was Nafion® perfluorosulfonic acid. It has been previously demonstrated to be effective in creating nanoscale porous domains inside carbon nanofibers [34]. In this work, its addition to PAN (sample labelled $Nafion_2$-PAN_8-CNF) resulted in the formation of thin carbon fibers (average fiber diameter of 180 nm) (Figures 1d and 2d). The CNF webs also had an obviously a higher density of fibers (compare Figure 1a,d), which is beneficial for improved conductivity and mechanical stability. The fibers have a well-developed internal porosity, as shown by TEM analysis of the cross-sections, Figure 3d. A homogeneous distribution of pores with a size of 4 nm can be observed within the CNF structure.

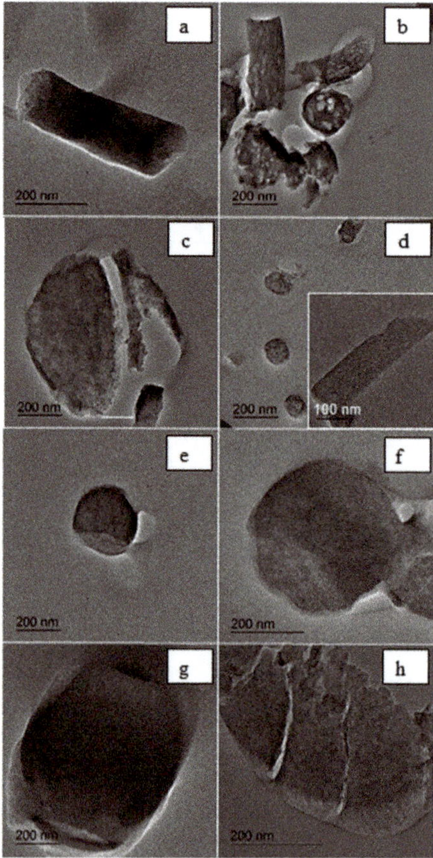

Figure 3. TEM micrographs of: (**a**) PAN_{10}-CNFs; (**b**) $15kPMMA_2$-PAN_8-CNF; (**c**) PVP_5-PAN_{10}-CNF; (**d**) $Nafion_2$-PAN_8-CNF; (**e**) Zn_1-PAN_{10}-CNF; (**f**) Zn_3-PAN_{10}-CNF; (**g**) Zn_5-PAN_{10}-CNF; (**h**) Zn_7-PAN_{10}-CNF.

Finally, the inorganic template precursor $ZnCl_2$ was investigated. The presence of $ZnCl_2$ may allow the formation of pores during the ensuing stabilization-carbonization steps in air and in Ar with the following mechanism: during the stabilization step, hydrated chloride hydrolyzes and forms an oxychloride, from which zinc oxide forms creating microporosity by etching carbon atoms [23]. FE-SEM characterization of Figure 1e–h showed that increasing the $ZnCl_2$ concentration in the electrospun solution from 1 wt % to 7 wt % (samples Zn_1-PAN_{10}-CNF, Zn_3-PAN_{10}-CNF, Zn_5-PAN_{10}-CNF, Zn_7-PAN_{10}-CNF) the diameter of the carbon fibers increased tremendously from 300 to 700 nm (Figure 2e–h). It is known that solution conductivity affects fiber uniformity. The high conductivity of the $PAN/ZnCl_2$ (52:1) solution was reported to lead to instability in the electrospinning process with the formation of large fibers and bundles as a cotton-like 3D deposit [48]. However, the reverse has also been observed: increasing the conductivity through the addition of salt may produce finer, more uniform fibers, resulting in an increased elongational force exerted on the fiber jet [49]. Indeed, other reports on the electrospinning of $ZnCl_2/PAN$ solutions showed that the average diameters of the obtained CNFs gradually decrease from 350 to 200 nm with increasing zinc chloride content from 1 wt % to 5 wt % [23].

Another particular feature of CNFs derived from electrospun $ZnCl_2$/PAN solutions is the formation of progressively larger fiber bundles with increased Zn salt concentration (see Figure 1g,h in particular). This phenomenon was already observed for PAN/polystyrene mixtures, where the template not only acted as sacrificial decomposable phase, but also controlled the formation of these architectures [13]. In the TEM micrographs of the cross-sections of Zn_1-PAN_{10}-CNF, Zn_3-PAN_{10}-CNF, Zn_5-PAN_{10}-CNF, Zn_7-PAN_{10}-CNF samples (Figure 3e–h), no pores are visible inside the fibers.

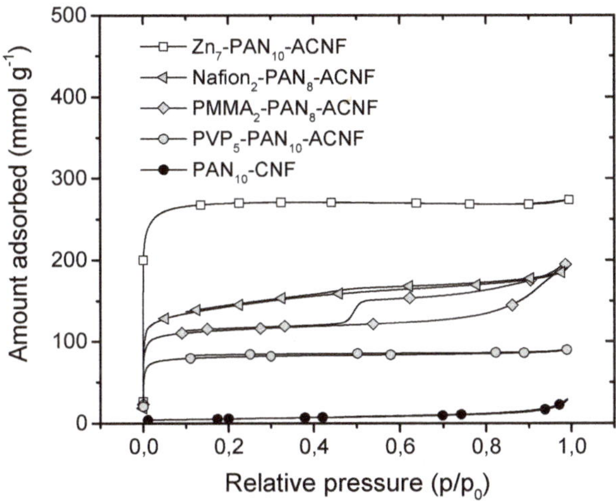

Figure 4. Selected N_2 adsorption-desorption isotherms of carbonized and NH_3-activated CNF webs (ACNF). The carbonized web obtained from PAN-only is also shown as a reference (PAN_{10}-CNF).

Except for the latter Zn-PAN-CNF samples, all other CNFs derived from polymer/PAN solutions containing the sacrificial polymers presented pores visible by TEM in the mesoscopic range, in agreement with previous works [6,14,29,30,34]. The nitrogen adsorption/desorption isotherms of all the investigated CNFs showed however very low S_{BET} values (below 20 $m^2 \cdot g^{-1}$, see * in column 2 of Table 2), almost unchanged compared to the reference PAN-CNFs (Table 2). This result, apparently in contrast with the micrographs of Figure 3 showing pores of different size in the prepared CNFs and in contrast with previous reports using the same templates and similar synthesis conditions [6,14,28–30,34], demonstrates that the obtained porosity is closed, and not accessible from outside the CNFs. This observation may be due to the partial collapse of the percolating porous network formed during the thermal removal of the template or porogen [34]. As a result, free-standing porous fiber webs are obtained, with closed porosity and low surface area. The latter is assigned only to the outer surface area of plain and smooth CNFs with diameters in the range 150–700 nm, depending on the porogen used. The appealing approach of using a soft or hard template added during the preparation step of electrospinning seems therefore ineffective in creating open porosity. The conditions necessary to reach an open porosity are shown later in this study.

Graphitic character and the related electrical conductivity of the carbon based electrode materials are relevant properties for electrochemical applications. To investigate them, Raman spectroscopy and the four-electrode method were used to characterize the carbon nanofiber networks prepared in this work. In particular, the modification of this properties upon the different steps of preparation was studied. The Raman spectra of all CNFs prepared with different porogens (Figure S2) present two intense and broad bands, the so-called D band at 1357 cm^{-1} ascribed to defects and disorder in the graphitic structure and the so-called G band at 1560 cm^{-1} corresponding to the in-plane vibration of sp^2-bonded carbon in graphite. The spectra were fitted with more contributions, namely the D4 band, ascribed to sp^3-carbon (1180 cm^{-1}), the D3 band (1500 cm^{-1}) associated with an amorphous sp^2 carbon bonded in the graphitic phase and the D2 band (1580 cm^{-1}) corresponding to graphitic lattices in the structure (Figure S3) [50]. The relative intensities of the D and G bands (I_D/I_G) as well as the relative areas (A_D/A_G) are often used to estimate the degree of graphitization of carbon materials [39,51]. Lower I_D/I_G ratios indicate higher levels of crystalline sp^2-carbon [52,53]. The I_D/I_G and A_D/A_G values obtained after deconvolution of the spectra of all CNFs derived from the carbonization of porogen/PAN solutions are summarized in Table 3. It is evident that the addition of polymer or inorganic salt porogen did not significantly influence the graphitization of the carbon fibers. Similar I_D/I_G ratios were obtained for reference PAN_{10}-CNFs and the other CNF webs. Their values between 1.8 and 2.7 demonstrate that the fibers are composed by disordered carbon (due to the relatively low carbonization temperature of 1000 °C) with local graphite inclusions (turbostratic domains) already evidenced for electrospun CNFs [24,39,47,54].

Table 2. Textural properties of carbonized (CNFs) and ammonia-activated (ACNFs) carbon nanofibers.

Fiber Precursor	CNFs	ACNFs				
	S_{BET}, $m^2 \cdot g^{-1}$	S_{BET}, $m^2 \cdot g^{-1}$	C_{BET}	V_{meso}, $cm^3 \cdot g^{-1}$	V_{micro}, $cm^3 \cdot g^{-1}$	d_{pore}, nm
PAN_{10}	20	n/a*	-	-	-	-
$15kPMMA_2$-PAN_8	35	450	3034	0.0599	0.1243	2.9
$120kPMMA_2$-PAN_8	n/a*	645	2410	0.0877	0.1669	1.9
$15kPMMA_4$-PAN_6	n/a*	360	2927	-	0.1006	0.5
$120kPMMA_4$-PAN_6	n/a*	410	3324	0.0083	0.1177	0.8
PVP_5-PAN_{10}	3	325	2012	-	0.0941	1.8
$Nafion_2$-PAN_8	n/a*	535	1993	0.1166	0.1327	2.4
Zn_1-PAN_{10}	n/a*	680	1986	0.0791	0.1814	1.6
Zn_3-PAN_{10}	n/a*	570	1707	0.0721	0.1614	2.5
Zn_5-PAN_{10}	n/a*	865	1791	-	0.1982	1.5
Zn_7-PAN_{10}	n/a*	1083	3558	0.0980	0.3098	1.4

* S_{BET} <3 $m^2 \cdot g^{-1}$

Table 3. Relative areas and intensities of D and G bands (I_D/I_G, A_D/A_G) in the Raman spectra after carbonization in Ar (CNF) and after a subsequent activation in ammonia (ACNF).

Fiber Precursor	CNF		ACNF	
	$A_{D/G}$	$I_{D/G}$	$A_{D/G}$	$I_{D/G}$
PAN_{10}	3.01	1.83	2.95	1.86
$15kPMMA_2$-PAN_8	4.73	2.45	3.06	1.88
$120kPMMA_2$-PAN_8	4.59	2.02	5.00	2.39
$15kPMMA_4$-PAN_6	3.24	2.20	3.78	2.08
$120kPMMA_4$-PAN_4	3.14	1.86	3.76	2.02
PVP_5-PAN_{10}	4.69	2.09	3.21	2.28
$Nafion_2$-PAN_8	5.34	2.77	2.52	2.33
Zn_1-PAN_{10}	4.40	2.43	4.65	2.49
Zn_3-PAN_{10}	6.34	2.40	9.61	2.56
Zn_5-PAN_{10}	4.24	2.57	2.30	1.88
Zn_7-PAN_{10}	6.54	2.39	3.06	1.89

The electrical conductivities of the CNFs prepared by electrospinning, determined directly on the self-standing CNF webs, are consistent with the partial graphitic character evidenced by Raman spectroscopy. Except for the fibers prepared using zinc chloride as porogen precursor, which show significantly lower conductivity, all CNF networks present similar conductivity values around 6–9 $S \cdot cm^{-1}$ (Table 4), which is in agreement with previous results obtained on PAN based electrospun materials [10,47,55].

Table 4. In-plane electrical conductivity and elemental content of all CNF webs and selected ACNF webs.

Fiber Precursor	CNF				ACNF			
	Conductivity S·cm^{-1}	C wt %	N wt %	O Wt %	Conductivity S·cm^{-1}	C wt %	N wt %	O wt %
PAN_{10}	6.6	66.8	5.6	17.5	3.8	78.4	7.0	7.8
$15kPMMA_2$-PAN_8	8.8	75.8	5.5	11.1	1.9	77.8	5.8	6.7
PVP_5-PAN_{10}	9.7	76.2	2.2	10.9	4.8	78.8	7.0	8.2
$Nafion_2$-PAN_8	7.0	74.3	5.5	16.0	2.4	70.3	5.5	9.6
Zn_1-PAN_{10}	1.1	70.3	5.2	14.6	-	-	-	-
Zn_3-PAN_{10}	0.8	71.5	4.6	17.1	-	-	-	-
Zn_5-PAN_{10}	3.3	61.2	6.8	20.9	-	-	-	-
Zn_7-PAN_{10}	0.8	68.1	2.5	19.2	0.3	71.3	6.1	11.3

The carbon fibers prepared from $ZnCl_2$-PAN_{10} fibers demonstrated very low conductivity, between 3.3 and 0.8 S·cm^{-1}. The graphitic structure being the same for all the samples as indicated by the Raman study, this decrease in conductivity from the pristine CNFs may be ascribed to the microstructure of the $ZnCl_2$-PAN-CNFs in which, due to $ZnCl_2$ removal, electron paths are likely disrupted. Furthermore, its particular morphology with fibers forming large bundles can affect the inter-fiber/inter-bundle connection (decreased number of connection points) and thus the electron transport.

3.2. Characterization of Porogen-PAN Fibrous Webs After Carbonization and NH$_3$ Activation

The post-treatment with ammonia at high temperature is a conventional approach to creating porosity in carbon-based materials by etching, which can also introduce nitrogen functionalities on the surface [56]. The mechanism for the formation of both pores and nitrogen groups is based on the reaction of ammonia with carbon. This gasification reaction continuously occurs during NH$_3$ pyrolysis and react with surface carbon atoms forming volatile compounds such as HCN [57,58]. When the reaction between ammonia and carbon occurs at different rates on the surface (due to carbon structure heterogeneity), increased porosity is obtained. When the ammonia pyrolysis is stopped, nitrogen atoms that were reacting with carbon surface atoms at that moment remain on the surface, with different environments including amino, cyanide, pyrrolic, pyridinic and quaternary nitrogen groups [59,60].

We first subjected the reference electrospun carbon nanofibers PAN_{10}-CNF web to treatment under pure flowing ammonia at 900 °C [61]. The resulting activated nanofibers (ACNFs), labelled PAN10-ACNFs, did not present significant modification of their average diameter (200 nm) nor of their porosity, still not visible by TEM cross-sectional observations (Figure S4). As a consequence, specific surface area determined from nitrogen adsorption/desorption isotherms remained low (Table 2). Due to the low surface area obtained with the porogen/PAN template method after carbonization in Ar, as described previously, and the low surface area obtained with ammonia activation on pure PAN-derived CNFs, the combination of the two approaches was then investigated as a last attempt to develop high specific area. All CNF webs that had been prepared by porogen/PAN templating and carbonized in Ar (labelled CNF) were then further activated in flowing ammonia gas in the same conditions as for PAN_{10}-ACNF.

N$_2$ adsorption-desorption measurements were performed to characterize their porosity and specific surface area (Table 2 and Figure 4). Figure 4 clearly shows that ACNF webs adsorbed a significant amount of N$_2$, the three polymer porogens resulting in intermediate adsorbed volumes and the $ZnCl_2$ porogen resulting in the highest adsorbed volumes. The isotherms of the different ACNF webs are of type I and/or type II according to IUPAC classification, which indicates that fibers presented an overall microporous structure, as well as mesopores. With Nafion® as a porogen, significant hysteresis of type H4 is observed during desorption, closing at P/P$_0$ = 0.45. This is assigned to mesopores with a bottleneck shape, which may be related to the existence of Nafion® polymer aggregates in the electrospinning solution [62,63], while other polymers were fully dissolved. In the case of microporous structures, the BET energetic constant (C_{BET}, related to the energy of adsorption of the first layer of N_2 adsorbate on the carbon surface) is expected to be higher than the usual value expected for mesoporous carbon structure as seen in Table 2 [64]. The values of the specific surface area, meso and micropore volume as well as average pore diameter of all ACNF webs are summarized in Table 2. The specific surface area of all ACNF webs ranged from 325 to 1083 $m^2 \cdot g^{-1}$, the maximum value corresponding to $ZnCl_7$-PAN_{10}-ACNF, while the average pore size was comprised between 0.5 and 2.9 nm.

The carbon fibers derived from PMMA-PAN solutions presented surface areas between 360 and 645 $m^2 \cdot g^{-1}$. The specific area increased by increasing the molecular weight of the sacrificial polymer from 15,000 to 120,000 $g \cdot mol^{-1}$ and by decreasing the PMMA:PAN ratio from 2:3 to 1:4. The carbon fibers derived from the PVP-PAN solution with a PVP:PAN ratio of 1:2 presented a surface area of 325 $m^2 \cdot g^{-1}$. This is comparable to the BET value obtained with PMMA in high ratio to PAN (2:3), and may also be increased if the PVP:PAN ratio had been further optimized (lowered). The carbon fibers derived from Nafion®-PAN solution with Nafion®:PAN ratio of 1:4 presented a surface area of 535 $m^2 \cdot g^{-1}$, also comparable to the S_{BET} value obtained with PMMA:PAN ratio of 1:4 (450 $m^2 \cdot g^{-1}$ and 645 $m^2 \cdot g^{-1}$, depending on PMMA molecular weight). The carbon fibers obtained from $ZnCl_2$-PAN solutions showed a S_{BET} value of 680 $m^2 \cdot g^{-1}$ already at a low $ZnCl_2$:PAN ratio of 1:10, further increasing with increasing amounts of $ZnCl_2$: PAN ratio from 3:10 to 7:10. Due to the high value of 1083 $m^2 \cdot g^{-1}$ measured for the carbon fibers obtained using a $ZnCl_2$:PAN ratio of 7:10 and for the regular ammonia activation duration of 15 min at 900 °C, the effect of the duration of activation was performed for this sample, at the same temperature of 900 °C. It is noted that the ammonia pyrolysis was performed in flash mode [65], (sample heated from room temperature to 900 °C in the range of 1–1.5 min), which allows very precise control of the pyrolysis duration down to 5 min. To stop the pyrolysis, the quartz tube was immediately removed from the split hinge oven. The results are depicted in Figure 5. The specific area was already high after only 5 min of ammonia activation, with slightly increased values for increased durations up to 20 min. The optimal duration was considered to be 15 min, corresponding to the maximum compromise between high developed area and high mechanical resistance. A longer treatment of 20 min led to an increased fragility of the carbon web, limiting its utilization as a self-standing electrode.

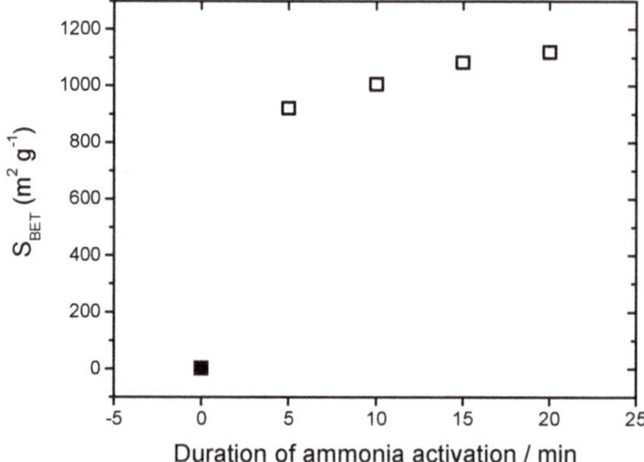

Figure 5. Effect of the duration of ammonia activation at 900 °C on the mass-specific surface area (S_{BET}) of Zn_7-PAN_{10}-CNF.

Raman spectra recorded on activated fibrous webs were very similar to those obtained with carbonized samples (Figure S5). The calculated I_D/I_G ratios (Table 3) were practically unchanged upon activation in ammonia, demonstrating that the degree of graphitization was not significantly impacted by the etching treatment.

As already mentioned, ammonia activation usually leads to (additional) nitrogen doping of the carbon [56,59]. In order to evaluate the introduction of nitrogen sites in the prepared ACNF webs, elemental analysis was performed before and after the NH_3 treatment, for selected CNF webs (Table 4). The nitrogen amount in the CNF derived from PAN only was around 5.6 wt % after carbonization (PAN_{10} row in Table 4, CNF column), in agreement with previous reports and with the carbonization temperature of 900 °C. It is known that higher pyrolysis temperatures result in lower nitrogen content [15]. For the reference PAN_{10}-CNF, the nitrogen amount slightly increased to 7.0% after ammonia activation, while a decrease in O wt % was observed. The latter effect is explained by selective etching of surface O-containing groups by ammonia. For PVP_5-PAN_{10} and Zn_7-PAN_{10}, an approximately three-fold increase in nitrogen amount was observed from CNF to ACNF webs, accompanied by a reduction in O wt % [59]. All ACNF webs however had comparable N wt % amounts, in the range of 6.1 to 7.0 wt %.

Nitrogen doping, in particular with ammonia (highly basic nitrogen groups), is interesting for some electrochemical applications such as supercapacitors or oxygen-reducing sites (in alkaline electrolytes in particular). The etching of some carbon masses during ammonia activation results in carbon fibers with increased internal porosity and also possibly weakened inter-fiber connections, which may impact the mechanical stability of self-standing fiber webs and also their electrical conductivity. The electrical conductivity of selected carbon fiber webs (one selected for each type of porogen) was halved after the ammonia activation (Table 4). For example, it dropped from 6.6 to 3.8 S·cm^{-1} for the reference web derived from PAN only, and from 9.7 to 4.8 S·cm^{-1} for PVP_5-PAN_{10}-CNF and PVP_5-PAN_{10}-ACNF, respectively. Interestingly, the fact that the conductivity decreased significantly even for PAN_{10}-ACNF while such fibers have no open porosity, indicates that the decreased in-plane conductivity cannot be assigned to the formation of open pores in the fibers. The decreased conductivity of carbon fiber webs after ammonia activation may be due to the nitrogen doping itself (decreasing the intrinsic conductivity of carbon) or weakened electrical contact at the nodes of the fibrous web (possibly due to etching of carbon at the nodes). As already mentioned earlier for the series of Zn_7-PAN_{10}-CNF fibers subjected to various duration of activation in ammonia, activation longer than 20 min still resulted in a self-standing web but with greatly increased fragility, which is directly related to the quality of the nodes of the webs.

All these results lead to the conclusion that the open porosity in the PAN-derived CNFs resulted from a two-step approach involving the combination of a template route with a chemical activation step. The porogens incorporated in the electrospun PAN solution led first to carbon fibers with closed pores after the carbonization in Ar. The closed pores could be opened and interconnected within the fiber upon a post-treatment in ammonia atmosphere at high temperature. Other examples of etching in combination with the use of porogens have been reported (e.g., water etching-assisted templating) [31].

The purpose of this work was the rationalization of an effective strategy to obtain CNF webs with both high porosity between fibers and high specific area by introducing porosity inside the fibers for application in electrochemical energy conversion and storage devices. Having this objective in mind, it is also important to maintain the electrical conductivity of the self-standing CNF webs to sufficiently high values. Typically, to be not be limited by the conduction of electrons across an electrode, the electric conductivity must be about 100 times higher than the electrolyte conductivity. The results indicate that the approach combining the introduction of templates and chemical activation led to open micro- and mesoporosity inside the fibers, while the electric conductivity was approximately halved during the ammonia activation step. This decrease in conductivity may be due to the removal of a fraction of the carbon from the bulk of the CNFs during ammonia activation and/or the fragilization of the interfibrous connection. In addition, when the electrical conductivity of self-standing CNF ammonia-activated webs is plotted against their BET area, a negative linear correlation is observed (Figure 6). The identification of such a correlation is useful to select the best compromise between conductivity and BET area, which will depend on the exact electrochemical application and range of typical current densities produced by an electrode (in the order of increasing current density: batteries < fuel cells < supercapacitors). With current densities <10 mA·cm^{-2}, the reported conductivities [66] are not a limiting factor in batteries, and thus any of the ACNF webs could be selected. With current densities in the range of 1–2 A·cm^{-2}, fuel cells are more demanding in terms of electron conductivity in the electrodes. By comparison, the effective proton conductivity in typical active layer of PEMFC is only in the range of 1–2 S·m^{-1}. Therefore, if the electronic conductivity through the active layer is >20 S·m^{-1} (0.2 S·cm^{-1}), it should not be a factor limiting the porous electrode performance [67–70]. All present ACNF webs could therefore be applicable for self-standing electrodes in, e.g., PEMFCs. Lastly, in electrochemical supercapacitors the instantaneous current density can reach extremely high values on the order of 10 A·cm^{-2} due to the non-faradaic process resulting in higher electric conductivity requirements. Moreover, the high electric conductivity requirement is accompanied by the high BET area requirement [71], which are shown in Figure 6 to be antagonistic.

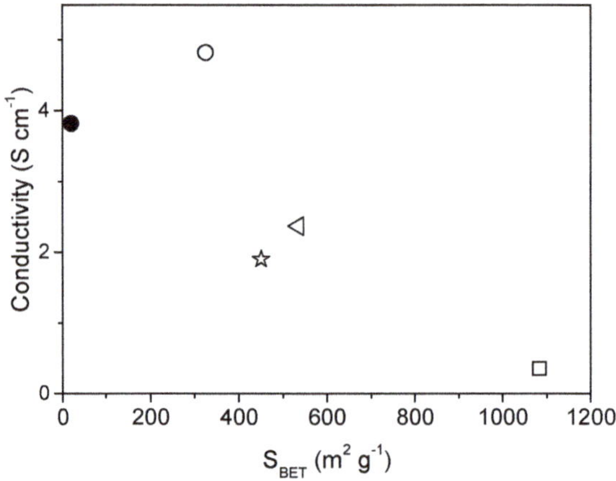

Figure 6. Electrical conductivity versus specific surface area of the ACNF fiber webs. ●—PAN$_{10}$-ACNF; ○—PVP$_5$-PAN$_{10}$-ACNF; ◁—Nafion$_2$-PAN$_8$-ACNF; ☆—15kPMMA$_2$-PAN$_8$-ACNF; □—Zn$_7$-PAN$_{10}$-ACNF.

4. Conclusions

- The templating approach of PAN fibers with polymer or inorganic porogens (PMMA, Nafion®, PVP, ZnCl$_2$) resulted in CNFs with closed porosity after carbonization in Ar.
- Subsequent activation in ammonia at 900 °C opened this porosity, resulting in specific surface area in the range of 325–1083 m^2·g^{-1}.
- Ammonia activation systematically decreased the electric conductivity of the ACNF webs by a factor of approximately two to three.
- A negative linear correlation between electric conductivity of ammonia-activated ACNF webs and their BET area is revealed.

Supplementary Materials: The following are available online at http://www.mdpi.com/2571-9637/2/1/13/s1, Figure S1. TEM micrographs of PAN CNFs prepared from PAN and PMMA after carbonization and activation; Figure S2. Raman spectra of template based CNFs after carbonization; Figure S3. Example of deconvolution of a Raman spectrum of CNFs (PAN$_{10}$-CNF) by using Gaussian and Lorentzian curves (Voigt); **Figure S4**. FE-SEM (left side) and cross-section TEM (right side) micrographs of PAN$_{10}$-ACNF fibers; Figure S5. Raman spectra of CNFs after activation.

Author Contributions: Conceptualization, S.C and F.J.; methodology, S.C., F.J. and D.J.; formal analysis, S.Y.; investigation, S.Y.; resources, S.C., F.J. and D.J.; writing—original draft preparation, S.Y and S.C.; writing—review and editing, F.J and D.J.; supervision, S.C and F.J.; project administration, F.J.; funding acquisition, F.J. and S.C.

Funding: The research leading to these results has received funding from the French National Research Agency under the CAT^{2}CAT contract (ANR-16-CE05-0007). SC acknowledges the financial support from the European

Research Council under the European Union's Seventh Framework Programme (FP/2007-2013) / ERC Grant Agreement n. 306682 and the French IUF.

Acknowledgments: Authors thank Nicolas Donzel for his help with nitrogen adsorption/desorption analysis and electrical conductivity measurements.

Conflicts of Interest: The authors declare no conflict of interest.

References

1. Aricò, A.S.; Bruce, P.; Scrosati, B.; Tarascon, J.-M.; van Schalkwijk, W. Nanostructured materials for advanced energy conversion and storage devices. *Nat. Mater.* **2005**, *4*, 366–377. [CrossRef] [PubMed]
2. Du, L.; Shao, Y.; Sun, J.; Yin, G.; Liu, J.; Wang, Y. Advanced catalyst supports for PEM fuel cell cathodes. *Nano Energy* **2016**, *29*, 314–322. [CrossRef]
3. Cavaliere, S.; Subianto, S.; Savych, I.; Jones, D.J.; Rozière, J. Electrospinning: Designed architectures for energy conversion and storage devices. *Energy Environ. Sci.* **2011**, *4*, 4761–4785. [CrossRef]
4. Peng, S.; Li, L.; Kong Yoong, J.L.; Tian, L.; Srinivasan, M.; Adams, S.; Ramakrishna, S. Electrospun carbon nanofibers and their hybrid composites as advanced materials for energy conversion and storage. *Nano Energy* **2016**, *22*, 361–395. [CrossRef]
5. Liu, Y.; Fan, L.Z.; Jiao, L. Graphene highly scattered in porous carbon nanofibers: A binder-free and high-performance anode for sodium-ion batteries. *J. Mater. Chem. A* **2017**, *5*, 1698–1705. [CrossRef]
6. Peng, Y.; Lo, C. Electrospun porous carbon nanofibers as lithium ion battery anodes. *J. Solid State Electrochem.* **2015**, *19*, 3401–3410. [CrossRef]
7. Li, X.; Zhao, Y.; Bai, Y.; Zhao, X.; Wang, R.; Huang, Y.; Liang, Q.; Huang, Z. A Non-Woven Network of Porous Nitrogen-doping Carbon Nanofibers as a Binder-free Electrode for Supercapacitors. *Electrochim. Acta* **2017**, *230*, 445–453. [CrossRef]
8. Gopalakrishnan, A.; Sahatiya, P.; Badhulika, S. Template-Assisted Electrospinning of Bubbled Carbon Nanofibers as Binder-Free Electrodes for High-Performance Supercapacitors. *ChemElectroChem* **2018**, *5*, 531–539. [CrossRef]
9. Liu, D.; Zhang, X.; Sun, Z.; You, T. Free-standing nitrogen-doped carbon nanofiber films as highly efficient electrocatalysts for oxygen reduction. *Nanoscale* **2013**, *5*, 9528–9531. [CrossRef] [PubMed]
10. Ercolano, G.; Farina, F.; Cavaliere, S.; Jones, D.J.; Rozière, J. Towards ultrathin Pt films on nanofibres by surface-limited electrodeposition for electrocatalytic applications. *J. Mater. Chem. A* **2017**, *5*, 3974–3980. [CrossRef]
11. Kayarkatte, M.K.; Delikaya, Ö.; Roth, C. Freestanding Catalyst Layers: A Novel Electrode Fabrication Technique for PEM Fuel Cells via Electrospinning. *ChemElectroChem* **2017**, *4*, 404–411. [CrossRef]
12. Chan, S.; Jankovic, J.; Susac, D.; Saha, M.S.; Tam, M.; Yang, H.; Ko, F. Electrospun carbon nanofiber catalyst layers for polymer electrolyte membrane fuel cells: Structure and performance. *J. Power Sources* **2018**, *392*, 239–250. [CrossRef]
13. Sun, J.; Zeng, L.; Jiang, H.R.; Chao, C.Y.H.; Zhao, T.S. Formation of electrodes by self-assembling porous carbon fibers into bundles for vanadium redox flow batteries. *J. Power Sources* **2018**, *405*, 106–113. [CrossRef]
14. Cao, Y.; Lu, H.; Hong, Q.; Bai, J.; Wang, J.; Li, X. Co decorated N-doped porous carbon nanofibers as a free-standing cathode for Li-O$_2$ battery: Emphasis on seamlessly continuously hierarchical 3D nano-architecture networks. *J. Power Sources* **2017**, *368*, 78–87. [CrossRef]
15. Liu, C.-K.; Lai, K.; Liu, W.; Yao, M.; Sun, R.-J. Preparation of carbon nanofibres through electrospinning and thermal treatment. *Polym. Int.* **2009**, *58*, 1341–1349. [CrossRef]
16. Nan, D.; Wang, J.-G.; Huang, Z.-H.; Wang, L.; Shen, W.; Kang, F. Highly porous carbon nanofibers from electrospun polyimide/SiO$_2$ hybrids as an improved anode for lithium-ion batteries. *Electrochem. Commun.* **2013**, *34*, 52–55. [CrossRef]
17. Zhang, L.; Jiang, Y.; Wang, L.; Zhang, C.; Liu, S. Hierarchical porous carbon nanofibers as binder-free electrode for high-performance supercapacitor. *Electrochim. Acta* **2016**, *196*, 189–196. [CrossRef]
18. Alothman, Z.A. A review: Fundamental aspects of silicate mesoporous materials. *Materials* **2012**, *5*, 2874–2902. [CrossRef]

19. Ma, C.; Cao, E.; Li, J.; Fan, Q.; Wu, L.; Song, Y.; Shi, J. Synthesis of mesoporous ribbon-shaped graphitic carbon nanofibers with superior performance as efficient supercapacitor electrodes. *Electrochim. Acta* **2018**, *292*, 364–373. [CrossRef]

20. Chen, Y.; Lu, Z.; Zhou, L.; Mai, Y.-W.; Huang, H. In situ formation of hollow graphitic carbon nanospheres in electrospun amorphous carbon nanofibers for high-performance Li-based batteries. *Nanoscale* **2012**, *4*, 6800. [CrossRef] [PubMed]

21. Xie, W.; Khan, S.; Rojas, O.J.; Parsons, G.N. Control of Micro- and Mesopores in Carbon Nanofibers and Hollow Carbon Nanofibers Derived from Cellulose Diacetate via Vapor Phase Infiltration of Diethyl Zinc. *ACS Sustain. Chem. Eng.* **2018**, *6*, 13844–13853. [CrossRef]

22. Ma, C.; Li, Z.; Li, J.; Fan, Q.; Wu, L.; Shi, J.; Song, Y. Lignin-based hierarchical porous carbon nanofiber films with superior performance in supercapacitors. *Appl. Surf. Sci.* **2018**, *456*, 568–576. [CrossRef]

23. Kim, C.; Ngoc, B.T.N.; Yang, K.S.; Kojima, M.; Kim, Y.A.; Kim, Y.J.; Endo, M.; Yang, S.C. Self-Sustained Thin Webs Consisting of Porous Carbon Nanofibers for Supercapacitors via the Electrospinning of Polyacrylonitrile Solutions Containing Zinc Chloride. *Adv. Mater.* **2007**, *19*, 2341–2346. [CrossRef]

24. Yin, D.; Han, C.; Bo, X.; Liu, J.; Guo, L. Prussian blue analogues derived iron-cobalt alloy embedded in nitrogen-doped porous carbon nanofibers for efficient oxygen reduction reaction in both alkaline and acidic solutions. *J. Colloid Interface Sci.* **2019**, *533*, 578–587. [CrossRef] [PubMed]

25. Zhang, W.; Yao, X.; Zhou, S.; Li, X.; Li, L.; Yu, Z.; Gu, L. ZIF-8/ZIF-67-Derived Co-N_x-Embedded 1D Porous Carbon Nanofibers with Graphitic Carbon-Encased Co Nanoparticles as an Efficient Bifunctional Electrocatalyst. *Small* **2018**, *14*, 1800423. [CrossRef] [PubMed]

26. Yao, Y.; Liu, P.; Li, X.; Zeng, S.; Lan, T.; Huang, H.; Zeng, X.; Zou, J. Nitrogen-doped graphitic hierarchically porous carbon nanofibers obtained: Via bimetallic-coordination organic framework modification and their application in supercapacitors. *Dalton Trans.* **2018**, *47*, 7316–7326. [CrossRef] [PubMed]

27. Wang, C.; Kaneti, Y.V.; Bando, Y.; Lin, J.; Liu, C.; Li, J.; Yamauchi, Y. Metal-organic framework-derived one-dimensional porous or hollow carbon-based nanofibers for energy storage and conversion. *Mater. Horiz.* **2018**, *5*, 394–407. [CrossRef]

28. Zhang, W.; Miao, W.; Liu, X.; Li, L.; Yu, Z.; Zhang, Q. High-rate and ultralong-stable potassium-ion batteries based on antimony-nanoparticles encapsulated in nitrogen and phosphorus co-doped mesoporous carbon nanofibers as an anode material. *J. Alloys Compd.* **2018**, *769*, 141–148. [CrossRef]

29. Jeong, J.; Choun, M.; Lee, J. Tree-Bark-Shaped N-Doped Porous Carbon Anode for Hydrazine Fuel Cells. *Angew. Chem. Int. Ed.* **2017**, *56*, 13513–13516. [CrossRef] [PubMed]

30. Zhao, X.; Xiong, P.; Meng, J.; Liang, Y.; Wang, J.; Xu, Y. High rate and long cycle life porous carbon nanofiber paper anodes for potassium-ion batteries. *J. Mater. Chem. A* **2017**, *5*, 19237–19244. [CrossRef]

31. An, G.-H.; Koo, B.-R.; Ahn, H.-J. Activated mesoporous carbon nanofibers fabricated using water etching-assisted templating for high-performance electrochemical capacitors. *Phys. Chem. Chem. Phys.* **2016**, *18*, 6587–6594. [CrossRef] [PubMed]

32. He, T.; Fu, Y.; Meng, X.; Yu, X.; Wang, X. A novel strategy for the high performance supercapacitor based on polyacrylonitrile-derived porous nanofibers as electrode and separator in ionic liquid electrolyte. *Electrochim. Acta* **2018**, *282*, 97–104. [CrossRef]

33. Yang, D.S.; Chaudhari, S.; Rajesh, K.P.; Yu, J.S. Preparation of nitrogen-doped porous carbon nanofibers and the effect of porosity, electrical conductivity, and nitrogen content on their oxygen reduction performance. *ChemCatChem* **2014**, *6*, 1236–1244. [CrossRef]

34. Tran, C.; Kalra, V. Fabrication of porous carbon nanofibers with adjustable pore sizes as electrodes for supercapacitors. *J. Power Sources* **2013**, *235*, 289–296. [CrossRef]

35. Wang, W.; Wang, H.; Wang, H.; Jin, X.; Li, J.; Zhu, Z. Electrospinning preparation of a large surface area, hierarchically porous, and interconnected carbon nanofibrous network using polysulfone as a sacrificial polymer for high performance supercapacitors. *RSC Adv.* **2018**, *8*, 28480–28486. [CrossRef]

36. Liu, J.; Xiong, Z.; Wang, S.; Cai, W.; Yang, J.; Zhang, H. Structure and electrochemistry comparison of electrospun porous carbon nanofibers for capacitive deionization. *Electrochim. Acta* **2016**, *210*, 171–180. [CrossRef]

37. Ji, L.; Zhang, X. Fabrication of porous carbon nanofibers and their application as anode materials for rechargeable lithium-ion batteries. *Nanotechnology* **2009**, *20*, 155705. [CrossRef] [PubMed]

38. Zhang, H.; Xie, Z.; Wang, Y.; Shang, X.; Nie, P.; Liu, J. Electrospun polyacrylonitrile/β-cyclodextrin based porous carbon nanofiber self-supporting electrode for capacitive deionization. *RSC Adv.* **2017**, *7*, 55224–55231. [CrossRef]

39. Le, T.H.; Tian, H.; Cheng, J.; Huang, Z.H.; Kang, F.; Yang, Y. High performance lithium-ion capacitors based on scalable surface carved multi hierarchical construction electrospun carbon fibers. *Carbon N. Y.* **2018**, *138*, 325–336. [CrossRef]

40. Heo, Y.-J.; Lee, H.I.; Lee, J.W.; Park, M.; Rhee, K.Y.; Park, S.-J. Optimization of the pore structure of PAN-based carbon fibers for enhanced supercapacitor performances via electrospinning. *Compos. Part B Eng.* **2019**, *161*, 10–17. [CrossRef]

41. Lee, H.-M.; Kim, H.-G.; Kang, S.-J.; Park, S.-J.; An, K.-H.; Kim, B.-J. Effects of pore structures on electrochemical behaviors of polyacrylonitrile (PAN)-based activated carbon nanofibers. *J. Ind. Eng. Chem.* **2015**, *21*, 736–740. [CrossRef]

42. Kim, C.H.; Kim, B.-H. Zinc oxide/activated carbon nanofiber composites for high-performance supercapacitor electrodes. *J. Power Sources* **2015**, *274*, 512–520. [CrossRef]

43. Kim, C.; Yang, K.S. Electrochemical properties of carbon nanofiber web as an electrode for supercapacitor prepared by electrospinning. *Appl. Phys. Lett.* **2003**, *83*, 1216–1218. [CrossRef]

44. Zhang, T.; Xiao, B.; Zhou, P.; Xia, L.; Wen, G.; Zhang, H. Porous-carbon-nanotube decorated carbon nanofibers with effective microwave absorption properties. *Nanotechnology* **2017**, *28*, 355708. [CrossRef] [PubMed]

45. Fitzer, E.; Frohs, W.; Heine, M. Optimization of stabilization and carbonization treatment of PAN fibres and structural characterization of the resulting carbon fibres. *Carbon N. Y.* **1986**, *24*, 387–395. [CrossRef]

46. Moon, S.; Farris, R.J. Strong electrospun nanometer-diameter polyacrylonitrile carbon fiber yarns. *Carbon N. Y.* **2009**, *47*, 2829–2839. [CrossRef]

47. Savych, I.; Bernard D'Arbigny, J.; Subianto, S.; Cavaliere, S.; Jones, D.J.; Rozière, J. On the effect of non-carbon nanostructured supports on the stability of Pt nanoparticles during voltage cycling: A study of TiO_2 nanofibres. *J. Power Sources* **2014**, *257*, 147–155. [CrossRef]

48. Heikkila, P. Electrospinning of polyacrylonitrile (PAN) solution: Effect of conductive additive and filler on the process. *eXPRESS Polym. Lett.* **2009**, *3*, 437–445. [CrossRef]

49. Subianto, S.; Cornu, D.; Cavaliere, S. Fundamentals of Electrospinning. In *Electrospinning for Advanced Energy and Environmental Applications*; Cavaliere, S., Ed.; CRC Press: Boca Raton, FL, USA, 2015; pp. 1–27.

50. Karacan, I.; Erzurumluoğlu, L. The effect of carbonization temperature on the structure and properties of carbon fibers prepared from poly(m-phenylene isophthalamide) precursor. *Fibers Polym.* **2015**, *16*, 1629–1645. [CrossRef]

51. Shi, R.; Han, C.; Xu, X.; Qin, X.; Xu, L.; Li, H.; Li, J.; Wong, C.P.; Li, B. Electrospun N-Doped Hierarchical Porous Carbon Nanofiber with Improved Degree of Graphitization for High-Performance Lithium Ion Capacitor. *Chem. A Eur. J.* **2018**, *24*, 10460–10467. [CrossRef] [PubMed]

52. Cançado, L.G.; Jorio, A.; Ferreira, E.H.M.; Stavale, F.; Achete, C.A.; Capaz, R.B.; Moutinho, M.V.O.; Lombardo, A.; Kulmala, T.S.; Ferrari, A.C. Quantifying Defects in Graphene via Raman Spectroscopy at Different Excitation Energies. *Nano Lett.* **2011**, *11*, 3190–3196. [CrossRef] [PubMed]

53. Cançado, L.G.; Gomes da Silva, M.; Martins Ferreira, E.H.; Hof, F.; Kampioti, K.; Huang, K.; Pénicaud, A.; Alberto Achete, C.; Capaz, R.B.; Jorio, A. Disentangling contributions of point and line defects in the Raman spectra of graphene-related materials. *2D Mater.* **2017**, *4*, 025039. [CrossRef]

54. Zhang, Z.; Li, X.; Wang, C.; Fu, S.; Liu, Y.; Shao, C. Polyacrylonitrile and carbon nanofibers with controllable nanoporous structures by electrospinning. *Macromol. Mater. Eng.* **2009**, *294*, 673–678. [CrossRef]

55. Inagaki, M.; Yang, Y.; Kang, F. Carbon nanofibers prepared via electrospinning. *Adv. Mater.* **2012**, *24*, 2547–2566. [CrossRef] [PubMed]

56. Atamny, F.; Blöcker, J.; Dübotzky, A.; Kurt, H.; Timpe, O.; Loose, G.; Mahdi, W.; Schlögl, R. Surface chemistry of carbon: Activation of molecular oxygen. *Mol. Phys.* **1992**, *76*, 851–886. [CrossRef]

57. Jaouen, F.; Serventi, A.M.; Lefèvre, M.; Dodelet, J.; Bertrand, P. Non-Noble Electrocatalysts for O 2 Reduction: How Does Heat Treatment Affect Their Activity and Structure? Part II. Structural Changes Observed by Electron Microscopy, Raman, and Mass Spectroscopy. *J. Phys. Chem. C* **2007**, *111*, 5971–5976. [CrossRef]

58. Jaouen, F.; Dodelet, J. Non-Noble Electrocatalysts for O_2 Reduction: How Does Heat Treatment Affect Their Activity and Structure? Part I. Model for Carbon Black Gasification by NH 3: Parametric Calibration and Electrochemical Validation. *J. Phys. Chem. C* **2007**, *111*, 5963–5970. [CrossRef]

59. Shen, W.; Fan, W. Nitrogen-containing porous carbons: Synthesis and application. *J. Mater. Chem. A* **2013**, *1*, 999. [CrossRef]

60. Artyushkova, K.; Kiefer, B.; Halevi, B.; Knop-Gericke, A.; Schlogl, R.; Atanassov, P. Density functional theory calculations of XPS binding energy shift for nitrogen-containing graphene-like structures. *Chem. Commun.* **2013**, *49*, 2539. [CrossRef] [PubMed]

61. Nan, D.; Huang, Z.H.; Lv, R.; Yang, L.; Wang, J.G.; Shen, W.; Lin, Y.; Yu, X.; Ye, L.; Sun, H.; et al. Nitrogen-enriched electrospun porous carbon nanofiber networks as high-performance free-standing electrode materials. *J. Mater. Chem. A* **2014**, *2*, 19678–19684. [CrossRef]

62. Subianto, S.; Cavaliere, S.; Jones, D.J.; Rozière, J. Effect of side-chain length on the electrospinning of perfluorosulfonic acid ionomers. *J. Polym. Sci. Part A Polym. Chem.* **2013**, *51*, 118–128. [CrossRef]

63. Ballengee, J.B.; Pintauro, P.N. Morphological Control of Electrospun Nafion Nanofiber Mats. *J. Electrochem. Soc.* **2011**, *158*, B568–B572. [CrossRef]

64. Brunauer, S.; Emmett, P.H.; Teller, E. Adsorption of Gases in Multimolecular Layers. *J. Am. Chem. Soc.* **1938**, *60*, 309–319. [CrossRef]

65. Jaouen, F.; Lefèvre, M.; Dodelet, J.-P.; Cai, M. Heat-Treated Fe/N/C Catalysts for O2 Electroreduction: Are Active Sites Hosted in Micropores? *J. Phys. Chem. B* **2006**, *110*, 5553–5558. [CrossRef] [PubMed]

66. Fehse, M.; Cavaliere, S.; Lippens, P.E.; Savych, I.; Iadecola, A.; Monconduit, L.; Jones, D.J.; Rozie, J.; Fischer, F.; Tessier, C.; et al. Nb-Doped TiO_2 Nanofibers for Lithium Ion Batteries. *J. Phys. Chem. C* **2013**, *117*, 13827–13835. [CrossRef]

67. Liu, Y.; Murphy, M.W.; Baker, D.R.; Gu, W.; Ji, C.; Jorne, J.; Gasteiger, H.A. Proton Conduction and Oxygen Reduction Kinetics in PEM Fuel Cell Cathodes: Effects of Ionomer-to-Carbon Ratio and Relative Humidity. *J. Electrochem. Soc.* **2009**, *156*, B970. [CrossRef]

68. Liu, Y.; Ji, C.; Gu, W.; Jorne, J.; Gasteiger, H.A. Effects of Catalyst Carbon Support on Proton Conduction and Cathode Performance in PEM Fuel Cells. *J. Electrochem. Soc.* **2011**, *158*, B614. [CrossRef]

69. Baghalha, M.; Eikerling, M.; Stumper, J.; Harvey, D.; Eikerling, M. Modeling the effect of low carbon conductivity of the cathode catalyst layer on PEM fuel cell performance. *ECS Trans.* **2010**, *28*, 113–123.

70. Savych, I.; Subianto, S.; Nabil, Y.; Cavaliere, S.; Jones, D.; Rozière, J. Negligible degradation upon in situ voltage cycling of a PEMFC using an electrospun niobium-doped tin oxide supported Pt cathode. *Phys. Chem. Chem. Phys.* **2015**, *17*, 16970–16976. [CrossRef] [PubMed]

71. Lu, X.; Wang, C.; Favier, F.; Pinna, N. Electrospun Nanomaterials for Supercapacitor Electrodes: Designed Architectures and Electrochemical Performance. *Adv. Energy Mater.* **2017**, *7*, 1601301. [CrossRef]

 *surfaces*

Article

Potential Dependent Structure and Stability of Cu(111) in Neutral Phosphate Electrolyte

Yvonne Grunder [1], Jack Beane [1], Adam Kolodziej [2], Christopher A. Lucas [1] and Paramaconi Rodriguez [2,*]

[1] Department of Physics, University of Liverpool, Liverpool L69 7ZE, UK; Yvonne.Grunder@liverpool.ac.uk (Y.G.); sgjbeane@student.liverpool.ac.uk (J.B.); clucas@liverpool.ac.uk (C.A.L.)

[2] School of Chemistry, University of Birmingham. Birmingham B15 2TT, UK; AXK595@student.bham.ac.uk

* Correspondence: p.b.rodriguez@bham.ac.uk

Received: 30 January 2019; Accepted: 20 February 2019; Published: 24 February 2019

Abstract: Copper and copper oxide electrode surfaces are suitable for the electrochemical reduction of CO_2 and produce a range of products, with the product selectivity being strongly influenced by the surface structure of the copper electrode. In this paper, we present in-situ surface X-ray diffraction studies on Cu(111) electrodes in neutral phosphate buffered electrolyte solution. The underlying mechanism of the phosphate adsorption and deprotonation of the (di)-hydrogen phosphate is accompanied by a roughening of the copper surface. A change in morphology of the copper surface induced by a roughening process caused by the formation of a mixed copper–oxygen layer could also be observed. The stability of the Cu(111) surface and the change of morphology upon potential cycling strongly depends on the preparation method and history of the electrode. The presence of copper islands on the surface of the Cu(111) electrode leads to irreversible changes in surface morphology via a 3D Cu growth mechanism.

Keywords: Cu(111); electrochemical interface; in-situ X-ray diffraction

---

1. Introduction

Compared to many metal catalysts, copper surfaces have been proven to electrochemically convert CO_2 to high value and energy-dense products, such as methane ethylene, formic acid, methanol and ethanol amongst others. However, the efficiency and the selectivity are far from optimal and the parameters controlling these factors are not-fully understood. Differences in reactivity and selectivity have been variously ascribed to surface area, particle size, surface structure and roughness and the electrolyte composition [1–6]. More recently, the effect of the role of the oxygen content of the copper catalyst and the oxidation state of the copper on the electrocatalytic activity and selectivity have illustrated the high complexity of the system. Oxide-derived Cu catalysts have shown high-selectivity towards the formation of C_2 products [7–10]. Such selectivity has been attributed to changes in the surface structure, including roughness and defects, with active sites being generated during the reduction pretreatment of the Cu-oxide catalyst. However, more recently, in addition to the structural factors, the presence of Cu^+ and residual subsurface oxygen has been suggested to affect the product selectivity [5,11,12]. Mistry et al. demonstrated, via operando XAFS, the presence of Cu^+ species and subsurface oxygen during the carbon dioxide reduction reaction [5]. In another report, LeDuff et al. [7] implemented a pulse sequence between reduction potentials where the carbon dioxide reduction reaction (CO2RR) takes place (<-0.5 V) and a potential in the region between -0.2 and -0.35 V vs. RHE where the co-adsorption of OH and other anions takes place [13]. They concluded that the positive potential of the pulse has a significant effect in the catalytic activity and selectivity of the copper single-crystal electrodes towards the CO2RR. This was associated to the adsorbed species, presumably

OH or anions, at the surface of the Cu single crystal electrodes which prevent any irreversible damage or changes to the Cu surface structure [7].

In this paper, we present the in-situ characterization by surface X-ray diffraction of the Cu(111) electrode at positive potentials with the aim to determine the composition and structure of the electrochemical interface. The key to understanding the stability and reactivity of the Cu(111) electrode is to control the surface morphology during the preparation of the clean surface. Depending on the details of surface preparation, two different Cu(111) surfaces can be obtained; although macroscopically rather rough, one surface is completely stable during potential cycling that involves considerable modification of the surface structure. The other surface gives rise to X-ray scattering features consistent with the presence of twinned Cu nano-islands and, although this surface exhibits a similar potential-dependent restructuring, it is not stable during potential cycling. The direct correlation of structural stability, only available via in-situ structural characterization, and reactivity is vital to understanding structure–reactivity relationships, especially in studies of more active metal electrodes [14,15].

2. Materials and Methods

The Cu(111) single crystal working electrode (MaTeck, miscut < 0.1°) was prepared by electropolishing for 10 s in 70 % orthophosphoric acid at 2 V against a high-surface area copper mesh. The crystal was then rinsed in ultra-pure water, covered with the electrolyte solution and transferred to the electrochemical cell. A copper wire was used as counter electrode and an Ag/AgCl electrode was used for the reference electrode. All potentials are quoted against this reference electrode. The experimental procedure for the surface X-ray diffraction experiments followed that of similar studies reported previously [16,17]. Surface X-ray diffraction measurements were carried out on the I07 beamline at Diamond Light Source (Harwell Science and Innovation Campus, Fermi Ave, Didcot OX11 0DE, Oxford, UK) [18], with a monochromatic beam of 25 keV X-rays. Beam defining slits were 0.5 mm × 0.5 mm and the beam size at the sample was estimated to be 200 μm × 300 μm (vertical × horizontal). The sample was mounted on a (2 + 3) circle diffractometer [19]. A 2D Pilatus Dectris 1 M detector (DECTRIS Ltd., Baden-Daettwil, Switzerland) was used for the data acquisition. For the measurement of crystal truncation rods, scans along the Q_z (surface normal) direction at specific points in reciprocal space were recorded. Background correction and standard instrumental corrections were applied to the dataset to be able to model the intensity distribution [20,21]. Errors on the individual data points were a combination of the statistical error and an estimated 10% systematic error. The model of X-ray diffraction was obtained using a Python (Python Software Foundation, Beaverton, OR, USA) program and the integrated lmfit leastsquare fitting routine, which returned the best values for the relaxation, coverage and β-factor. The Cu(111) surface has a close packed structure with a hexagonal unit cell, where the surface normal is along the $(0, 0, L)_{hex}$ direction and the surface plane contains the $(H, 0, 0)_{hex}$ and $(0, K, 0)_{hex}$ vectors, which are separated by 60°. The Miller indices H, K, and L have units of $a^* = b^* = 4\pi/\sqrt{3}a_{NN}$ and $c^* = 2\pi/\sqrt{6}a_{NN}$, where the nearest neighbour for copper is $a_{NN} = 2.556$ Å.

3. Results

The Cu(111) surfaces were characterised in a phosphate buffered electrolyte by surface X-ray diffraction and with electrochemical methods.

3.1. Electrochemical Characterization

The Cu(111) electrode was characterised in phosphate buffer solution of concentration equal to 0.01 M, 0.05 M and 0.1 M at different pH values of 6 and 8 with cyclic voltammetry at different scan rates (Figure 1A). In the anodic scan, a shoulder (A1) at approximately −0.65 V and a peak (A2) approximately −0.5 V can be observed at both pHs; however, at lower pH, the second anodic peak is more defined. At pH = 6, the anodic peak shifts towards higher potentials with decreasing phosphate concentration (Figure 1B). The total charge associated to the electrochemical process shows equal charge in both the anodic and cathodic scan of 52–60 $\mu C/cm^2$, which corresponds to ~0.2 electrons per surface atom. The charge values were found to be independent of the pH and phosphate concentration (Figure 1C,D).

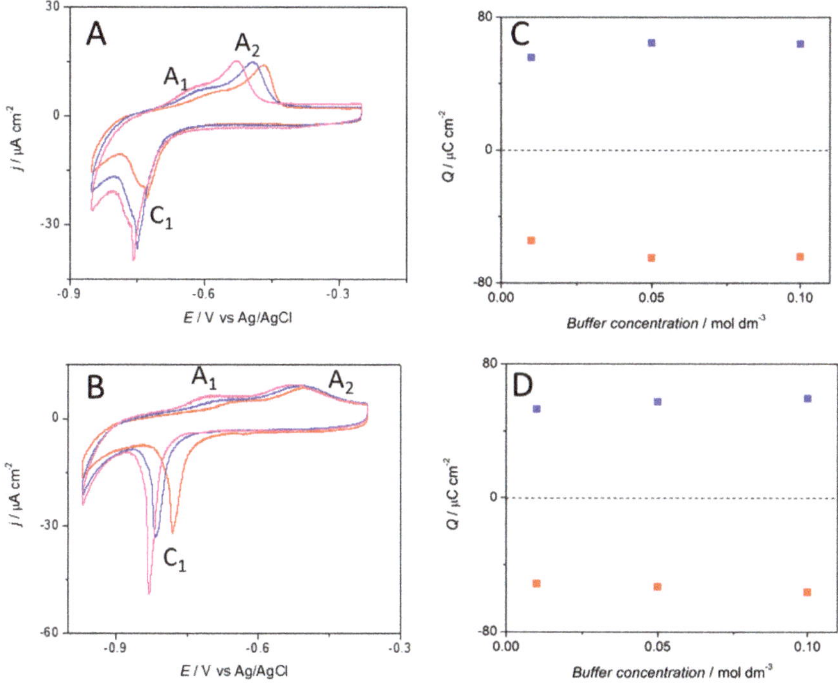

Figure 1. Cyclic voltammetry profiles of a Cu(111) electrode in a phosphate buffer solution at (**A**) pH = 6 and (**B**) pH = 8 with phosphate concentration equal to 0.01 M (red line), 0.05 M (blue line) and 0.1 M (magenta line). Scan rate ϑ = 0.02 V s^{-1}. (**C,D**) Charge involved in the oxidative (■) and reductive (■) process at the Cu(111) electrode as a function of the phosphate buffer concentration for pH = 6 and pH = 8 respectively.

3.2. Structural Characterization

The Cu(111) surface was investigated in phosphate buffer solution of concentration 0.1 M at pH = 8 with in-situ surface X-ray diffraction. Throughout the experiment, we noted the difference in the quality/morphology of the starting surface, which resulted specifically in a difference in reversibility of adsorption processes and stability. The differences in surface morphology are described in more detail below. Prior to any prolonged potential cycling resulting in excessive time spent at potentials where phosphate is adsorbed, all surfaces exhibited a rather similar change in structure as the potential was cycled. Figure 2 shows the X-ray intensities measured at the reciprocal space positions (0, 0, 1.4) and (0, 1, 0.5), shown as dotted and solid lines, respectively, and are representative of the potential dependent behaviour. The non-specular (0, 1, 0.5) position is sensitive to any changes in the atomic structure of the Cu(111) surface or to the adsorption of species into well-defined Cu adsorption sites, whereas the specular (0, 0, 1.4) position is sensitive to any changes in the surface normal electron density distribution including the electrolyte side of the interface. The results indicate that there is a clear structural change at the interface, most likely involving restructuring of the Cu surface itself and this exhibits considerable hysteresis consistent with significant rearrangement of surface atoms.

Following the X-ray voltammetry (XRV) measurements, such as those shown in Figure 2, the Cu(111) surface structures were characterised while changing the potential in steps of 0.1 V over the range from −1.0 V to −0.5 V. For each potential, a rocking scan was measured at the (0, 1, 0.5) position and the integrated intensity obtained from this measurement is also shown in Figure 2 as black squares. It can be seen that these lie halfway in the hysteresis loop. The surface was characterised by detailed crystal truncation rod (CTR) measurements at potentials in the order −1.0 V, −0.9 V, −0.8 V, −0.7 V, −0.6 and −0.5 V, after which the potential was stepped back to −0.8 V to check the reproducibility and stability of the surface structure. The exact potential history of the experiment is shown in Figure 3 together with the integrated intensity and the peak width at the (0, 1, 0.5) position. It can be seen that the peak width did not change throughout the course of the experiment, indicating a stable surface, i.e., there was no change in surface morphology or domain size (which is inversely proportional to the width of the peak in the rocking scan). The integrated intensity changed with the applied potential as expected from the results shown in Figure 2. This change was reversible, as indicated by the two datasets recorded at −0.8 V, before and after the potential cycle, which gave the same structural parameters within the experimental error.

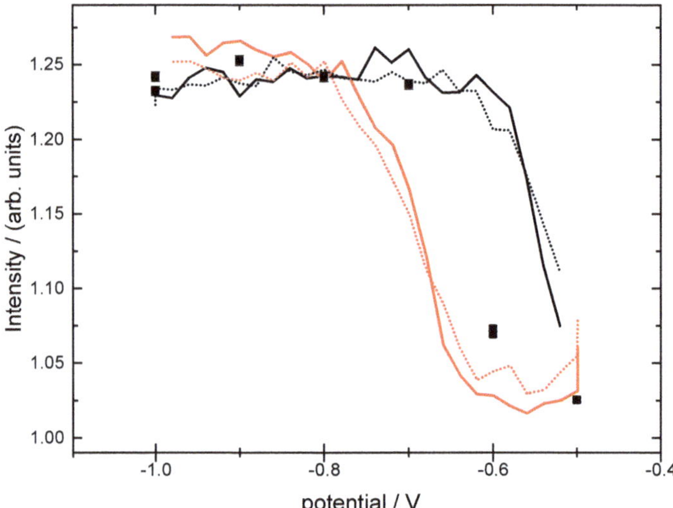

Figure 2. The potential dependence of the X-ray intensities measured at the non-specular, (0, 1, 0.5), (solid line) and specular, (0, 0, 1.4), anti-Bragg positions (sweep rate = 2 mV/s). In addition, the intensity obtained from measuring the integrated intensity at selected potentials under static conditions is also shown (black squares).

The CTR data and the best fit to the data obtained through a least-square fitting procedure for potentials of −0.5 V and −0.8 V are shown in Figure 4. The structural model used to fit the data allowed variation in the coverage of the topmost copper layer, the relaxation of the two topmost copper layers and also included a β-roughness factor which is typically used to model surface roughness in CTR measurements [22]. For the data measured at positive potentials it was also necessary to include an adsorbed oxygen layer with the oxygen atoms occupying *fcc*/*hcp* three-fold hollow Cu sites. The oxygen coverage and distance to the topmost Cu layer was allowed to vary to obtain the best fit. The obtained parameters for the best fits to the data at the different potentials are summarised in Figure 5. At potentials positive of −0.7 V, the coverage of the topmost copper layer started decreasing and the oxygen coverage increased, together with an increased surface roughness. The inclusion of the oxygen layer in the structural model resulted in a decrease of the reduced $\chi2$ (indicating the goodness of fit in the least squares method) from 3.3 to 1.8 for the fit to the data recorded at −0.5 V. In-plane scans were measured at all potentials at L = 0.4 along the high symmetry <1, 0, L>, <0, 1, L> and <1, 1, L> directions but did not show any additional superstructure peaks. Thus, we conclude that there is no specific ordering of the phosphate or any oxygenated species into a commensurate ordered adsorbate or copper oxide layer, as such ordering would give rise to additional scattering features that would be detected in the scans along the high symmetry directions. Previous works, in gas phase and computational studies, suggest the formation of hexagonal or quasi-hexagonal structures on the Cu(111) surface upon oxygen adsorption, structures which can be viewed as the initial layer of a Cu_2O (111) film. The model proposed by Platzman et al. suggests a three-step oxidation mechanism comprising the formation of a Cu_2O layer, followed by the formation of a metastable overlayer of $Cu(OH)_2$ and finally the transformation of this metastable overlayer phase into CuO layer [23]. Our results in electrochemical environment do not show any of these ordered structures in the potential range under study.

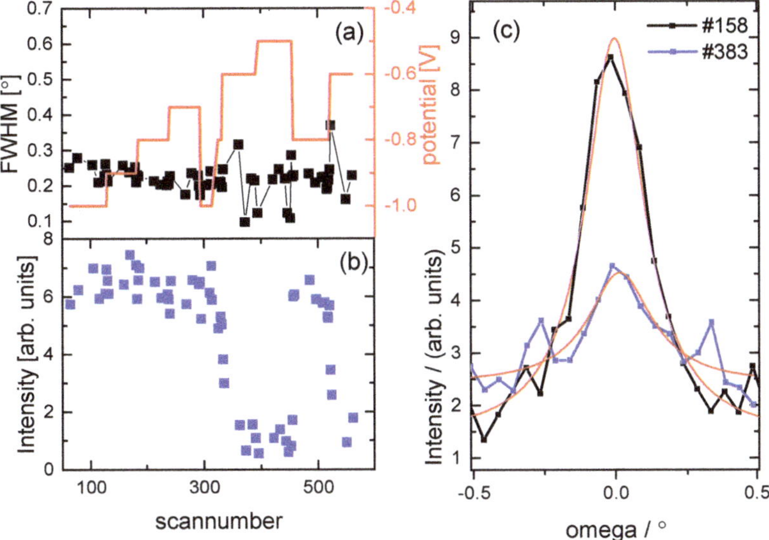

Figure 3. The surface stability has been monitored throughout the course of the experiment. (**a**) The full width half maximum (FWHM) and integrated intensity obtained from rocking scans at the (0, 1, 0.5) position are shown as function of the scan number during the experiment. The potential is also shown in red. The peak width is stable, indicating a stable surface with a domain size of ~100 nm. (**b**) The integrated intensity is changing reversibly with the applied potential. (**c**) The peak profiles of the rocking scan and the fit of the peak with a Lorentzian line shape are shown for scan numbers 158 and 383 at a potential of −0.9 V and −0.6 V, respectively.

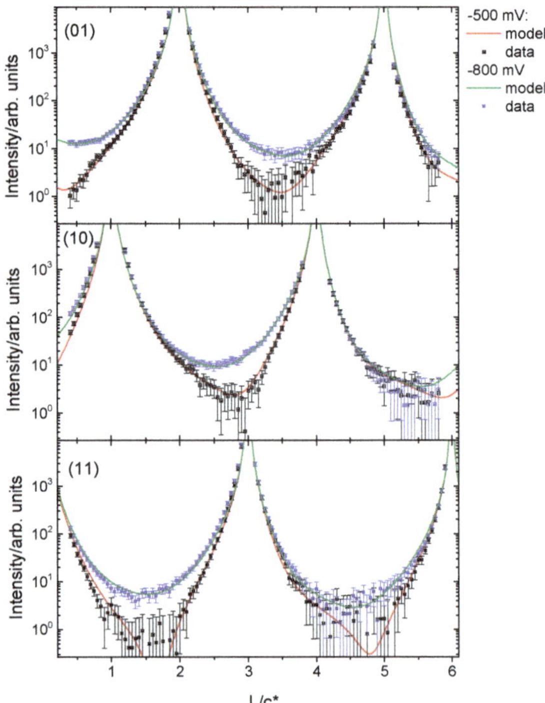

Figure 4. CTR data measured at −0.8 V (blue symbols) and −0.5 V (black symbols) together with the calculated fits to the data (solid lines) according to the structural model described in the text. At −0.5 V, an overall decrease in intensity at the surface sensitive positions can be observed. This has been modelled with an increased roughness and adsorption of an oxygenated species into the surface Cu layer. The change in intensity induced by the potential is fully reversible.

As noted above, the morphology of the surface depends crucially on the surface preparation and transfer into the electrochemical system. For the above case following the electropolishing of the Cu crystal, the transfer process to the X-ray electrochemical cell was smooth and the working electrode was contacted and held at −0.8 V as soon as the crystal was in the cell. Any exposure to oxygen during the preparation or not applying a potential in the oxygen reduction region immediately upon immersion into the cell resulted in a different surface morphology than for the surface described above. Previous work in gas phase has shown that at room temperature the oxidation of Cu(111) proceeds through the epitaxial growth of copper oxide islands [24,25]. It has been shown that, in gas phase, the growth of oxides on Cu(111) depends on the oxygen pressure and temperature and it can follows three possible processes: (i) growth from step edges; (ii) interrace growth from vacancy islands; and (iii) growth of on-terrace oxide [24,25]. The CTRs in this case showed additional peaks at (0, 1, 1), (0, 1, 4), (1, 0, 2) and (1, 0, 5) (Figure 6), which arose from stacking faults (twinning) induced by the nucleation of copper atoms into *hcp* sites. The CTR data in Figure 6 were best modelled by including additional Lorentzian peaks to represent the twinned Cu, rather than a smooth surface with stacking faults, and indicate that the peaks arose due to nano-crystalline copper islands on the surface. Due to the presence of the additional peaks resulting from the nano-crystalline copper at the surface, the CTRs could be reproduced reliably with a very simple model including surface roughness and Cu surface relaxation only, i.e., for this more complex surface, the fitting was not sensitive to the inclusion of an oxygen layer. The surface roughness and relaxation rely on the overall intensity change across the whole range of *L* and the intensity distribution close to the Bragg peaks, respectively. We were not able to reliably include any coverage of individual atomic layers or oxygen atoms, which changed the intensity close to the anti-Bragg position as the intensity was dominated by the Lorentzian peaks arising from the nano-crystalline copper. Although these parameters, which are summarized in Figure 7, are only indicative, a similar trend in the potential dependent relaxation and roughness was observed when stepping the potential slowly positively. Surprisingly, at −0.8 V, modelling of the CTR data indicated a much lower surface roughness parameter, *β*, than the data shown in Figure 4. We attribute this to the presence of the nano-clustered copper islands acting as nucleation sites for copper ad-atoms, thus decreasing the overall surface roughness. The changes observed during the potential steps were, however, not reversible. Upon stepping back to the negative potential, an increase in the surface roughness, as indicated by the red symbols in Figure 7, could be observed. This increase in surface roughness was accompanied by a decrease in surface domain size and an increase in the size of the nano-crystalline copper domains present on the surface. This was shown by the rocking scan measurements at the (0, 1, 0.5) position (Figure 8), which, as for the data shown in Figure 3, also indicated the potential history of this sample. A gradual increase in the width of the rocking scan, from 0.2° to 0.4°, was observed, corresponding to a domain size decreasing from ~100 nm to ~50 nm. Correspondingly, the height D of the nano-clusters could be estimated from the width of the peaks observed in the CTR data in Figure 6, D = 2 π/ΔQ$_z$, and was found to grow in the course of the experiment from 20 to 40 nm. Thus, it can be concluded that this Cu surface—with Cu islands present on the surface—was unstable during potential cycling.

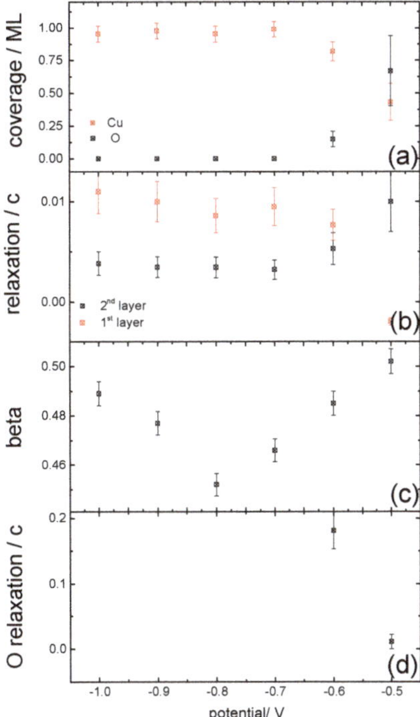

Figure 5. The structural parameters obtained from modelling the CTR data are plotted as a function of potential: (**a**) the coverage of the surface Cu layer and the adsorbed oxygen layer; (**b**) the relaxation of the top two copper layers; (**c**) the roughness as modelled through a β-roughness model; and (**d**) the Cu-O layer separation.

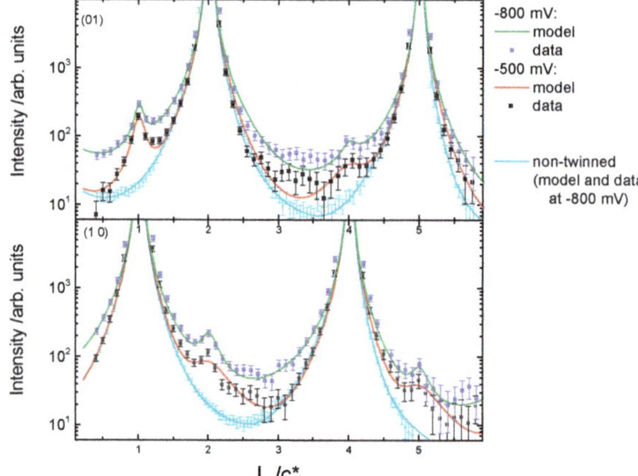

Figure 6. CTR data measured from a different Cu surface preparation at -0.8 V (blue symbols) and -0.5 V (black symbols) together with the calculated fits to the data (solid lines) according to the structural model described in the text. In comparison to the CTR data presented in Figure 4, additional peaks arising from Cu nano-islands can be observed.

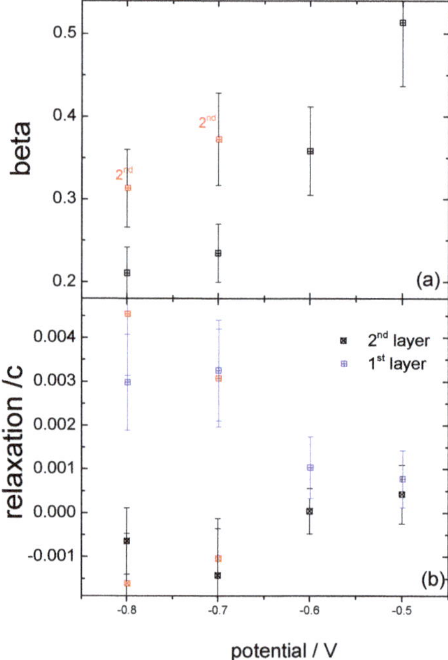

Figure 7. The structural parameters obtained from modelling the CTR data in Figure 6 are plotted as a function of potential: (**a**) the β-roughness factor; and (**b**) the relaxation of the two topmost atomic Cu layers. The data obtained after having stepped the potential to the positive limit are shown in red symbols. Although the potential-induced change in structure is similar, the roughness of the surface has increased.

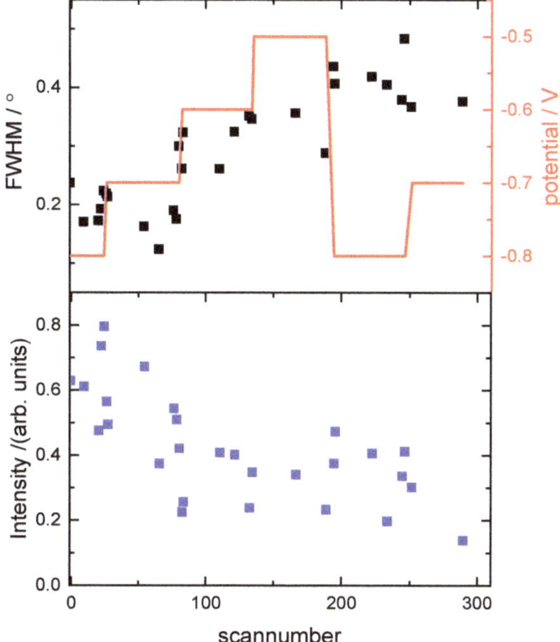

Figure 8. The surface stability of the Cu sample, which showed additional peaks arising from the presence of Cu nano-crystals, i.e., equivalent to the data shown in Figure 3: (a) the FWHM; and (b) the integrated intensity obtained from rocking scans at the (0, 1, 0.5) position as a function of the scan number during the experiment. The potential is also shown in red. The peak width is increasing and the integrated intensity is not recovering after having held the potential positively, indicating a change in surface morphology with increased roughness and smaller domain size during the course of the experiment.

4. Discussion

In this paper, we present a structural investigation of the Cu(111) electrode surface in phosphate-containing neutral electrolyte solution. We have found two different surface behaviours depending on the initial morphology and defect density of the surface observed directly after preparation and transfer into the X-ray electrochemical cell. In both cases, however, a clear structural transition with potential was found at -0.6 V, which coincides with a current peak in the anodic scan. The discussion of the structural results is based on the dataset presented in Figures 4–6, which gave the most concise insight into the underlying surface processes. At potentials negative of the observed transition, the coverage of the copper is constant and no potential dependent relaxation is observed. An increase in roughness can, however, be observed for increasingly cathodic potentials. The structural transition occurring at potentials >-0.7 V can be attributed to an increased surface roughness combined with adsorption of an oxygenated species, followed by the formation of a mixed copper-oxygen layer at positive potentials. This transition is accompanied by an inward relaxation of the surface Cu layer and a slight outward relaxation of the second atomic Cu layer.

Cyclic voltammograms were recorded at two different pH values and different phosphate concentrations to get an estimate of the charge transfer and possible coverage of an anionic layer and/or oxygenated species. For all electrolyte solutions, an anodic peak with a shoulder could be observed indicating a slow two-step adsorption process. The charge under the anodic and cathodic current peaks in the cyclic voltammetry corresponded to 0.2 electron transferred per surface atom. The in situ X-ray structural analysis of the Cu(111) surface at -0.5 V gave an oxygen coverage $\theta_O = 0.6$. It is known that phosphate species have a tetrahedral structure with a P-O bond length of the order $d_{P-O} \sim 1.5$–1.6 Å, thus the corresponding O-O distance is $d_{O-O} \sim 2.44$–2.62 Å [26–28]. The atomic distance of the copper atoms at the Cu(111) surface is 2.556 Å, which is similar to the O-O distance in the phosphate, thus making the adsorption of phosphate or hydrogen phosphate with three oxygen atoms sitting in the same adsorption site likely. This suggests that the adsorbate species at -0.5 V can be either hydrogen phosphate or phosphate and not dihydrogen phosphate. The exact nature of the phosphate species and the adsorption process is, however, not fully understood. From the structural analysis at -0.6 V, it is clear that the difference between the data recorded at the two most positive potentials is not just due to a change in coverage of the adsorbed oxygenated species. The change in relaxation of the topmost copper layer and the change in the distance of the oxygen atoms to the Cu surface indicate an intermediate structure at this potential. This is also evident from the XRV and potentiostatic measurements presented in Figure 2.

To further elucidate the adsorption process, the peak positions in the cyclic voltammetry measured in 0.1 M phosphate solution at pH = 6 and pH = 8 are shown as a function of the logarithm of the scan rate in Figure 9. The cathodic peak and the first anodic peak (corresponding to the shoulder in the cyclic voltammogram) did not show a scan rate dependence, indicating a fast adsorption process, whereas the second anodic peaks showed a slope of approximately 0.12 V/log (scan rate/(V/s)), independent of the pH of the electrolyte, indicating is a slow process. The rate determining step in the adsorption process is thus associated with the second anodic peak. The surface X-ray diffraction data show that at the corresponding potentials a substantial rearrangement of the surface atomic structure took place. The X-ray voltammetry obtained in-situ shows a large hysteresis consistent with a slow mass transfer process. Static CTR measurement indicate the incorporation of oxygen/phosphate into the Cu surface resulting in mixed surface copper-oxide and increased surface roughness in this potential region. Structural rearrangements of the surface Cu layer with similarly slow kinetics have been observed in alkaline electrolytes [29,30]. In that case, the slow kinetics were associated with formation of a single Cu(I) oxide layer and a decrease of the Cu atomic density by 30% [31].

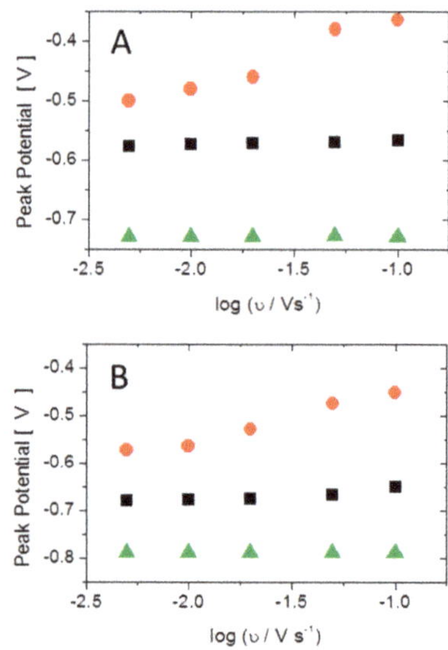

Figure 9. Peak potentials obtained from cyclic voltammograms are shown as function of the log of the scan rate: (**A**) pH = 6; and (**B**) pH = 8. The black and red symbols correspond to the first and second anodic peaks, respectively. The green symbols correspond to the cathodic peak.

With the acid dissociation constant $pKa(H_2PO_4^-/HPO_4^{2-}) = 7.2$, at pH = 8 a proportion of 6% $H_2PO_4^-$ and 94% HPO_4^- is expected, thus mostly hydrogen phosphate. At pH = 6 a proportion of 87% $H_2PO_4^-$ and 13% HPO_4^{2-} is expected in the electrolyte solution. On gold and platinum surfaces, the adsorbed phosphate species has been found to be pH- and potential-dependent [32,33]. In contrast, on copper surfaces, phosphate dihydrogen was found to adsorb for acidic electrolyte solutions (pH < 5) but not in alkaline solution (pH > 9) [34]. Although we cannot totally exclude that dihydrogen is adsorbing in the first step on the Cu surface, the fact that the two adsorption steps were observed at pH = 6 and pH = 8 in the electrochemical data and did not show a difference in the kinetics of the processes, suggest that the adsorption process occurred through a similar mechanism. In addition, a more covalent bonding between the oxygen of the phosphate and copper was found, in comparison to silver and gold [34], and this could lead to the atomic rearrangement and slow mass transport observed in the X-ray measurements.

By combining the structural and electrochemical results, we propose the following adsorption mechanism: The first peak observed in the cyclic voltammograms corresponds to adsorption of hydrogen phosphate to the surface (either directly or through the deprotonation of dihydrogen phosphate depending on the pH). This is supported by the structural data where a small amount of oxygen was found to be specifically adsorbed into three-fold hollow sites on the Cu surface. The second anodic peak is associated with the deprotonation of the hydrogen phosphate and the adsorption of the phosphate anion which forms a strong bond with the Cu surface leading to an atomic rearrangement involving the incorporation of three oxygen atoms from the phosphate into the surface Cu layer. This process is totally reversible, as also confirmed by the electrochemical characterization which gives the same charge under the peak in the cyclic voltammogram for the anodic and cathodic process.

The elucidation of the adsorption process helps in understanding the differences in stability of the Cu(111) electrodes during adsorption process and the dependence on the details of Cu surface preparation. The in-situ X-ray diffraction data enable three main properties of the surface contributing to the morphology to be distinguished: (a) the domain size of the atomic terraces present at the surface; (b) the presence of Cu nano-crystals at the surface; and (c) the overall atomic roughness, described through the β-model and hereafter referred to as rugosity. We have shown that, even though the Cu(111) surfaces exhibit identical terrace lengths, the rugosity of the surface at terrace level might be different. The Cu(111) surface with large rugosity shows a reversible adsorption process of phosphates and oxygen incorporation into the crystalline structure. On the other hand, the Cu(111) surface with copper nano-crystals present on the surface showed degradation upon phosphate adsorption/desorption driven by the potential cycling. Surprisingly the rugosity was larger on the surface showing reversible adsorption, the reversible nature of the adsorption/desorption being evident from the stable intensity at the anti-Bragg position of the CTRs. The oxygen induced reconstruction of the Cu(111) surface at room temperature have shown the formation of a disordered surface overlayer, with O and Cu atoms at different heights [35]. It has been proposed that the formation of this disorder overlayer favours the oxygen diffusion on Cu (111) leading to fast nucleation of a large number of oxide islands. We propose a similar process in electrochemical media through the oxygens of the adsorbed phosphates.

As the adsorption of phosphate increases the rugosity of the surface, we associate the differences in stability of the two surfaces to the presence of surface defects acting as nucleation sites for Cu ad-atoms created during the phosphate incorporation into the surface Cu layer. Depending on the surface morphology, these atoms can be incorporated into the nano-crystals present on the surface, thus leading to further roughening and decrease in the surface domain size, i.e., the Cu surface is unstable. On a Cu electrode surface with high rugosity, no single nucleation sites stand out and the formation of nano-crystals is prevented. This is a more stable Cu electrode surface exhibiting a reversible phosphate adsorption process.

5. Conclusions

In this paper, we present electrochemical and in-situ surface X-ray diffraction measurements of phosphate adsorption onto Cu(111) electrodes. The underlying mechanism of the phosphate adsorption and deprotonation of the (di)-hydrogen phosphate was accompanied by a roughening of the Cu surface. We report for the first time the roughening of the Cu surface through the formation of a mixed copper–oxygen layer, where the oxygen from an adsorbed phosphate species was incorporated into the surface Cu layer. The stability of the Cu(111) surface and the change of rugosity upon potential cycling strongly depend on the preparation method and history of the electrode. It was shown that the presence of Cu islands on the Cu(111) surface leads to a 3D nucleation and growth mechanism which causes irreversible changes in surface morphology. The results demonstrate the importance of controlling the surface preparation of copper catalysts, as this determines the stability of the catalyst under operation conditions for the electrochemical CO_2 reduction reaction. The incorporation of oxygen into the metal surface from the adsorbed phosphate anion is a process that could also be relevant to the study of similar oxoanions, such as sulphate, and their role in the stability of electrocatalysts during oxidation processes.

Author Contributions: P.R. and Y.G. conceived the experiments. All authors contributed to the experimental measurements. Y.G., J.B. and C.A.L. performed the analysis of the in-situ X-ray measurements. All authors contributed to the discussion, writing and revision of the manuscript. All authors have given approval to the final version of the manuscript.

Funding: This research received no external funding.

Acknowledgments: We thank the Diamond Light Source for supplying the beamtime necessary to conduct the measurements presented in this paper. In particular we would like to thank our local contacts, Jonathan Rawle and Hadeel Hussain, and the Diamond I07 team for their support during the experiment. AK acknowledges the University of Birmingham for financial support through Ph.D. scholarships at the School of Chemistry. P.R. acknowledges the University of Birmingham for financial support through the Birmingham fellowship program. Y.G. acknowledges the Royal Society for funding through a Royal Society Research Fellowship.

References

1. Larrazábal, G.O.; Martín, A.J.; Pérez-Ramírez, J. Building blocks for high performance in electrocatalytic CO_2 reduction: materials, optimization strategies, and device engineering. *J. Phys. Chem. Lett.* **2017**, *8*, 3933–3944.
2. Arán-Ais, R.M.; Gao, D.; Roldan cuenya, B. structure- and electrolyte-sensitivity in CO_2 electroreduction. *Acc. Chem. Res.* **2018**, *51*, 2906–2917. [CrossRef] [PubMed]
3. Engstfeld, A.K.; Maagaard, T.; Horch, S.; Chorkendorff, I.; Stephens, I.E.L. Polycrystalline and Single-Crystal Cu Electrodes: Influence of Experimental Conditions on the Electrochemical Properties in Alkaline Media. *Chem. A Eur. J.* **2018**, *24*, 17743–17755. [CrossRef] [PubMed]
4. Grosse, P.; Gao, D.; Scholten, F.; Sinev, I.; Mistry, H.; Roldan Cuenya, B. Dynamic Changes in the Structure, Chemical State and Catalytic Selectivity of Cu Nanocubes during CO_2 Electroreduction: Size and Support Effects. *Angew. Chem.* **2018**, *130*, 6300–6305. [CrossRef]
5. Mistry, H.; Varela, A.S.; Bonifacio, C.S.; Zegkinoglou, I.; Sinev, I.; Choi, Y.-W.; Kisslinger, K.; Stach, E.A.; Yang, J.C.; Strasser, P.; et al. Highly selective plasma-activated copper catalysts for carbon dioxide reduction to ethylene. *Nat. Commun.* **2016**, *7*, 12123. [CrossRef] [PubMed]
6. Resasco, J.; Chen, L.D.; Clark, E.; Tsai, C.; Hahn, C.; Jaramillo, T.F.; Chan, K.; Bell, A.T. Promoter Effects of Alkali Metal Cations on the Electrochemical Reduction of Carbon Dioxide. *J. Am. Chem. Soc.* **2017**, *139*, 11277–11287. [CrossRef] [PubMed]
7. Le Duff, C.S.; Lawrence, M.J.; Rodriguez, P. Role of the Adsorbed Oxygen Species in the Selective Electrochemical Reduction of CO_2 to Alcohols and Carbonyls on Copper Electrodes. *Angew. Chem. Int. Ed.* **2017**, *56*, 12919–12924. [CrossRef] [PubMed]

8.   Kas, R.; Kortlever, R.; Milbrat, A.; Koper, M.T.M.; Mul, G.; Baltrusaitis, J. Electrochemical CO_2 reduction on Cu_2O-derived copper nanoparticles: controlling the catalytic selectivity of hydrocarbons. *Phys. Chem. Chem. Phys.* **2014**, *16*, 12194–12201. [CrossRef] [PubMed]

9.   Mandal, L.; Yang, K.R.; Motapothula, M.R.; Ren, D.; Lobaccaro, P.; Patra, A.; Sherburne, M.; Batista, V.S.; Yeo, B.S.; Ager, J.W.; et al. Investigating the Role of Copper Oxide in Electrochemical CO_2 Reduction in Real Time. *ACS Appl. Mater. Interfaces* **2018**, *10*, 8574–8584. [CrossRef] [PubMed]

10.   Verdaguer-Casadevall, A.; Li, C.W.; Johansson, T.P.; Scott, S.B.; McKeown, J.T.; Kumar, M.; Stephens, I.E.L.; Kanan, M.W.; Chorkendorff, I. Probing the Active Surface Sites for CO Reduction on Oxide-Derived Copper Electrocatalysts. *J. Am. Chem. Soc.* **2015**, *137*, 9808–9811. [CrossRef] [PubMed]

11.   Lee, S.; Kim, D.; Lee, J. Electrocatalytic Production of C3-C4 Compounds by Conversion of CO_2 on a Chloride-Induced Bi-Phasic Cu_2O-Cu Catalyst. *Angew. Chem. Int. Ed.* **2015**, *54*, 14701–14705. [CrossRef] [PubMed]

12.   Eilert, A.; Cavalca, F.; Roberts, F.S.; Osterwalder, J.; Liu, C.; Favaro, M.; Crumlin, E.J.; Ogasawara, H.; Friebel, D.; Pettersson, L.G.M.; et al. Subsurface Oxygen in Oxide-Derived Copper Electrocatalysts for Carbon Dioxide Reduction. *J. Phys. Chem. Lett.* **2017**, *8*, 285–290. [CrossRef] [PubMed]

13.   Friebel, D.; Broekmann, P.; Wandelt, K. Electrochemical in situ STM study of a Cu(111) electrode in neutral sulfate containing electrolyte. *Phys. Status Solidi* **2004**, *201*, 861–869. [CrossRef]

14.   Lucas, C.A.; Markovic, N.M. Structure Relationships in Electrochemical Reactions. In *Encyclopedia of Electrochemistry*; Wiley-VCH Verlag GmbH & Co. KGaA: Weinheim, Germany, 2007; ISBN 9783527610426.

15.   Gründer, Y.; Lucas, C.A. Surface X-ray diffraction studies of single crystal electrocatalysts. *Nano Energy* **2016**, *29*, 378–393. [CrossRef]

16.   Lucas, C.A.; Thompson, P.; Gründer, Y.; Markovic, N.M. The structure of the electrochemical double layer: Ag(111) in alkaline electrolyte. *Electrochem. Commun.* **2011**, *13*, 1205–1208. [CrossRef]

17.   Sisson, N.; Gründer, Y.; Lucas, C.A. Structure and Stability of Underpotentially Deposited Ag on Au(111) in Alkaline Electrolyte. *J. Phys. Chem. C* **2016**, *120*, 16100–16109. [CrossRef]

18.   Nicklin, C.; Arnold, T.; Rawle, J.; Warne, A. Diamond beamline I07: A beamline for surface and interface diffraction. *J. Synchrotron Radiat.* **2016**, *23*, 1245–1253. [CrossRef] [PubMed]

19.   Vlieg, E. A (2 + 3)-Type Surface Diffractometer: Mergence of the z-Axis and (2 + 2)-Type Geometries. *J. Appl. Crystallogr.* **1998**, *31*, 198–203. [CrossRef]

20.   Schlepütz, C.M.; Herger, R.; Willmott, P.R.; Patterson, B.D.; Bunk, O.; Brönnimann, C.; Henrich, B.; Hülsen, G.; Eikenberry, E.F. Improved data acquisition in grazing-incidence X-ray scattering experiments using a pixel detector. *Acta Crystallogr. Sect. A Found. Crystallogr.* **2005**, *61*, 418–425. [CrossRef] [PubMed]

21.   Schlepütz, C.M.; Mariager, S.O.; Pauli, S.A.; Feidenhansl, R.; Willmott, P.R. Angle calculations for a (2+3)-type diffractometer: Focus on area detectors. *J. Appl. Crystallogr.* **2011**, *44*, 73–83. [CrossRef]

22.   Robinson, I.K. Crystal truncation rods and surface roughness. *Phys. Rev. B* **1986**, *33*, 3830–3836. [CrossRef]

23.   Platzman, I.; Brener, R.; Haick, H.; Tannenbaum, R. Oxidation of Polycrystalline Copper Thin Films at Ambient Conditions. *J. Phys. Chem. C* **2008**, *112*, 1101–1108. [CrossRef]

24.   Pérez León, C.; Sürgers, C.; Löhneysen, H. Formation of copper oxide surface structures via pulse injection of air onto Cu(111) surfaces. *Phys. Rev. B* **2012**, *85*, 035434.

25.   Matsumoto, T.; Bennett, R.A.; Stone, P.; Yamada, T.; Domen, K.; Bowker, M. Scanning tunneling microscopy studies of oxygen adsorption on Cu(1 1 1). *Surf. Sci.* **2001**, *471*, 225–245. [CrossRef]

26.   Gamoke, B.; Neff, D.; Simons, J. Nature of PO Bonds in Phosphates. *J. Phys. Chem. A* **2009**, *113*, 5677–5684. [CrossRef] [PubMed]

27.   Pye, C.C.; Rudolph, W.W. An ab Initio, Infrared, and Raman Investigation of Phosphate Ion Hydration. *J. Phys. Chem. A* **2003**, *107*, 8746–8755. [CrossRef]

28.   Rose, J.; Flank, A.M.; Masion, A.; Bottero, J.Y.; Elmerich, P. Nucleation and Growth Mechanisms of Fe Oxyhydroxide in the Presence of PO4 Ions. 2. P K-Edge EXAFS Study. *Langmuir* **1997**, *13*, 1827–1834. [CrossRef]

29.   Maurice, V.; Strehblow, H.H.; Marcus, P. In situ STM study of the initial stages of oxidation of Cu(111) in aqueous solution. *Surf. Sci.* **2000**, *458*, 185–194. [CrossRef]

30.   Kunze, J.; Maurice, V.; Klein, L.H.; Strehblow, H.H.; Marcus, P. In situ scanning tunneling microscopy study of the anodic oxidation of Cu(111) in 0.1 M NaOH. *J. Phys. Chem. B* **2001**, *105*, 4263–4269. [CrossRef]

31. Kunze, J.; Maurice, V.; Klein, L.H.; Strehblow, H.-H.; Marcus, P. In situ STM study of the anodic oxidation of Cu(0 0 1) in 0.1 M NaOH. *J. Electroanal. Chem.* **2003**, *554–555*, 113–125. [CrossRef]

32. Yaguchi, M.; Uchida, T.; Motobayashi, K.; Osawa, M. Speciation of Adsorbed Phosphate at Gold Electrodes: A Combined Surface-Enhanced Infrared Absorption Spectroscopy and DFT Study. *J. Phys. Chem. Lett.* **2016**, *7*, 3097–3102. [CrossRef] [PubMed]

33. Gisbert, R.; García, G.; Koper, M.T.M. Adsorption of phosphate species on poly-oriented Pt and Pt(1 1 1) electrodes over a wide range of pH. *Electrochim. Acta* **2010**, *55*, 7961–7968. [CrossRef]

34. Niaura, G.; Gaigalas, A.K.; Vilker, V.L. Surface-Enhanced Raman Spectroscopy of Phosphate Anions: Adsorption on Silver, Gold, and Copper Electrodes. *J. Phys. Chem. B* **1997**, *101*, 9250–9262. [CrossRef]

35. Jensen, F.; Besenbacher, F.; Lægsgaard, E.; Stensgaard, I. Oxidation of Cu(111): two new oxygen induced reconstructions. *Surf. Sci. Lett.* **1991**, *259*, L774–L780.

Article

Interface Science Using Ambient Pressure Hard X-ray Photoelectron Spectroscopy

Marco Favaro [1,*], **Fatwa Firdaus Abdi** [1], **Ethan Jon Crumlin** [2], **Zhi Liu** [2,3], **Roel van de Krol** [1] and **David Edward Starr** [1,*]

[1] Helmholtz-Zentrum Berlin für Materialien und Energie GmbH, Institute for Solar Fuels, Hahn-Meitner-Platz 1, D-14109 Berlin, Germany; fatwa.abdi@helmholtz-berlin.de (F.F.A.); roel.vandekrol@helmholtz-berlin.de (R.v.d.K.)

[2] Advanced Light Source, Lawrence Berkeley National Laboratory, 1 Cyclotron Rd, Berkeley, CA 94720, USA; ejcrumlin@lbl.gov (E.J.C.); liuzhi@shanghaitech.edu.cn (Z.L.)

[3] School of Physical Science and Technology, ShanghaiTech University, Pudong, Shanghai 201210, China

* Correspondence: marco.favaro@helmholtz-berlin.de (M.F.); david.starr@helmholtz-berlin.de (D.E.S.); Tel.: +49-(0)30-8062-42835 (M.F.); +49-(0)30-8062-42320 (D.E.S.)

Received: 18 December 2018; Accepted: 24 January 2019; Published: 28 January 2019

Abstract: The development of novel in situ/operando spectroscopic tools has provided the opportunity for a molecular level understanding of solid/liquid interfaces. Ambient pressure photoelectron spectroscopy using hard X-rays is an excellent interface characterization tool, due to its ability to interrogate simultaneously the chemical composition and built-in electrical potentials, in situ. In this work, we briefly describe the "dip and pull" method, which is currently used as a way to investigate in situ solid/liquid interfaces. By simulating photoelectron intensities from a functionalized TiO_2 surface buried by a nanometric-thin layer of water, we obtain the optimal photon energy range that provides the greatest sensitivity to the interface. We also study the evolution of the functionalized TiO_2 surface chemical composition and correlated band-bending with a change in the electrolyte pH from 7 to 14. Our results provide general information about the optimal experimental conditions for characterizing the solid/liquid interface using the "dip and pull" method, and the unique possibilities offered by this technique.

Keywords: in situ ambient pressure XPS; hard X rays; photoelectron simulations; solid/liquid interface; TiO_2; APTES

---

1. Introduction

Molecular-level processes occurring at solid/liquid interfaces are intriguing not only from a fundamental physical-chemical perspective, but also from a practical perspective since they are an essential part of (photo)electrochemical systems, which are key to future renewable energy storage technologies. This provides a powerful driving force for the development of novel in situ/operando characterization tools that can directly probe the interface [1–7]. For a complete understanding of the solid/liquid interface, it is mandatory to investigate the properties of the solid bulk, the bulk liquid, and the thin interface layer between the solid and the liquid phases [8]. This constitutes the most important part of the solid/liquid junction where specific adsorption of ions, charge transfer dynamics and electrical potential formation occur.

On the other hand, the measurement of the interface layer properties is a challenging task. First, it requires operando or in situ experimental capabilities in order to capture the true makeup of the interfacial layer under operating or realistic working conditions, respectively [1,8,9]. Secondly, to characterize the true interfacial properties, one should use an experimental probe that limits the perturbation of the junction itself. Therefore, several spectroscopic methods based on *photon in/photon out* or *photon in/electron out* approaches have been applied to investigate solid/liquid interfaces. Among them, synchrotron-based techniques such as surface X-ray diffraction, X-ray absorption/emission and photoelectron spectroscopy, and "laboratory-based" techniques, such as infrared, Raman and non-linear optical spectroscopies have been recently developed [10].

In this context, X-ray photoelectron spectroscopy (XPS) stands as an excellent characterization tool, since it offers elemental and chemical sensitivity, simultaneously measuring local built-in electrical potentials via the detection of rigid photoelectron kinetic energy shifts in both core and valence levels [11–17]. Due to the high vapor pressure of many liquids of interest, differential pumping schemes have to be used in photoelectron analyzers to minimize the elastic and inelastic scattering of electrons in the gas phase above the liquid side of the junction [18–20]. In addition, small sample-to-analyzer aperture working distances (WDs) must be used, for the same purpose [18–20]. A reasonable trade-off needs to be found between limiting the electron scattering by the gas phase molecules and keeping the pressure at the sample surface above 90–95% of the nominal pressure in the chamber. Usual WDs (at which the analyzer focus is optimized) are about the diameter of the aperture itself [18–20]. Modern state-of-the-art electron analyzers are capable to operate at pressures of and above 30 mbar (the vapor tension of water at room temperature) and at high photoelectron kinetic energies (KEs, up to 12 keV) [11]. The extension of AP-XPS to high photon energies (i.e., high photoelectron KEs) has two main advantages: the reduced photoelectrons/gas molecules scattering provides higher signal intensity at higher gas pressure (i), whereas the high photoelectron KEs lead to increased photoelectron inelastic mean free paths (IMFPs) in the liquid (ii). The latter enables the investigation of solid/liquid junctions through liquid layers with thickness on the same order of the IMFP (ii).

Two additional challenges exist for the AP-XPS investigation of solid/liquid electrolyte interfaces. The first is the preparation of stable liquid films thick enough to be representative of a realistic solid/liquid interface, but thin enough to allow photoelectrons ejected from the interfacial region to penetrate and emerge from the liquid on their path to the photoelectron analyzer. The second is the tuning of the experimental conditions (detection geometry, photon energy etc.) to enhance the photoelectron signal coming from the interface.

In this work, we briefly describe the so-called "dip and pull" method that can be used as a way to investigate the solid/liquid interface in situ by coupling it with ambient pressure hard X-ray photoelectron spectroscopy (AP-HAXPES) using X-ray energies between 2.0 and 10.0 keV [21].

By using a model system of a solid/liquid interface and numerical simulations, we estimate the optimal photon energy range for the signal coming from the interface. The simulations have been generated using the Simulation of Electron Spectra for Surface Analysis (SESSA) software [22]. Our model system is a TiO_2 surface functionalized with 1 monolayer equivalent (1 MLE) of 3-aminopropyl triethoxysilane (APTES). Alkylsilanes adsorb at the surface of transition metal oxides [23] as self-limiting monolayer systems, thereby constituting ideal interfaces. Furthermore, APTES contains one nitrogen and silicon atom, whose corresponding 1s photoionization yield was monitored as a function of the photon energy. The difference in binding energy (BE) between these two core levels (N 1s BE ~ 400 eV, Si 1s BE ~ 1850 eV) will be used to highlight the different photoionization yield trends with the photon energy.

In the last part of this work, we report on the evolution of the APTES-functionalized TiO_2 surface chemistry and correlated band-bending by systematically changing the pH of the aqueous electrolyte from 7 to 14. This was performed at beamline 9.3.1 at the Advanced Light Source (ALS, Lawrence Berkeley National Laboratory, Berkeley, U.S.A.) [11]. Our results provide general information about the experimental conditions that allow for an optimal characterization of the solid/liquid interface using the "dip and pull" method and AP-HAXPES, showing at the same time the capabilities offered by this technique to address fundamental questions in energy materials and conversion research.

2. Materials and Methods

2.1. Numerical Simulation Details

The simulations reported in this work have been generated using the Simulation of Electron Spectra for Surface Analysis (SESSA) software (National Institute of Standards and Technology, Gaithersburg, MD, USA.) [22].

The APTES thickness corresponding to 1 monolayer equivalent (MLE) was found as follows, using SESSA simulations. First, 1 ML of TiO_2 having a thickness of 3.57 Å (the relaxed Ti-Ti distance along the rutile (110) direction [24]) was supported on a planar Cu substrate. Then, the APTES layer thickness deposited atop the TiO_2 film was systematically varied until the differential cross section-normalized integrated photoelectron intensities of Ti 2p and Si 1s were identical. This leads to an identical concentration of Si atoms with respect to the Ti atoms contained within 1 ML of TiO_2 (for practical reasons chosen along the rutile (110) direction). Therefore, the thickness of APTES corresponding to 1 MLE on rutile TiO_2(110) was found to be 12.7 Å.

Note that the differential cross sections $(d\sigma/d\Omega)_{nl}$ (the probability for photoelectrons to be emitted into the solid angle $d\Omega$) were computed in SESSA using the first order nondipole approximation, in which nondipole effects are treated in a perturbative manner. Within the Cooper parametrization, the differential cross section $(d\sigma/d\Omega)_{nl}$ for 100% linearly polarized radiation [25] can be expressed as follows:

$$(d\sigma/d\Omega)_{nl} = (\sigma_{nl}/4\pi) \times \{[1 + \frac{1}{2}\beta_{nl} \times (3\cos^2\alpha - 1)] + [\sin\alpha\cos\varphi \times (\delta + \gamma\cdot\cos^2\alpha)]\} \qquad (1)$$

σ_{nl} is the total photoionization cross section (TPCS) for a core level described by a principal and angular quantum number n and l, β_{nl} is the corresponding asymmetry parameter describing the anisotropy of the photoelectron angular distribution, α is the angle formed between the photoelectron direction k and the polarization vector ε, φ is the angle formed between the radiation Poynting vector S (parallel to the propagation direction) and the plane containing k and ε (see Figure 1 for the definition of the aforementioned angles). Since the simulations have been performed considering radiation from a synchrotron (horizontal photon beam) and bending magnet as source (electric field polarization vector laying on the orbit plane, see Figure 1), the angle φ is equal to 0° for all α values (in our simulations, $\alpha = 15°$ as implemented at BL 9.3.1, ALS). The term within the first square parentheses represents the dipole contribution, whereas the second term is the parametrization of the first order nondipole effects, which are non-negligible for photon energies above 2000 eV [25]. Within the nonrelativistic single-electron approximation, the γ parameter represents the major correction term corresponding to electric dipole-quadrupole interference. The magnetic dipole term, δ, can only be present if core-relaxation occurs [25]. δ and γ, as well as σ_{nl} and β_{nl}, are explicit functions of the photon energy. Finally, the take-off angle θ (angle between the photoelectron direction k and the normal n to the sample surface) was kept to 0° for all the simulations (normal emission detection, NE).

The atomic density (number of emitters per unit volume) of TiO_2 and APTES used in the simulations were equal to 9.568×10^{22} cm^{-3} and 9.522×10^{22} cm^{-3}, respectively (the APTES stoichiometry used in the simulations was $SiNO_3C_9H_{23}$). The energy gap between the valence and conduction bands of TiO_2 was set to 3.03 eV [26]. The energy gap between the HOMO and LUMO of APTES was determined from the fluorescence spectrum of APTES, which exhibits an intense band at about 438 nm [27]. The related HOMO-LUMO energy gap was therefore estimated to be equal to 2.83 eV.

To complete the model of the solid/liquid interface, a liquid water layer with a thickness of 20 nm was placed atop the sample surface, as schematically reported in Figure 1. This value has been taken as a realistic representation of the liquid layer thickness typically obtained during "dip and pull" experiments. The water atomic density and band gap used in the SESSA simulations were 1.003×10^{23} cm^{-3} and 6.9 eV [28], respectively.

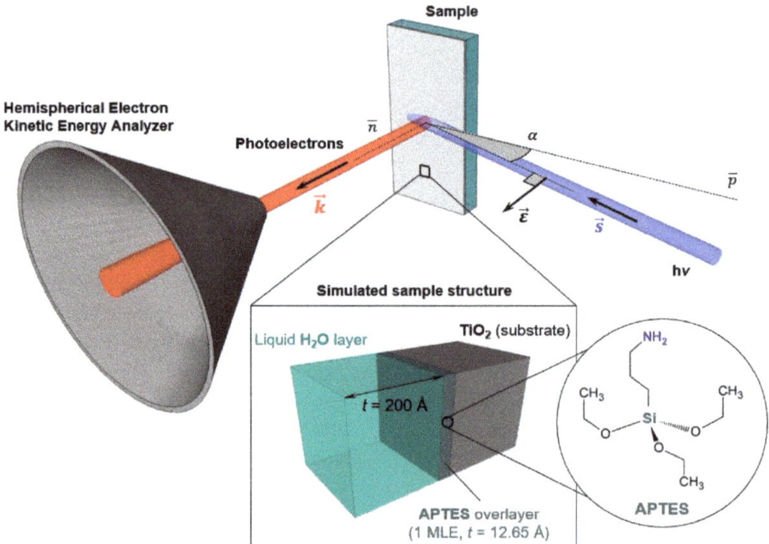

Figure 1. Schematization of the ambient pressure hard X-ray photoelectron spectroscopy (AP-HAXPES) experimental geometry (implemented in Simulation of Electron Spectra for Surface Analysis (SESSA) calculations) and 3-aminopropyl triethoxysilane (APTES)-functionalized TiO_2 model system used in this work.

2.2. Sample Preparation

The TiO_2 precursor solution was made by mixing titanium-isopropoxide (Alfa-Aesar, Haverhill, MA, USA.) with glacial acetic acid (Sigma-Aldrich, Saint Louis, MO, USA.) in anhydrous ethanol. 1 mL of Ti-isopropoxide was added to 5 mL of anhydrous ethanol followed by adding varying amounts of glacial acetic under vigorous stirring to give mole ratios ranging from 1:1 to 1:2, Ti-isopropoxide:acetic acid. All solutions resulted in TiO_2 films of similar quality. The precursor solution was spin-coated onto FTO/glass substrates in four 100 μL aliquots at 2000 rpm for 120 s. Prior to spin coating the substrates were sonicated in a 1% solution of Triton X-100 (Sigma-Aldrich, Saint Louis, MO, USA.) in de-ionized water, DI water, and a 1:1 solution of acetone:absolute ethanol, followed by a final sonication in DI water. The spin-coated films were sintered at 500 °C for 12 h.

The TiO_2 coated substrates were functionalized with APTES by placing the dried (300 °C for 15 min) substrates into a solution of 0.25 mL APTES in 100 mL of anhydrous toluene under Ar(g) at 80 °C for 5 h. Following reaction, the APTES/TiO_2 coated substrates were dried at 70 °C for 30 min and packed into vacuum sealed bags in a N_2(g) filled glove box.

2.3. In Situ AP-HAXPES Measurements and Beamline 9.3.1 Experimental Details

The source for beamline 9.3.1 at the Advanced Light Source (ALS) of Lawrence Berkeley National Laboratory is a bending magnet. A Si(111) double crystal monochromator (DCM) provides an energy range between 2000 and 6000 eV (denoted as "hard" throughout the text to be consistent with the acronym HAXPES). The minimum X-ray spot size on the sample is 0.7 mm (v) × 1.0 mm (h). All spectra were taken with a photon energy of 4000 eV, at room temperature, and in NE (see Figure 1 for a schematic of the experimental geometry). The pressure in the experimental chamber was kept at the equilibrium vapor pressure of the aqueous electrolyte solution at room temperature. To limit evaporation from the electrochemical cell a large volume (∼500 mL) of outgassed pure water was introduced in the analysis chamber.

The pass energy of the Scienta analyzer (R4000 HiPP-2, Scienta, Uppsala, Sweden) was set to 200 eV. A step size of 100 meV and a dwell time of 300 ms was used. Under these conditions, the total resolution (X-ray and electron analyzer) was equal to about 550 meV at 4.0 keV, determined by measuring the spectral broadening of a gold Fermi edge taken on a clean polycrystalline surface. Binding energy (BE) scale calibration was done by using the Au $4f_{7/2}$ (BE = 84.00 eV) photoelectron peak position taken on a clean gold polycrystalline surface as reference values measured under all the experimental conditions. During operation, the sample and the electron energy analyzer were commonly grounded. Spectral fitting was carried out using a Doniach-Šunjić shape for the Au 4f photoelectron peaks, whereas symmetric pseudo-Voigt functions (G/L ratio ranging from 85/15 to 75/25) were used to fit the Si 1s, O 1s, Ti 2p and N 1s core levels (after Shirley background subtraction). During the fitting procedure, the Shirley background was optimized together with the spectral components. Finally, chi-square (χ^2) minimization was ensured by the use of a nonlinear least-squares routine, with increased stability over simplex minimization.

2.4. "Dip and Pull" Method and in Situ Measurements

Prior to its introduction into the experimental chamber, the aqueous solutions at different pH (see Table 1) were outgassed for at least 30 min at low pressure (between 10 and 20 Torr) in a dedicated offline chamber. After degassing, the electrolyte solution was placed into the AP-HAXPES analysis chamber, where the pressure was carefully decreased to just below equilibrium vapor pressure of the electrolyte at room temperature (the chamber was pumped by a MVP 035-2 diaphragm pump, Pfeiffer Vacuum, Aßlar, Germany). Then, the pumping via the diaphragm pump was stopped and the pressure drifted up to the equilibrium vapor pressure of the used electrolyte solution at room temperature (Table 1). At this point, the only active pumping was provided by the aperture of the HEA nozzle that separates the high pressure analysis chamber from the differentially-pumped electrostatic lens system (aperture diameter: 300 μm). To create the solid/liquid electrolyte interface, the sample was immersed deeply into the aqueous solution. It was then slowly extracted from the liquid by raising the sample manipulator at a constant rate (100 μm s^{-1}). Following this procedure, a thin layer of aqueous electrolyte film was formed on the sample surface. The latter was then positioned at the intersection of the X-ray beam and the focal point of the hemispherical electron analyzer (HEA), thereby allowing AP-HAXPES measurements of the solid/liquid interface.

Table 1. Concentration and pH values for the different aqueous solutions used in this study. The water vapor partial pressure above pure water and KOH aqueous solutions at different concentration has been determined using the Bridgeman-Aldrich [29] and the Balej equation [29], respectively.

Aqueous Solution	KOH Concentration [M = mol L^{-1}]	pH	H$_2$O Vapor Partial Pressure at r.t. (= 298 K) [mbar]
mQ H$_2$O	–	7	31.67
KOH 10^{-4} M	10^{-4}	10	31.41
KOH 10^{-2} M	10^{-2}	12	31.39
KOH 1.0 M	1.0	14	30.29

3. Results and Discussion

3.1. The Challenge of Preparing a Realistic Solid/Liquid Interface

The main challenge in the investigation of solid/liquid interfaces using a *photon in/electron out* technique lies in the preparation of the interface itself. The electron detection implies that the electrons ejected from the interfacial region need to penetrate through condensed phases whose thickness must be less than two to three times the electron inelastic mean free path (IMFP). Although the use of hard X-rays allows obtaining high photoelectron KEs (and therefore IMFPs), a photoelectron having a KE spanning from 2000 eV to 10000 eV is characterized by an IMFP on the order of tens of nanometers. This poses severe limitations and constraints in the preparation of the solid/liquid interface.

Currently, two different preparation and investigation approaches are used, which differ from each other for the side of the interface that is used for the X-ray incidence and electron detection [9].

Photoelectron spectroscopy can be conducted from the solid side of the solid/liquid interface using few layer graphene membranes supporting the solid phase [30,31]. Although this technique is appealing for the fact that gases or liquids can be flown through the system thereby providing facile mass transport, it limits the investigation to thin solid films since the photo-emitted electrons must travel through the solid phase/graphene membrane to reach the photoelectron analyzer. For instance, a photoelectron travelling in Au with a KE of 500 and 5000 eV is characterized by an IMFP of about 0.8 and 4.2 nm, respectively. Using the definition of the information depth as three times the IMFP (i.e., the depth in the material at which 95% of the ejected photoelectrons are inelastically scattered in their path toward the surface), we find that the maximum probed thickness of Au achievable with soft (hard) X-rays is equal to about 2 nm (13 nm).

A second approach, using X-ray incidence and electron detection from the liquid side, requires the formation of a thin liquid layer atop the solid surface [9]. This approach has the advantage of being able to investigate a broad range of solid materials of arbitrary thickness. This is particularly important for photoelectrochemical (PEC) interfaces, due to the diffusion length of the excited charge carriers (λ_{CC}) that can span from tens to hundreds of nm. This poses a constraint on the thickness (t) of the investigated semiconductor, where $t \sim \lambda_{CC}$. In addition, such approach is applicable to fundamental investigations at interfaces such as the probing of the electrical potential distribution simultaneously within the solid (i.e., band-bending) and the liquid side of the junction (the double/diffuse layer) [13,17].

Similar to the previous discussion, the maximum thickness of the liquid layer must be three times the photoelectron IMFP to allow the photoelectrons ejected from the interface to penetrate the liquid film, to cross the liquid/gas interface and finally to travel through the gas phase on their path toward the electron analyzer. Therefore, for photoelectrons traveling in water with a KE of 500 eV and 5000 eV, the maximum water thickness must be equal to about 6 nm and 38 nm, respectively.

The preparation of liquid layers characterized by a thickness of the order of tens of nanometer and that are stable for the duration of the measurements (often several hours) is not straightforward and requires the development of novel techniques. As schematically reported in Figure 2, three experimental procedures have been identified to obtain such "free-surface" liquid layers: the "emersion technique" (recently called "dip and pull", Figure 2a), "the tilted sample" procedure (Figure 2b) and the "offset droplet" method (Figure 2c).

The "dip and pull" method was developed from the early works on the extended liquid meniscus of Bockris and Cahan [32], Siegbahn [33] and in particular by Hansen and Kolb [34–36]. It obtains nanometer-thick liquid layers by partially extracting the sample from the liquid solution under controlled conditions (i.e., equilibrium between the liquid and its vapor at a given temperature, constant rate of pulling). This method will be discussed in detail in the next Section 3.1.1.

In "the tilted sample" technique developed by Kötz and coworkers [37], the liquid is contained in a pocket in which the sample is immersed with a tilt angle. This tilt creates a region at the boundary between the liquid free surface and the sample that can be positioned at the focal point of the analyzer and irradiated by the X-ray beam. The authors were able to detect the photoelectron spectra of the elements of the investigated ionic liquid (1-ethyl-3-methylimidazolium tetrafluoroborate) and of the Pt sample simultaneously in one survey spectrum, using the Al K_α emission line as excitation source [37]. Similar to what has been described above, it is possible to create a stable evaporation/condensation equilibrium by dosing water in the analysis chamber to match the water vapor pressure above the liquid contained in the pocket, at the given temperature of the experiment.

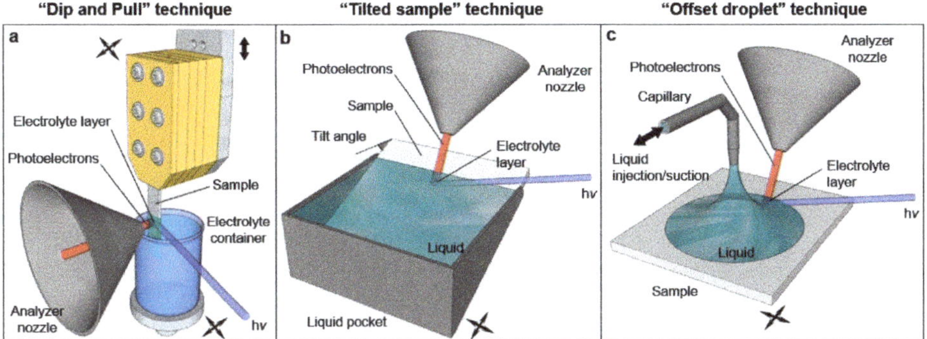

Figure 2. Schematization of the "dip and pull" (**a**), "tilted sample" (**b**) and "offset droplet" techniques (**c**) that are successfully used to prepare free-surface liquid layers thin enough to probe the solid/liquid interface using photoelectron spectroscopy.

Recently, Walton and coworkers developed instead an alternative way by using a fine capillary (100 µm internal diameter) to inject and control a droplet deposited in situ on the sample surface [38]. Since an offset exists between the position of the capillary and the measurement spot, the authors named this technique as the "offset droplet" method. The capillary is connected with an external HPLC pump via a liquid feedthrough. By applying a fixed flow rate to the pump, the liquid can be directly introduced into the analysis chamber and deposited on the sample surface. After an initial transient in which the pressure in the analysis chamber will equilibrate to the liquid vapor tension, a steady-state condition can be reached by choosing an appropriate flow to counterbalance the evaporation rate. By rigidly moving the sample and the capillary, the droplet edge can be positioned at the intersection of the photon beam and the analyzer focus, enabling therefore photoemission experiments at the solid/liquid interface.

It is worth mentioning that the three aforementioned techniques do not allow a fine control of the liquid layer thickness, although it is possible to "select" the suitable thickness for the experimental purposes. With the "dip and pull" technique, it is possible to change the height of the measurement spot by changing the positions of the sample and liquid container with respect to the X-ray beam-analyzer focus intersection. The same procedure can be applied to the "tilted sample" and "offset droplet" techniques, by changing the tilt angle and the liquid injection/suction rate, respectively. The described procedure holds exclusively for wettable surfaces, where the contact angle ψ at the liquid meniscus is smaller than 90° (indicating relatively intense interactions between the solid and the liquid). This poses a severe restriction on the type of samples/surfaces accessible using these techniques, due to the fact that no nanometric-thick continuous liquid layer can be formed on non-wettable surfaces ($\psi \geq 90°$). A practical example of this drawback will be given in Section 3.3.

3.1.1. The "Dip and Pull" Method

Figure 3 reports a schematization of the "dip and pull" procedure. The sample is first placed in front of the analyzer nozzle, aligning its bottom edge with the nozzle aperture (Figure 3a). The sample is then lowered till its bottom edge enters in contact with the liquid free surface in the electrolyte beaker (Figure 3b). By measuring these two positions using the sample manipulator encoder, it is possible to get the value d, which is the distance between the nozzle aperture and the free surface of the electrolyte (typically some millimeters). The sample is then immersed into the liquid by the same length, plus an additional portion Δ (Figure 3c). The sample is then retracted from the liquid by a distance d, as reported in Figure 3d. In this manner, the point P that was previously placed at the liquid meniscus (liquid/gas interface) is now aligned with the nozzle aperture, thereby facilitating the search of the optimal liquid layer thickness. By scanning the region around P (changing the height of the measurement spot as reported above) it is possible to find the suitable layer thickness for the experiment (in the case of AP-HAXPES the film thickness can span from few to some tens of nm).

The liquid film thickness depends on a number of factors and their corresponding interplays: on the wettability of the solid surface for a given liquid (i), the height of the measurement spot above the free surface of the bulk liquid (ii), the presence of an eventual temperature gradient between the two positions (iii), and the solvent evaporation rate (iv). To counterbalance water evaporation from both the liquid layer and the container, two options can be used: either dosing water in the analysis chamber through a valve connected with a heated external reservoir, or by introducing a second (larger) water volume within the analysis chamber. Both solutions work at pressures close to the water vapor pressure at room temperature (100% relative humidity, RH), thus generating a dynamic equilibrium at the thin liquid layer where the evaporation and condensation rates are the same. Then, by moving the sample (and the liquid container accordingly), the interface can be positioned at the intersection of the incident X-ray beam and the focal point of the electron analyzer, thereby enabling in situ photoemission measurements of the solid/liquid interface.

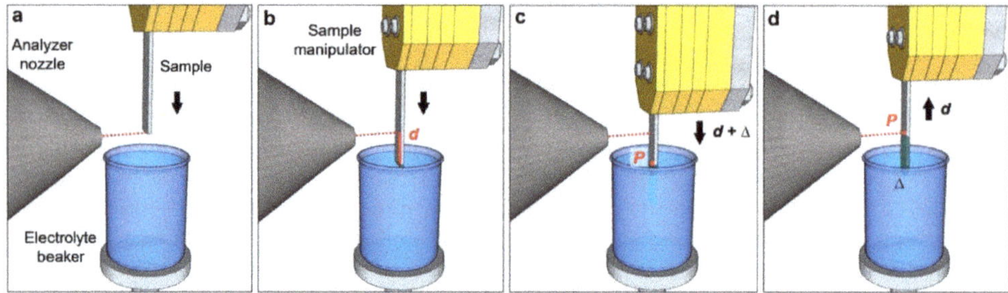

Figure 3. Schematization of the "dip and pull" method coupled with AP-HAXPES experiments. See text for the detailed explanation of the experimental procedure. The pressure inside the analysis chamber is generally close to the water vapor pressure at room temperature, thus enabling to work at about 100% relative humidity.

The "dip and pull" method can be also utilized to perform in situ (photo)electrochemical and photoemission measurements. Two additional electrodes can be mounted on the sample holder, leading to an effective three-electrode electrochemical cell [11–17]. The bottom part of the electrodes (Δ), being kept in the bulk electrolyte, provides the electrochemical continuity to apply a potential to the thin electrolyte layer on the sample (working electrode) surface, thus enabling the investigation of solid/liquid electrified interfaces [11–17]. A drawback of this method (and, in general, of the "free-surface" liquid layer approach) is constituted by the mass transport limitation in the electrolyte layer, since the latter is effectively static in the direction parallel to the solid surface (with the exception of liquid flow due to eventual differences in the temperature between the electrolyte reservoir and the measurement spot above the liquid meniscus, causing thermo-capillary or Bénard–Marangoni convection). The mass transport limitations have been experimentally addressed in our recent works [15,39]. We have demonstrated that the liquid layer undergoes instability for faradaic reactions involving consumption of the electrolyte, such as during the oxygen evolution reaction (OER) in 1.0 M KOH aqueous solution [39]. Holding the potential at the Pt working electrode at +1.93 V vs. RHE (reversible hydrogen electrode), we observed the loss of potential control within the liquid layer in less than 2 h from the beginning of the experiment [39]. The loss of anodic polarization was assessed by the deviation of the binding energy (BE) shift of the O 1s liquid phase water (LPW) component and K 2p core levels from the applied OER potential, as a function of the observation time [39]. The following mechanism is likely to occur: first, the hydroxyl anions are depleted from the thin liquid layer due to the ongoing oxidation to molecular oxygen. Second, the consequent decrease over time of the pH within the liquid layer leads to an increasing *IR* drop, which is responsible for the observed BE deviation from the applied potential. Interestingly, after the loss of potential control occurred, we observed also an important decrease of the LPW signal mirrored by the K 2p intensity, thereby indicating a progressive thinning of the liquid layer. To estimate the diffusion time scale of the hydroxyl groups from the liquid meniscus through the liquid layer, we can use Fick's first law of diffusion (Equation (2)). The distance z of the liquid meniscus from the AP-HAXPES measurement position is typically about 0.8 mm, and the diffusion coefficient D of hydroxyl anions in water at infinite dilution at room temperature is equal to 5.30×10^{-5} cm^2 s^{-1} [40].

$$J_{\parallel} = n/(\Sigma \times \Delta t) = -D \, (dC/dz)_{\parallel} \tag{2}$$

$J_{\parallel}$ represents the diffusional flux parallel to the sample surface, which is the amount of hydroxyl groups (n, in mol) flowing through the liquid layer cross-section (Σ) within the time Δt, whereas $(dC/dz)_{\parallel}$ is the 1-dimensional concentration gradient within the liquid layer (parallel to the sample surface) between the measurement spot and the liquid meniscus. First, we take into account a liquid layer thickness of 20 nm, which is a typical value for the "dip and pull" technique, and 0.7 mm as the lateral dimension of the sample. The cross-section Σ of the electrolyte film is therefore 0.7 cm $\times$ 20×10^{-7} cm = 1.4×10^{-6} cm^2. Secondly, let us define a complete depletion of hydroxyl within the liquid layer after a time Δt. This sets the concentration gradient $(dC/dz)_{\parallel}$ to be linear between the measurement spot and liquid meniscus, where the OH$^-$ concentration (C) is nominally equal to 1.0 M (1 mol L^{-1}). Therefore, $(dC/dz)_{\parallel}$ = 1 mol L^{-1}/0.8 cm = 10^{-3} mol cm^{-3}/0.8 cm = 1.25×10^{-3} mol cm^{-4}. We can then determine the flux J of hydroxyl through the cross section Σ per unit time: $J_{\parallel} = n/(\Sigma \times \Delta t) = C \times \Sigma \times z/\Sigma \times \Delta t = C \times z/\Delta t = 10^{-3}$ mol cm^{-3} $\times$ 0.8 cm/Δt = 8.0×10^{-4} mol cm^{-2}/Δt. Using Equation (2) we can calculate Δt, which is the time needed by the hydroxyl groups to diffuse from the liquid meniscus to replenish the solution volume within the thin electrolyte layer. Δt is found to be about 12×10^3 s (~3.3 h). This limitation in the ionic diffusion rate is confirmed by a recent study of Shavorskiy et al., where the authors observe a significant *IR* drop in the liquid film at the hematite/liquid electrolyte interface under PEC conditions (for applied potentials above ~1.2 V vs. RHE) [17]. This mass transport limitation has also an important effect on the achievable current densities within the liquid layer. In a recent study, we masked the bottom of a Pt sample immersed in KOH 1.0 M aqueous electrolyte and compared to an unmasked electrode [15]. We determined that a bulk (unmasked) current density of about 1.0 mA cm^{-2} under OER conditions corresponds to about 0.3 mA cm^{-2} in the emersed part of the Pt surface [15]. This observation was confirmed by the polarization resistance (R$_p$) for the masked and unmasked configurations, determined using electrochemical impedance spectroscopy (EIS) measurements. In line with the current densities results, the ratio between the two configuration R$_p$ (unmasked to masked) was equal to 3.37 [15]. It is important to highlight that it was not possible to completely avoid the contribution arising from the macroscopic liquid meniscus between the sample and the liquid free surface. Therefore, the actual current density at the AP-HAXPES measuring spot might be even lower. As a general remark, the interplay between the applied overpotential and the limitations in the mass transport kinetics through the liquid layer plays a crucial role during "dip and pull" experiments, and needs to be evaluated case by case (depending on the nature of the solid surface, the catalytic activity of the (photo)electrocatalyst etc.). Moreover, the electrode potential at the measuring spot must be corrected for the actual pH value (since an eventual concentration gradient leads to a pH gradient as well within the liquid layer). This can be performed by calculating the electrolyte/water ratio using the corresponding photoionization core levels (such as K 2p and O 1s for a KOH aqueous solution, as reported in reference [15]).

The lowering of the reaction kinetics within the liquid layer due to the mass transport limitations can be exploited to investigate the evolution of the interfacial properties on a time scale accessible by AP-HAXPES (seconds). For instance, in a recent work we were able to monitor the light-induced formation of a bismuth phosphate (BiPO$_4$) layer atop a polycrystalline bismuth vanadate (BiVO$_4$) surface, working at the half-cell open circuit potential and in a phosphate-containing electrolyte layer with a thickness ranging between 24 and 32 nm [41]. We found that the $BiPO_4$ formation was reversible upon restoring dark conditions. The $BiPO_4$ formation and dissolution kinetics have been characterized by fitting the temporal trend of the phosphate/water liquid phase (LPW) intensity ratio under visible light and dark conditions, respectively (the spectral components have been determined from the O 1s core level spectra acquired under light and dark conditions as a function of time). The retrieved time constants of the $BiPO_4$ formation and dissolution were found to be equal to 321 ± 61 s and 498 ± 89 s, respectively [41].

Finally, a general challenge closely related to the use of the aforementioned methods concerns the enhancement of the photoelectron signal coming from the narrow interfacial region atop the solid and buried by the liquid side of the junction. In the next section, using a model solid/liquid junction and numerical simulations, we will address this point and assess the optimal X-ray energy that allows the enhancement of the interface photoelectron signal.

3.2. Optimization of the AP-HAXPES Experimental Conditions

The simulations have been generated using the SESSA software [22] (see "Materials and Methods" section for the simulation details). Our model system is constituted by a TiO_2 surface functionalized with 1 MLE of APTES, characterized by a thickness of 12.7 Å. To complete our model of the solid/liquid interface, a water layer with a thickness of 20 nm was placed atop the sample surface, as schematically reported in Figure 1. This value has been taken as a realistic representation of the liquid layer thickness typically obtained during "dip and pull" experiments and accessible by HAXPES [11–17].

The use of HAXPES in combination with AP experiments leads to several advantages:

1. The high photon energy (and correspondingly the photoelectron KE) drastically decreases the secondary electron emission cross sections. This has a beneficial effect in limiting the radiation-induced damage suffered by the sample and the correlated radiolysis of water [42];
2. The inelastic scattering between photoelectrons and gas molecules decreases as a result of the large energy difference existing between the X-ray energy and the rotovibrational/electronic excitations and ionization thresholds in the gas molecules (typically falling within the UV-VIS and soft X-ray regions, respectively, for most gases of interest such as O_2, N_2, H_2O, CO, CO_2, gaseous hydrocarbons and alcohols);
3. The elastic scattering between photoelectrons and gas molecules, which is responsible for smearing out the photoelectron angular distribution, is less pronounced due to the generally increasing forward focusing effect as the excitation energy increases [43];
4. To maximize the overall photoelectron yield in HAXPES, quasi-normal emission detection is typically coupled with X-ray grazing incidence ($\alpha \leq 5°$) [44–47]. This has the advantage of minimizing the secondary (inelastic) electron background [46];
5. The high photoelectron KEs lead to increased photoelectron IMFPs in the liquid and thus enable the investigation of solid/liquid junctions through liquid layers whose thickness is of the same order of the photoelectron information depth.

On the other hand, the large probing depth reduces the relative contribution of the interface region with respect to the overall detection (which includes the bulk liquid and solid signals). In addition, the total photoionization cross sections (TPCSs) decrease with increasing photon energy, thereby further reducing the signal coming from the interface (within our model, the signal coming from the APTES overlayer whose thickness is comparable to typical electrical double layer thickness in solution). Figure 4a,b clearly illustrate the scenario, reporting the photoelectron IMFP (right axes) and TPCS (left axes) for Si 1s and N 1s core levels (from the APTES overlayer) as a function of photon energy. The IMFP is calculated for photoelectrons traveling in pure water. The IMFP (Λ_e) and TPCS (σ) trends have been fitted using the following relations (Equations (3) and (4), respectively):

$$\Lambda_e = C \times (h\nu)^p \tag{3}$$

$$\sigma = \sigma_0 \times \exp[-\{(h\nu - h\nu_0)/\tau\}] \tag{4}$$

C, p, σ_0 and τ are fitting parameters, whereas $h\nu$ denotes the photon energy. Note that we have chosen to report the TPCS instead of the differential PCS (DPCS) in order to provide a general discussion about the photon energy dependency of the PCS, since the DPCS strongly depends on the specific experimental geometry and utilized multipole expansion truncation (see "Materials and Methods" section for details). The TPCSs have been determined by interpolation of the TPCS values reported by Yeh and Lindau [48], whereas the electron IMFPs have been computed using the Tanuma–Powell–Penn (TPP-2M) predictive equation [49]. Table 2 reports the determined p, τ and σ_0 values for both analyzed core levels, which will be used later to rationalize the observed behavior of the Si and N 1s core level photoelectron intensity trend as a function of the photon energy.

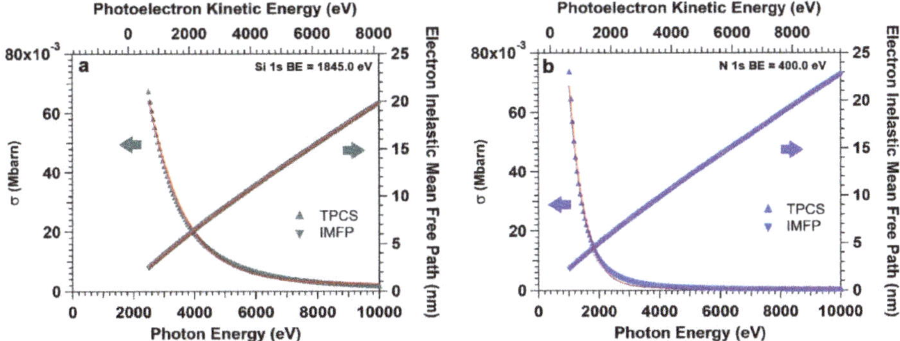

Figure 4. Electron inelastic mean free path (IMFP, right axes) and total photoionization cross section (TPCS, left axes) as a function of the photon energy, for Si 1s (**a**) and N 1s (**b**) core levels. The photoelectron kinetic energy scale has been generated from the photon energy values, using Einstein's energy conservation law and a binding energy of 1845.0 eV and 400.0 eV for Si 1s and N 1s core levels, respectively. The IMFPs are determined for photoelectrons travelling through water. Red curves are fits according to Equations (3) and (4).

Table 2. p, τ, and σ_0 fitting parameters describing the functional dependency of the electron inelastic mean free path (IMFP) and total photoionization cross section (TPCS) on the photon energy, for Si 1s and N 1s core levels. Note that the cross section unit is expressed in barn (1 barn = $1 \cdot 10^{-24}$ cm^{-2}).

Core level	Nominal Binding Energy [eV]	p	τ [eV]	$h\nu_0$ [eV]	σ_0 [Mbarn]
Si 1s	1845.0	0.843 ± 0.01	1387 ± 18	2500	0.062 (at 2500 eV)
N 1s	400.0	0.850 ± 0.01	622 ± 7	1000	0.071 (at 1000 eV)

Therefore, it is necessary to find the optimal photon energy that enables to selectively enhance the signal coming from the interfacial region. In addition, as Figure 4a,b and Table 2 show, this evaluation needs to be performed for the different core level spectra involved in the investigation, since they are characterized by different cross sections with different photon energy dependency.

Let us start by quantitatively defining the photoelectron intensity for a given core level characterized by principal and angular quantum numbers n and l, respectively. The photoelectron intensity I_{nl} in vacuum, normalized by the incident X-ray flux integrated over the irradiated sample volume and the analyzer acceptance solid angle over the surface, is given by Equation (5):

$$I_{nl} = \int_0^\infty \rho(x, y, z) \times (d\sigma/d\Omega)_{nl} \times \exp[-z/(\Lambda_e \times \cos\theta)] \, dxdydz \tag{5}$$

ρ(x, y, z) is the number of emitters for unit volume (atomic density), z is the photoelectron path in the material measured along the surface normal, Λ_e is the IMFP, $(d\sigma/d\Omega)_{nl}$ is the DPCS, and θ is the take-off angle formed between the photoelectron propagation direction and the surface normal. We need now to solve the integral reported in Equation (5) for core levels belonging to the interface (in our case, the APTES overlayer) by systematically changing the photon energy, in order to find the energy value that maximizes I_{nl}. This is given by the trade-off between the increase in the probing depth and the DPCS (TPCS) lowering at increasing photon energies. To this end, we have performed a series of SESSA simulations keeping the detection geometry fixed ($\alpha = 15°$, $\theta = 0°$) and varying the photon energy of the incoming radiation (considered as 100% linearly polarized in the orbit (horizontal) plane) with a step of 50 eV. Note that the simulated detection geometry is the same to the one experimentally available at BL 9.3.1, ALS. The results of these simulations are reported in Figure 5a,b for the Si and N 1s core level photoelectron intensity (left axes), respectively.

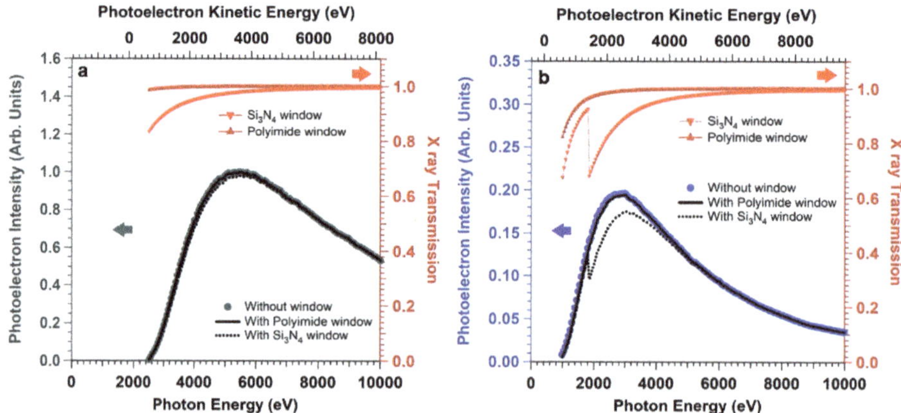

Figure 5. Si 1s (**a**) and N 1s (**b**) photoelectron intensity evolution as a function of the photon energy, with and without the X-ray window placed between the X-ray source and the sample. The two elements belong to the APTES overlayer buried by 20 nm of water. For a better comparison, the reported photoelectron intensities with no simulated absorber are normalized by the maximum value obtained at 5390eV for the Si 1s core level. The right axes of Figure **a**,**b** report the X-ray transmission through a Si_3N_4 and polyimide (Kapton) window. For both materials, the simulated window thickness is 500 nm, which is a common value for X-ray windows used during AP-HAXPES experiments. To simulate the effect of the X-ray window on the photoelectron response, the X-ray transmission curves have been convoluted with the photoelectron intensity trends obtained in absence of an X-ray absorber.

The intensity of the Si 1s core level spectrum (BE = 1850 eV) initially increases for photon energy values between 2500 and 4500 eV (Figure 5a). This is due to the fact that the increasing photoelectron KE leads to an increase of the IMFP, which balances the decay of TPCS (since the photon energy moves away from the Si KLL threshold, centered between 1800 and 1900 eV). As the photon energy further increases, the exponential decay of the TPCS starts to dominate over the other terms in Equation (5), with the intensity curve leveling off and eventually reaching a maximum at a photon energy of 5390 eV (the intensity curve has been fitted using a lognormal function in order to accurately characterized the maximum). At this energy, the TPCS is about $e^{-2.0}$ the value at 2500 eV ($\sigma_0 = 0.062$ Mbarn), whereas the probed depth (defined as the depth from which 95% of the emitted electrons are inelastically scattered, i.e., $3 \cdot \lambda_e$) is about 40 nm (that is two times the thickness of the water layer). For higher energies, the TPCS dominates and a monotone decrease of the photoelectron intensity can be observed in line with the exponential decay of the TPCS. The intensity trend of the N 1s core level spectrum as a function of the photon energy (Figure 5b) follows a similar qualitative trend, however shifted in photon energy (photoelectron KE) due to the lower energy threshold of the KLL transition (at around 400 eV). The maximum of the intensity curve is reached at a photon energy of about 2850 eV, for which the TPCS is equal to about $e^{-3.0}$ the value at 1000 eV (σ_0 = 0.071 Mbarn). The information depth at 2890 eV is equal to about 25 nm.

We can therefore rationalize the observed dependencies drawing some general conclusions:

1. For core levels whose electronic excitation energies (and therefore BEs) fall within the soft X-ray region (below 1000 eV), the TPCS decreases rapidly as the photon energy increases within the hard X-ray region (above 2000 eV). For core levels characterized by higher electronic excitation energies (approaching or within the hard X-ray region), the corresponding TPCS decreases with a lower rate with the increase of the photon energy;

2. The IMFP follows the same functional dependency with the photon energy for both type of core levels;

3. The combination of the two aforementioned points lead to the following phenomenology: the maximum of the photoelectron intensity of different core levels belonging to the interface region lies at different photon energies. For soft X-ray core levels, the trade-off between relatively high KEs and the fast decay of TPCS leads to probed depths at the curve maximum essentially matching the electrolyte overlayer thickness. For hard X-ray core levels, instead, the slower decay of the TPCS with the photon energy leads to information depth at the maximum of the photoelectron intensity curve higher than the water overlayer thickness. In addition, the intensity curve for hard X-ray core levels is characterized by a broader spectral range compared to that for soft X-ray spectra. In our case, the full-width at half-maximum (FWHM) of the Si 1s intensity curve is about 6500 eV, whereas the same for the N 1s core level is characterized by a FWHM of about 4150 eV.

Furthermore, we simulated the effect of introducing an X-ray absorber between the X-ray source and the sample, which is typically done in order to seal the X-ray source (beamline or anode source) from the high pressure environment. Figure 5 reports the X-ray transmission (right axes) through a 500 nm-thick Si_3N_4 and Kapton window (the calculations of the X-ray transmission as a function of the photon energy (1000 eV–10000 eV) and fixed incidence angle α at the window ($\alpha = 90°$) have been performed using the database of the Center for X-ray Optics (CXRO) of the Lawrence Berkeley National Laboratory (Berkeley, U.S.A) [50]). We can observe that working with hard X-rays (photon energy above 2000 eV) and Kapton windows keeps the X-ray transmission close to unity even for relatively thick windows (X-ray transmission above 95 %). On the other hand, for Si_3N_4 windows the Si KLL absorption edge between 1800 and 1900 eV decreases the transmission above 2000 eV (by about 30%, 20%, and 10% at a photon energy of 2000, 2500, and 3100 eV, respectively). The Si and N 1s photoelectron intensities have been then convoluted with the simulated X-ray transmission. We can observe that the Si 1s intensity trend (Figure 5a) is basically not influenced by the presence of the window (either Si_3N_4 or Kapton), with an intensity decrease at the maximum (hv = 5390 eV) of about 2% in the case of a Si_3N_4 absorber. On the other hand, for the soft X-ray N 1s core level (Figure 5b), the presence of the Si_3N_4 window leads to an appreciable loss of the photoelectron intensity at the curve maximum (by about 15 % compared to the simulation performed without absorber). Moreover, the maximum of the intensity curve is blue-shifted (~250 eV) by the presence of the Si_3N_4 window compared to the curve obtained with no absorber. In line with the findings obtained for the Si 1s core level intensity curve, the presence of the 500 nm-thick Kapton window does not induce a shift of the N 1s core level intensity curve, with a loss of signal at the maximum less than 2% compared to the calculation performed without absorber. We want to highlight that the presence of a differentially-pumped aperture (pin-hole) instead of an X-ray window will lead to the same results obtained for no X-ray absorber, since the former implies a simple photon energy-constant lowering of the photon flux at the sample.

Overall, this analysis elucidates the importance of tuning the photon energy for different core levels belonging to the interface region buried by a nanometric-thick electrolyte layer, in order to keep the optimal core level "brightness" and enhance the investigation of the solid/liquid interface. Moreover, it shows that another important parameter to keep in mind for enhancing the photoelectron intensity is the choice of the X-ray window, in particular for soft X-ray core levels.

3.3. Evolution of the Physical/Chemical Properties of the Solid/Liquid Interface as a Function of the Electrolyte pH

In this section, we will show a practical example of the potentials offered by coupling the "dip and pull" method with AP-HAXPES. We investigated a sol-gel spin-coated TiO_2 surface functionalized with ~1 MLE of APTES, using a fixed photon energy of 4000 eV. This value is close to the average (4120 eV) of the two energies retrieved above to enhance the signal intensity from Si and N 1s core levels within the APTES overlayer. In addition, the photon flux at the sample at photon energies of 2850 eV and 5390 eV at BL 9.3.1 at the Advanced Light Source was not optimal due to experimental limitations of the beamline optics. We studied the solid/liquid interface formation by investigating the surface in its pristine conditions (high vacuum, HV, about 10^{-6} mbar) (i), by exposing it to ~70% RH at room temperature (hydrated conditions, HC) (ii), and finally after dipping the surface in pure water and different aqueous electrolytes (iii), changing the pH from 7 to 14 (see also Table 1). The dipping was performed in about 70% RH environment at room temperature (p_{water} ~ 21 mbar).

It is important to note that the experimental BE values measured in this work for light elements such as O and N are slightly higher than those reported in the literature for similar systems. This might be due to recoil effects when momentum is transferred from the ejected photoelectron to the emitting atom. Recoil is present in all photoemission processes and its effects are non-negligible for high photoelectron KEs and light elements [51,52]. At a photon energy of 4000 eV the calibration performed on gold using the Au $4f_{7/2}$ core level signal (KE ~ 3916 eV) is essentially not influenced by the recoil (the corresponding loss is about 10 meV). On the other hand, the energy loss for the O 1s (KE ~ 3470 eV) core level spectrum is about 120 meV.

To monitor the formation of the liquid electrolyte layer on the surface, we acquired the O 1s core level spectrum at all the aforementioned conditions, as reported in Figure 6a. At pristine conditions, the O 1s spectrum exhibits an intense peak centered at a BE of about 530.7 eV, attributable to the TiO_2 lattice oxygen (O^{2-}) [53,54]. A minor component can be observed at about 533.1 eV, most likely due to molecularly adsorbed water ($H_2O_{ads.}$) on Ti^{4+} sites [53] and on the APTES terminal $-NH_2$ groups upon exposure to air [55]. It is in fact suggested by Meroni et al. in a recent work that APTES chemisorption on TiO_2 mainly occurs through the formation of one or two $Si-O-Ti$ bonds involving the Si headgroup [56], whereas adsorption via the terminal amino group seems instead considerably more labile. Therefore, the Lewis base-character of the amine group might induce water adsorption at the nitrogen atom through hydrogen bond formation with water molecules.

At HC and for the dipping experiments in the different aqueous electrolytes, we can observe the presence of two new spectral features in the O 1s core level spectrum. The first, centered at a BE between 533.6 and 534.0 eV is attributable to liquid phase water (LPW) due to the formation of the liquid layer, whereas the second peak is associated to the gas phase water (GPW) (in the BE range of 536.3 and 536.6 eV) [53]. The shift in BE of these two components is typically associated to work function changes at the sample surface [57] due to the different explored conditions.

In addition, passing from the pristine conditions to HC and for the dipping experiments in the different aqueous electrolytes, it is possible to observe the development of a third feature centered at a BE of about 531.8 eV. The origin of such a spectral component, under the particular conditions of the experiment, is not trivial and it can be due to a number of different causes. First, the observed BE is typical of surface hydroxyl groups generated from the well-known dissociative adsorption of water for pressures above ~ 10^{-3} mbar [53]. Second, oxygen-containing carbon compounds from background contamination can also produce a peak at this BE, particularly after filling the chamber with water vapor for several minutes at relatively high pressures. The contamination originates most likely from environmental CO_2 and hydrocarbons containing CO_x groups. This is typically caused by the displacement of molecules from the chamber walls upon exposure to water vapor at or above the mbar pressure range or from molecules contained in the water vapor source itself [53]. Third, the formation of non-volatile carbonates during the preparation of the alkaline electrolytes in ambient conditions ($CO_2 + 2\ KOH \rightarrow K_2CO_3 + H_2O$) can lead to the "deposition" of carbonates on the sample surface during the "dip and pull" procedure. The reported BE for carbonate groups (531.9 eV [58]) is also consistent with our findings. The "deposition" of carbonates at the highest investigated pH might be the main cause of the important intensity increase of the spectral component observed when passing from pH 12 to pH 14.

We want now to focus the reader's attention on the photoelectron intensity evolution of the LPW component during the experiment. Figure 6b,c report the LPW to O^{2-} ratio and the corresponding water layer thickness estimation, respectively. To do so, SESSA simulations have been performed using the same sample structure reported in Figure 1. By changing the water overlayer thickness, the simulated LPW/O^{2-} ratio was adjusted to match the experimentally-determined value for each condition. Passing from the first dipping in pure water to the third in KOH 10^{-2} M (pH 12), the LPW/O^{2-} ratio changes from about 0.3 to 0.5, which corresponds to an electrolyte layer thickness between 3 and 4 nm. Within the experimental uncertainty, these are similar thicknesses found at HC, after exposing the sample to about 21 mbar of water (~ 2.5 nm). This means that a thin layer of water condensed on the surface at HC (~ 70% RH), and that the thickness of the liquid layer did not depend on the successive dipping procedures in pure water (pH 7) and in KOH 10^{-4} and 10^{-2} M aqueous solutions (pH 10 and 12, respectively). Differently, at pH 14 (KOH 1.0 M) the LPW/O^{2-} ratio increases up to ~2.5 (Figure 6b), which corresponds to an electrolyte layer thickness of about 13 nm (Figure 6c).

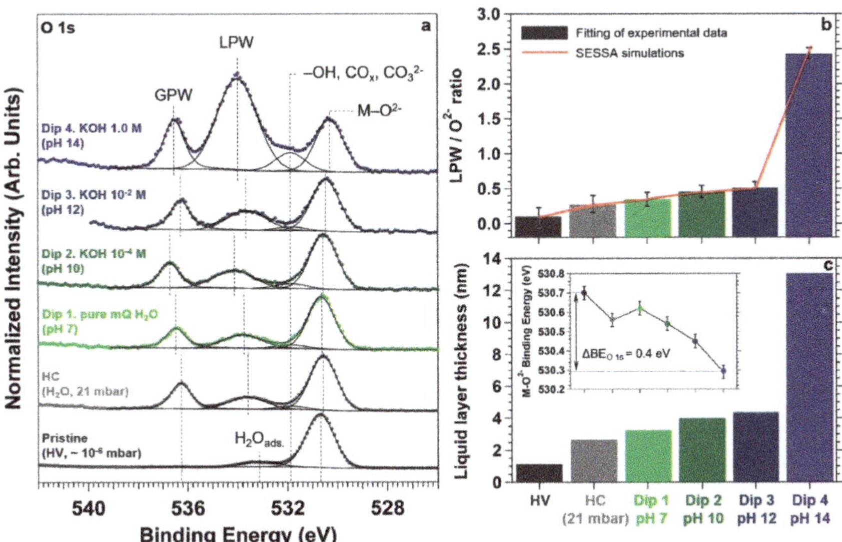

Figure 6. (a) O 1s core level spectra and corresponding multi-peak fitting for the explored experimental conditions (the spectra are normalized by the intensity of the M–O^{2-} component); (b): ratio between the liquid phase water (LPW) and the lattice oxygen (O^{2-}) component determined via the multi-peak fitting procedure; (c): corresponding estimation of the liquid layer thickness, performed using the intensity attenuation of the O^{2-} component. The inset reported in Figure c shows the BE negative shift of the O^{2-} component throughout the experiment, pointing to an upward band bending in TiO_2 as consequence of the deprotonation of the surface –OH groups. The experiment was conducted at room temperature.

To find the reason for the observed phenomenology, we use the Henderson–Hasselbalch equation to estimate the ratio between the unprotonated (–NH$_2$) and protonated (–NH$_3^+$) amine groups as a function of the electrolyte pH (–NH$_2$/–NH$_3^+$ = 10^{pH-pKa}). This procedure allows us to qualitatively assess the net charge at the surface under the different explored conditions, thereby inferring about the interaction between the functionalized charged surface and the liquid water. Using a value of the APTES amine group acid dissociation constant (pKa) of 10.6 as reported by Notsu et al. [59], we find a –NH$_2$/–NH$_3^+$ ratio of 2.5 $\cdot$ 10^{-4}, 0.25, 25, and 2500 at pH 7, 10, 12, and 14, respectively. This means that for pH values below (above) the pKa, the amine group is present on the surface mainly in its protonated (unprotonated) form. This is qualitatively demonstrated by the N 1s core level spectra reported in Figure 7a for the pristine conditions, hydrated conditions and after dipping the sample in the KOH 1.0 M solution (pH 14). The spectrum acquired on the pristine surface exhibits a clear asymmetry toward high BEs, and can be fitted using two spectral components centered at a BE of 400.4 and 401.7 eV (with a FWHM of 1.8 and 2.4 eV, respectively). Taking into account a positive 140 meV shift of the BE due to the recoil loss, the identified BEs for these components are in line with previous studies [56] and can be associated with carbamate (–NHC(=O)OH) and –NH$_3^+$ moieties present at the APTES overlayer, respectively. Interestingly, within the detection limit of the technique (about 1 at.%), we do not observe the presence of unprotonated amine groups (which should generate a peak at a BE of 399.6 eV [56]). This might be due to the fact that the –NH$_2$ groups readily reacted with environmental CO$_2$ to form carbamates upon exposing the sample to ambient conditions, after the APTES functionalization of the TiO$_2$ surface [60]. Exposing the sample to ~ 70% RH at room temperature leads to an increase of the –NH$_3^+$ surface concentration (in agreement with the Henderson–Hasselbalch equation for pH 7), accompanied by an upward BE shift of both spectral components (by about 0.3 eV). After dipping the sample in the KOH 1.0 M solution (pH 14), it is possible to observe the disappearance of the protonated amine component as predicted by the Henderson–Hasselbalch equation. Under these conditions, the N 1s core level peak could be fitted using only one spectral component, characterized by a FWHM of 2.2 eV and centered at a BE of about 400.0 eV. This value is about 0.4 eV lower than the BE reported above for the carbamate moieties present on the pristine surface (400.4 eV). We will give a plausible explanation for such a shift later at the end of this section.

We can then conclude that for pH values below the –NH$_2$ pKa (7 and 10), the presence of the protonated amine groups, although positively charged, prevents the formation of a stable and relatively thick water layer leading to the observed partial non-wetting behavior. Above the –NH$_2$ pKa (pH 12 and 14), no protonated –NH$_3^+$ groups are essentially present on the surface, while the majority of the free (unprotonated) amine groups are converted into carbamate moieties by the nucleophilic CO$_2$ or CO$_3^{2-}$ attack (keep in mind that the O 1s revealed an enhanced CO$_3^{2-}$ photoelectron signal at pH 14, most likely due to the high availability within the liquid layer of non-volatile carbonates formed during the preparation of the solution). In this case, the absence of formal charges at the surface might weaken its interaction with water. On the other hand, at pH 14 we observe a different behavior, in which the surface shows a higher hydrophilicity character. It is reported that the surface hydroxy groups on TiO$_2$ possess acidic character [61] and can therefore easily donate H (Brønsted acid) in strong alkaline conditions [61,62]. Therefore, the deprotonation reaction occurring at the solid/liquid interface (–OH + H$_2$O → –O$^-$ + H$_3$O$^+$) at pH 14 generates a surface negative charge that might be responsible for the stabilization of a thicker liquid layer compared to those observed at lower pH values.

The development of a negative surface charge driven by the pH increase leads also to a second effect. In Figure 6a it is possible to qualitatively observe a negative BE shift of the O^{2-} component. The inset reported in Figure 6c shows the O^{2-} BE trend as a function of the different explored conditions. The downward BE shift is equal to 0.4 eV passing from pristine conditions to the final dip in pH 14, and it is attributable to an upward band-bending in TiO$_2$ (formation of a depletion layer at the solid/liquid interface). This is confirmed by the observation of the same negative BE shift when acquiring the Ti 2p core level spectrum, as reported in Figure 7b for the same experimental conditions. Passing from pristine to pH 14 conditions, the BE measured on the Ti 2p$_{3/2}$ spectrum is equal to 459.3 and 458.9 eV, respectively. The observed negative 0.4 eV shift is thus in agreement with the previous findings. It is worth noting that the photoelectrons generated from the ionization of the O 1s and Ti $2p_{3/2}$ core levels have similar KEs (the difference being about 70 eV), which leads to the same information depth at a photon energy of 4000 eV (~17 nm). This means that the information carried by the BE shift is generated by the spectral summation within the depletion layer, convoluted with the exponential decay of the photoelectron intensity. In addition, we want to highlight that no appreciable FWHM change is observed on the lattice oxygen spectral component or in the Ti $2p_{3/2}$ spectrum, passing from pristine conditions to pH 14. This might be due to the difference between the information depth (using O 1s and Ti $2p_{3/2}$ core levels) and the thickness of the space charge region (SCR). Using a concentration of donors of 1×10^{18} cm^{-3} for intrinsic *n*-doped TiO$_2$ [63] and a band bending potential of 0.4 V as determined experimentally (assuming flat bands at the pristine conditions), the space charge region thickness can be estimated to be about 45 nm [64].

We can make use of the adsorption of the APTES molecules at the TiO$_2$ top-most layer to probe the electrical potential value at the upper part of the band-bending potential distribution, by using a core level spectrum from an element belonging to the APTES overlayer (e.g., Si and N 1s). In the case of the Si 1s spectrum (Figure 7c), the negative BE shift between the pristine and the pH 14 conditions is considerably larger than that observed for O 1s and Ti $2p_{3/2}$, being equal to ~1.1 eV. This value represents therefore the maximum band bending in the depletion layer of TiO$_2$.

On the other hand, as already introduced above, for the N 1s core level spectrum we observe a negative shift of the carbamate group BE of about 0.4 eV, passing from pristine to pH 14 conditions. This could be generated by two causes. First, the nitrogen atom is not "directly" adsorbed at the TiO$_2$ top-most layer, differently than the silicon atom, but it is screened by three alkyl units from the latter. Therefore, the presence of four single covalent bonds (Si–C, C–C, C–C and C–N) between the silicon and the nitrogen atom might lower the potential experienced in the latter. Second, as schematically reported in Figure 7d, the nitrogen atom is dangling from the surface into the liquid layer, thereby experiencing the electrical potential drop in solution. The Debye length k^{-1} for a monovalent ion in an aqueous solution at room temperature is related to the concentration C of the electrolyte by $k^{-1} = 3.04 \times C^{-1/2}$ (C is expressed in mol L^{-1}, or M, and k in Å). This means that for a 1.0 M solution, the Debye length is equal to about 0.3 nm. Defining the thickness of the Gouy–Chapman diffuse layer as three times the Debye length (i.e., at which distance the potential at the sample surface is ∼95% screened in solution), we find that such a thickness is equal to 0.9 nm. We performed a molecular dynamics simulation of the adsorption of one APTES molecule on a rutile TiO$_2$(110) surface through the formation of one Ti–O–Si bond (Figure 7d). To further confirm this, we used an UFF force field and a convergence cut off of 10^{-6}. The simulated distance between the nitrogen and the ideal plane containing the-top most layer oxygen atoms is equal to about 0.7 nm, which is similar to the thickness of the diffuse layer calculated above. This means that the nitrogen atom might experience the local potential at the slip plane (that is, at the ideal plane between the diffuse layer and the bulk solution). The two described effects might then account for the different BE shift measured on the N 1s core level spectra with respect to the value measured using the Si 1s core level.

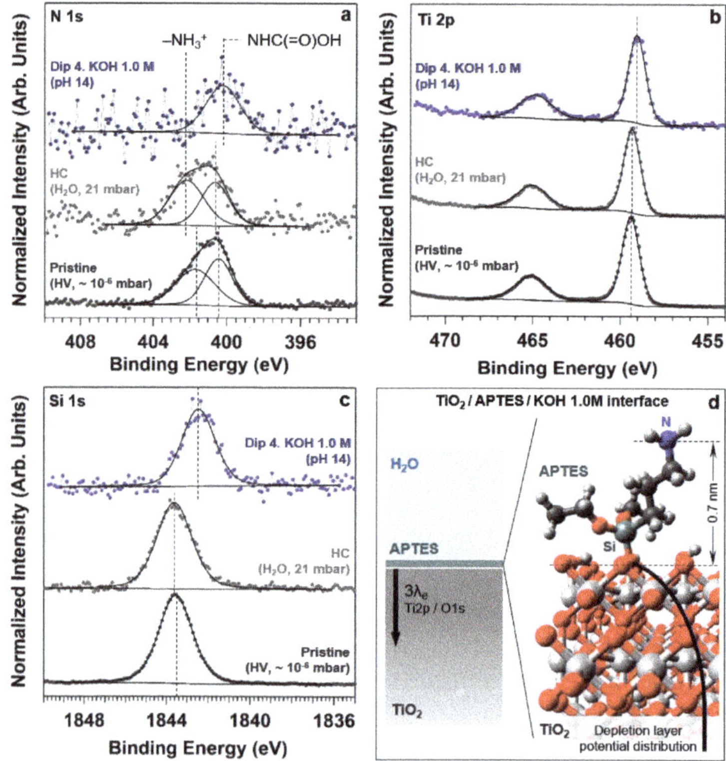

Figure 7. N 1s (**a**), Ti 2p (**b**) and Si 1s (**c**) core level spectra acquired at the pristine conditions, hydrated conditions (HC) and after dipping the sample in the KOH 1.0 M solution (pH 14). (**d**) Relaxed structure obtained from the molecular dynamics simulation of the adsorption of one APTES molecule on a rutile $TiO_2(110)$ surface through the formation of one Ti–O–Si bond (white: titanium; light grey: hydrogen; dark grey: carbon; light green: silicon; red: oxygen; blue: nitrogen).

4. Conclusions

In conclusion, we have described the "dip and pull" procedure for the generation of nanometric-thick layers of liquid electrolyte on solid surfaces. The prepared solid/liquid interfaces can be then investigated using AP-XPS. The use of photon energies within the hard X-ray range (above 2000 eV) allows to probe solid surfaces buried by liquid layers as thick as tens of nanometers, enabling the extension of the parameter space and the possibility to study a wide variety of systems, electrolytes, and electrolyte concentrations.

We discussed the advantages and current drawbacks of using the static "dip and pull" method and related techniques. The main limitation of such techniques is the mass transport in the direction parallel to the interface, which sets restrictions on the current densities achievable in the liquid layer. On the other hand, the lowering of the reaction kinetics within the liquid layer opens up the possibility of performing time-resolved studies on time scales accessible by AP-HAXPES. In addition, this technique is appealing for the investigation of fundamental properties of the solid/liquid interface, such as electrical double layer structure and its modulation by electrical potential and illumination transients, specific ion adsorptions and charge transfers across the interface. This is demonstrated by the recent development of AP-XPS and AP-HAXPES instruments utilizing the "dip and pull" method, at the Advanced Light Source (U.S.A.), Paul Scherrer Institute (Switzerland), MAXIV (Sweden), and BESSY II (Helmholtz-Zentrum Berlin, Germany).

By using a model system of a solid/liquid organic/inorganic hybrid interface and numerical simulations, we retrieved the optimal photon energy range that enhances the signal coming from the interface when coupling the "dip and pull" method with AP-HAXPES. This is an important parameter that needs to be tuned for each investigated system, due to the fact that the use of hard X-rays to induce photoemission implies relatively low photoelectron intensities due to the low cross sections.

Finally, we presented a practical application of the technique, investigating the APTES-functionalized TiO_2 polycrystalline surface in KOH aqueous electrolytes at different concentrations. Our experiments show that it is possible to monitor in situ the physical/chemical changes of the solid/liquid interface induced by the variation of the electrolyte pH. Our findings open up the possibility to optimize the solid/liquid interface in order to study simultaneously the potential profile in semiconductors (band-bending) and in the liquid side of the junction (double + diffuse layer), using local molecular probes directly functionalizing the semiconductor surface.

We think that the detailed molecular level comprehension of the solid/liquid interface properties will advance materials and process research in many key-technological fields, such as (photo)electrocatalysis, corrosion protection, water remediation, and CO_2 capture and conversion.

Author Contributions: Conceptualization, M.F. and D.E.S.; Formal analysis, M.F.; Investigation, M.F., F.F.A. and D.E.S.; Writing—original draft, M.F.; Writing—review & editing, M.F., F.F.A., E.J.C., Z.L., R.v.d.K. and D.E.S.

Funding: This work was supported by the German Federal Ministry of Education and Research (BMBF project "JointLab—Grundlagen elektrochemischer Phasengrenzen", GEP, #13XP5023C). This research was partially conducted at the Advanced Light Source of the Lawrence Berkeley National Laboratory, a DOE Office of Science User Facility under contract no. DE-AC02-05CH11231. Z.L. was supported by the National Natural Science Foundation of China (11227902).

Conflicts of Interest: The authors declare no conflict of interest.

References

1. Salmeron, M.; Schlogl, R. Ambient pressure photoelectron spectroscopy: A new tool for surface science and nanotechnology. *Surf. Sci. Rep.* **2008**, *63*, 169–199. [CrossRef]

2. Crumlin, E.J.; Bluhm, H.; Liu, Z. In situ investigation of electrochemical devices using ambient pressure photoelectron spectroscopy. *J. Electr. Spectr. Relat. Phenom.* **2013**, *190*, 84–92. [CrossRef]

3. Liu, X.; Yang, W.; Liu, Z. Recent progress on synchrotron-based in-situ soft X-ray spectroscopy for energy materials. *Adv. Mater.* **2014**, *26*, 7710–7729. [CrossRef]

4. Crumlin, E.J.; Liu, Z.; Bluhm, H.; Yang, W.; Guo, J.; Hussain, Z. X-ray spectroscopy of energy materials under in situ/operando conditions. *J. Elect. Spectr. Relat. Phenom.* **2015**, *200*, 264–273. [CrossRef]

5. Starr, D.E.; Liu, Z.; Hävecker, M.; Knop-Gericke, A.; Bluhm, H. Investigation of solid/vapor interfaces using ambient pressure X-ray photoelectron spectroscopy. *Chem. Soc. Rev.* **2013**, *42*, 5833–5857. [CrossRef] [PubMed]

6. Lewerenz, H.-J.; Lichterman, M.F.; Richter, M.H.; Crumlin, E.J.; Hu, S.; Axnanda, S.; Favaro, M.; Drisdell, W.; Hussain, Z.; Brunschwig, B.S.; et al. Operando analyses of solar fuels light absorbers and catalysts. *Electrochim. Acta* **2016**, *211*, 711–719. [CrossRef]

7. Roy, K.; Artiglia, L.; van Bokhoven, J.A. Ambient pressure photoelectron spectroscopy: Opportunities in catalysis from solids to liquids and introducing time resolution. *Chem. Cat. Chem.* **2018**, *10*, 666–682. [CrossRef]

8. Liu, Z.; Bluhm, H. Liquid/solid interfaces studied by ambient pressure HAXPES. In *Hard X-ray Photoelectron Spectroscopy (HAXPES)*, 1st ed.; Woicik, J., Ed.; Springer: Cham, Switzerland, 2016; pp. 447–466, ISBN 978-3-319-24043-5.

9. Starr, D.E.; Favaro, M.; Abdi, F.F.; Bluhm, D.; Crumlin, E.J.; van de Krol, R. Combined soft and hard X-ray ambient pressure photoelectron spectroscopy studies of semiconductor/electrolyte interfaces. *J. Elect. Spectr. Relat. Phenom.* **2017**, *221*, 106–115. [CrossRef]

10. Zaera, F. Surface chemistry at the liquid/solid interface. *Surf. Sci.* **2011**, *605*, 1141–1145. [CrossRef]

11. Axnanda, S.; Crumlin, E.J.; Mao, B.; Rani, S.; Chang, R.; Karlsson, P.G.; Edwards, M.O.M.; Lundqvist, M.; Moberg, R.; Ross, P.N.; et al. Using "tender" X-ray ambient pressure X-ray photoelectron spectroscopy as a direct probe of solid-liquid interface. *Sci. Rep.* **2015**, *5*, 9788. [CrossRef]

12. Karslıoğlu, O.; Nemšák, S.; Zegkinoglou, I.; Shavorskiy, A.; Hartl, M.; Salmassi, F.; Gullikson, E.M.; Ng, M.L.; Rameshan, C.; Rude, B.; et al. Aqueous solution/metal interfaces investigated in operando by photoelectron spectroscopy. *Faraday Discuss.* **2015**, *180*, 35–53. [CrossRef] [PubMed]

13. Favaro, M.; Jeong, B.; Ross, P.N.; Yano, J.; Hussain, Z.; Liu, Z.; Crumlin, E.J. Unravelling the electrochemical double layer by direct probing of the solid/liquid interface. *Nat. Commun.* **2016**, *7*, 12695. [CrossRef] [PubMed]

14. Favaro, M.; Drisdell, W.S.; Marcus, M.A.; Gregoire, J.M.; Crumlin, E.J.; Haber, J.A.; Yano, J. An operando investigation of (Ni–Fe–Co–Ce)O$_x$ system as highly efficient electrocatalyst for oxygen evolution reaction. *ACS Catal.* **2017**, *7*, 1248–1258. [CrossRef]

15. Favaro, M.; Valero-Vidal, C.; Eichhorn, J.; Toma, F.M.; Ross, P.N.; Yano, J.; Liu, Z.; Crumlin, E.J. Elucidating the alkaline oxygen evolution reaction mechanism on platinum. *J. Mater. Chem. A* **2017**, *5*, 11634–11643. [CrossRef]

16. Favaro, M.; Yang, J.; Nappini, S.; Magnano, E.; Toma, F.M.; Crumlin, E.J.; Yano, J.; Sharp, I.D. Understanding the oxygen evolution reaction mechanism on CoO$_x$ using operando ambient-pressure X-ray photoelectron spectroscopy. *J. Am. Chem. Soc.* **2017**, *139*, 8960–8970. [CrossRef] [PubMed]

17. Shavorskiy, A.; Ye, X.; Karslıoğlu, O.; Poletayev, A.D.; Hartl, M.; Zegkinoglou, I.; Trotochaud, L.; Nemšák, S.; Schneider, C.M.; Crumlin, E.J.; et al. Direct mapping of band positions in doped and undoped hematite during photoelectrochemical water splitting. *J. Phys. Chem. Lett.* **2017**, *8*, 5579–5586. [CrossRef] [PubMed]

18. Ogletree, D.F.; Bluhm, H.; Hebenstreit, E.D.; Salmeron, M. Photoelectron spectroscopy under ambient pressure and temperature conditions. *Nucl. Instrum. Meth. Phys. Res. A* **2009**, *601*, 151–160. [CrossRef]

19. Bluhm, H. Photoelectron spectroscopy of surfaces under humid conditions. *J. Elect. Spectr. Relat. Phenom.* **2010**, *177*, 71–84. [CrossRef]

20. Kahk, J.M.; Villar-Garcia, I.J.; Grechy, L.; Bruce, P.J.K.; Vincent, P.E.; Eriksson, S.K.; Rensmo, H.; Hahlin, M.; Åhlund, J.; Edwards, M.O.M.; et al. A study of the pressure profiles near the first pumping aperture in a high pressure photoelectron spectrometer. *J. Elect. Spectr. Relat. Phenom.* **2015**, *205*, 57–65. [CrossRef]

21. Fadley, C.S. X-ray photoelectron spectroscopy: From origins to future directions. *Nucl. Instrum. Meth. Phys. Res. A* **2009**, *601*, 8–31. [CrossRef]

22. Smekal, W.; Werner, W.S.M.; Powell, C.J. Simulation of electron spectra for surface analysis (SESSA): A novel software tool for quantitative auger-electron spectroscopy and X-ray photoelectron spectroscopy. *Surf. Interface Anal.* **2005**, *37*, 1059–1067. [CrossRef]

23. Barlow, S.M.; Raval, R. Complex organic molecules at metal surfaces: Bonding, organisation and chirality. *Surf. Sci. Rep.* **2003**, *50*, 201–341. [CrossRef]

24. Abrahams, S.C.; Bernstein, J.L. Rutile: Normal probability plot analysis and accurate measurement of crystal structure. *J. Chem. Phys.* **1971**, *55*, 3206. [CrossRef]

25. Cooper, J.W. Multipole corrections to the angular distribution of photoelectrons at low energies. *Phys. Rev. A* **1990**, *42*, 6942. [CrossRef] [PubMed]

26. Pascual, J.; Camassel, J.; Mathieu, H. Resolved quadrupolar transition in TiO$_2$. *Phys. Rev. Lett.* **1977**, *39*, 1490. [CrossRef]

27. Saravanan, P.; Jayamoorthy, K.; Ananda Kumar, S. Binding interaction and morphology studies of 3-aminopropyltriethoxysilane treated nanoparticles. In *Eco-Friendly Nano-Hybrid Materials for Advanced Engineering Applications*, 1st ed.; Ananda Kumar, S., Ed.; CRC Press: Boca Raton, FL, USA, 2017; pp. 263–280, ISBN 978-1-77188-295-8.

28. Do Couto, P.C. Electronically Excited Water Aggregates and the Adiabatic Band Gap of Water. *J. Chem. Phys.* **2007**, *126*, 014509. [CrossRef] [PubMed]

29. Balej, J. Water vapour partial pressures and water activities in potassium and sodium hydroxide solutions over wide concentration and temperature ranges. *Int. J. Hydrogen Energy* **1985**, *10*, 233–243. [CrossRef]

30. Velasco-Vélez, J.J.; Pfeifer, V.; Hävecker, M.; Wang, R.; Centeno, A.; Zurutuza, A.; Algara-Siller, G.; Stotz, E.; Skorupska, K.; Teschner, D.; et al. Atmospheric pressure X-ray photoelectron spectroscopy apparatus: Bridging the pressure gap. *Rev. Sci. Instrum.* **2016**, *87*, 053121. [CrossRef] [PubMed]

31. Velasco-Vélez, J.J.; Pfeifer, V.; Hävecker, M.; Weatherup, R.S.; Arrigo, R.; Chuang, C.-H.; Stotz, E.; Weinberg, G.; Salmeron, M.; Schlögl, R.; et al. Photoelectron spectroscopy at the graphene-liquid interface reveals the electronic structure of an electrodeposited cobalt/graphene electrocatalyst. *Angew. Chem. Int. Ed.* **2015**, *54*, 14554–14558. [CrossRef]

32. Bockris, J.; Cahan, B.D. Effect of a finite-contact-angel meniscus on kinetics in porous electrode systems. *J. Phys. Chem.* **1969**, *50*, 1307–1324. [CrossRef]

33. Siegbahn, H. Electron spectroscopy for chemical analysis of liquids and solutions. *J. Phys. Chem.* **1985**, *89*, 897–909. [CrossRef]

34. Kolb, D.M.; Hansen, W.N. Electroreflectance spectra of emersed metal electrodes. *Surf. Sci.* **1979**, *79*, 205–211. [CrossRef]

35. Hansen, W.N. Electrode resistance and the emersed double layer. *Surf. Sci.* **1980**, *101*, 109–122. [CrossRef]

36. Kolb, D.M.; Rath, D.L.; Wille, R.; Hansen, W.N. An ESCA study on the electrochemical double layer of emersed electrodes. *Ber. Bunsenges. Phys. Chem.* **1983**, *87*, 1108. [CrossRef]

37. Weingarth, D.; Foelske-Schmitz, A.; Wokaun, A.; Kötz, R. In situ electrochemical XPS study of the Pt/[EMIM][BF$_4$] system. *Electrochem. Commun.* **2011**, *13*, 619–622. [CrossRef]

38. Booth, S.G.; Tripathi, A.M.; Strashnov, I.; Dryfe, R.A.W.; Walton, A.S. The offset droplet: A new methodology for studying the solid/water interface using X-ray photoelectron spectroscopy. *J. Phys. Condens. Matter* **2017**, *29*, 454001. [CrossRef] [PubMed]

39. Stoerzinger, K.A.; Favaro, M.; Ross, P.N.; Hussain, Z.; Liu, Z.; Yano, J.; Crumlin, E.J. Stabilizing the meniscus for operando characterization of platinum during the electrolyte-consuming alkaline oxygen evolution reaction. *Top. Catal.* **2018**, *61*, 2152–2160. [CrossRef]

40. Atkins, P.; Paula, J.D. *Physical Chemistry*, 7th ed.; Oxford University Press: Oxford, UK, 2006; p. 1104, ISBN 9780198700722.

41. Favaro, M.; Abdi, F.F.; Lamers, M.; Crumlin, E.J.; Liu, Z.; van de Krol, R.; Starr, D.E. Light-induced surface reactions at the bismuth vanadate/potassium phosphate interface. *J. Phys. Chem. B* **2018**, *122*, 801–809. [CrossRef]

42. Weatherup, R.S.; Wu, C.H.; Escudero, C.; Pérez-Dieste, V.; Salmeron, M.B. Environment-dependent radiation damage in atmospheric pressure X-ray spectroscopy. *J. Phys. Chem. B* **2018**, *122*, 737–744. [CrossRef]

43. Fadley, C.S. Hard X-ray photoemission: An overview and future perspective. In *Hard X-ray Photoelectron Spectroscopy (HAXPES)*, 1st ed.; Woicik, J., Ed.; Springer: Cham, Switzerland, 2016; pp. 1–34, ISBN 978-3-319-24043-5.

44. Gorgoi, M.; Svensson, S.; Schäfers, F.; Öhrwall, G.; Mertin, M.; Bressler, P.; Karis, O.; Siegbahn, H.; Sandell, A.; Rensmo, H.; et al. The high kinetic energy photoelectron spectroscopy facility at BESSY: Progress and first results. *Nucl. Instrum. Meth. Phys. Res. A* **2009**, *601*, 48–53. [CrossRef]

45. Kobayashi, K. Hard X-ray photoemission spectroscopy. *Nucl. Instrum. Meth. Phys. Res. A* **2009**, *601*, 32–47. [CrossRef]

46. Strocov, V.N. Optimization of the X-ray incidence angle in photoelectron spectrometers. *J. Synchrotron Rad.* **2013**, *20*, 517–521. [CrossRef] [PubMed]

47. Rueff, J.-P.; Rault, J.E.; Ablett, J.M.; Utsumi, Y.; Céolin, D. HAXPES for materials science at the GALAXIES beamline. *arXiv* **2018**, arXiv:1807.07522.

48. Yeh, J.-J.; Lindau, I. Atomic Subshell Photoionization Cross Sections and Asymmetry Parameters: 1 < Z < 103. *Atom. Data Nucl. Data Tables* **1985**, *32*, 1–155. [CrossRef]

49. Tanuma, S.; Powell, C.J.; Penn, D.R. Calculation of electron inelastic mean free paths (IMFPs) VII. Reliability of the TPP-2M IMFP predictive equation. *Surf. Interface Anal.* **2003**, *35*, 268–275. [CrossRef]

50. Henke, B.L.; Gullikson, E.M.; Davis, J.C. X-ray interactions: Photoabsorption, scattering, transmission, and reflection at E = 50–30,000 eV, Z = 1–92. *Atom. Data Nucl. Data Tables* **1993**, *54*, 181–342. [CrossRef]

51. Weiland, C.; Rumaiz, A.K.; Pianetta, P.; Woicik, J.C. Recent applications of hard X-ray photoelectron spectroscopy. *J. Vac. Sci. Technol. A* **2016**, *34*, 030801. [CrossRef]

52. Takata, Y.; Kayanuma, Y.; Yabashi, M.; Tamasaku, K.; Nishino, Y.; Miwa, D.; Harada, Y.; Horiba, K.; Shin, S.; Tanaka, S.; et al. Recoil effects of photoelectrons in a solid. *Phys. Rev. B Condens. Matter Mater. Phys.* **2007**, *75*, 233404. [CrossRef]

53. Ketteler, G.; Yamamoto, S.; Bluhm, H.; Andersson, K.; Starr, D.E.; Ogletree, D.F.; Ogasawara, H.; Nilsson, A.; Salmeron, M. The Nature of water nucleation sites on TiO_2(110) surfaces revealed by ambient pressure X-ray photoelectron spectroscopy. *J. Phys. Chem. C* **2007**, *111*, 8278–8282. [CrossRef]

54. Head, A.R.; Johansson, N.; Niu, Y.; Snezhkova, O.; Chaudhary, S.; Schnadt, J.; Bluhm, H.; Chen, C.; Avila, J.; Asensio, M.-C. In situ characterization of the deposition of anatase TiO_2 on rutile TiO_2(110). *J. Vac. Sci. Technol. A* **2018**, *36*, 02D405. [CrossRef]

55. Ukaji, E.; Furusawa, T.; Sato, M.; Suzuki, N. The effect of surface modification with silane coupling agent on suppressing the photo-catalytic activity of fine TiO_2 particles as inorganic UV filter. *Appl. Surf. Sci.* **2007**, *254*, 563–569. [CrossRef]

56. Meroni, D.; Lo Presti, L.; Di Liberto, G.; Ceotto, M.; Acres, R.G.; Prince, K.C.; Bellani, R.; Soliveri, G.; Ardizzone, S. A close look at the structure of the TiO_2-APTES interface in hybrid nanomaterials and its degradation pathway: An experimental and theoretical study. *J. Phys. Chem. C* **2017**, *121*, 430–440. [CrossRef] [PubMed]

57. Axnanda, S.; Scheele, M.; Crumlin, E.J.; Mao, B.; Chang, R.; Rani, S.; Faiz, M.; Wang, S.; Alivisatos, A.P.; Liu, Z. Direct work function measurement by gas phase photoelectron spectroscopy and its application on PbS nanoparticles. *Nano Lett.* **2013**, *13*, 6176–6182. [CrossRef] [PubMed]

58. Deng, X.; Verdaguer, A.; Herranz, T.; Weis, C.; Bluhm, H.; Salmeron, M. Surface chemistry of Cu in the presence of CO_2 and H_2O. *Langmuir* **2008**, *24*, 9474–9478. [CrossRef] [PubMed]

59. Notsu, H.; Fukazawa, T.; Tatsuma, T.; Tryk, D.A.; Fujishima, A. Hydroxyl groups on boron-doped diamond electrodes and their modification with a silane coupling agent. *Electrochem. Solid-State Lett.* **2001**, *4*, H1–H3. [CrossRef]

60. Muüller, K.; Lu, D.; Senanayake, S.D.; Starr, D.E. Monoethanolamine adsorption on TiO_2(110): Bonding, structure, and implications for use as a model solid-supported CO_2 capture material. *J. Phys. Chem. C* **2014**, *118*, 1576–1586. [CrossRef]

61. Ohtani, B.; Okugawa, Y.; Nishimoto, S.I.; Kagiya, T. Photocatalytic activity of TiO_2 powders suspended in aqueous silver nitrate solution. Correlation with pH-dependent surface structures. *J. Phys. Chem.* **1987**, *91*, 3550–3555. [CrossRef]

62. Sham, T.K.; Lazarus, M.S. X-ray photoelectron spectroscopy (XPS) studies of clean and hydrated TiO_2 (rutile) surfaces. *Chem. Phys. Lett.* **1979**, *68*, 426–432. [CrossRef]

63. Mattioli, G.; Alippi, P.; Filippone, F.; Caminiti, R.; Bonapasta, A.A. Deep versus shallow behavior of intrinsic defects in rutile and anatase TiO_2 polymorphs. *J. Phys. Chem. C* **2010**, *114*, 21694–21704. [CrossRef]

64. Van de Krol, R. Principles of Photoelectrochemical Cells. In *Photoelectrochemical Hydrogen Production*, 1st ed.; Van de Krol, R., Grätzel, M., Eds.; Springer: New York, NY, USA, 2012; pp. 13–68, ISBN 978-1-4614-1379-0.

Article

Oxygen Reduction Reaction on Polycrystalline Platinum: On the Activity Enhancing Effect of Polyvinylidene Difluoride

Alessandro Zana [1], Gustav K. H. Wiberg [1,2] and Matthias Arenz [1,*]

[1] Department of Chemistry and Biochemistry, University of Bern, 3006 Bern, Switzerland; alessandro.zana@dcb.unibe.ch (A.Z.); gustav.wiberg@gmail.com (G.K.H.W.)
[2] Department of Physical Sciences, Harold Washington College, Chicago, IL 60601, USA
* Correspondence: matthias.arenz@dcb.unibe.ch; Tel. +41-31631-5384

Received: 20 December 2018; Accepted: 21 January 2019; Published: 24 January 2019

Abstract: There have been several reports concerning the performance improving properties of additives, such as polyvinylidene difluoride (PVDF), to the membrane or electrocatalyst layer of proton exchange membrane fuel cells (PEMFC). However, it is not clear if the observed performance enhancement is due to kinetic, mass transport, or anion blocking effects of the PVDF. In a previous investigation using a thin-film rotating disk electrode (RDE) approach (of decreased complexity as compared to membrane electrode assembly (MEA) tests), a performance increase for the oxygen reduction reaction (ORR) could be confirmed. However, even in RDE measurements, reactant mass transport in the catalyst layer cannot be neglected. Therefore, in the present study, the influence of PVDF is re-examined by coating polycrystalline bulk Pt electrodes by PVDF and measuring ORR activity. The results on polycrystalline bulk Pt indicate that the effects of PVDF on the reaction kinetics and anion adsorption are limited, and that the observed performance increase on high surface area Pt/C most likely is due to an erroneous estimation of the electrochemical active surface area (ECSA) from CO stripping and H_{upd}.

Keywords: ORR; Platinum; PVDF; PEMFC

---

1. Introduction

Efficient energy converters, such as the proton exchange membrane fuel cell (PEMFC), are integral parts in an environmentally friendly hydrogen economy. Although the PEMFC technology has been around for several decades, it still suffers from relatively high costs [1]. With the aim of making the fuel cell technology more economical, significant effort has been invested in finding cheaper catalysts and fabrication techniques, but also in optimizing properties of the solid membrane electrolyte [2–6]. There are several reports stating that the properties of the solid membrane electrolyte can be improved by including different additives. For example, a performance enhancement of PEMFCs by including additives such as polyvinylidene difluoride (PVDF) into the membrane has been reported [7].

If such studies are performed in single cell membrane electrode assembly (MEA) tests, it is very difficult to analyze the origin of observed performance increases. This is due to the complexity of the MEA system and the influence of multiple structural aspects on its performance. Therefore, the effect of PVDF additions to the membrane is not yet fully clarified. In PEMFC research, a thin-film rotating disk electrode (RDE) approach is often applied to reduce the system complexity and to study electrocatalyst layers in a more fundamental fashion [8]. The previous work of our research group used such an approach to mix PVDF into the ink of a high surface area catalyst. The electrochemical properties of catalyst layers prepared by casting the ink on a glassy carbon disk were compared with respect to PVDF free films [9]. The specific oxygen reduction reaction (ORR) activity established by

RDE in phosphoric acid electrolytes was found to improve when the catalyst layer contained PVDF. Further work, however, showed that even in a RDE setup, the apparent (measured) specific ORR activity is extremely sensitive to the film quality, which in turn depends on ink properties, such as pH [10]. Thus, the observed activity increase can be due to intrinsic changes, such as oxygen solubility and suppression of anion adsorption, or it might be due to improved mesoscopic film properties enhancing the local reactant mass transport. To investigate the possible catalytic enhancement due to PVDF further, in the presented work we therefore investigated thin PVDF films on a flat surface of polycrystalline bulk Pt. This system excludes any influence of catalyst film properties, and therefore focusses on the interface between catalyst and electrolyte.

2. Experiment

2.1. Sample Preparation

To prepare PVDF films onto a Pt surface of an RDE electrode, two different amounts of PVDF (Polysciences Inc. Warrington, PA, USA) were dissolved in Tetrahydrofuran (THF) to obtain two solutions containing 0.1 mM and 0.1 M of PVDF, respectively. The electrode tip for the PVDF coating was a polycrystalline Pt disk inserted in a cylinder of polytetrafluoroethylene (PTFE), in such a manner that only a circular area of the platinum is exposed. The Pt tip was sonicated in acetone, ethanol, isopropanol, and MilliQ water, repeating each step three times. Following the sonication, the Pt tip was dip coated into one of the two PVDF solutions for 30 seconds and pulled out of the electrolyte with the surface facing down. The PDVF coated Pt electrode tip was left to dry in air in a closed glass beaker. After drying, the mirror-like shine of the bare Pt surface was lost and the surface exhibited a pale milky color. In the following, PVDF 1 will refer to the polycrystalline Pt surface dip coated in the solution containing 0.1 mM PVDF, and PVDF 2 will refer to the Pt surface dip coated in the PVDF solution containing 0.1 M PVDF. After the drying process, the coated Pt electrode was introduced in an Ar-saturated 0.05 M H_2SO_4 electrolyte solution for electrochemical characterization.

2.2. Electrochemical Characterization

The electrochemical characterization was performed by RDE according to the procedure described in [11,12]. In brief, the electrochemical measurements were conducted in a three- compartment Teflon cell, using an RDE system (Radiometer Analytical, Europe Hach Lange GmbH *Headquarter* Willstätterstr. 11, D-40549 Düsseldorf, Germany). A saturated calomel electrode (SCE) and a Pt mesh have been used as reference and counter electrodes, respectively. The working electrode (WE) was a polycrystalline Pt disk ($\varnothing$ = 5 mm, A = 0.196 cm^2). All the potentials in the paper are expressed with respect to the reversible hydrogen electrode (RHE) potential experimentally determined for each measurement series. The 0.05 M H_2SO_4 aqueous electrolyte solution was prepared using Millipore Milli Q water (> 18.2 MΩcm, TOC < 5ppb) and 96% H_2SO_4 Suprapur (Sigma-Aldrich). The oxygen reduction rate and specific activity (SA) are typically calculated from the positive-going polarization curves recorded in oxygen saturated electrolytes at a scan rate of 50 mVs^{-1} and 1600 rpm rotation rate. In all cases, a background subtraction was performed to remove any contributions from capacitive and surface oxidation processes according to [13]. The iR-drop was measured and compensated according to the methodology used in [13,14]. In brief, the solution resistance was recorded online with the potentiostat (ECi-200 Nordic Electrochemistry Aps c/o Skelvangsvej 43, DK-8920 Randers NV, Denmark) by superimposing a 5 kHz, 5 mV AC signal, and compensated for by an analogue positive feedback scheme. The effective solution resistance was around 2Ω [15]. The electrochemical active surface area (ECSA) was determined by H_{upd} integration and CO stripping experiments described in [12,13]. In all the measurements the potential was cycled in the potential window between 0.05 and 1.05 V_{RHE}.

3. Results and Discussion

3.1. Surface Characterization in Ar Saturated Electrolyte

In Figure 1, the cyclic voltammograms (CVs) of the bare Pt surface and the two different PVDF films on polycrystalline Pt recorded in Ar saturated electrolyte solution are shown. The two different dip coatings were made from solutions containing 0.1 M and 0.1 mM of PVDF solution, respectively. The voltammetric profiles of polycrystalline Pt are generally divided into three electrode potential regions: (i) the hydrogen under potential deposition (H_{upd}) region (0.05–0.3 V_{RHE}), (ii) the double layer region (0.3–0.7 V_{RHE}), and (iii) the potential region of OH_{ad} and O_{ad} formation and reduction (>0.7 V_{RHE}) [16]. The effect of PVDF can be scrutinized by comparing the different CVs in the respective potential regions. Starting with the H_{upd} region, two characteristic peaks of polycrystalline Pt in contact with a H_2SO_4 electrolyte solution are present for all samples. However, the H_{upd} currents decrease according to the amount of deposited PVDF (concentration of PVDF in the dipping solution). The formation and reduction oxygenated species exhibit the same trend, i.e., the intensity of the peaks decreases as more PVDF is deposited. In contrast, the double layer potential region remains similar for all the samples. Only a small increase in the pseudo-capacity is discernible for the Pt surfaces covered by PVDF, which might be interpreted by a minor surface roughening or by a small increase in the concentration of water dipoles close to the electrode surface. However, at the same time, the presence of PVDF clearly reduces the amount of H_{upd} and OH_{ad} species adsorbed onto the Pt surface. Further, the analysis of the H_{upd} peaks is a common method to determine the ECSA used for normalizing the measured activity of a Pt electrode. Integration of the H_{upd} peaks and the resulting charge, $Q_{H_{upd}}$, can be related to the charge of a freshly flame annealed polycrystalline Pt surface, which has a $Q_{H_{upd}}$ per area unit 210 $\mu C\ cm^{-2}$ [17]. From this analysis, we can determine the loss in ECSA due to the PVDF coverage to 27 % and 43 % for PVDF 1 and PVDF 2, respectively.

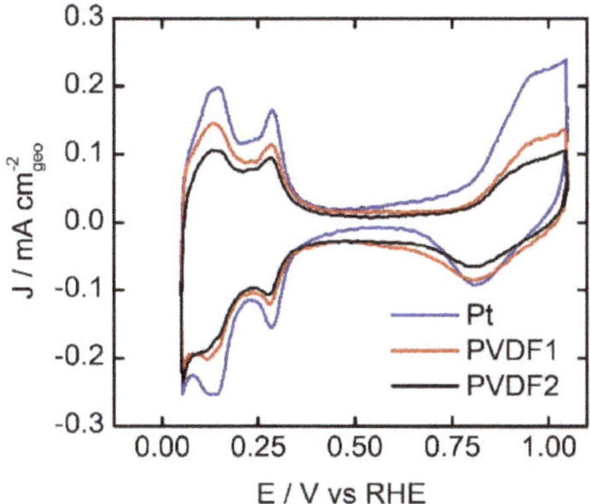

Figure 1. Cyclic voltammograms recorded in Ar saturated 0.05 M H_2SO_4 electrolyte solution for bare and polyvinylidene difluoride (PVDF) modified Pt. The scan rate was 100 mVs^{-1} and the measurements were recorded at room temperature. PVDF 1 stands for the film prepared by dip coating the Pt electrode into 0.1 mM PVDF solution, PVDF 2 for the film prepared by dip coating into 0.1 M PVDF solution. The roughness factors (Rf), calculated by H_{upd} for bare Pt, PVDF1, and PVDF2, are: 1.65, 1.20, 0.94. This corresponds to an ECSA loss with respect to the bare polycrystalline Pt electrode of 27 % and 43 % for PVDF 1 and PVDF 2, respectively.

3.2. CO-Stripping Measurements

CO-stripping is an alternative technique to determine the ECSA [11]. A monolayer of CO is formed on Pt from a CO saturated electrolyte solution, keeping the electrode at a low fixed potential (typically around 50 mV$_{RHE}$). Thereafter, the dissolved CO is displaced from the electrolyte solution by saturating it with Ar, and the CO adlayer is oxidized in a positive-going voltage scan. As CO oxidation is a two electron process, the surface area A_{CO}, relating to the CO oxidation charge (Q_{CO}), is calculated by [17]:

$$A_{CO} = \frac{Q_{CO}}{2 \times 196 \; \mu C \; cm^{-2}} \tag{1}$$

In Figure 2, we show the CO stripping profiles for the bare surface and the two PVDF coated polycrystalline Pt surfaces. At the start of the stripping sweep starting from 0.05 V_{RHE}, the CO covered Pt electrodes show no features, such as the H_{upd}, until the characteristic CO oxidation peak at around 0.8V_{RHE} is observed [18]. The presence of the PVDF coating clearly reduces the amount of stripping charge and at the same time the oxidation onset, as well as the peak position of the maximum being shifted towards higher potentials. Calculating the ECSA loss from Q_{CO}, we determine losses of 17 % and 64%, respectively (with respect to the roughness factor of the bare Pt surface). This means that the ECSA loss calculated by H_{upd} and CO-oxidation led to comparable values, as seen above.

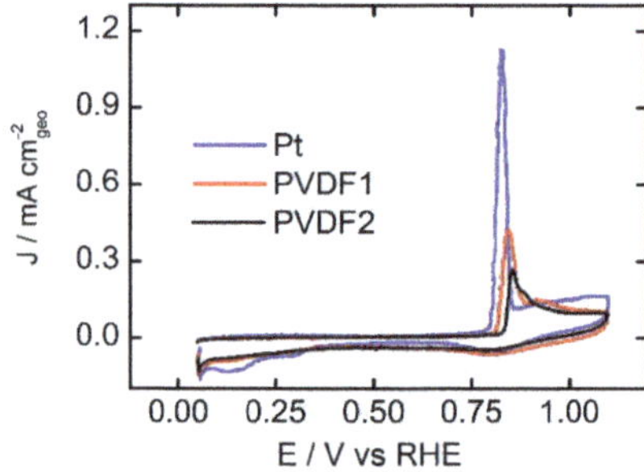

Figure 2. CO stripping curves recorded for polycrystalline Pt and the two PVDF modified surfaces at room temperature in Ar saturated 0.05 M H_2SO_4 electrolyte solution. The scan rate was 50 mVs^{-1}. PVDF 1 stands for the film prepared by dip coating the Pt electrode into 0.1 mM PVDF solution, PVDF 2 for the film prepared by dip coating into 0.1 M PVDF solution. The roughness factors (Rf) calculated by CO stripping for the bare Pt surface, PVDF1 and PVDF2 are: 1.40, 1.17, 0.51. This corresponds to an ECSA loss with respect to the bare polycrystalline Pt electrode of 17 % and 64 % for PVDF 1 and PVDF 2, respectively.

3.3. Oxygen Reduction

In order to determine the ORR activity, polarization curves of the bare and PVDF coated Pt surfaces were recorded in an oxygen saturated 0.05 M H_2SO_4 electrolyte solution, applying different rotations rates. The resulting background corrected data and their analysis are summarized in Figure 3.

In Figure 3a, the ORR polarization curves of the three samples are separated into positive- and negative-going curves and compared with each other. Based on the current profiles, two distinct potential regions can be observed: the kinetic potential region at an electrode potential E > 0.8 V_{RHE}, and the hydro dynamic mass transport limited potential region at E < 0.6 V_{RHE}. The latter is easily identified by its strong rotation rate dependency (not shown). The influence of the thick PVDF coating (PVDF 2) can clearly be seen in the mass transport limited potential region, as well as in the kinetic region by a significant decrease in current. Furthermore, all curves exhibit a hysteresis, i.e., a difference in currents between the positive- and negative-going potential scan is observed. This hysteresis is a typical feature in RDE measurements and can be interpreted due to the fact that the steady state coverage of oxygenated species is not instantaneously achieved on Pt during potential sweeps when coming from the reduced or oxidized surface (lower or higher potentials). Thus, the observed ORR rate is artificially enhanced [19,20] in the positive-going polarization curves, whereas it may be too low in the negative-going polarization curves. This is explicitly shown in Figure 3b in a logarithmic Tafel plot of the kinetic current densities normalized to the geometric electrode area, i.e., J_K. Note that the curves have been corrected for diffusion effects by the use of the Koutecký-Levich equation. In the positive-going potential sweeps, significantly higher kinetic ORR rates are determined than in the negative going potential sweeps. However, in the Tafel plot, the curves of the different kinetic currents form almost parallel lines, indicating that the same ORR reaction pathway is valid for all investigated surfaces. Furthermore, while larger amounts of PVDF inhibit the observed kinetic ORR current, the kinetic ORR currents on the bare Pt surface and PVDF 1 almost overlap. This means that although the ECSA on PVDF 1 is reduced, and the kinetic current is not affected.

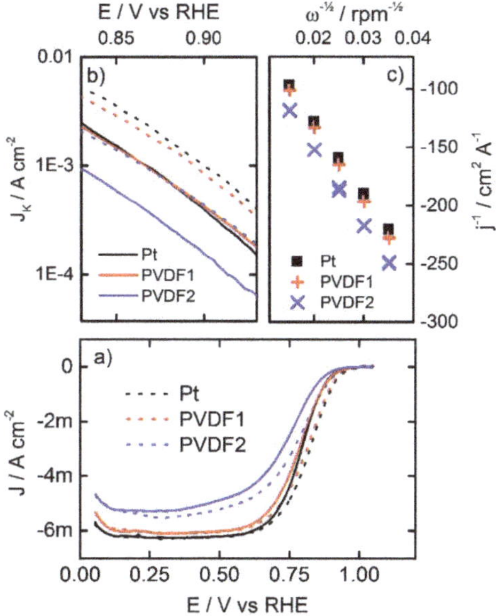

Figure 3. (**a**) Oxygen reduction reaction (ORR) polarization curves recorded for bare polycrystalline Pt and PVDF modified Pt surfaces. Dotted and solid lines indicate the positive- and negative-going potential sweeps, respectively. The scan rate was 50 mVs^{-1} and the rotation rate 1600 rpm. The electrolyte was 0.05 M H$_2$SO$_4$ aqueous solution. (**b**) Tafel plot of the kinetic current. (**c**) Koutecky-Levich plot of the currents at 0.35 V$_{RHE}$.

As we consider the electrode during the negative-going potential sweep to be closer to the steady state condition, we chose the negative going potential sweeps for further analysis. In this analysis, we studied the hydrodynamic response of the system by changing of the convection (rotation rate of the RDE). In Figure 3c, the inverse of the measured current density is plotted versus $\omega^{\frac{1}{2}}$, in a so-called Koutecký-Levich plot based on:

$$J_m^{-1} = J_k^{-1} + J_{DL}^{-1} = J_k^{-1} + (B_L(\omega))^{-1} \tag{2}$$

where J_m is the measured current density, J_k the kinetic current density, J_{DL} the diffusion limited current density, B_L the Levich B factor, and ω the rotation rate.

It can be seen that the data points in the Koutecký-Levich plot form parallel lines, which is a clear indication that the hydrodynamic properties and the number of electrons transferred per oxygen molecule are not affected by the PVDF [21]. Thus, we conclude that the PVDF coating does not favor the formation of H$_2$O$_2$. The negative shift of the polarization curves of PVDF 2 can instead be related to a reduced interfacial current. The reduced interfacial current can be the result of a 3D PVDF film that reduces the diffusion to the electrode, or that a large part of the surface is blocked for the reaction by the PVDF, or a combination of both [22].

So far, we have only analyzed the kinetic current densities normalized to the geometric area, as is often done when working with bulk metal electrodes. To obtain catalytic turnover rates, or the equivalent in electrocatalysis, the (surface area) specific activity (SA), the kinetic current needs to be normalized to the ECSA of the catalyst. The determination of the ECSA of a catalyst is an essential task that can be challenging, in particular when working with catalysts for which no defined procedure is yet established [23]. Even though this is not the case for Pt based catalysts, where H_{upd} and CO stripping are standard procedures for the ECSA determination, the ECSA determination is not trivial. The procedure implies that the catalytic reactions with H^+ and CO and the ORR probe the same type of active site on the catalyst surface. In addition, their accuracy is limited. In particular, the charge related to the H_{upd} process is sensitive to the lower potential limit of the CV, where H_{upd} overlaps with the hydrogen evolution reaction. Furthermore, it has to be mentioned that PVDF retains a dipole moment that could interact with polar molecules or protons, and thus alter the adsorption process on a PVDF coated surface. On the contrary, O_2 does not retain any dipole moment, and low amounts of PVDF are not expected to affect the ORR process. Deviations in the ECSA between determination by H_{upd} and CO stripping, as shown here, are therefore not surprising, and can be considered within the accuracy of the method (usually around 10–15 %).

In Figure 4, comparing the J_K with the normalized SA (normalized with the CO stripping charge) for Pt, PVDF 1, and PVDF 2, a complex behavior becomes apparent. In the positive-going potential scan (which is in general analyzed in RDE measurements), J_K decreases with the PVDF content. However, this decrease is "overcompensated" by the loss in ECSA. This leads to a somewhat higher SA, especially for PVDF 2. However, in the negative-going potential scan—which resembles most steady state conditions—an increase in SA is observed for PVDF 1, but almost no increase in SA is observed for PVDF 2. Taking all results into account, these findings indicate that the ORR kinetics are neither significantly inhibited nor enhanced by the PVDF films. Improvement in SA seems mostly related to the fact that the surface area determination is "more affected" by the PVDF than the ORR itself, which might be related to the fact that for the ECSA determination, different probe molecules are used.

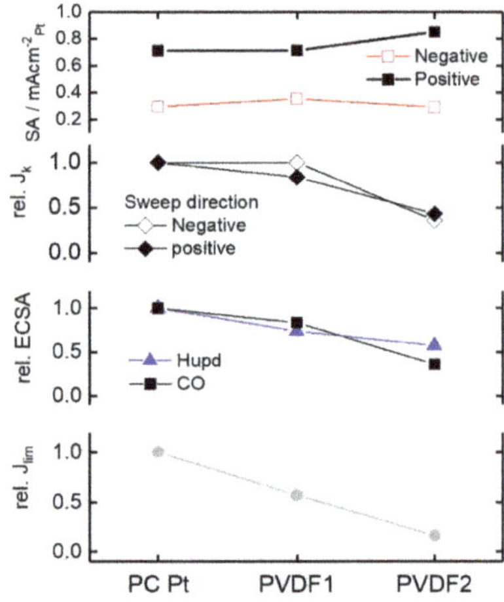

Figure 4. Summary of results. Specific activity (SA) values at 0.9 V_{RHE} for bare and PVDF modified Pt in the positive and negative potential scan, respectively. Relative (normalized to bare Pt) current density J_K; relative ECSA based on H_{upd} and CO stripping charge; relative limiting current density J_{lim} for bare and PVDF modified Pt.

It was not an arbitrary choice to use H_2SO_4 as a supporting electrolyte. H_2SO_4 is known to be a "strongly adsorbing" electrolyte, and decreased ORR kinetics in H_2SO_4 as compared to a "non-absorbing" electrolyte, i.e. $HClO_4$, are observed. This is attributed to the site blocking nature of the HSO_4^- and SO_4^{2-} anions. In literature reporting a positive influence of PVDF on the ORR activity, it was speculated that PVDF might positively influence (inhibit) anion absorption. However, in the presented work we never observed an increase in J_k for a PVDF coated Pt surface, and thus conclude that there is no positive influence of PVDF on anion absorption, as a diminished anion absorption would result in a decreased overpotential of the ORR. The activity enhancing effect of PVDF that has been previously observed for high surface area catalysts (i.e. Pt nanoparticles supported on a carbon support (Pt/C)), therefore, is most likely due to different reasons—a combination of reduced active surface area and improved film properties that has been seen to strongly influence the apparent ORR activity of Pt/C in RDE measurements [10].

Lastly, it is seen that the diffusion limited current J_{lim} is affected by the amount of PVDF used to cast the Pt surface. This could be the result of two different processes: a change in the reaction pathway of the ORR, or the formation of a 3D diffusion barrier. As the Levich analysis shows a straight line with constant B values (Fig. 3) for all the samples, a change in reaction pathway has to be discarded. We therefore conclude that the main reason for the reduction of J_k is the formation of a permeable 3D layer, which partially blocks the active catalyst surface.

4. Conclusions

We have examined ORR on polycrystalline Pt with different amounts of PVDF deposited. The experimental results indicate that PVDF forms a 3D film which disturbs the mass transport of oxygen to the Pt surface. The PVDF coating also exhibits an active site-blocking effect, as both the ECSA and the apparent kinetic activity are reduced for thick PVDF layers. However, for a thin PVDF layer, the kinetic region overlaps with the one measured for bare Pt, even though the ECSA is decreased. Thus, the SA exhibits a slight increase. However, this increase in SA is associated with the loss in ECSA rather than a real kinetic enhancement effect. Depending on the scan direction, the same can be observed for thick PVDF films. Here, the kinetic current is reduced, but this reduction is overcompensated by the ECSA loss. Thus, we conclude that PVDF does not affect the ORR kinetics in 0.05 M H_2SO_4. The reduction in ECSA with even low PVDF coverage might be related to a difference between the interaction of the probe molecules (H^+ and CO) with the PVDF covered surface and the reactant. Thus, PVDF used in small amounts inhibits the adsorption of CO and H^+ without strongly blocking the ORR kinetics. This leads to the "artificial" increase in SA. Therefore, we can conclude that the addition of PVDF to Pt bulk and Pt-based catalysts does not affect the kinetics of ORR. Previously reported performance enhancing properties might be due the reduced ECSA, as well as other factors, like improved film quality.

Author Contributions: Conceptualization, A.Z. and G.K.H.W.; Data curation, A.Z.; Formal analysis, A.Z. and G.K.H.W.; Supervision, M.A.; Writing—original draft, A.Z. and G.K.H.W.; Writing—review & editing, M.A.

Conflicts of Interest: The authors declare no conflict of interest.

References

1. Debe, M.K. Electrocatalyst approaches and challenges for automotive fuel cells. *Nature* **2012**, *486*, 43–51. [CrossRef] [PubMed]
2. Gasteiger, H.A.; Kocha, S.S.; Sompalli, B.; Wagner, F.T. Activity benchmarks and requirements for Pt, Pt-alloy, and non-Pt oxygen reduction catalysts for PEMFCs. *Appl. Catal. B-Environ.* **2005**, *56*, 9–35. [CrossRef]
3. Frey, T.; Linardi, M. Effects of membrane electrode assembly preparation on the polymer electrolyte membrane fuel cell performance. *Electrochim. Acta* **2004**, *50*, 99–105. [CrossRef]
4. Shahgaldi, S.; Ghasemi, M.; Wan Daud, W.R.; Yaakob, Z.; Sedighi, M.; Alam, J.; Ismail, A.F. Performance enhancement of microbial fuel cell by PVDF/Nafion nanofibre composite proton exchange membrane. *Fuel Process. Technol.* **2014**, *124*, 290–295. [CrossRef]
5. Li, H.-B.; Shi, W.-Y.; Zhang, Y.-F.; Liu, D.-Q.; Liu, X.-F. Effects of Additives on the Morphology and Performance of PPTA/PVDF in Situ Blend UF Membrane. *Polymers* **2014**, *6*, 1846–1861. [CrossRef]

6. Mu, S.; Tian, M. Optimization of perfluorosulfonic acid ionomer loadings in catalyst layers of proton exchange membrane fuel cells. *Electrochim. Acta* **2012**, *60*, 437–442. [CrossRef]

7. Brodt, M.; Wycisk, R.; Dale, N.; Pintauro, P. Power Output and Durability of Electrospun Fuel Cell Fiber Cathodes with PVDF and Nafion/PVDF Binders. *J. Electrochem. Soc.* **2016**, *163*, F401–F410. [CrossRef]

8. Schmidt, T.; Gasteiger, H.; Stäb, G.; Urban, P.; Kolb, D.; Behm, R. Characterization of high-surface-area electrocatalysts using a rotating disk electrode configuration. *J. Electrochem. Soc.* **1998**, *145*, 2354–2358. [CrossRef]

9. Holst-Olesen, K.; Nesselberger, M.; Perchthaler, M.; Hacker, V.; Arenz, M. Activity inhibition and its mitigation in high temperature proton exchange membrane fuel cells: The role of phosphoric acid, ammonium trifluoromethanesulfonate, and polyvinylidene difluoride. *J. Power Sources* **2014**, *272*, 1072–1077. [CrossRef]

10. Inaba, M.; Quinson, J.; Arenz, M. pH matters: The influence of the catalyst ink on the oxygen reduction activity determined in thin film rotating disk electrode measurements. *J. Power Sources* **2017**, *353*, 19–27. [CrossRef]

11. Mayrhofer, K.J.J.; Strmcnik, D.; Blizanac, B.B.; Stamenkovic, V.; Arenz, M.; Markovic, N.M. Measurement of oxygen reduction activities via the rotating disc electrode method: From Pt model surfaces to carbon-supported high surface area catalysts. *Electrochim. Acta* **2008**, *53*, 3181–3188. [CrossRef]

12. Zana, A.; Speder, J.; Roefzaad, M.; Altmann, L.; Bäumer, M.; Arenz, M. Probing Degradation by IL-TEM: The Influence of Stress Test Conditions on the Degradation Mechanism. *J. Electrochem. Soc.* **2013**, *160*, F608–F615. [CrossRef]

13. Nesselberger, M.; Ashton, S.; Meier, J.C.; Katsounaros, I.; Mayrhofer, K.J.J.; Arenz, M. The Particle Size Effect on the Oxygen Reduction Reaction Activity of Pt Catalysts: Influence of Electrolyte and Relation to Single Crystal Models. *J. Am. Chem. Soc.* **2011**, *133*, 17428–17433. [CrossRef] [PubMed]

14. Wiberg, G.K.H. The Development of a State-of-the-Art Experimental Setup Demonstrated by the Investigation of Fuel Cell Reactions in Alkaline Electrolyte. Ph.D. Thesis, Technical Universtity of Munich, Munich, Germany, 2010.

15. Deng, Y.-J.; Wiberg, G.K.H.; Zana, A.; Arenz, M. On the oxygen reduction reaction in phosphoric acid electrolyte: Evidence of significantly increased inhibition at steady state conditions. *Electrochim. Acta* **2016**, *204*, 78–83. [CrossRef]

16. Deng, Y.-J.; Arenz, M.; Wiberg, G.K. Equilibrium coverage of OH ad in correlation with platinum catalyzed fuel cell reactions in HClO 4. *Electrochem. Commun.* **2015**, *53*, 41–44. [CrossRef]

17. Trasatti, S.; Petrii, O. Real surface area measurements in electrochemistry. *Pure Appl. Chem.* **1991**, *63*, 711–734. [CrossRef]

18. Arenz, M.; Mayrhofer, K.J.J.; Stamenkovic, V.; Blizanac, B.B.; Tomoyuki, T.; Ross, P.N.; Markovic, N.M. The effect of the particle size on the kinetics of CO electrooxidation on high surface area Pt catalysts. *J. Am. Chem. Soc.* **2005**, *127*, 6819–6829. [CrossRef]

19. Wiberg, G.K.H.; Arenz, M. Establishing the potential dependent equilibrium oxide coverage on platinum in alkaline solution and its influence on the oxygen reduction. *J. Power Sources* **2012**, *217*, 262–267. [CrossRef]

20. Hodnik, N.; Baldizzone, C.; Cherevko, S.; Zeradjanin, A.; Mayrhofer, K.J. The Effect of the Voltage Scan Rate on the Determination of the Oxygen Reduction Activity of Pt/C Fuel Cell Catalyst. *Electrocatalysis* **2015**, *6*, 1–5. [CrossRef]

21. Bard, A.J.; Faulkner, L.R. *Electrochemical Methods Fundamentals and Applications*, 2nd ed.; Wiley: New York, NY, USA, 2001; p. 833.

22. Jinnouchi, R.; Kudo, K.; Kitano, N.; Morimoto, Y. Molecular Dynamics Simulations on O2 Permeation through Nafion Ionomer on Platinum Surface. *Electrochim. Acta* **2016**, *188*, 767–776. [CrossRef]

23. Wiberg, G.K.; Mayrhofer, K.J.; Arenz, M. Investigation of the oxygen reduction activity on silver—A rotating disc electrode study. *Fuel Cells* **2010**, *10*, 575–581. [CrossRef]

*surfaces*

Article

Surface Investigation on Electrochemically Deposited Lead on Gold

Alicja Szczepanska [1,2]**, Gary Wan** [1,2]**, Mattia Cattelan** [3,*]**, Neil A. Fox** [3] **and Natasa Vasiljevic** [1,2]

[1]   School of Physics, HH Wills Physics Laboratory, University of Bristol, Tyndall Avenue, Bristol BS8 1TL, UK; alicja.szczepanska@bristol.ac.uk (A.S.); gary.wan@bristol.ac.uk (G.W.); N.Vasiljevic@bristol.ac.uk (N.V.)
[2]   Bristol Centre for Functional Nanomaterials, University of Bristol, Tyndall Avenue, Bristol BS8 1TL, UK
[3]   School of Chemistry, Cantocks Close, Bristol BS8 1TS, UK; Neil.Fox@bristol.ac.uk
*   Correspondence: mattia.cattelan@bristol.ac.uk; Tel.: +44-117-394004

Received: 17 December 2018; Accepted: 14 January 2019; Published: 17 January 2019

Abstract: Electrodeposition of Pb on Au has been of interest for the variety of surface phenomena such as the UnderPotential Deposition (UPD) and surface alloying. Here, we examined the interface between the electrodeposited Pb film on Au, using surface sensitive techniques such as X-ray Photoelectron Spectroscopy (XPS), Ultraviolet Photoelectron Spectroscopy (UPS), Energy-Filtered Photoemission Electron Microscopy (EF-PEEM) and Work Function (WF) mapping. The initially electrodeposited Pb overlayer (~4 ML equivalent thickness) was transferred from the electrochemical cell to the UHV system. The deposited Pb layer was subjected to Argon sputtering cycles to remove oxide formed due to air exposure and gradually thinned down to a monolayer level. Surface science acquisitions showed the existence of a mixed oxide/metallic Pb overlayer at the monolayer level that transformed to a metallic Pb upon high temperature annealing (380 °C for 1 h) and measured changes of the electronic interaction that can be explained by Pb/Au surface alloy formation. The results show the electronic interaction between metallic Pb and Au is different from the interaction of Au with the PbO and Pb/PbO mixed layer; the oxide interface is less strained so the surface stress driven mixing between Au is not favored. The work illustrates applications of highly surface sensitive methods in the characterization of the surface alloy systems that can be extended to other complex and ultrathin mixed-metallic systems (designed or spontaneously formed).

Keywords: Lead OPD; surface alloy; XPS; UPS; EF-PEEM

---

1. Introduction

Electrodeposition of Pb on Au is a well-known system characterized by Stranski-Krastanov growth [1]. The electrodeposition of Pb on Au starts by the formation of the epitaxial monolayer positive from the equilibrium potential for bulk deposition known as Underpotential Deposition (UPD), followed by the nucleation of 3D islands in the OverPotential Deposition (OPD) region [2,3].

The UPD of Pb Monolayer (ML) on Au (111) has been considered as one of the model UPD systems and has been widely studied in the past few decades with a range of electrochemical and surface science techniques [2,4–11]. An incommensurate, electrocompressed closely packed hexagonal structure of a complete Pb UPD ML has been analyzed by a range of techniques including Scanning Tunneling Microscopy (STM), Atomic Force Microscopy (AFM), ex situ Low Energy Electron Diffraction (LEED) and in situ x-ray diffraction [8,12–16].

The UPD of Pb on Au has been used in a wide range of applications including measurements of surface area/roughness, surface structure characterization, chemical composition measurements, electrocatalysis, sensing as well as a mediation of the epitaxial thin film growth. For example, Pb UPD on Au has been used for the determination of surface areas, surface structures and crystallographic orientation of nanostructures such as gold nanorods [17], controlled-shape nanoparticles [17,18] and nanoporous gold electrodes [19]. The Au surfaces modified by a Pb UPD layer have shown to enhance the electrocatalytic activity and selectivity of oxygen reduction reactions [5,20,21], methanol oxidation [22] and glucose oxidation reactions [23,24]. Moreover, the Pb UPD layers have been successfully employed as mediators and manipulators of thin film growth kinetics in electrochemical methods such as the Surfactant-Mediated Growth (SMG) [25,26] and Defect Mediated Growth (DMG) [27,28] of epitaxial films of Ag, Co and Cu. In the recently developed Surface Limited Redox Replacement (SLRR) method, the UPD monolayer of Pb has been exploited as a sacrificial layer for the deposition of 2D epitaxial noble metal films, such as Pt, Ag, Au [26,29–36], nano-alloys [37–40] and nanoporous fuel-cell electrodes [41]. The SLRR method led to the development of a special class of highly active Pt-ML electrocatalysts for oxygen reduction reaction [42] and has become a tool for the design of model systems for fundamental electrocatalytic studies of structure-activity relationship [38] and catalytic stability [43]. The most recent advances include different ideas and concepts for a so-called electroless (e-less) deposition of Pb monolayer that were integrated into a deposition method called "electroless Atomic Layer Deposition" (e-less ALD) [44].

Pb on Au system is characterized by a large atomic size mismatch of 21%. Thus, it is an ideal model for studies of the surface-stress evolution and the related phenomena. This is an immiscible system, i.e., Pb and Au do not form bulk alloys. However, surface alloying has been reported in both electrochemical [6,7,45–48] and ultrahigh vacuum [49,50] environments. Several early studies argued that there was no evidence of surface alloy formation in this system [2,8,14,15]. However, later investigations, by in situ STM and surface stress measurements, showed the surface transformations that can be associated with the surface alloy formation/dissolution [6,7,46–48]. The in situ STM studies showed roughening of the initially flat Au substrate (by vacancy and ad-islands formation) after a Pb UPD layer dissolution [6,7,45,48]. The compressive stress evolution during monolayer formation of the magnitude of 1.2 N/m [48] featured a very narrow tensile drop within the main Pb UPD peak, corresponding to the formation of a high coverage Pb layer [47,48].

While most electrodeposition studies have been focused on Pb UPD, the bulk or so-called OPD for this system has not been studied so extensively [48]. The morphology of Pb OPD on Au studied by a combination of in situ STM [48] and ex situ AFM [3], have shown dependence on the deposition potential and specific ion adsorption (i.e. nature of ions in the deposition solution [3]). In the ionic liquid solution, the nucleation process of Pb leads to the formation of triangular-shaped vacancies on the Au (111), followed by the creation of large, well-defined 3D crystals of Pb [51]. The compressive stress evolution during 3D island creation as a result from isolated Pb island nucleation and growth has been studied in combination with the in situ Oblique-Incidence Reflectivity Difference (OIRD) [48]. The most intriguing was the observation of the surface alloy formation during OPD deposition that seems to be surface stress related and localized around the edge areas of 3D Pb crystals [48].

To better understand the phenomenon of surface alloy formation of Pb ML on Au, use of state-of-the-art characterization methods is required. In our work, we present results from X-ray Photoelectron Spectroscopy (XPS), Ultraviolet Photoelectron Spectroscopy (UPS), Energy-Filtered Photoemission Electron Microscopy (EF-PEEM) and Work Function (WF) mapping. XPS is a technique generally used to analyze the chemical composition of both bulk samples and surface alloys [52–55]. The UPS is a method that is highly surface sensitive and has been used to characterize a broad range of surface alloys, such as Au/Ni, Co/Cu(001) [52], Pt-bimetallic systems such as Pt_xM (M=Ni, Co, Fe, Ti, V) [53,56,57]. EF-PEEM and WF mapping are methods that can provide information about surface orientation and segregation [55,58,59]; also WF provides the information about surface activity [54]. These techniques are particularly useful due to their high surface sensitivity, ability to detect electronic changes from chemical environments at the Pb/Au interface and imaging capabilities at the nanoscale.

Studying Pb UPD on Au surface alloying is difficult as the UPD layer stability without the potential control can be compromised by exposure to air/oxygen [60]. Also, the bulk electrodeposited thicker Pb films inevitably oxidize in the air (i.e. forming of oxides). Therefore, the approach used in our work to study Pb ML level films was based on the deposition of bulk (OPD deposition) of equivalent 4 ML thickness films. Upon transfer from the electrochemical cell, PbO was removed gradually from the surface by cycles of Ar-sputtering and gentle annealing until the layer was thinned down to the desired ML level. Using a combination of XPS, UPS, EF-PEEM and WF mapping we then explored the structure of the layer.

2. Materials and Methods

Electrochemical measurements were performed in a standard three-electrode cell with a BioLogic VSP potentiostat (Bio-Logic Science Instruments SAS, Seyssinet-Pariset, France) and controlled by built-in software.

A solution containing 1 mM $Pb(ClO_4)_2$ and 100mM $HClO_4$ was prepared by mixing high purity chemicals ($PbCO_3$, 99.999%, Alfa Aesar, Heysham, UK and 70% $HClO_4$, 99.999%, Sigma Aldrich, Steinheim, Germany). Before the experiment, the solution was deaerated with nitrogen gas (Oxygen Free Nitrogen, CAS: 7727-37-9, BOC, Manchester, UK) for 30 min. The oxygen free environment was maintained during the experiment with nitrogen flow above the solution.

Pb and Pt wires were used as pseudo Reference (RE) and Counter (CE) Electrodes respectively. Prior to experiments, the electrodes were prepared by cleaning in HNO_3 ($\geq$65%, Sigma Aldrich, Steinheim, Germany) for 10 s, rinsed with Milli-Q water (Millipore/Merck, Darmstadt, Germany) and dried with nitrogen. The Pt wire was then flame annealed by a butane torch. All the potentials will be reported with respect to the Pb pseudo RE (0.64 V vs Standard Hydrogen Electrode)

A vacuum evaporated Au thin film (250 nm) deposited on a glass slide (with a 4 nm Ti adhesion layer) was used as a Working Electrode (WE). Prior to experiments, the Au sample was cleaned with H_2SO_4 (95.0–97.0%, Sigma Aldrich, Steinheim, Germany) for 10 min, rinsed with Milli-Q water and dried with nitrogen. To obtain the Au (111) orientation, the sample was flame annealed for 2 min using a butane torch and rinsed with Milli-Q water.

The sample was then immersed in the deaerated solution with the half of its exposed surface area above the solution. The part of the substrate remained deposit-free, for the sake of comparison between the Pb covered and bare substrate. The deposition of Pb film was done by Linear Sweep Voltammetry (LSV) with a scan rate of 10 mV/s starting from the bare Au surface at a potential of 0.9 V to −0.1 V. The LSV potential range covered the Pb UPD ML deposition (between 0.9 V and 0.0 V) followed by the OPD growth (from 0.0V to −0.1 V). The potential of −0.1 V was then held constant for 20 s allowing for further (excess) overpotential Pb deposition. The sample was then removed from the cell, rinsed with Milli-Q water and dried with nitrogen gas. The equivalent thickness of the deposited Pb film of 4 ML was calculated by integration of the deposition current normalized by a charge of 305 $\mu C/cm^2$ corresponding to an 'ideal' ML of Pb (111).

The prepared sample was immediately transferred into the Bristol NanoESCA Facility's Ultra High Vacuum (UHV) chambers and heated overnight at 150 °C to clean the surface from physisorbed contaminants without desorbing and altering the deposited Pb. Other surface cleaning procedures involved 0.5 kV argon sputtering (at 45° to the sample normal) for 5 min and 2 min 30 s (for a total of 7 min 30 s), and a final 1 h anneal at 380 °C. All the surface analysis techniques were performed on both the Au and Pb deposited sides of the sample before proceeding to the next procedure. Throughout the experiments, the sample was kept in UHV chambers.

High resolution XPS was carried out using a monochromatic Al k_α (hv = 1486.7 eV) excitation source, with the analyzer at 45° to the sample normal. A Pass Energy (PE) of 6 eV was used for detailed photoemission lines, giving an overall energy resolution of about 300 meV obtained by fitting the Fermi edge of the polycrystalline Au sample using a sigmoid function; PE of 50 eV was used for surveys. Angle resolved XPS acquisitions were performed using a non-monochromatic Al k_α source and a PE of 50 eV.

The NanoESCA II EF-PEEM was used for WF mapping employing a non-monochromatic Hg source (hv $\approx$ 5.2 eV) with an overall energy resolution of about 140 meV and lateral resolution of about 150 nm. UPS was performed with monochromatic He-I (hv = 21.2 eV) and He-II (hv = 40.8 eV) with an overall energy resolution of about 140 meV. EF-PEEM was also used for chemical imaging on Pb 5*d* photoemission lines using a monochromated He-II (hv = 40.8 eV) source; a lateral resolution of about 500 nm and energy resolution of 0.9 eV was used to improve the signal-to-noise ratio.

3. Results

3.1. Core-Level Spectroscopy

XPS is a surface sensitive technique that provides insight on the elemental composition and chemical environment of a sample. The presence of Pb on the Au surface was identified and quantified using XPS shown in Figure 1.

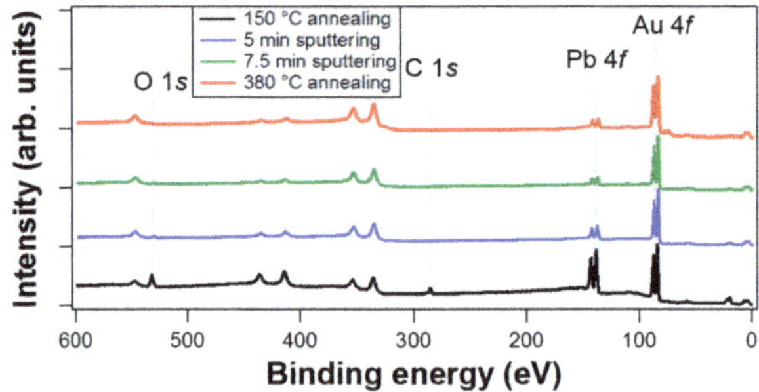

Figure 1. XPS surveys of the Pb deposited on Au after 150 °C annealing (black), after 5 min of sputtering (blue), after 7 min 30 s of sputtering (green) and after 380 °C annealing (red). The spectra are shifted vertically for clarity.

XPS was carried out on the Pb deposited side of the sample after each heating and sputtering treatment stage, the XPS surveys are shown in Figure 1. After the first 150 °C annealing step, the XP spectrum showed the presence of Pb and Au, as well as carbon and oxygen from air contaminants and oxidation. 5 min sputtering was used to gently clean the surface [59,61], successfully removing the carbon contamination and significantly reducing the oxygen signal as well. However, the Pb quantity strongly decreased (see Table 1) and the remaining oxygen was predicted to be from oxidized Pb. Further argon sputtering for an extra 2 min 30 s did not appear to preferentially remove Pb oxide over metallic Pb, while annealing at 380 °C succeeded in removing the oxygen.

Table 1. Tabulated key results obtained from XPS after each surface treatment step.

Treatment	Pb 4$f_{7/2}$ Metal BE	Pb 4$f_{7/2}$ Oxide BE	Area Pb Oxide/Area Pb Metal	O/Pb (at. %)	Overlayer Coverage
150 °C anneal.	136.9 eV	138.3 eV	12.0	1.7	11.4 Å (3.7 ML)
5 min sputter.	136.9 eV	138.1 eV	4.5	0.9	5.5 Å (1.8 ML)
7.5 min sputter.	136.9 eV	138.0 eV	3.5	0.8	3.0 Å (1.0 ML)
380 °C anneal.	136.8 eV	/	/	/	2.5 Å (0.9 ML)

To investigate the presence of Pb oxide, detailed scans of the Pb 4*f* and O 1*s* photoemission lines were acquired, which are shown in Figure 2. After the initial 150 °C annealing step, the fitted Pb 4*f* peaks showed contributions from higher and lower binding energy (BE) doublets, which are from oxidized and metallic Pb, respectively. The higher BE of the oxide doublets was due to the higher electronegativity of oxygen with respect to Pb [62]. The O 1*s* peak also showed two components after the first annealing step, where the lower BE component was expected to be from Pb oxide, and the higher component to be mainly from oxygen on carbon contaminants [62,63]. This was further confirmed when 5 min of argon sputtering removed carbon, and significantly decreased the higher BE component of the O 1s peak.

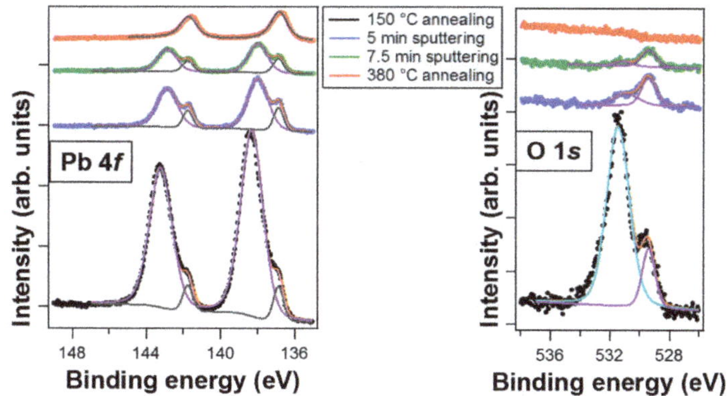

Figure 2. Detailed XP spectra of the Pb 4f (**left**) and O 1s (**right**) emission lines of the Pb deposited on Au after different treatments, normalized by Au 4f areas. The acquired data are represented by the dotted lines while the peak deconvolutions are drawn as thin lines. The black, blue, green and red dotted lines represent the data after each treatment procedure, the same as in Figure 1. For Pb 4f the grey, purple and orange lines represent the fitted peaks for Pb metallic, Pb oxide and total fit respectively. For O 1s, the light blue, purple and orange lines represent the fitted peaks for oxygen bonded with carbon, with Pb and total fit respectively.

Table 1 lists the fitted BEs for the Pb-metal and Pb-oxide Pb $4f_{7/2}$ peaks, and the stoichiometry of Pb and O which were calculated by considering the Pb 4f area of oxidized Pb and the overall peak of O; the peak areas ratios were normalized by their sensitivity factors.

The stoichiometry of O and Pb were not reliable for the first measurement after the 150 °C annealing due to carbon and oxygen surface contamination. But after sputtering, the carbon contamination was removed and the O/Pb ratio after 5 and 7 min 30 s was found to be relatively constant at 0.9 and 0.8 respectively, indicating that the layer after sputtering was mainly composed of PbO. The BE of the Pb $4f_{7/2}$ oxide decreased significantly with the sputtering procedure from 138.3 eV to 138.0 eV, probably due to the formation of suboxides and damage to the surface due to Ar implantation (see also WF histograms discussion below).

Table 1 also reports the Pb overlayer estimated thicknesses. For the thickest layers of Pb (after 150 °C annealing and after 5 min sputtering), the coverages have been determined using angle resolved XPS as conducted in Ref. [64] approximating a homogeneous PbO layer. The coverage estimation by angle resolved acquisitions on much thinner layers (after 7 min 30 s of sputtering and 380 °C annealing) is not reliable due to the low signal of the overlayers and their complex morphology. Their coverages have therefore been extrapolated from the Au 4f /Pb 4f area ratio, calibrated with the overlayer thickness of the first two acquisitions. The monolayer coverages were obtained using the lattice parameter of PbO [65] after annealing at 150 °C, 5 and 7 min 30 s of sputtering and using the distance between atoms in bulk Pb (111) after the 380 °C anneal.

Comparing the ratio of Pb oxide to metallic in the Pb 4f photoemission lines after the 150 °C anneal, we observed a ratio of 12.0 which then dropped to 4.5 and 3.5 after the 5 min and 7 min 30 s of sputtering, respectively (see Table 1). The significant ratio difference after the first sputtering confirms that not all the Pb layer was homogenously oxidized. This was indirect evidence of Stranski- Krastanov growth, where the Pb islands, formed after the first monolayer, strongly oxidized when exposed to air due to their high surface areas and probably maintained only a very small core of metallic Pb. The first mild sputtering removed about half of the Pb layer (see Table 1), and the remaining layer was still partially oxidized but with the presence of more metallic Pb. Interestingly the oxide/metal ratio did not change drastically upon a further 2 min 30 s sputtering. This proves, along with a similar O/Pb stoichiometry, that the layer compositions at 1.8 ML and 1 ML of PbO are similar and that the oxides do not form a uniform passivating layer on the top of the Pb metal. On the contrary, the oxidation from air exposure created oxide particles all over the surface and therefore affected even the 1st Pb ML in contact with Au [16,66], see also EF-PEEM discussion below.

The final annealing was then performed at 380 °C to remove the Pb oxide without also desorbing the metal Pb [67]. From Figure 2, the Pb oxide and oxygen emission lines disappeared after the 380 °C anneal, but more interestingly, the metallic Pb peak increased significantly. This indicates that the annealing did not simply desorb Pb oxide from the Au surface, but rather removed some of the oxygen from the Pb oxide, probably facilitated by the Pb/Au interface. The calculated layer coverage after 7 min 30 s was estimated to be about 1 ML, this value is preserved after the 380 °C anneal (see Table 1), confirming the transformation of the mixed oxide-metal layer to an only metallic homogeneous overlayer. This latter finding is confirmed by WF maps, and EF-PEEM acquisition reported below.

3.2. Valence Band Photoemission and Photoemission Microscopy

UPS (using He-I and He-II) was also performed on the sample after every annealing and sputtering procedure (Figure 3), allowing for a much higher surface sensitive characterization due to the lower photoelectron energies with respect to the XPS [68]. He-II (hv = 40.8 eV) is even more surface sensitive than He-I and allows for emission from the Pb 5*d* levels [69].

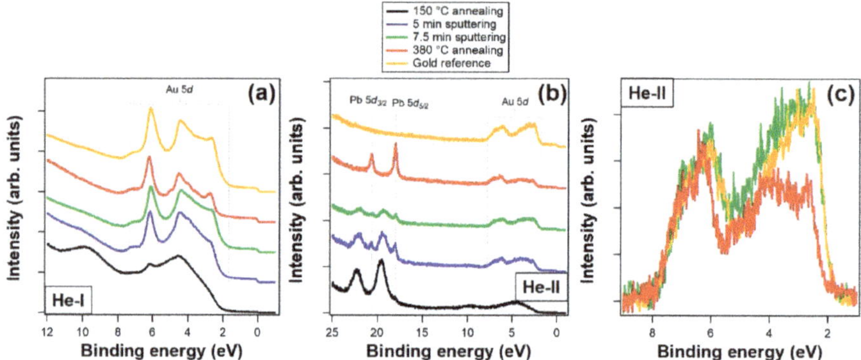

Figure 3. UP spectra of Pb on Au after the different treatments, using (**a**) He-I and (**b,c**) He-II excitation. (**c**) shows a magnified, normalized and background subtracted region of the Au 5*d* bands after 7 min 30 s sputtering and annealing at 380 °C. UPS of the clean Au side of the sample is also plotted as a reference on all graphs. The spectra are indicated in black after 150 °C annealing, blue after 5 min of sputtering, green after 7 min 30 s of sputtering, red after 380 °C annealing and gold for the Au reference.

After the first annealing step (black lines), both UP spectra showed indistinct Au 5*d* bands, located between 2 and 7 eV [70,71], but showed two broad features at about at 4.5 and 9.8 eV, as seen in Figure 3a,b. Two distinct photoemission lines appeared at around 19.5 eV and 22.3 eV BE for the He-II spectrum of the sample. These correspond to the Pb 5*d* lines and have distinctly higher BEs than the 18.0 eV and 20.6 eV of metallic Pb [69] confirming the presence of Pb oxide in agreement with the XPS findings. The significantly higher oxide/metal ratio visible in Figure 3b with respect to the XPS was due to the much higher surface sensitivity of He-II UPS than XPS.

The 5 min of sputtering significantly sharpened the Au 5*d* band peaks in both the He-I and He-II UP spectra, while the Pb oxide peaks decreased and the metallic Pb peaks grew in intensity, indicating that metallic Pb was initially buried under the Pb oxide and carbon contamination. Further sputtering of 2 min 30 s reduced both the metallic and oxide peak areas. As found with XPS, the oxide/metal ratio visible in the Pb 5*d* region showed that between the 5 min and 7 min 30 s sputtering the layer compositions were very similar. Also, the presence of both oxide and metal on the surface after 7 min 30 s of sputtering showed the presence of oxide in contact with Au, further indicating that the metal in contact with Au oxidized during air exposure. Also confirmed by XPS, the final annealing at 380 °C completely removed the O from Pb and increased the metallic Pb signal at the oxide's expense.

Figure 3c shows an enlarged view of the He-II UP spectra of the sample after the 7 min 30 s sputtering and 380 °C annealing along with undeposited Au region on the sample. The spectra were normalized to the Fermi edge and subtracted of a Shirley background for easier comparison. For the sample after 380 °C annealing, a clear shift in the Au 5*d* bands to higher BEs and shrinkage were observed, confirming electronic interactions between metallic Pb and Au. On the other hand, comparing the Au 5*d* bands after 7 min 30 s sputtering, the shift and shrinkage were minimal and possibly caused by the overlapping of the broad oxide features at 4.5 eV. Interestingly He-I acquisitions for the 380 °C annealed sample only showed a minimal shift in BE because the Au signal also originated from deeper atom layers, He-II was necessary for the extreme surface sensitivity sampling, almost exclusively at the Pb/Au interface.

To visualize the surface structure at the nanoscale, the low energy emission cutoffs from EF-PEEM images were used to generate WF maps by performing an error function fitting at each pixel in the field of view. Figure 4 shows the WF maps of the same position of the sample after 5 min of sputtering, where the Pb signal was clear and without surface contamination, and after the final 380 °C annealing stage. The WF was roughly 3.95 eV and 4.15 eV respectively, in agreement with the literature where Pb oxide has a lower WF than metallic Pb [72,73], considering also that the high extractor voltage used in the PEEM reduces WF due to the Schottky effect [58].

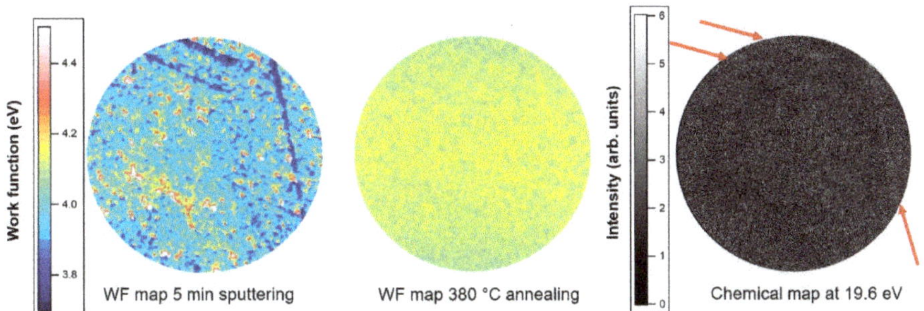

Figure 4. WF maps of the Pb on Au after the 5 min sputtering (**left**) and 380 °C annealing (**center**) steps. On the (**right**), a normalized EF-PEEM image at 19.6 eV BE for Pb after 5 min of sputtering. The arrows are a guide to the eye highlighting some features in the image. The images were taken in the same area, with a field of view of 17.2 μm.

The WF map after 5 min sputtering showed a surface dominated by circular domains of about 300 nm in diameter, with a few stripes of low WF. EF-PEEM and WF maps obtained on the undeposited Au side (not reported here) also showed similar domains and stripes, indicating that they came from the substrate structure and that Pb OPD did not induce any changes in the general surface morphology as observed with the selected lateral resolution (150 nm).

EF-PEEM was performed using the He-II excitation on the Pb 5$d_{5/2}$ emission line, providing a direct mapping of oxidized Pb at 19.6 eV. After 5 min sputtering, although very weak, some stripes of higher intensity can be seen on the chemical map (indicated by red arrows). These features are in the same position as the low WF stripes indicating that the low WF was due to Pb oxide. Notably, the chemical map shows also that the oxide was not only localized on the stripes, but it was distributed on all the sample's surface. The poor lateral resolution (500 nm) and the low signal made it difficult to see its exact position, but it could be roughly related to the zone with WF < 4.1 eV in the WF map after 5 min of sputtering reported in Figure 4.

The WF contrast on the Pb side disappeared after annealing to 380 °C suggesting that the stripes were initially covered with Pb oxide but reduced to Pb after the annealing, confirming the transformation of part of the oxide into the metal as found in the XPS analysis. This finding is further confirmed by the fact that the chemical map at 18.0 eV (Pb $5d_{5/2}$ metallic) after annealing at 380 °C (not reported here) was homogeneous. Within our lateral resolution (150 nm), we could notice that the annealing process did not change the surface morphology.

Histograms of all the WF maps were plotted in Figure 5 to show the distribution of WF across the surface after each treatment step.

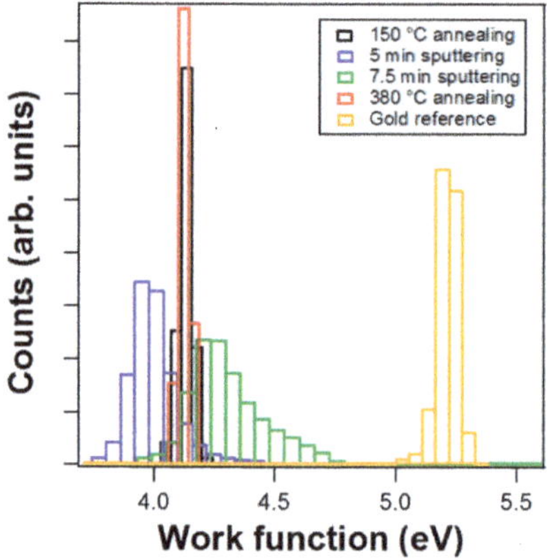

Figure 5. WF distributions after each treatment step plotted as a histogram after 150 °C annealing (black), after 5 min of sputtering (blue), after 7 min 30 s of sputtering (green), after 380 °C annealing (red) and for the Au reference (golden). All the distributions were taken from the same area other than the Au reference, which was done on the undeposited half of the sample.

The WF varied throughout each step, where the sample at the first step (black) was carbon contaminated, then (blue) was cleaned, but further sputtering (green) increased the damage to the sample surface leading to surface roughening and broadening of the WF histogram; this broadening can also be related to the formation of suboxides as discussed above in the XPS section. With the final annealing (red), only metallic Pb remained on the surface with a very narrow WF distribution. The WF of the Pb film remained lower than WF of the polycrystalline Au [74] throughout each step, indicating that the OPD fully covered Au and the treatments did not expose the underlying Au, confirming the XPS coverage calculation in Table 1.

4. Discussion

With the combination of multiple surface sensitive techniques, we were able to determine the general structure of the Pb deposited on Au by OPD, its oxidation from exposure to atmosphere, the transformation of the surface with sputtering and annealing, and then the changes and shifts in the electronic bands from interaction between Au and Pb.

With XPS, we were able to determine the Pb oxidation species after sputtering to be PbO. The amount of deposited Pb was also quantified and found to be as expected from the OPD charge. Contaminants from the solution used in the electrochemical deposition were also not detected. XPS and UPS also gave an indirect insight into the nanostructure of the materials, confirming Stranski-Krastanov growth and that air oxidation reached the 1st Pb strained ML.

Moreover, we observed that the Pb layer thickness after 7 min 30 s sputtering and the 380 °C of annealing did not change. This might be an indication of type-I alloying, which agrees with the in situ STM results published by our group [48]. In type-I alloys, the surface adatoms are found to be diffusing on the surface and are highly mobile, due to the kinetics of alloying and repulsion [75]. Indeed, we have found that the Pb was able to migrate over to the clean Au side of the sample (over 2 mm migration) after annealing to 380 °C.

One particularly interesting finding is the interaction between Au and the metallic Pb from the UPS in Figure 3c (red curve). The shift to higher BE in the Au 5*d* bands could be a result of differential relaxation, exchange and correlation effects [76] due to the presence of Pb in the Au lattice [77]. Shifts in the 5*d* band have also been reported for Au alloys [70] and were explained by charge transfer and electron *d*-band repulsion where the band narrows [78]. Due to the alloying, the Au 5*d* peaks narrowed, and the spin-orbit splitting decreased and overall shifted to higher BEs, as observed with Au alloyed with Cd [77]. Comparing our findings to these reports we can assume we formed a Pb/Au surface alloy after the annealing at 380 °C.

From our UPS investigation we also showed that, in the case of partially oxidized Pb, the interaction with Au was negligible. The interfacial strain at the interface due to the Pb oxidation was lower than in the case of Pb metal. The Pb oxide affected the possibility of Pb metal to form a surface alloy surrounding it with oxide (see Figure 4) and eliminated the stress at the interface, which can explain why the Pb metal, present after only sputtering treatments, did not interact with Au. The high diffusion rate of Pb has been observed during annealing at high temperature. It is possible that the Pb-Au bond is much weaker at elevated temperatures compared to that at room temperature and the fast diffusion of Pb on the Au surface suggests almost 'lattice gas' behavior. In the presence of the metallic Pb, the system relaxes by minimizing the elastic energy through surface alloying. This is typical for immiscible and highly strained systems such as ours [79,80].

The high temperature annealing, removing the oxide and transforming it into metal, is the only treatment after which the metal enters in direct contact with the Au surface and forms a surface alloy. On the other hand, this treatment allows faster diffusion of Pb during which the alloy formation is not favorable. In future papers, we will explore these and similar materials at different temperatures to investigate the interplay between the surface alloying and surface interaction (ad-atom diffusion).

Author Contributions: Data curation, M.C.; Funding acquisition, N.A.F.; Investigation, A.S., G.W. and M.C.; Methodology, A.S., N.V., G.W. and M.C.; Supervision, M.C., N.A.F. and N.V.; Validation, M.C. and N.V.; Visualization, M.C.; Writing—original draft, A.S., G.W. and M.C.; Writing—review & editing, A.S., M.C., N.A.F. and N.V.

Funding: This research was partially funded by EPSRC Strategic Equipment, Grant number EP/K035746/1 and EP/M000605/1; Ph.D. studentships are funded through BCFN: EPSRC (EP/L016648/1), and Renewtec Technologies, Al Hamad Group.

Acknowledgments: The authors acknowledge the Bristol NanoESCA Facility (EPSRC Strategic Equipment Grant EP/K035746/1 and EP/M000605/1). A.S. and G.W. acknowledge Ph.D. studentships funded through BCFN: EPSRC (EP/L016648/1), and Renewtec Technologies, Al Hamad Group, respectively.

Conflicts of Interest: The authors declare no conflict of interest.

References

1. Obretenov, W.; Schmidt, U.; Lorenz, W.J.; Staikov, G.; Budevski, E.; Carnal, D.; Müller, U.; Siegenthaler, H.; Schmidt, E. Underpotential Deposition and Electrocrystallization of Metals An Atomic View by Scanning Tunneling Microscopy. *J. Electrochem. Soc.* **1993**, *140*, 692–703. [CrossRef]

2. Hamelin, A.; Lipkowski, J. Underpotential deposition of lead on gold single crystal faces: Part II. General discussion. *J. Electroanal. Chem. Interfacial Electrochem.* **1984**, *171*, 317–330. [CrossRef]

3. Wu, G.Y.; Schwarzacher, W. The effect of halide additives on the electrodeposition of Pb on polycrystalline Au. *J. Electroanal. Chem.* **2009**, *629*, 164–168. [CrossRef]

4. Adzic, R.; Yeager, E.; Cahan, B. Optical and Electrochemical Studies of Underpotential Deposition of Lead on Gold Evaporated and Single-Crystal Electrodes. *J. Electrochem. Soc.* **1974**, *121*, 474–484. [CrossRef]

5. Alvarez-Rizatti, M.; Jüttner, K. Electrocatalysis of oxygen reduction by UPD of lead on gold single-crystal surfaces. *J. Electroanal. Chem. Interfacial Electrochem.* **1983**, *144*, 351–363. [CrossRef]
6. Green, M.P.; Hanson, K.J.; Carr, R.; Lindau, I. STM Observations of the Underpotential Deposition and Stripping of Pb on Au(111) under Potential Sweep Conditions. *J. Electrochem. Soc.* **1990**, *137*, 3493–3498. [CrossRef]
7. Green, M.P.; Hanson, K.J. Alloy formation in an electrodeposited monolayer. *Surf. Sci.* **1991**, *259*, L743–L749. [CrossRef]
8. Toney, M.F.; Gordon, J.G.; Samant, M.G.; Borges, G.L.; Melroy, O.R. In-Situ Atomic Structure of Underpotentially Deposited Monolayers of Pb and Tl on Au(111) and Ag(111): A Surface X-ray Scattering Study. *J. Phys. Chem.* **1995**, *99*, 4733–4744. [CrossRef]
9. Lorenz, W.J.; Staikov, G. 2D and 3D thin film formation and growth mechanisms in metal electrocrystallization—An atomistic view by in situ STM. *Surf. Sci.* **1995**, *335*, 32–43. [CrossRef]
10. Herrero, E.; Buller, L.J.; Abruña, H.D. Underpotential Deposition at Single Crystal Surfaces of Au, Pt, Ag and Other Materials. *Chem. Rev.* **2001**, *101*, 1897–1930. [CrossRef]
11. Hsieh, S.-J.; Gewirth, A.A. Poisoning the catalytic reduction of peroxide on Pb underpotential deposition modified Au surfaces with iodine. *Surf. Sci.* **2002**, *498*, 147–160. [CrossRef]
12. Ganon, J.P.; Clavilier, J. Electrochemical adsorption of lead and bismuth at gold single crystal surfaces with vicinal (111) orientations. I. *Surf. Sci.* **1984**, *145*, 487–518. [CrossRef]
13. Samant, M.G.; Toney, M.F.; Borges, G.L.; Blum, L.; Melroy, O.R. Grazing incidence x-ray diffraction of lead monolayers at a silver (111) and gold (111) electrode/electrolyte interface. *J. Phys. Chem.* **1988**, *92*, 220–225. [CrossRef]
14. Tao, N.J.; Pan, J.; Li, Y.; Oden, P.I.; DeRose, J.A.; Lindsay, S.M. Initial stage of underpotential deposition of Pb on reconstructed and unreconstructed Au(111). *Surf. Sci.* **1992**, *271*, L338–L344. [CrossRef]
15. Chen, C.H.; Washburn, N.; Gewirth, A.A. In situ atomic force microscope study of lead underpotential deposition on gold (111): Structural properties of the catalytically active phase. *J. Phys. Chem.* **1993**, *97*, 9754–9760. [CrossRef]
16. Crepaldi, A.; Pons, S.; Frantzeskakis, E.; Calleja, F.; Etzkorn, M.; Seitsonen, A.P.; Kern, K.; Brune, H.; Grioni, M. Combined ARPES and STM study of Pb/Au(111) Moiré structure: One overlayer, two symmetries. *Phys. Rev. B* **2013**, *87*. [CrossRef]
17. Hernández, J.; Solla-Gullón, J.; Herrero, E.; Aldaz, A.; Feliu, J.M. Characterization of the Surface Structure of Gold Nanoparticles and Nanorods Using Structure Sensitive Reactions. *J. Phys. Chem. B* **2005**, *109*, 12651–12654. [CrossRef]
18. Jeyabharathi, C.; Zander, M.; Scholz, F. Underpotential deposition of lead on quasi-spherical and faceted gold nanoparticles. *J. Electroanal. Chem.* **2017**. [CrossRef]
19. Liu, Y.; Bliznakov, S.; Dimitrov, N. Comprehensive Study of the Application of a Pb Underpotential Deposition-Assisted Method for Surface Area Measurement of Metallic Nanoporous Materials. *J. Phys. Chem. C* **2009**, *113*, 12362–12372. [CrossRef]
20. Adžić, R.R.; Tripkovic, A.V.; Markovic, N.M. Oxygen reduction on electrode surfaces modified by foreign metal ad-atoms: Lead ad-atoms on gold. *J. Electroanal. Chem.* **1980**, *114*, 37–51. [CrossRef]
21. Wang, Y.; Laborda, E.; Plowman, B.J.; Tschulik, K.; Ward, K.R.; Palgrave, R.G.; Damm, C.; Compton, R.G. The strong catalytic effect of Pb(II) on the oxygen reduction reaction on 5 nm gold nanoparticles. *Phys. Chem. Chem. Phys.* **2014**, *16*, 3200–3208. [CrossRef] [PubMed]
22. Hernández, J.; Solla-Gullón, J.; Herrero, E.; Aldaz, A.; Feliu, J.M. Methanol oxidation on gold nanoparticles in alkaline media: Unusual electrocatalytic activity. *Electrochim. Acta* **2006**, *52*, 1662–1669. [CrossRef]
23. Chen, Y.; Milenkovic, S.; Hassel, A.W. {110}-Terminated Square-Shaped Gold Nanoplates and Their Electrochemical Surface Reactivity. *ChemElectroChem* **2017**, *4*, 557–564. [CrossRef]
24. Yu, L.; Akolkar, R. Lead underpotential deposition for the surface characterization of silver ad-atom modified gold electrocatalysts for glucose oxidation. *J. Electroanal. Chem.* **2017**, *792*, 61–65. [CrossRef]
25. Brankovic, S.R.; Dimitrov, N.; Sieradzki, K. Surfactant Mediated Electrochemical Deposition of Ag on Au(111). *Electrochem. Solid State Lett.* **1999**, *2*, 443–445. [CrossRef]
26. Wu, D.; Solanki, D.J.; Joi, A.; Dordi, Y.; Dole, N.; Litvnov, D.; Brankovic, S.R. Pb Monolayer Mediated Thin Film Growth of Cu and Co: Exploring Different Concepts. *J. Electrochem. Soc.* **2019**, *1666*, D3013–D3021. [CrossRef]

27. Sieradzki, K.; Dimitrov, N. Electrochemical Defect-Mediated Thin-Film Growth. *Science* **1999**, *284*, 138. [CrossRef]

28. Hwang, S.; Oh, I.; Kwak, J. Electrodeposition of Epitaxial Cu(111) Thin Films on Au(111) Using Defect-Mediated Growth. *J. Am. Chem. Soc.* **2001**, *123*, 7176–7177. [CrossRef]

29. Viyannalage, L.T.; Vasilic, R.; Dimitrov, N. Epitaxial Growth of Cu on Au(111) and Ag(111) by Surface Limited Redox ReplacementAn Electrochemical and STM Study. *J. Phys. Chem. C* **2007**, *111*, 4036–4041. [CrossRef]

30. Fayette, M.; Liu, Y.; Bertrand, D.; Nutariya, J.; Vasiljevic, N.; Dimitrov, N. From Au to Pt via Surface Limited Redox Replacement of Pb UPD in One-Cell Configuration. *Langmuir* **2011**, *27*, 5650–5658. [CrossRef]

31. Jayaraju, N.; Viravapandian, D.; Kim, G.Y.; Banga, D.; Stickney, J.L. Electrochemical Atomic Layer Deposition (E-ALD) of Pt Nanofilms Using SLRR Cycles. *J. Electrochem. Soc.* **2012**, *159*, D616–D622. [CrossRef]

32. Ambrozik, S.; Rawlings, B.; Vasiljevic, N.; Dimitrov, N. Metal deposition via electroless surface limited redox replacement. *Electrochem. Commun.* **2014**, *44*, 19–22. [CrossRef]

33. Dimitrov, N. Recent Advances in the Growth of Metals, Alloys, and Multilayers by Surface Limited Redox Replacement (SLRR) Based Approaches. *Electrochim. Acta* **2016**, *209*, 599–622. [CrossRef]

34. Al Amri, Z.; Mercer, M.P.; Vasiljevic, N. Surface Limited Redox Replacement Deposition of Platinum Ultrathin Films on Gold: Thickness and Structure Dependent Activity towards the Carbon Monoxide and Formic Acid Oxidation reactions. *Electrochim. Acta* **2016**, *210*, 520–529. [CrossRef]

35. Brankovic, S.R. Fundamentals of Metal Deposition via Surface Limited Redox Replacement of Underpotentially-Deposited Monolayer. *Electrochem. Soc. Interface* **2018**, *27*, 57–63. [CrossRef]

36. Brankovic, S.R.; Wang, J.X.; Adžić, R.R. Metal monolayer deposition by replacement of metal adlayers on electrode surfaces. *Surf. Sci.* **2001**, *474*, L173–L179. [CrossRef]

37. Mercer, M.P.; Plana, D.; Fermiotan, D.J.; Morgan, D.; Vasiljevic, N. Growth of epitaxial Pt1-xPbx alloys by surface limited redox replacement and study of their adsorption properties. *Langmuir* **2015**, *31*, 10904–10912. [CrossRef]

38. Vasiljevic, N. Electrodeposition of Pt-bimetallic Model Systems for Electrocatalysis and Electrochemical Surface Science. *Electrochem. Soc. Interface* **2018**, *27*, 71–75. [CrossRef]

39. Zhang, J.; Vukmirovic, M.B.; Sasaki, K.; Nilekar, A.U.; Mavrikakis, M.; Adzic, R.R. Mixed-Metal Pt Monolayer Electrocatalysts for Enhanced Oxygen Reduction Kinetics. *J. Am. Chem Soc.* **2005**, *127*, 12480–12481. [CrossRef]

40. Bromberg, L.; Fayette, M.; Martens, B.; Luo, Z.P.; Wang, Y.; Xu, D.; Zhang, J.; Fang, J.; Dimitrov, N. Catalytic Performance Comparison of Shape-Dependent Nanocrystals and Oriented Ultrathin Films of Pt4Cu Alloy in the Formic Acid Oxidation Process. *Electrocatalysis* **2012**, 1–13. [CrossRef]

41. McCurry, D.; Kamundi, M.; Fayette, M.; Wafula, F.; Dimitrov, N. All Electrochemical Fabrication of a Platinized Nanoporous Au Thin-Film Catalyst. *ACS Appl. Mater. Interfaces* **2011**, *3*, 4459–4468. [CrossRef] [PubMed]

42. Adzic, R.R.; Zhang, J.; Sasaki, K.; Vukmirovic, M.B.; Shao, M.; Wang, J.X.; Nilekar, A.U.; Mavrikakis, M.; Valerio, J.A.; Uribe, F. Platinum Monolayer Fuel Cell Electrocatalysts. *Top. Catal.* **2007**, *46*, 249–262. [CrossRef]

43. Fayette, M.; Nutariya, J.; Vasiljevic, N.; Dimitrov, N. A Study of Pt Dissolution during Formic Acid Oxidation. *ACS Catalysis* **2013**, *3*, 1709–1718. [CrossRef]

44. Wu, D.; Solanki, D.J.; Ramirez, J.L.; Yang, W.; Joi, A.; Dordi, Y.; Dole, N.; Brankovic, S.R. Electroless Deposition of Pb Monolayer: A New Process and Application to Surface Selective Atomic Layer Deposition. *Langmuir* **2018**, *34*, 11384–11394. [CrossRef] [PubMed]

45. Green, M.; Hanson, K.; Scherson, D.; Xing, X.; Richter, M.; Ross, P.; Carr, R.; Lindau, I. In situ scanning tunneling microscopy studies of the underpotential deposition of lead on gold (111). *J. Phys. Chem.* **1989**, *93*, 2181–2184. [CrossRef]

46. Stafford, G.R.; Bertocci, U. In Situ Stress and Nanogravimetric Measurements During Underpotential Deposition of Pb on (111)-Textured Au. *J. Phys. Chem. C* **2007**, *111*, 17580–17586. [CrossRef]

47. Shin, J.W.; Bertocci, U.; Stafford, G.R. Stress Response to Surface Alloying and Dealloying during Underpotential Deposition of Pb on (111)-Textured Au. *J. Phys. Chem. C* **2010**, *114*, 7926–7932. [CrossRef]

48. Nutariya, J.; Velleuer, J.; Schwarzacher, W.; Vasiljevic, N. Surface alloying/dealloying in Pb/Au (111) system. *ECS Trans.* **2010**, *28*, 15–25. [CrossRef]

49. Biberian, J.P.; Rhead, G.E. Spontaneous alloying of a gold substrate with lead monolayers. *J. Phys. F Met. Phys.* **1973**, *3*, 675–682. [CrossRef]

50. Perdereau, J.; Biberian, J.P.; Rhead, G.E. Adsorption and surface alloying of lead monolayers on (111) and (110) faces of gold. *J. Phys. F Met. Phys.* **1974**, *4*, 798. [CrossRef]

51. Wang, F.-X.; Pan, G.-B.; Liu, Y.-D.; Xiao, Y. Pb deposition onto Au(111) from acidic chloroaluminate ionic liquid. *Chem. Phys. Lett.* **2010**, *488*, 112–115. [CrossRef]

52. Kim, S.K.; Kim, J.S.; Han, J.Y.; Seo, J.M.; Lee, C.K.; Hong, S.C. Surface alloying of a Co film on the Cu(001) surface. *Surf. Sci.* **2000**, *453*, 47–58. [CrossRef]

53. Stamenkovic, V.R.; Mun, B.S.; Mayrhofer, K.J.J.; Ross, P.N.; Markovic, N.M. Effect of Surface Composition on Electronic Structure, Stability, and Electrocatalytic Properties of Pt-Transition Metal Alloys: Pt-Skin versus Pt-Skeleton Surfaces. *J. Am. Chem. Soc.* **2006**, *128*, 8813–8819. [CrossRef] [PubMed]

54. Tang, Q.; Zhang, H.; Meng, Y.; He, B.; Yu, L. Dissolution Engineering of Platinum Alloy Counter Electrodes in Dye-Sensitized Solar Cells. *Angew. Chem. Int. Ed.* **2015**, *54*, 11448–11452. [CrossRef] [PubMed]

55. Ali-Löytty, H.; Hannula, M.; Honkanen, M.; Östman, K.; Lahtonen, K.; Valden, M. Grain orientation dependent Nb–Ti microalloying mediated surface segregation on ferritic stainless steel. *Corros. Sci.* **2016**, *112*, 204–213. [CrossRef]

56. Stamenkovic, V.R.; Mun, B.S.; Arenz, M.; Mayrhofer, K.J.J.; Lucas, C.A.; Wang, G.; Ross, P.N.; Markovic, N.M. Trends in electrocatalysis on extended and nanoscale Pt-bimetallic alloy surfaces. *Nat. Mater.* **2007**, *6*, 241. [CrossRef] [PubMed]

57. Mun, B.S.; Watanabe, M.; Rossi, M.; Stamenkovic, V.R.; Markovic, N.M.; Ross, P.N. A study of electronic structures of Pt3M (M=Ti,V,Cr,Fe,Co,Ni) polycrystalline alloys with valence-band photoemission spectroscopy. *J. Chem. Phys.* **2005**, *123*, 204717. [CrossRef]

58. Renault, O.; Brochier, R.; Roule, A.; Haumesser, P.H.; Krömker, B.; Funnemann, D. Work-function imaging of oriented copper grains by photoemission. *Surf. Interface Anal.* **2006**, *38*, 375–377. [CrossRef]

59. Tiwari, D.; Cattelan, M.; Harniman, R.L.; Sarua, A.; Abbas, A.; Bowers, J.W.; Fox, N.A.; Fermin, D.J. Mapping Shunting Paths at the Surface of Cu2ZnSn(S,Se)4 Films via Energy-Filtered Photoemission Microscopy. *iScience* **2018**, *9*, 36–46. [CrossRef]

60. Vasilic, R.; Vasiljevic, N.; Dimitrov, N. Open Circuit Stability of Underpotentially Deposited Pb Monolayer on Cu(111). *J. Elecatroanal. Chem.* **2005**, *580*, 203–212. [CrossRef]

61. Tiwari, D.; Cattelan, M.; Harniman, R.L.; Sarua, A.; Fox, N.; Koehler, T.; Klenk, R.; Fermin, D.J. Impact of Sb and Na Doping on the Surface Electronic Landscape of Cu2ZnSnS4 Thin Films. *ACS Energy Lett.* **2018**, 2977–2982. [CrossRef]

62. Kim, K.S.; O'Leary, T.J.; Winograd, N. X-ray photoelectron spectra of lead oxides. *Anal. Chem.* **1973**, *45*, 2214–2218. [CrossRef]

63. Desimoni, E.; Casella, G.I.; Morone, A.; Salvi, A.M. XPS determination of oxygen-containing functional groups on carbon-fibre surfaces and the cleaning of these surfaces. *Surf. Interface Anal.* **1990**, *15*, 627–634. [CrossRef]

64. Fadley, C.S. Angle-resolved x-ray photoelectron spectroscopy. *Prog. Surf. Sci.* **1984**, *16*, 275–388. [CrossRef]

65. Light, T.B.; Eldridge, J.M.; Matthews, J.W.; Greiner, J.H. Structure of thin lead oxide layers as determined by x−ray diffraction. *J. Appl. Phys.* **1975**, *46*, 1489–1492. [CrossRef]

66. Brunt, T.A.; Rayment, T.; O'Shea, S.J.; Welland, M.E. Measuring the Surface Stresses in an Electrochemically Deposited Metal Monolayer: Pb on Au(111). *Langmuir* **1996**, *12*, 5942–5946. [CrossRef]

67. Vurens, G.H.; Salmeron, M.; Somorjai, G.A. The preparation of thin ordered transition metal oxide films on metal single crystals for surface science studies. *Prog. Surf. Sci.* **1989**, *32*, 333–360. [CrossRef]

68. Seah, M.P.; Dench, W.A. Quantitative electron spectroscopy of surfaces: A standard data base for electron inelastic mean free paths in solids. *Surf. Interface Anal.* **1979**, *1*, 2–11. [CrossRef]

69. Nyholm, R.; Berndtsson, A.; Martensson, N. Core level binding energies for the elements Hf to Bi (Z=72-83). *J. Phys. C* **1980**, *13*, L1091. [CrossRef]

70. Nicholson, J.A.; Riley, J.D.; Leckey, R.C.G.; Jenkin, J.G.; Liesegang, J.; Azoulay, J. Ultraviolet photoelectron spectroscopy of the valence bands of some Au alloys. *Phys. Rev. B* **1978**, *18*, 2561–2567. [CrossRef]

71. Eastman, D.E.; Grobman, W.D. Photoemission Energy Distributions for Au from 10 to 40 eV Using Synchrotron Radiation. *Phys. Rev. Lett.* **1972**, *28*, 1327–1330. [CrossRef]

72. Michaelson, H.B. The work function of the elements and its periodicity. *J. Appl. Phys.* **1977**, *48*, 4729–4733. [CrossRef]

73. Evans, S.; Thomas, J.M. Electronic structure of the oxides of lead. Part 1.—A study using X-ray and ultraviolet photoelectron spectroscopy of the oxidation of polycrystalline lead. *J. Chem. Soc. Faraday Trans.* **1975**, *71*, 313–328. [CrossRef]
74. Uda, M.; Nakamura, A.; Yamamoto, T.; Fujimoto, Y. Work function of polycrystalline Ag, Au and Al. *J. Electron. Spectros. Relat. Phenom.* **1998**, *88*, 643–648. [CrossRef]
75. Hoster, H. *Surface Alloys*; Wiley-VCH Verlag GmbH & Co.: Weinheim, Germany, 2014; Volume 3, pp. 329–367.
76. Pantelides, S.T.; Mickish, D.J.; Kunz, A.B. Correlation effects in energy-band theory. *Phys. Rev. B* **1974**, *10*, 2602–2613. [CrossRef]
77. Shevchik, N.J. Evolution of energy bands from the Au 5d level in Cd-Au alloys. *J. Phys. F* **1975**, *5*, 1860. [CrossRef]
78. Moruzzi, V.L.; Williams, A.R.; Janak, J.F. Band narrowing and charge transfer in six intermetallic compounds. *Phys. Rev. B* **1974**, *10*, 4856–4862. [CrossRef]
79. Tersoff, J. Surface-Confined Alloy Formation in Immiscible Systems. *Phys. Rev. Lett.* **1984**, *74*, 434–437. [CrossRef] [PubMed]
80. Besenbacher, F.; Nielsen, L.P.; Sprunger, P.T. Chapter 6 Surface alloying in heteroepitaxial metal-on-metal growth. In *The Chemical Physics of Solid Surfaces*; King, D.A., Woodruff, D.P., Eds.; Elsevier: Amsterdam, The Netherlands, 1997; Volume 8, pp. 207–257.

Article

Co$_3$O$_4$ Nanopetals on Si as Photoanodes for the Oxidation of Organics

Leonardo Girardi [1], Luca Bardini [1], Niccolò Michieli [2], Boris Kalinic [2], Chiara Maurizio [2], Gian Andrea Rizzi [1,*] and Giovanni Mattei [2]

[1] Department of Chemical Sciences, University of Padova, via Marzolo 1, 35121 Padova, Italy; leonardo.girardi@phd.unipd.it (L.G.); luca.bardini@unipd.it (L.B.)
[2] Physics and Astronomy Department, University of Padova, via Marzolo 6, 35121 Padova, Italy; niccolotomaso.michieli@unipd.it (N.M.); boris.kalinic@unipd.it (B.K.); chiara.maurizio@unipd.it (C.M.); giovanni.mattei@unipd.it (G.M.)
* Correspondence: gianandrea.rizzi@unipd.it; Tel.: +39-049-827-5722

Received: 13 November 2018; Accepted: 9 January 2019; Published: 11 January 2019

Abstract: Cobalt oxide nanopetals were grown on silicon electrodes by heat-treating metallic cobalt films deposited by DC magnetron sputtering. We show that cobalt oxide, with this peculiar nanostructure, is active towards the photo-electrochemical oxidation of water as well as of organic molecules, and that its electrochemical properties are directly linked to the structure of its surface. The formation of Co$_3$O$_4$ nanopetals, induced by oxidizing annealing at 300 °C, considerably improves the performance of the material with respect to simple cobalt oxide films. Photocurrent measurements and electrochemical impedance are used to explain the behavior of the different structures and to highlight their potential application in water remediation technologies.

Keywords: cobalt oxide; water oxidation; photo-electrochemistry; hydroxyl radical; electro-oxidation

---

1. Introduction

The photoelectrochemical oxidation of water is a major topic in modern chemical engineering; key technologies, such as the production of hydrogen and oxygen as renewable fuels [1] and the quantification [2] and abatement of organic pollutants [3,4] in the field of water remediation, in fact, depend on it. The cornerstone for the development of scalable Photoelectrochemical Water Oxidation (PEC-WO) processes is to find cost-effective, active, and durable semiconducting photo-electrodes able to sustain oxygen evolution reaction (OER) for long periods of time. To this end, silicon is a preferred material, as it possesses a suitable band gap and is readily available. The amount of oxygen produced by Si photoelectrodes, however, is usually small. More importantly, the performances of silicon photoanodes decay rapidly due to corrosion. To overcome these problems, Si-based water-splitting electrodes require the use of a thin (210 nm) protective coating; the thickness of the coating should be precisely controlled, as a layer which is too thick is reported to hamper the performance of the underlying Si [5,6]. The metals commonly used for this purpose are Ru [7], Ni [8], NiO$_x$ [9], Ir [10], Cu [11], CuO$_x$ [12], MnO$_x$ [13], Co, and CoO$_x$ [14].

Cobalt oxide, in particular, has been shown to be effective as a protective layer in a variety of conditions [15,16], as it is particularly stable in the strongly basic conditions employed for the electro-oxidation of water. The use of Co$_3$O$_4$ thin films as passivating layers was also explored by Yang et al. [17], who obtained a stable photocurrent as high as 17 mA/cm^{-2} in 1 M potassium hydroxide (KOH; pH = 13.6). On the other hand, thanks to its absorption in the visible range, cobalt oxide has also been used as a visible light sensitizer to increase the activity of wide band-gap photoanodes at longer wavelengths [18–20].

In this work, we show that cobalt oxide is an active participant in the electrochemical oxidation of water under visible light irradiation, and that it can be effectively used as a stand-alone photoanode. The thermal oxidation of a Co film, deposited by Direct Current (DC) magnetron sputtering on n-Si(100), forms p-type Co_3O_4 layers. These systems are electrodes with a stable anodic photocurrent in both neutral Na_2SO_4 and NaOH solutions. Moreover, we demonstrate that the controlled nano-structuring of the cobalt oxide structure produces significant improvements in photoelectrode performance. We further apply these findings to the oxidation of various organic molecules, and show that the photocurrent increases proportionally with the organic content of the solution [21,22], indicating that cobalt oxide photoanodes can be a promising material for applications in water remediation, as well as a greener, more efficient alternative for the determination of chemical oxygen demand (COD), with respect to commonly used dichromate methods [23]. In particular, we will show that different thermal treatments of the Co layer give rise to different nanostructures characterized by different electrical and sensing properties.

2. Materials and Methods

2.1. Sample Preparation

The substrates used were n-type Si(100) wafer, preliminarily washed in H_2SO_4:H_2O_2 at 90 °C. The native silicon oxide passivation layer, whose typical thickness is about 2 nm [24], was not removed prior to deposition. To this respect, it is worth mentioning that there are several examples in the literature of Si-based photoanodes where the presence of the native silica layer was found to improve the photo-electrochemical performances; some authors suggested that the silica layer may act as an adhesive between the substrate and the deposited film [6]. 50 nm-thick Co films were deposited on the Si substrates by DC magnetron sputtering, with a deposition rate of 1 Å/s. The deposited films were annealed in O_2 flux at different temperature values, as detailed in Table 1.

Table 1. Details of the post-deposition annealing performed in pure oxygen flux.

Sample Name	Treatment Conditions
PETAL	4 h at 300 °C
HYBRID	4 h at 300 °C + 1 h at 450 °C
NOPETAL	1 h at 450 °C

2.2. Structural and Morphological Characterizations

The thickness of the deposited films was measured by Atomic Force Microscopy (AFM) using a NT-MDT PRO Solver microscope (NT-MDT, Moscow, Russia).

Scanning Electron Microscopy (SEM) measurements were carried out with a Zeiss Sigma HDF Field-Emission Scanning Electron Microscopy (FE-SEM), equipped with InLens, Secondary Electrons (SE), and Backscattered Electrons (BSE) detectors for the imaging (Zeiss, Jena, Germany).

Grazing Incidence X-ray Diffraction (GIXRD) spectra were collected with an X'Pert Pro diffractometer (incidence angle = 1 deg) using the Cu K_α radiation. The GIXRD analysis was performed by using MAUD software (University of Trento, Italy) [25].

Surface composition was determined by using X-ray Photoelectron Spectroscopy (XPS) measurements, performed on a custom-built Ultra High Vacuum (UHV) chamber (base pressure = 5×10^{-10} mbar) equipped with a non-monochromatized, double-anode X-ray source (Omicron DAR-400, Scienta-Omicron GmbH, Uppsala, Sweden), a hemispherical electron analyzer (Omicron EA-125, Scienta-Omicron GmbH, Uppsala, Sweden), and a 5-channeltron detector assembly. The electron analyzer had an acceptance angle of $\pm 4°$, and the diameter of the analyzed area was 3 mm. The spectra were acquired with Mg K_α radiation.

2.3. Electrochemical Measurements

Electrochemical measurements to test the activity were performed either in 0.1 M sodium sulphate or in 0.1 M sodium hydroxide in milliQ water. The measurements were made with Autolab PGSTAT204 potentiostat (Metrohm, Utrecht, The Netherlands) in a Teflon PhotoElectroChemical (PEC) cell. A Pt coil wire and Ag/AgCl electrode were used as the counter electrode and reference electrode, respectively. The samples were mounted outside the cell and kept in position by an O-ring seal. The electrical contact was obtained by a metal tip which was firmly pressed onto the back of the Si wafer by a spring. All samples were illuminated from the front. Photoelectrochemical experiments were carried out with a white LED with an intensity of about 100 mW/cm^2. The light intensity was measured using a photodiode at the same distance from the light source as the sample; the light was passing through the electrolyte solution, and the quartz window mounted on the cell. The light intensity was controlled by the optical bench (Metrohm-Autolab) coupled to the Autolab PGSTAT204 instrument (Metrohm, Utrecht, The Netherlands). All Linear Sweep Voltammetry (LSV) measures have been obtained with a scan rate of 5 mV/s. Electrochemical Impedance Spectroscopy (EIS) data were obtained under illumination, and in the dark condition, the voltage amplitude for EIS measurements was ±10 mV with the frequency range set from 10^5 Hz to 10^0 Hz, performing 50 points with logarithmic distribution.

3. Results and Discussion

3.1. Structure and Activity

In Figure 1a–c, the FE-SEM top-view images of the three electrodes are shown. The heating process of the Co metallic film at 300 °C for 4 h (sample labeled PETAL) induces the growth of nanopetals from the layer underneath (Figure 1a). The petals' surface, ≈100–200 nm wide, is preferentially normal to the original film surface, and the petal thickness is limited to about 10 nm. These structures are similar to others which have previously been observed [26]. The petal formation is dependent on the annealing temperature. Indeed, a similar treatment performed at 450 °C (sample labeled NOPETAL, Figure 1c) induces the formation of a rough surface characterized by relatively large grains and with no nanopetals. If the PETAL sample is further treated at 450 °C (sample labeled HYBRID), the nanopetals are less visible (Figure 1b), and the overall surface morphology is intermediate between that of sample PETAL and of sample NOPETAL. The GIXRD patterns of the three samples (Figure 1d) are all very similar and show the signature of Co_3O_4 nanocrystals, whose average size is about 25 nm for the sample PETAL, 40 nm for the HYBRID one, and 50 nm for the NOPETAL one.

The different surface morphology has a direct impact on the electrochemical properties of the material towards the oxidation of water. A sketch of the electronic band structure of the prepared electrodes is shown in Figure 1e, where the n-Si substrate and the Co_3O_4 layer (p-type semiconductor, as shown below) form a "tunneling" p-n junction with an interposed SiO_x dielectric barrier [27]. An approximate position of the Valence Band (VB) and Conduction Band (CB) edges of the materials, before the junction is formed, can be easily derived from Pearson absolute electronegativity values [28], where the band-gap values are the ones reported in the literature for Si(100), native SiO_2 [29], and Co_3O_4 [30]. After Fermi-level alignment, the approximate position of the CB and VB edges can be obtained for the sample PETAL from the Mott-Schottky (MS) plots (see SI Figure S2, curve acquired at 2000 Hz) using the formulas [31–33]: $E_{cb}(V) = V_{fb} + 0.1$ for Si(100) and $E_{vb}(V) \approx V_{fb}$ for Co_3O_4, where V_{fb} is the flat-band potential obtained from the MS plots (V_{fb} = 0.95 V vs. RHE for Si and V_{fb} = 1.66 V vs. RHE for Co_3O_4 at 2000 Hz). Different values of V_{fb} can be obtained in the case of samples NOPETAL and HYBRID due their different nanostructuration [34]. The sketch of Figure 1e indicates that, after the junction has formed, there is an electron migration from Si to Co_3O_4 that shifts the VB edge of Co_3O_4 slightly above the VB of Si(100).

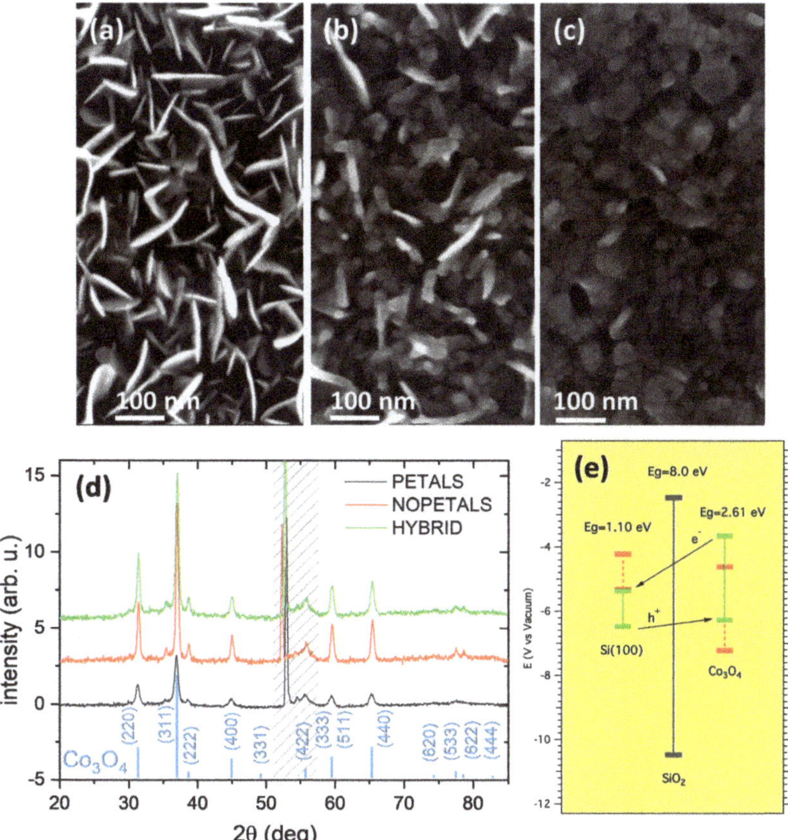

Figure 1. (a–c) FE-SEM top-view images of the samples PETAL (**a**), HYBRID (**b**), and NOPETAL (**c**). (**d**) Corresponding Grazing Incidence X-ray Diffraction (GIXRD) patterns. The Co_3O_4 reflections are marked (ICSD-63164). In the dashed region, a spurious diffraction signal from the Si substrate is visible (the X-ray penetration depth is about 80 nm). (**e**) Sketch of the electronic energy bands of the prepared electrodes. The theoretical positions of edges CB and VB, obtained from Pearson's electronegativity values, are in red. The approximate positions of CB and VB edges, after the junction is formed, obtained from Mott-Schottky plots (SI. Figure S2) are indicated by green bars.

Electrochemical measurements were carried out either in 0.1 M NaOH or in Na_2SO_4, the former being the ideal condition for water oxidation, the latter being more relevant for sensing and water remediation. In fact, in NaOH electrolytes, the difference in the photocurrent values after the addition of small volumes of different analytes (EtOH, glucose, etc.) is less evident, since it is overshadowed by the already high photocurrent value due to hydroxyls oxidation. As shown in Figure 2a, the samples HYBRID and NOPETAL show similar current outputs with respect to water oxidation, despite the different surface morphology. The current for the sample PETAL, on the other hand, is at least three times higher in sodium sulphate, and almost two times higher in alkaline conditions, with a concomitant shift of the onset potential (SI Figure S1).

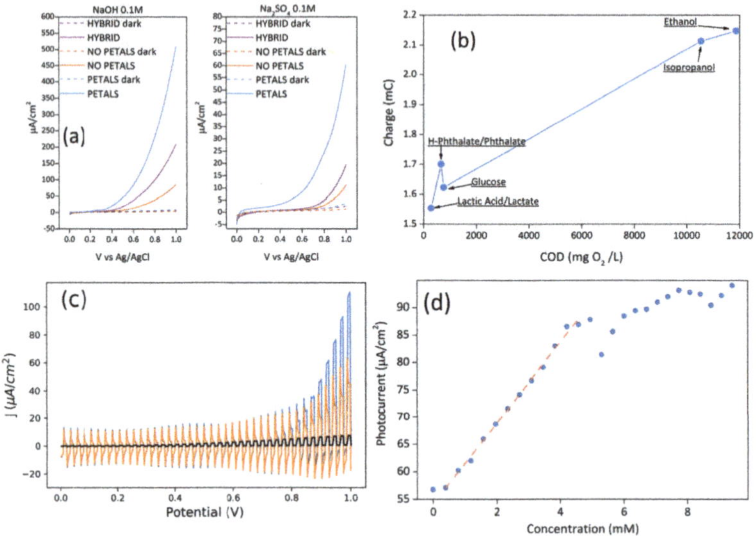

Figure 2. (**a**) LSV in 0.1 M Na$_2$SO$_4$ and 0.1 M NaOH of the three samples in dark and illuminated conditions; (**b**) charge values at 1V vs. Ag/AgCl, from chronoamperometries (pH = 6), as a function of theoretical Chemical Oxygen Demand (COD) of the sample PETAL with different organic molecules in a 0.1 M sodium sulphate solution; (**c**) LSV with chopped illumination of the sample PETAL in sodium sulphate before (orange) and after (blue) the addition of glucose. The behavior of the bare silicon electrode is also shown for reference (black curve); (**d**) photocurrent dependence of sample PETAL at 1 V vs. Ag/AgCl on the concentration of glucose.

For the sample PETAL, if a source of carbon like EtOH, MeOH, or glucose is added to the sodium sulphate solution, a clear enhancement of the photocurrent is seen (Figure 2c), indicating that the mineralization process is the decomposition of organic substances and can be described with the reaction [2]: $C_yH_mO_jN_kX_q + (2y - j)H_2O \rightarrow yCO_2 + q\,X^- + kNH_3 + (4y - 2j + m - 3k)H^+ + (4y - 2j + m - 3k - q)e^-$.

The process is partly due to electron capture by holes on the Co$_3$O$_4$ surface (the other reaction that leads to the mineralization to CO$_2$ involves the reaction with hydroxyl radicals generated during the oxidation of water). The electrode response to chopped illumination is characterized by "spike and overshoot" photocurrent transients. We attribute the photocurrent spikes of Figure 2c (blue and orange curves) to a high degree of electron and hole recombination on the surface, and to the formation of surface states (trap-states). This is a known behavior which is described, for instance, in [35,36]. When the light is turned on, holes generated in the space-charge region are swept rapidly towards the semiconductor electrolyte junction. Due to the slow kinetics of the four-hole oxidation of water to molecular oxygen, the concentration of holes increases considerably at the interface, until the rate of arrival of holes is balanced in the steady state by the rates of charge transfer and recombination. Since surface recombination leads to a flux of electrons towards the surface, the resulting photocurrent transient is the sum of the hole and electron contributions. A careful analysis of Figure 2c also shows that as soon as a hole scavenger such as glucose is added, the cathodic spikes are strongly attenuated.

It is worth reminding that, quite often, the COD value is obtained by measuring the total charge due to the photocurrent after the degradation to CO$_2$ is complete (see Figure 3a), and its value is expressed as mg O$_2$/L, according to the equation [2]:

$$\mathrm{COD(O_2\ mg/L)} = \frac{nC}{4} \times 32000 = \frac{Q}{4FV} \times 32000 = K \times Q \tag{1}$$

where n is the number of electrons, C is the concentration of substances, Q is the charge passed, V is the volume of the solution, and F is the Faraday constant. We report in Figure 2b the total charge due to the mineralization process for five different substances, as a function of theoretical COD (mg O_2/L). This graph suggests that the charge is directly correlated to the theoretical COD values—in other words, the higher the theoretical COD value, the higher the charge.

To better highlight the proportionality between the photocurrent/charge and organic content, as well as to verify the concentration range to obtain a linear response, further tests were conducted with glucose as a proof-of-concept carbon source, using the PETAL sample as an electrode. Since the current arising from the oxidation of water is lower in sodium sulfate, the current generated by the oxidation process of the organic molecule can be better highlighted. For this reason, the quantification of different concentrations of glucose was reported in a sodium sulfate solution (Figure 2d).

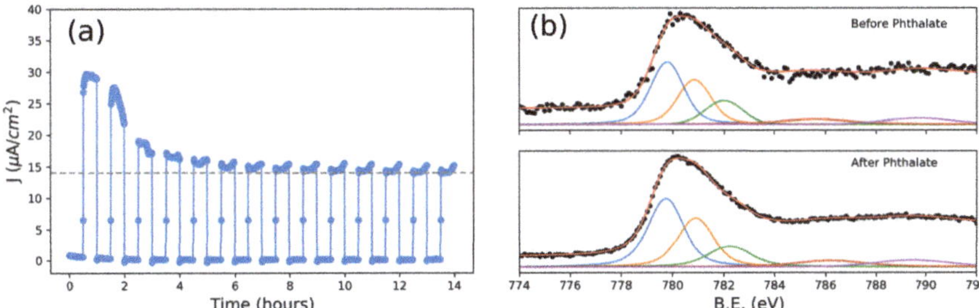

Figure 3. (**a**) Photocurrent at 0.95 V vs. Ag/AgCl of the layer PETAL over time in a 3 mM phthalate, 0.1 M sodium sulphate solution (pH = 6); (**b,c**) fitted X-ray photoelectron spectroscopy (XPS) spectra [37] of the Co $2p_{3/2}$ orbitals taken before (**b**) and after (**c**) 14 h of phthalate oxidation.

All samples show a linear response with glucose concentration, with the sample PETAL showing the widest linear range of 0.4–4.2 mM (see SI, Table S1), a sensitivity of 7.5 $\pm$ 0.1 μA/cm^{2}mM^{-1}, a limit of detection (LOD) of 2.3 $\pm$ 0.2 μM, and a limit of quantification (LOQ) of 7.7 $\pm$ 0.6 μM (The limit of detection (LOD) and limit of quantification (LOQ) were calculated from the standard deviation (σ) of the linear part of LSV measurements (Figure 2a) and from the slope (b) of the calibration curve (Figure 2d). For LOD, we considered $\Delta j = 3\sigma$ as the minimum value of Δj to detect glucose (LOD = 3σ/b), while for LOQ, we considered it to be $\Delta j = 10\sigma$ (LOD = 10σ/b). The error was calculated using the propagation formula). Sensitivity and linear range are comparable with some of the values found in the literature for metal oxide nanostructures [38].

Considering a previous report which demonstrates that a passivating film with a thickness of 10 nm is enough to significantly hamper the photoelectro-catalytic activity of silicon [6], we can safely state that the surface of the Co_3O_4 nanostructured layer (whose average thickness after annealing is about 80 nm) is the only active species in the present scenario. This observation is supported by Mott-Schottky plots (see SI, Figure S2), which show a positive slope (n-type behavior of the silicon wafer) in the dark, but a negative slope (p-type behavior of Co_3O_4) under illumination, showing that cobalt oxide dominates the photoelectrochemical behavior of the electrode. In this case, the completely different shape of the MS plots under illumination is probably due to inhomogeneities of the electrical field and formation of surface states in the layer where the photocurrent is built up, as recently described in a paper by Kirchartz et al. [39].

To further probe the performance of our nanostructured cobalt oxide layers, phthalate was chosen as the next target molecule. Phthalates form a class of extremely common plasticizers and pollutants, which, due to their large diffusion, cause serious concerns for human health [40]. A long-term test showed that after more than 10 h, the photocurrent had gradually decreased to the value of the clean solution, containing only Na_2SO_4. In this case, it was possible to calculate the value of COD according to Equation 1, and the proportionality constant K was found to be 1.594×10^{-4} O_2 mg/L $\times$ C, a value similar to the ones found in the literature [2], indicating that the mineralization process is mainly due to interaction with hydroxyl radicals. Moreover, X-ray photoelectron spectra (XPS), taken on an as-prepared electrode and on the electrode after the long-term test, showed no significant changes in the 2p peaks of cobalt; in particular, satellites which can be related to Co(II) oxides/hydroxides are absent [41] (Figure 3b), highlighting the excellent stability of the cobalt oxide surface even after prolonged operation.

3.2. Mechanistic Analysis

To fully characterize the effect of the surface structure of Co_3O_4 on its electrochemical properties, glucose was chosen for further analyses. In the literature, the oxidation of glucose is said to be caused by the presence of Co^{3+} and Co^{4+} surface states according to the following reaction sequence [42,43]:

$$Co_3O_4 + OH^- + H_2O \leftrightarrow 3CoOOH + e^- \text{ (0.3 V vs. Ag/AgCl)} \tag{2}$$

$$CoOOH + OH^- \leftrightarrow CoO_2 + H_2O + e^- \text{ (0.6 V vs. Ag/AgCl)} \tag{3}$$

$$2CoO_2 + C_6H_{12}O_6 \text{ (glucose)} \rightarrow 2CoOOH + C_6H_{10}O_6 \text{ (gluconolactone)} \tag{4}$$

In the present case, cyclic voltammetries, acquired in NaOH electrolyte, as in the case of Co_3O_4 nanofibers [44], (CVs, SI Figure S3) never clearly showed the presence of oxidation peaks at about 0.3 V and 0.6 V vs. Ag/AgCl. We attribute this behavior to a much lower surface area of our samples compared with the examples found in the literature, where Co_3O_4 nano-structrures characterized by a very high surface area were used [45]. Moreover, it is likely that a different mechanism was involved in our case, since an enhancement in the anodic (photo)current appeared only when the sample surface was illuminated.

The effect of light is likely a combination of the population of the conduction band of the silicon substrate, which allows the current to tunnel through the $Co_3O_4/SiO_2/Si$ junction, as well as the creation of active cobalt species on the surface of the electrode. The electrochemical impedance spectra (EIS) reported in Figure 4 and performed in the dark and under illumination with a white LED in Na_2SO_4 electrolyte, also highlight the presence of two different contributions to the electrochemical properties of the electrodes. EIS experiments were performed at two different overpotentials in order to investigate the behavior of the polarized electrode both with (1 V) and without (0.1 V) the presence of electrochemical reactions.

In the dark, significant differences were already apparent between the NOPETAL and the PETAL layers. At low bias (100 mV vs. Ag/AgCl), the PETAL layer has two contributions of comparable magnitude, one in the low frequency range and one in the high frequency one. The NOPETAL layer shows a major contribution at low frequency, which largely overshadows the high-frequency semi-circle. At low bias, the same behavior is present even when the sample is illuminated. For both samples PETAL and HYBRID, on the other hand, there is a stark decrease in the low-frequency contribution when the LED is turned on. This information can be taken as an indication that, for samples PETAL and HYBRID, the lower-frequency part of the spectrum is dominated by the contribution of surface states. For sample PETAL in particular, a description with three time constants can also be proposed; in this scenario, the second time constant (roughly in the range 10 kHz–150 Hz) disappears when the sample is illuminated. Such an interpretation would suggest the second contribution being due to the space charge layer.

At a bias of 1 V in the dark, the low-frequency contribution remains dominant for both NOPETAL and HYBRID layers; it is significant that, with the increase in bias, the impedance of the high-frequency component for the sample NOPETAL increases. Conversely, in the case of the

PETAL layer, the high-frequency contribution remains relatively unvaried, while there is an increase in the mid/low frequency semi-circle, a behavior consistent with the application of a bias at a blocking electrode-solution interface. These results indicate that the structuring of the Co_3O_4 layer in sample PETAL is also beneficial for the electrical connection with the underlying silicon substrate.

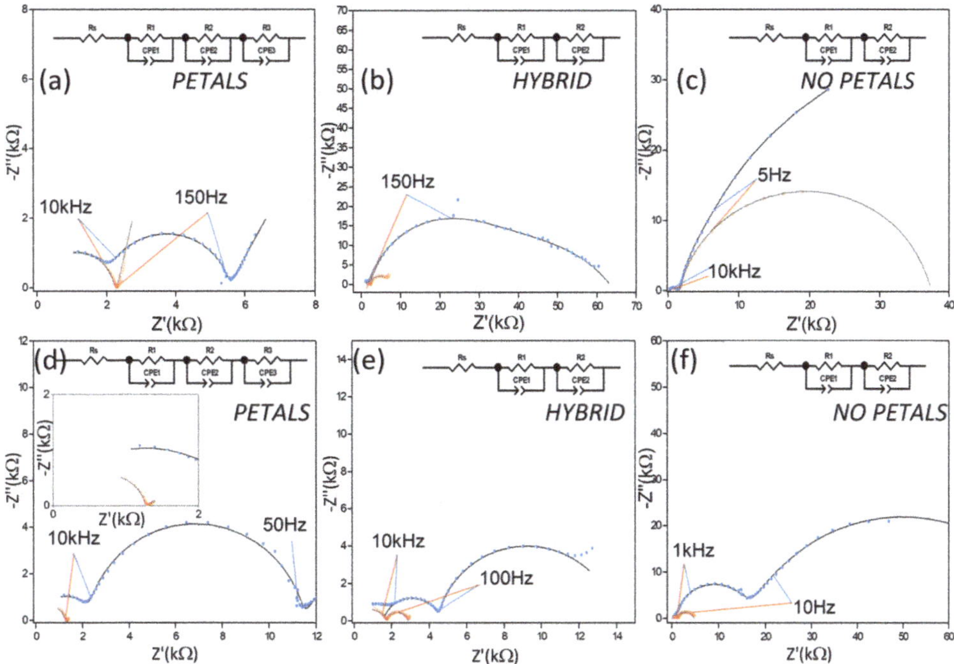

Figure 4. Comparison of Nyquist impedance spectra for the PETAL , NOPETAL , and HYBRID layers. (**a–c**) Spectra taken at 0.1 V; (**d–f**) spectra taken at 1 V. In all figures, spectra recorded in the dark are presented in blue, while spectra recorded under illumination are presented in orange. All spectra have been acquired in 0.1 M Na_2SO_4 electrolyte. Solid lines were obtained by fitting the EIS spectra with the EC (equivalent circuit) represented in the upper part of each panel. CPE indicates a constant phase element, Rs indicates the solution resistance. The inset of Figure (**d**) shows an enlargement of the fitting results at low impedance values.

When the electrodes are illuminated at 1 V, Nyquist plots indicate a similar overall decrease in impedance across the entire frequency range for all samples. For samples NOPETAL and HYBRID , a significant impedance response is also present at lower frequencies; a closer inspection of the behavior of the phase angle reveals significant differences in the behavior of the three surface structures in this region (Figure 5). While the PETAL layer shows near-ideal resistive behavior across the frequency spectrum up to 100 Hz, a residual RC contribution at low frequency remains for the NOPETAL and HYBRID layers. We also note that, upon illumination, the phase behavior, as a function of the frequency, is intermediate between the ones of the HYBRID and PETAL layers, corresponding to the fact that the surface morphology is also in between the two (see Figure 1a–c).The overall picture suggests that the formation of surface states (i.e., CoOOH or CoO_2) is linked to the presence of the petal-like structures. The time constant visible in Figure 5 with light on, at around 50 Hz, can then be linked to the charge transfer resistance between the electrode and the species in the solution. For sample NOPETAL , there is significant resistance probably given by the fact that the reaction mechanism largely does not directly involve the formation of surface states on the cobalt oxide surfaces. At the opposite extreme, sample

PETAL does not have significant contributions in this region due to the direct involvement of Co_3O_4 surface states in the reaction, which form a low impedance pathway for electrons. Sample HYBRID lies in between; as can be seen in Figure 1, some petal-like structures can be spotted in sample HYBRID, but most of them were destroyed by the thermal treatment.

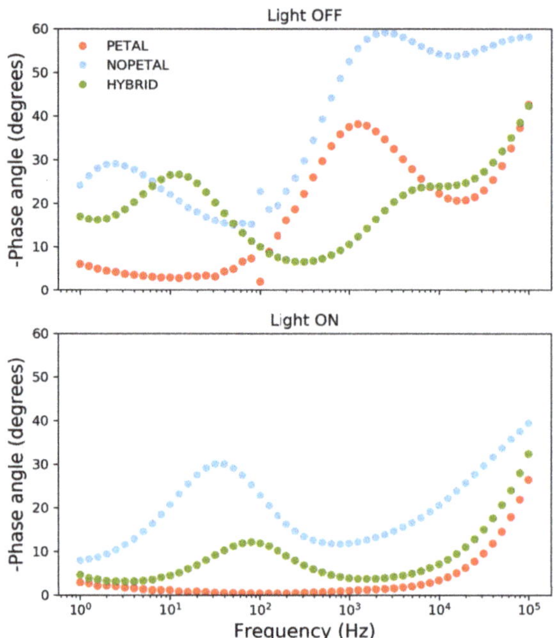

Figure 5. Phase angle spectra recorded at 1 V in 0.1 M sodium sulphate for samples PETAL (red), NOPETAL (blue), and HYBRID (green).

This qualitative description, implying that the existence of surface states is linked to the presence of petal-like structures, is confirmed by fitting the impedance data by an equivalent circuit (EC), containing, in the case of sample PETAL, three RQ elements corresponding to the Co_3O_4 interface with a SiO_x/Si, Co_3O_4 space charge region (characterized by the presence of surface states) and the double layer at the interface with the electrolyte [46]. In the case of samples HYBRID and NOPETAL, the EC consists in only two RQ (Q = CPE) elements, relatable to the $Co_3O_4/SiO_x/Si$ interface and Helmholtz layer, confirming the fact that the contribution to the overall process of surface states in these samples is negligible. A table containing the calculated resistances and CPE values, in the three cases, is reported in the Supplementary Materials (Table S2).

The significantly different behavior of the sample PETAL is then to be assigned to a particularly favorable interplay of surface area and the creation of active surface states, showing that nanostructuring is not merely morphological, but has important implications on the electronic structure of the Co_3O_4 at the interface with the solution. It should be noted that one further element which could contribute to the improved performance of the PETAL sample is the change in the wettability of the surface, which can play an important role when gas-evolving reactions are involved [47]. Further experiments are needed to clarify this point. The possibility of combining the effect of surface structuring, as described herein, with the creation of oxygen vacancies through surface doping [48] makes cobalt oxide photoelectrodes a promising new substrate for the engineering of simple, stable, and effective photoanodes.

4. Conclusions

Nanostructured cobalt oxide layers obtained by oxidation of a 50 nm Co film deposited on 100 n-Si by DC-magnetron sputtering have been produced. The choice of the annealing parameters allowed for the production of a Co_3O_4 layer decorated with nanopetals. This system can form a versatile and durable photoanode, and has been tested as a sensor for different water-soluble organic molecules in a sodium sulphate solution, obtaining, for glucose, a sensitivity of 7.5 ± 0.1 mA cm^{-2}M^{-1}. This nanostructured electrode can also be effective for water remediation. Indeed, we have shown that under positive bias and visible light illumination, the phthalate gets completely mineralized after about 6 h of electrochemical work. Moreover, the composition of the Co_3O_4 nanopetal layer after repeated sensing experiments and electrochemical work under the application of illumination and bias is virtually identical to that of the as-grown electrode, as shown by XPS analysis. Different layer morphologies can be obtained by specific thermal treatments which remarkably influence the electrical properties of these electrodes, as proven by EIS measurements and, consequently, their sensing properties. We believe that this nanostructured electrode, formed by Co_3O_4 nanopetals on Si, can be quite easily engineered to produce sensing devices or larger-area electrodes for water remediation applications.

Supplementary Materials: The following are available at http://www.mdpi.com/2571-9637/2/1/4/s1, Figure S1: Linear sweep voltammetry (5 mV/s) under illumination of the PETAL sample in different electrolyte solutions; Figure S2: Mott-Schottky plots of the prepared samples in the dark and under illumination in Na_2SO_4; Figure S3: Cyclic voltammetry performed on sample PETALS, in 0.1 M NaOH and 5 mM glucose with scan rate of 40 mV/s in the dark and with illumination; Table S1: Sensing in Sodium Sulphate 0.1 M at 1 V vs. Ag/AgCl; Table S2:Equivalent Circuit parameters.

Author Contributions: Investigation, L.G., L.B.; investigation, writing–original draft preparation, G.A.R.; conceptualization, investigation, writing–review and editing: N.M., B.K., C.M.; writing–review and editing: G.M.

Funding: This research was partially funded by Università degli Studi di Padova, Physics and Astronomy Department, grant number BIRD178923/17. Authors gratefully acknowledge the Italian Minister of University (MIUR) for financial support to the SMARTNESS (Solar driven chemistry: new materials for photo- and electrocatalysis) financed thorough the PRIN 2015K7FZLH.

Acknowledgments: We thank Alessandro Michieletto for helpful discussions.

Conflicts of Interest: The authors declare no conflict of interest. The founding sponsors had no role in the design of the study; in the collection, analyses, or interpretation of data; in the writing of the manuscript, and in the decision to publish the results.

Abbreviations

The following abbreviations are used in this manuscript:

UHV	Ultra High Vacuum
FE-SEM	Field Emission- Scanning Electron Microscopy
GIXRD	Grazing Incidence X-Ray Diffraction
VB	Valence Band
CB	Conduction Band
EIS	Electrochemical Impedance Spectroscopy
PEC	PhotoElectroChemical
COD	Chemical Oxygen Demand
LSV	Linear Sweep Voltammetry
XPS	X-Ray Photoelectron Spectroscopy

References

1. Cesar, I.; Sivula, K.; Kay, A.; Zboril, R.; Grätzel, M. Influence of Feature Size, Film Thickness, and Silicon Doping on the Performance of Nanostructured Hematite Photoanodes for Solar Water Splitting. *J. Phys. Chem. C* **2009**, *113*, 772–782. [CrossRef]
2. Zhang, J.; Zhou, B.; Zheng, Q.; Li, J.; Bai, J.; Liu, Y.; Cai, W. Photoelectrocatalytic COD determination method using highly ordered TiO$_2$ nanotube array. *Water Res.* **2009**, *43*, 1986–1992. [CrossRef] [PubMed]
3. Ganiyu, S.O.; Zhou, M.; Martínez-Huitle, C.A. Heterogeneous electro-Fenton and photoelectro-Fenton processes: A critical review of fundamental principles and application for water/wastewater treatment. *Appl. Catal. B Environ.* **2018**, *235*, 103–129. [CrossRef]
4. Garcia-Segura, S.; Brillas, E. Applied photoelectrocatalysis on the degradation of organic pollutants in wastewaters. *J. Photochem. Photobiol. C Photochem. Rev.* **2017**, *31*, 1–35. [CrossRef]
5. Shi, Y.; Han, T.; Gimbert-Suriñach, C.; Song, X.; Lanza, M.; Llobet, A. Substitution of native silicon oxide by titanium in Ni-coated silicon photoanodes for water splitting solar cells. *J. Mater. Chem. A* **2017**, *5*, 1996–2003. [CrossRef]
6. Han, T.; Shi, Y.; Song, X.; Mio, A.; Valenti, L.; Hui, F.; Privitera, S.; Lombardo, S.; Lanza, M. Ageing mechanisms of highly active and stable nickel-coated silicon photoanodes for water splitting. *J. Mater. Chem. A* **2016**, *4*, 8053–8060. [CrossRef]
7. Li, L.; Duan, L.; Xu, Y.; Gorlov, M.; Hagfeldt, A.; Sun, L. A photoelectrochemical device for visible light driven water splitting by a molecular ruthenium catalyst assembled on dye-sensitized nanostructured TiO$_2$. *Chem. Commun.* **2010**, *46*, 7307–7309. [CrossRef]
8. Kenney, M.J.; Gong, M.; Li, Y.; Wu, J.Z.; Feng, J.; Lanza, M.; Dai, H. High-Performance Silicon Photoanodes Passivated with Ultrathin Nickel Films for Water Oxidation. *Science* **2013**, *342*, 836–840. [CrossRef]
9. Sun, K.; McDowell, M.T.; Nielander, A.C.; Hu, S.; Shaner, M.R.; Yang, F.; Brunschwig, B.S.; Lewis, N.S. Stable Solar-Driven Water Oxidation to O$_2$(g) by Ni-Oxide-Coated Silicon Photoanodes. *J. Phys. Chem. Lett.* **2015**, *6*, 592–598. [CrossRef]
10. Chen, Y.W.; Prange, J.D.; Duehnen, S.; Park, Y.; Gunji, M.; Chidsey, C.E.D.; McIntyre, P.C. Atomic layer-deposited tunnel oxide stabilizes silicon photoanodes for water oxidation. *Nat. Mater.* **2011**, *10*, 539–544. [CrossRef]
11. Du, J.; Chen, Z.; Ye, S.; Wiley, B.J.; Meyer, T.J. Copper as a Robust and Transparent Electrocatalyst for Water Oxidation. *Angew. Chem. Int. Ed.* **2015**, *54*, 2073–2078. [CrossRef] [PubMed]
12. Yu, F.; Li, F.; Zhang, B.; Li, H.; Sun, L. Efficient Electrocatalytic Water Oxidation by a Copper Oxide Thin Film in Borate Buffer. *ACS Catal.* **2015**, *5*, 627–630. [CrossRef]
13. Strandwitz, N.C.; Comstock, D.J.; Grimm, R.L.; Nichols-Nielander, A.C.; Elam, J.; Lewis, N.S. Photoelectrochemical Behavior of n-type Si(100) Electrodes Coated with Thin Films of Manganese Oxide Grown by Atomic Layer Deposition. *J. Phys. Chem. C* **2013**, *117*, 4931–4936. [CrossRef]
14. Artero, V.; Chavarot-Kerlidou, M.; Fontecave, M. Splitting Water with Cobalt. *Angew. Chem. Int. Ed.* **2011**, *50*, 7238–7266. [CrossRef] [PubMed]
15. Kosmala, T.; Calvillo, L.; Agnoli, S.; Granozzi, G. Enhancing the Oxygen Electroreduction Activity through Electron Tunnelling: CoO$_x$ Ultrathin Films on Pd(100). *ACS Catal.* **2018**, *8*, 2343–2352. [CrossRef]
16. Bae, D.; Mei, B.; Frydendal, R.; Pedersen, T.; Seger, B.; Hansen, O.; Vesborg, P.C.K.; Chorkendorff, I. Back-Illuminated Si-Based Photoanode with Nickel Cobalt Oxide Catalytic Protection Layer. *ChemElectroChem* **2016**, *3*, 1546–1552. [CrossRef]
17. Yang, J.; Walczak, K.; Anzenberg, E.; Toma, F.M.; Yuan, G.; Beeman, J.; Schwartzberg, A.; Lin, Y.; Hettick, M.; Javey, A.; et al. Efficient and Sustained Photoelectrochemical Water Oxidation by Cobalt Oxide/Silicon Photoanodes with Nanotextured Interfaces. *J. Am. Chem. Soc.* **2014**, *136*, 6191–6194. [CrossRef]
18. Ramakrishnan, V.; Kim, H.; Park, J.; Yang, B. Cobalt oxide nanoparticles on TiO$_2$ nanorod/FTO as a photoanode with enhanced visible light sensitization. *RSC Adv.* **2016**, *6*, 9789–9795. [CrossRef]
19. Maeda, K.; Ishimaki, K.; Tokunaga, Y.; Lu, D.; Eguchi, M. Modification of Wide-Band-Gap Oxide Semiconductors with Cobalt Hydroxide Nanoclusters for Visible-Light Water Oxidation. *Angew. Chem. Int. Ed.* **2016**, *55*, 8309–8313. [CrossRef]

20. Maeda, K.; Ishimaki, K.; Okazaki, M.; Kanazawa, T.; Lu, D.; Nozawa, S.; Kato, H.; Kakihana, M. Cobalt Oxide Nanoclusters on Rutile Titania as Bifunctional Units for Water Oxidation Catalysis and Visible Light Absorption: Understanding the Structure—Activity Relationship. *ACS Appl. Mater. Interfaces* **2017**, *9*, 6114–6122. [CrossRef]

21. Wang, C.; Wu, J.; Wang, P.; Ao, Y.; Hou, J.; Qian, J. Photoelectrocatalytic determination of chemical oxygen demand under visible light using Cu_2O-loaded TiO_2 nanotube arrays electrode. *Sens. Actuators B Chem.* **2013**, *181*, 1–8. [CrossRef]

22. Zhang, Z.; Chang, X.; Chen, A. Determination of chemical oxygen demand based on photoelectrocatalysis of nanoporous TiO_2 electrodes. *Sens. Actuators B Chem.* **2016**, *223*, 664–670. [CrossRef]

23. Hejzlar, J.; Kopáček, J. Determination of low chemical oxygen demand values in water by the dichromate semi-micro method. *Analyst* **1990**, *115*, 1463–1467. [CrossRef]

24. Raider, S.; Flitsch, R.; Palmer, M. Oxide Growth on Etched Silicon in Air at Room Temperature. *J. Electrochem. Soc.* **1975**, *122*, 413–418. [CrossRef]

25. Lutterotti, L.; Chateigner, D.; Ferrari, S.; Ricote, J. Texture, Residual Stress and Structural Analysis of Thin Films Using a Combined X-Ray Analysis. *Thin Solid Films* **2004**, *450*, 34–41. [CrossRef]

26. Yu, T.; Zhu, Y.; Xu, X.; Shen, Z.; Chen, P.; Lim, C.T.; Thong, J.L.; Sow, C.H. Controlled Growth and Field-Emission Properties of Cobalt Oxide Nanowalls. *Adv. Mater.* **2005**, *17*, 1595–1599. [CrossRef]

27. Xie, X.; Chung, H.; Sow, C.; Wee, A. Oxide growth and its dielectrical properties on alkylsilated native-SiO_2/Si surface. *Chem. Phys. Lett.* **2004**, *388*, 446–451. [CrossRef]

28. Xu, Y.; Schoonen, M.A. The absolute energy positions of conduction and valence bands of selected semiconducting minerals. *Am. Mineral.* **2000**, *85*, 543. [CrossRef]

29. Grządziel, L.; Krzywiecki, M.; Peisert, H.; Chassé, T.; Szuber, J. Photoemission study of the Si(111)-native SiO_2/copper phthalocyanine (CuPc) ultra-thin film interface. *Org. Electron.* **2012**, *13*, 1873–1880. [CrossRef]

30. Qiao, L.; Xiao, H.Y.; Meyer, H.M.; Sun, J.N.; Rouleau, C.M.; Puretzky, A.A.; Geohegan, D.B.; Ivanov, I.N.; Yoon, M.; Weber, W.J.; Biegalski, M.D. Nature of the band gap and origin of the electro-/photo-activity of Co_3O_4. *J. Mater. Chem. C* **2013**, *1*, 4628–4633. [CrossRef]

31. Guo, W.; Chemelewski, W.D.; Mabayoje, O.; Xiao, P.; Zhang, Y.; Mullins, C.B. Synthesis and Characterization of CuV_2O_6 and $Cu_2V_2O_7$: Two Photoanode Candidates for Photoelectrochemical Water Oxidation. *J. Phys. Chem. C* **2015**, *119*, 27220–27227. [CrossRef]

32. Hsu, Y.K.; Yu, C.H.; Chen, Y.C.; Lin, Y.G. Synthesis of novel Cu_2O micro/nanostructural photocathode for solar water splitting. *Electrochim. Acta* **2013**, *105*, 62–68. [CrossRef]

33. Cardon, F.; Gomes, W.P. On the determination of the flat-band potential of a semiconductor in contact with a metal or an electrolyte from the Mott-Schottky plot. *J. Phys. D Appl. Phys.* **1978**, *11*, L63. [CrossRef]

34. Muñoz, A. Semiconducting properties of self-organized TiO_2 nanotubes. *Electrochim. Acta* **2007**, *52*, 4167–4176. [CrossRef]

35. Dunn, H.K.; Feckl, J.M.; Müller, A.; Fattakhova-Rohlfing, D.; Morehead, S.G.; Roos, J.; Peter, L.M.; Scheu, C.; Bein, T. Tin doping speeds up hole transfer during light-driven water oxidation at hematite photoanodes. *Phys. Chem. Chem. Phys.* **2014**, *16*, 24610–24620. [CrossRef] [PubMed]

36. Berger, T.; Monllor-Satoca, D.; Jankulovska, M.; Lana-Villarreal, T.; Gómez, R. The Electrochemistry of Nanostructured Titanium Dioxide Electrodes. *ChemPhysChem* **2012**, *13*, 2824–2875. [CrossRef]

37. Chuang, T.; Brundle, C.; Rice, D. Interpretation of the X-ray photoemission spectra of cobalt oxides and cobalt oxide surfaces. *Surf. Sci.* **1976**, *59*, 413–429. [CrossRef]

38. Dhara, K.; Mahapatra, D.R. Electrochemical nonenzymatic sensing of glucose using advanced nanomaterials. *Microchim. Acta* **2017**, *185*, 49. [CrossRef]

39. Zonno, I.; Martinez-Otero, A.; Hebig, J.C.; Kirchartz, T. Understanding Mott-Schottky Measurements under Illumination in Organic Bulk Heterojunction Solar Cells. *Phys. Rev. Appl.* **2017**, *7*, 034018. [CrossRef]

40. Zarean, M.; Keikha, M.; Poursafa, P.; Khalighinejad, P.; Amin, M.; Kelishadi, R. A systematic review on the adverse health effects of di-2-ethylhexyl phthalate. *Environ. Sci. Pollut. Res.* **2016**, *23*, 24642–24693. [CrossRef]

41. Biesinger, M.C.; Payne, B.P.; Grosvenor, A.P.; Lau, L.W.; Gerson, A.R.; Smart, R.S. Resolving surface chemical states in XPS analysis of first row transition metals, oxides and hydroxides: Cr, Mn, Fe, Co and Ni. *Appl. Surf. Sci.* **2011**, *257*, 2717–2730. [CrossRef]

42. Fan, S.; Zhao, M.; Ding, L.; Liang, J.; Chen, J.; Li, Y.; Chen, S. Synthesis of 3D hierarchical porous Co_3O_4 film by eggshell membrane for non-enzymatic glucose detection. *J. Electroanal. Chem.* **2016**, *775*, 52–57. [CrossRef]

43. Li, M.; Han, C.; Zhang, Y.; Bo, X.; Guo, L. Facile synthesis of ultrafine Co_3O_4 nanocrystals embedded carbon matrices with specific skeletal structures as efficient non-enzymatic glucose sensors. *Anal. Chim. Acta* **2015**, *861*, 25–35. [CrossRef] [PubMed]

44. Ding, Y.; Wang, Y.; Su, L.; Bellagamba, M.; Zhang, H.; Lei, Y. Electrospun Co_3O_4 nanofibers for sensitive and selective glucose detection. *Biosens. Bioelectron.* **2010**, *26*, 542–548. [CrossRef]

45. George, G.; Anandhan, S. A comparative study on the physico–chemical properties of sol–gel electrospun cobalt oxide nanofibres from two different polymeric binders. *RSC Adv.* **2015**, *5*, 81429–81437. [CrossRef]

46. Lopes, T.; Andrade, L.; Ribeiro, H.A.; Mendes, A. Characterization of photoelectrochemical cells for water splitting by electrochemical impedance spectroscopy. *Int. J. Hydrog. Energy* **2010**, *35*, 11601–11608. [CrossRef]

47. Xu, W.; Lu, Z.; Sun, X.; Jiang, L.; Duan, X. Superwetting Electrodes for Gas-Involving Electrocatalysis. *Acc. Chem. Res.* **2018**, *51*, 1590–1598. [CrossRef]

48. Wang, S.; He, T.; Yun, J.H.; Hu, Y.; Xiao, M.; Du, A.; Wang, L. New Iron-Cobalt Oxide Catalysts Promoting $BiVO_4$ Films for Photoelectrochemical Water Splitting. *Adv Funct. Mater.* **2018**, *28*, 1802685. [CrossRef]

Article

Potential Driven Non-Reactive Phase Transitions of Ordered Porphyrin Molecules on Iodine-Modified Au(100): An Electrochemical Scanning Tunneling Microscopy (EC-STM) Study

Tomasz Kosmala [1,2,*], Matías Blanco [1], Gaetano Granozzi [1] and Klaus Wandelt [2,3,*]

1   Department of Chemical Sciences and INSTM Unit, University of Padova, Via F. Marzolo 1,
    35131 Padova, Italy; matias.blancofernandez@unipd.it (M.B.); gaetano.granozzi@unipd.it (G.G.)
2   Institute of Physical and Theoretical Chemistry, Bonn University, Wegeler-Strasse 12,
    D-53115 Bonn, Germany
3   Institute of Experimental Physics, Plaza Maxa Borna 9, 50-204 Wroclaw, Poland
*   Correspondence: tomasz.kosmala@unipd.it (T.K.); k.wandelt@uni-bonn.de (K.W.)

Received: 3 July 2018; Accepted: 24 July 2018; Published: 25 July 2018

Abstract: The modelling of long-range ordered nanostructures is still a major issue for the scientific community. In this work, the self-assembly of redox-active tetra(N-methyl-4-pyridyl)-porphyrin cations (H_2TMPyP) on an iodine-modified Au(100) electrode surface has been studied by means of Cyclic Voltammetry (CV) and in-situ Electrochemical Scanning Tunneling Microscopy (EC-STM) with submolecular resolution. While the CV measurements enable conclusions about the charge state of the organic species, in particular, the potentio-dynamic in situ STM results provide new insights into the self-assembly phenomena at the solid-liquid interface. In this work, we concentrate on the regime of positive electrode potentials in which the adsorbed molecules are not reduced yet. In this potential regime, the spontaneous adsorption of the H_2TMPyP molecules on the anion precovered surface yields the formation of up to five different potential-dependent long-range ordered porphyrin phases. Potentio-dynamic STM measurements, as a function of the applied electrode potential, show that the existing ordered phases are the result of a combination of van der Waals and electrostatic interactions.

Keywords: porphyrins; self-assembly; surface nanostructures; in situ EC-STM; metal-electrolyte interface; potential-dependent structures; combined non-covalent control

---

1. Introduction

One of the major challenges in the development of supramolecular nanoarchitectures is to find effective and flexible methods in order to create nanostructures with long range order [1,2]. In recent years, supramolecular self-assembly attracted much attention by the scientific community: as a well-known bottom-up method, molecular self-assembly on surfaces is a simple and fast tool to build nanoscale structures, which can be readily tuned by the on-purpose synthesis of appropriate molecular building blocks, and, thereby, the design of the noncovalent interactions between the molecules and the substrate and the molecules themselves [3–7]. Moreover, to direct the ordering process of the supramolecular nanoarchitectures, one can take advantage of well-defined single crystal surfaces, which can serve as electronic and/or geometric templates due to the operation of specific adsorbate-substrate interactions. In this case, the resulting supramolecular structures may be dictated by the symmetry and periodicity of the substrate surface. In contrast to the assembly of supramolecular architectures in three-dimensional (3D) bulk environments, e.g., solutions, where assembly is governed only by specific and directional intermolecular interactions, a complex interplay between adsorbate-adsorbate and adsorbate-substrate interactions controls the two-dimensional (2D)-phase formation behavior when surfaces come into play [8–10]. Furthermore, if these molecular building-blocks are deposited electrochemically in the form of ions, the electrochemical (EC) potential becomes a very useful control parameter, which allows for influencing the self-assembly process [11–13]. In the end, the capability of many organic molecules to spontaneously self-assemble into molecular monolayers on suitable solid substrates may lead to functional surfaces with broad applications in electronic devices with nanometer dimensions, "green energy" catalysts, or novel light-stimulated sensors [14,15].

Among other examples, porphyrins, due to the wide range of potential applications of this class of molecules, e.g., in cancer therapy [16,17], catalysis [18,19], and sensing [20,21], have become prototypical systems for the fabrication and design of supramolecular nanoarchitectures and the characterization of their formation mechanism [22,23]. As an example from the point of applications, tetra (*N*-methyl-4-pyridyl)-porphyrin *p*-tolylsulfonate molecules (H$_2$TMPyP, Figure 1) were proven to be good sensors for the detection of benzene and heavy metal ions (Hg^{2+}, Pb^{2+}, Cd^{2+}) in water [24]. From self assembling studies in ultra-high-vacuum (UHV), it is known that adsorbed porphyrins on bare metal surfaces lie flat due to their large conjugated π-electron systems, which have the tendency to maximize the π-bonding with the surface [25]. In the EC environment, however, it is expected that electrostatic interactions of the porphyrin cations with the charged metallic electrode surface will also have a strong influence. Moreover, when compared to an adsorbed porphyrin layer in UHV, the electrode surface in contact with an electrolyte may be modified by the presence of specifically adsorbed anions from the supporting electrolyte. The influence of these preadsorbed anions on the deposition of organic cations depends obviously very much on their charge state. In the present work the surface is modified by iodide anions, which are known to be almost uncharged, so that, the iodine-porphyrin interactions are mainly dispersive in nature, i.e., of van der Waals type [26–28]. Actually, self-assembly of H$_2$TMPyP has been already investigated on various iodine modified metallic surfaces, such as Au(111) [26,27], Pt(100), Ag(111) [29,30], Cu(111), and Cu(100) [11,28]. To the best of our knowledge, studies on the self-assembly of porphyrins on the iodine precovered Au(100) surface have not been reported yet. Here, we present the results of combined Cyclic Voltammetry (CV) and in situ Scanning Tunneling Microscopy (STM) investigations. While the CV measurements enable conclusions about the charge state of the adsorbed molecules, in particular, potentio-dynamic STM measurements provide insight into their structural self-assembly on the surface. In this work, we concentrate on the "non-reactive" potential regime, i.e., the potential regime in which the molecules retain their oxidation state upon adsorption. Already in this restricted regime at positive potentials up to five different potential-dependent long-range ordered phases of the H$_2$TMPyP molecules are found on the iodine-modified Au(100) electrode. The different observed phases are thus not a consequence of the variation of the redox-state of the molecules themselves, but only of the electrode potential (and the concomitant change in iodine coverage). The potential induced modification of the non-covalent interactions, namely van der Waals and electrostatic forces between the assembled molecules and the substrate, is the only driving force for the observed phase transitions between the five porphyrin structures in their constant oxidized form ([H$_2$TMPyP(0)]$^{+4}$).

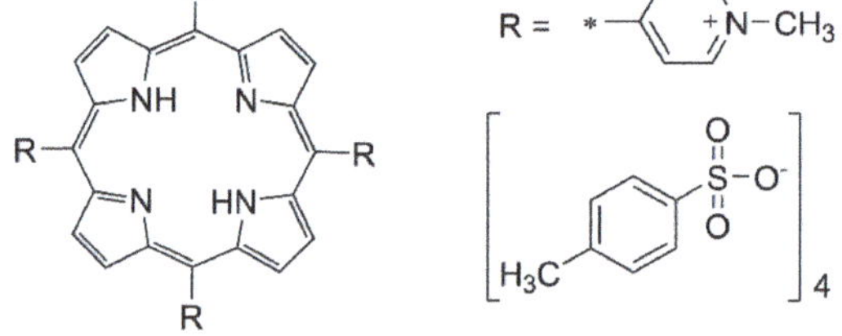

Figure 1. H$_2$TMPyP molecular structure.

2. Materials and Methods

The results that are presented in this work were carried out using a home build in-situ EC scanning tunneling microscope at constant current mode described in detail by Wilms et al. [31]. All of the electrolytes were prepared by using deionized water from a Millipore®-Pure (Merck, Burlington, MA, USA) water system (with a specific resistance of 18 MΩ·cm and a residual amount of organic impurities in the ppb regime) and purged with suprapure argon gas for several hours before use. Chemicals used (purchased from Sigma-Aldrich, Saint Louis, MO, USA) have the highest commercially available quality level and were used without further purification. The STM tips were electrochemically etched from a 0.25 mm tungsten wire in 2 M KOH solution and subsequently rinsed with high purity water, dried, and coated by passing the tip through a drop of hot glue.

The Au(100) electrode (MaTeck Company, Juelich, Germany) was annealed before each measurement in order to clean and smooth the surface. For this purpose, the sample was located on a Ceran® plate and was annealed by a butane-oxygen gas flame for three minutes up to faint red glow (600–700 °C). Subsequently, the crystal was cooled down to room temperature (about 15 min) in an atmosphere of argon. Then, it was heated again to a red glow and cooled for six minutes. To prevent the re-contamination of the surface after the cleaning procedure, the crystal was covered by a protective drop of Millipore®-water and immediately mounted into the EC cell of the EC-STM.

Cyclic Voltammetry (CV) and STM measurements were first performed in the pure supporting aqueous 5 mM H_2SO_4 + 1 mM KI electrolyte. In this electrolyte, the iodine-modified Au(100)-surface shows several phase transitions, starting from a $\left(\sqrt{2} \times p\sqrt{2}\right)$ at negative potentials, over a $\left(\sqrt{2} \times 2\sqrt{2}\right)$ to a *pseudo-hex-rot* structure at positive potentials. Both the $\left(\sqrt{2} \times p\sqrt{2}\right)$ and the *pseudo-hex-rot* structure exhibit the phenomenon of reversible electro-compression [32–34]. For the adsorption experiments the supporting electrolyte in the cell was replaced by a solution containing sulfate, iodide, and H_2TMPyP porphyrin molecules (5 mM H_2SO_4 + 1 mM KI + 0.01 mM H_2TMPyP). The potential of the gold electrode was controlled with respect to a Pt/PtI quasi-reference electrode, whose potential is related to that of the reversible hydrogen electrode (RHE) by the relation: RHE = Pt/PtI + 580 mV.

The base vectors that were employed for the description of the ordered structures of the various in the Results and Discussion section follow the common code:

- $\vec{a_x}$ for the base vector of the Au(100) lattice ($x = 1,2$);
- $\vec{b_x}$ and $\vec{b'_x}$ for the base vectors of the iodine lattice ($x = 1,2$); and,
- $\vec{c_x}, \vec{d_x}, \vec{e_x}, \vec{f_x}$ and $\vec{g_x}$ for the base vectors of the porphyrin phases P_I–P_V ($x = 1,2$).

3. Results and Discussion

3.1. Cyclic Voltammetry and H$_2$TMPyP Structures

The structural investigations of the organic layers that are presented in this work have exclusively been carried out under "non-reactive" conditions, i.e., in the regime where no redox-processes of the porphyrin molecules take place, as revealed by the survey cyclic voltammogram (CV) of the overall system shown in Figure 2a (see the red indicated region in Figure 2a). The first surface reduction/re-oxidation step of the molecules is indicated by the peak couple P_{red}/P_{ox}, as verified by the following EC-STM measurements. Also, the adsorption/desorption of the iodine anions is marked in the CV traces. As can be seen, the iodine desorption and the hydrogen evolution reaction (HER) are retarded due to the presence of the molecules. Hence, we focus and place emphasis exclusively on the structures of the [H$_2$TMPyP(0)]$^{4+}$ adlayer and their changes merely driven by changes of the electrode potential on the positive side of P_{red}/P_{ox}.

It is well known that iodide anions adsorb specifically on the Au(100) surface, and even low coverages stabilize the unreconstructed Au(100)(1 $\times$ 1) structure of the substrate. At high coverages the iodide anions form highly ordered layers, whose structure is potential dependent in terms of electro-compression or electro-expansion [32–34]. However, within the electrode potential range examined here, iodide forms a commensurate *pseudo-hex-rot* iodine structure on Au(100) [32–34], and our results indicate a *pseudo-hex-rot* iodine structure that is rotated by about 2.6° with respect to the substrate $\sqrt{2}$ direction (Figure 2b), and consequently exhibits a periodic variation of the iodine binding sites, resulting in the observed height, i.e., brightness, modulation. A ball model of this phase acquired at an electrode potential of −150 mV vs. Pt/PtI is shown in Figure 2c, and it includes the unit cell.

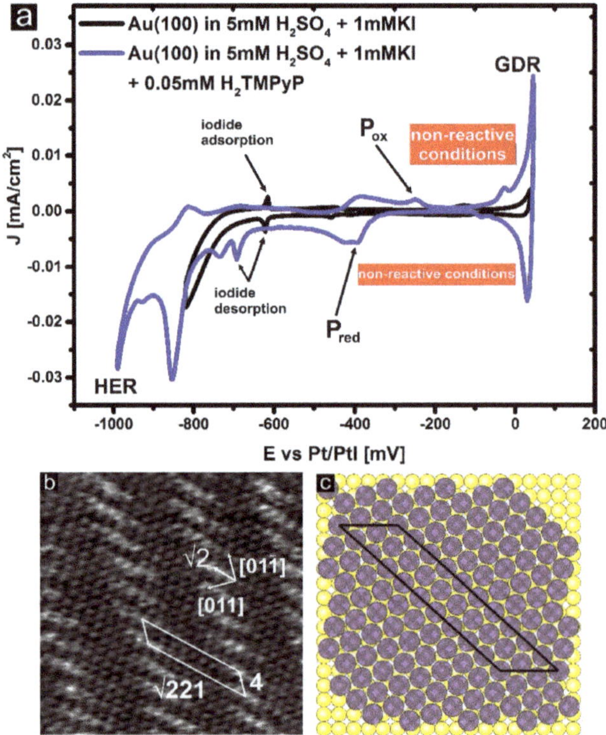

Figure 2. (**a**) Cyclic voltammograms of the Au(100) surface in 5 mM H_2SO_4 + 1 mM KI electrolyte solution (black curve) and in the H_2TMPyP containing electrolyte (blue curve, 5 mM H_2SO_4 + 1 mM KI + 0.05 mM H_2TMPyP), dE/dt = 10 mV·s^{-1}; GDR = gold dissolution reaction, HER = hydrogen evolution reaction, and the red bars indicate *non-reactive* regimes; (**b**) high-resolution in-situ Scanning Tunneling Microscopy (STM) image of the *pseudo-hex-rot* iodine structure obtained from a Au(100) surface in 5 mM H_2SO_4 + 1 mM KI solution; image parameters: 13.73 nm × 13.73 nm, I_t = 1 nA, U_b = −93 mV, E = −150 mV vs. Pt/PtI; (**c**) ball-model of the (4 × $\sqrt{221}$) iodine structure shown in (**b**).

Exposure of this iodine-modified Au(100) surface to the H_2TMPyP molecules containing 1 mM KI + 5 mM H_2SO_4 solution at potentials between the oxidative gold dissolution reaction (GDR) and the first reduction step of the porphyrin species (P_{red}) leads to spontaneous adsorption and the subsequent lateral ordering of H_2TMPyP molecules at the surface. The solution contains actually $[H_2TMPyP]^{4+}$ cations, but for the sake of ease we simply refer to H_2TMPyP molecules throughout this paper. The accompanying p-tolylsulfonate- as well as the SO_4^{2-}-anions from the supporting electrolyte adsorb weaker than iodide anions and therefore do not displace the latter [34].

Figure 3a presents a highly ordered layer of H_2TMPyP at an electrode potential -200 mV vs. Pt/PtI. Each bright dot represents one porphyrin molecule. Close-ups of this layer (Figure 3b–d) reveal that the molecules are self-assembled into differently ordered domains of molecular rows covering the atomically flat terraces. In addition, step edges (panel a) are oriented parallel to and decorated by molecular rows (see white and yellow lines in Figure 3a). A careful examination of Figure 3a–d clearly indicates the coexistence of five different porphyrin phases (P_I, P_{II}, P_{III}, P_{IV}, and P_V). Within each phase, translational domains (Figure 3b,c molecular rows of the domain highlighted by yellow solid lines) and the respective domain boundaries between them (Figure 3b,c yellow dashed lines) are observed.

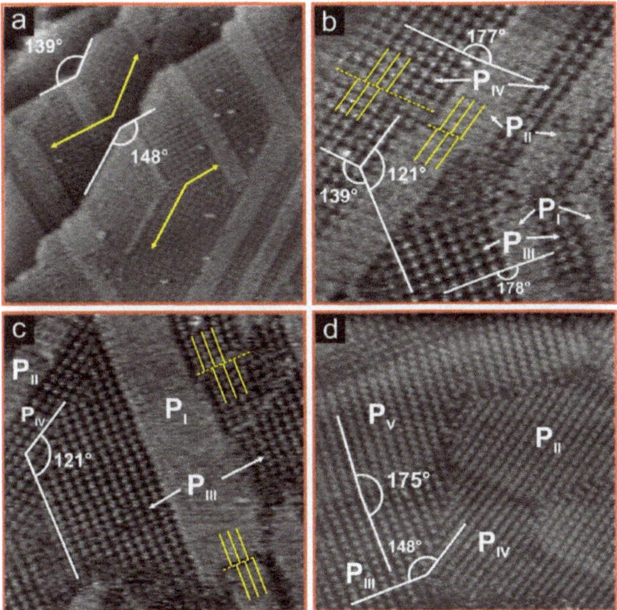

Figure 3. Electrochemical Scanning Tunneling Microscopy (EC-STM) images of self-assembled H_2TMPyP molecules on an iodine-modified Au(100) electrode in the *non-reactive* regime. P_I–P_V refer to structurally different phases. (**a**) 83.55 nm × 83.55 nm, $I_t = 1$ nA, $U_b = -320$ mV, E = -200 mV vs. Pt/PtI; (**b**) 42.35 nm × 42.35 nm, $I_t = 1$ nA, $U_b = -472$ mV, E = -200 mV vs. Pt/PtI; (**c**) 56.47 nm × 56.47 nm, $I_t = 1$ nA, $U_b = -491$ mV, E= -200 mV vs. Pt/PtI; (**d**) 56.47 nm × 56.47 nm, $I_t = 1$ nA, $U_b = -303$ mV, E = -200 mV vs. Pt/PtI.

In order to shed light on these porphyrin phases, STM images with atomic resolution were registered (Figure 4a–d). These images clearly reveal that the organic macrocycles are lying flat on the substrate due to their large molecular π-electron system and the tendency to maximize the π-bonding to the surface, as reported before [11,26–28]. The STM images show that each flat-lying porphyrin molecule can be recognized as a square-shaped motif with the characteristic four additional lobes placed at the four corners of each square (see inset and blue molecules in Figure 4a). The center-to-center distance measured diagonally across one molecule between the corner spots is 1.2 ± 0.1 nm, which is consistent with the distance between two diagonally located pyridinium units [26]. Furthermore, a detailed analysis of images, like in Figure 4, yields the characteristic lattice parameters, like vectors, rotation angles, molecules per unit cell, and surface coverage of the five different phases (P_I–P_V). These structural data are collected in Table 1.

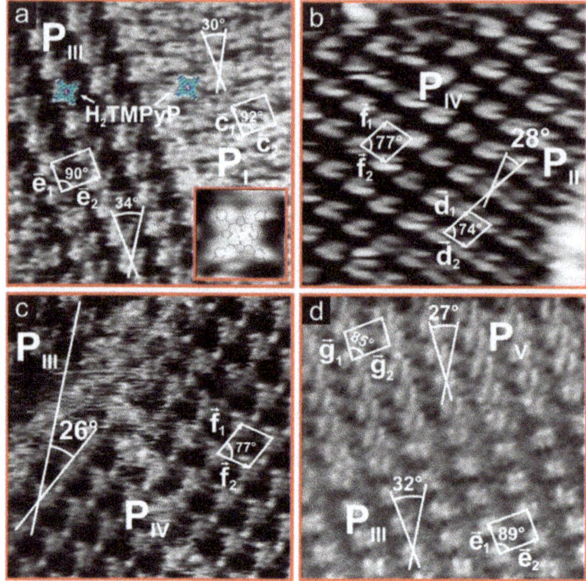

Figure 4. High-resolution EC-STM images of the ordered H_2TMPyP adlayer on an iodine-modified Au(100) electrode in the *non-reactive* regime. (**a**) 14.12 nm × 14.12 nm, I_t = 1 nA, U_b = −472 mV, E = −200 mV vs. Pt/PtI. Size of inset: 1.94 nm × 1.94 nm; (**b**) 14.15 nm × 14.15 nm, I_t = 1 nA, U_b = −291 mV, E = −200 mV vs. Pt/PtI; (**c**) 14.12 nm × 14.12 nm, I_t = 1 nA, U_b = −472 mV, E = −200 mV vs. Pt/PtI; (**d**) 14.12 nm × 14.12 nm, I_t = 1 nA, U_b = −303 mV, E= −200 mV vs. Pt/PtI.

Table 1. Structural data of the five different self-assembled, potential dependent H_2TMPyP structures P_I–P_V on an iodine-modified Au(100) electrode surface as detected by in situ STM in the non-reactive potential regime.

Parameter	P_I	P_{II}	P_{III}	P_{IV}	P_V
lattice constants [±0.05 nm]	c_1 = 1.61 c_2 = 1.80	d_1 = 1.94 d_2 = 1.56	e_1 = 1.61 e_2 = 2.06	f_1 = 1.94 f_2 = 1.76	g_1 = 1.46 g_2 = 2.06
angle [±1°]	92°	74°	89°	77°	85°
molecule rotation [α ±1°]	+30° vs. $\vec{c_1}$	−28° vs. $\vec{d_1}$	+32° vs. $\vec{e_1}$	+26° vs. $\vec{f_1}$	+27° vs. $\vec{g_1}$
surface concentration [molecules/cm²]	3.453×10^{13}	3.437×10^{13}	3.016×10^{13}	3.014×10^{13}	3.333×10^{13}

Based on these experimental results of the symmetry and orientation of the H_2TMPyP molecules, in Figure 5 we propose a schematic model of the ordered H_2TMPyP porphyrin molecules for each phase. It is important to emphasize that the value of the surface coverage of phase P_V lies in the middle of those of the other phases. Furthermore, P_V was observed rather rarely in comparison to phases P_I–P_{IV}: it only appears when the adsorption process started at an electrode potential of ca. −200 mV vs. Pt/PtI, and it vanishes altogether at those potential values where the phases P_I–P_{IV} were observed within more extended regimes of the electrode potential (*vide infra*). This suggests that P_V is less stable than P_I–P_{IV} and it is observable only under specific conditions in this very narrow electrode potential window, e.g., after adsorption at −200 mV vs. Pt/PtI and within a narrow range of electrode potential.

As mentioned previously, within the potential window investigated here, the porphyrin molecules do not undergo any redox process. Hence, the observed phase transitions are *not* triggered by a variation of the chemical state of the porphyrin due to an electron transfer process; but it can only be traced back to changes of the interactions between the porphyrin molecules and the substrate [11,35,36].

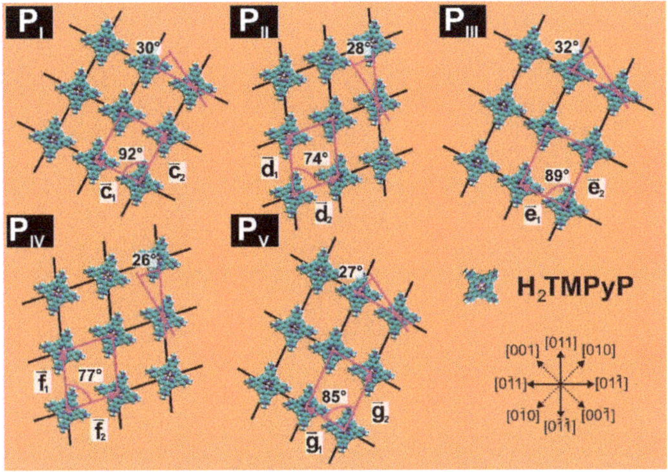

Figure 5. Schematic models of the five ordered H$_2$TMPyP molecular phases in the *non-reactive* regime on I/Au(100), with the denoted vectors of the unit cells, the angles between them, and the angle of rotation of individual molecules off the molecular row direction (see Table 1).

3.2. Electrode Potential Stability of H$_2$TMPyP Phases on I/Au(100)

In order to check the stability and existence of phases P$_I$–P$_V$ as a function of potential, and whether there is a competition (i.e., a difference in stability) between them expressed by their different equilibrium coverages, we also performed potentio-dynamic STM studies. These potentio-dynamic STM measurements were started at −200 mV vs. Pt/PtI (see Figure 2) with a highly ordered porphyrin layer comprising the first four phases P$_I$–P$_{IV}$, as shown in Figure 6a). The following images in panels b–i of Figure 6 are a selection out of a series of 170 successive images and are taken at the indicated potentials. They are largely registered at the same surface position (indicated by the red arrow in Figure 6a–i). However, due to thermal drift this marker slowly drifted out of the scanned area so that the last three images 6j–l were recorded around a new point of the same area (marked with a light blue arrow). From the starting point (Figure 6a reflects the very initial state, which was allowed to equilibrated for 210 s before the series of 170 images was started with Figure 6b), the electrode potential was first scanned in the cathodic direction. The areas covered by P$_I$ and P$_{II}$ are observed to shrink while those of P$_{III}$ and P$_{IV}$ start to grow. When the potential reaches −300 mV vs. Pt/PtI, within less than 5 min, the phases P$_I$ and P$_{II}$ are finally completely replaced by P$_{III}$ and P$_{IV}$ (see Figure 6c,d). Moreover, new bright spots appear in the images (see white arrows in Figure 6c), which represent the growth of a completely new ordered phase. A height profile measured along the short orange line in Figure 6d and shown in the inset yields a height difference between the bright spots and the dark region between them of around 0.3 nm. This value approximately matches the thickness of one flat lying porphyrin molecule [37].

These observations suggest that, when approaching the reactive regime (below −300 mV vs. Pt/PtI), the new bright spots represent adsorption of molecules in the second layer. This is supported by lowering of the electrode potential even further to −300 mV and −400 mV, which leads to an increase of the bi-layer coverage and the creation of new phases under these reactive conditions, named P$_{VI}$ and P$_{VII}$ (Figure 6d–f). Both monolayer phases P$_{III}$ and P$_{IV}$ phases are finally completely replaced by the bi-layer when the potential reaches the maximum of peak P$_{red}$ at −395 mV vs. Pt/PtI, which represents the first reduction step of the porphyrin molecules. Thus, the appearance of the bi-layer in the STM images in coincidence with the first reduction peak in the CV clearly marks the cathodic limit of the *non-reactive* monolayer adsorption regime. The full analysis of the EC-STM data under reactive conditions will be the subject of a forthcoming paper. Here, we concentrate on changes on the surface upon sweeping the potential back into the positive direction. No changes where noticed until the electrode potential reached again -300 mV vs. Pt/PtI. At this potential, re-entering in the *non-reactive* regime, a disordering of the lateral structure at the domain boundaries as well as desorption of second layer porphyrin molecules starts to take place (see yellow arrows in Figure 6g). At the potential -250 mV vs. Pt/PtI (corresponding to the anodic peak P_{ox}) the second layer has almost completely disappeared, and at a potential of ca. -200 mV vs. Pt/PtI, represented by Figure 6i, phases P_I and P_{II} have reappeared on the surface. Changing the electrode potential even further in anodic direction leads to growth of P_I and P_{II} at the expense of P_{III} and P_{IV} (see Figure 6j,k), and finally a complete replacement of the latter two phases. Moreover, at a potential of -100 mV vs. Pt/PtI the GDR starts at the step-edges as revealed by the course of the step edge in Figure 6k. After reaching the electrode potential of -50 mV vs. Pt/PtI where the GDR proceeds very quickly (see Figure 6l), P_{II} completely disappears from the surface, while P_I is still clearly observable.

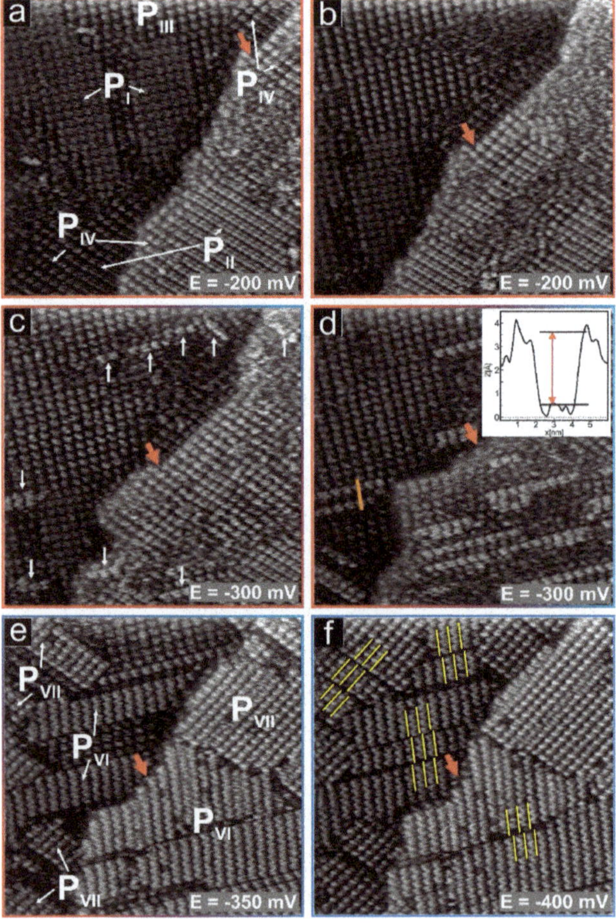

Figure 6. *Cont.*

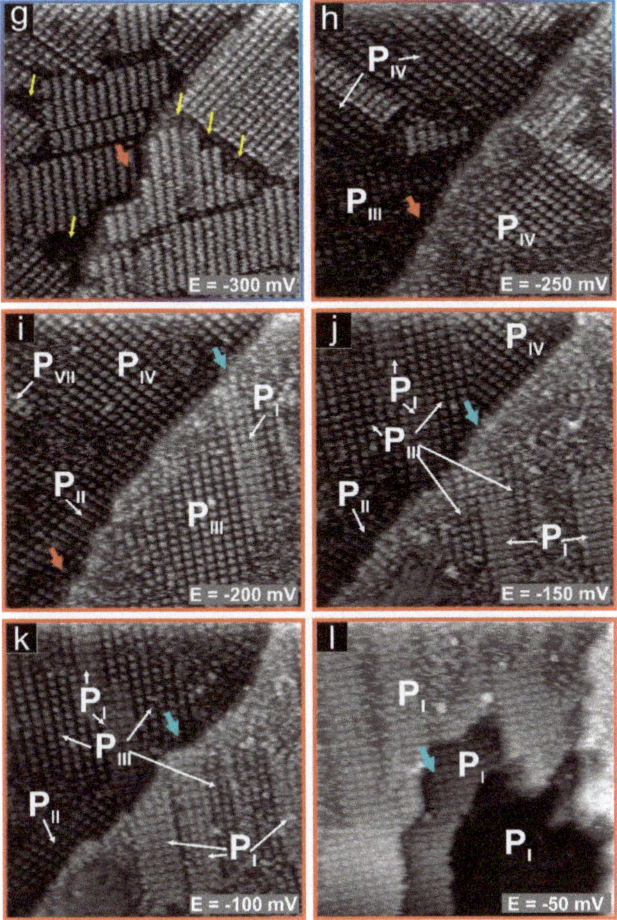

Figure 6. Potential induced phase transition of molecular H_2TMPyP adlayer on an I/Au(100) electrode surface, STM series: 50.32 nm × 50.32 nm; (**a**) I_t = 1 nA, U_b = −291 mV; (**b**) I_t = 1 nA, U_b = −267 mV; (**c**) I_t = 1 nA, U_b = −267 mV; (**d**) I_t = 1 nA, U_b = −267 mV; (**e**) I_t = 1 nA, U_b = −208 mV; (**f**) I_t = 1 nA, U_b = −208 Mv; (**g**) I_t = 1 nA, U_b = −208 mV; (**h**) I_t = 1 nA, U_b = −208 mV; (**i**) I_t = 1 nA, U_b = −250 mV; (**j**) I_t = 1 nA, U_b = −477 mV; (**k**) I_t = 1 nA, U_b = −461 mV; (**l**) I_t = 1 nA, U_b = −461 mV. For explanation of the different colored arrows see the text.

Figure 7 represents a diagram that indicates the stable potential and coexistence regimes of the five phases P_I–P_V. The above potentio-dynamic STM measurements reveal a competition between these phases, i.e., their relative stability as a function of electrode potential, as manifested by their different surface concentrations. Furthermore, from a correlation of the structural changes with the relative coverage of each phase, we arrive at a strict relationship between the potential regime of each phase and its surface concentration: according to Table 1, P_I has the highest surface density (3.453×10^{13} molecules/cm^2) and its existence regime is at the most positive potentials in comparison to all other phases. Next, P_{II} with a lower density (3.437×10^{13} molecules/cm^2) than P_I occurs at lower potentials. Finally, P_{III} and P_{IV} with the lowest surface concentrations among all five monolayer phases (of 3.016×10^{13} molecules/cm^2 and 3.006×10^{13} molecules/cm^2, respectively) have their existence regimes at the most negative potentials.

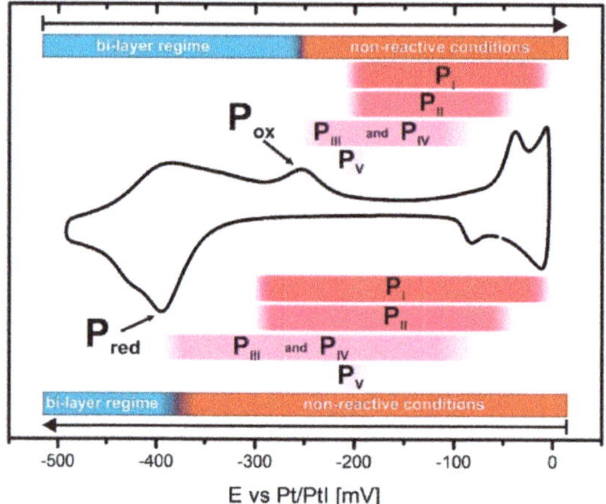

Figure 7. Diagram of the stability regimes of the P$_I$–P$_V$ monolayer phases of adsorbed H$_2$TMPyP on an iodine-modified Au(100) electrode surface.

Therefore, in order to get more insights and to understand the relations between the applied electrode potential and the different surface concentration of each porphyrin phase, a set of STM images was taken, which provides a direct correlation between the structure of the different organic phases and that of the underlying iodine and gold lattice.

3.3. Substrate-Adlayer Structure Correlations

3.3.1. Phases P$_I$, P$_{III}$

Owing to the overlap of the existence regimes, it is possible to record a set of STM images that comprise both phases P$_I$ and P$_{III}$ (14.12 nm × 14.12 nm, Figure 8a,b and Figure 8d,e) within the same surface area. It is well known that by changing the tunneling conditions (namely tunneling bias) a resonant tunneling may be achieved [38,39]. At high bias voltage (e.g., U$_b$ = −416 mV in Figure 8a, and U$_b$ = −472 mV in the lower half part of Figure 8d), almost all electrons flow via resonant tunneling through quantized energy levels that are defined by the molecular orbitals or the band generated by the periodic arrangement of the molecules. Thus, tunneling through these states emphasizes the contribution of the organic molecules in the respective STM images. On the other hand, at lower bias voltage (e.g., U$_b$ = −9 mV in the upper-left part of Figure 8d and left part of Figure 8e), the normal tunneling process dominates the tunneling current, where electrons tunnel through the potential well without interacting with the localized discrete molecular levels. Most importantly, in some cases at an intermediate bias voltage it is possible that both tunneling channels through the potential well and the localized discrete levels contribute to the image. Thus, images that are registered at this intermediate bias voltage comprise both types of features that can be associated either with the substrate or the adsorbate (see e.g., Figure 8b at U$_b$ = −13 mV, and its Reverse Fast Fourier Transform (RFFT) in Figure 8c in which we can observe a signal from the porphyrin molecules as well as from the iodine lattice; and the upper-right part of Figure 8d and right part of Figure 8e at U$_b$ = −9 mV, with the RFFT presented in Figure 8f performed on the right part of Figure 8e, in which contributions from gold and the porphyrin lattice are observed). Such images enable a direct correlation between the structure of the organic overlayer and that of the underlying iodine and the supporting gold lattice. Moreover, the *simultaneous* observation of features from both the molecules on the one hand, and iodine or gold underneath on the other hand, excludes that the observed bias dependence of the images is due to a physical removal of the molecules from the substrate by the tip at low bias voltages (Figure 8b,e).

This approach circumvents any problem with drift during image taking because both contributions, that from the molecules and that from the substrate, are included in the same image.

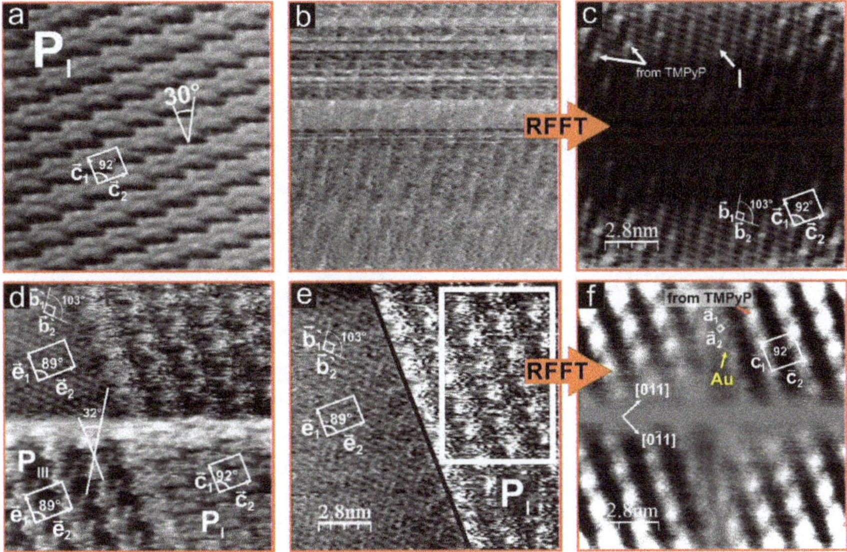

Figure 8. Correlation of the structures of the P_I and P_{III} monolayer phases of adsorbed H_2TMPyP molecules with the underlying iodine and gold layer, (**a**) P_I porphyrin layer at high tunneling bias, 14.12 nm × 14.12 nm, I_t = 1 nA, U_b = −416 mV, E = −200 mV vs. Pt/PtI; (**b**) Iodine layer underneath P_I at low tunneling bias, 14.12 nm × 14.12 nm, I_t = 1 nA, U_b = −13 mV, E = −200 mV vs. Pt/PtI; (**c**) a Reverse Fast Fourier Transform (RFFT) of the STM image (**b**); (**d**) upper part: Iodine layer underneath P_{III} at low tunneling bias: U_b = −9 mV, lower part: P_{III} porphyrin layer at high tunneling bias: U_b = −472 mV, 14.12 nm × 14.12 nm, I_t = 1 nA, E = −200 mV vs. Pt/PtI; (**e**) left part: Iodine layer underneath P_I at low tunneling bias: U_b = −9 mV, right part: combination of the Porphyrin layer and the substrate layer at low tunneling bias: U_b = −9 mV, 14.12 nm × 14.12 nm, I_t = 1 nA, E = −200 mV vs. Pt/PtI; (**f**) a Reverse Fast Fourier Transform (IFFT) of the left part of STM image (**e**).

Therefore, after close analysis and superimposition of the signals that are described above from the porphyrin-, iodine-, and Au-structure presented in Figure 8, it is possible to relate the structure of the molecular phases P_I and P_{III} to those of the iodine and the gold lattice underneath (see Figure 9). The iodine layer under both molecular phases was found to be a *pseudo-hex-rot* phase, but when considering that this iodine phase is electro-compressible (the iodine lattice becomes more compressed with increasing electrode potential), a differently compressed *pseudo-hex-rot* iodine phase should exist, and was actually found, under the two H_2TMPyP phases. With the set of images from Figure 8 in hand, it is thus possible to devise structure models of both phases P_I and P_{III}, and the corresponding data are summarized in Table 2.

Starting with P_I (Figure 9), we detect that in the direction of the vectors $\overline{b_1}$ every nineteenth and in the direction of the vector $\overline{b_2}$ every seventh iodine atom in the respective iodine row occupies a site on top of a gold atom. On the other hand, the molecular rows of the H_2TMPyP layer of P_I are not aligned with any high symmetry direction of the substrate, neither of the iodine lattice nor of the Au(100) surface. However, when considering the symmetry axis of individual molecules that are centred atop of an iodine atom, we observe that this axis is rotated by 30° off the row direction (see Table 1) and coincides with close-packed iodine rows running in the [0 7 35] direction of the substrate, respectively.

On the basis of this model the surface coverage of the porphyrin adlayer is calculated to be 0.0286 monolayers (ML) relative to the density of the *gold* layer.

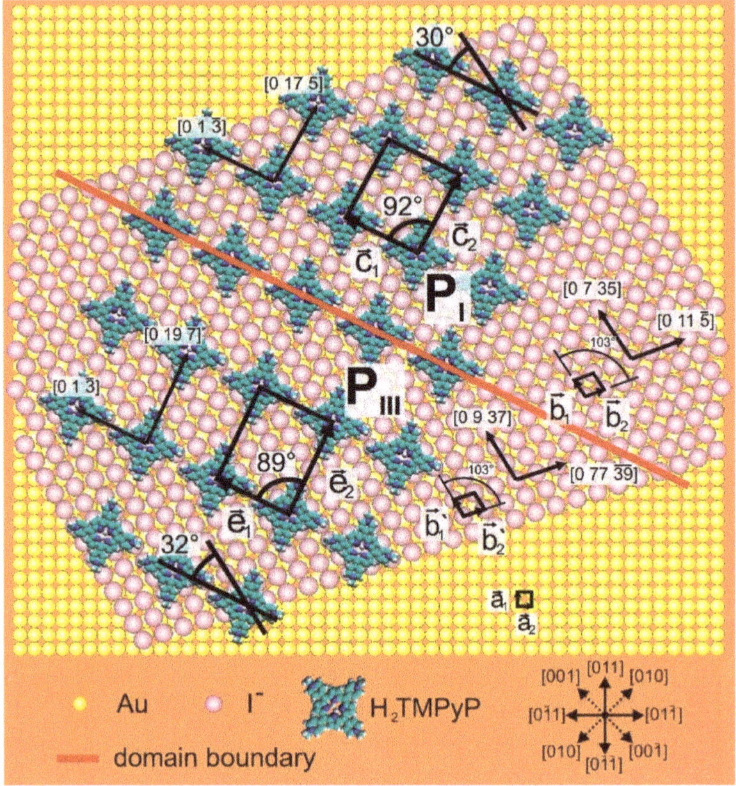

Figure 9. Structure models of the phases P_I and P_{III} on an iodine modified Au(100) surface.

In turn, for phase P_{III}, the iodine rows follow the direction of the vectors $\overline{b_1'}$ and $\overline{b_2'}$ instead, which in these directions leads to periodic atop binding sites of an iodine atom on a gold atom every twenty-first and forty-first iodine, respectively. The symmetry axis of individual molecules is rotated by 32° off the row direction (see Table 1), which coincides with the [0 9 37] substrate direction in this case. As a consequence, the iodine coverage under P_{III} (0.250 ML) is slightly lower than that under P_I (0.0286 ML). This lower iodine coverage is a consequence of the electro-expansion of the *pseudo-hex-rot* phase with decreasing electrode potential (0.52 ML under P_I vs. 0.50 ML under P_{III}, see Table 2).

Table 2. Data correlating the structures of the phases P_I, P_{III} and P_{IV} of self-assembled H_2TMPyP molecules on an iodine modified Au(100) electrode with those of the respective iodine layer and the Au(100) substrate underneath.

Parameter		P_I	P_{III}	P_{IV}
Au lattice parameters	vectors direction		$\overline{a_1} = [0\,1\,1]$ $\overline{a_2} = [0\,1\,\overline{1}]$	
	lattice constant		$a_1 = 0.289$ nm $a_2 = 0.289$ nm	
Iodine lattice parameters	vectors direction	$\overline{b_1} = [0\,7\,35]$ $\overline{b_2} = [0\,11\,\overline{5}]$	$\overline{b'_1} = [0\,9\,37]$ $\overline{b'_2} = [0\,77\,\overline{39}]$	$\overline{b'_1} = [0\,9\,37]$ $\overline{b'_2} = [0\,77\,\overline{39}]$
	lattice constant	$b_1 = 0.404$ nm $b_2 = 0.410$ nm	$b'_1 = 0.388$ nm $b'_2 = 0.439$ nm	$b'_1 = 0.388$ nm $b'_2 = 0.439$ nm
	Θ^*	0.52 ML	0.50 ML	0.50 ML
	Matrix with respect to Au(100)	$\begin{pmatrix} \vec{b_1} \\ \vec{b_2} \end{pmatrix} = \begin{pmatrix} 3 & -2 \\ 3 & 8 \end{pmatrix} \begin{pmatrix} \vec{a_1} \\ \vec{a_2} \end{pmatrix}$	$\begin{pmatrix} \vec{b'_1} \\ \vec{b'_2} \end{pmatrix} = \begin{pmatrix} 11.5 & -7 \\ 19 & 58 \end{pmatrix} \begin{pmatrix} \vec{a_1} \\ \vec{a_2} \end{pmatrix}$	$\begin{pmatrix} \vec{b'_1} \\ \vec{b'_2} \end{pmatrix} = \begin{pmatrix} 11.5 & -7 \\ 19 & 58 \end{pmatrix} \begin{pmatrix} \vec{a_1} \\ \vec{a_2} \end{pmatrix}$
Porphyrins lattice parameters	vectors direction	$\overline{c_1} = [0\,1\,\overline{3}]$ $\overline{c_2} = [0\,17\,5]$	$\overline{e_1} = [0\,1\,\overline{3}]$ $\overline{e_2} = [0\,19\,7]$	$\overline{f_1} = [0\,61\,73]$ $\overline{f_2} = [0\,77\,\overline{39}]$
	Θ^*	0.0286 ML	0.0250 ML	0.0250 ML
	Matrix with respect to iodine	$\begin{pmatrix} \vec{c_1} \\ \vec{c_2} \end{pmatrix} = \begin{pmatrix} 3 & -2 \\ 3 & 4 \end{pmatrix} \begin{pmatrix} \vec{b_1} \\ \vec{b_2} \end{pmatrix}$	$\begin{pmatrix} \vec{e_1} \\ \vec{e_2} \end{pmatrix} = \begin{pmatrix} 3 & -2 \\ 4 & 4 \end{pmatrix} \begin{pmatrix} \vec{b'_1} \\ \vec{b'_2} \end{pmatrix}$	$\begin{pmatrix} \vec{f_1} \\ f_2 \end{pmatrix} = \begin{pmatrix} 5 & 2 \\ 0 & 4 \end{pmatrix} \begin{pmatrix} \vec{b'_1} \\ \vec{b'_2} \end{pmatrix}$
	Matrix with respect to Au(100)	$\begin{pmatrix} \vec{c_1} \\ \vec{c_2} \end{pmatrix} = \begin{pmatrix} 2.5 & -5 \\ 5.5 & 3 \end{pmatrix} \begin{pmatrix} \vec{a_1} \\ \vec{a_2} \end{pmatrix}$	$\begin{pmatrix} \vec{e_1} \\ \vec{e_2} \end{pmatrix} = \begin{pmatrix} 2.5 & -5 \\ 6.5 & 3 \end{pmatrix} \begin{pmatrix} \vec{a_1} \\ \vec{a_2} \end{pmatrix}$	$\begin{pmatrix} 5\vec{f_1} \\ 5\vec{f_2} \end{pmatrix} = \begin{pmatrix} 33.5 & -3 \\ 9.5 & 29 \end{pmatrix} \begin{pmatrix} \vec{a_1} \\ \vec{a_2} \end{pmatrix}$

* Surface coverage relative to the density of the gold layer.

Even though the determined differences in coverage of P_I and P_{III}, and, even more so, for the respective underlying iodine structures, may appear small, there is no doubt that they are real, because they are obtained from the coexisting phases in the same image of the same measurement.

Thus, P_I with a monolayer coverage of 0.0286 ML is situated on an iodine layer of coverage 0.52 ML, while P_{III} of coverage 0.0250 ML is found on an iodine underlayer of coverage 0.50 ML at lower potentials. In both phases, the direction of closest-packed molecular rows do not coincide with any high symmetry direction of neither the iodine nor the gold lattice underneath, but all of the molecules are centred on an iodine atom, and a symmetry axis of the molecules is aligned with the direction of a close-packed iodine row. Since both phases, P_I and P_{III}, occur within the *non-reactive* potential regime, and thus, retain their charge state, only differences in density and atomic arrangement of the iodine underlayer can be taken responsible for the differences between P_I and P_{III}, namely their different lattice parameters, and, in particular, their different atomic density.

Two arguments can be put forward to explain the higher coverage of P_I when compared to P_{III}. Firstly, the denser iodine anion layer underneath P_I (at higher potential) may correspond to a higher density of negative charge, which attracts more molecular cations. This argument, however, may not be valid because among the halides iodine binds most covalently, in particular the higher its coverage (at higher potentials) is. Iodine layers on noble metals, like Au or Pt, were found to be hydrophobic and almost uncharged [26]. The same conclusion was drawn from studies with iodine adsorption on Cu(100) [40,41]. Secondly, the denser the iodine layer the more effectively it screens the positive charge of the metal substrate, and thereby, reduces the electrostatic repulsion between the electrode and the molecular cations, and, instead, relatively strengthens van-der-Waals interactions between the molecules and the covalently bound iodine layer. Even though this dominance of van-der-Waals interactions has been favoured in the literature, our results indicate a combination of van-der-Waals and electrostatic forces. While the simultaneous increase of iodine- and H_2TMPyP-coverage with increasing electrode potential supports the notion of dominant van-der-Waals bonding of the molecules to the surface, the match of the symmetry direction of the individual molecules with the nearly rectangular iodine lattice underneath points to a contribution of electrostatic interactions, namely interlayer attraction between the positively charged *N*-methyl-4-pyridyl-ligands and the iodine lattice underneath, and intralayer repulsion between these ligands of the neighbouring molecules. Therefore, these latter electrostatic forces are actually responsible for the lateral order formation.

3.3.2. Phases P_{III}, P_{IV}

A similar comparative analysis can be done with phases P_{III} and P_{IV} based on images, as shown in Figure 10. According to Table 1, these two phases have essentially equal surface coverage and they are stable in the same potential range, as revealed by the potentio-dynamic STM measurements shown in Figure 6. In order to achieve a structural correlation between the molecular adlayer and the substrate underneath, we performed again bias dependent STM measurements (Figure 10a–c). As listed in Table 2, the *pseudo-hex-rot* structure of the iodine structure on the substrate is the same under both phases, and is described by the vectors $\overline{b}'_1$ and $\overline{b}'_2$ with directions of $[0\,9\,37]$ and $[0\,77\,\overline{39}]$, respectively. In these directions, every twenty-first and forty-first iodine atom, respectively, occupies an atop binding site on a gold atom, and the iodine coverage is 0.50 ML (see Table 2). While the structural correlation of phase P_{III} with the iodine lattice underneath has already been described in the previous section, molecular rows of phase P_{IV} described by $\overline{f}_1$ and $\overline{f}_2$ with directions $[0\,61\,73]$ and $[0\,77\,\overline{39}]$, respectively, run parallel to the iodine $\overline{b}'_2$ direction, but, interestingly, have the same orientation of the molecular symmetry axis as in P_{III} (coinciding with the $[0\,9\,37]$ direction of close packed iodine rows), as shown in Figure 11a. Therefore, if both unit cells of P_{III} (blue) and P_{IV} (red) are superimposed (as shown in Figure 11b) a new coincidence lattice $(\overline{S}_1,\overline{S}_2)$ (green in Figure 11b) becomes apparent. The vector $\overline{S}_1$ of this structure is aligned with the densely packed iodine rows in the direction $\begin{bmatrix} 0 & 9 & 37 \end{bmatrix}$. The new (green) unit cell contains four molecules from P_{III} and four molecules from P_{IV}, plus one shared additional molecule (the green atom can be either blue or red in the model) every eleventh iodine atom underneath. The fact that both phases P_{III} and P_{IV} have the same coverage (0.025 ML), means that both phases adsorb with equal probability on the surface at this potential regime.

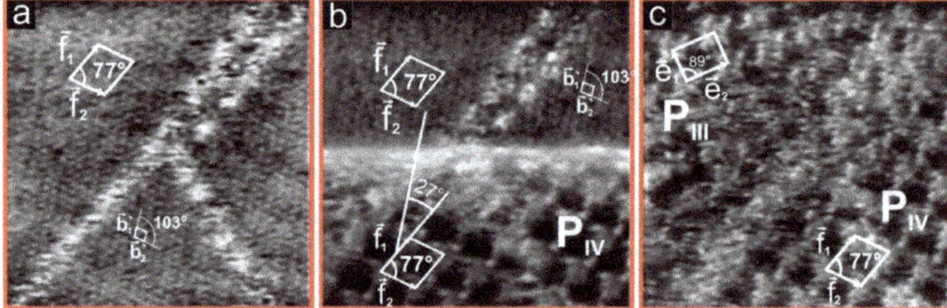

Figure 10. Structure correlation between the H$_2$TMPyP layer of P_{III} and P_{IV} and the underlying iodine and gold layer, (**a**) Iodine layer underneath at low tunneling bias, 14.12 nm × 14.12 nm, I_t = 1 nA, U_b = −17 mV, E = −200 mV vs. Pt/PtI; (**b**) upper part: Iodine layer underneath at low tunneling bias: U_b = −17 mV, lower part: Porphyrin layer at high tunneling bias: U_b = −472 mV, 14.12 nm × 14.12 nm, I_t = 1 nA, E = −200 mV vs. Pt/PtI; (**c**) Porphyrin layer at high tunneling bias, 14.12 nm × 14.12 nm, I_t = 1 nA, U_b = −472 mV, E = −200 mV vs. Pt/PtI.

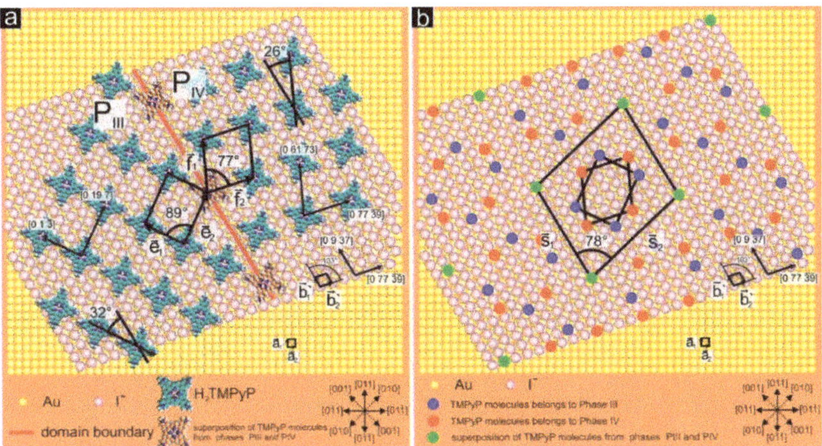

Figure 11. Structural model of P$_{III}$ and P$_{IV}$ on iodine modified Au(100) surface (**a**); structure model of superstructure created by superimposing of Phase III and IV (**b**).

4. Conclusions

In this study, we have investigated in situ the self-assembly of H$_2$TMPyP molecules on an iodine modified Au(100) electrode surface by EC-STM methods exclusively in the *non-reactive* regime, i.e., in the potential regime in which the molecules retain their oxidized [H$_2$TMPyP(0)]$^{4+}$ state throughout. The exposure of the *pseudo-hex-rot* iodine modified Au(100) surface to the porphyrin containing solution results in the spontaneous adsorption and lateral ordering of the molecules on the surface, with the organic macrocycles lying flat on the substrate. Up to five different porphyrin phases (P$_I$–P$_V$) were detected on the surface of the substrate, whose stabilities, surface concentrations, and existence regimes were demonstrated yet to be potential dependent. In fact, potentio-dynamic STM measurements revealed that P$_I$ and P$_{II}$, stable structures at more positive potentials, transform into phases P$_{III}$ and P$_{IV}$ by applying more negative electrode potential, while P$_V$ was found to be stable and detected in a very narrow potential window only. Therefore, there is a competition between the different phases, resulting in a strict relationship between their respective surface concentrations and their existence regimes as a function of the applied electrode potential.

In addition, by applying adequate imaging conditions, i.e., bias voltages, STM images were obtained, which enable a correlation between the structures of the porphyrin adlayer and of the underlying iodine and gold lattices. The lattice parameters regarding the relative orientation of the iodine underlayer and the organic overlayer for P$_I$ (at more positive potentials) on the one hand, and for P$_{III}$ and P$_{IV}$ (at more negative potentials) on the other hand were analysed. The results reveal that the observed phase transitions are accompanied by a change of the underlying electro-compressible iodine lattice, in particular, its density, as a function of the electrode potential. Since iodide anions are largely discharged upon adsorption and essentially bound covalently as neutral iodine atoms, it has been suggested in the literature [27,29,30] that porphyrin molecules on iodine modified metal surfaces are predominantly controlled by van der Waals forces. However, our results about potential dependent lattice coincidences and molecular orientations strongly support the notion that not only the van-der-Waals forces are responsible for the self-assembly of the porphyrin molecules on the polarized substrate, but that also electrostatic interactions between the molecules and the iodine modified metal substrate, as well as between the molecules themselves, play a decisive role in the 2D ordering process.

Author Contributions: All authors have made a substantial contribution to the work, and approved it for publication.

Funding: This work has in part been funded by the collaborative research centre 624 of the German Science Foundation.

Acknowledgments: M.B. gratefully acknowledges the Clarin CoFund Program postdoctoral fellowship (ACA17-29), funded by Gobierno del Principado de Asturias and Marie Curie Actions (grant 600196).

Conflicts of Interest: The authors declare no conflict of interest.

References

1. Madueno, R.; Räisänen, M.T.; Silien, C.; Buck, M. Functionalizing hydrogen-bonded surface networks with self-assembled monolayers. *Nature* **2008**, *454*, 618–621. [CrossRef] [PubMed]
2. Li, Z.; Wandlowski, T. Structure formation and annealing of isophthalic acid at the electrochemical Au(111)-electrolyte interface. *J. Phys. Chem. C* **2009**, *113*, 7821–7825. [CrossRef]
3. Barth, J.V.; Weckesser, J.; Lin, N.; Dmitriev, A.; Kern, K. Supramolecular architectures and nanostructures at metal surfaces. *Appl. Phys. A Mater. Sci. Process.* **2003**, *76*, 645–652. [CrossRef]
4. Dmitriev, A.; Lin, N.; Weckesser, J.; Barth, J.V.; Kern, K. Supramolecular assemblies of trimesic acid on a Cu(100) surface. *J. Phys. Chem. B* **2002**, *106*, 6907–6912. [CrossRef]
5. Whitesides, G.M. Self-Assembly at All Scales. *Science* **2002**, *295*, 2418–2421. [CrossRef] [PubMed]
6. Desiraju, G.R. Chemistry beyond the molecule. *Nature* **2001**, *412*, 397–400. [CrossRef] [PubMed]
7. Sosa-Vargas, L.; Kim, E.; Attias, A.J. Beyond "decorative" 2D supramolecular self-assembly: Strategies towards functional surfaces for nanotechnology. *Mater. Horiz.* **2017**, *4*, 570–583. [CrossRef]
8. Safarowsky, C.; Rang, A.; Schalley, C.A.; Wandelt, K.; Broekmann, P. Formation of 2D supramolecular architectures at electrochemical solid/liquid interfaces. *Electrochim. Acta* **2005**, *50*, 4257–4268. [CrossRef]
9. Ciesielski, A.; Palma, C.A.; Bonini, M.; Samori, P. Towards supramolecular engineering of functional nanomaterials: Pre-programming multi-component 2D self-assembly at solid-liquid interfaces. *Adv. Mater.* **2010**, *22*, 3506–3520. [CrossRef] [PubMed]
10. Gottfried, J.M. Surface chemistry of porphyrins and phthalocyanines. *Surf. Sci. Rep.* **2015**, *70*, 259–379. [CrossRef]
11. Hai, N.T.M.; Gasparovic, B.; Wandelt, K.; Broekmann, P. Phase transition in ordered porphyrin layers on iodide modified Cu(111): An EC-STM study. *Surf. Sci.* **2007**, *601*, 2597–2602. [CrossRef]
12. Phan, T.H.; Kosmala, T.; Wandelt, K. Potential dependence of self-assembled porphyrin layers on a Cu(111) electrode surface: In-Situ STM study. *Surf. Sci.* **2015**, *631*, 207–212. [CrossRef]
13. Madry, B.; Morawski, I.; Kosmala, T.; Wandelt, K.; Nowicki, M. Porphyrin Layers at Cu/Au(111)–Electrolyte Interfaces: In Situ EC-STM Study. *Top. Catal.* **2018**, 1–15. [CrossRef]
14. Barth, J.V.; Costantini, G.; Kern, K. Engineering atomic and molecular nanostructures at surfaces. *Nature* **2005**, *437*, 671–679. [CrossRef] [PubMed]
15. Castner, D.G.; Ratner, B.D. Biomedical surface science: Foundations to frontiers. *Surf. Sci.* **2002**, *500*, 28–60. [CrossRef]
16. Lang, K.; Mosinger, J.; Wagnerová, D.M. Photophysical properties of porphyrinoid sensitizers non-covalently bound to host molecules; models for photodynamic therapy. *Coord. Chem. Rev.* **2004**, *248*, 321–350. [CrossRef]
17. Vicente, M.G. Porphyrin-based sensitizers in the detection and treatment of cancer: Recent progress. *Curr. Med. Chem. Anti-Cancer Agents* **2001**, *1*, 175–194. [CrossRef] [PubMed]
18. Lei, J.; Ju, H.; Ikeda, O. Supramolecular assembly of porphyrin bound DNA and its catalytic behavior for nitric oxide reduction. *Electrochim. Acta* **2004**, *49*, 2453–2460. [CrossRef]
19. Ema, T.; Miyazaki, Y.; Shimonishi, J.; Maeda, C.; Hasegawa, J.Y. Bifunctional porphyrin catalysts for the synthesis of cyclic carbonates from epoxides and CO_2: Structural optimization and mechanistic study. *J. Am. Chem. Soc.* **2014**, *136*, 15270–15279. [CrossRef] [PubMed]
20. Purrello, R.; Gurrieri, S.; Lauceri, R. Porphyrin assemblies as chemical sensors. *Coord. Chem. Rev.* **1999**, *190–192*, 683–706. [CrossRef]
21. Hirano, T.; Yui, T.; Okazaki, K.I.; Kajino, T.; Fukushima, Y.; Inoue, H.; Torimoto, T.; Takagi, K. Photo-Induced Electron Migrations in the Nano-Cavities of Mesoporous Silica Sensitized by a Cationic Porphyrin Dye. *J. Nanosci. Nanotechnol.* **2009**, *9*, 495–500. [CrossRef] [PubMed]

22. Lehn, J.-M. Supramolecular Chemistry—Scope and Perspectives Molecules, Supermolecules, and Molecular Devices (Nobel Lecture). *Angew. Chem. Int. Ed. Engl.* **1988**, *27*, 89–112. [CrossRef]

23. Phan, T.H.; Wandelt, K. Molecular self-assembly at metal-electrolyte interfaces. *Int. J. Mol. Sci.* **2013**, *14*, 4498–4524. [CrossRef] [PubMed]

24. Pansu, R.B.; Delmarre, D.; Me, R. Heavy metal ions detection in solution, in sol ± gel and with grafted porphyrin monolayers. *Photochem. Photobiol.* **1999**, *124*, 23–28.

25. He, Y.; Ye, T.; Borguet, E. Porphyrin self-assembly at electrochemical interfaces: Role of potential modulated surface mobility. *J. Am. Chem. Soc.* **2002**, *124*, 11964–11970. [CrossRef] [PubMed]

26. Kunitake, M.; Batina, N.; Itaya, K. Self-Organized Porphyrin Array on Iodine-Modified Au(111) in Electrolyte Solutions: In Situ Scanning Tunneling Microscopy Study. *Langmuir* **1995**, *11*, 2337–2340. [CrossRef]

27. Kunitake, M.; Akiba, U.; Batina, N.; Itaya, K. Structures and dynamic formation processes of porphyrin adlayers on iodine-modified Au(111) in solution: In Situ STM study. *Langmuir* **1997**, *13*, 1607–1615. [CrossRef]

28. Hai, N.T.M. Preparation and characterization of copper-iodide thin films and organic supramolecular layers at copper/electrolyte interfaces. Ph.D. Dissertation, University of Bonn, Bonn, Germany, 2007.

29. Ogaki, K.; Batina, N.; Kunitake, M.; Itaya, K. In Situ Scanning Tunneling Microscopy of Ordering Processes of Adsorbed Porphyrin on Iodine-Modified Ag(111). *J. Phys. Chem.* **1996**, *100*, 7185–7190. [CrossRef]

30. Sashikata, K.; Sugata, T.; Sugimasa, M.; Itaya, K. In Situ Scanning Tunneling Microscopy Observation of a Porphyrin Adlayer on an Iodine-Modified Pt(100) Electrode. *Langmuir* **1998**, *14*, 2896–2902. [CrossRef]

31. Wilms, M.; Kruft, M.; Bermes, G.; Wandelt, K. A new and sophisticated electrochemical scanning tunneling microscope design for the investigation of potentiodynamic processes. *Rev. Sci. Instrum.* **1999**, *70*, 3641. [CrossRef]

32. Gao, X.; Edens, G.J.; Liu, F.-C.; Weaver, M.J.; Hamelin, A. Sensitivity of Electrochemical Adlayer Structure to the Metal Crystallographic Orientation: Potential-Dependent Iodide Adsorption on Au(100) in Comparison with Other Low-Index Surfaces. *J. Phys. Chem.* **1994**, *98*, 8086–8095. [CrossRef]

33. Valenzuela-Benavides, J.; Herrera-Zaldívar, M. Structural transitions of chemisorbed iodine on Au(1 0 0): A STM and LEED study. *Surf. Sci.* **2005**, *592*, 150–158. [CrossRef]

34. Magnussen, O.M. Ordered anion adlayers on metal electrode surfaces. *Chem. Rev.* **2002**, *102*, 679–725. [CrossRef] [PubMed]

35. Su, G.-J.; Zhang, H.-M.; Wan, L.-J.; Bai, C.-L.; Wandlowski, T. Potential-Induced Phase Transition of Trimesic Acid Adlayer on Au(111). *J. Phys. Chem. B* **2004**, *108*, 1931–1937. [CrossRef]

36. Cunha, F.; Jin, Q.; Tao, N.J.; Li, C.-Z. Structural phase transition in self-assembled 1,10′ phenanthroline monolayer on Au(111). *Surf. Sci.* **1997**, *389*, 19–28. [CrossRef]

37. Bai, Y. Photoelectron Spectroscopic Investigations of Porphyrins and Phthalocyanines on Ag(111) and Au (111): Adsorption and Reactivity. Friedrich-Alexander-Universität Erlangen-Nürnberg. 2010. Available online: http://opus4.kobv.de/opus4-fau/frontdoor/deliver/index/docId/1146/file/YunBaiDissertation.pdf (accessed on 13 May 2016).

38. Mizutani, W.; Shigeno, M.; Ono, M.; Kajimura, K. Voltage-dependent scanning tunneling microscopy images of liquid crystals on graphite. *Appl. Phys. Lett.* **1990**, *56*, 1974–1976. [CrossRef]

39. Claypool, C.L.; Faglioni, F.; Goddard, W.A.; Gray, H.B.; Lewis, N.S.; Marcus, R.A. Source of Image Contrast in STM Images of Functionalized Alkanes on Graphite: A Systematic Functional Group Approach. *J. Phys. Chem. B* **1997**, *101*, 5978–5995. [CrossRef]

40. Huemann, S.; Hai, N.T.M.; Broekmann, P.; Wandelt, K.; Zajonz, H.; Dosch, H.; Renner, F. X-ray diffraction and STM study of reactive surfaces under electrochemical control: Cl and I on Cu(100). *J. Phys. Chem. B* **2006**, *110*, 24955–24963. [CrossRef] [PubMed]

41. Röefzaad, M. Ordnungsphänomene redox-aktiver Moleküle auf Elektrodenoberflächen unter reaktiven Bedingungen. Ph.D. Thesis, University of Bonn, Bonn, Germany, 2011.

Article

Methanol Oxidation on Graphenic-Supported Platinum Catalysts

Gladys Arteaga, Luis M. Rivera-Gavidia, Sthephanie J. Martínez, Rubén Rizo, Elena Pastor * and Gonzalo García *

Área de Química Física, Departamento de Química, Facultad de Ciencias, Universidad de La Laguna (ULL), Instituto de Materiales y Nanotecnología (IMN), 38200 La Laguna, Tenerife, Spain; alu0100846713@ull.edu.es (G.A.); mlriviera@gmail.com (L.M.R.-G.); sthephanie94@hotmail.com (S.J.M.); rjrizo57@gmail.com (R.R.)
* Correspondence: epastor@ull.edu.es (E.P.); ggarcia@ull.edu.es (G.G.);
  Tel.: +34-922-318-071 (E.P.); +34-922-318-032 (G.G.)

Received: 19 November 2018; Accepted: 29 December 2018; Published: 6 January 2019

Abstract: Graphene oxide (GO), reduced graphene oxide by thermal treatment (rGO-TT), nitrogen-modified rGO (N-rGO), and carbon Vulcan were synthesized and employed in the current work as catalyst support for Pt nanoparticles, to study their properties and impact toward the methanol oxidation reaction (MOR) in sulfuric acid medium. Several physicochemical techniques, such as X-ray photoelectron spectroscopy (XPS), X-ray powder diffraction (XRD), Transmission electron microscopy (TEM), energy-dispersive X-ray spectroscopy (EDX), Raman, and elemental analysis were employed to characterize the novel materials, while potentiodynamic and potentiostatic methods were used to study catalytic performance toward the methanol oxidation reaction in acidic medium. The main results indicate a high influence of the support on the surface electronic state of the catalyst, and consequently the catalytic performance toward the MOR is modified. Accordingly, Pt/N-rGO and Pt/rGO-TT show the lowest and the highest catalytic performance toward the MOR, respectively.

Keywords: electrocatalysis; catalysts; methanol oxidation reaction; graphene; DMFC; Pt

---

1. Introduction

The search for alternatives to the use of fossil fuels, and the study and development of new green technologies for energy supply is an imperative necessity in order to fight global warming. Direct methanol fuel cells (DMFC) are an example of these energy-conversion devices [1–8]. They are able to supply energy from the electro-oxidation of a cheap resource like methanol, resulting not only in the reduction of release of harmful emissions, but also in the supply of high energy densities at low temperatures. The design of the catalysts to be used in DMFC systems is of outstanding importance. Many studies have confirmed that platinum is the most active metal when employed as a catalyst for the methanol oxidation reaction (MOR) [9–11].

However, the scarcity of this metal in the world and its high cost hinder the commercialization of such systems. In order to reduce as much as possible the amount of the catalytic material, the synthesis of Pt nanoparticles with high surface area/volume ratios, as well as the use of catalyst supports for a good dispersion of the metallic nanoparticles, is essential [12–14]. Furthermore, a high electrical conductivity of the catalyst support is the key factor for electrochemical reactions like the MOR. In this sense, graphene has emerged as a new-generation catalyst support, because of its excellent electrical conductivity and high surface area [15–18]. Many works have described chemical exfoliation methods of graphite to obtain so-called graphene oxide (GO) as a catalyst support. However, the conductivity of such materials is lower than the highly desirable two-dimensional graphene, because of the high amount of oxygen groups on the surface. To fulfill conductivity problems, a great diversity of physical and chemical methods have been described in the literature for GO reduction to obtain reduced GO (rGO), with higher conductivity than GO itself [15–18]. On the other hand, it has been demonstrated that the surface chemistry of the catalytic support plays an essential role in the activity of the catalyst toward the MOR [19–23]. For example, the modification of carbon materials with heteroatoms, like nitrogen or sulfur, causes an electronic modulation of the carbonaceous lattice, resulting in a change in the electrocatalytic properties toward the MOR and other reactions of interest [15,18,23–25].

In the present paper, the synthesis of reduced graphene oxide (rGO) by a thermal treatment (rGO-TT) and nitrogen-modified rGO by the caffeine route (N-rGO) is reported, both of which are employed as supports for Pt nanoparticles. A deep physicochemical characterization of the novel catalysts was performed, and their catalytic performance toward the MOR was studied and compared with that of Vulcan XC72R supported Pt catalyst (Pt/C).

2. Materials and Methods

Sulfuric acid (p.a.; Merck, Darmstadt, Germany), methanol (p.a.; Merck, Darmstadt, Germany), potassium permanganate (>99.8%; Sigma-Aldrich, Saint Louis, MO, USA), caffeine (>99%; Sigma-Aldrich, Saint Louis, MO, USA), Vulcan XC-72R (Cabot, Boston, MA, USA), graphite (>99.8%; Sigma-Aldrich, Saint Louis, MO, USA), hydrogen peroxide (30% v/v; Foret, Barcelona, Spain), hydrogen (5% H_2/95% N_2; Air Liquide, Tenerife, Spain), (8 wt % $H_2PtCl_6 \cdot 6H_2O$, Sigma-Aldrich, Saint Louis, MO, USA), Nafion solution (5 wt %; Sigma-Aldrich, Saint Louis, MO, USA), Ar (99.999%, Air Liquide, Tenerife, Spain), CO (99.997%, Air Liquide, Tenerife, Spain), and water (18.2 M $\Omega \cdot cm^{-1}$; Milli-Q, Millipore, Burlington, VT, USA) were purchased and then used for the synthesis of the graphene-based materials and the preparation of electrolyte solutions.

2.1. Synthesis of Graphene Oxide

Graphene oxide (GO) was prepared by following a modified Hummers method [15]. Briefly, 1 g of the graphite powder was mixed with 30 mL of concentrated H_2SO_4 cooled in an ice bath. Then, 3.5 g of $KMnO_4$ are slowly added while being stirred and cooled continuously. After removal from the ice bath, the mixture was diluted with Milli-Q water under stirring for 1 h at 35 °C. Then, the solution was heated up to 95–98 °C over 30 min. Next, 200 mL of ultrapure water was gradually introduced, followed by 1.25 mL of 30% v/v H_2O_2, and stirring was maintained for 30 min. Finally, the dispersion was centrifuged with Milli-Q water, until a pH of 7 was achieved in the supernatant liquid. The material was dried using an oven at 60 °C overnight.

2.2. Synthesis of Reduced Graphene Oxide Materials

In order to produce N-rGO, an adequate amount (4 mmol) of reducing agent (caffeine) was ultrasonically dispersed in Milli-Q water and mixed with a GO aqueous dispersion (0.015 g mL^{-1}). The final dispersion was placed into a Teflon-lined autoclave and heated at 160 °C for 10 h. Then, N-rGO was washed by centrifugation, employing Milli-Q water, and transferred to an oven at 60 °C for 24 h.

The rGO-TT was obtained by reduction of GO through a heat treatment under a reductive atmosphere (5% H_2/95% N_2 at 100 mL/min) in a tubular furnace. The thermal treatment comprised a ramp temperature of 5 °C/min from room temperature, up to 200 °C. This temperature was maintained for 30 min; finally, a ramp temperature of 5 °C/min was applied up to 450 °C where it was preserved for 1 h.

2.3. Synthesis of Pt/C, Pt/N-rGO, and Pt/rGO Catalysts

Platinum nanoparticles supported on carbon black Vulcan XC-72R, N-rGO, and rGO-TT were synthesized following the formic acid method [6]. Briefly, catalyst support was dispersed in 2 M formic acid solution under sonication. Then, the mixture was heated at 80 °C under stirring, and an appropriate amount of metal precursor (H_2PtCl_6) was employed to obtain theoretical platinum loading of 20 wt % in the catalyst. The suspension was filtered, washed with ultrapure water, and dried at 60 °C overnight.

2.4. Physicochemical Characterization

Powder XRD spectra were acquired from X'Pert PRO X-ray diffractometer (PANalytical, Madrid, Spain) to determine the crystal structure. CuKα radiation (λ = 1.5405 Å) was employed, and 2θ data were collected from 20° to 100° with a scanning rate of 0.04 s^{-1}. Crystalline phases were identified by comparing the experimental diffraction patterns with the Joint Committee on Powder Diffraction Standards (JCPDS).

Elemental analysis with an experimental error close to 0.04% was performed using an Elemental Analyzer CNHS FLASH EA 1112 (Thermo Scientific, Madrid, Spain).

Raman spectra were collected employing a RENISHAW confocal Raman microscope, model inVia (RENISHAW, Gloucestershire, United Kingdom), with a green laser (λ = 532 nm) in the 100 to 3200 cm^{-1} range.

Transmission electron microscopy (TEM) images were obtained from a HRTEM JEOL JEM 2100 operating (JEOL Ltd., Tokyo, Japan) at 200 eV. Obtained images were used to evaluate the morphology and agglomeration degree, as well as to calculate the average Pt nanoparticle sizes.

Chemical composition of the catalysts was determined by energy dispersive X-ray analysis (EDX, Oxford 6699 ATW, Oxford, Oxfordshire, UK), and X-ray photoelectron spectroscopy (XPS) data were obtained with a SPECS GmbH customized system ($SPECS^{TM}$, Berlin, Germany) for surface analysis, equipped with a non-monochromatic X-ray source XR 50 and a hemispherical energy analyzer (PHOIBOS 1509MCD). Photoelectrons were excited by using MgKα line (1253.6 eV) operating at 200 W/12 kV. Powder samples were attached onto Cu foil and were placed first in the pre-treatment chamber at room temperature for several hours, before being transferred to the analysis chamber. The XPS data were signal averaged for a large enough number of scans to get a good signal/noise ratio, at increments of 0.1 eV and fixed pass energy of 25 eV. High-resolution spectra envelopes were obtained by curve fitting synthetic peak components, using the software XPSpeak. The raw data were used with no preliminary smoothing. Symmetric Gaussian—Lorentzian product functions were employed to approximate the line shapes of the fitting components. Binding energies were calibrated relative to the C 1s peak from the graphitic carbon at 284.6 eV.

2.5. Electrochemical Characterization

All electrochemical experiments were carried out at room temperature using a three-electrode glass cell, a carbon rod as a counter electrode and a reversible hydrogen electrode (RHE) as a reference electrode. All the potentials in this work were given against the RHE.

The working electrode consisted of a certain amount of the catalyst deposited as a thin layer over a glassy carbon disc (0.196 cm^2). The ink was prepared by mixing 2 mg of the catalyst powder, 15 μL of Nafion solution, and 500 μL of Milli-Q water.

Electrochemical measurements were performed with a PC-controlled Autolab PGSTAT30 potentiostat-galvanostat (Metrohm, Herisau, Switzerland) in 0.5 M H_2SO_4 electrolyte solution. A total of 2 M CH_3OH + 0.5 M H_2SO_4 solution was used to study the methanol oxidation reaction (MOR). Ar was employed to deoxygenate all solutions, as well as CO for CO stripping experiments. The last experiments were recorded at 0.02 V s^{-1} after bubbling CO through the cell for 10 min while keeping the electrode at 0.10 V, followed by Ar purging for 30 min to completely remove the excess CO. CO stripping voltammograms were recorded, by first scanning negatively until 0.05 V so that the entire hydrogen region was probed, and then scanning positively up to 1.2 V. Methanol cyclic voltammograms (CVs) were recorded between 0.05 and 1.0 V at a scan rate of 0.02 V s^{-1}, while current transients were obtained by stepping the potential from 0.05 V to the final oxidation potential ($0.50 < E_f < 0.90$ V). Tafel plots were calculated from stationary currents achieved at 300 s during the chronoamperometry experiments. Currents were normalized by the electroactive surface area (ESA) calculated from CO stripping experiments, by using a conversion factor of 240 $\mu C\ cm^{-2}$ (12.6 cm^2 for Pt/C, 8.2 cm^2 for Pt/rGO-TT and 9.1 cm^2 for Pt/N-rGO).

3. Results and Discussion

3.1. Physicochemical Characterization

Figure 1 displays the X-ray diffractograms for graphite, carbon Vulcan, GO, rGO-TT, N-rGO, Pt/C, Pt/rGO-TT, and Pt/N-rGO. Graphite depicts a sharp diffraction peak at 24.6°, which is related to the (002) plane, and a small diffraction peak at 54° associated to the (004) facet [15]. The GO formation from the oxidation of graphite is detected by the broad diffraction peak at 10.7°, which is related to the (101) diffraction plane and by the absence of the diffraction peaks associated to graphite [15]. Additionally, Table 1 shows the expansion of the C–C interplanar spacing from 0.34 nm (graphite) to 0.82 nm (GO), due to the growth of oxygen functional groups between the graphitic layers [15].

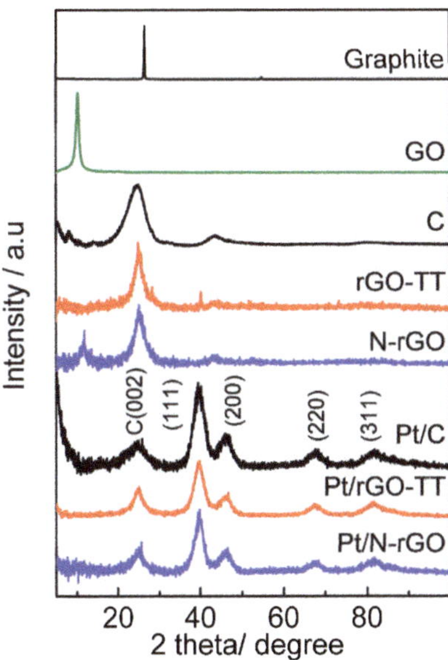

Figure 1. XRD patterns of graphite, graphene oxide (GO), Vulcan carbon (C), reduced graphene oxide by thermal treatment (rGO-TT), nitrogen-modified reduced graphene oxide (N-rGO), Pt/C, Pt/rGO-TT, and Pt/N-rGO.

Table 1. X-ray diffraction (XRD) parameters for all studied materials (d_{hkl} = interplanar spacing).

Catalyst	Position$_{hkl}$ 2θ/°	d_{hkl}/nm	Number of Layers	Pt Crystallite Size/nm
Graphite	26.5 C(002)	0.34 C(002)	121	
GO	10.7 C(002)	0.82 C(002)	11	
Vulcan	24.7 C(002)	0.36 C(002)	6	-
rGO-TT	25.2 C(002)	0.35 C(002)	12	
N-rGO	25.3 C(002)	0.35 C(002)	9	
Pt/C	67.4 Pt(220)	0.23 Pt(111)	-	2.3
Pt/rGO-TT	67.5 Pt(220)	0.23 Pt(111)	-	2.4
Pt/N-rGO	67.6 Pt(220)	0.23 Pt(111)	-	2.0

On the other hand, Vulcan carbon, rGO-TT, and N-rGO reveal similar interplanar spacing (Table 1) and X-ray patterns (Figure 1), which indicate the restoration of the graphite C–C lattice and the elimination of oxygen functional groups between the graphitic layers in rGO materials. Also, a small diffraction peak at ca. 45° associated to the (100) plane is observed at the three carbon supports. The number of graphene layers that was estimated from the Debye–Scherrer equation and the Bragg law is also included in Table 1. The success of GO, rGO-TT, and N-rGO formation can be easily perceived by the diminution of the graphene layers in comparison to graphite. In this context, a graphite behavior was suggested for materials with more than 12 graphene layers, and therefore, a graphenic-like behavior is expected for all materials of the current work [26]. On the other hand, it is also important to note that N-rGO reveals a small diffraction peak at 10.7° associated to GO, which suggests a partial reduction of GO when caffeine is employed as a reducing agent [15].

Additionally, Figure 1 depicts X-ray diffractograms of Pt/C, Pt/rGO-TT, and Pt/N-rGO, in which the typical face centered cubic (*fcc*) structure of Pt is discerned at $2\theta = 40°, 46°, 68°$, and $82°$, associated to the subsequent planes: (111), (200), (220), and (311) (JPCDS 00-004-0802). It is remarkable that the absence of the diffraction peak is at 10.7° for Pt/N-rGO, which suggests a reduction of the oxygen functional groups by the formic acid that is employed to reduce the metal precursor. In addition, it is important to highlight the absence of Pt crystalline oxides at all catalysts. However, the presence of Pt amorphous oxides should not be ruled out, since the oxygen content in graphenic-based materials is high (see Table 2). The most important crystallographic parameters of Pt-based catalysts are depicted in Table 1, in which similar interplanar spacing for all catalysts, smaller crystallite size for Pt/N-rGO than Pt/C and Pt/rGO-TT, and a small but visible lattice contraction in graphenic-based catalysts are discerned.

Table 2. Composition acquired by energy dispersive X-ray analysis (EDX) analysis and * elemental analysis for all studied materials.

Composition (% Weight)			
Material	**C**	**O**	**N ***
Graphite	80.0	20.0	
Carbon	96.5	3.5	-
GO	48.0	52.0	-
rGO-TT	75.3	24.7	-
N-rGO	73.0	21.0	6.0
Pt/C	96.4	3.6	-
Pt/rGO-TT	86.9	13.1	-
Pt/N-rGO	78.3	15.7	6.0

Scanning electron spectroscopy (SEM) coupled with energy dispersive X-ray spectroscopy (EDX) was employed to study the morphology, distribution, surface topography, and semi-quantitative bulk composition of the catalysts. The amount of Pt incorporated into each sample was close to the nominal value (19.0 ± 1.5 wt %). Figure 2 displays SEM images in the microscale for all materials synthesized. Pt/C reveals a smooth and homogeneous surface, while a granular and porous structure is observed at Pt/rGO-TT and Pt/N-rGO.

Figure 2. Scanning electronic microscopy (SEM) images for Pt/C (**left panel**), Pt/rGO-TT (**middle panel**), and Pt/N-rGO (**right panel**).

Table 2 discloses a strong increment of the oxygen content after GO production, which is reduced in the first stage after rGO-TT and N-rGO formation, and in the second stage after Pt deposition. The last stage is in agreement with the X-ray diffraction pattern of N-rGO, as the small peak associated with oxide species disappears after the addition of formic acid. It is noteworthy that even after Pt deposition, both graphenic catalysts show a higher amount of oxygen than carbon-supported Pt.

Raman spectroscopy (spectra are not shown) was employed to study the structural changes, the disorder degree, and the crystallographic size in the basal plane (L_a) of graphenic materials. With this end, the following equation was employed [27,28]:

$$L_a(\text{nm}) = \left(2.4 \times 10^{-10}\right) \cdot \lambda_l^4 \cdot \left(\frac{I_D}{I_G}\right)^{-1} \tag{1}$$

where λ_l is the excitation wavelength of the laser (532 nm), I_D is the peak intensity at 1360 cm^{-1} (associated to sp^3 carbon domains), and I_G is the peak intensity at 1580 cm^{-1} (related to sp^2 bonds into the graphitic grid). The main results indicate an increment of the disorder (I_D/I_G ~ 0.9) compared to graphite (I_D/I_G ~ 0.15), and the similar L_a ~ 20 nm for Vulcan carbon and all graphenic materials. The latter is the expected result, since structural disorder by edges, finite size, and carbon hybridization by heteroatom–C bond formation is usually reported in this type of materials [15,18].

Morphology, particle size, and agglomeration degree were studied by transmission electron microscopy (TEM). Figure 3 shows TEM images of Pt/C, Pt/rGO-TT, and Pt/N-rGO catalysts. Meanwhile, Figures S1–S3 depict high-resolution TEM images for all electrocatalysts employed in the current work. Spherical Pt nanoparticles are homogeneously dispersed on carbon sheets; however, the agglomeration degree increases with the amount of oxygen in the catalyst in the subsequent way: Pt/C < Pt/rGO-TT < Pt/N-rGO. Therefore, Pt nanoparticles seem to nucleate and grow referentially at surface oxygenated sites. Regarding the particle sizes, Figure 3 includes histograms for each catalyst, and shows increments in the following way: Pt/ N-rGO (3.14 ± 0.50 nm) < Pt/C (3.28 ± 0.25 nm) < Pt/rGO-TT (3.32 ± 0.43 nm). The last is in agreement with the trend observed for the crystallite size values calculated by XRD, although the values are slightly lower than those achieved by TEM, which is expected (i.e., crystallite size ≤ grain size ≤ particle size).

X-ray photoelectron spectroscopy (XPS) was carried out to identify and quantify the surface composition and the chemical state of the elements (see Table 3). Deconvoluted spectra for Pt, C, O, and N are presented in Figures 4 and 5. The Pt 4f signal (Figure 4) is resolved into three doublets in all electrocatalysts. The most intense Pt 4f$_{7/2}$ component at lower binding energy is attributed to metallic Pt^0, while the second doublet is assigned to Pt^{2+}, and the third doublet at higher binding energy is related to the higher oxidation state of Pt^{4+}. According to Table 3, there is a peak shift of metallic Pt to a lower binding energy for graphenic-based supports. The latter indicates a charge transfer from graphenic-based supports to platinum. However, Pt^{2+} species (i.e., PtO) increase to the detriment of metallic Pt at Pt/N-rGO material.

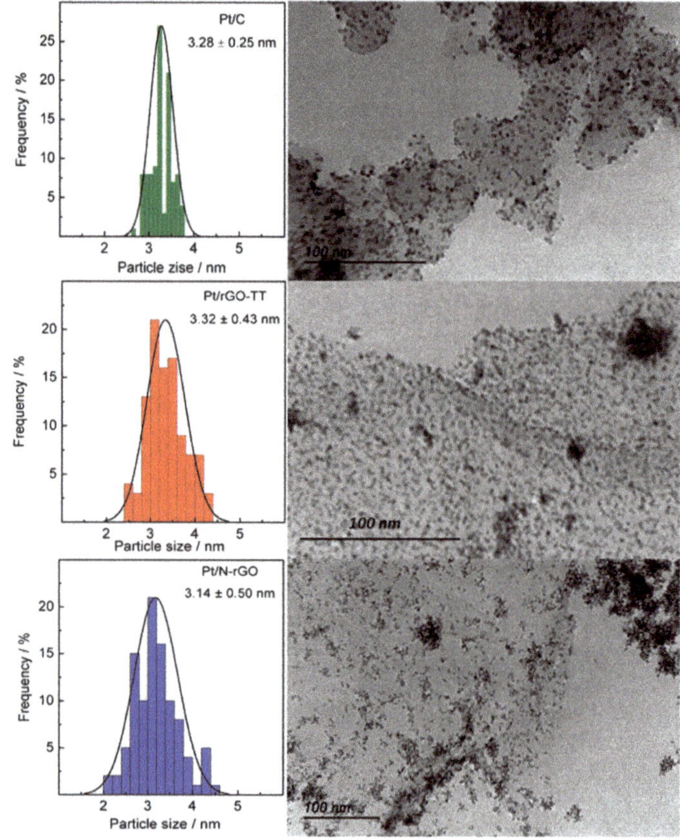

Figure 3. Transmission electron micrographs and particle size distribution histograms of the catalysts.

Table 3. X-ray photoelectron spectroscopy (XPS) analysis of the catalysts.

Elements / Catalysts	Pt4f		C1s		O1s	
	Pt^0	71.5 (62)	C–C	284.6 (81)	Pt oxides	530 (5)
	Pt^{2+}	72.4 (31)	C–N	-	O=C	531.1 (10)
Pt/C	Pt^{4+}	73.5 (7)	C–OH	285.8 (14)	OH–C	532.8 (72)
			C=O	287 (4)	O–C=O	534.1 (13)
			O–C=O	289.1 (1)	H_2O	-
Pt/rGO-TT	Pt^0	71.2 (63)	C–C	284.6 (68)	Pt oxides	530 (3)

Table 3. *Cont.*

Elements / Catalysts	Pt4f		C1s		O1s		N1s	
Pt/rGO-TT	Pt^{2+}	72.3 (30)	C–N	-	O=C	531.2 (23)		
	Pt^{4+}	74.3 (7)	C–OH	285.8 (15)	OH–C	532.9 (58)		
			C=O	287 (10)	O–C=O	533.9 (9)		
			O–C=O	288.7 (7)	H$_2$O	535.2 (7)		
Pt/N-rGO	Pt^{0}	71.2 (49)	C–C	284.5 (52)	Pt oxides	530.1 (4)		
	Pt^{2+}	72.1 (42)	C–N	285.3 (14)	O=C	531.2 (33)		
	Pt^{4+}	74.0 (10)	C–OH	285.9 (12)	OH–C	532.7 (48)	Pyridinic	398.7 (29)
			C=O	287.1 (12)	O–C=O	534 (12)	Pyrrolic	400.2 (49)
			O–C=O	288.5 (10)	H$_2$O	535 (2)	Quaternary	401.2 (22)
			π–π	290.0 (2)	-	-	-	-

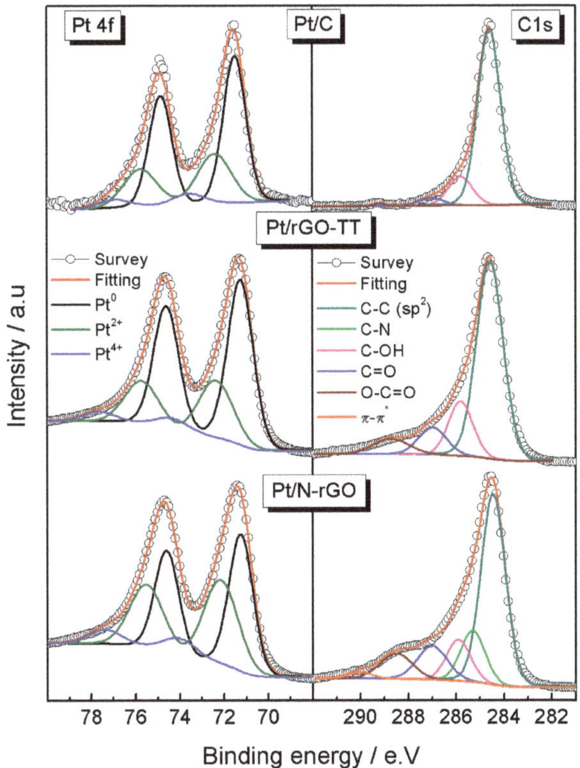

Figure 4. Pt 4f (**left panels**) and C1s (**right panels**) XPS spectra of the catalysts.

The C1s signal provides detailed information about the presence of oxygen and nitrogen species bonded to carbon atoms (Figure 4). As can be seen, the carbon signal is mainly composed by graphitic carbon (C1), followed by different oxygenated species at higher binding energy values, such as C–OH (C3), C=O (C4), and carboxylic bonds (C5) [29]. In addition to these signals, Pt/NrGO reveals an additional component at ca. 285.3 (C2) ascribed to nitrogen species bonded to carbon atoms. Indeed, the strong N1s signal (Figure 5) confirms that the last statement since the nitrogen species, such as pyridinic, pyrrolic, and N-quaternary, are clearly identified. Also, a small but visible signal at ca. 289 eV ascribed to a π–π* shakeup satellite for Pt/Vulcan and Pt/NrGO catalysts is observed [30].

Interestingly, the O1s signal (Figure 5) reveals different surface oxygenated species, which strongly depend on the substrate. In agreement with the C1s signal, C=O (O2:~531.05 ± 0.05 eV) and C–OH (O3:~533.5 ± 0.5 eV) species are the main components for all materials [29]. Signals related to carboxyl groups (O4) and platinum oxides (O1) appear in minor proportion, while intercalated water (O5) is only perceived at graphenic materials. On the other hand, Pt/NrGO shows similar intensities for C–OH and C=O species, and the most intense component is associated to C–O species. All XPS information is summarized in Table 3.

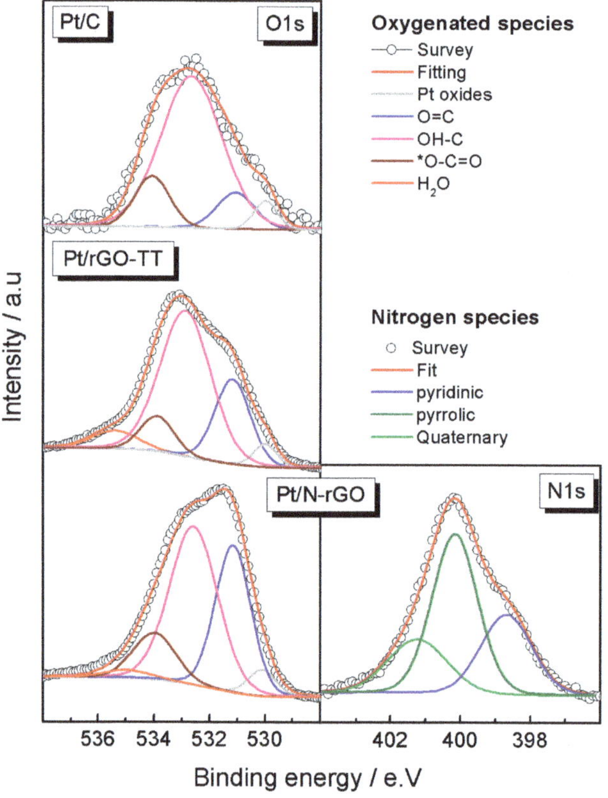

Figure 5. O1s (**left panels**) and N1s (**right panel**) XPS spectra of the catalysts.

3.2. Electrochemical Characterization

CO stripping experiments were performed to (i) characterize the catalyst surface; (ii) achieve catalytic activity towards CO oxidation, which is one of the most important catalyst poisons; and (iii) acquire the electroactive surface area (ESA). Figure 6 shows a similar CO oxidation peak value (CO_P = 0.83 V) for all catalysts. Noticeable, Pt/N-rGO reveals a broader CO oxidation peak, with an anodic tail that ends at 1.2 V. Additionally, the onset potential for the CO oxidation reaction is 0.70 V for Pt/rGO-TT, 0.72 V for Pt/C, and 0.75 V for Pt/N-rGO. These parameters indicate a higher CO tolerance for Pt/rGO-TT than for Pt/C, and the even lower one for Pt/N-rGO. Also, an increase of the double-layer capacitive current is observed at graphenic-based materials, especially relevant for Pt/N-rGO. The latter results and the low catalytic activity towards the CO oxidation on Pt/N-rGO may be ascribed to the high amount of surface-oxygenated species, i.e., PtO and PtO_2. Regarding the ESA, the values increase in the following way: Pt/rGO-TT (8.2 cm^2) < Pt/N-rGO (9.1 cm^2) < Pt/C (12.6 cm^2). The last is in agreement with TEM analysis, in which high Pt agglomeration degree was detected at graphenic-based catalysts.

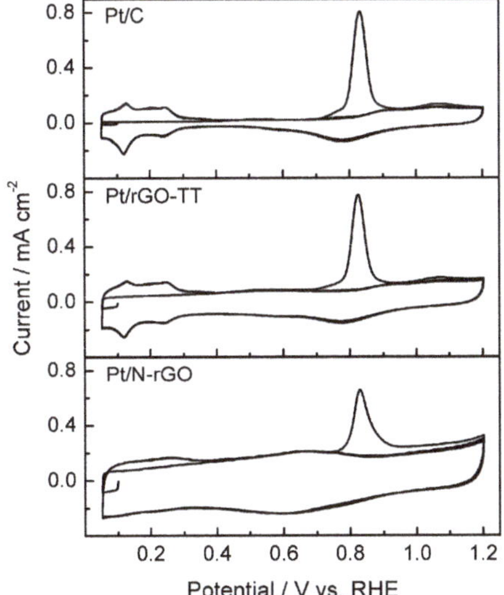

Figure 6. CO stripping voltammograms recorded at Pt/C, Pt/rGO-TT, and Pt/N-rGO catalysts in 0.5 M H_2SO_4. E_{ad} = 0.1 V; scan rate = 0.02 V s^{-1}.

The catalytic performance toward the methanol oxidation reaction (MOR) was evaluated by potentiodynamic (Figure 7) and potentiostatic (Figure 8) experiments. Figure 7 shows the lowest performance for Pt/N-rGO toward the MOR, which is in disagreement with what has been reported by other authors [31,32]. This different catalytic performance may be ascribed to the elevated amount of Pt surface oxide species (PtO and PtO_2 observed by XPS analysis) by using the caffeine route for the synthesis of the N-rGO, which inhibits the methanol adsorption step (see below).

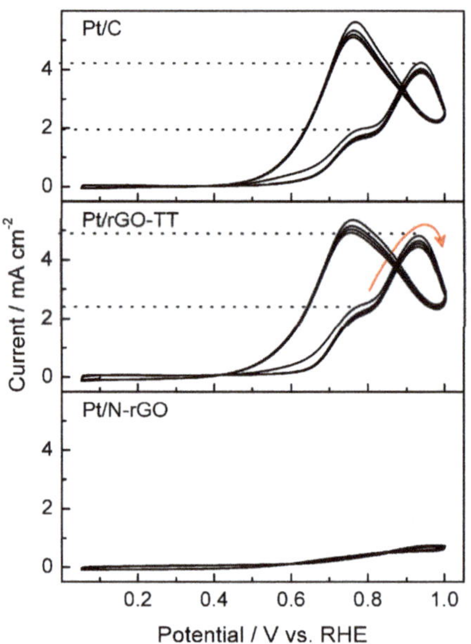

Figure 7. Methanol oxidation recorded for Pt/C, Pt/rGO-TT, and Pt/N-rGO catalysts in 2 M CH$_3$OH + 0.5 M H$_2$SO$_4$. Scan rate = 0.02 V s^{-1}.

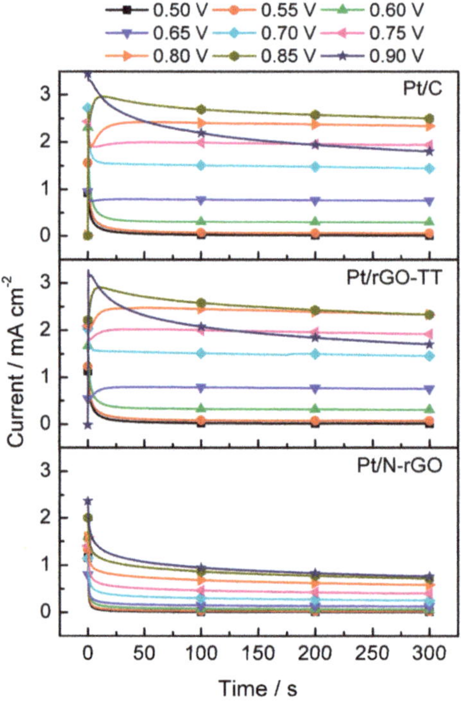

Figure 8. Current transients for methanol oxidation on Pt/C, Pt/rGO-TT, and Pt/N-rGO catalysts in 2 M CH$_3$OH + 0.5 M H$_2$SO$_4$. E_i = 0.05 V, 0.50 V $\leq$ E_f $\leq$ 0.90 V.

Additionally, similar voltammetric profiles with comparable onset potentials (~0.5 V) and analogous anodic ($E_{Pa} \sim 0.93$ V) and cathodic ($E_{Pc} \sim 0.76$ V) peak potentials are observed for Pt/rGO-TT and Pt/C. However, a close inspection of Figure 7 reveals higher anodic currents at Pt/rGO-TT than Pt/C during the positive sweep potential. Indeed, the pre-peak at ca. 0.77 V reveals higher current density at Pt/rGO-TT (2.4 mA cm^{-2}) than at Pt/C (2.0 mA cm^{-2}). The current peak during the anodic sweep (j_{Pa}) is 4.9 mA cm^{-2} at Pt/rGO-TT and 4.2 mA cm^{-2} at Pt/C, while the current peak during the cathodic sweep (j_{Pc}) is 5.4 mA cm^{-2} at Pt/rGO-TT and 5.6 mA cm^{-2} at Pt/C. The j_{Pa}/j_{Pc} ratio is usually employed to estimate the tolerance of the catalyst to adsorbed reaction intermediates (mainly CO$_{ad}$). Pt/C shows the lowest j_{Pa}/j_{Pc} value and appears as the catalyst with the highest tolerance, which is not in agreement with the CO stripping experiments in Figure 6. However, a recent work states that the j_{Pc} is not related to a residual intermediate of the MOR, and the j_{Pa}/j_{Pc} ratio is associated with the oxophilicity degree of the catalyst [33]. Thus, higher j_{Pa}/j_{Pc} value indicates higher oxophilicity degree, and consequently, Pt/rGO-TT appears to be the material with the highest oxophilicity. The last is in agreement with the physicochemical characterization, as well as with the CO stripping experiments, in which Pt/rGO-TT reveals the highest CO tolerance.

Figure 7 depicts current transients achieved from an initial potential of 0.05 V, in which methanol does not oxidize, to a final potential (0.50 $\leq$ E $\leq$ 0.90 V) where the MOR takes place. In agreement with potentiodynamic experiments, Pt/N-rGO develops the lowest performance toward the MOR in all potential ranges studied. In the same way, Pt/rGO-TT and Pt/C reveal similar better performance toward the alcohol oxidation. Interestingly, Pt/rGO-TT and Pt/C develops inflection points at short times at the same potential range (0.65 $\leq$ E $\leq$ 0.75), where the onset potential for the CO removal (Figure 5) and the pre-peak at the potentiodynamic experiments in presence of methanol is observed (Figure 6). This result suggests a change of rate-determining step (RDS), which may be attributed to an enhancement of the water dissociation reaction (i.e., OH$_{ad}$ formation) necessary to remove CO from the catalyst surface [34,35]. Additionally, it is observed that at E $\geq$ 0.85 V (anodic peak during the positive sweep potential, Figure 6), the current decay is faster at Pt/rGO-TT and Pt/C, which is related to the low performance of Pt oxide surface toward the MOR [36].

Further characterization of the catalytic activity of the synthesized materials was carried out by a Tafel plot analysis. A Tafel slope is usually employed to determine the rate-determining step (RDS), and therefore, the reaction mechanism may be inferred from Tafel slope values. Figure 9 depicts a Tafel plot obtained from the current transients of methanol oxidation for the three catalysts (Figure 8) at different final potentials (0.50 $\leq$ E $\leq$ 0.90 V) in an acidic medium. Pt/C shows a Tafel slope of 120 mV dec^{-1}, while Pt/rGO-TT and Pt/C reveal a Tafel slope of 75 mV dec^{-1}. In order to clarify the operating reaction mechanism on the catalysts involved in the current work, the main steps for MOR are briefly described.

First there is the electrosorption and dehydrogenation of methanol onto a suitable surface. The complete electrosorption and dehydrogenation steps yield to adsorbed carbon monoxide (CO$_{ad}$), otherwise formaldehyde and formic acid may be produced [37]:

$$H_3OH \rightleftharpoons CH_3OH_{ad} \tag{2}$$

$$CH_3OH_{ad} \rightarrow CH_2OH_{ad} + H^+ + e^- \tag{3}$$

$$CH_2OH_{ad} \rightarrow CHOH_{ad} + H^+ + e^- \tag{4}$$

$$CH_2OH_{ad} \rightarrow CH_2O + H^+ + e^- \tag{5}$$

$$CH_2O \rightleftharpoons CH_2O_{ad} \tag{6}$$

$$CH_2O_{ad} \rightarrow CHO_{ad} + H^+ + e^- \tag{7}$$

$$CHO_{ad} + H_2O \rightarrow HCOOH + H^+ + e^- \tag{8}$$

$$CHO_{ad} \rightarrow CO_{ad} + H^+ + e^- \tag{9}$$

Second, there is the electrooxidation of CO_{ad}. Water dissociation on the catalyst surface (formation of adsorbed OH) is needed to perform the CO_{ad} oxidation to CO_2, and consequently liberate the catalyst surface [31,38,39]:

$$H_2O \rightleftharpoons OH_{ad} + H^+ + e^- \tag{10}$$

$$COad + OH_{ad} \rightarrow COOH_{ad} \tag{11}$$

$$COOH_{ad} \rightarrow CO_2 + H^+ + e^- \tag{12}$$

A Tafel slope of 120 mV dec^{-1} suggests that the rate-determining step is related to the first electrochemical step, and therefore the rate of MOR at Pt/N-rGO is limited by the "electrosorption and dehydrogenation step", i.e., Reaction 3. The latter is in complete agreement with the low current developed by the Pt/N-rGO materials during the MOR, and should be ascribed to a non-suitable surface for methanol dehydrogenation, which is comprised of strongly bonded oxygen species onto the surface, i.e., PtO and PtO$_2$ [36]. On the other hand, a Tafel slope close to 60 mV dec^{-1} (Pt/C and Pt/rGO-TT) indicates a chemical reaction after an electrochemical reaction as the RDS, i.e., the surface diffusion of CO_{ad} toward the active site (reaction 11). Both materials revealed similar catalytic performance toward the MOR, although a slight enhancement of the CO tolerance at Pt/rGO-TT was perceived. The last seems to be related to electronic effects (charge transfer from rGO-TT to Pt) and to a higher amount of C–OH species onto the graphenic-based catalyst (bifunctional effect).

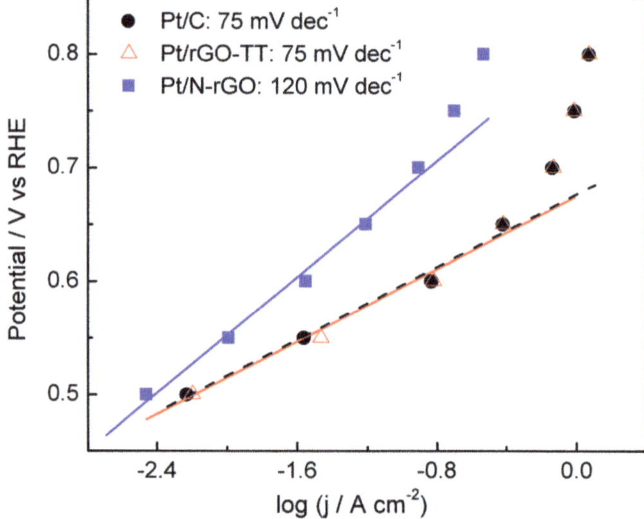

Figure 9. Tafel plots obtained at 300 s from Figure 8.

4. Conclusions

Graphene oxide (GO), reduced graphene oxide (rGO-TT), nitrogen-modified rGO (N-rGO), and carbon-supported Pt materials using these supports were successfully synthesized and fully characterized. Results have shown that the support conditions the state of oxidation of metal nanoparticles at the surface of the catalysts, which control the activity of a specific electrochemical reaction. Indeed, a small addition of nitrogen into the catalyst support induces a change of the rate-determining step from a chemical reaction after an electrochemical reaction to the first electron. Thus, Pt/N-rGO displays the worst performance toward the methanol oxidation reaction, while

Pt/rGO-TT not only enhances the catalytic activity toward the alcohol oxidation, but also toward the most important catalyst poison.

Supplementary Materials: The following are available online at http://www.mdpi.com/2571-9637/2/1/2/s1, Figure S1: TEM image for Pt/C electrocatalyst, Figure S2. TEM image for Pt/rGO-TT electrocatalyst, Figure S3. TEM image for Pt/N-rGO electrocatalyst.

Author Contributions: Data curation, G.A. and S.J.M.; Formal analysis, L.M.R.-G.; Writing – original draft, R.R.; Writing – review & editing, E.P. and G.G.

Acknowledgments: The Spanish Ministry of Economy and Competitiveness (MINECO) has supported this work under project ENE2017-83976-C2-2-R (co-funded by FEDER). L.M.R. thanks the ACIISI for his PhD fellowships. G.G. acknowledges the Viera y Clavijo program (ACIISI and ULL) for financial support.

Conflicts of Interest: The authors declare no conflict of interest.

References

1. Guillén-Villafuerte, O.; Guil-López, R.; Nieto, E.; García, G.; Rodríguez, J.L.; Pastor, E.; Fierro, J.L.G. Electrocatalytic Performance of Different Mo-Phases Obtained during the Preparation of Innovative Pt-MoC Catalysts for DMFC Anode. *Int. J. Hydrog. Energy* **2012**, *37*, 7171–7179. [CrossRef]
2. Li, M.; Adzic, R.R. Low-Platinum-Content Electrocatalysts for Methanol and Ethanol Electrooxidation. In *Lecture Notes in Energy*; Shao, M., Ed.; Springer: London, UK, 2013; Volume 9, pp. 1–25.
3. Pérez-Rodríguez, S.; Corengia, M.; García, G.; Zinola, C.F.; Lázaro, M.J.; Pastor, E. Gas Diffusion Electrodes for Methanol Electrooxidation Studied by a New DEMS Configuration: Influence of the Diffusion Layer. *Int. J. Hydrog. Energy* **2012**, *37*, 7141–7151. [CrossRef]
4. Huang, L.; Zhang, X.; Wang, Q.; Han, Y.; Fang, Y.; Dong, S. Shape-Control of Pt–Ru Nanocrystals: Tuning Surface Structure for Enhanced Electrocatalytic Methanol Oxidation. *J. Am. Chem. Soc.* **2018**, *140*, 1142–1147. [CrossRef] [PubMed]
5. Roca-Ayats, M.; García, G.; Peña, M.A.; Martínez-Huerta, M.V. Titanium Carbide and Carbonitride Electrocatalyst Supports: Modifying Pt-Ti Interface Properties by Electrochemical Potential Cycling. *J. Mater. Chem. A* **2014**, *2*, 18786–18790. [CrossRef]
6. Calderón, J.C.; García, G.; Querejeta, A.; Alcaide, F.; Calvillo, L.; Lázaro, M.J.; Rodríguez, J.L.; Pastor, E. Carbon Monoxide and Methanol Oxidations on Carbon Nanofibers Supported Pt-Ru Electrodes at Different Temperatures. *Electrochim. Acta* **2015**, *186*, 359–368. [CrossRef]
7. Flórez-Montaño, J.; García, G.; Rodríguez, J.L.; Pastor, E.; Cappellari, P.; Planes, G.A. On the Design of Pt Based Catalysts. Combining Porous Architecture with Surface Modification by Sn for Electrocatalytic Activity Enhancement. *J. Power Sources* **2015**, *282*, 34–44. [CrossRef]
8. Rizo, R.; García, G.; Pastor, E. Methanol Oxidation on Bimetallic Electrode Surfaces. In *Encyclopedia of Interfacial Chemistry: Surface Science and Electrochemistry*; Wandelt, K., Ed.; Elsevier: Amsterdam, The Netherlands, 2018; Volume 5, pp. 719–729.
9. Anitha, V.C.; Zazpe, R.; Krbal, M.; Yoo, J.; Sopha, H.; Prikryl, J.; Cha, G.; Slang, S.; Schmuki, P.; Macak, J.M. Anodic TiO 2 Nanotubes Decorated by Pt Nanoparticles Using ALD: An Efficient Electrocatalyst for Methanol Oxidation. *J. Catal.* **2018**, *365*, 86–93. [CrossRef]
10. Wasmus, S.; Küwer, A. Methanol Oxidation and Direct Methanol Fuel Cells: A Selective Review. *J. Electroanal. Chem.* **1999**, *461*, 14–31. [CrossRef]
11. Planes, G. a.; García, G.; Pastor, E. High Performance Mesoporous Pt Electrode for Methanol Electrooxidation. A DEMS Study. *Electrochem. Commun.* **2007**, *9*, 839–844. [CrossRef]
12. Martínez Huerta, M.V.; García, G. Fabrication of Electro-Catalytic Nano-Particles and Applications to Proton Exchange Membrane Fuel Cells. In *Micro & Nano-Engineering of Fuel Cells*; Leung, D.Y.C., Xuan, J., Eds.; Sustainable Energy Developments; CRC Press: London, UK, 2015; pp. 95–129.
13. Rizo, R.; Sebastián, D.; Rodríguez, J.L.; Lázaro, M.J.; Pastor, E. Influence of the Nature of the Carbon Support on the Activity of Pt/C Catalysts for Ethanol and Carbon Monoxide Oxidation. *J. Catal.* **2017**, *348*, 22–28. [CrossRef]

14. Rizo, R.; Arán-Ais, R.M.; Padgett, E.; Muller, D.A.; Lázaro, M.J.; Solla-Gullón, J.; Feliu, J.M.; Pastor, E.; Abruña, H.D. Pt-Richcore/Sn-Richsubsurface/Ptskin Nanocubes As Highly Active and Stable Electrocatalysts for the Ethanol Oxidation Reaction. *J. Am. Chem. Soc.* **2018**, *140*, 3791–3797. [CrossRef] [PubMed]

15. Rivera, L.M.; Fajardo, S.; Arévalo, M.D.C.; García, G.; Pastor, E. S- and N-Doped Graphene Nanomaterials for the Oxygen Reduction Reaction. *Catalysts* **2017**, *7*, 278. [CrossRef]

16. Zhu, Y.; Chen, G.; Zhong, Y.; Zhou, W.; Shao, Z. Rationally Designed Hierarchically Structured Tungsten Nitride and Nitrogen-Rich Graphene-Like Carbon Nanocomposite as Efficient Hydrogen Evolution Electrocatalyst. *Adv. Sci.* **2018**, *5*, 1700603. [CrossRef] [PubMed]

17. Liang, H.; Li, C.; Chen, T.; Cui, L.; Han, J.; Peng, Z.; Liu, J. Facile Preparation of Three-Dimensional Co 1-x S/Sulfur and Nitrogen-Codoped Graphene/Carbon Foam for Highly Efficient Oxygen Reduction Reaction. *J. Power Sources* **2018**, *378*, 699–706. [CrossRef]

18. Rivera, L.M.; García, G.; Pastor, E. Novel Graphene Materials for the Oxygen Reduction Reaction. *Curr. Opin. Electrochem.* **2018**, *9*, 233–239. [CrossRef]

19. Sun, M.; Liu, H.; Liu, Y.; Qu, J.; Li, J. Graphene-based transition metal oxide nanocomposites for the oxygen reduction reaction. *Nanoscale* **2015**, *7*, 1250–1269. [CrossRef] [PubMed]

20. Chen, D.; Feng, H.; Li, J. Graphene Oxide: Preparation, Functionalization, and Electrochemical Applications. *Chem. Rev.* **2012**, *112*, 6027–6053. [CrossRef] [PubMed]

21. Chen, D.; Tang, L.; Li, J. Graphene-based materials in electrochemistry. *Chem. Soc. Rev.* **2010**, *39*, 3157–3180. [CrossRef]

22. Chetty, R.; Kundu, S.; Xia, W.; Bron, M.; Schuhmann, W.; Chirila, V.; Brandl, W.; Reinecke, T.; Muhler, M. PtRu Nanoparticles Supported on Nitrogen-Doped Multiwalled Carbon Nanotubes as Catalyst for Methanol Electrooxidation. *Electrochim. Acta* **2009**, *54*, 4208–4215. [CrossRef]

23. Maiyalagan, T. Synthesis and Electro-Catalytic Activity of Methanol Oxidation on Nitrogen Containing Carbon Nanotubes Supported Pt Electrodes. *Appl. Catal. B Environ.* **2008**, *80*, 286–295. [CrossRef]

24. Sebastián, D.; Nieto-Monge, M.J.; Pérez-Rodríguez, S.; Pastor, E.; Lázaro, M.J. Nitrogen Doped Ordered Mesoporous Carbon as Support of PtRu Nanoparticles for Methanol Electro-Oxidation. *Energies* **2018**, *11*, 831. [CrossRef]

25. Antonietti, M.; Oschatz, M. The Concept of "Noble, Heteroatom-Doped Carbons," Their Directed Synthesis by Electronic Band Control of Carbonization, and Applications in Catalysis and Energy Materials. *Adv. Mater.* **2018**, *30*, 1706836. [CrossRef] [PubMed]

26. Ambrosi, A.; Chua, C.K.; Latiff, N.M.; Loo, A.H.; Wong, C.H.A.; Eng, A.Y.S.; Bonanni, A.; Pumera, M. Graphene and Its Electrochemistry—An Update. *Chem. Soc. Rev.* **2016**, *45*, 2458–2493. [CrossRef] [PubMed]

27. Zickler, G.A.; Smarsly, B.; Gierlinger, N.; Peterlik, H.; Paris, O. A Reconsideration of the Relationship between the Crystallite Size La of Carbons Determined by X-Ray Diffraction and Raman Spectroscopy. *Carbon N. Y.* **2006**, *44*, 3239–3246. [CrossRef]

28. Cançado, L.G.; Takai, K.; Enoki, T.; Endo, M.; Kim, Y.A.; Mizusaki, H.; Jorio, A.; Coelho, L.N.; Magalhães-Paniago, R.; Pimenta, M.A. General Equation for the Determination of the Crystallite Size La of Nanographite by Raman Spectroscopy. *Appl. Phys. Lett.* **2006**, *88*, 163106. [CrossRef]

29. Wagner, C.; Riggs, W.M.; Davis, L.E.; Moulder, J.F.; Muilenberg, G.E. *Handbook of X-ray Photoelectron Spectroscopy*; Muilenberg, G.E., Ed.; Perkin-Elmer Corporation: Eden Prairie, MN, USA, 1979.

30. Ganguly, A.; Sharma, S.; Papakonstantinou, P.; Hamilton, J. Probing the Thermal Deoxygenation of Graphene Oxide Using High-Resolution In Situ X-Ray-Based Spectroscopies. *J. Phys. Chem. C* **2011**, *115*, 17009–17019. [CrossRef]

31. Liu, D.; Li, L.; You, T. Superior catalytic performances of platinum nanoparticles loaded nitrogen-doped graphene toward methanol oxidation and hydrogen evolution reaction. *J. Colloid Interface Sci.* **2017**, *487*, 330–335. [CrossRef]

32. Xiong, B.; Zhou, Y.; Zhao, Y.; Wang, J.; Chen, X.; O'Hayre, R.; Shao, Z. The use of nitrogen-doped graphene supporting Pt nanoparticles as a catalyst for methanol electrocatalytic oxidation. *Carbon N. Y.* **2013**, *52*, 181–192. [CrossRef]

33. Chung, D.Y.; Lee, K.J.; Sung, Y.E. Methanol Electro-Oxidation on the Pt Surface: Revisiting the Cyclic Voltammetry Interpretation. *J. Phys. Chem. C* **2016**, *120*, 9028–9035. [CrossRef]

34. García, G.; Koper, M.T.M. Carbon Monoxide Oxidation on Pt Single Crystal Electrodes: Understanding the Catalysis for Low Temperature Fuel Cells. *ChemPhysChem* **2011**, *12*, 2064–2072. [CrossRef]

35. García, G. Correlation between CO Oxidation and H Adsorption/Desorption on Pt Surfaces in a Wide PH Range: The Role of Alkali Cations. *ChemElectroChem* **2017**, *4*, 459–462. [CrossRef]

36. Iwasita, T. Electrocatalysis of Methanol Oxidation. *Electrochim. Acta* **2002**, *47*, 3663–3674. [CrossRef]

37. Guillén-Villafuerte, O.; García, G.; Guil-López, R.; Nieto, E.; Rodríguez, J.L.; Fierro, J.L.G.; Pastor, E. Carbon monoxide and methanol oxidations on Pt/X@MoO$_3$/C (X = Mo$_2$C, MoO$_2$, Mo^0) electrodes at different temperatures. *J. Power Sources* **2013**, *231*, 163–172. [CrossRef]

38. Gisbert, R.; García, G.; Koper, M.T.M. Oxidation of carbon monoxide on poly-oriented and single-crystalline platinum electrodes over a wide range of pH. *Electrochim. Acta* **2011**, *56*, 2443–2449. [CrossRef]

39. Gilman, S. The Mechanism of Electrochemical Oxidation of Carbon Monoxide and Methanol on Platinum. II. The "Reactant-Pair" Mechanism for Electrochemical Oxidation of Carbon Monoxide and Methanol1. *J. Phys. Chem.* **1964**, *68*, 70–80. [CrossRef]

Article

Electrochemical Behavior of Pt–Ru Catalysts Supported on Graphitized Ordered Mesoporous Carbons toward CO and Methanol Oxidation

Juan Carlos Calderón Gómez [1,*], Verónica Celorrio [2,3], Laura Calvillo [4], David Sebastián [1], Rafael Moliner [1] and María Jesús Lázaro Elorri [1,*]

[1]   Instituto de Carboquímica (CSIC), Miguel Luesma Castán 4, 50018 Zaragoza, Spain; dsebastian@icb.csic.es (D.S.); rmoliner@icb.csic.es (R.M.)

[2]   UK Catalysis Hub, Research Complex at Harwell, RAL, Oxford OX11 0FA, UK; v.celorrio@ucl.ac.uk

[3]   Kathleen Lonsdale Building, Department of Chemistry, University College London, Gordon Street, London WC1H 0AJ, UK

[4]   Dipartimento di Scienze Chimiche, Università di Padova, Via Marzolo 1, 35131 Padova, Italy; laura.calvillolamana@unipd.it

*   Correspondence: jccalderon@icb.csic.es (J.C.C.G.); mlazaro@icb.csic.es (M.J.L.E.); Tel.: +34-976-733-977 (J.C.C.G. & M.J.L.E.)

Received: 29 November 2018; Accepted: 27 December 2018; Published: 3 January 2019

Abstract: In this work, graphitized ordered mesoporous carbons (gCMK-3) were employed as support for Pt and Pt–Ru nanoparticles synthesized by different reduction methods. The catalysts displayed metal contents and Pt:Ru atomic ratios close to 20 wt % and 1:1, respectively. A comparison of the physical parameters of Pt and Pt–Ru catalysts demonstrated that Ru enters into the Pt crystal structure, with well-dispersed nanoparticles on the carbon support. The Pt catalysts exhibited similar surface oxide composition, whereas a variable content of surface Pt and Ru oxides was found for the Pt–Ru catalysts. As expected, the Pt–Ru catalysts showed low CO oxidation onset and peak potentials, which were attributed to the high relative abundances of both metallic Pt and Ru oxides. All the studied catalysts exhibited higher maximum current densities than those observed for the commercial Pt and Pt–Ru catalysts, although the current–time curves at 0.6 V vs. reversible hydrogen electrode (RHE) demonstrated a slightly higher stationary current density in the case of the Pt/C commercial catalyst compared with Pt nanoparticles supported on gCMK-3s. However, the stationary currents obtained from the Pt–Ru/gCMK-3 catalysts surpassed those of the commercial Pt–Ru material, suggesting the suitability of the prepared catalysts as anodes for these devices.

Keywords: Pt–Ru catalysts; ordered mesoporous carbons; graphitization; CO oxidation; methanol oxidation; direct methanol fuel cells

1. Introduction

A quite interesting carbon material used as support for electrodes in direct methanol fuel cells (DMFCs) is the ordered mesoporous carbon (OMC). This material is synthesized by nanocasting using a silica template, which is impregnated with a carbon-based resin that, after carbonization and silica removal, results in an OMC with a 3D interconnected porous structure, high surface area, and rich content of mesopores [1], which promote the formation of uniformly distributed nanoparticles on the carbon surface [2]. Moreover, it is possible to modify its surface chemistry by creating surface functional groups that act as anchoring sites and enhance the electronic transfer between the catalytic nanoparticles and the support. The challenge is to maintain the ordered structure during the functionalization treatments in order to preserve the mesoporous structure as well as the high surface area [3,4], which has been demonstrated to be crucial for the performance of OMC-supported Pt nanoparticles [5,6].

Despite the mentioned outstanding properties of OMCs, their low electrical conductivity hinders their use in fuel cells. The relatively low temperatures employed during the carbonization stage seem to be the cause for the poor formation of extensive graphite layers [7,8]. To avoid this drawback, graphitization of OMCs has been suggested by means of different procedures, such as the use of aromatic and polyaromatic hydrocarbons as carbon precursors during the silica template impregnation, as they can produce more graphitic domains after carbonization. Some of these organic molecules could be benzene, naphthalene, anthracene, and pyrene [9]. A similar procedure indicated that Fe phthalocyanines can also generate graphitic planes, but this methodology is expensive due to the high cost of these molecules, even when low temperatures are used for carbonization [10]. Postsynthesis heat treatments are accepted as a useful way to obtain graphitized OMCs, although part of the ordered structure can be altered, broadening the pore size distribution [11]. Some of these treatments can be performed at temperatures lower than 1000 °C in the presence of transition metal salts [12–14] or increasing the temperature until 1500 °C in N_2 or Ar atmosphere, avoiding the use of these metal compounds.

Graphitized OMCs have been employed to support Pt nanoparticles, finding enhanced stationary current densities for the oxidation of methanol compared with other Pt catalysts supported on carbon nanofibers, carbon nanocoils, and carbon black [6]. Pd nanoparticles have also been supported on these materials to test their viability as anode catalysts in formic acid fuel cells. The current densities associated with the oxidation of formic acid on these catalysts were higher than those observed on a Pd catalyst supported on carbon black [15]. Regarding their use as cathode supports, Rivera et al. prepared Pt_2CrCo alloys supported on graphitized OMCs to assess the oxygen reduction reaction (ORR) on them. The authors obtained materials with outstanding activity toward the ORR, even in the presence of methanol, considering the crossover of this fuel between the anode and the cathode in DMFCs [16].

In this work, the activity of Pt and Pt–Ru catalysts supported on different graphitized ordered mesoporous carbon (gCMK-3) was studied in order to determine their behavior toward the CO and methanol electro-oxidation. It is important to highlight that the activity of Pt–Ru catalysts supported on graphitized ordered mesoporous carbons toward the mentioned molecules have not been reported in the literature, whereas the state of the art related to the behavior and performance of Pt catalysts supported on graphitized ordered mesoporous carbons toward the methanol electrochemical oxidation is in an incipient state. The CMK-3 was prepared using different TEOS/P123 mass ratios and subsequently treated at 1500 °C in Ar atmosphere to induce the formation of graphitic planes [6,15,16]. The Pt and Pt–Ru catalysts were prepared employing different reducing agents to verify their influence on the physical and electrochemical behavior of the catalysts. The gCMK-3-supported catalysts displayed higher CO tolerance and improved methanol oxidation current densities than those observed for commercial Pt and Pt–Ru catalysts supported on carbon black from E-TEK, which were used as reference to evaluate the results displayed for these catalysts.

2. Materials and Methods

2.1. Synthesis of Graphitized Ordered Mesoporous Carbon (gCMK)

CMK-3 was prepared by an incipient wetness impregnation procedure as described in Reference [15]. Briefly, the SBA-15 silica (Sigma-Aldrich, Madrid, Spain) was impregnated with the carbon precursor (furan resin, Huttenes-Albertus, Düsseldorf, Germany). The volume of furan resin used in the synthesis was equal to the silica porous volume (~1 cm^3 g^{-1}). After letting the material dry, it was subjected to a thermal treatment in nitrogen atmosphere at 700 °C for 2 h, and the silica was then removed using HF (40%, Fluka, Bucharest, Romania). Two silica templates with different textural properties were used for the synthesis. They were prepared by following the procedure described in

Reference [17] and using a TEOS/P123 (Sigma-Aldrich, Madrid, Spain) mass ratio (R) equal to 2 and 8 (SBA-15/R2 and SBA-15/R8, respectively). The carbon supports derived from these silica materials were labeled as CMK-3-R2 and CMK-3-R8 [17,18]. Finally, the CMK-3 materials were heat-treated at 1500 °C in a graphite electrical furnace for 1 h under argon flow in order to obtain the graphitized material (gCMK-3) [15].

2.2. Synthesis of gCMK-3-Supported Pt and Pt–Ru Catalysts

Pt and Pt–Ru catalysts supported on gCMK-3s were synthesized employing three different procedures described below. In all methods, a solution of the metal precursor salts (H_2PtCl_6 8% w/w solution and $RuCl_3$, 99.999%, Sigma-Aldrich, Madrid, Spain) was prepared in water or ethylene glycol (Sigma-Aldrich, Madrid, Spain) in order to obtain materials with a Pt:Ru atomic ratio of 1:1.

Method 1 (sodium borohydride method, BM): gCMK-3 was dispersed in water, and the solution containing the metal precursors was then slowly added under vigorous stirring. Finally, a 26.4 mM sodium borohydride (Sigma-Aldrich, Madrid, Spain) solution was added dropwise under sonication [3].

Method 2 (formic acid method, FAM): gCMK-3 was dispersed in a 2 M formic acid (Merck, Darmstadt, Germany) solution and then heated at 80 °C. Afterward, the metal precursor solution was slowly added and kept at this temperature for 12 h [4].

Method 3 (ethylene glycol method, EG): the metal precursors and the carbon support were dispersed in EG, and the pH was adjusted to 11 using a NaOH (Sigma-Aldrich, Madrid, Spain) solution in EG. The dispersion was heated at 195 °C for 2 h, quickly cooled down to room temperature, and the pH adjusted to 1 with HCl (Merck, Darmstadt, Germany) [19].

Finally, all the materials were filtered, washed, and dried at 60 °C. The catalyst were labeled as Pt/gCMK-3-RX-Y or Pt–Ru/gCMK-3-RX-Y, with X = 2, 8 and Y = BM, FAM, EG.

2.3. Physicochemical Characterization of the Catalysts

Energy dispersive X-ray (EDX) was employed to determine the metal content of the synthesized catalysts using a scanning electron microscope (SEM) Hitachi S-3400N (Hitachi, Düsseldorf, Germany) coupled to a Röntec XFlash analyzer (RÖNTEC GmbH, Berlin, Germany). This instrument is equipped with a Be window and a Si(Li) detector (Hitachi, Düsseldorf, Germany) operating at 15 keV.

X-ray diffraction (XRD) analyses were carried out to study the crystalline properties of the catalysts. XRD patterns were obtained by means of a Bruker AXS D8 Advance diffractometer (Bruker, Karlsruhe, Deutschland), with a θ-θ configuration and a Cu-Kα radiation (λ = 0.154 nm). The scans were achieved between 0° and 100° for 2θ values. The (220) peak located at 2θ $\approx$ 70° and the Scherrer's equation were used to calculate the crystallite size [20].

X-ray photoelectron spectroscopy (XPS) analysis was performed in order to determine the chemical state and the surface composition of Pt and Ru in the catalysts. An ESCAPlus Omicron spectrometer (Omicron, Houston, TX, USA) equipped with a concentric hemispherical analyzer, seven channeltron detectors, and an Al/Mg monochromated X-ray source (Kα = 1253.6 eV) was employed. This equipment was operated at 15 kV and 15 mA. The XPS spectra were recorded at pressures lower than 8×10^{-9} mbar. Data processing was performed with the CASAXPS® software (SurfaceSpectra, Manchester, UK), fitting the experimental data with Gaussian–Lorentzian curves.

In order to determine the particle size and the dispersion of the nanoparticles on the gCMK-3s, transmission electron microscopy (TEM) analyses were made using a 200 kV JEOL-2000 FXII transmission electron microscope (JEOL, Peabody, MA, USA). The catalysts were dropped from an ethanol suspension on a carbon grid. The images were taken with a MultiScan CCD Gatan 694 (Gatan, Pleasanton, CA, USA) camera and treated with the ImageJ® software (National Institutes for Health, Bethesda, MD, USA).

2.4. Electrochemical Characterization

A three-electrode cell connected to an AUTOLAB NS85630 modular equipment (Metrohm, Herisau, Switzerland) was employed to assess the electrochemical activity of the synthesized materials. A reversible hydrogen electrode (RHE) placed inside a Luggin capillary was used as the reference electrode, and a glassy carbon bar was used as the counter electrode. A catalyst-modified glassy carbon disk was used as the working electrode. The catalyst inks were prepared with 2.0 mg of catalyst, 15 mL of Nafion® (5 wt %, Sigma-Aldrich, Madrid, Spain) and 500 mL of ultrapure water. Afterward, a 60-μL aliquot was deposited and dried on the glassy carbon disk. All the aqueous solutions were prepared with high resistivity deoxygenated 18.2 MΩ cm H_2O. 0.5 M H_2SO_4 (95%–97%, Merck, Darmstadt, Germany) was used as the supporting electrolyte. For the methanol electrochemical characterization, 2.0 M methanol (99%, Merck, Darmstadt, Germany) solutions in the supporting electrolyte were employed. CO activity was studied by bubbling CO (99.999%, Air Liquide, Madrid, Spain) into the electrochemical cell for 10 min at 0.2 V vs. RHE and 20 °C in order to form a CO monolayer on the deposited catalyst; then, nitrogen (MicroGeN2, GasLab, Fremont, CA, USA) was bubbled for 20 min to remove the dissolved CO from the supporting electrolyte. Two potential scans at 0.020 V s^{-1} between 0.05 V and 1.0 V vs. RHE were applied. All the experiments were performed at 20 °C, while all the current densities presented in this paper have been normalized by the electroactive area calculated from the charge associated with the oxidation of an adsorbed CO monolayer on Pt (420 μC cm^{-2} $_{Pt}$).

The results obtained from the physical and the electrochemical characterization were compared against those exhibited by the commercial Pt and Pt–Ru catalysts supported on carbon black from E-TEK (DeNora, Milan, Italy). These catalysts have a metal content close to 20 wt% and a Pt:Ru ratio near 1:1.

3. Results

3.1. Physical Characterization

Table 1 summarizes the metal content of the different studied catalysts. In all the cases, metal contents around 20 wt% were observed, agreeing with the nominal values. Pt:Ru atomic ratios were close to 1:1, except in the case of the catalyst prepared by the EG method, which exhibited a higher Pt atomic content (66%).

Figures 1 and 2 depict the XRD patterns for the prepared Pt and Pt–Ru catalysts, respectively. All the catalysts presented the typical (110), (200), and (220) peaks for the face-centered cubic structure of Pt located at 2θ of 39.75°, 46.31°, and 67.48°, as those observed for the commercial Pt/C and Pt–Ru/C. On the other hand, the peaks for the Pt–Ru materials exhibited a shift toward higher 2θ values with respect to those of the Pt catalysts (see Figure 2), which is attributed to the structure contraction due to the incorporation of Ru into the fcc Pt crystalline structure [21–23]. This fact was corroborated by the smaller lattice parameter calculated for the Pt–Ru materials compared with the Pt values (see Table 1). Furthermore, a peak around 25° in the XRD patterns of the Pt and Pt–Ru commercial catalysts was detected, which corresponds to the (002) reflection of the graphite basal planes present in the carbon black acting as support for these materials.

Table 1. Physical characterization of the catalysts supported on graphitized ordered mesoporous carbons (gCMK-3s) and the commercial Pt/C E-TEK. Metal surface area: SA (m^2 g^{-1}) = 6 × 10^3/ρd, with d as the average crystallite size (nm) and ρ as the alloy density (g cm^{-3}). ρ was determined by considering that ρ_{Pt-Ru}(g cm^{-3}) = ρ_{Pt} X$^m_{Pt}$ + ρ_{Ru} X$^m_{Ru}$, with ρ_{Pt} = 21.4 g cm^{-3} and ρ_{Ru} = 12.3 g cm^{-3} and X$^m_{Pt}$ and X$^m_{Ru}$ as the weight percentage of Pt and Ru, respectively.

Catalyst	Metal Content, wt %	Pt:Ru Atomic Ratio	Crystallite and Particle Size, nm		Lattice Parameter, Å	Metal Surface Area, m^2 g^{-1}
			XRD	TEM		
Pt/gCMK-3-R2-BM	20	—	5.9	3.9 ± 1.3	3.923	47
Pt/gCMK-3-R2-FAM	22	—	3.7	3.2 ± 1.0	3.926	76
Pt/gCMK-3-R8-BM	16	—	5.6	4.1 ± 1.4	3.920	50
Pt/gCMK-3-R8-EG	24	—	6.2	5.9 ± 1.4	3.919	45
Pt/gCMK-3-R8-FAM	23	—	4.1	4.9 ± 1.6	3.920	68
Pt/C E-TEK	20	—	3.0	—	3.921	93
Pt–Ru/gCMK-3-R2-FAM	26	53:47	6.6	5.2 ± 2.3	3.896	49
Pt–Ru/gCMK-3-R8-EG	22	67:33	3.6	6.7 ± 2.8	3.921	85
Pt–Ru/gCMK-3-R8-FAM	24	55:45	5.6	3.9 ± 1.3	3.898	57
Pt–Ru/C E-TEK	20	45:55	4.4	—	3.866	76

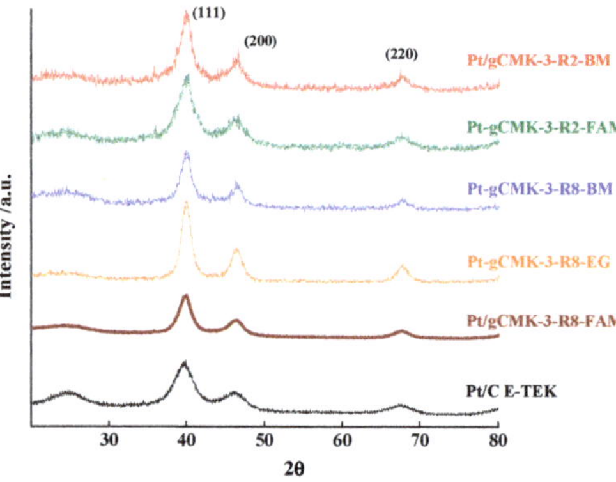

Figure 1. XRD patterns of the Pt catalysts supported on gCMK-3s and the commercial Pt/C from E-TEK.

The crystallite sizes were calculated from the (220) reflection, and the values are reported in Table 1 [21]. For the Pt catalysts, the smallest crystallite sizes were obtained for the FAM method and the largest ones for the EG, whereas the opposite was observed for the Pt–Ru materials. This result is in agreement with that reported by Calderón et al. for Pt–Ru catalysts supported on carbon nanofibers and prepared by different synthesis routes [21]. In general, the addition of Ru led to an increase in the crystallite size and thus the metal surface areas (SA) agreeing with the expression described in Table 1. As expected, SA diminished with the increase in the crystallite size, with Pt/gCMK-3-R8-EG displaying the lowest value in the group of the Pt/gCMK-3 catalysts. In the case of Pt–Ru materials, Pt–Ru/gCMK-3-R2-FAM presented lower value in the SA as a consequence of the largest crystallite diameter determined for this material.

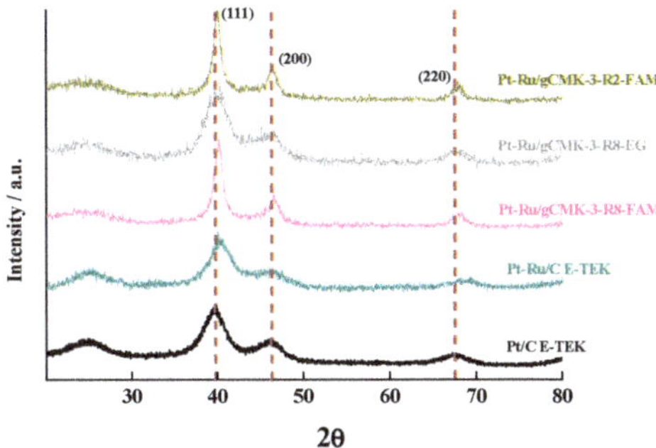

Figure 2. XRD patterns of the Pt–Ru catalysts supported on gCMK-3s and the commercial Pt/C and Pt–Ru/C from E-TEK.

Figure 3 shows the TEM images obtained for the Pt/gCMK-3 catalysts, which demonstrate well-dispersed nanoparticles on the carbon support, although some agglomerates can be observed in some cases, such as Pt/gCMK-3-R2-BM and Pt/gCMK-3-R2-EG. Pt/gCMK-3-R2-FAM showed the smallest particle size, while Pt/gCMK-3-R8-EG exhibited the highest one, following the trend observed for the crystallite size values determined by XRD. For the Pt–Ru catalysts (Figure 4), larger particles were observed, especially for Pt–Ru/gCMK-3-R2-FAM and Pt–Ru/gCMK-3-R8-EG. These materials displayed a wide distribution of diameters between 1 and 12 nm, although the catalyst Pt–Ru/gCMK-R8-FAM showed a narrower distribution and the smallest particle size among the synthesized Pt–Ru catalysts.

The difference in the magnitude between the crystallite and the particle diameters can be explained as follows: In XRD, an average of the crystallite size is obtained as a consequence of the X-ray beam passing through the crystallites, even if they are agglomerated. During the TEM analysis, a counting and measuring process is performed, which considers dispersed and nonagglomerated crystalline and amorphous particles. Therefore, smaller values are determined [24].

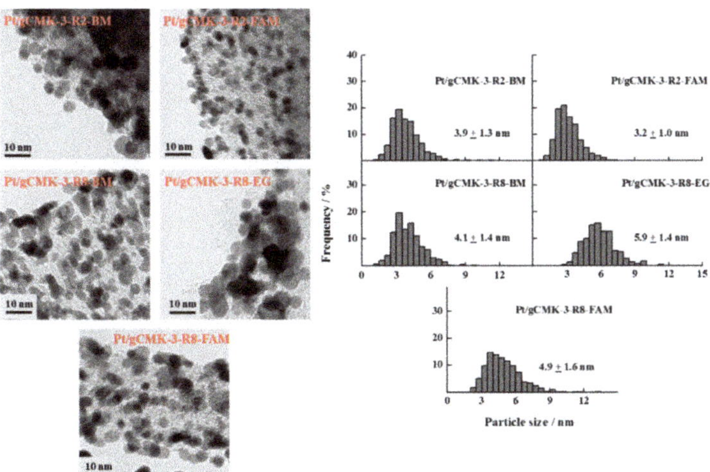

Figure 3. TEM images and particle size distributions of the synthesized Pt/gCMK-3 catalysts.

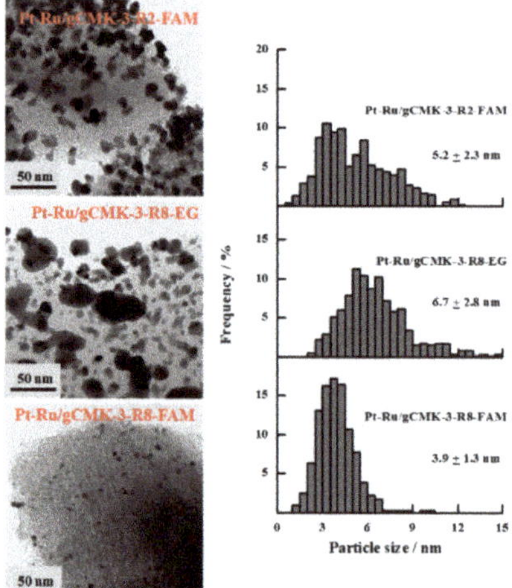

Figure 4. TEM images and particle size distributions of the synthesized Pt–Ru/gCMK-3 catalysts.

XPS analysis was carried out to identify the chemical and electronic states of the metals present in the catalysts as well as their relative surface abundance, and the results are reported in Table 2. From the deconvolution of the binding energy spectra of Pt 4f for Pt/gCMK-3 and Pt–Ru/gCMK-3 catalysts, three pairs of Pt peaks were observed, which correspond to the three oxidation states of this metal (see Figure S1 in the Supplementary Information). The transitions generated in the 4f 7/2 and 4f 5/2 orbitals of Pt (0) are described by the peaks close to 71 and 75 eV, while the second and third pair of peaks, located at 73–76 and 75–79 eV, are assigned to Pt (II) and Pt (IV), respectively. In the case of the catalysts Pt/gCMK-3, a similar surface composition of the nanoparticles was observed as all of them displayed a Pt (0) relative surface abundance around 60%. The oxidized states Pt (II) and Pt (IV) exhibited similar abundances between them (close to 20%), indicating that neither the synthesis routes nor the carbon support affect the surface composition of these catalysts. However, in the case of the Pt–Ru/gCMK-3 catalysts (spectra depicted in Figure S2 in the Supplementary Information), several differences were observed in the relative abundance of the Pt surface oxidation states, which are attributed to the entrance of Ru into the Pt lattice that induces changes in the electronic structure and interactions of this metal [25–27]. The catalysts supported on gCMK-3-R8 displayed a higher abundance of Pt (0), a result that matched with the trend observed for the Pt catalysts supported on gCMK-3-R8, which displayed a slightly higher abundance of the Pt (0) oxidation state compared with that observed for the Pt nanoparticles supported on gCMK-R2. Regarding the catalyst Pt–Ru/gCMK-3-R2-FAM, it is possible to conclude that, on this carbon material, the presence of Ru induced the formation of nanoparticles with a high surface content of Pt oxides.

Table 2. Electronic and composition parameters of the catalysts supported on gCMK-3s obtained from the XPS data.

Catalyst	Chemical State	Relative Area, %	Binding Energy, eV
Pt/gCMK-3-R2-BM			*Pt 4f 7/2*
	Metallic Pt	59.4	71.4
	Pt^{2+}	21.7	72.6
	Pt^{4+}	18.9	75.3
Pt/gCMK-3-R2-FAM			*Pt 4f 7/2*
	Metallic Pt	60.9	71.4
	Pt^{2+}	20.6	72.6
	Pt^{4+}	18.5	75.3
Pt/gCMK-3-R8-BM			*Pt 4f 7/2*
	Metallic Pt	61.1	71.4
	Pt^{2+}	19.5	72.6
	Pt^{4+}	19.4	75.3
Pt/gCMK-3-R8-EG			*Pt 4f 7/2*
	Metallic Pt	61.7	71.4
	Pt^{2+}	20.1	72.6
	Pt^{4+}	18.2	75.4
Pt/gCMK-3-R8-FAM			*Pt 4f 7/2*
	Metallic Pt	62.1	71.5
	Pt^{2+}	18.8	72.7
	Pt^{4+}	19.1	75.5
Pt–Ru/gCMK-3-R2-FAM			*Pt 4f 7/2*
	Metallic Pt	28.7	71.4
	Pt^{2+}	17.4	72.6
	Pt^{4+}	53.9	74.2
			Ru 3p 3/2
	Metallic Ru	14.9	461.0
	Ru^{4+}	27.9	463.2
	Ru^{4+} hydrate	57.2	465.4
Pt–Ru/gCMK-3-R8-EG			*Pt 4f 7/2*
	Metallic Pt	59.1	71.4
	Pt^{2+}	22.2	72.7
	Pt^{4+}	18.7	75.7
			Ru 3p 3/2
	Metallic Ru	43.6	461.6
	Ru^{4+}	18.6	463.2
	Ru^{4+} hydrate	37.8	465.3
Pt–Ru/gCMK-3-R8-FAM			*Pt 4f 7/2*
	Metallic Pt	63.5	71.6
	Pt^{2+}	19.2	72.9
	Pt^{4+}	17.28	75.4
			Ru 3p 3/2
	Metallic Ru	18.9	461.1
	Ru^{4+}	40.7	463.3
	Ru^{4+} hydrate	40.4	465.1

Table 2 also reports the results related to the XPS analysis of Ru and its signals. The studied transitions correspond to the 3p orbitals located between 450 and 494 eV (see Figure S3 in the Supplementary Information). A doublet attributed to 3p 1/2 and 3p 3/2 transitions from Ru (0) around 461 and 483 eV was determined, besides two doublets around 463-485 and 465-487, assigned to the transitions of Ru (IV) and hydrated Ru (IV), respectively. The deconvolution analysis suggests an influence of the synthesis method on the Ru oxides surface abundance as the catalysts reduced with formic acid (FAM) exhibited higher Ru (IV) abundances than that observed for the catalyst reduced with ethylene glycol. This means that the obtaining of different relative abundances for the Ru oxidation states depends on the employed reducing agent, with formic acid leading a less complete reduction of Ru and ethylene glycol promoting a major reduction of Ru.

3.2. CO Stripping

The activity of the catalysts toward the carbon monoxide electro-oxidation was evaluated, and the results are depicted in Figures 5 and 6. Pt/gCMK-3-R8 samples exhibited lower onset potential (values between 0.52–0.61-V) than those supported on gCMK-3-R2 (0.63–0.67 V), indicating that linearly adsorbed CO is promoted on these materials due to the high content of metallic Pt in these catalysts, which was detected by XPS (see Table 2) [28]. From this analysis, it is possible to conclude that CO oxidation starts at low potentials on catalysts with a high metallic Pt abundance. Regarding the CO peak potentials, all the Pt samples exhibited a main oxidation peak value around 0.80–0.83 V vs. RHE, similar to that observed for the commercial Pt/C. However, Pt/gCMK-3-R8-EG and Pt/gCMK-3-R8-FAM displayed an additional peak at 0.71 V vs. RHE, indicating a higher ability of the materials supported on gCMK-3-R8 to oxidize the adsorbed CO and thus suggesting an improved tolerance of these materials toward this intermediate formed during the methanol oxidation [21].

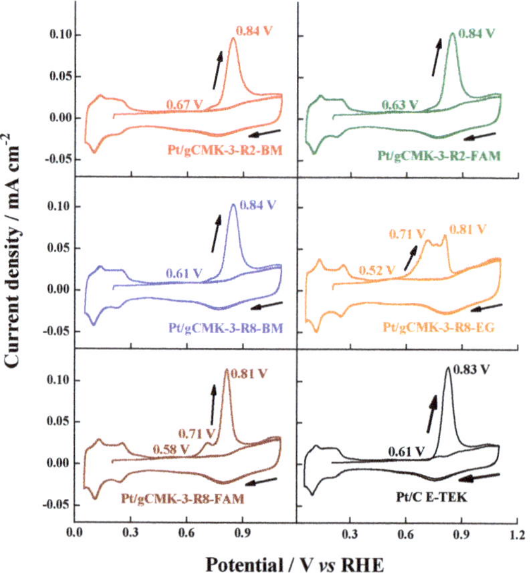

Figure 5. Cyclic voltammograms for the CO_{ad} oxidation in 0.5 M H_2SO_4 aqueous solution on the Pt/gCMK-3 catalysts. Experiments were performed at 0.020 V s^{-1} and 20 °C. CO adsorption potential: 0.2 V vs. reversible hydrogen electrode (RHE).

As expected, the Pt–Ru catalysts oxidized CO at lower potentials than the Pt catalysts (Figure 6) due to the presence of Ru, which helps to oxidize the CO adsorbed on Pt [21]. These values were between 0.43 and 0.47 V compared with 0.50 V detected for the commercial Pt–Ru/C from E-TEK. For the Pt–Ru samples, the main oxidation peak was attained at around 0.54–0.57 V vs. RHE, slightly lower than the commercial one (0.60 V). Only the Pt–Ru/gCMK-3-R8-FAM sample showed an additional peak at lower potential (0.49 V), which could be interpreted as enrichment of Pt and Ru oxides on the surface of the catalyst as this catalyst displayed the highest metallic Pt abundance and also a high presence of Ru oxides [21,29]. This argument is also valid for explaining the positive CO onset and peak potential values observed for the catalyst Pt–Ru/gCMK-3-R8-EG, which also exhibited the lowest Ru oxides abundance values (see Table 2).

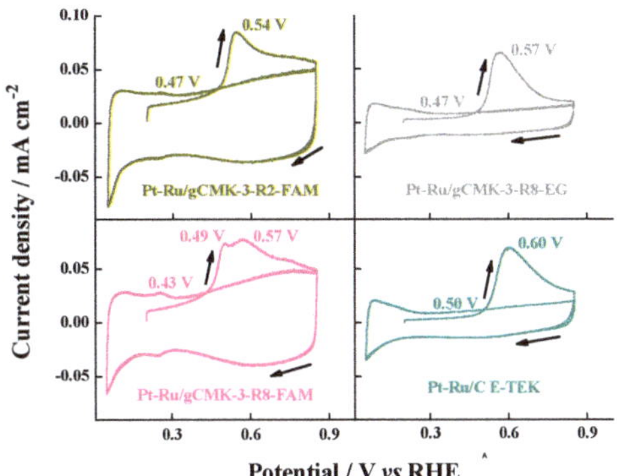

Figure 6. Cyclic voltammograms for the CO_{ad} oxidation in 0.5 M H_2SO_4 aqueous solution on the Pt–Ru/gCMK-3 catalysts. Experiments were performed at 0.020 V s^{-1} and 20 °C. CO adsorption potential: 0.2 V vs. RHE.

These results suggest a positive effect of the use of gCMK-3 carbons on the CO oxidation reaction, which could be attributed to its enhanced diffusion through the gCMK-3 ordered mesoporous structure and a good electronic transfer between the nanoparticles and the support, as already proposed by Calvillo et al. for a Pt catalyst supported on gCMK-3. The mentioned authors reported a significant decrease in these values compared with a commercial Pt/C catalyst from E-TEK, a result explained by the changes in the electronic structure of the metal due to its interaction with the carbon support [6]. On the other hand, the graphitization of the carbon supports did not improve the catalytic activity toward the CO electrochemical oxidation as more positive potential values for the oxidation of this adsorbate were observed compared with those reported for catalysts supported on non-graphitized CMK materials [30]. At the same time, this result matches those found in Reference [6], which were attributed to a lack of surface oxygen groups, which play a key role in the oxidation of this adsorbate.

3.3. Methanol Oxidation

Figure 7 shows the behavior of the Pt catalysts supported on gCMK-3s toward the methanol oxidation. The catalysts supported on gCMKs developed significantly higher current densities than the commercial Pt/C. Moreover, the magnitude of the current densities were higher than those reported by Salgado et al. for the oxidation of this fuel on non-graphitized CMK-3-supported Pt and Pt–Ru catalysts [30]. Our results match with those reported in Reference [6], which can be explained by the enhanced electrical conductivity of gCMK-3, which promotes the obtaining of increased current densities for the methanol oxidation.

All the samples showed a methanol oxidation current hysteresis, which has recently been explained by a change in the rate determining step, from water dissociation to the methanol dehydrogenation step. In the forward scan, oxygenated species are adsorbed on the electrode, changing the surface nature of the electrode and thus generating a peak in the backward scan at different potentials attributed to the oxidation of fresh methanol adsorbed on the Pt surface [31,32]. Among the studied catalysts, only Pt/gCMK-R8-BM demonstrated a high oxophilicity as the peak current in the forward scan, I_f, was higher than that of the backward scan, I_b [32], while the catalyst supported on gCMK-3-R2 and Pt/gCMK-R8-EG and Pt/gCMK-R8-FAM developed a higher peak current during the backward scan, indicating lower oxophilicity of these materials that improved the adsorption of methanol on the oxidized Pt surface, allowing its oxidation on oxidized Pt active sites [31].

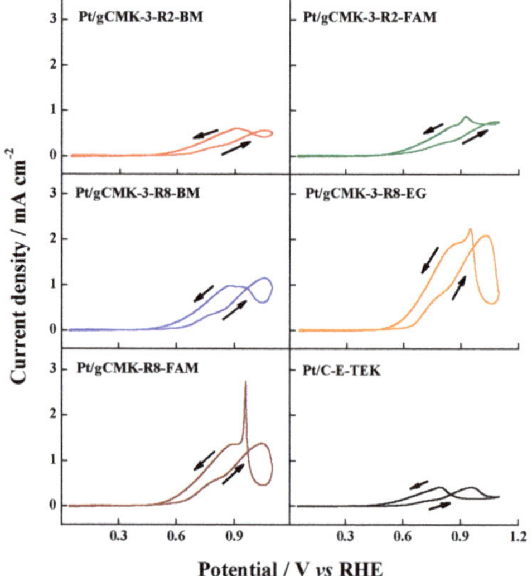

Figure 7. Cyclic voltammograms in 2 M CH$_3$OH + 0.5 M H$_2$SO$_4$ aqueous solution on the Pt/gCMK-3 catalysts. Experiments were performed at 0.020 V s^{-1} and 20 °C.

Similar results were observed for the Pt–Ru catalysts. The catalysts reduced with formic acid showed higher current densities than the commercial Pt–Ru/C and the catalyst reduced with ethylene glycol (see Figure 8). The best performance was obtained with Pt–Ru/gCMK-3-R8-FAM, which can again be related to the high content of Ru oxides in these catalysts, which promote the CO oxidation [21].

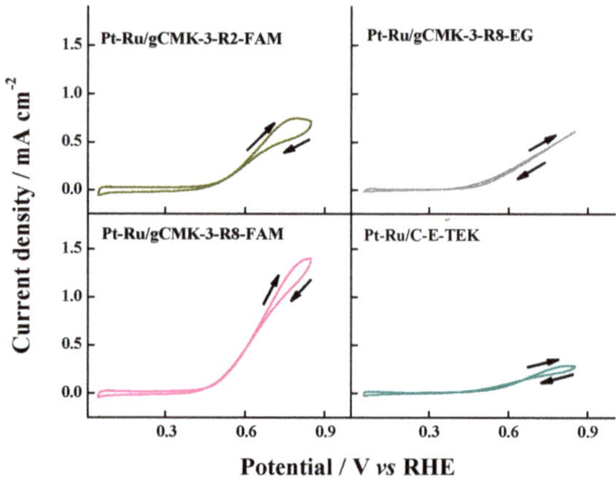

Figure 8. Cyclic voltammograms in 2 M CH$_3$OH + 0.5 M H$_2$SO$_4$ aqueous solution on the Pt–Ru/gCMK-3 catalysts. Experiments were performed at 0.020 V s^{-1} and 20 °C.

Figure 9a depicts the *j–t* curves for the methanol oxidation on the Pt and Pt–Ru catalysts supported on graphitized ordered mesoporous carbons recorded at 0.6 V vs. RHE, a typical DMFC work potential. In the case of the Pt catalysts, the commercial one presented the highest current densities, a fact explained by an improved reoxidation of the intermediate species formed during the methanol oxidation, which could be favored by the presence of the micropores of the carbon black, which delays their diffusion and allows their reoxidation [33]. Nevertheless, the difference in the stationary current densities among the commercial catalyst and the gCMK-3-supported catalyst with the highest performance (Pt/gCMK-3-R8-EG) was around 11–12 $\mu A\ cm^{-2}$, low values that allowed us to conclude that the use of Pt/gCMK-3 catalysts could still be an alternative as anodes in DMFCs. Among the synthesized catalysts, no significant differences in the magnitude of the stationary current densities were found, except in the case of the catalyst Pt/gCMK-3-R2-BM, which displayed the lowest current density (0.0279 $mA\ cm^{-2}$).

The behavior of the stationary current densities for the Pt–Ru catalysts was different to that observed for the Pt materials, as can be seen in Figure 9b. The synthesized catalysts developed higher current densities than those observed for the commercial Pt–Ru/C catalyst from E-TEK and the studied Pt catalysts, making evident the effect of Ru and their surface oxides determined by XPS analysis on the enhancement of the methanol oxidation stationary current densities obtained on these catalysts, with Pt–Ru/gCMK-3-R8-FAM being the catalyst displaying the highest stationary current density, as previously observed in the cyclic voltammetry experiments.

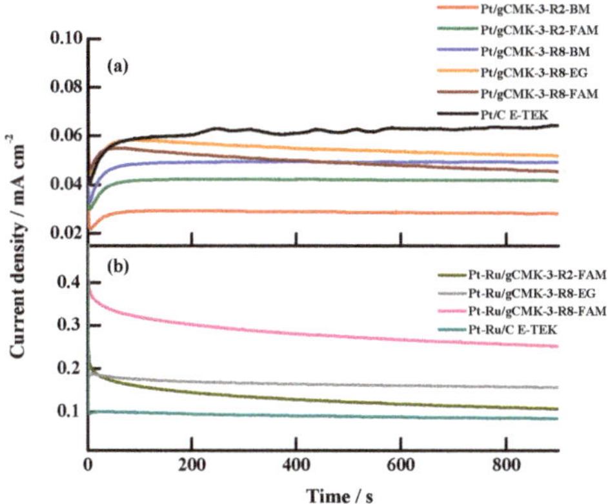

Figure 9. *j–t* curves in 2 M CH_3OH + 0.5 M H_2SO_4 aqueous solution at 0.60 V vs. RHE on the Pt/gCMK-3 (**a**) and Pt–Ru/gCMK-3 (**b**) catalysts.

Considering the results presented here, it is possible to establish a comparison between the performance of the catalysts presented in this work and those reported for materials supported on ordered mesoporous carbons. In the case of Pt catalysts supported on ordered mesoporous carbons, Su et al. prepared materials using sodium borohydride as reducing agent, obtaining catalysts that achieved current densities close to 1.6 $mA\ cm^{-2}$, surpassing the values generated for a commercial Pt/C catalyst from E-TEK [34]. The authors explained this high performance by the improved transport of methanol toward the catalytic nanoparticles through the pores of the support. As mentioned above, Salgado et al. [30] synthesized Pt catalysts supported on ordered mesoporous carbons, which displayed CO oxidation peak potentials and methanol oxidation current densities comparable to those presented in this work. The improved diffusion of the electroactive species through the carbon support was stated as the main cause for the remarkable performance of these catalysts. It is important to highlight that the results presented here are quite similar to those described in Reference [6], although in this case, the Pt nanoparticles were supported on graphitized ordered mesoporous carbons. This fact corroborates that the materials studied here possess a catalytic activity that follows the trend of the catalysts already reported in the literature.

The behavior Pt–Ru catalysts supported on ordered mesoporous carbons toward the methanol oxidation have also been studied. Salgado et al. [30] reported the activity of Pt–Ru catalysts supported on non-graphitized ordered mesoporous carbons, which displayed lower methanol oxidation current densities than those reported by Calvillo et al. [6] as well as that obtained in the present study. The increased current densities obtained on graphitized CMK-3 can be explained by the high conductivity of these materials achieved after the heat treatment applied on them. Other OMCs have been demonstrated to play a key role in the increase in the performance of the as-supported nanoparticles. Maiyalagan et al. compared the activity of highly stable Pt–Ru nanoparticles supported on ordered mesoporous carbons and carbon blacks, finding higher current densities for the methanol oxidation on the OMC-supported materials [35]. Nitrogen doping of ordered mesoporous carbons has also been suggested as an alternative to improve the performance of Pt–Ru catalysts, generating methanol oxidation current densities higher than 6 mA cm^{-2} in acidic and alkaline media, which were attributed to both the presence of nitrogen as dopant and the increase in the porosity of the carbon supports as a consequence of the treatment employed to dope the ordered mesoporous carbons [36]. From the comparisons presented here, it is possible to suggest that the Pt and Pt–Ru catalysts supported on graphitized CMK-3 are suitable materials to be employed as anode catalysts in DMFCs. However, further studies are necessary to fully understand and characterize the mechanism of methanol oxidation on these catalysts and therefore finding the effect of carbon support and composition of the catalysts on this reaction. Studies at different temperatures can be useful to determine the rate determining step on the different catalysts, along with differential electrochemical mass spectrometry to identify the onset potentials for both CO and methanol oxidations.

4. Conclusions

Pt and Pt–Ru catalysts supported on graphitized ordered mesoporous carbons (gCMK-3) were synthesized using sodium borohydride, formic acid, and ethylene glycol as reducing agents. Metal contents close to 20 wt% and Pt:Ru atomic ratios close to 1:1 were obtained with well-dispersed metal nanoparticles on the carbon materials and a narrow particle size distribution around 5 nm. The Pt/gCMK-3 catalysts displayed similarities in terms of the surface relative abundance for Pt (0), Pt (II), and Pt (IV) oxidation states, with values near 60%, 20%, and 20%, respectively. On the other hand, the Pt–Ru/gCMK-3 materials exhibited a higher abundance of surface Pt and Ru oxides. As expected, the Pt–Ru materials were the most tolerant to CO as they were able to oxidize this adsorbate at the lowest potentials, an effect attributed to the presence of Ru oxides present on the surface of the catalyst. In agreement with the CO oxidation results, the Pt–Ru catalysts, in particular the one supported on gCMK-3-R8, were the most active ones toward the methanol oxidation. Pt/gCMK-3 catalysts displayed oxidation current densities with some hysteresis, indicating the adsorption of methanol on oxidized Pt to react on this surface. Moreover, the stationary current densities at 0.6 V developed by the Pt–Ru/gCMK-3 catalysts were higher than those of the Pt materials, suggesting a profitable effect of this metal and their oxides on these catalysts. The results found in this research indicate that the studied catalysts can be good candidates to be used as anodes in direct methanol fuel cells.

Supplementary Materials: The following are available online at http://www.mdpi.com/2571-9637/2/1/1/s1, Figure S1: Pt 4f XPS spectra for the synthesized Pt/gCMK-3 catalysts. Red line: Pt (0), blue line: Pt (II), green line: Pt (IV), black dotted line: fitted curve, black full line: spectrum. Figure S2: Pt 4f XPS spectra for the synthesized Pt–Ru/gCMK-3 catalysts. Red line: Pt (0), blue line: Pt (II), green line: Pt (IV), black dotted line: fitted curve, black full line: spectrum. Figure S3: Ru 3p XPS spectra for the synthesized Pt–Ru/gCMK-3 catalysts. Red line: Ru (0), blue line: Ru (IV), green line: Ru (IV) hydrated, black dotted line: fitted curve, black full line: spectrum.

Author Contributions: J.C.C.G. performed the experiments and wrote the paper. V.C. and L.C. synthesized the carbon supports, prepared the catalysts, and performed the physical characterization (EDX, XRD, TEM). D.S. performed the XPS experiments and analysis. R.M. and M.J.L.E. contributed reagents/materials/analysis tools. All the authors conceived and designed the experiments and revised the manuscript.

Funding: This work was carried out with the support of the Conversion Group of Gobierno de Aragon T06 and FEDER.

Acknowledgments: V.C. kindly thanks the UK Catalysis Hub for resources and support provided via the membership of the UK Catalysis Hub Consortium and funded by EPSRC (EPSRC grants EP/K014706/1 and EP/K014714/1).

Conflicts of Interest: The authors declare no conflict of interest.

References

1. Lu, A.; Schüth, F. Nanocasting: A Versatile strategy for creating nanostructured porous materials. *Adv. Mater.* **2006**, *18*, 1793–1805. [CrossRef]
2. Joo, S.H.; Choi, S.J.; Oh, I.; Kwak, J.; Liu, Z.; Terasaki, O.; Ryoo, R. Ordered nanoporous arrays of carbon supporting high dispersions of platinum nanoparticles. *Nature* **2001**, *412*, 169–172. [CrossRef] [PubMed]
3. Calvillo, L.; Lázaro, M.J.; García-Bordejé, E.; Moliner, R.; Cabot, P.L.; Esparbé, I.; Pastor, E.; Quintana, J.J. Platinum supported on functionalized ordered mesoporous carbon as electrocatalyst for direct methanol fuel cells. *J. Power Sources* **2007**, *169*, 59–64. [CrossRef]
4. Salgado, J.R.C.; Quintana, J.J.; Calvillo, L.; Lázaro, M.J.; Cabot, P.L.; Esparbé, I.; Pastor, E. Carbon monoxide and methanol oxidation at platinum catalysts supported on ordered mesoporous carbon: The influence of functionalization of the support. *Phys. Chem. Chem. Phys.* **2008**, *10*, 6796–6806. [CrossRef] [PubMed]
5. Lei, Z.; An, L.; Dang, L.; Zhao, M.; Shi, J.; Bai, S.; Cao, Y. Highly dispersed platinum supported on nitrogen-containing ordered mesoporous carbon for methanol electrochemical oxidation. *Microporous Mesoporous Mater.* **2009**, *119*, 30–38. [CrossRef]
6. Calvillo, L.; Celorrio, V.; Moliner, R.; García, A.B.; Camean, I.; Lázaro, M.J. Comparative study of Pt catalysts supported on different high conductive carbon materials for methanol and ethanol oxidation. *Electrochim. Acta* **2013**, *102*, 19–27. [CrossRef]
7. Eftekhari, A.; Fan, Z. Ordered mesoporous carbon and its applications for electrochemical energy storage and conversion. *Mater. Chem. Front.* **2017**, *1*, 1001–1027. [CrossRef]
8. Fulvio, P.F.; Mayes, R.T.; Wang, X.; Mahurin, S.M.; Bauer, J.C.; Presser, V.; McDonough, J.; Gogotsi, Y.; Dai, S. "Brick-and-mortar" self-assembly approach to graphitic mesoporous carbon nanocomposites. *Adv. Funct. Mater.* **2011**, *21*, 2208–2215. [CrossRef]
9. Kim, C.H.; Lee, D.K.; Pinnavaia, T.J. Graphitic mesostructured carbon prepared from aromatic precursors. *Langmuir* **2004**, *20*, 5157–5159. [CrossRef]
10. Lee, K.; Ji, X.; Rault, M.; Nazar, L. Simple synthesis of graphitic ordered mesoporous carbon materials by a solid-state method using metal phthalocyanines. *Angew. Chem.* **2009**, *48*, 5661–5665. [CrossRef]
11. Fuertes, A.B.; Alvarez, S. Graphitic mesoporous carbons synthesised through mesostructured silica templates. *Carbon* **2004**, *42*, 3049–3055. [CrossRef]
12. Tang, J.; Wang, T.; Sun, X.; Guo, Y.; Xue, H.; Guo, H.; Liu, M.; Zhang, X.; He, J. Effect of transition metal on catalytic graphitization of ordered mesoporous carbon and Pt/metal oxide synergistic electrocatalytic performance. *Microporous Mesoporous Mater.* **2013**, *177*, 105–112. [CrossRef]
13. Nettelroth, D.; Schwarz, H.C.; Burblies, N.; Guschanski, N.; Behrens, P. Catalytic graphitization of ordered mesoporous carbon CMK-3 with iron oxide catalysts: Evaluation of different synthesis pathways. *Phys. Status Solidi A* **2016**, *213*, 1395–1402. [CrossRef]
14. Sultana, K.N.; Fadhel, A.L.; Deshmane, V.G.; Ilias, S. Novel method for synthesis of electrocatalyst via catalytic graphitization of ordered mesoporous carbon for PEMFC application. *Sep. Sci. Technol.* **2018**, *53*, 1948–1956. [CrossRef]
15. Celorrio, V.; Sebastián, D.; Calvillo, L.; García, A.B.; Fermín, D.J.; Lázaro, M.J. Influence of thermal treatments on the stability of Pd nanoparticles supported on graphitised ordered mesoporous carbons. *Int. J. Hydrog. Energy* **2016**, *41*, 19570–19578. [CrossRef]
16. Rivera Gavidia, L.M.; García, G.; Celorrio, V.; Lázaro, M.J.; Pastor, E. Methanol tolerant Pt_2CrCo catalysts supported on ordered mesoporous carbon for the cathode of DMFC. *Int. J. Hydrog. Energy* **2016**, *41*, 19645–19655. [CrossRef]
17. Calvillo, L.; Celorrio, V.; Moliner, R.; Cabot, P.L.; Esparbé, I.; Lázaro, M.J. Control of textural properties of ordered mesoporous materials. *Microporous Mesoporous Mater.* **2008**, *116*, 292–298. [CrossRef]
18. Lázaro, M.J.; Calvillo, L.; Bordejé, E.G.; Moliner, R.; Juan, R.; Ruiz, C.R. Functionalization of ordered mesoporous carbons synthesized with SBA-15 silica as template. *Microporous Mesoporous Mater.* **2007**, *103*, 158–165. [CrossRef]
19. Wang, X.; Hsing, I.M. Surfactant stabilized Pt and Pt alloy electrocatalyst for polymer electrolyte fuel cells. *Electrochim. Acta* **2002**, *47*, 2981–2987. [CrossRef]
20. Warren, B.E. *X-ray Diffraction*, 1st ed.; Addison-Wesley: Reading, UK, 1969; pp. 27–40. ISBN 0486663175.

21. Calderón, J.C.; García, G.; Calvillo, L.; Rodríguez, J.L.; Lázaro, M.J.; Pastor, E. Electrochemical oxidation of CO and methanol on Pt–Ru catalysts supported on carbon nanofibers: The influence of synthesis method. *Appl. Catal. B-Environ.* **2015**, *165*, 676–686. [CrossRef]

22. Calderón, J.C.; Mahata, N.; Pereira, M.F.R.; Figueiredo, J.L.; Fernandes, V.R.; Rangel, C.M.; Calvillo, L.; Lázaro, M.J.; Pastor, E. Pt-Ru catalysts supported on carbon xerogels for PEM fuel cells. *Int. J. Hydrog. Energy* **2012**, *37*, 7200–7211. [CrossRef]

23. Sebastián, D.; Calderón, J.C.; González-Expósito, J.A.; Pastor, E.; Martínez-Huerta, M.V.; Suelves, I.; Moliner, R.; Lázaro, M.J. Influence of carbon nanofiber properties as electrocatalyst support on the electrochemical performance for PEM fuel cells. *Int. J. Hydrog. Energy* **2010**, *35*, 9934–9942. [CrossRef]

24. Calderón, J.C.; Ndzuzo, L.; Bladergroen, B.J.; Pasupathi, S. Oxygen reduction reaction on Pt-Pd catalysts supported on carbon xerogels: Effect of the synthesis method. *Int. J. Hydrog. Energy* **2018**, *43*, 16881–16896. [CrossRef]

25. Yang, G.; Sun, Y.; Lv, P.; Zhen, F.; Cao, X.; Chen, X.; Wang, Z.; Yuan, Z.; Kong, X. Preparation of Pt-Ru/C as an oxygen-reduction electrocatalyst in microbial fuel cells for wastewater treatment. *Catalysts* **2016**, *6*, 150. [CrossRef]

26. Ye, F.; Liu, H.; Hu, W.; Zhong, J.; Chen, Y.; Cao, H.; Yang, J. Heterogeneous Au-Pt nanostructures with enhanced catalytic activity toward oxygen reduction. *Dalton Trans.* **2012**, *41*, 2898–2903. [CrossRef] [PubMed]

27. Qu, J.; Liu, H.; Ye, F.; Hu, W.; Yang, J. Cage-bell structured Au-Pt nanomaterials with enhanced electrocatalytic activity toward oxygen reduction. *Int. J. Hydrog. Energy* **2012**, *37*, 13191–13199. [CrossRef]

28. Kim, G.J.; Kwon, D.W.; Hong, S.C. Effect of Pt particle size and valence state on the performance of Pt/TiO$_2$ catalysts for CO oxidation at room temperature. *J. Phys. Chem. C* **2016**, *120*, 17996–18004. [CrossRef]

29. Jeon, T.Y.; Lee, K.S.; Yoo, S.J.; Cho, Y.H.; Kang, S.H.; Sung, Y.E. Effect of surface segregation on the methanol oxidation reaction in carbon-supported Pt−Ru alloy nanoparticles. *Langmuir* **2010**, *26*, 9123–9129. [CrossRef]

30. Salgado, J.R.C.; Alcaide, F.; Álvarez, G.; Calvillo, L.; Lázaro, M.J.; Pastor, E. Pt–Ru electrocatalysts supported on ordered mesoporous carbon for direct methanol fuel cell. *J. Power Sources* **2010**, *195*, 4022–4029. [CrossRef]

31. Chung, D.Y.; Lee, K.J.; Sung, Y.E. Methanol electro-oxidation on the Pt surface: Revisiting the cyclic voltammetry interpretation. *J. Phys. Chem. C* **2016**, *120*, 9028–9035. [CrossRef]

32. Fu, X.; Zhao, Z.; Wan, C.; Wang, Y.; Fan, Z.; Song, F.; Cao, B.; Li, M.; Xue, W.; Huang, Y.; et al. Ultrathin wavy Rh nanowires as highly effective electrocatalysts for methanol oxidation reaction with ultrahigh ECSA. *Nano Res.* **2019**, *12*, 211–215. [CrossRef]

33. Calderón, J.C.; García, G.; Querejeta, A.; Alcaide, F.; Calvillo, L.; Lázaro, M.J.; Rodríguez, J.L.; Pastor, E. Carbon monoxide and methanol oxidations on carbon nanofibers supported Pt–Ru electrodes at different temperatures. *Electrochim. Acta* **2015**, *186*, 359–368. [CrossRef]

34. Su, F.; Zeng, J.; Bao, X.; Yu, Y.; Lee, J.Y.; Zhao, X.S. Preparation and characterization of highly ordered graphitic mesoporous carbon as a Pt catalyst support for direct methanol fuel cells. *Chem. Mater.* **2005**, *17*, 3960–3967. [CrossRef]

35. Maiyalagan, T.; Alaje, T.O.; Scott, K. Highly stable Pt-Ru nanoparticles supported on three-dimensional cubic ordered mesoporous carbon (Pt-Ru/CMK-8) as promising electrocatalysts for methanol oxidation. *J. Phys. Chem. C* **2012**, *116*, 2630–2638. [CrossRef]

36. Sebastián, D.; Nieto-Monge, M.J.; Pérez-Rodríguez, S.; Pastor, E.; Lázaro, M.J. Nitrogen doped ordered mesoporous carbon as support of PtRu nanoparticles for methanol electro-oxidation. *Energies* **2018**, *11*, 831. [CrossRef]

surfaces

MDPI

Article

Ir-Ni Bimetallic OER Catalysts Prepared by Controlled Ni Electrodeposition on Ir$_{poly}$ and Ir(111)

Ebru Özer [1], Ilya Sinev [2], Andrea M. Mingers [3], Jorge Araujo [1], Thomas Kropp [4],
Manos Mavrikakis [4], Karl J. J. Mayrhofer [3], Beatriz Roldan Cuenya [2] and Peter Strasser [1,5,*]

[1] The Electrochemical Energy, Catalysis and Materials Science Laboratory, Department of Chemistry,
Technical University Berlin, Straße des 17. Juni 124, Berlin 10623, Germany;
ebru.oezer@campus.tu-berlin.de (E.Ö.); ferreiradearaujo@campus.tu-berlin.de (J.A.)

[2] Fritz Haber Institute of the Max Planck Society, Department of Interface Science, Faradayweg 4-6,
Berlin 14195, Germany; ilya.sinev@rub.de (I.S.); roldan@fhi-berlin.mpg.de (B.R.C.)

[3] Max-Planck-Institut für Eisenforschung GmbH, Max-Planck-Strasse 1, Düsseldorf 40237, Germany;
mingers@mpie.de (A.M.M.); k.mayrhofer@fz-juelich.de (K.J.J.M.)

[4] University of Wisconsin-Madison, 1415 Engineering Drive, Madison WI 53706, USA;
tekropp@wisc.edu (T.K.); emavrikakis@wisc.edu (M.M.)

[5] Ertl Center for Electrochemistry and Catalysis, Gwangju Institute of Science and Technology,
Gwangju 500-712, Korea

[*] Correspondence: pstrasser@tu-berlin.de; Tel.: +49-030-314-29542

Received: 26 November 2018; Accepted: 13 December 2018; Published: 14 December 2018

Abstract: The alteration of electrocatalytic surfaces with adatoms lead to structural and electronic modifications promoting adsorption, desorption, and reactive processes. This study explores the potentiostatic electrodeposition process of Ni onto polycrystalline Ir (Ir$_{poly}$) and assesses the electrocatalytic properties of the resulting bimetallic surfaces. The electrodeposition resulted in bimetallic Ni overlayer (OL) structures and in combination with controlled thermal post-deposition annealing in bimetallic near-surface alloys (NSA). The catalytic oxygen evolution reaction (OER) activity of these two different Ni-modified catalysts is assessed and compared to a pristine, unmodified Ir$_{poly}$. An overlayer of Ni on Ir$_{poly}$ showed superior performance in both acidic and alkaline milieu. The reductive annealing of the OL produced a NSA of Ni, which demonstrated enhanced stability in an acidic environment. The remarkable activity and stability improvement of Ir by Ni modification makes both systems efficient electrocatalysts for water oxidation. The roughness factor of Ir$_{poly}$ is also reported. With the amount of deposited Ni determined by inductively coupled plasma mass spectrometry (ICP-MS) and a degree of coverage (monolayer) in the dependence of deposition potential is established. The density functional theory (DFT) assisted evaluation of H adsorption on Ir$_{poly}$ enables determination of the preferred Ni deposition sites on the three low-index surfaces (111), (110), and (100).

Keywords: oxygen evolution reaction; iridium; nickel; electrodeposition; model catalyst; water oxidation; CO oxidation; DFT; hydrogen adsorption

---

1. Introduction

Iridium is one of the most important electrode materials for basic surface science and industrial applications especially those in the field of electrochemistry due to its enhanced catalytic properties in water splitting electrolyzers [1,2]. The efficiency of electrolyzers and, thus, the catalytic properties are highly dependent on the interface between the electrode surface and the electrolyte. However, the main impediments to an effective application of water electrolysis as a sustainable method for hydrogen production are the high potential losses at the anode side and the high Ir material costs.

In this context, studies on low index surfaces of single crystals offer a meaningful opportunity for a fundamental understanding and, based on it, the rational design of enhanced electrocatalytic materials. A wealth of information on structural and electronic effects can emerge from these studies particularly in the context of electrochemical deposition of metal adsorbates at a sub-monolayer coverage including underpotential deposition (UPD). Potential-controlled deposition results in the growth of a new two-dimensional phase, which is believed to have electronic properties that differ significantly from those of the respective bulk materials. In addition, the electronic structure of the substrate itself and, thus, the interfacial reactivity is altered by the metal adatoms [3,4]. The over-layer and near-surface alloy systems, therefore, offer a great variability in optimizing the binding energies to reaction intermediates and adsorption properties [5–9]. The deposition of metal adatoms on foreign metal substrates provides attractive systems with highly controllable surface coverages by using the electrode potential and deposition time. In particular, the resulting structure and surface properties of the new electrochemical materials are dependent on the variable surface coverage and, thereby, allow for a precise and reproducible control of the coverage-dependent properties. Due to its practical importance in surface chemistry and electrocatalysis, a number of ex situ and in situ techniques such as cyclic voltammetry (CV), scanning tunneling microscopy (STM), X-ray photoelectron spectroscopy (XPS), and low-energy electron diffraction (LEED) have been applied to augment the process of deposition at the molecular and atomic level [10].

In this study, the deposition of Ni on polycrystalline Ir and well-characterized Ir(111) electrodes in both sulfate-containing electrolytes and phosphate-buffered solutions will be presented employing experimental techniques. In situ cyclic current-potential curves are used to detect the redox potentials of the electroactive species or rather the potential range in which deposition takes place. Ex situ XPS data provides information on the chemical composition of the surface. In addition, a more quantitative trace metal analysis was performed with inductively coupled plasma-mass spectrometry (ICP-MS) to determine the adsorbed metal concentration based on high sensitivity and detection capability. Surface sensitive CO stripping measurements provide valuable insights into the morphological and structural composition of the deposit. Theoretically calculated hydrogen binding energies were taken into consideration. The Faraday charge efficiencies (FE) towards oxygen evolution of the modified electrodes were measured with differential electrochemical mass spectrometry (DEMS) and it was confirmed that the entire charge is transferred to facilitate the oxygen evolution reaction (OER).

In this case, a majority of the work has been carried out on polycrystalline Ir electrodes to study the multi-faceted Ni deposition process and to establish profound procedures and protocols for the precise control of surface coverage and structural composition. The intent is to present a broad overview and to provide detailed observations of the deposition process of Ni on Ir using not only electrochemical methods but also knowledge obtained from various surface sensitive techniques. The following studies and conclusions, therefore, present the first step in the electrodeposition of Ni layers on Ir electrodes and their electrochemical properties in the oxygen evolution reaction (OER). The results of this research expand the fundamental knowledge about the deposition process on polycrystalline Ir and show initial studies on Ir(111) single crystal electrodes under well-defined conditions of cleanliness thanks to a custom-made inductive and electrochemical crystal chamber [11].

2. Materials and Methods

2.1. Substrates

Ir(111) Single Crystal and Polycrystalline Ir (Ir$_{poly}$)

Ir(111) single crystal electrodes with a diameter of 10 mm and a precision of orientation of $\leq 0.1°$ (MaTecK GmbH, Jülich, Germany) were applied as working electrodes for the study of surface modification. To mount the cylinder into the inductive and electrochemical setup in which the surface was refined thermally, two lateral holes with a diameter of 0.5 mm were machined into the shell surface through which loops made of Ir wire ($\varnothing$ = 0.25 mm, 99.9%; Alfa Aesar, Germany) run. The detailed description of mounting can be found elsewhere [11]. A heating protocol consisting of several heating cycles per heating phase is applied until the determined electrochemical surface characterization is consistent with a perfectly clean and well-ordered (111) single-crystalline Ir surface [12].

2.2. Electrochemical Characterization

2.2.1. Ni Sweep Voltammetry

The electrodeposition of Ni on polycrystalline Ir was conducted in a standard rotating disk electrode (RDE) setup. Measurements on Ir(111) single crystals were performed in a custom-made setup [11].

Initially, the Ni electrodeposition was studied in 0.05 M H_2SO_4 (diluted from 95% H_2SO_4, Suprapure, Carl Roth, Germany) and in a phosphate buffer solution of pH 6.85 (0.1 M KH_2PO_4 and 0.1 M K_2HPO_4, 1:1 mixture by volume, Merck, Germany) using the cyclic voltammetry (CV) technique. The electrodes were transferred into the degassed solution containing 2 mM $NiSO_4$. A voltammetric scan was then performed from the immersion potential of 1.0 V_{RHE} for Ir_{poly} and 0.8 V_{RHE} for Ir(111) to the negative potential limit of 0.05 V_{RHE} at a scan rate of 50 mV s^{-1}. Under the same conditions, CVs of the electrodes were recorded in the absence of Ni^{2+} ions to locate the Ni adsorption-desorption region. These solutions contained 2 mM Na_2SO_4 to have the same anion concentration. To trace conductivity changes of the electrode, impedance spectroscopy was performed to determine the ohmic drop of the electrode. In both electrochemical measurements, the voltammetric scan ended at the positive potential limit to ensure the oxidation of all deposited Ni. All electrolytes were purged with either Ar (Ir(111) in the inductive setup) or N_2 (Ir_{poly} in the RDE) for at least 15 min to remove dissolved oxygen prior to use.

2.2.2. Controlled-Potential Technique

The potential range of Ni deposition on Ir_{poly} in phosphate buffer was examined more thoroughly. A dependence of the deposited Ni amount on Ir_{poly} as a function of deposition potential was established. The total metal content of the deposition process was determined at six potentials between 0.05 and 0.5 V_{RHE} using ICP-MS. The given potential window was determined in the Ni sweep voltammetry experiments. The electrodes were immersed under a potential control at the deposition potential and the respective potential was held for a time of 10 min. Subsequently, the coverage of Ni was determined by ICP-MS after dissolving all Ni by immersing the electrode in 5 mL 0.1 M HNO_3 for one hour. Beforehand, the electrodes were rinsed thoroughly with milli-q water.

2.2.3. Preparation of Ni Overlayers (OLs) and Ni (near-) Surface Alloys (NSAs)

To produce an overlayer of Ni (Ni/Ir OL), the electrode was immersed at the desired deposition potential (E_{dep}) into a phosphate buffer solution of pH 6.85 (0.1 M KH_2PO_4, 0.1 M K_2HPO_4) containing 2 mM $NiSO_4$ and held for 10 min. The Ni/Ir_{poly} OL electrode catalyst for further electrochemical testing was produced at a deposition potential of 0.05 V_{RHE}.

For the preparation of a Ni/Ir_{poly} near surface alloy (NSA), the Ni/Ir_{poly} OL electrode (E_{dep} = 0.05 V_{RHE}) was subsequently annealed for 1 min in 4% H_2/Ar at 400 °C in a custom-made inductive setup [11]. The Ni-modified catalysts were then tested for OER activity in 0.1 M KOH and 0.05 M H_2SO_4.

2.2.4. Electrochemical OER Protocol at RDE

Two types of OER protocols were used to test the performance of unmodified and Ni-modified Ir_{poly} electrodes at the RDE. First, the OER activities of Ir_{poly}, Ni/Ir_{poly} OL, and Ni/Ir_{poly} NSA were tested in both 0.05 M H_2SO_4 and 01 M KOH. In a separate protocol, the stability of the Ni/Ir_{poly} OL catalyst was subject to greater stress. The catalyst was first treated in 0.05 M H_2SO_4 before continuing the OER measurement in 0.1 M KOH. In between measurements, the electrode was rinsed thoroughly with milli-q water.

The electrolytes were prepared by dilution of 95% H_2SO_4 (Suprapure, Carl Roth, Germany) with ultrapure water. Purified Fe-free 0.1 M KOH was prepared according to the method reported by Trotochaud et al. [13].

For both procedures, the immersion of the electrodes at 1.00 V_{RHE} was followed by 10 polarization curves into the potential region of OER at a scan rate of 5 mV s^{-1} from the immersion potential up until a current of 0.785 mA (current density of 1 mA cm^{-2}) was reached. The measured currents are normalized to the geometric area of the electrodes to provide comparable current densities (j_{geo}). The geometric surface area of the electrodes is 0.785 cm^2 corresponding to the diameter of 10 mm. After 10 consecutive cycles, electrochemical impedance spectroscopy (PEIS) was measured to obtain the ohmic resistance. The respective potentials are referred to as E_{iR}.

2.2.5. CO Stripping Protocol

The CO stripping measurements were carried out in a rotating disk (RDE) electrode setup with a reversible hydrogen electrode (RHE) as a reference electrode and a Pt-mesh as a counter electrode. The measurements were recorded with a Bio-Logic SP-200 potentiostat (Bio-Logic, France). The CO stripping measurements were performed on an unmodified Ir$_{poly}$ and two Ni-modified Ir$_{poly}$ electrodes in purified Fe-free 0.1 M KOH (semiconductor grade, 99.99% trace metals basis, Merck, Germany). Ni/Ir$_{poly}$ OL and Ni/Ir$_{poly}$ NSA electrodes were prepared as described in Section 2.2.2.

2.3. Theoretical Characterization

Density Functional Theory (DFT) Calculated H-Ir Binding Energies

Theoretical H-Ir binding energies were calculated using the projector augmented wave (PAW) method [14,15] as implemented in the Vienna ab initio simulation package (VASP) [16,17]. Exchange-correlation energies were obtained by using the functional by Perdew, Burke, and Ernzerhof (PBE) [18]. A plane wave cutoff of 600 eV was used. Structure optimizations were performed until total energies were converged to 10^{-6} eV and forces acting on the relaxed ions were below 0.02 eV Å^{-1}.

Surface models were obtained by cutting bulk iridium (lattice constant a = 3.877 Å) in (100), (110), and (111) orientation. The resulting slab models consist of five atomic layers separated by a 12 Å vacuum layer. The lowest two atomic layers are frozen to simulate the bulk. For Ir(100) and Ir(111), the Brillouin zone was sampled at 6 × 6 × 1 k points while 4 × 4 × 1 k points were used for the larger Ir(110) model.

2.4. Physicochemical Characterization

2.4.1. Inductively Coupled Plasma Mass Spectrometry (ICP-MS)

ICP-MS of the HNO$_3$ aliquots taken to dissolve the electrodeposited nickel was measured with a NexION 300 X (PerkinElmer®, Shelton, CT, USA) equipped with a concentric PTFE Nebulizer-PFA-ST (PerkinElmer®, Shelton, CT, USA). and a spray chamber.

2.4.2. Atomic Force Microscopy (AFM)

The AFM images were done in a tapping mode using a Bruker MultiMode 8 microscope (Bruker, Berlin, Germany) on the samples after transport in Ar.

2.4.3. X-ray Photoelectron Spectroscopy (XPS)

The pristine and Ni-modified Ir(111) surfaces were analyzed by X-ray photoelectron spectroscopy in an ultra-high vacuum (UHV) setup equipped with Phoibos 150-MCD 9 hemispherical analyzer (SPECS GmbH, Berlin, Germany). A monochromatic Al Kα X-ray source (hv = 1486.5 eV) at 12 kV of anodic voltage and 300 W power was used to perform the characterization. For calibration of the binding energy, the C1s signal of graphitic-like carbon (285 eV) was used. For peak deconvolution, Gaussian-Lorentzian product functions and a Shirley background were applied in the Casa XPS software.

2.5. Mass Spectrometry

Differential Electrochemical Mass Spectrometry (DEMS)

Differential electrochemical mass spectrometry (DEMS) was acquired in a custom-made dual thin-layer electrochemical flow cell based on the concept reported elsewhere [19,20]. A Prisma quadrupole mass spectrometer (QMS 200, Pfeiffer-Vacuum, Germany) equipped with two turbomolecular pumps (HiPace 80, Pfeiffer, Germany) operating under 10^{-6} mbar were used to detect any volatile products. The instrument allows a continuous flow of the electrolyte of 10 μL s^{-1} through the hydrophobic membrane to the mass spectrum ion source. We used the hydrophobic PTFE membrane with a pore size of 30 nm and a thickness of 150 μm (Cobetter, Cat. No. PF-003HS) as the separation interface for the electrolyte and vacuum. A high mass transport rate is guaranteed due to the low volume of electrolyte flowing over the electrode surface as well a stable ionic current for the electrochemical process.

The electrochemistry was controlled using a BioLogic potentiostat, a leak-free Ag/AgCl as reference electrode (Warner Instruments, USA), and a Pt mesh as a counter electrode. Cyclic voltammograms in 0.1 M KOH and 0.05 M H_2SO_4 were recorded in a potential range of 0.6 and 1.6 V_{RHE} (prior to iR-compensation) at a scan rate of 50 mV s^{-1}. The flow cell setup was used for evaluating the faradaic charge efficiencies toward oxygen evolution for unmodified and Ni-modified polycrystalline Ir electrodes. The preparation of the Ni-modified electrodes is described above. In the first step, the ionic current baseline was recorded in a steady-state measurement to obtain a catalyst-specific calibration factor K^*. A constant purge of N_2 gas over the electrodes' surface was done during integration of the electrode with the flow cell system. In addition, during DEMS experiments, the purge N_2 was kept over the non-electrolyte contact areas of the electrode to ensure that the electrochemistry is running in an oxygen-free environment. DEMS measurements were monitorized with control product formation species with m/z values of 2, 16, 18, 28, 30, 36, 32, 34, 44 during the CV scan. The previously mentioned catalyst-specific calibration factor K_j^* was applied for quantification of the faradaic efficiency (FE) of O_2 (m/z 32). Chronopotentiometric steps were applied in the linear region of the ionic current baseline during the steady-state measurement protocol. In Equation (1), the relation between the mass spectrometric ion current ($i_{MS,j}$) and the faradaic current (i_F) to the volatile product j is given as:

$$K_j^* = \frac{i_{MS,j} \cdot z_j}{i_F}. \tag{1}$$

In this case, z_j is the number of transferred electrons per molecule of volatile product j. The division of the integrated mass spectrometric ion current ($Q_{MS,j}$) by the integrated capacitive current-corrected, faradaic current (Q_F) allows for the determination of the faradaic efficiency (FE) (charge selectivity) according to:

$$FE(\%) = \frac{Q_{MS,j} \cdot z_j}{Q_F \cdot K_j^*} \cdot 100 = \frac{Q_{F,j}^{DEMS}}{Q_F} \cdot 100. \tag{2}$$

In the analysis, only anodic faradaic currents are taken into account since solely the anodic processes of molecular oxygen evolution is of interest. The limits of integration were given by the detection signal of oxygen evolution (m/z 32) and the initial baseline level. For the different samples, the limits varied due to distinct onset potentials for oxygen evolution.

3. Results and Discussion

3.1. Roughness Factor (RF) of Ir$_{poly}$

For the subsequent evaluation of the accurate amount of deposited Ni on Ir$_{poly}$ (which will be transferred into fractions of a single monolayer), the following calculations aim at determining the surface roughness of a polycrystalline Ir electrode. Equation (3) establishes a relationship between the double-layer charging capacitive current density j_c and the double-layer capacitance C_{dl} (per geometric area) [21].

$$j_c = v \cdot C_{dl}. \tag{3}$$

For Ir$_{poly}$, the potential window between 0.4 and 0.5 V$_{RHE}$ lacks any faradaic process (see Figure S1) and was, therefore, chosen to be the electrochemical double-layer capacitance. The roughness factor (RF) of Ir$_{poly}$ is then calculated from the double-layer capacitance according to the Equation (4) below.

$$RF = \frac{C_{dl}}{C_S}. \tag{4}$$

C_S in Equation (4) is the electrochemical double-layer capacitance of a smooth and planar electrode surface of the same material measured under the same conditions. These measurements have been conducted for a single crystal Ir(111) electrode. The double-layer capacitance measurements for Ir$_{poly}$ and Ir(111) are depicted in Figure 1. Given the slopes of both electrode surfaces, a roughness factor of 4.3 of the polycrystalline Ir electrode can be defined according to the following calculation.

$$RF = \frac{148.6 \; \mu F}{34.3 \; \mu F} = 4.33. \tag{5}$$

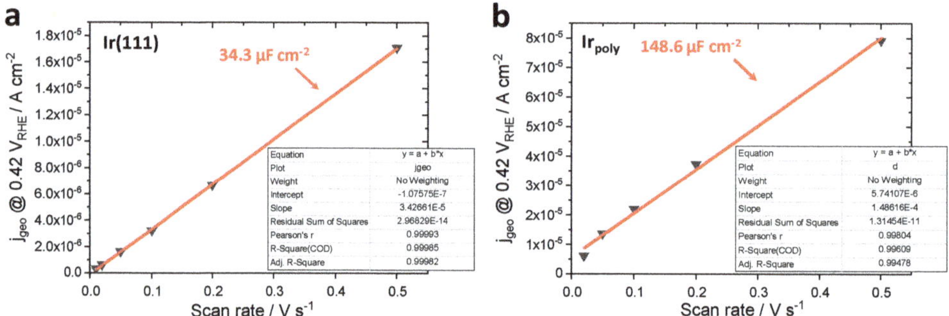

Figure 1. Double-layer capacitance measurements of (**a**) Ir(111) and (**b**) Ir$_{poly}$ displaying the anodic charging current densities measured at 0.42 V$_{RHE}$.

For calculating the mass of Ni needed to produce a monolayer on a polycrystalline Ir cylinder with a diameter of 10 mm, it is legitimate to assume that, on an Ir polycrystal, only a (111) face exists as the terraces of a polycrystal, which are presumed to consist of the three low-index surfaces (111), (110), and (100) nearly to the same extent.

In Table 1, the individual steps of our estimation are summarized. In the first step, the atoms per unit area are determined by dividing the number of Ir atoms per (111) face in the unit cell by the surface area of the (111) face. By using the Avogadro constant N_A, the amount of substance per surface area is introduced. Given a geometric projected surface area of 0.785 cm^2, the molar or atomic amount of Ni needed to form a complete monolayer is obtained. Analogously, the mass of Ni per monolayer can be obtained by introducing the molar mass of Ni (M = 58.68 g mol^{-1}). With a given RF of 4.33, the estimated Ni amount to produce a monolayer on Ir$_{poly}$ is 0.52 µg.

Table 1. Calculations for determining the mass needed to produce a monolayer of Ni on a polycrystalline Ir cylinder with a diameter of 0.785 cm^2. (Lattice parameter Ir: a = 3.8394 Å, according to PDF: 00-006-0598).

Steps	Calculation
surface area calculation	$\sqrt{3/4} \cdot a^2$
surface area/10^{-15} cm^2	1.277
atoms per (111) face in unit cell	2
atoms per unit area/10^{15} cm^{-2}	1.566
amount of substance per surface area/nmol cm^{-2}	2.6
amount of substance for monolayer/nmol ML^{-1}	2.04
mass of Ni per monolayer/µg ML^{-1}	0.12
mass of Ni per monolayer for Ir$_{poly}$/µg ML^{-1}	0.52

3.2. Surface Chemical Characterization

Ni deposition was conducted and analyzed on polycrystalline Ir (Ir$_{poly}$) and single-crystalline Ir(111) electrodes. The experiments with Ir$_{poly}$ have served to gain a better understanding of the deposition process and for determining the influence of adsorbed Ni on the surface and the resulting catalytic properties.

3.2.1. Ni Deposition on Ir$_{poly}$ and Ir(111) by Sweep Voltammetry

The Ni deposition process on a polycrystalline Ir electrode was first studied by using the cyclic voltammetry technique in order to determine the potential area in which Ni deposition occurs. Figure 2a depicts the cyclic voltammogram of a polished and electrochemically cleaned polycrystalline Ir cylinder in a phosphate buffer solution with and without 2 mM NiSO$_4$. The solution without Ni^{2+} contains 2 mM Na$_2$SO$_4$ to guarantee the same concentration of anions in the electrolyte and to make the measurements comparable.

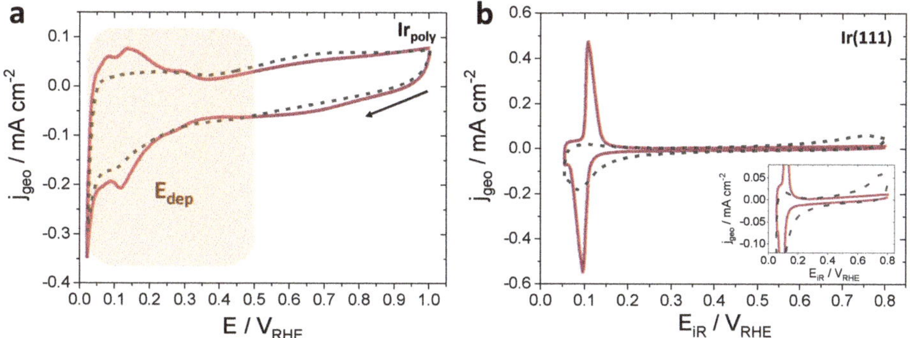

Figure 2. Cyclic voltammograms of (**a**) a polycrystalline Ir electrode and (**b**) a pristine Ir(111) in a phosphate-buffered solution (pH = 6.85) containing either 2 mM Na$_2$SO$_4$ (solid lines) or 2 mM NiSO$_4$ (dashed lines) at a scan rate of v = 0.05 V s^{-1}. The electrodes were immersed at a potential of 1.0 V$_{RHE}$ and 0.8 V$_{RHE}$ for Ir$_{poly}$ and Ir(111), respectively.

In the presence of Ni, the hydrogen adsorption-desorption region is reduced significantly, which suggests a blocking of Ir sites by Ni ad-atoms. By contrast, in the negative going sweep, there is only a slight distinction in the total electric charge between the cyclic voltammograms in the potential area of 0.5 to 0.0 V$_{RHE}$, which implies that Ni electrodeposition on Ir is not recognizable as strongly defined deposition peaks. It cannot be determined with absolute certainty if oxygen desorption in the Ni-free solution (at ~+0.7 V$_{RHE}$) is shifted toward lower potentials (at ~+0.65 V$_{RHE}$) due to the so-called ligand effect [22], which is triggered by adsorbed Ni adatoms or if deposition of Ni occurs on distinct adsorption sites, which are merged into one broad redox wave. In the positive going sweep, desorption of Ni ad-atoms begins beyond the hydrogen underpotential deposition (HUPD) region and attains its maximum value at around +0.65 V_{RHE}.

Figure 2b displays the cyclic voltammogram of a pristine Ir(111) single crystal electrode in phosphate buffer of pH 6.85 containing 2 mM $NiSO_4$. A cyclic voltammogram of the same electrode in the absence of Ni is also shown for comparison. Both voltammograms were recorded in the potential window of 0.05–0.8 V_{RHE}, which is a potential range adequate for a sufficient characterization of the Ir(111) surface [12]. The CV of Ir(111) in phosphate buffer without Ni is mostly featureless and only displays the HUPD redox peak at ~+0.1 V_{RHE}. The significant difference between both voltammograms is located in the HUPD region. The characteristic HUPD peak of the (111) face is mostly masked in the presence of Ni. The voltammogram is presumably dominated by a broad deposition and stripping region of deposited Ni between 0.50–0.05 V_{RHE} and 0.3–0.8 V_{RHE}, respectively. The peak in the positive going sweep at ~+0.1 V_{RHE} can either be ascribed to stripping off Ni or to a hydrogen desorption charge due to remaining hydrogen adsorption on Ir(111) sites or both. The latter case presupposes that not all Ir sites are being blocked by Ni adatoms during deposition. This might be a kinetically controlled problem if the deposition process by cyclic voltammetry requires some time to form a complete Ni layer. Free Ir sites may be explained by the deposition of Ni on existing Ni adatoms.

In the negative going sweep, the charge is found to be higher ($\approx$375.9 μC cm^{-2}) than the sum of the charges in the positive going sweep. From this observation, it seems that Ni deposition at a single crystalline Ir(111) is not entirely reversible, which means that Ni is not completely stripped off the surface in the anodic scan up to 0.8 V_{RHE}.

3.2.2. Ni Deposition on Ir_{poly} by the Controlled-Potential Technique

The formation of a Ni layer is likely a process that requires some time since the cyclic voltammogram in Figure 2a does not display defined Ni redox peaks. Hence, the cycling technique may not be the most suitable approach. On this account, the potential range E_{dep} in Figure 2a was used to establish a relationship between Ni coverage and the applied deposition potential (E_{dep}). Therefore, the electrode was immersed at the desired potential under the potential control and the potential was held for 10 min. The deposited amount of Ni was determined chemically by ICP-MS. The process was conducted for six distinct deposition potentials. In Figure 3, the effect of the deposition potential E_{dep} on the amount of adsorbed Ni is shown. The obtained values from ICP-MS and the resultant extent of Ni monolayer formation on polycrystalline Ir are presented in Table S1 As described earlier, the analyzed amount of Ni with ICP-MS was translated into a degree of coverage (ML) by assuming that (111) facets are prevalent on a polycrystalline Ir surface.

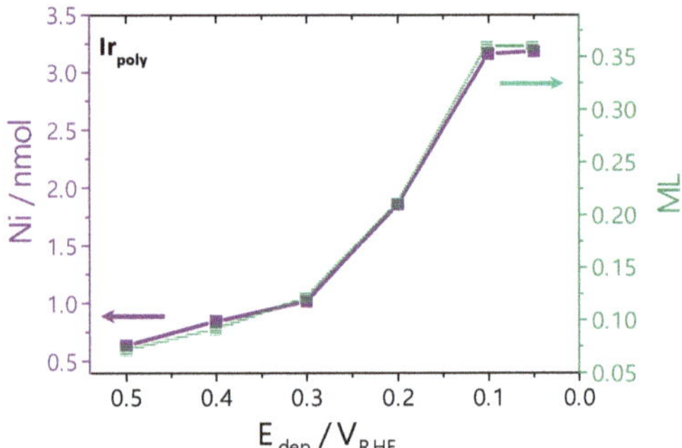

Figure 3. Relationship between amount of deposited Ni (in nmol, left axis) on polycrystalline Ir electrode and deposition potential E_{dep}. The electrode was held for 10 min at each potential in a phosphate-buffered solution (pH = 6.85) containing 2 mM $NiSO_4$. The amount of deposited Ni was analyzed by ICP-MS. The right axis (green) gives the determined value of the monolayer (ML) formation.

At first, a slight increase in Ni coverage and a decreasing deposition potential is observed. The linear rise becomes larger until the Ni amount approaches a limiting value at a potential of $\leq$+0.1 V_{RHE}. Since the surface of a polycrystalline Ir electrode is not well-defined, the calculated values of Ni ML coverage provide approximate guidance for the relationship between absolute molar Ni amount and deposition potential. The assumption of a prevalent (111) facet on Ir_{poly} is thereby the most severe assumption as a polycrystalline surface generally consists of a number of different types of facets. Some of these orientations may be less suitable or have less potential for Ni deposition. However, with an estimated average roughness factor of 4.33 for the polycrystalline Ir electrode, a maximum coverage of ~0.36 ML was obtained.

3.2.3. Surface Chemical State of Ni/Ir OL

The X-ray photoemission spectra of the Ni-modified single and polycrystalline Ir electrodes are presented in Figure 4.

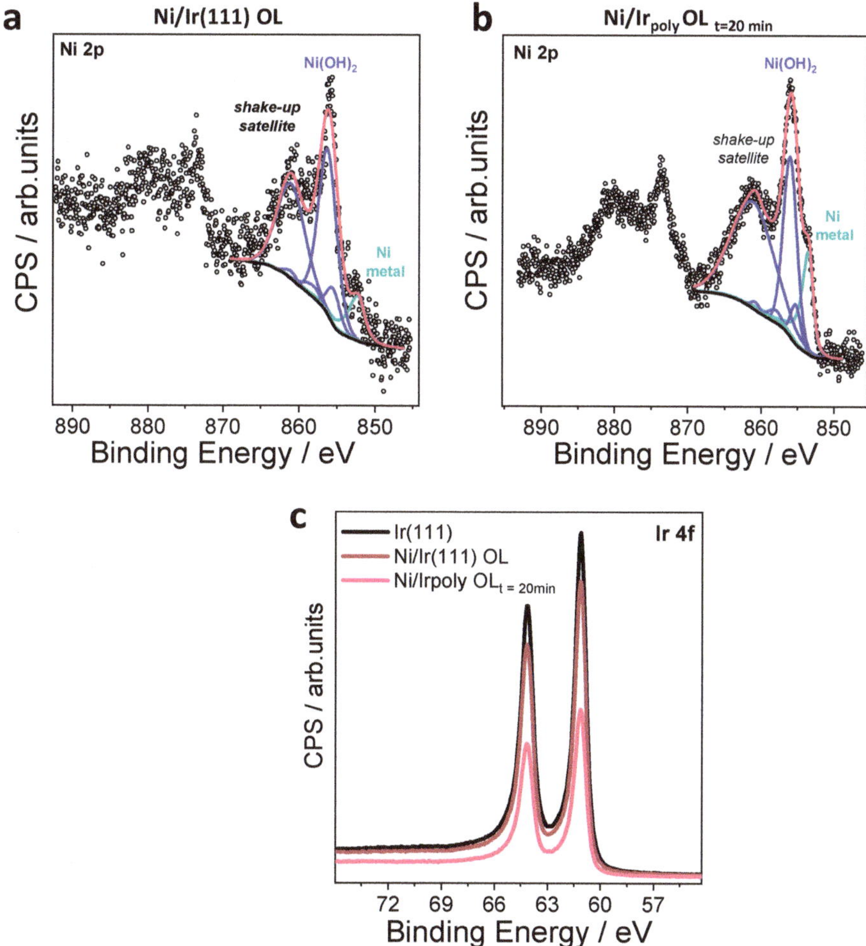

Figure 4. XP spectra of (**a**) Ni-modified Ir(111) single crystal (Ni/Ir(111) OL) and (**b**) a Ni-modified Ir polycrystal (Ni/Ir_{poly} $OL_{t=20min}$). (**c**) XP spectra from the Ir 4f core level of Ir(111), Ni/Ir(111) OL, and Ni/Ir_{poly} $OL_{t=20min}$. The deposition potential was 0.05 V_{RHE}. The deposition time for the formation of the over layer on Ir_{poly} was twice as long as for Ir(111).

The deposition was performed at a potential of 0.05 V_{RHE} to form an over layer (OL). The Ni over layer on polycrystalline Ir (Ni/Ir$_{poly}$ OL$_{t=20min}$) was formed by deposition for 20 min, which is twice as long as for the formation of an OL on Ir(111). For comparison, the XP core level spectra of the well-defined and unchanged Ir(111) single crystal are given in Figure S2. In this case, the survey spectrum over the entire energy range (Figure S2a, top view) is characterized by intense photoemission peaks distinctive for Ir and weak O1s and C1s lines. The high-resolution spectra in Figure S2b (4 panes) excludes the presence of Ni in the sample. During transfer of the single crystal into the UHV chamber and until vacuum conditions were reached, the sample was shortly exposed to a laboratory atmosphere. The detected carbon and oxygen signals can, therefore, be assigned to chemisorbed species during the loading process. After extended exposure time, a Ni-free near the surface region could be affirmed.

The binding energy and line shape of Ir(111) and Ir$_{poly}$ in the Ni-modified samples in Figure 4c are in accordance with the pure Ir spectrum. Thus, the deposition process does not affect the Ir state. This observation is, moreover, consistent with images of the topography of the Ni-modified Ir(111) surface analyzed by AFM. The AFM images and their respective height profiles are presented in Figure S3 with increasing magnifications from Figure S3a–c. The Ni/Ir(111) OL surface was found to exhibit well-resolved flat terraces vertically separated by ~1.53 Å tall steps, which is coherent with single or double atomic steps. The Ni 2p region of all prepared samples in Figure 4a,b indicate the prevalence of Ni^{2+} species with their main peaks at ~857 and 874 eV and their typical shake-up satellite peaks at ~863 and 880 eV. The intensity of the Ni lines increase for longer deposition times while the underlying Ir lines decrease in intensity (see Figure 4c), which implies higher Ni coverages. In Ni/Ir$_{poly}$ OL$_{t=20min}$, the Ni content determined by XPS was three times greater than on the Ni-modified single crystalline Ir sample (Ni/Ir(111) OL). The bright spots in the AFM images correspond with very high probability to contamination by particles of dirt since these measurements have been performed in air.

3.3. Structure-Activity Correlations

3.3.1. Oxygen Evolution Reaction (OER) and Faraday Efficiencies (FE)

The electrochemical OER activity of Ni-modified Ir$_{poly}$ electrodes have been tested using the rotating disk electrode (RDE) technique in both acidic and alkaline environments. The respective OER polarization curves are depicted for the first cycle in Figure 5a,c and are compared to the pristine and unmodified Ir$_{poly}$ catalyst. The 10th OER cycle of the electrode materials is depicted (Figure S4a,b) to monitor the stability of the catalysts. The Ni/Ir$_{poly}$ NSA sample was formed by reductive annealing of Ni/Ir$_{poly}$ OL for 1 min and shows a very weak pronounced Ni(OH)$_2$/NiOOH redox peak in 0.1 M KOH in the 10th cycle (see Figure S4b). The reductive treatment of the over layer for 2 min was sufficient to produce a near-surface alloy of Ni in the subsurface region [11]. Thus, a reduced time was chosen to design a surface in which Ni is both at the surface and in the subsurface of the catalyst.

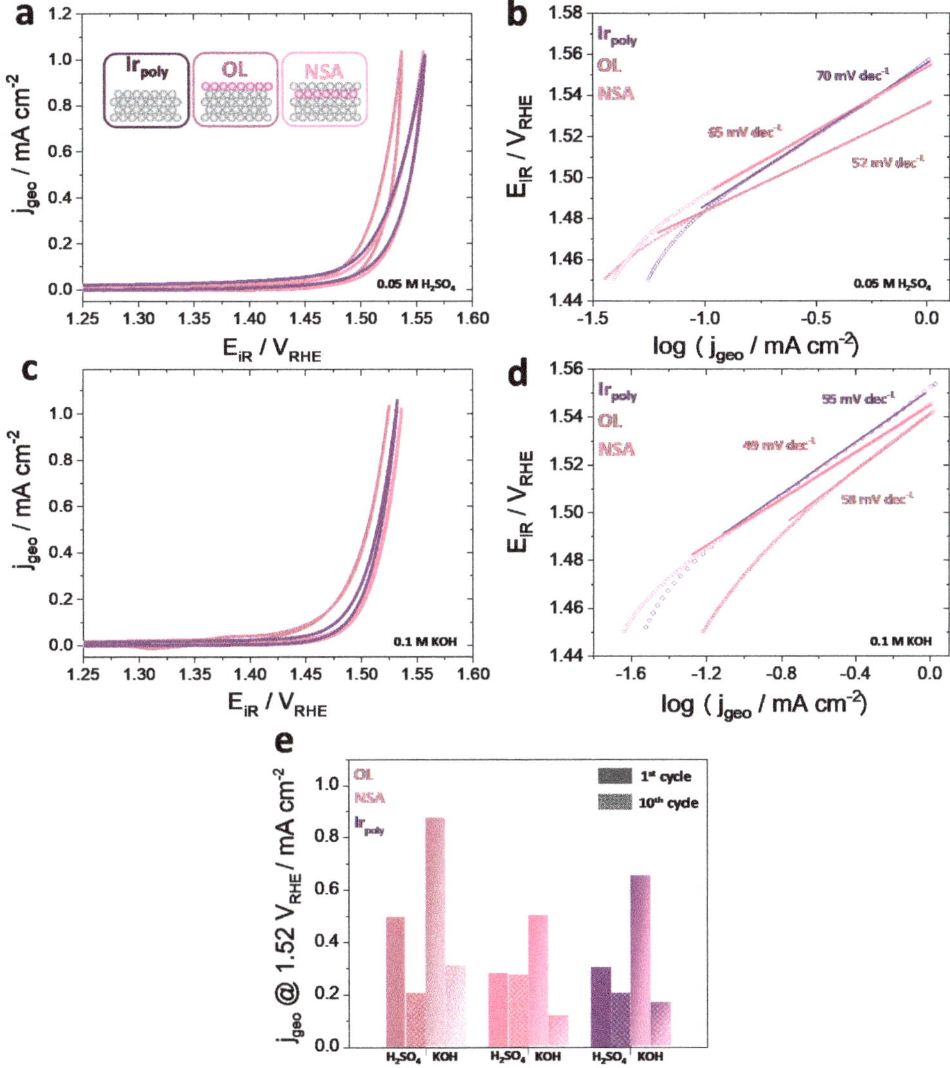

Figure 5. Electrocatalytic OER activities of Ni-modified Ir electrodes compared to pristine polycrystalline Ir at scan rates of 5 mV s^{-1}. (**a**) 1st OER cycle in 0.05 M H$_2$SO$_4$ and (**b**) corresponding Tafel plots (with ohmic-drop correction) of Ir$_{poly}$, Ni/Ir$_{poly}$ OL, and Ni/Ir$_{poly}$ NSA. (**c**) 1st OER cycle in 0.1 M KOH and (**d**) corresponding Tafel plots of all catalysts. (**e**) Current densities of electrodes measured at 1.53 V$_{RHE}$ normalized to the electrode's geometric surface area of 0.785 cm^{-2} for the 1st and 10th cycle. The 10th cycle is depicted in Figure S4 in the Supplementary Information.

As can be seen from Figure 5a,c, the geometric OER activity of the Ir$_{poly}$ reference catalyst was improved by the addition of Ni in the form of an over layer (OL) in both electrolyte systems. The superior catalytic performance of Ni/I$_{rpoly}$ OL in an acidic solution is not only displayed via the higher current density of the cyclic voltammogram (CV) curves but is also reflected by its lower Tafel slope at higher potentials of ~52 mV dec^{-1} (Figure 5b), which is significantly smaller than the slope of the unmodified (~70 mV dec^{-1}) and NSA (~65 mV dec^{-1}) electrocatalysts. Ni atoms in the subsurface layer (NSA) formed by the annealing of the OL do not seem to have an enhanced effect on the OER activity in acidic and alkaline media. In acidic solution, the OER performance of the NSA is almost identical to the unmodified Ir_{poly}. These observations indicate that the performance of the modified material is strongly influenced by the relative position of the second component in the electrode material. In order to correlate the observed reactivity with the structure of the catalyst, the surface composition of a polycrystalline surface has to be considered in detail. Given the high roughness factor of the polycrystalline Ir surface (due to the grain boundaries), the Ni/Ir_{poly} OL might have a similar surface structure to an Ir-Ni surface alloy in which Ni atoms are adjacent to Ir surface atoms. The result is an optimization of the binding energy towards critical reaction intermediates of the proposed oxygen evolution mechanism such as O, OH, and OOH.

In an alkaline environment, the mechanism of the OER has been proposed to occur via the following steps [23–26]: (*M* represents the catalytic active site).

$$M + OH^- \rightarrow M - OH + e^- \tag{6}$$

$$M - OH + OH^- \rightarrow M - O + H_2O + e^- \tag{7}$$

$$M - O + OH^- \rightarrow M - OOH + e^- \tag{8}$$

$$M - OOH + OH^- \rightarrow M - OO + H_2O + e^- \tag{9}$$

$$M - OO \rightarrow O_2. \tag{10}$$

Annealing the OL system in a reductive atmosphere significantly reduces the amount of Ni adatoms at the surface. Since the characteristic $Ni(OH)_2/NiOOH$ redox peak can be recognized very weakly in the cyclic voltammogram of the 10th curve in 0.1 M KOH (see Figure 5c), Ni is still present at the surface albeit to a considerably smaller extent. The electronic effect of underlying Ni atoms on surface Ir atoms seems to have a slightly negative impact on the binding energy of reaction intermediates to Ir. However, the activity behavior of the three examined catalysts is somehow similar in KOH (see Figure 5c) whereby the pristine Ir_{poly} is slightly more active than the NSA. The Tafel slopes of the electrocatalysts differ significantly in both examined electrolyte solutions. In acidic solution, the OL system exhibits the lowest slope followed by the NSA and the unmodified catalyst. Under alkaline conditions, the order of the Tafel slopes is as follows: NSA > Ir_{poly} > OL. The analysis of the experimentally observed Tafel slopes is a powerful tool for determining the reaction steps that are rate-limiting at the reaction interface. It must be kept in mind that these parameters are greatly influenced by the concentration of the surface species formed during the reaction. In an acidic medium, the Tafel slopes of the catalysts differ far more than in alkaline milieu, which suggests a greater influence of Ni on the reaction mechanism. In 0.1 M KOH, the Tafel slopes lie in the range of $\sim$60 mV dec^{-1}. This predicts a slow chemical step to be rate-limiting [27]. The above presented reaction mechanism (6)–(10) proposes electron-transfer steps that determine the overall kinetics. However, the reaction step in Equation (7) can also be divided into an electron transfer process subsequent to a proton transfer step as:

$$M - OH + OH^- \rightarrow M - O^- + H_2O \tag{11}$$

$$M - O^- \rightarrow M - O + e^-. \tag{12}$$

Thus, if the rate-determining step involves the purely chemical step in Equation (11), a Tafel slope of 60 mV dec^{-1} is predicted.

The electrode processes were also measured in a dual-thin layer flow cell setup in combination with differential electrochemical mass spectroscopy (DEMS). The respective Faraday efficiencies (FE) and faradaic current densities are given in Figure S5. In this case, the same order of activity performance was measured in the alkaline solution (OL > Ir_{poly} > NSA). The superior OER performance of the Ni/Ir_{poly} OL catalyst can be deduced to surface enrichment of Ir atoms promoting positive synergies between Ir and Ni atoms. The increase in activity with Ir-Ni binary systems has been observed in various studies in alkaline and acidic electrolytes [23,28–30]. The higher activity of Ni/Ir_{poly} OL in acidic electrolyte may be assigned to an increase of surface OH groups initiated by leaching off Ni adatoms from the surface [23]. At these high potentials, partial oxidation of the catalyst surface cannot be precluded. The oxygen atoms, which were previously bound to Ni, will be protonated to become electroneutral. As OH is considered to act as a crucial intermediate in the oxygen evolution mechanism, a higher fraction of OH at the surface will have an advantageous effect on the reactivity.

As can be seen from Figure 5e, all Ni-modified catalyst systems display a higher OER reactivity in 0.1 M KOH compared to the acidic milieu, which might be partly attributed to the already higher activity of polycrystalline Ir in alkaline media when compared to the acidic substance. During repetitive cycling, the OER performance is declining for all catalysts in both systems (compare Figure S4a,b). In H_2SO_4, the OL system shows the highest drop in activity, which is most likely due to dissolution of Ni surface atoms. After 10 cycles into the potential regions of OER, the activities of the OL and NSA in an acidic solution approach each other, or in other words, the OL catalyst approaches the more stable NSA system. The relative position of Ni is critical for the stability of the catalyst material. The addition of Ni in the form of an over layer significantly enhances the electrocatalytic activity in comparison with an unmodified reference system. However, the stability of the Ni OL system is not given in acidic media. By contrast, Ni in the subsurface layer of Ir_{poly} seems to positively affect the stability of the respective catalyst material since it shows the lowest drop in current density during the 10 cycles.

The OER activity of polycrystalline Ir has been enhanced in both an acidic and an alkaline solution with the addition of Ni at the topmost layer (OL) using the controlled-potential technique. Based on the significantly higher OER performance of the OL by contrast with the NSA and unmodified Ir electrode, the Ni/Ir_{poly} OL system was investigated in a second OER protocol. In Figure 6, the OER polarization curves of two Ni OL catalyst systems in 0.1 M KOH are depicted. The electrochemical performance of each sample was tested by 10 consecutive cycles. However, both samples differ in their electrochemical pretreatment. The Ni-modified Ir_{poly} electrode OL_{di} was tested in KOH directly after the process of Ni deposition. The electrochemical activity of the catalyst OL_{pr} was initially recorded in 0.05 M H_2SO_4 before it was transferred into the alkaline setup.

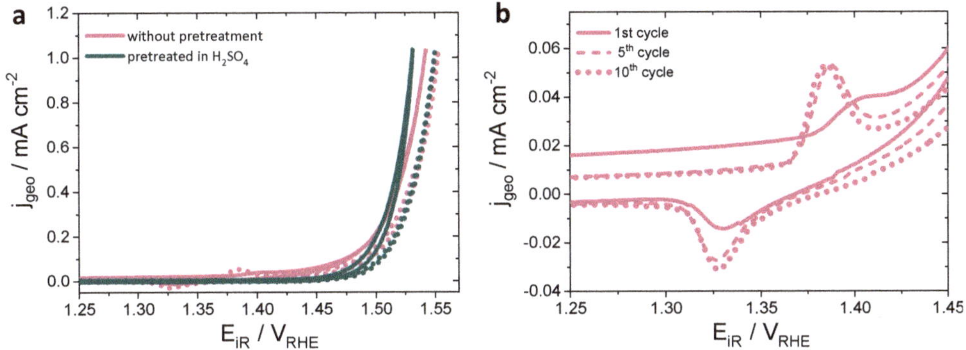

Figure 6. (**a**) OER polarization curves of Ni/Ir_{poly} OL in N_2-saturated 0.1 M KOH without (lilac) and with (green) electrochemical OER pre-treatment in 0.05 M H_2SO_4. Scan rate $\nu = 5$ mV s^{-1}. 1st (solid lines) and 10th cycle (dotted lines) included. The over layer (OL) was formed by electrodeposition of Ni at 0.05 V_{RHE} for 10 min. (**b**) Cathodic shift of the Ni(OH)$_2$/NiOOH redox peak of the OL implies destabilization of Ni^{2+}.

For the OL_{di} catalyst, the characteristic Ni(OH)$_2$/NiOOH redox transition [31,32] at ~+1.35 V_{RHE} (see Figure 6b) is clearly visible and grows with an increasing cycle number. A slight cathodic shift of the precatalytic Ni$^{2+/3+}$ redox wave is observed with an increasing cycle number, which indicates a destabilization of Ni^{2+} under OER catalytic potential. The redox charge obtained by integration of the area under the cathodic wave from the last scan amounts to 137.15 µC (with consideration of geometric surface area), which is equivalent to 1.42 nmol redox active Ni atoms. Taking into consideration that the Ni amount determined by ICP-MS for deposition at 0.05 V_{RHE} was 3.178 nmol (see Table S1), this result would imply that not all Ni sites in the Ni/Ir_{poly} OL system contribute equally to the OER current. The pre-treated OL_{pr} catalyst shows no sign of a $Ni(OH)_2$/NiOOH redox peak, which clearly confirms the instability of a Ni OL in acidic media. The characteristic redox peak of the OL_{di}, however, demonstrates the successful implementation of the electrochemical deposition of Ni on the polycrystalline Ir electrode and forms the basis of further atomic structure modifications. The prior OER treatment of the OL catalyst in sulfuric acid seems to activate the catalyst by ~31%. The enhanced electrochemical property of the pre-treated catalyst might be attributed to an increased presence of electrophilic oxygen species [33–35]. Pfeifer et al. identified the reactive O^{I-} species by characteristic near-edge X-ray absorption fine structure (NEXAFS) features in Ir(III/IV) oxohydroxides. The dissolution of Ni adatoms along with the introduction of Ir vacancies in highly acidic solution and the accompanying formation of electronic defects in the cationic and anionic framework might lead to the formation of highly electrophilic oxygen atoms. Since these oxygen atoms are highly susceptible to a nucleophilic attack, they are suggested to facilitate the O-O bond formation, which is a key step during electrocatalytic OER. Another conclusion that can be drawn from the higher activity of the OL_{pr} catalyst is roughening of the surface by repetitive cycling.

The selective leaching of a sacrificial component is a promising approach to design alloy systems with enhanced activity and stability towards the OER. In the work of Seitz et al., a highly active IrO_x/$SrIrO_3$ electrocatalyst was formed by leaching strontium in the near-surface during electrochemical testing from the surface of $SrIrO_3$ thin-films [36]. In comparison to known IrO_x catalyst materials, these IrO_x surface layers induced by Sr deficiencies showed significantly higher OER activities, which supports the theory of increasing active sites by selective leaching of sacrificial components from surface alloys.

The observed order of activity of the Ir_{poly} catalysts measured in KOH in the RDE setup could be confirmed with an electrochemical flow cell setup, which directly interfaced to a mass spectrometer (MS). In Figure S5a, the OL system displays the highest measured current density at 1.56 V_{RHE} followed by Ir_{poly} and the NSA system, which are quite close to one another. The subsequent OER measurements in acidic solution slightly differ in the sequence of activity. These were measured directly after the catalysts were treated in alkaline solution. The unmodified polycrystal experiences a drastic loss in current density. The NSA catalyst system, however, offers a greater stability and activity compared to Ir_{poly}. The enhanced activity behavior of the NSA system was also observed in the 10th cycle directly measured in acidic solution (see Figure S4a). In Figure 5e, almost no activity loss is observed in acidic solution. Within the context of the measuring accuracy, the faraday efficiency (FE) of O_2 evolution in Figure S5b reaches ~100% for Ir_{poly} and for the OL catalyst in both electrolytes and in KOH, respectively. The FE over 100% for the OL catalyst in acidic solution is significant and may be deduced to Ni dissolution in the highly corrosive environment, which simultaneously produces additional oxygen. For both Ni-modified NSA catalysts, the FE amounts to ~104% within the precision of measurement. It should be noted that the faradaic efficiencies estimated using the calibration factor K^* are assuming 100% efficiency in the steady state protocol and, therefore, the OER efficiencies reflect processes during non-equilibrium conditions.

3.3.2. CO Stripping Voltammetry

The surface-sensitive electrochemical oxidation of adsorbed CO on pristine Ir_{poly} and Ni-modified Ir_{poly} electrodes has been studied in alkaline solution (0.1 M KOH) using stripping voltammetry. In this scenario, the Ni/Ir_{poly} NSA catalyst was prepared by annealing the OL for only 1 min in the reductive atmosphere. Due to the reduced exposure to the reductive treatment, the characteristic $Ni(OH)_2$/NiOOH feature is still visible in the cyclic voltammogram. The CO oxidation mechanism is examined in the alkaline electrolyte in order to integrate the Ni electrochemistry and to identify if CO adsorption is favored in the presence of Ni since the (sub)surface layers of Ni on Ir have not been stable in an acidic environment. Furthermore, the concept of OER in alkaline media may be of interest from the catalytic point-of-view since the pretreatment of the Ni/Ir_{poly} OL catalyst in H_2SO_4 seems to have a beneficial effect in the subsequent activity measurement in KOH.

Figure 7 compares the anodic stripping voltammetry observed for a CO adlayer on the three distinct electrode surfaces in 0.1 M KOH solution at 50 mV s^{-1}, which is followed by the second sweep cycle between the potentials of 0.05 and 1.40 V_{RHE}. The presence of a CO surface adlayer can be confirmed since, for each electrocatalyst, a CO stripping peak was recorded. The voltammetric profiles further validate the complete electrooxidation of the adsorbed CO in the first anodic sweep since the second cycle produces a voltammogram displaying HUPD peaks characteristic for Ir.

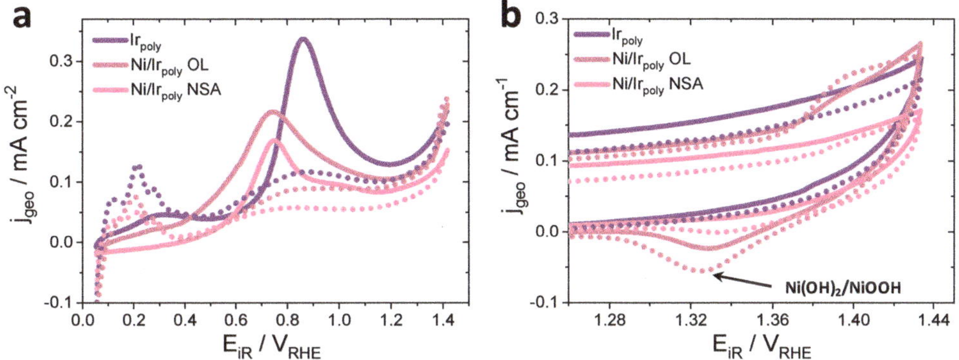

Figure 7. CO stripping (solid line) and the subsequent cyclic voltammogram (dotted line) for Ir_{poly}, Ni/Ir_{poly} OL, and Ni/Ir_{poly} NSA measured in 0.1 M KOH. (**a**) Anodic curves over the entire potential range and (**b**) anodic and cathodic curves in a potential range of 1.25–1.44 V_{RHE}.

It can be noticed that there is a clear distinction between the Ni-modified electrocatalysts and the unmodified, pristine polycrystalline Ir electrocatalyst with regard to the stripping peak shape and position in Figure 7a. The Ni/Ir_{poly} OL and Ni/Ir_{poly} NSA electrodes produced a stripping peak at much lower potentials than the pure Ir_{poly}, which suggests that the oxidation of surface CO to CO_2 is promoted by the presence of Ni. In accordance with the enhanced CO oxidation of the Ni-modified catalysts, the onset of OH adsorption on the OL and NSA starts at more negative potentials than on pristine Ir_{poly} [37]. The oxidation of Ir sites is favored, or in other words, the adsorption of HO at free metallic sites is highly favored in the presence of Ni. An additional voltammetric feature of the CO-loaded Ir polycrystal is a broad peak at ~+0.3 V_{RHE} (Figure 7a), which is only weakly defined for the OL and no longer present for the NSA. The possible origin of this peak will be discussed below.

To gain a quantitative understanding of the effect of Ni modification on the CO oxidation activity in alkaline electrolyte, the CO stripping profiles were deconvoluted into multiple peaks. In order to place hidden peaks, a linear background-correction was performed to the CO stripping peaks and, subsequently, the individual peaks were fitted. The respective CO stripping curves are presented in Figure 8 for Ir_{poly}, Ni/Ir_{poly} OL, and Ni/Ir_{poly} NSA catalysts. At least two peaks can sufficiently fit all profiles.

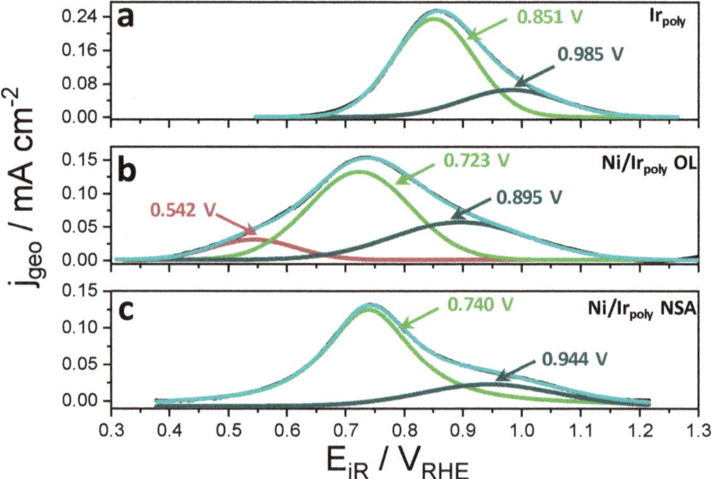

Figure 8. CO stripping curves for (**a**) the unmodified polycrystalline Ir electrode and the Ni-modified (**b**) Ir_{poly} OL and (**c**) Ir_{poly} NSA catalysts obtained at 50 mV s^{-1} in N_2-saturated 0.1 M KOH (purified Fe free) at a rotation of 1000 rpm. A linear-background correction was performed through baseline subtraction for the anodic CO stripping voltammetry in Figure 7a between 0.4 and 1.2 V_{RHE}. The peaks were fit by Gaussian functions and the individual contributions are shown in colors. The cumulative curve is given in turquoise.

The major component is centered at ~+0.85 V_{RHE} for Ir_{poly} and is shifted toward more cathodic potentials for Ni/Ir_{poly} OL and Ni/Ir_{poly} NSA. The effect of a more facile CO oxidation with respect to the unmodified Ir_{poly} is quite likely due to a Ni richer surface, which has been observed in earlier studies [38–40]. The hypothesis is further reinforced by the fact that no Ni was removed by the CO stripping procedure since the characteristic $Ni(OH)_2/NiOOH$ redox couple was observed for both Ni-modified catalysts in Figure 8b after CO was stripped off the surface. The early onset of CO oxidation for the OL and NSA is initiated by highly hydrophilic Ni-rich areas forming Ni(oxo)hydroxides, which further react with adsorbed CO on neighboring Ir sites as early as ~+0.72 V_{RHE} for Ni/Ir_{poly} OL. The most likely reaction scheme to the CO oxidation process in alkaline solution can be assigned to a Langmuir-Hinshelwood mechanism involving the competitive adsorption of the reactants CO and OH [41].

$$CO_{ads} + OH_{ads} \rightarrow COOH_{ads} \tag{13}$$

$$COOH_{ads} \rightarrow CO_2 + H^+ + e^-. \tag{14}$$

The oxidative removal of adsorbed CO on Ir in the presence of Ni is presumed to proceed according to the following equations.

$$Ni_{ads} + H_2O \rightarrow Ni_{ads} \cdots OH_{ads} + H^+ + e^- \tag{15}$$

$$Ni_{ads} \cdots OH_{ads} + CO_{ads} \rightarrow Ni_{ads} + CO_2 + H^+ + e^- \tag{16}$$

In the CO stripping voltammetry of the pure Ir polycrystal in Figure 7a, an anodic redox wave is visible at ~+0.3 V_{RHE}. The origin of this peak cannot be adequately verified. However, a rather small 'pre-peak' observed by Spendelow et al. [41] and García et al. [42] was attributed to oxidation of CO at step sites. In the study of the mechanism and kinetics of CO oxidation on stepped surfaces by in situ infrared spectroscopy, Lebedeva et al. found that the step trough is the most active site for CO oxidation [43]. After deposition of Ni, this peak is weakly identifiable, which suggests that Ni adatoms adsorb to some extent on Ir step sites. After annealing, however, the peak disappeared probably due to healing of a majority of defect sites on the polycrystalline Ir surface.

The CO stripping peaks in Figure 8 all consist of a small peak or shoulder at higher potentials (dark green fit) arising very likely from the adsorption of HO. Ir oxidation is prevented until the oxidative removal of adsorbed CO. On the Ni-modified catalysts, OH adsorption is more favored and, thereby, shifted to lower potentials as discussed earlier. The main peak positioned at ~+0.85 V_{RHE} for Ir_{poly}, ~+0.72 V_{RHE} for Ni/Ir_{poly} OL, and ~+0.74 V_{RHE} for Ni/Ir_{poly} NSA can be attributed to CO oxidation at terrace sites of Ir [44]. Based on the fitted data, the cathodic shift of the CO stripping peak is slightly more pronounced for Ni/Ir_{poly} OL (~+0.72 V_{RHE}) than for Ni/Ir_{poly} NSA (~+0.74 V_{RHE}). This finding can be interpreted under consideration of the CO oxidation charge of both Ni-modified catalysts. The CO stripping charge is frequently used to determine the electrochemical active surface area (ECSA), which is required to assess the stability and activity properties of electro-catalytic materials [38,45–47]. The baseline-corrected deconvoluted peaks (light green) were integrated for each individual catalyst. As expected, the net CO stripping charge of Ir_{poly} was found to be the highest with 803 $\mu C\ cm^{-2}$. The electrodeposition of Ni on the surface of Ir reduces the surface area available to CO, as Ir sites are blocked by Ni adatoms. A net charge of 569 $\mu C\ cm^{-2}$ can be assigned to the OL catalyst. The CO charge for the NSA is slightly reduced to 529 $\mu C\ cm^{-2}$. However, in addition to purely geometric blocking, electronic effects due to the Ni atoms need to be taken into consideration as well when discussing the CO coverages. Since the outer layer of the NSA in this study consists of both Ir and Ni atoms, the lower CO coverage on the NSA must be deduced to the modified electronic properties of the Ir surface atoms due to Ni atoms in the subsurface layer. A lowering of the chemisorption energy leads to a weaker adsorbate binding. A similar behavior has been reported by Bandarenka et al. in the study of the CO oxidation on alloys of Cu/Pt (111) [48]. It is, therefore, feasible that the electronic effect of Ni in the subsurface region on Ir either has a negative effect for the oxidation of CO or does not compensate for the reduced amount of surface Ni atoms. Consequently, the peak position of the main CO stripping peak is not only a function of the pure composition but depends strongly on the specific position of the non-noble metal. Apparent from Figure 8b, the CO stripping curve of the OL seems to consist of a second feature at ~+0.54 V_{RHE}, which might be deduced to an OH adsorption on Ni adatoms prior to the facilitated electrochemical oxidation of CO. Even after intense fitting, this rather small peak was not found on the NSA most likely due to the comparatively small amount of Ni surface atoms. However, the absence of this peak on the NSA system strengthens the observed higher activity of the OL towards CO oxidation.

In the foregoing discussion, a distinction between distinct active Ir sites has not been addressed yet. The behavior of hydrogen on the polycrystalline Ir surface, however, may provide important information on the deposition process of Ni. In Figure 7a, there are three distinct redox features in the anodic branch of the HUPD region of the pristine Ir_{poly} in the second scan after the complete oxidation of adsorbed CO. These redox peaks indicate distinct Ir sites at which oxidation of UPD hydrogen occurs. The well-defined HUPD peaks can be assigned to the three low-index crystallographic facets (111), (110), and (100) of Ir. A thorough theoretical study of the adsorption/deposition of hydrogen on various close-packed facets of noble and non-noble metals including the face-centered cubic (fcc) facets of Ir(111), Ir(110), and Ir(100) was performed. The energetically most favored adsorption sites for Ir(111) and Ir(100) were adopted by Ferrin et al. [49]. The binding energies (BE) for the three distinct low-index surfaces along with the coordination numbers (CN) of surface Ir atoms and calculated geometries for surface hydrogen (Site) are given in Table 2.

Table 2. DFT-calculated binding energies of hydrogen and H-Ir bond distances on the three low-index surfaces of Ir.

Face	CN	Site	H BE/eV	H-Ir Bond Distance/pm
(100)	6	bridging	−2.91	2 × 181
(111)	9	atop	−2.73	159
(110)	7	bridging	−2.74	2 × 181

The binding energies were calculated relative to the hydrogen atom in the gas phase. The computational details reveal that the Ir(100) surface (−2.91 eV) binds hydrogen the strongest and is followed by Ir(111) (−2.73 eV) and Ir(110) (−2.74 eV). Based on these calculations, Ir(100) binds hydrogen ~0.20 eV more strongly than the Ir(111) and (110) surfaces. The difference in binding energy between those facets can be deduced to the hydrogen-metal bond distances and binding geometries. It has been reported that, for Ir(111), adsorption on the top site [50,51] is favored while, on Ir(100), the bridge site [52] is more favorable in terms of energy. Due to the distinct geometries of the adsorption configuration for surface hydrogen, the H-Ir bond distance on (100) is larger (1.81 Å) than on (111) (1.59 Å). The very similar hydrogen binding energies of the (111) and (110) sites hamper the assignment of the HUPD peaks. For this, an additional DFT study on Ni binding energies was performed revealing that Ni binds the strongest on the Ir(110) face (−4.55 eV), less strongly on the Ir(100) face (−4.47 eV), and the weakest on Ir(111) (−3.76 eV). Given the stronger binding of Ni to the (110) sites, we assume the first peak to originate from HUPD on (110) sites. The combined results from the DFT studies are presented in Figure 9.

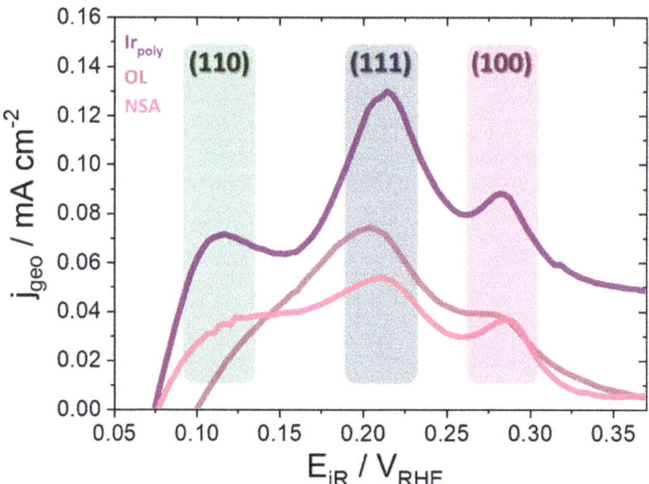

Figure 9. The second cycle of the CO stripping CV (from Figure 7) displaying the HUPD peaks of the three distinct catalyst systems. The allocation of the HUPD peaks are based on DFT calculations of the hydrogen binding energy on the three low-index surfaces (111), (110), and (100) of Ir.

As the HUPD peak attributed to the (110) sites is significantly reduced by forming an over layer of Ni, it appears that Ni adsorbs rather on (110) sites and that (100) sites are less favored. For the NSA catalyst system, the (110) peak is slightly visible whereas the (100) is more pronounced due to Ni in the subsurface layers. The HUPD charge is the lowest for the NSA catalyst (0.10 mC cm^{-2}). For pristine Ir$_{poly}$ and Ni/Ir$_{poly}$ OL, the HUPD charge was determined to be 0.20 mC cm^{-2} and 0.13 mC cm^{-2}, respectively. The given order was also observed for the CO charge and might provide important indications for the activity of the Ni-modified catalysts. The smaller HUPD charge of the NSA catalyst in comparison to the system in which Ni is present at the surface might be set into relation with its lower activity performance toward catalyzing the oxygen evolution.

4. Conclusions

The analysis of the three distinct catalyst systems with respect to their electrocatalytic performance towards oxygen evolution revealed that the Ni-modified polycrystalline Ir electrodes display distinct electrochemical properties. Incorporating Ni in the form of an over layer significantly reduces the overpotential to evolve oxygen in both the acidic and the alkaline solution. The promoting effect of Ni in IrNi bimetallic catalysts has been observed in various studies [23,29,35]. In acidic milieu, the higher activity might be assigned to the sacrificial leaching of Ni leading to a higher content of hydroxyl groups (OH). The Ni-depleted active catalyst structure is expected to be composed of oxygen atoms protonated from the electrolyte to form surface hydroxyl groups, which presumably act as crucial intermediates in the OER mechanism. The beneficial effect adding Ni to the OER activity, however, is no longer given to the same extent if Ni is present in the subsurface layer of the electrode. Thus, we can conclude that Ni in the NSA configuration is not able to perform as a sacrificial catalyst component as effectively as Ni in the OL configuration. Furthermore, the electronic effect of Ni on Ir might exercise a negative effect on the binding energy of Ir towards reaction intermediates in the OER. However, it cannot be neglected that the reductive treatment, which formed the NSA, led to partial healing of the Ir surface with a smaller density of defect and step sites. Since the environment of Ir was significantly altered, so was its activity. In alkaline media, the highly hydrophilic Ni adatoms are sufficient to achieve the desired effect of increasing the OER performance of the catalyst material. The performance of the OL system increases in KOH after it was pretreated in acid electrolyte beforehand. This observation confirms the hypothesis of Ni dissolution in acid. It appears as if the preceding OER treatment in 0.05 M H_2SO_4 has an advantageous impact on the electrocatalytic properties in KOH and can be deduced to the previously mentioned leaching of Ni in an acidic solution, which raises the concentration of surface OH atoms. The performance during OER of the catalysts including the shift of CO oxidation peaks is not only a function of the pure composition of the material but does strongly depend on the relative position of the metals to each other. However, it must be kept in mind that the interpretation of the peaks with regard to the CO stripping measurements especially of the shoulders is quite complex due to the presence of surface defects.

Supplementary Materials: The following are available online at http://www.mdpi.com/2571-9637/1/1/13/s1. Figure S1. Double-layer capacitance measurements displaying cyclic voltammograms of Ir_{poly} measured in a potential window in which no faradic currents occur. Following scan rates were adjusted: 500 mV s^{-1}, 200 mV s^{-1}, 100 mV s^{-1}, 50 mV s^{-1}, 20 mV s^{-1} and 10 mV s^{-1}; Figure S2. X-ray photoelectron spectroscopy measurements of (a) Ir(111) over the entire energy range. (b) High-resolution spectra of Ir(111) displaying the Ir4f, C1s, O1s and Ni 2p spectra. (c) XP spectra of Ni/Ir_{poly} $OL_{t=20min}$ over the entire energy range; Figure S3. (a) Atomic force microscopy (AFM) images of Ni/Ir(111) OL with respective height profiles. (b) Magnification of a terrace assembly of (a) and (c) higher magnification. Scale bar is 400 nm. Arrows give the direction of the calculated height profile; Figure S4. 10th cycle of the electrocatalytic oxygen evolution reaction (OER) activities of Ni-modified Ir electrodes compared to pristine polycrystalline Ir at scan rates of 5 mV s^{-1} in (a) 0.05 M H_2SO_4 and (b) in 0.1 M KOH of Ir_{poly}, Ni/Ir_{poly} OL and Ni/Ir_{poly} NSA. Current densities of the electrodes are normalized to the electrodes' geometric surface area of 0.785 cm^{-2}; Figure S5. Electrocatalytic OER activities of Ni-modified electrocatalysts and pristine Ir_{poly} measured by differential electrochemical mass spectrometry (DEMS) in a dual-thin layer flow cell setup at a scan rate of 10 mV s^{-1}. (a) Faradaic current densities measured at an OER overpotential of 300 mV (1.53 V_{RHE}) normalized to the electrodes' geometric surface area of 0.785 cm^{-2}. The electrocatalytic OER activity in H_2SO_4 was measured directly after the treatment in KOH. (b) Faraday efficiency of oxygen evolution (FE_{O2}) determined for the catalysts; Table S1. Calculated amount of adsorbed Ni on a polycrystalline Ir electrode in dependence of deposition potential Edep. The coverage of Ni (ML) was determined from ICP MS based on the assumption of a predominant (111) face on Irpoly.

Author Contributions: Conceptualization, E.Ö and P.S. Methodology, E.Ö. Validation, E.Ö. Formal analysis, E.Ö., J.A., A.M.M., I.S., and T.K. Investigation, E.Ö., J.A., and I.S. Resources, P.S., K.M., M.M., and B.R.C. Data analysis, E.Ö., J.A., I.S., and T.K. Writing—original draft preparation, E.Ö. Writing—review and editing, E.Ö., P.S., B.R.C., K.J.J.M., T.K., J.A., and A.M.M. Visualization, E.Ö. and P.S. Supervision, P.S. Project administration, E.Ö.

Funding: This research was funded by the Deutsche Forschungsgemeinschaft under the priority program SPP1613, grant number STR 596/3-1. The U.S. Department of Energy (DOE) through the Office of Basic Energy Sciences (BES) by grant DE-FG02-05ER15731 supported work at UW–Madison. Computational work was performed using supercomputing resources at the Center for Nanoscale Materials (CNM) at the Argonne National Laboratory under contract number DE-AC02-06CH11357 and at the National Energy Research Scientific Computing Center (NERSC) under contract number DE-AC02-05CH11231. Thomas Kropp is grateful for partial financial support by the Alexander von Humboldt Foundation.

Acknowledgments: We thank Benjamin Paul and Zarina Pawolek for helpful discussions and help with the set up.

Conflicts of Interest: The authors declare no conflict of interest.

References

1. Cherevko, S.; Geiger, S.; Kasian, O.; Kulyk, N.; Grote, J.-P.; Savan, A.; Shrestha, B.R.; Merzlikin, S.; Breitbach, B.; Ludwig, A.; et al. Oxygen and hydrogen evolution reactions on Ru, RuO_2, Ir, and IrO_2 thin film electrodes in acidic and alkaline electrolytes: A comparative study on activity and stability. *Catal. Today* **2016**, *262*, 170–180. [CrossRef]

2. Kasian, O.; Geiger, S.; Stock, P.; Polymeros, G.; Breitbach, B.; Savan, A.; Ludwig, A.; Cherevko, S.; Mayrhofer, K.J.J. On the Origin of the Improved Ruthenium Stability in RuO_2–IrO_2 Mixed Oxides. *J. Electrochem. Soc.* **2016**, *163*, F3099–F3104. [CrossRef]

3. Schmeisser, D.; Jacobi, K. Photoelectron spectroscopy of nickel on zinc oxide in the monolayer and submonolayer range. *Surf. Sci.* **1979**, *88*, 138–152. [CrossRef]

4. Gaebler, W.; Jacobi, K.; Ranke, W. The structure and electronic properties of thin palladium films on zinc oxide studied by AES and UPS. *Surf. Sci.* **1978**, *75*, 355–367. [CrossRef]

5. Greeley, J.; Mavrikakis, M. Near-surface alloys for hydrogen fuel cell applications. *Catal. Today* **2006**, *111*, 52–58. [CrossRef]

6. Greeley, J.; Mavrikakis, M. Alloy catalysts designed from first principles. *Nat. Mater.* **2004**, *3*, 810–815. [CrossRef] [PubMed]

7. Greeley, J.; Mavrikakis, M. Surface and Subsurface Hydrogen: Adsorption Properties on Transition Metals and Near-Surface Alloys. *J. Phys. Chem. B* **2005**, *109*, 3460–3471. [CrossRef] [PubMed]

8. Herron, J.A.; Mavrikakis, M. On the composition of bimetallic near-surface alloys in the presence of oxygen and carbon monoxide. *Catal. Commun.* **2014**, *52*, 65–71. [CrossRef]

9. Kandoi, S.; Ferrin, P.A.; Mavrikakis, M. Hydrogen on and in Selected Overlayer Near-Surface Alloys and the Effect of Subsurface Hydrogen on the Reactivity of Alloy Surfaces. *Top. Catal.* **2010**, *53*, 384–392. [CrossRef]

10. Kim, C.; Möller, T.; Schmidt, J.; Thomas, A.; Strasser, P. Suppression of Competing Reaction Channels by Pb Adatom Decoration of Catalytically Active Cu Surfaces During CO_2 Electroreduction. *ACS Catal.* under review.

11. Özer, E.; Paul, B.; Spöri, C.; Strasser, P. Coupled Inductive Annealing-Electrochemical Setup for Controlled Preparation and Characterization of Alloy Crystal Surface Electrodes. *Small Methods* **2018**, 1800232. [CrossRef]

12. Özer, E.; Spöri, C.; Reier, T.; Strasser, P. Iridium(1 1 1), Iridium(1 1 0), and Ruthenium(0 0 0 1) Single Crystals as Model Catalysts for the Oxygen Evolution Reaction: Insights into the Electrochemical Oxide Formation and Electrocatalytic Activity. *ChemCatChem* **2016**, *9*, 597–603. [CrossRef]

13. Trotochaud, L.; Young, S.L.; Ranney, J.K.; Boettcher, S.W. Nickel–Iron Oxyhydroxide Oxygen-Evolution Electrocatalysts: The Role of Intentional and Incidental Iron Incorporation. *J. Am. Chem. Soc.* **2014**, *136*, 6744–6753. [CrossRef] [PubMed]

14. Blöchl, P.E. Projector augmented-wave method. *Phys. Rev. B* **1994**, *50*, 17953–17979. [CrossRef]

15. Kresse, G.; Joubert, D. From ultrasoft pseudopotentials to the projector augmented-wave method. *Phys. Rev. B* **1999**, *59*, 1758–1775. [CrossRef]

16. Kresse, G.; Furthmüller, J. Efficient iterative schemes for ab initio total-energy calculations using a plane-wave basis set. *Phys. Rev. B* **1996**, *54*, 11169–11186. [CrossRef]

17. Kresse, G.; Furthmüller, J. Efficiency of ab-initio total energy calculations for metals and semiconductors using a plane-wave basis set. *Comput. Mater. Sci.* **1996**, *6*, 15–50. [CrossRef]

18. Perdew, J.P.; Burke, K.; Ernzerhof, M. Generalized Gradient Approximation Made Simple. *Phys. Rev. Lett.* **1996**, *77*, 3865–3868. [CrossRef]

19. Jusys, Z.; Massong, H.; Baltruschat, H. A New Approach for Simultaneous DEMS and EQCM: Electro-oxidation of Adsorbed CO on Pt and Pt-Ru. *J. Electrochem. Soc.* **1999**, *146*, 1093–1098. [CrossRef]

20. Wonders, A.H.; Housmans, T.H.M.; Rosca, V.; Koper, M.T.M. On-line mass spectrometry system for measurements at single-crystal electrodes in hanging meniscus configuration. *J. Appl. Electrochem.* **2006**, *36*, 1215–1221. [CrossRef]

21. McCrory, C.C.L.; Jung, S.; Peters, J.C.; Jaramillo, T.F. Benchmarking Heterogeneous Electrocatalysts for the Oxygen Evolution Reaction. *J. Am. Chem. Soc.* **2013**, *135*, 16977–16987. [CrossRef] [PubMed]

22. Bligaard, T.; Nørskov, J.K. Ligand effects in heterogeneous catalysis and electrochemistry. *Electrochim. Acta* **2007**, *52*, 5512–5516. [CrossRef]

23. Reier, T.; Pawolek, Z.; Cherevko, S.; Bruns, M.; Jones, T.; Teschner, D.; Selve, S.; Bergmann, A.; Nong, H.N.; Schlögl, R.; et al. Molecular Insight in Structure and Activity of Highly Efficient, Low-Ir Ir–Ni Oxide Catalysts for Electrochemical Water Splitting (OER). *J. Am. Chem. Soc.* **2015**, *137*, 13031–13040. [CrossRef]

24. Rossmeisl, J.; Qu, Z.W.; Zhu, H.; Kroes, G.J.; Nørskov, J.K. Electrolysis of water on oxide surfaces. *J. Electroanal. Chem.* **2007**, *607*, 83–89. [CrossRef]

25. Sanchez Casalongue Hernan, G.; Ng May, L.; Kaya, S.; Friebel, D.; Ogasawara, H.; Nilsson, A. In Situ Observation of Surface Species on Iridium Oxide Nanoparticles during the Oxygen Evolution Reaction. *Angew. Chem. Int. Ed.* **2014**, *53*, 7169–7172. [CrossRef] [PubMed]

26. Favaro, M.; Valero-Vidal, C.; Eichhorn, J.; Toma, F.M.; Ross, P.N.; Yano, J.; Liu, Z.; Crumlin, E.J. Elucidating the alkaline oxygen evolution reaction mechanism on platinum. *J. Mater. Chem. A* **2017**, *5*, 11634–11643. [CrossRef]

27. Doyle, R.L.; Lyons, M.E.G. The Oxygen Evolution Reaction: Mechanistic Concepts and Catalyst Design. In *Photoelectrochemical Solar Fuel Production: From Basic Principles to Advanced Devices*, 1st ed.; Giménez, S., Bisquert, J., Eds.; Springer International Publishing: Cham, Switzerland, 2016; pp. 41–104, ISBN 978-3-319-29641-8.

28. Gong, L.; Ren, D.; Deng, Y.; Yeo, B.S. Efficient and Stable Evolution of Oxygen Using Pulse-Electrodeposited Ir/Ni Oxide Catalyst in Fe-Spiked KOH Electrolyte. *ACS Appl. Mater. Interfaces* **2016**, *8*, 15985–15990. [CrossRef]

29. Nong, H.N.; Oh, H.S.; Reier, T.; Willinger, E.; Willinger, M.G.; Petkov, V.; Teschner, D.; Strasser, P. Oxide-Supported IrNiO$_x$ Core–Shell Particles as Efficient, Cost-Effective, and Stable Catalysts for Electrochemical Water Splitting. *Angew. Chem.* **2015**, *127*, 3018–3022. [CrossRef]

30. Wang, C.; Sui, Y.; Xu, M.; Liu, C.; Xiao, G.; Zou, B. Synthesis of Ni–Ir Nanocages with Improved Electrocatalytic Performance for the Oxygen Evolution Reaction. *ACS Sustain. Chem. Eng.* **2017**, *5*, 9787–9792. [CrossRef]

31. Bode, H.; Dehmelt, K.; Witte, J. Zur kenntnis der nickelhydroxidelektrode—I.Über das nickel (II)-hydroxidhydrat. *Electrochim. Acta* **1966**, *11*, 1079–1087. [CrossRef]

32. Corrigan, D.A.; Knight, S.L. Electrochemical and Spectroscopic Evidence on the Participation of Quadrivalent Nickel in the Nickel Hydroxide Redox Reaction. *J. Electrochem. Soc.* **1989**, *136*, 613–619. [CrossRef]

33. Pfeifer, V.; Jones, T.E.; Velasco Velez, J.J.; Arrigo, R.; Piccinin, S.; Havecker, M.; Knop-Gericke, A.; Schlögl, R. In situ observation of reactive oxygen species forming on oxygen-evolving iridium surfaces. *Chem. Sci.* **2017**, *8*, 2143–2149. [CrossRef] [PubMed]

34. Pfeifer, V.; Jones, T.E.; Velasco Velez, J.J.; Massue, C.; Greiner, M.T.; Arrigo, R.; Teschner, D.; Girgsdies, F.; Scherzer, M.; Allan, J.; et al. The electronic structure of iridium oxide electrodes active in water splitting. *Phys. Chem. Chem. Phys.* **2016**, *18*, 2292–2296. [CrossRef] [PubMed]

35. Nong, H.N.; Reier, T.; Oh, H.-S.; Gliech, M.; Paciok, P.; Vu, T.H.T.; Teschner, D.; Heggen, M.; Petkov, V.; Schlögl, R.; et al. A unique oxygen ligand environment facilitates water oxidation in hole-doped IrNiOx core–shell electrocatalysts. *Nat. Catal.* **2018**. [CrossRef]

36. Seitz, L.C.; Dickens, C.F.; Nishio, K.; Hikita, Y.; Montoya, J.; Doyle, A.; Kirk, C.; Vojvodic, A.; Hwang, H.Y.; Norskov, J.K.; et al. A highly active and stable IrO$_x$/SrIrO$_3$ catalyst for the oxygen evolution reaction. *Science* **2016**, *353*, 1011–1014. [CrossRef] [PubMed]

37. Cui, C.; Ahmadi, M.; Behafarid, F.; Gan, L.; Neumann, M.; Heggen, M.; Cuenya, B.R.; Strasser, P. Shape-selected bimetallic nanoparticle electrocatalysts: Evolution of their atomic-scale structure, chemical composition, and electrochemical reactivity under various chemical environments. *Faraday Discuss.* **2013**, *162*, 91–112. [CrossRef] [PubMed]

38. Rudi, S.; Cui, C.; Gan, L.; Strasser, P. Comparative study of the electrocatalytically active surface areas (ECSAs) of Pt alloy nanoparticles evaluated by Hupd and CO-stripping voltammetry. *Electrocatalysis* **2014**, *5*, 408–418. [CrossRef]

39. Beermann, V.; Gocyla, M.; Kühl, S.; Padgett, E.; Schmies, H.; Goerlin, M.; Erini, N.; Shviro, M.; Heggen, M.; Dunin-Borkowski, R.E.; et al. Tuning the Electrocatalytic Oxygen Reduction Reaction Activity and Stability of Shape-Controlled Pt–Ni Nanoparticles by Thermal Annealing—Elucidating the Surface Atomic Structural and Compositional Changes. *J. Am. Chem. Soc.* **2017**, *139*, 16536–16547. [CrossRef]

40. Chatenet, M.; Soldo-Olivier, Y.; Chaînet, E.; Faure, R. Understanding CO-stripping mechanism from Ni_{UPD}/Pt(110) in view of the measured nickel formal partial charge number upon underpotential deposition on platinum surfaces in sulphate media. *Electrochim. Acta* **2007**, *53*, 369–376. [CrossRef]

41. Spendelow, J.S.; Goodpaster, J.D.; Kenis, P.J.A.; Wieckowski, A. Mechanism of CO Oxidation on Pt(111) in Alkaline Media. *J. Phys. Chem. B* **2006**, *110*, 9545–9555. [CrossRef]

42. García, G.; Koper, M.T.M. Stripping voltammetry of carbon monoxide oxidation on stepped platinum single-crystal electrodes in alkaline solution. *Phys. Chem. Chem. Phys.* **2008**, *10*, 3802–3811. [CrossRef] [PubMed]

43. Lebedeva, N.P.; Rodes, A.; Feliu, J.M.; Koper, M.T.M.; van Santen, R.A. Role of Crystalline Defects in Electrocatalysis: CO Adsorption and Oxidation on Stepped Platinum Electrodes As Studied by in situ Infrared Spectroscopy. *J. Phys. Chem. B* **2002**, *106*, 9863–9872. [CrossRef]

44. López-Cudero, A.; Cuesta, A.; Gutiérrez, C. Potential dependence of the saturation CO coverage of Pt electrodes: The origin of the pre-peak in CO-stripping voltammograms. Part 1: Pt(111). *J. Electroanal. Chem.* **2005**, *579*, 1–12. [CrossRef]

45. Garrick, T.R.; Moylan, T.E.; Carpenter, M.K.; Kongkanand, A. Editors' Choice—Electrochemically Active Surface Area Measurement of Aged Pt Alloy Catalysts in PEM Fuel Cells by CO Stripping. *J. Electrochem. Soc.* **2017**, *164*, F55–F59. [CrossRef]

46. Hasché, F.; Oezaslan, M.; Strasser, P. Activity, stability and degradation of multi walled carbon nanotube (MWCNT) supported Pt fuel cell electrocatalysts. *Phys. Chem. Chem. Phys.* **2010**, *12*, 15251–15258. [CrossRef] [PubMed]

47. Li, G.; Anderson, L.; Chen, Y.; Pan, M.; Abel Chuang, P.-Y. New insights into evaluating catalyst activity and stability for oxygen evolution reactions in alkaline media. *Sustain. Energy Fuels* **2018**, *2*, 237–251. [CrossRef]

48. Bandarenka Aliaksandr, S.; Varela Ana, S.; Karamad, M.; Calle-Vallejo, F.; Bech, L.; Perez-Alonso Francisco, J.; Rossmeisl, J.; Stephens Ifan, E.L.; Chorkendorff, I. Design of an Active Site towards Optimal Electrocatalysis: Overlayers, Surface Alloys and Near-Surface Alloys of Cu/Pt(111). *Angew. Chem. Int. Ed.* **2012**, *51*, 11845–11848. [CrossRef]

49. Ferrin, P.; Kandoi, S.; Nilekar, A.U.; Mavrikakis, M. Hydrogen adsorption, absorption and diffusion on and in transition metal surfaces: A DFT study. *Surf. Sci.* **2012**, *606*, 679–689. [CrossRef]

50. Hagedorn, C.J.; Weiss, M.J.; Weinberg, W.H. Dissociative chemisorption of hydrogen on Ir(111): Evidence for terminal site adsorption. *Phys. Rev. B* **1999**, *60*, R14016–R14018. [CrossRef]

51. Krekelberg, W.P.; Greeley, J.; Mavrikakis, M. Atomic and Molecular Adsorption on Ir(111). *J. Phys. Chem. B* **2004**, *108*, 987–994. [CrossRef]

52. Lerch, D.; Klein, A.; Schmidt, A.; Müller, S.; Hammer, L.; Heinz, K.; Weinert, M. Unusual adsorption site of hydrogen on the unreconstructed Ir(100) surface. *Phys. Rev. B* **2006**, *73*, 075430. [CrossRef]

Article

Metallic Iridium Thin-Films as Model Catalysts for the Electrochemical Oxygen Evolution Reaction (OER)—Morphology and Activity

Ebru Özer [1,†], Zarina Pawolek [1,†], Stefanie Kühl [1], Hong Nhan Nong [2], Benjamin Paul [1], Sören Selve [3], Camillo Spöri [1], Cornelius Bernitzky [1] and Peter Strasser [1,4,*]

[1] The Electrochemical Energy, Catalysis and Materials Science Laboratory, Department of Chemistry, Technical University Berlin, Straße des 17. Juni 124, D-10623 Berlin, Germany; ebru.oezer@campus.tu-berlin.de (E.Ö.); zarina@mailbox.tu-berlin.de (Z.P.); stefanie.kuehl@tu-berlin.de (S.K.); benjamin.paul@tu-berlin.de (B.P.); camillo.spoeri@tu-berlin.de (C.S.); cornelius.bernitzky@campus.tu-berlin.de (C.B.)

[2] Max Planck Institute for Chemical Energy Conversion, Department of Heterogeneous Reactions, Stiftstr. 34, D-45470 Mülheim, Germany; hongnhan.nong@gmail.com

[3] Zentraleinrichtung Elektronenmikroskopie, Technische Universität Berlin, D-10623 Berlin, Germany; soeren.selve@tu-berlin.de

[4] Ertl Center for Electrochemistry and Catalysis, Gwangju Institute of Science and Technology, Gwangju 500-712, Korea

* Correspondence: p.strasser@tu-berlin.de; Tel.: +49-(0)30-314-29542

† These author contributed equally to this work.

Received: 3 November 2018; Accepted: 5 December 2018; Published: 6 December 2018

Abstract: Iridium (Ir) oxide is known to be one of the best electrocatalysts for the oxygen evolution reaction (OER) in acidic media. Ir oxide-based materials are thus of great scientific interest in current research on electrochemical energy conversion. In the present study, we applied Ir metal films as model systems for electrochemical water splitting, obtained by inductive heating in a custom-made setup using two different synthesis approaches. X-ray photoelectron spectroscopy (XPS) and selected area electron diffraction (SAED) confirmed that all films were consistently metallic. The effects of reductive heating time of calcined and uncalcined Ir acetate films on OER activity were investigated using a rotating disk electrode (RDE) setup. The morphology of all films was determined by scanning electron microscopy (SEM). The films directly reduced from the acetate precursor exhibited a strong variability of their morphology and electrochemical properties depending on heating time. The additional oxidation step prior to reductive heating accelerates the final structure formation.

Keywords: oxygen evolution reaction; water splitting; iridium; thin-films; spin-coating; model systems; electrocatalysts

1. Introduction

Water electrolysis as a non-polluting way of producing hydrogen is considered as one of the key technologies for an affordable and sustainable energy cycle [1]. Proton exchange membrane (PEM) electrolyzers are the most promising candidates for the supply of high purity hydrogen, because of their operation at high current densities and low gas crossover rates [2]. Among electrocatalysts applied in acidic media of water electrolysis, Ir oxide represents the state-of-the-art oxygen evolution catalyst with excellent activity and stability [3–5]. As a noble metal, it is less sensitive to corrosion in acidic media and further combines a high activity and stability.

The conditions under which the oxide electrodes are formed have a significant influence on the oxygen evolution reaction (OER) performance. In relation to thermally prepared oxides, this comprises the applied temperature [4], gas atmosphere [6], and particularly the precursor [6], which in turn has a significant effect on the morphology. These oxides were reported to have a higher stability, but lower activity in contrast to electrochemically oxidized Ir [7,8]. Accordingly, further investigation of electrochemically formed oxides on metallic Ir surfaces is of great scientific interest. In order to understand the role of morphology and structure of these materials, an adequate model system of well-defined Ir metal thin films is required. The fundamental research with model systems can elucidate the structure–activity–stability challenges of OER materials and, therefore, enable the systematic development and design of enhanced catalysts for industrial applications [9,10]. Given the high sensitivity of the OER kinetics towards the catalyst's preparation, the variability in either film thickness or film composition in a model system should be kept to a minimum, which facilitates future performance comparisons and performance enhancements.

Here, we present a study of novel Ir metal films on Ti substrate obtained by a precursor-based synthesis route with subsequent reductive heating in a custom-made inductive chamber. The highly accurate temperature and time regulation in the inductive setup guarantees the formation of homogeneous metallic Ir thin films. Design details and capabilities of the setup can be found elsewhere [11]. Herein, the metallic Ir film was obtained by reduction of two distinct Ir species, namely Ir acetate and Ir oxide. The latter was formed from the acetate by an additional calcination step prior to reduction in order to study the templating effect of Ir oxide on the resulting catalyst film. Three different heating times were applied during synthesis to monitor possible sintering processes during heat treatment. In the setup, the morphology of the metallic films can be varied under controlled heating conditions. The catalytically active Ir oxide species was obtained by subsequent electrochemical oxidation. It is well accepted that the synthetic oxidative route from metallic Ir to Ir oxide, that is, either electrochemical or thermal oxidation, has a great influence on the OER performance. To our knowledge, the influence of the synthesis route of the initial metallic Ir, however, has never been taken into account. The present work addresses this issue. It presents a comparative investigation of electrochemically oxidized Ir oxide species, which originate from distinct Ir metal thin films with distinct morphology. In this way, we were able to influence the stability and activity of the catalytic systems to a certain degree by the different arrangement and environment of the active sites.

A wide range of physical and electrochemical characterization, including cyclic voltammetry, stepped potential voltammetry (SPV), X-ray photoelectron spectroscopy, transmission electron microscopy (TEM)/selected area electron diffraction (SAED), and scanning electron microscopy (SEM)/energy-dispersive X-ray spectroscopy measurements (EDX), was performed on the as-prepared and the electrochemically oxidized films to uncover links between structural and catalytic properties.

2. Materials and Methods

As-prepared (AP) iridium acetate films (Chempur, Karlsruhe, Germany) on titanium were prepared via spin coating as described by the authors of [10]. From here, two different synthesis routes were applied. In the direct reduction route, the films were directly reduced in a custom-made inductive cell directly from the acetate in 4% H_2/Ar (purity 5N, 99.999%, Linde, Munich, Germany) at 400 °C, while in the oxide route, the acetate films were annealed at 450 °C for 15 min prior to the reduction in the inductive cell. Samples will be referred to as Ac*t* (samples reduced directly from the acetate) and Ox*t* (samples reduced from the oxide), where *t* is replaced by the time (in min) that the sample was exposed to reduction in the inductive cell. For each synthesis route and each heating time, several samples have been produced to ensure the reproducibility and to provide sufficient samples for the various physical analysis methods, so not every characterization was made with the same sample.

Electrochemical activity and stability measurements were performed at room temperature in 0.05 M H_2SO_4 (diluted from 95% H_2SO_4, Suprapure, Carl Roth, Germany) using a potentiostat (SP-240, BioLogic, Seyssinet-Pariset, France) and a common three-electrode rotating disk electrode (RDE) setup. A platinum mesh was used as counter electrode and a mercury/mercury sulfate electrode (Ametek, Berwyn, PA, USA, sat., calibrated against reversible hydrogen electrode, RHE) was used as reference electrode. All potentials were converted into RHE scale (referred to as V_{RHE}). Nitrogen was used for degassing. The electrodes were immersed in the electrolyte at a potential of 0.4 V_{RHE}. First, for surface characterization, cyclic voltammetry was performed in different ranges below the OER onset. The lower turning potential was 0.05 V_{RHE}, the upper turning potential was varied between 0.8 and 1.2 V_{RHE} in 0.2 V steps (five cycles each), followed by 50 cycles between 0.05 and 1.4 V_{RHE} for electrochemical oxidation (samples referred to as _CV). After every 10th cycle of the latter, three cycles in a range of 0.4 to 1.4 V_{RHE} were performed to obtain the anodic charge. A scan rate of 50 mV s^{-1} was applied. The anodic charge of the last scan was used for charge normalization of the current density curves obtained from the following activity measurements. A rotation speed of 1600 rpm was applied to the RDE working electrode and chronoamperometries (CA) were performed for 15 min at different potentials between 1.534 and 1.784 V_{RHE}, resulting in a stepped potential voltammetry (SPV) curve. In this potential range, adequate currents were obtained in pre-measurements. At each potential, electrochemical impedance spectroscopy was measured to obtain the ohmic resistance. After the last potential, the electrolyte was degassed; the rotation of the RDE was turned off; and again, three cycles at 50 m V s^{-1} in the potential range of 0.4 to 1.4 V_{RHE} were conducted. The protocol was performed twice, and the respective samples are referred to as _OER.

For electrochemical stability testing, the potential steps in the OER protocol were replaced by one long chronoamperometry at circa 1.734 V_{RHE} (without correction of ohmic resistance) for 20 h.

Scanning electron microscopy (SEM) images were obtained using a JEOL 7401F instrument (Tokyo, Japan) with an accelerating voltage of 10 kV. Corresponding EDX analysis was established by a Bruker XFLASH 4010 (Bruker, Billerica, MA, USA) at an accelerating voltage of 6 keV. The EDX was normalized by the sample current, which was measured in a Faraday cup if not stated otherwise.

XRD patterns were obtained by grazing-incident X-ray diffraction (GIXRD) measurement in a Bruker D8 Advance diffractometer (Bruker AXS). The analysis was performed at a grazing incidence of 1° with a Goebel mirror, a Cu Kα source, a 0.23° secondary Soller, and a scintillation counter as detector. The identification of the individual phases in the measured diffractograms was enabled with reference patterns. Rietveld refinement with the software TOPAS (Bruker AXS) allowed a clear definition of the individual phases present in the sample. A quantitative analysis, however, is not given with Rietveld refinement, as it is primarily aimed at examining powder samples. The structure and composition of film samples thus do not allow an exact determination of concentrations as the individual phases might occupy distant layers, which consequently affects the intensity of the incident beam.

For transmission electron microscopy (TEM), Cu grids with a 200 mesh and a holey carbon layer by Quantifoil (Großlöbichau, Germany) were coated with the dispersed catalyst. The measurements were performed at a FEI Tecnai G^{2}20 S-TWIN transmission electron microscope (Hillsboro, OR, USA) with a LaB$_6$ cathode operated at an accelerating voltage of 200 kV. The SAED evaluation was performed by PASAD [12,13] plugin in Digital Micrograph™, and the obtained reciprocal distances were transferred to Cu Kα scale to obtain a calculated diffraction pattern.

For the investigation of a cross section of Ir film on Ti, a TEM lamella was prepared by focused ion beam (FIB). A FEI Helios NanoLab 600 workstation equipped with (1) an Omniprobe lift-out system, (2) a Pt & W gas injector system, and (3) a scanning electron microscope (SEM) detector. High resolution TEM (HRTEM) was performed using an FEI Titan 80-300 TEM electron microscope with a high brightness X-FEG electron source and a Cs corrector for the objective lens. The microscope was operated at 200 kV.

Laboratory-based X-ray photoelectron spectroscopy (XPS) measurements were carried out at room temperature in an ultrahigh vacuum (UHV) setup using a monochromatic Al Kα source at 1486.6 eV (SPECS, Berlin, Germany). The XP spectra were analyzed using CasaXPS software. Fitting of O 1s spectra was performed using three singlets corresponding to lattice O (at ~ 530.0 eV), OH (at ~ 531.4 eV), adsorbed water, and organic O (at ~533.0 eV), respectively. The Ir 4f spectra were fitted with one component (for Ac10_AP and Ox10_AP samples, corresponding to Ir^0 species) or three components (for the electrochemically treated samples Ac10_CV, Ac10_OER, Ox10_CV, and Ox10_OER, corresponding to Ir^0, Ir^{IV}, and Ir^{III} species). The Ir 4f 7/2 positions of the Ir^0, Ir^{IV}, and Ir^{III} species were located at ~60.9 eV, ~61.7 eV, and ~62.3 eV, respectively. The peak models consist of either a doublet (Ir^0 component), a doublet and three satellite peaks (Ir^{IV} component), or a doublet and two satellite peaks (Ir^{III} component) [14]. The peak models (including asymmetry parameter) were derived from the fits of the Ir 4f XPS of the polycrystalline metallic Ir standard (for the Ir^0 component) or of a rutile-type IrO_2 standard sample (for the Ir^{IV} component) [15].

3. Results and Discussion

In the study of the morphology, structure, and catalytic activity of supported metallic Ir films, two different synthesis routes were established. The synthesis of the 'Ox' films was based on the reductive treatment of Ir oxide films, which were prepared according to a previously reported precursor-based spin coating method [10]. The 'Ac' films, however, skipped the calcination step and were directly reduced from the spin coated acetate film.

3.1. Morphology of Ir Metal Thin-Film Catalysts

3.1.1. Scanning Electron Microscopy (SEM)

In order to investigate the morphology of the Ox and Ac films, micrographs of the samples before (_AP) and after OER (_OER) measurements were obtained using a scanning electron microscope (SEM). The respective results are displayed in Figure 1. The Ox_AP films (Figure 1, second row) display a homogeneously cracked surface with clear porosity. Their morphology (Ox10, Ox30, and Ox60) does not change significantly with heating time, in contrast to the morphology of the Ac_AP film series. The surface of the Ac films (Figure 1, third row) displays larger cracks compared with that of the Ox films, and in addition, they exhibit a larger crack size distribution. Thus, a clear tendency towards a more cracked surface with a longer annealing time was observed for the Ac_AP films. The morphological restructuring (and probable sintering process) produces a highly cracked, 'sponge-like', and more fibrous morphology. It can be assumed that in contrast to the Ac samples, the calcination step in the Ox route creates a defined and crystalline rutile structure [10] from the beginning, which consequently leads to a faster formation of the final metallic Ir structure through reductive heating. After OER measurements, the Ac_OER samples (Figure 1, fourth row) exhibit a swollen and filled surface with rounded edges on the cracks. A similar effect is only weakly visible for the Ox films (Figure 1, first row) as a result of their less pronounced cracks.

3.1.2. Energy-Dispersive X-ray Spectroscopy Measurements (EDX)

EDX measurements performed on the _AP samples during SEM investigation in Figure S1 (in the Supplementary Information) show the atomic percentage (at%) ratio of O to Ir (not normalized by sample current) depending on the reductive heating time. The results reveal a significantly higher amount of O in the Ox samples than in the Ac samples, regardless of the heating time. This indicates the presence of an O species in the Ox films that cannot be removed by reductive heating under applied conditions, or that occurs while handling the samples under air, respectively. This species might either originate from a small fraction of (re)oxidized Ir in the film; physically adsorbed or trapped O within the pores of the film; or, most likely, a Ti oxide layer formed during the calcination step [10] and not being reduced at applied reduction conditions.

The position of O within the film, however, cannot be obtained by a single EDX measurement. For this purpose, several EDX measurements have been conducted along the radius of an Ox10_AP sample and IrO_x film on Ti (before reductive heating). The results are given in Figure 2. Here, the intensity (standardized to the sample current and time) of each element (left y-axis) along with the calculated O/Ir count ratio (right y-axis) is plotted against the radius of the film.

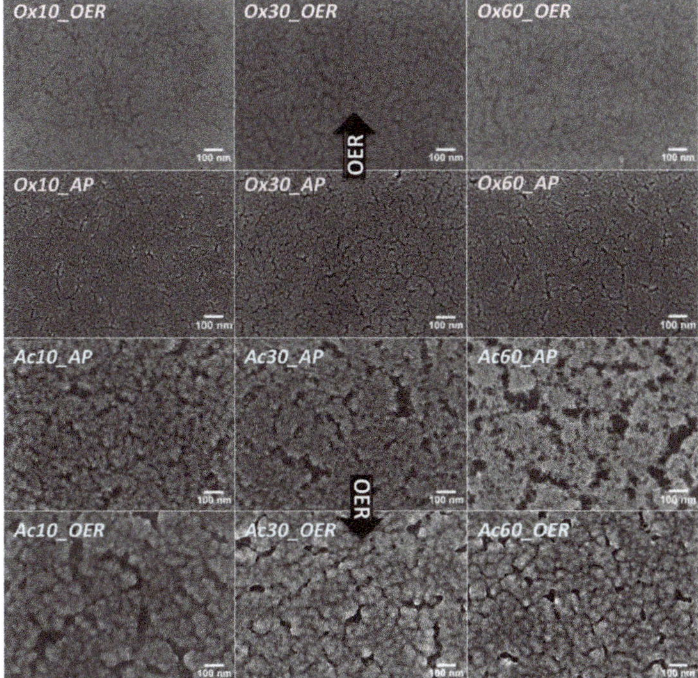

Figure 1. Scanning electron microscopy (SEM) images of Ox (top) and Ac (bottom) film series before (_AP, second and third row) and after oxygen evolution reaction (OER) (_OER, first and fourth row). The cracks in the Ac films enlarge with increasing heating time. The weakly cracked morphology of the Ox films does not show any change depending on the reductive heating time.

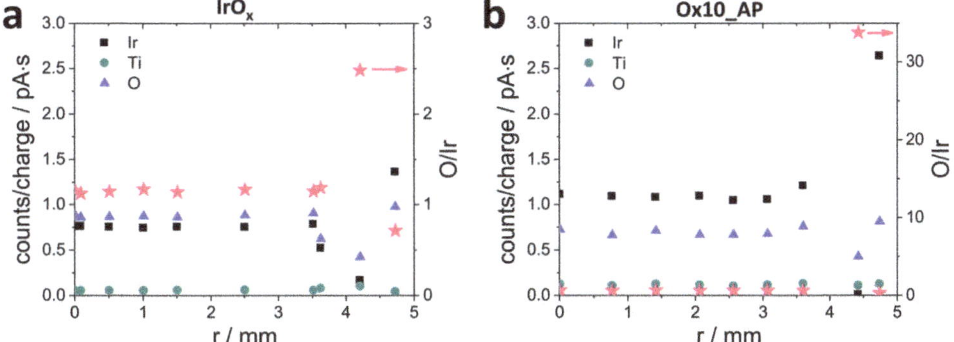

Figure 2. Counts per charge of the three main elements detected in energy-dispersive X-ray spectroscopy measurements (EDX) of (**a**) IrO_x and (**b**) Ox10_AP. The different intensities are caused by different film thicknesses, mainly due to border effects during spin coating.

Within a radius of 3.5 mm from the center, both samples show no variation in film thickness. Closer to the edge, the film thickness is altered as a result of the surface tension/boundary effects of the precursor solution during the spin coating process. The uneven film thickness at the edge of the sample and the ability of Ir to partially shield radiation emitted from deeper layers may at least give a hint on the O distribution, considering two extreme cases. If O was (exclusively) homogeneously distributed within the Ir(oxide) film (regardless of the chemical nature of O), the O/Ir ratio would be independent of the film thickness. In other words, at spots with lower Ir counts (lower film thickness), the O counts should also be lower. If oxygen only existed within a Ti oxide layer underneath the film, the O/Ir ratio would increase with decreasing film thickness as a result of lower shielding by Ir. With respect to the IrO_x film on Ti (Figure 2a), both cases seem to apply here; a lower amount of Ir accompanied by a lower amount of O (lower film thickness, 3.6 and 4.3 mm from center) points to an oxidized Ir film. At the same time, a higher O/Ir ratio speaks for a Ti oxide or Ir/Ti mixed oxide layer underneath the Ir oxide film. The homogenous part of the Ox10_AP film (Figure 2b) exhibits a drastically lower O/Ir ratio compared with the oxide film (Figure 2a), so a considerable amount of O has been removed from the sample by reductive heating. Even though inductive heating has successfully reduced the film, the location of remaining O has not changed during reduction; areas with a lower Ir amount in Figure 2b (lower film thickness, 4.4 mm) exhibit a larger O/Ir count ratio, indicating a significant amount of O beneath the film incorporated within a TiO_x or Ir–Ti mixed oxide layer. At areas with a higher Ir amount (3.6 mm), the O total counts increases with Ir, pointing to a fraction of O existing inside the film. However, the metallic state of the as-prepared film, which was confirmed by XPS (as discussed later in the manuscript), contradicts that the O is chemically bound to Ir.

3.2. Structural Characterization of Ir Metal Thin-Film Catalysts

3.2.1. X-Ray Diffraction (XRD)

To examine if the crystallinity and atomic structure of the Ac and Ox films changes depending on the reductive heating time, a grazing-incident X-ray diffraction (GIXRD) analysis was conducted prior to OER measurements (Figure S2). For both preparation methods, two crystalline phases are detected in the as-prepared state, where the dominating reflections (35° and 40°, purple sphere) can be assigned to the hexagonal phase of the Ti substrate. The weak reflections (47° and 68°, red star), however, are assigned to metallic Ir phases, evidencing crystallinity of the films. The low intensity of the reflections is a result of a low amount of Ir, or rather, a low thickness of the film in contrast to the bulk Ti cylinder.

The Rietveld refinement technique was used to analyze the XRD patterns of the as-prepared samples (Figure S3). The analysis evidences the absence of a crystalline IrO_2 phase, but suggests the presence of a TiO_2 rutile phase to a very small extent. Still, the existence of an amorphous or very nanocrystalline Ir oxide phase cannot be ruled out. To achieve a reasonable Rietveld refinement, an additional hexagonal phase had to be incorporated into the pattern fitting. The fittings further incorporate a metallic Ti phase and a nanocrystalline phase of metallic Ir. This supplementary reflection will be referred to as TiO_x and can be assigned to a Ti phase with larger lattice spacing rather than a rutile-type TiO_2. The presence of this phase might be deduced as a result of inclusion of oxygen into the lattice of Ti prior to the annealing process.

3.2.2. Selected Area Electron Diffraction (SAED)

To investigate the structure of the films at the nanoscale, selected area electron diffraction (SAED) was performed on the Ac_AP and Ox_AP catalyst films, as well as on the electrochemically treated Ac10_OER and Ox10_OER samples. The SAED patterns are given in Figure S4, and the corresponding calculated diffraction patterns (Cu Kα scale) are shown in Figure 3a–d and were made from dispersion of scraped off films. The reciprocal diffraction patterns were converted back into a diffraction pattern on the Cu scale.

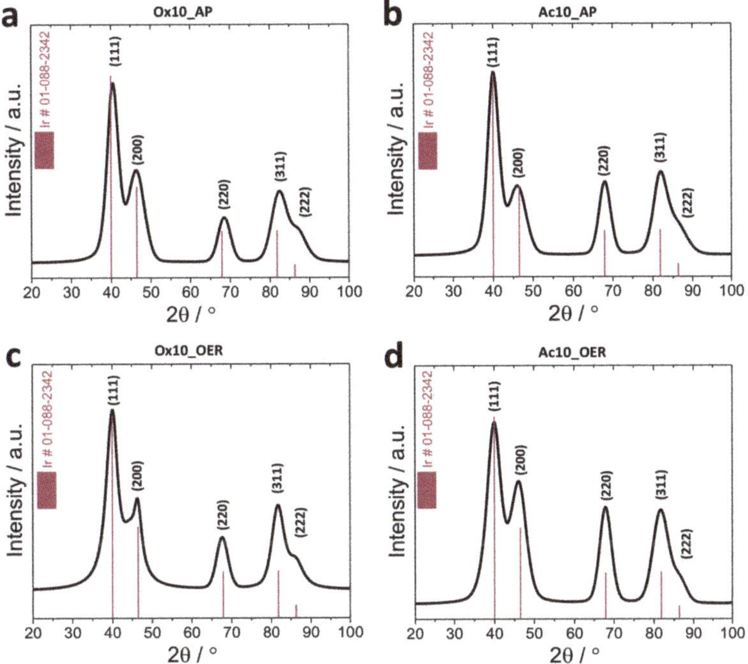

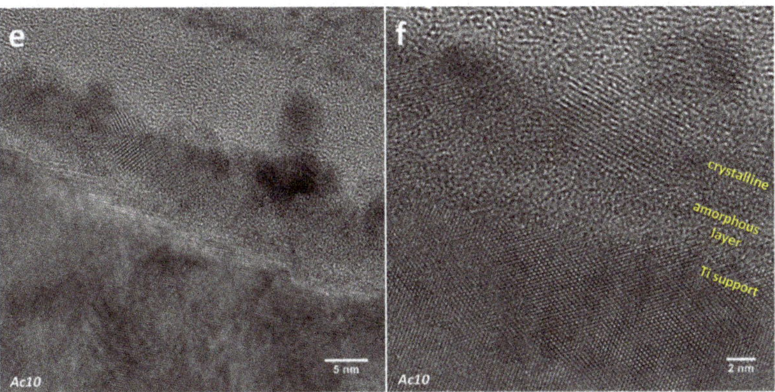

Figure 3. (**a–d**) Diffraction patterns evaluated from selected area electron diffraction (SAED) patterns (Figure S4) of Ox10 (a and c) and Ac10 films (b and d) before (_AP; a and b) and after electrochemical OER protocol (_OER; c and d). The patterns only show metallic Ir. (**e,f**) TEM cross-section of Ac10_AP film shows crystalline Ir particles on an amorphous interlayer.

For the calcined IrO_x film (prior to any reductive treatment), the clear diffraction pattern reveals a rutile-type phase corresponding to IrO_2 (Figure S5). According to SAED, after 10 min heating in the reducing atmosphere at 400 °C, the rutile-type reflections are replaced by a metallic Ir phase (Figure 3a), which is still present upon cycling into OER potential regions (Figure 3c). The same applies for the Ac10 catalyst, where the evaluation of the internal order of the catalyst films confirms no

IrO$_x$ phase at several positions of Ac10, as well as Ac10_OER after polarization into OER potential regions. Accordingly, SAED evaluation conclusively confirms that all investigated films are consistently metallic and that the oxidation observed by XPS analysis (see Section 3.3.3) is limited to near-surface Ir. Furthermore, the morphology of the thin-films was selectively studied on a FIB-prepared TEM cross-section of an Ac10 sample. The resulting images are displayed in Figure 3e and 3f. After 10 min of heating in the reducing atmosphere, the catalyst film including the crystalline particles and the amorphous layer is about 10 nm thick. The high-resolution TEM (HRTEM) images reveal the presence of an amorphous interlayer between the crystalline Ti substrate and crystalline Ir particles. The particles, which are believed to be caused by cracking of the film, cover the amorphous layer unevenly. These particles have also been found in the TEM images of the scraped film (Figure S6).

3.3. Surface Electrochemical Characterization

3.3.1. OER Activity

The electrochemical activity curves of the stepped potential voltammetry are depicted in Figure 4a. Ac and Ox samples show different electrochemical behaviors. A longer heating period of the Ac samples is accompanied with a slight decrease in the electrocatalytic activity. The distinct activities of the directly reduced films are likely the result of sintering or reorganization of the surface structure occurring during heating. The Ox samples, however, do not show remarkable differences between different heating times, which agrees with a stable morphology of the films observed by SEM (Figure 1, top row). This indicates the formation of a thermic stable structure, which is possibly benefited by the predefined structure of the former Ir oxide.

A comparison between the current density j_{geo} (Figure 4a) and the charge normalized current density j_{spec} (Figure 4b) also supports the prediction of a sintering process on the Ac films; the charge normalized curves of the Ac films approach each other in comparison with the unnormalized current density, indicating a decrease of active surface area by sintering. Accordingly, the decrease in electrocatalytic activity with longer heating time mainly results from the smaller active surface area as a result of sintering. On the other hand, the charge normalization barely changes the relative position of the current density curves of the Ox films, which is further proof of the stable morphology of these films. Compared with polycrystalline bulk iridium (measured on a polished Ir cylinder), all films possess higher charge normalized current densities. The slightly non-linear/bend behavior of the Tafel plot (Figure 4c) and the loss of current density during each potential step of the SPV (Figure S7) is an interplay between the approach of the steady-state activity after a sudden potential step and the dissolution of the electrochemically obtained iridium oxides, as was evidenced by Cherevko et al. [16].

3.3.2. OER Stability

The stability performance was investigated by CA at approximately +1.734 V_{RHE} in 0.05 M H_2SO_4; the results are shown in Figure S8. First, an abrupt drop of the current can be observed, assumed to be an approach to steady-state after quick potential change. The current then decreases slowly and almost linearly, presumably associated with the dissolution of oxidic Ir. After about 3 h, the current becomes almost constant. There are two reasonable explanations for this behavior. Either an equilibrium of formation and dissolution of the active iridium oxide species (oxidation and dissolution rates are the same) is maintained, or a more stable active species is present on the catalyst surface, as was reported by Li et al. [17].

To examine the stability of the catalysts, EDX was measured on all samples, which have undergone electrochemical treatment including OER, stability testing, and Faraday efficiency. Here, a loss of roughly 50 % Ir is observed, with an accuracy deviation of 10 % at most. This behavior agrees with the already mentioned lack of stability of electrochemically oxidized iridium.

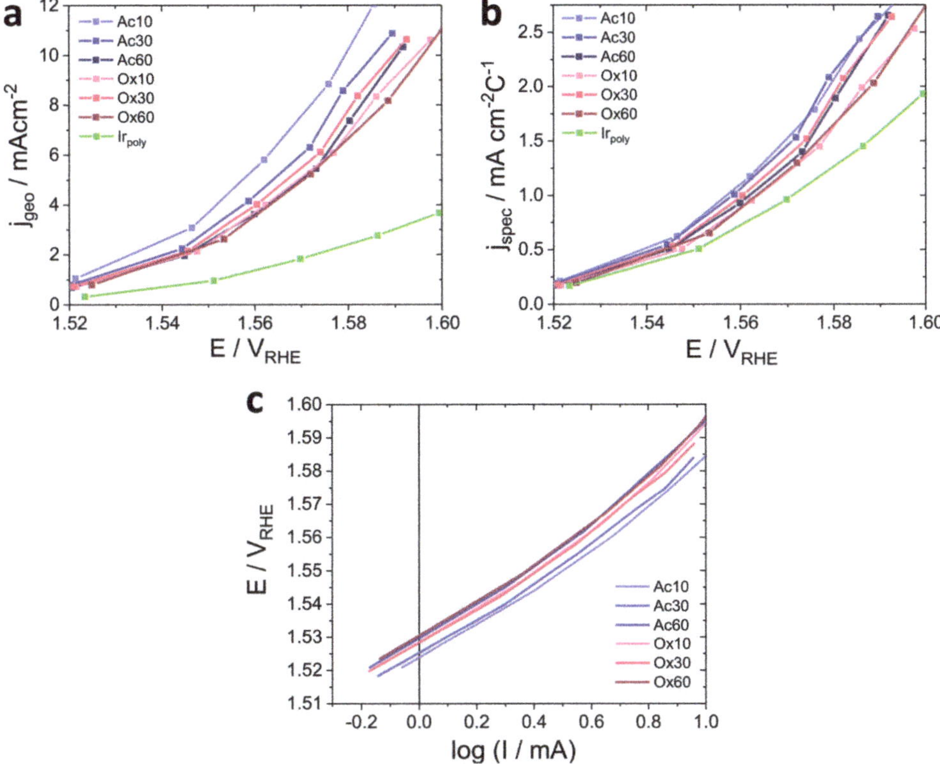

Figure 4. Electrochemical OER activity curves of Ac and Ox films obtained by stepped potential voltammetry. (**a**) Geometrical current densities in 0.05 M H_2SO_4. The shorter heated Ac films exhibit a slightly higher activity compared with the Ox films and longer heated Ac films. (**b**) Specific current density in 0.05 M H_2SO_4 based on the anodic charge obtained from the last scan (between 0.4 and 1.4 V_{RHE}) before SPV. Because of charge normalization, the activities of the Ac films no longer differ significantly from each other. The relation of the Ox films' activities to each other is not affected. (**c**) Tafel plot of the _OER films extrapolated from the stepped potential voltammetry (SPV) measurements.

3.3.3. X-Ray Photoelectron Spectroscopy (XPS)

The changes in the surface chemical state of Ir after activation and oxidation were traced with X-ray photoelectron spectroscopy (Figure 5).

The Ir 4f XP spectra in Figure 5a of the as-prepared Ac10 and Ox10 samples confirmed the presence of only metallic Ir in the XPS accessible area, indicating that the applied annealing step in H_2 is sufficient to reduce Ir on the surface to its metallic state. After the electrochemical treatments, the Ir 4f core-level spectra shifted to higher binding energy as a result of the formation of surface Ir oxide. The line shapes of the Ir 4f signals differ significantly depending on the potential range. Fits of the Ir 4f spectra revealed three different Ir species presented in these samples: metallic Ir^0, Ir^{IV}, and Ir^{III} species. The Ir species with a binding energy located at ~0.6 eV higher than that of the Ir^{IV} species was assigned as Ir^{III} in this work, although the true origin of this component (Ir^{III} or Ir^V) is still a controversy [14,18]. However, it is unlikely that Ir^V can exist in an Ir simple oxide under ex situ conditions.

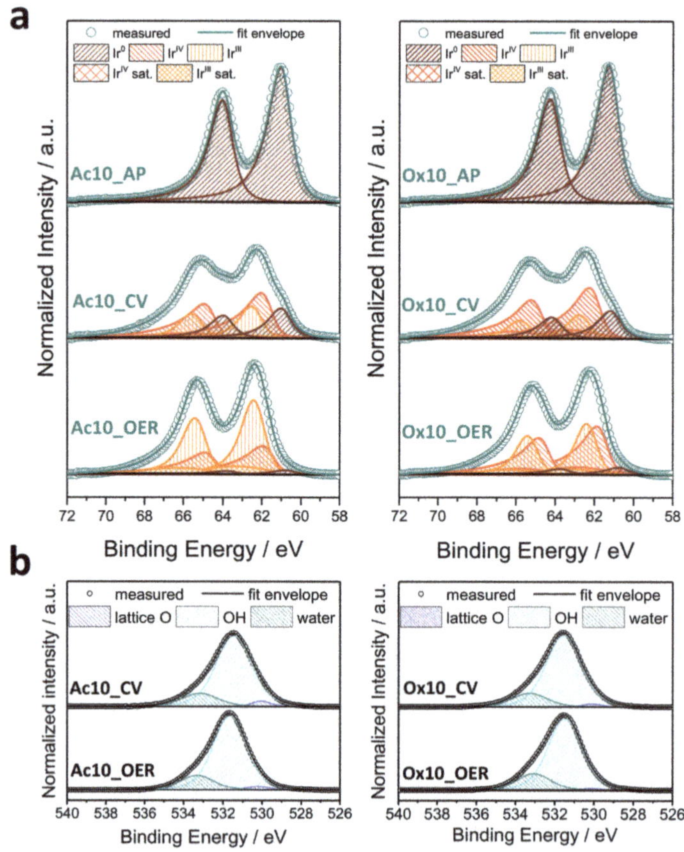

Figure 5. X-ray photoelectron spectroscopy of the Ac10 and Ox10 films at as-prepared states (_AP), after cyclic voltammetry between 0.05–1.4 V_{RHE} (_CV) and OER activity testing (_OER) with fitted components of (**a**) Ir 4f and (**b**) O 1s species. The as-prepared samples only show metallic Ir (Ir^0). Through potential cycling, the major part of metallic Ir was oxidized to Ir^{III} and Ir^{IV}. After stepped potential voltammetry with potentials up to 1.60 V_{RHE}, almost no Ir^0 is left and the ratio of Ir^{III} to Ir^{IV} is increased.

After cycling in the potential range of 0.05–1.4 V_{RHE}, both Ac10 and Ox10 films were oxidized to a major extent; however, there is still discernible contribution of metallic Ir to the line shapes. On closer inspection, the Ir4f XP spectra of both films differ in the Ir^{III}/Ir^{IV} species ratio, highlighting different oxidation behaviors. The Ac10_CV sample shows a higher Ir^{III}/Ir^{IV} ratio than the Ox10_CV sample, which might be the origin of its higher OER activity. In previous studies by Pfeifer et al. [14,19], electronic defects in the form of O 2p holes (O^{I-} species) were identified in amorphous IrO_x. These highly electrophilic O^{I-} species are created by iridium vacancies in the IrO_x framework and possess a great activity in catalyzing the OER. The formation of these species is in turn accompanied by the reduction of neighboring Ir^{IV} to Ir^{III} species, which appear at an Ir $4f_{7/2}$ binding energy of ~0.6 eV higher than that of the Ir^{IV} species, according to their density functional theory calculations. Drawing upon these findings, the enhanced activity of the Ac10 catalyst may be deduced to coordination defects in the amorphous morphology facilitating the O-O formation in the OER [19].

Upon exposing the catalysts to higher OER potentials (Ac10_OER and Ox10_OER), the metallic components diminished significantly to below 3 %. Two major oxidic components contribute to the Ir 4f spectrum. The Ir^{III} component in the Ac10_OER sample is now much higher compared with that of Ox10_OER, indicating that the dominant species is Ir^{III}. The higher specific OER activity based on charge normalization gives reason to believe that the O^{I-} species are present to a higher concentration on the Ac films. The difference in oxidation behavior of the Ac10 and Ox10 samples is plausibly explained by the difference in the morphology of the as-prepared metallic samples, as discussed previously. The Ac10 film with a highly cracked, sponge-like morphology is prone to defect formation during electrochemical treatment, in contrast to the Ox10 film with a smoother surface and more defined structure, as the starting point of the Ox film series is a rutile phase with Ir in the oxidation state +4.

The O 1s spectra in Figure 5b of the Ac catalysts prior to and after OER testing revealed three different oxygen species, which can be assigned to adsorbed water (~533.1 eV), lattice oxygen (~530.1 eV), and hydroxyl groups (~531.5 eV). All samples are distinguished by a high surface hydroxyl group and very small lattice oxygen contents, typical for electrochemically grown iridium oxide [20]. The O 1s XP spectra indicate that the oxide layers in the electrochemically treated samples are largely amorphous, which explains the absence of any crystalline IrO_x phase in the SAED of these samples, as discussed earlier.

4. Conclusions

In the study of the morphology, structure, and catalytic activity of supported metallic Ir films, we presented the successful synthesis of metallic Ir thin-films on Ti substrate using two different synthesis approaches. It was found that the catalyst structure was significantly influenced by the synthesis routes. We emphasize that the choice of the synthesis route to the initial metallic Ir strongly affects the performance of the subsequent electrochemically obtained Ir oxide. Further, these catalytic films exhibited an enhanced stability and activity in the OER compared with the surface of a polished, massive polycrystalline Ir cylinder.

Regardless of the heating time, the Ox route consistently resulted in the identical structure, creating a system insensitive against fluctuations in synthesis parameters. The Ox route might thus also be conducted in a common tube furnace, with a less sharp heating window (heating-up and cool down). In contrast, SEM and TEM revealed that the morphology of the Ir films synthesized over the Ac route exhibited a significant sensitivity towards the heating treatment. The application of a sharp inductive heating accompanied by a defined heating time thus enables one to control the progress of morphological restructuring, like sintering processes. These distinctive morphological properties were found to have a strong effect on the catalytic properties. A slight decrease of the OER activity with increasing heating period was observed. The difference between the catalytic properties of both catalyst types is reduced, as the Ac films approach the Ox samples with a longer heating time. Although this effect is not observable in the given scale of the morphology analysis, it cannot be excluded that structural similarities occur on a smaller scale. As expected, a complete oxidation of surface near Ir occurs during OER, while the majority of the film stays in its metallic state.

In the search for cost-effective materials catalyzing the OER, modifications of Ir with metallic additives are under broad investigation. On the basis of our results, we propose the Ox route as a promising synthesis approach for such bimetallic catalysts, as the final surface structure of the resulting products is quickly formed and is tolerant to different heating periods.

Supplementary Materials: The following are available online at http://www.mdpi.com/2571-9637/1/1/12/s1.

Author Contributions: Conceptualization, E.Ö. and Z.P.; methodology, E.Ö and Z.P.; validation, E.Ö, Z.P., and P.S.; formal analysis, E.Ö., Z.P., S.K., B.P., C.S., and S.S.; investigation, E.Ö., Z.P., S.K., H.N.N, B.P., S.S., and C.S.; resources, P.S., S.S., and C.B.; data curation, E.Ö, B.P, and S.K.; writing—original draft preparation, E.Ö. and Z.P.; writing—review and editing, E.Ö., Z.P., S.K., H.N.N, and C.S.; visualization, E.Ö. and Z.P.; supervision, P.S.; project administration, E.Ö., Z.P., B.P., and P.S.

Funding: This research was funded by Deutsche Forschungsgemeinschaft, grant number SPP1613.

Acknowledgments: We thank the Fritz-Haber-Institut of the Max-Planck-Gesellschaft for access to XPS and ZELMI for access to TEM/SAED and the focused ion beam preparation of a cross-section lamella.

Conflicts of Interest: The authors declare no conflict of interest. The funders had no role in the design of the study; in the collection, analyses, or interpretation of data; in the writing of the manuscript; or in the decision to publish the results.

References

1. Singh, S.; Jain, S.; Ps, V.; Tiwari, A.K.; Nouni, M.R.; Pandey, J.K.; Goel, S. Hydrogen: A sustainable fuel for future of the transport sector. *Renew. Sust. Energ. Rev.* **2015**, *51*, 623–633. [CrossRef]
2. Carmo, M.; Fritz, D.L.; Mergel, J.; Stolten, D. A comprehensive review on PEM water electrolysis. *Int. J. Hydrogen Energ.* **2013**, *38*, 4901–4934. [CrossRef]
3. McCrory, C.C.L.; Jung, S.; Peters, J.C.; Jaramillo, T.F. Benchmarking heterogeneous electrocatalysts for the oxygen evolution reaction. *J. Am. Chem. Soc.* **2013**, *135*, 16977–16987. [CrossRef] [PubMed]
4. Cherevko, S.; Reier, T.; Zeradjanin, A.R.; Pawolek, Z.; Strasser, P.; Mayrhofer, K.J.J. Stability of nanostructured iridium oxide electrocatalysts during oxygen evolution reaction in acidic environment. *Electrochem. Comm.* **2014**, *48*, 81–85. [CrossRef]
5. Reier, T.; Oezaslan, M.; Strasser, P. Electrocatalytic Oxygen Evolution Reaction (OER) on Ru, Ir, and Pt Catalysts: A Comparative Study of Nanoparticles and Bulk Materials. *ACS Catal.* **2012**, *2*, 1765–1772. [CrossRef]
6. De Oliveira-Sousa, A.; da Silva, M.A.S.; Machado, S.A.S.; Avaca, L.A.; de Lima-Neto, P. Influence of the preparation method on the morphological and electrochemical properties of Ti/IrO$_2$-coated electrodes. *Electrochim. Acta* **2000**, *45*, 4467–4473. [CrossRef]
7. Danilovic, N.; Subbaraman, R.; Chang, K.-C.; Chang, S.H.; Kang, Y.J.; Snyder, J.; Paulikas, A.P.; Strmcnik, D.; Kim, Y.-T.; Myers, D.; et al. Activity-Stability Trends for the Oxygen Evolution Reaction on Monometallic Oxides in Acidic Environments. *J. Phys. Chem. Lett.* **2014**, *5*, 2474–2478. [CrossRef] [PubMed]
8. Cherevko, S.; Geiger, S.; Kasian, O.; Mingers, A.; Mayrhofer, K.J.J. Oxygen evolution activity and stability of iridium in acidic media. Part 2.—Electrochemically grown hydrous iridium oxide. *J. Electroanal. Chem.* **2016**, *774*, 102–110.
9. Bandarenka, A.S.; Koper, M.T.M. Structural and electronic effects in heterogeneous electrocatalysis: Toward a rational design of electrocatalysts. *J. Catal.* **2013**, *308*, 11–22. [CrossRef]
10. Reier, T.; Teschner, D.; Lunkenbein, T.; Bergmann, A.; Selve, S.; Kraehnert, R.; Schlögl, R.; Strasser, P. Electrocatalytic Oxygen Evolution on Iridium Oxide: Uncovering Catalyst-Substrate Interactions and Active Iridium Oxide Species. *J. Electrochem. Soc.* **2014**, *161*, F876–F882. [CrossRef]
11. Özer, E.; Paul, B.; Spöri, C.; Strasser, P. Coupled Inductive Annealing-Electrochemical Setup for Controlled Preparation and Characterization of Alloy Crystal Surface Electrodes. *Small Methods* **2018**, *0*, 1800232. [CrossRef]
12. Gammer, C.; Mangler, C.; Rentenberger, C.; Karnthaler, H.P. Quantitative local profile analysis of nanomaterials by electron diffraction. *Scripta Mater.* **2010**, *63*, 312–315. [CrossRef]
13. PASAD. Available online: http://www.univie.ac.at/pasad/ (accessed on 18 May 2018).
14. Pfeifer, V.; Jones, T.E.; Velasco Velez, J.J.; Massue, C.; Greiner, M.T.; Arrigo, R.; Teschner, D.; Girgsdies, F.; Scherzer, M.; Allan, J.; et al. The electronic structure of iridium oxide electrodes active in water splitting. *Phys. Chem. Chem. Phys.* **2016**, *18*, 2292–2296. [CrossRef] [PubMed]
15. Reier, T.; Pawolek, Z.; Cherevko, S.; Bruns, M.; Jones, T.; Teschner, D.; Selve, S.; Bergmann, A.; Nong, H.N.; Schlögl, R.; et al. Molecular Insight in Structure and Activity of Highly Efficient, Low-Ir Ir–Ni Oxide Catalysts for Electrochemical Water Splitting (OER). *J. Am. Chem. Soc.* **2015**, *137*, 13031–13040. [CrossRef] [PubMed]
16. Cherevko, S.; Geiger, S.; Kasian, O.; Mingers, A.; Mayrhofer, K.J.J. Oxygen evolution activity and stability of iridium in acidic media. Part 1.—Metallic iridium. *J. Electroanal. Chem.* **2016**, *773*, 69–78. [CrossRef]
17. Li, T.; Kasian, O.; Cherevko, S.; Zhang, S.; Geiger, S.; Scheu, C.; Felfer, P.; Raabe, D.; Gault, B.; Mayrhofer, K.J.J. Atomic-scale insights into surface species of electrocatalysts in three dimensions. *Nat. Catal.* **2018**, *1*, 300–305. [CrossRef]
18. Sanchez Casalongue, H.G.; Ng, M.L.; Kaya, S.; Friebel, D.; Ogasawara, H.; Nilsson, A. In situ observation of surface species on iridium oxide nanoparticles during the oxygen evolution reaction. *Angew. Chem. Int. Ed.* **2014**, *53*, 7169–7172. [CrossRef] [PubMed]

19. Pfeifer, V.; Jones, T.E.; Velasco Vélez, J.J.; Arrigo, R.; Piccinin, S.; Hävecker, M.; Knop-Gericke, A.; Schlögl, R. *In situ* observation of reactive oxygen species forming on oxygen-evolving iridium surface. *Chem. Sci.* **2017**, *8*, 2143–2249. [CrossRef] [PubMed]
20. Saveleva, V.A.; Wang, L.; Teschner, D.; Jones, T.; Gago, A.S.; Friedrich, K.A.; Zafeiratos, S.; Schlögl, R.; Savinova, E.R. Operando Evidence for a Universal Oxygen Evolution Mechanism on Thermal and Electrochemical Iridium Oxides. *J. Phys. Chem. Lett.* **2018**, *9*, 3154–3160. [CrossRef] [PubMed]

Article

Diazonium Salts: Versatile Molecular Glues for Sticking Conductive Polymers to Flexible Electrodes

Momath Lo [1,2,3], **Rémi Pires** [3], **Karim Diaw** [1], **Diariatou Gningue-Sall** [1,*], **Mehmet A. Oturan** [2], **Jean-Jacques Aaron** [2,*] **and Mohamed M. Chehimi** [3,*]

[1]   Faculté des Sciences, Université Cheikh Anta Diop, BP 5005 Dakar-Fann, Sénégal;
    momath.lo@ucad.edu.sn (M.L.); karimdiaw@hotmail.com (K.D.)
[2]   Laboratoire Géomatériaux et Environnement, Université Paris-Est, 5 Bd. Descartes,
    77454 Marne-la-Vallée CEDEX 2, France; mehmet.oturan@u-pem.fr
[3]   Université Paris Est, CNRS, ICMPE (UMR 7182), 2-8 rue Henri Dunant, 94320 Thiais, France;
    pires@icmpe.cnrs.fr
*   Correspondence: diariatougningue.sall@ucad.edu.sn (D.G.-S.); jeanjacquesaaron@yahoo.fr (J.-J.A.);
    chehimi@icmpe.cnrs.fr (M.M.C.); Tel.: +33-149-781-160 (M.M.C.)

Received: 20 June 2018; Accepted: 22 July 2018; Published: 8 August 2018

Abstract: Adhesion of polymers to surfaces is of the upmost importance in timely applications such as protective coatings, biomaterials, sensors, new power sources and soft electronics. In this context, this work examines the role of molecular interactions in the adhesion of polypyrrole thin films to flexible Indium Tin Oxide (ITO) electrodes grafted with aryl layers from various diazonium salts, namely 4-carboxybenzenediazonium (ITO-CO_2H), 4-sulfonicbenzenediazonium (ITO-SO_3H), 4-*N,N*-dimethylbenzenediazonium (ITO-N(CH_3)$_2$), 4-aminobenzenediazonium (ITO-NH_2), 4-cyanobenzenediazonium (ITO-CN) and 4-*N*-phenylbenzenediazonium (ITO-NHPh). It was demonstrated that PPy thin layers were adherent to all aryl-modified surfaces, whereas adhesive failure was noted for bare ITO following simple solvent washing or sonication. Adhesion of polypyrrole was investigated in terms of hydrophilic/hydrophobic character of the underlying aryl layer as probed by contact angle measurements. It was found that sulfonic acid-doped polypyrrole (PPy-BSA) thin films were preferably deposited on the most hydrophobic surfaces. More importantly, the redox properties and electrochemical impedance of PPy were closely related to the hydrophobic character of the aryl layers. This work demonstrates that diazonium compounds are unique molecular glues for conductive polymers and permit to tune their interfacial properties. With robust, diazonium-based architectured interfaces, one can design high performance materials for e.g., sensors, printed soft electronics and flexible thermoelectrics.

Keywords: polypyrrole; diazonium salts; flexible ITO; adhesion; redox properties

1. Introduction

In the recent decades, conductive organic polymers have attracted a great deal of interest due to their salient features, namely electric, thermoelectric, electronic, optical, ion exchange, and reinforcing properties, to name but a few [1–3]. Particularly, polypyrrole (PPy) constitutes an excellent material for building electronic devices [4], because of its ease of preparation by a variety of chemical [5], electrochemical [6], biocatalytic [7], and radiation-induced [8,9] methods. In addition, it is possible to tune its electrical conductivity upon doping, copolymerization, blending and nanostructuring [10,11].

Several studies have been focused on the effects solvent nature [12], electrolyte [13], potential or current density [14,15], concentration of monomer [16], composition of working electrode (metals, glassy carbon, ITO) [17,18], on the electrochemical deposition and mechanical properties of PPy films. However, the performances as well as the long-term stability of polymer thin films is often limited by various factors, including poor adhesion to the substrate and migration of metal contamination from the substrate to the polymer [19]. To overcome these adhesion problems, coupling agents such as organosilanes and aryl layers from diazonium salts proved to efficiently anchor PPy [20–23] and other insulating polymers [24–26] to a large variety of materials surfaces. These coupling agents not only impart strong adhesive bonding to the conductive polymer top layer, but also limit or suppress the diffusion in the film of elements leached from the substrates [21].

While organosilanes are the main coupling agents used in adhesion science and technology [27–29], studies of diazonium salts as adhesion promoters remain sparse. This is due to the fact that the diazonium interface chemistry is still incipient. The interest of diazonium salts lies in their fast reaction with practically all surfaces, triggered by chemicals, reactive surfaces or radiations [30]. In addition, diazonium compounds are versatile as they impart functionalities to the materials surfaces depending on the nature of the substituent borne by the benzene ring in either *para-*, *meta-* or *ortho-* position of the diazonium group. In this way, one can design new surfaces possessing a plethora of properties such as hydrophilic/hydrophobic character [31], electrocatalytic activity [32], initiation of surface radical or oxidative polymerization [33–35], resistance to biofouling [36], sensing [37,38], reactivity towards polymer matrices [39,40].

However, to the best of our knowledge, no systematic study comparing the effect of several functional groups on the adhesion of diazonium-modified surfaces has been undertaken. Therefore, the objective of this work is to examine the propensity of a series of diazonium compounds to provide ultrathin adhesive layers to conductive polymer top coatings and to check to which extent the said aryl adhesive layers govern the physicochemical properties of the polymer top coatings. To illustrate the actual study, polypyrrole was selected as a reactive and functional polymer coating, and flexible ITO as a substrate which is generally difficult to coat with conductive polymers for obtaining long-term robust adhesion.

Herein, we investigated the electrografting of a series of diazonium compounds of the general formulae N_2-C_6H_4-R, with R in *para* position $=CO_2H$, SO_3H, $N(CH_3)_2$, NH_2, CN and NH-Ph. The hydrophobic/hydrophilic character of aryl-modified ITO electrodes was probed by means of contact angle measurements of water drops. The modified flexible ITO sheets further served as platforms for the electrodeposition of benzene sulfonic acid-doped polypyrrole (PPy-BSA), and the electrochemical properties were interrogated by cyclic voltammetry and electrochemical impedance spectroscopy. The electrochemical properties of PPy-BSA coatings will be discussed in terms of surface composition and wettability of the underlying flexible aryl-modified ITO sheets.

2. Materials and Methods

2.1. Reagents

4-Aminobenzoic acid, 4-aminobenzene sulfonic acid, *N,N*-dimethyl-1,4-phenylenediamine, 1,4-phenylenediamine, 4-aminobenzonitrile, *N*-phenyl-4-phenylenediamine, were purchased from Sigma Aldrich. Sodium nitrite ($\geq$99.0%, Sigma-Aldrich, St. Louis, MO, USA), chloric acid (37%), ferrocene (Sigma-Aldrich) and potassium chloride were used without further purification. Pyrrole (purity $\geq$ 98%, Aldrich) and benzene sulfonic acid were refrigerated prior to synthesis. All aqueous solutions were prepared using Milli-Q ultrapure water (MQ 18.2 MΩ·cm). ITO-coated polyethylene naphtalate (PEN) sheets (Peccel, Yokohama, Japan) were cut to 5 × 10 mm with scissors and served as working electrodes.

2.2. Diazonium Modification of Flexible ITO

The diazonium cations were generated in situ from the parent aromatic amines. The aromatic amines and the corresponding modified electrodes ITO-R (R = functional group) were the following: 4-aminobenzoic acid (ITO-CO_2H), 4-aminobenzene sulfonic acid (ITO-SO_3H), *N,N*-dimethyl-1,4-phenylenediamine (ITO-$N(CH_3)_2$), 1,4-phenylenediamine(ITO-NH_2), 4-aminobenzonitrile (ITO-CN),

N-phenyl-p-phenylenediamine (ITO-NH-Ph). The ITO surface was cleaned by sonication in a mixture of water/ethanol 80:20 (v/v) for 2 min. After sonication, the electrode was washed with DI water. The electrografting of aryl layers from the diazonium salts generated in situ was performed using a three-electrode system, including an ITO-coated polyethylene naphthalate (PEN) working electrode (area ~ 5 × 5 mm), a stainless steel grid as counter electrode and a Ag/AgCl (saturated KCl) reference electrode.

The diazonium salt was produced by mixing 0.1 mmole of aromatic amine (equivalent to 13.7, 17.3, 13.6, 10.8, 11.8, or 18.4 mg of 4-aminobenzoic acid, 4-aminobenzene sulfonic acid, N,N-dimethyl-1,4-phenylenediamine, 1,4-phenylenediamine, 4-aminobenzonitrile, or N-phenyl-4-phenylenediamine, respectively) with 90 mL of 0.5 M HCl in a glass bath for 30 min, before adding 10 mL of 10 mM $NaNO_2$ aqueous solution (this 10 mL solution contains 0.1 mmol $NaNO_2$). The resulting mixture was stirred for 1 h. The electrografting was then carried out in potentiostatic conditions (-0.8 V/SCE for 45 s), using a Biologic SP 150 potentiostat (Seyssinet-Pariset, France), computer-controlled with an EC-lab software. The blocking effect of the bare and modified ITO electrodes was investigated in a 1 mM $Fe(CN)^{3-/4-}$ redox solution by cyclic voltammetry (CV) in the -0.4 and 0.6 V/SCE range at a scan rate of 50 mV·s^{-1}. The relative electroactivity of the modified electrode, which is inversely related to its blocking effect properties, was quantified by CV in the presence of the redox probe species. The Parameter I_{rel} is defined as:

$$I_{rel} = \frac{Ipa\ with\ a\ film}{Ipa\ for\ a\ bare\ ITO} \times 100 \tag{1}$$

where Ipa is the intensity of the anodic peak current. Electrochemical impedance spectroscopy measurements were performed at open circuit potential in 1 mM $K_3Fe(CN)_6$ and 0.1 M KCl.

2.3. Preparation and Electrochemical Characterization of Polypyrrole Films on Diazonium-Modified ITO

Prior to the electropolymerization, diazonium-modified ITO electrodes were sonicated in a water/ethanol 80:20 (v/v) mixture for 2 min. The PPy on the ITO modified aryl layer was prepared by electropolymerization in a 0.01 M BSA + 0.1 M pyrrole aqueous solution (20 mL) by CV in a potential range of -1.0 to 1.2 V. The fresh PPy films were characterized by CV under the experimental conditions indicated in Section 3.

The impedance spectra were recorded in the 0.1 Hz to 100 kHz frequency range.

2.4. Surface Analysis

The X-ray photoelectron spectra (XPS) were recorded using a K Alpha apparatus (Thermo Fisher Scientific, East Grinsted, UK) fitted with a monochromatic Al X-ray source (hν = 1486.6 eV; spot size = 400 μm). A flood gun was used for static charge compensation.

An Easy Drop Krüss instrument (Hamburg, Germany) was used to determine the contact angles of water drops deposited on diazonium-modified ITO and on PPy samples.

3. Results and Discussion

3.1. Brief Description of the Strategy and the Objectives of the Work

Six diazonium compounds of the general formulae N_2-C_6H_4-R, with R = CO_2H, SO_3H, $N(CH_3)_2$, NH_2, CN and NH-Ph, were selected for the preparation of aryl layers on flexible ITO surfaces. Electrochemistry was selected as a means of attaching aryl monolayers, because it is a simple and straightforward method which can be conducted in aqueous medium. Diazonium compounds were generated in situ from the commercially-available parent aromatic amines; this adds up to the versatility of chemical surface engineering with diazonium compounds. The choice of the functional group in *para* position of the diazonium depends on two criteria: the electron-donor or electron-acceptor character of the functional group, on the one hand, and its propensity to govern the wetting of ITO surfaces, on the other hand.

Figure 1 depicts the chemical structures of the in situ generated diazonium cations (Upper panel) and the simple strategy adopted to attach PPy in two steps (Lower panel). Attachment of an aryl adhesive layer was followed by electrodeposition of PPy. It is to note that no physicochemical study will be provided on the adhesion of PPy on bare ITO, which is known to be very poor, as previously reported [22]. Instead, we will concentrate the present investigation on the diazonium susbstituent effect on the PPy adhesion and its redox behaviour.

Hereafter, the study is split into three main sections: (i) attachment and properties of aryl layers, (ii) electropolymerization and behaviour of the PPy topcoat, and (iii) adhesion aspects of PPy top coat to modified ITO sheets.

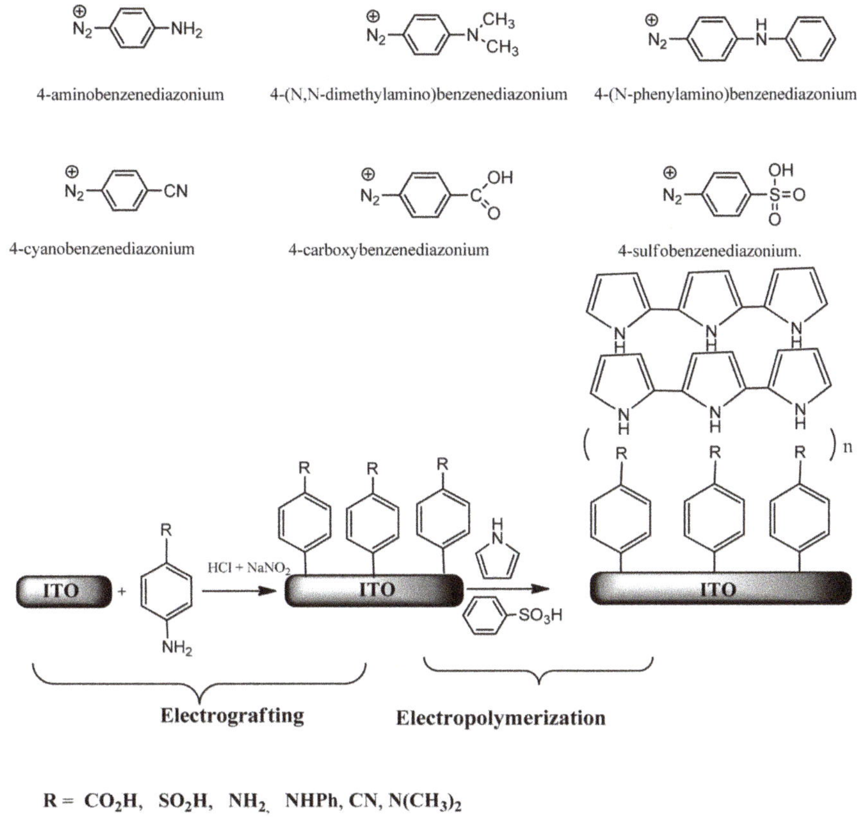

Figure 1. Upper panel: Chemical structures of the diazonium compounds under test. Lower panel: electrodeposition of benzene sulfonic acid-doped polypyrrole on diazonium-modified flexible ITO electrodes.

3.2. Electrochemical Modification of Flexible ITO Sheets with Diazonium Salts

Aryl layers were electrografted by chronoamperometry by setting the potential to -0.8 V/SCE for 45 s. This electrografting period of time was sufficient for good polypyrrole adhesion without inducing any major blocking effect that hampers the deposition of the polymer [22,41]. The blocking effect of the aryl layers was assessed using $Fe(CN)_6^{3-/4-}$, as depicted in Figure 2. The quasi-reversible redox wave obtained with bare ITO (Figure 2A(a), peaks at 0.25 and 0.1 V) suffered a decrease in the intensity for ITO-N(CH$_3$)$_2$ (Figure 2A(b), peaks at 0.38 and 0.1 V), but disappeared quasi-completely in the case of ITO-SO$_3$H (Figure 2A(c)), indicating a blocking effect of the aryl layer as reported by several authors [22,42–47]. The relative Fe(CN)$_6^{3-/4-}$ redox peak intensity (I$_{rel}$) decreased in the order ITO > ITO-N(CH$_3$)$_2$ > ITO-NH$_2$ > ITO-NH-Ph > ITO-CN > ITO-CO$_2$ H > ITO-SO$_3$H (see Table 1). The influence of the aryl grafting to ITO electrodes is thus straightforward since the response of Fe(CN)$_6^{3-/4-}$ is flattened for modified ITO plates compared to bare ITO strips. From measurements similar to those of Figure 2A, I$_{rel}$ values were computed and reported in Table 1 for all aryl layers. The ITO-N(CH$_3$)$_2$, ITO-NH-Ph and ITO-NH$_2$ surfaces exhibit a blocking effect for Fe(CN)$_6^{3-/4-}$ redox system reactions with I$_{rel}$ values of 49, 46.7, 23.6%, respectively. The highest blocking effects correspond to ITO-CN, ITO-CO$_2$H and ITO-SO$_3$H, with I$_{rel}$ values of 15.7, 11.2 and 3.9%, respectively. Under the same grafting conditions, the blocking effect appear to be more important for the ITO-SO$_3$H modified electrode [48].

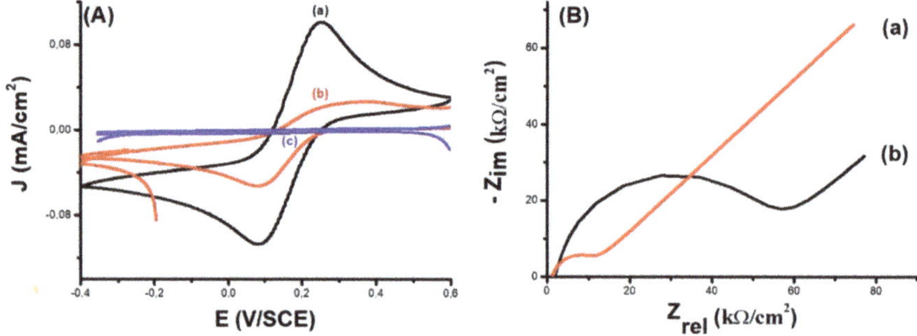

Figure 2. Electrochemical characterization of diazonium-modified ITO electrodes determined in aqueous solutions of 1 mM K$_3$Fe(CN)$_6$ and 0.1 M KCl. (**A**) Typical cyclic voltammograms (**a**) bare ITO, (**b**) ITO-N(CH$_3$)$_2$, and (**c**) ITO-SO$_3$H. (**B**) Electrochemical impedance spectroscopy plots recorded with (**a**) ITO-N(CH$_3$)$_2$, and (**b**) ITO-SO$_3$H.

Table 1. I_{rel}, R_{ct} and water drop contact angles (θ) obtained for aryl-modified ITO.

Surface	I_{rel} (%)	R_{ct} (kΩ/cm^2)	θ (°)
ITO-SO$_3$H	3.90	68.8	48.7 ± 0.5
ITO-CO$_2$H	11.2	47.3	70.7 ± 0.4
ITO-CN	15.7	44.0	81.2 ± 0.3
ITO-NH-Ph	23.6	9.32	103.1 ± 0.3
ITO-NH$_2$	46.6	8.97	116 ± 0.4
ITO-N(CH$_3$)$_2$	48.6	7.15	136.7 ± 0.1

The Nyquist plots for ITO-SO$_3$H and ITO-N(CH$_3$)$_2$ electrodes (Figure 2B) present typical semicircle shapes in the high-frequency domain, characteristic of an interfacial charge-transfer mechanism. The surfaces exhibit straight lines in the low-frequency regime, characteristic of a semi-infinite diffusion phenomenon. ITO-N(CH$_3$)$_2$ displays the lowest charge transfer resistance (R_{ct}), ca. 7.15 kΩ/cm^2 compared to 68.8 kΩ/cm^2 for ITO-SO$_3$H, a trend that is in agreement with previously-reported studies on glassy carbon electrodes [48]. Table 1 reports R_{ct} values for all diazonium-modified ITO surfaces and clearly indicates the general following trend: low values for electron-donor aryl groups (7.15–9.32 kΩ cm^2) and ~5–10-fold higher R_{ct} values (44.0–68.8 kΩ cm^2) for electron-acceptor aryl groups.

Figure 3 shows images of water drops deposited on the various ITO-R surfaces. For R = N(CH$_3$)$_2$, NH$_2$ and NH-Ph, the contact angles were found to be 136.7 ± 0.1, 116 ± 0.4 and 103.1 ± 0.3°, respectively. Taking into account the accepted rule of thumb, the water contact angle (θ) values above 90° indicated hydrophobic surfaces. In contrast, more wetting was observed for the electron-acceptor groups R = CN, COOH and SO_3H, the water contact angles of which are 81.2 ± 0.3, 70.7 ± 0.4 and $48.7 \pm 0.5°$, respectively. These surfaces can thus be ranked as relatively hydrophilic ($\theta < 90°$). As expected, the layers with more ionizable groups COOH and SO_3H produced the lowest contact angles and thus led to the highest hydrophilicity. It is thus possible to control the hydrophobic/hydrophilic character of flexible ITO surfaces by simply changing the radical R in *para* position of the diazonium group.

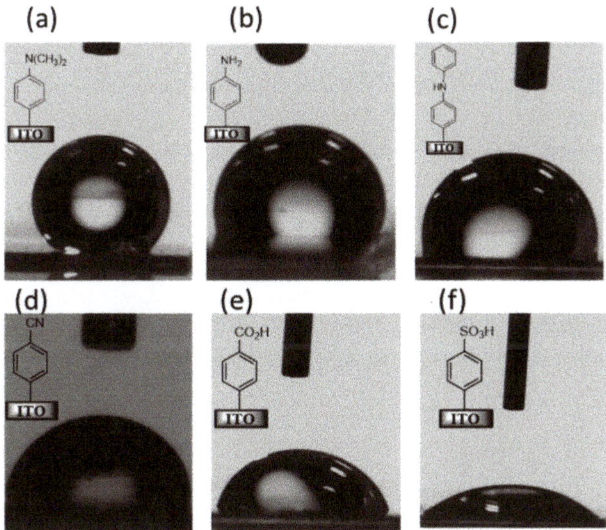

Figure 3. Contact angles of water drops gently deposited on diazonium-modified ITO: (**a**) ITO-N(CH$_3$)$_2$, (**b**) ITO-NH$_2$, (**c**) ITO-NH-Ph, (**d**) ITO-CN, (**e**) ITO-CO$_2$H, (**f**) ITO-SO$_3$H.

Figure 4 displays plots of the electrochemical characteristics of the aryl layers versus their wetting property, expressed by the water drop contact angle values. For the hydrophilic surfaces, the barrier to the $Fe(CN)_6^{3-/4-}$ probe is most probably due to the favourable water-aryl interfacial interactions. Under such wetting conditions, the probe molecules hardly reach the aryl-modified ITO surfaces and undergo redox reactions. Higher I_{rel} values indicate weaker barrier, and the probe interact more favorably with the surfaces compared to water.

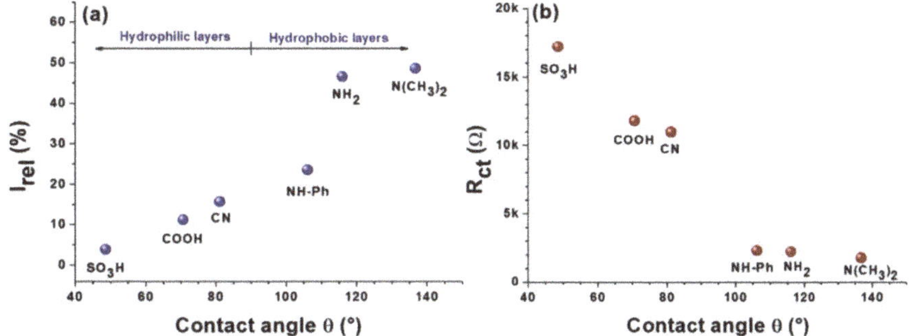

Figure 4. Correlations between the electrochemical and wetting properties of diazonium-modified flexible ITO electrodes: (**a**) I_{rel}-*vs*-water contact angle; and (**b**) R_{ct}-*vs*-water contact angle.

As far as R_{ct} is concerned, this physical property decreases for hydrophobic aryl-modified ITO surfaces ($\theta > 90°$). These results are due to the steric hindrance or the physical barrier role played by the free pair of the organic molecules [48].

All in one, the results can be discussed in terms of hydrophobic/hydrophilic effects [49]. Indeed, $Fe(CN)_6^{3-/4-}$ redox probe is relatively hydrophilic [49] and can rather interact with water instead of the aryl-modified surface. In addition, the hydrophilic aryl groups are negatively charged which may add up electrostatic repulsions at the ITO-aryl/probe interface.

3.3. Electrodeposition of Polypyrrole on ITO

3.3.1. Electrochemical Characterization of Poly(pyrrole-benzene sulfonic acid) Films on Aryl-Modified ITO

The aryl layers served as a primer for the electrodeposition of PPy-BSA. We achieved the electropolymerization on aryl-modified ITO strips by performing ten potential scanning cycles between -1 and 1.2 V/ECS at 100 mV·s^{-1} in an aqueous solution of 0.1 M of pyrrole and 0.01 M BSA (Figure 5). Table 2 reports the Ep$_{ox}$ and Ep$_{red}$ values for the electropolymerization process together with the oxidation peak density (Jp$_{ox}$). Highly uniform and adherent PPy-BSA films were easily obtained on all modified electrodes, regardless the nature of the diazonium compound. The films obtained on the hydrophobic surfaces (ITO-N(CH$_3$)$_2$, ITO-NH$_2$, ITO-NH-Ph) were much more conductive than those obtained on the hydrophilic surfaces (ITO-CO$_2$H, ITO-SO$_3$H, ITO-CN), as deduced from the corresponding high current in the CV curves.

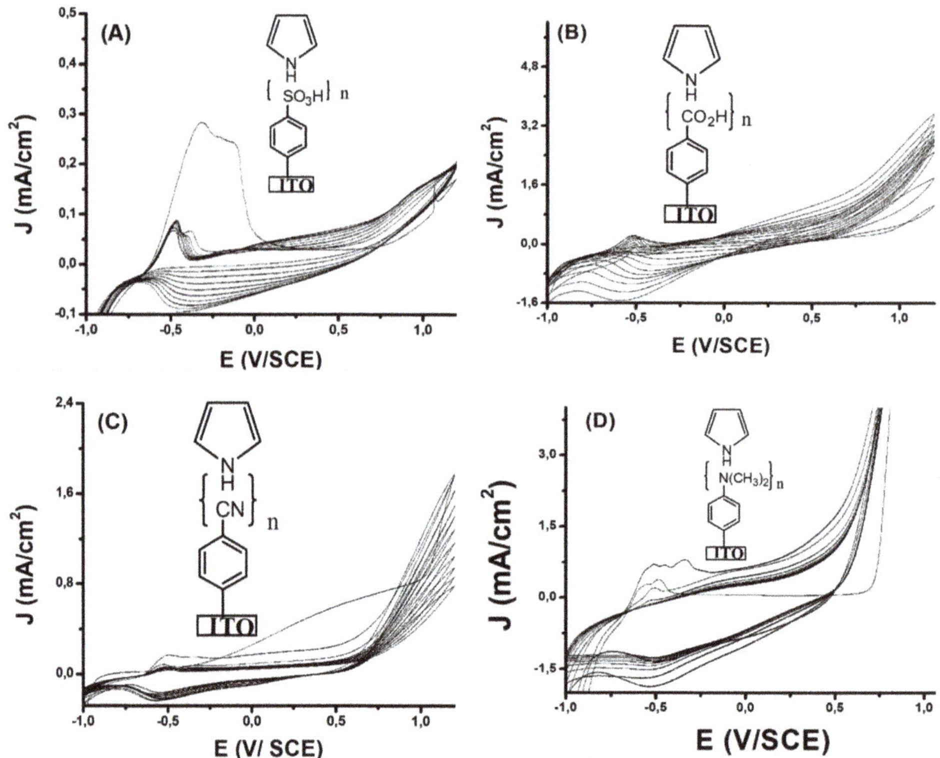

Figure 5. *Cont.*

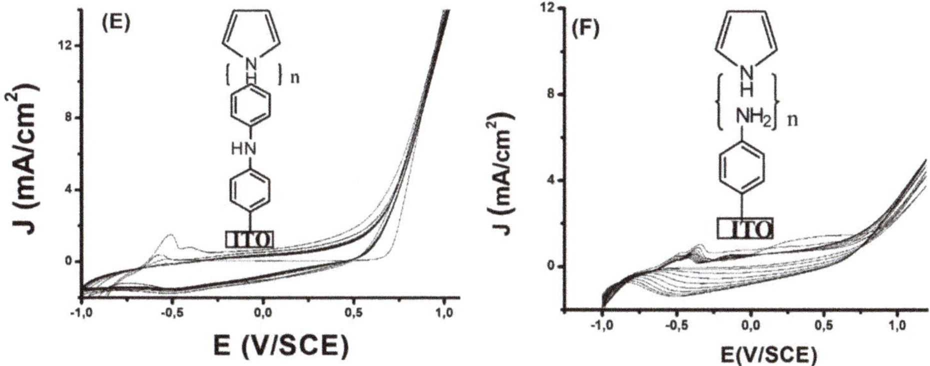

Figure 5. Electrosynthesis of PPy-BSA films by cyclic voltammetry on (**A**) ITO-SO$_3$H; (**B**) ITO-CO$_2$H; (**C**) ITO-CN; (**D**) ITO-N(CH$_3$)$_2$; (**E**) ITO-NH-Ph; and (**F**) ITO-NH$_2$.

Table 2. Electrochemical characteristics of electrodeposition of polypyrrole top layers: E_{ox}, E_{red} and Jp_{ox} values for the 6 ITO-aryl electrode materials.

Electrode Material	E_{ox} (V)	E_{red} (V)	Jp_{ox} (μA/cm^2)
ITO-N(CH$_3$)$_2$	0.25	−0.53	972
ITO-NH$_2$	0.27	−0.52	960
ITO-NH-Ph	0.2	−0.5	936
ITO-CN	0.25	−0.58	208
ITO-CO$_2$H	0.41	−0.59	908
ITO-SO$_3$H	0.24	−0.43	228

As a matter of fact, the oxidation peak is centered at 0.25, 0.27, and 0.2 V/SCE for ITO-N(CH$_3$)$_2$, ITO-NH$_2$ and ITO-NH-Ph electrodes, respectively. These peaks are positioned near those obtained on the hydrophilic surfaces ITO-CN and ITO-SO$_3$H (0.25 and 0.24 V/SCE, respectively). In contrast, on ITO-CO$_2$H the oxidation peak is centred at a higher potential value (0.41 V/SCE) and the reduction peak at −0.59 V/SCE. This is the largest potential difference for the electropolymerization of pyrrole on the actual series of diazonium-modified ITO electrodes. The density of Jp_{ox} of PPy-BSA on electrodes materials takes the maximum values for hydrophobic surface (972 μA/cm^2 on ITO-N(CH$_3$)$_2$), hence the highest conductivity achieved for electropolymerization of pyrrole on the actual electrode materials.

The results obtained so far show that thin films of PPy were preferentially deposited on hydrophobic surfaces, in line with the literature [50,51].

Figure 6 shows the blocking effect and EIS results of PPy films on all aryl-modified ITO electrodes. The blocking effect of PPy on all aryl layers was assessed with Fe(CN)$_6^{3-/4-}$, as depicted in Figure 6A. The quasi-reversible redox couple (located at 0.25 V and 0.1 V) obtained on the hydrophobic surfaces was characterized by a higher current, which accounted for a more conductive surface. It is worth to note that the PPy coating on ITO-N(CH$_3$)$_2$, ITO-NH$_2$, ITO-NH-Ph (Figure 6B) exhibited the lowest charge transfer resistance (0.15, 0.36, 0.6 kΩ/cm^2, respectively). Moreover, the R$_{ct}$ increased for PPy on ITO-SO$_3$H, ITO-CO$_2$H, and ITO-CN; the impedance curves were significantly broadened and the semi-circle diameters were about 28.59, 21.44 and 10 kΩ/cm^2 respectively. These results indicated that ITO-SO$_3$H had the largest obstruction, which resulted in reducing electron-transfer rate or increasing resistance to the flow of electrons. Furthermore, EIS showed that charge transfer resistance of ITO-N(CH$_3$)$_2$-PPy was much lower compared to others.

The results reported so far stress the paramount importance of the hydrophobic character of the substrate, here imparted by the aryl layers. As a matter of fact, PPy is relatively hydrophobic, and preferentially binds to hydrophobic surfaces [52]. Conversely, the strength of (macro) molecule-polypyrrole molecular interactions are maximized in water [53].

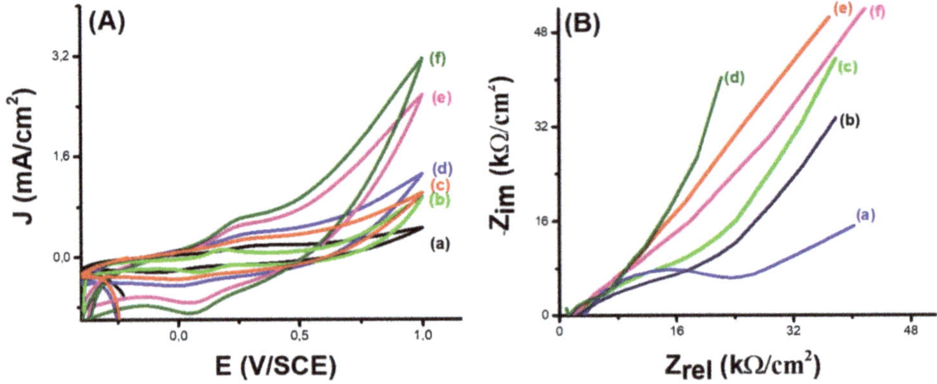

Figure 6. PPy-BSA film on (**a**) ITO-SO$_3$H, (**b**) ITO-CN, (**c**) ITO-CO$_2$H, (**d**) ITO-NH-Ph, (**e**) ITO-NH$_2$, (**f**) ITO-N(CH$_3$)$_2$. (**A**) Blocking effect of the PPy-ABS films on the ITO electrodes modified by the different diazonium cations. (**B**) Impedance measurements recorded for PPy-ABS coatings on modified ITO sheets.

3.3.2. XPS Characterization of PPy Top Coats and Reference Materials

Figure 7 displays survey regions of ITO, ITO-SO$_3$H, ITO-N(CH$_3$)$_2$, ITO-N(CH$_3$)$_2$-PPy samples in the 100–600 eV range. S2p, C1s, N1s, and In3d/Sn3d doublets are centered at 168, 285, 400, 445–453 and 487–496 eV, respectively.

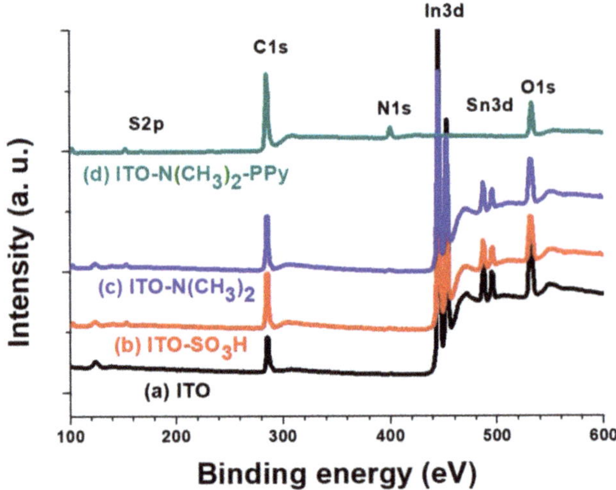

Figure 7. XPS survey regions of (**a**) ITO, (**b**) ITO-SO$_3$H, (**c**) ITO-N(CH$_3$)$_2$ and (**d**) ITO-N(CH$_3$)$_2$-PPy.

Bare ITO (Figure 7a) exhibits a quasi-horizontal background in the 160–285 eV spectral region and a sharp In3d doublet together with Sn3d peaks. Attachment of an aryl monolayer screened ITO (Figure 7b,c) therefore resulting in an increase of the relative C1s peak intensity, and in the appearance of an S2p peak at ~168 eV for ITO-SO$_3$H (Figure 7b) and of a N1s peak at 400 eV for ITO-N(CH$_3$)$_2$ (Figure 7c). After electropolymerization, the survey region resembled that of pure polypyrrole with effective screening of the characteristic In3d and Sn3d features from ITO (Figure 7d).

All apparent compositions (in at. %) are reported in Table 2 for the main elements. The carbon and nitrogen contents increased in the order ITO < ITO-aryl < ITO-aryl-PPy. Sulfur content is due to the sulfonate group from ITO-SO$_3$H, and to all PPy-BSA topcoats. This sulfur content is invariably higher for PPy-BSA than for ITO-SO$_3$H. In contrast, there is a clear decrease for the O at. %, as one shifts from ITO to ITO-aryl, then PPy. Indium and tin atoms are detected for bare and aryl-modified ITO surfaces only. After electropolymerization, In and Sn could not be detected since XPS is surface specific technique (about 12–15 nm depth) while PPy topcoat thickness is in the micrometer scale range (~3 μm, SEM image not shown). This is a much too large thickness for a topcoat to possibly detect the buried layers using XPS (10–12 nm sampling depth for organic coatings).

Table 3 provides also the In4d/In3d intensity ratio for bare and aryl-modified ITO sheets. Invariably, this ratio is higher for the modified ITO sheets. Indeed, an aryl layer screens ITO, but at a different extent for the In4d and In3d core electrons. The sampling depth of In4d is about 12.6 nm higher than that of In3d (10.8 nm) [47]. It follows that In3d peak attenuation is larger, hence In4d/In3d intensity ratio increases upon aryl grafting. All aryl layers have nanometer scale thickness well below 10.8 nm (the sampling depth of In3d).

Table 3. XPS-determined apparent surface chemical composition of ITO, ITO-aryl and ITO-aryl-PPy samples.

Materials	C	N	S	O	In	Sn	In4d/In3d
ITO	32.5	-	-	39.1	25.4	3.06	0.155
ITO-SO$_3$H	45.2	0.77	1.00	30.0	20.5	2.43	0.167
ITO-SO$_3$H-PPy	73.9	9.29	1.38	15.4	-	-	
ITO-COOH	58.6	1.68	-	25.1	11.3	3.34	0.174
ITO-COOH-PPy	78.7	10.5	1.13	9.71	-	-	
ITO-CN	71.4	4.91	-	16.8	5.88	1.04	0.179
ITO-CN-PPy	75.2	9.23	2.12	13.5	-	-	
ITO-NH-Ph	49.2	1.97	-	29.6	16.2	3.05	0.171
ITO-NH-Ph-PPy	75.5	9.83	1.71	12.9	-	-	
ITO-NH$_2$	42.9	1.31	-	33.1	16.2	6.55	0.159
ITO-NH$_2$-PPy	77.3	12.2	2.22	8.22	-	-	
ITO-N(CH$_3$)$_2$	52.7	1.13	-	28.2	15.5	2.48	0.181
ITO-N(CH$_3$)$_2$-PPy	76.5	6.94	1.18	15.4	-	-	

As the blocking effect can be due to the thickness or grafting density of the top layer, we were tempted to correlate I_{rel} (in Table 1) to I_{4d}/I_{3d} XPS intensity ration (in Table 3). The correlation is poor and the data corresponding to ITO-NH$_2$ (I_{4d}/I_{3d} = 0.159; I_{rel} = 46.6%) differs markedly from the trend given by the 5 other surfaces. As a matter of fact, Baranton and Bélanger [48] clearly indicated that blocking effect cannot be simply related to the film thickness or the surface coverage. Herein, we do confirm such an observation previously reported [48] and confirm that the hydrophilic/hydrophobic effects operate at the interface, which gives full credit to Downard and Prince [49]. It follows that barrier properties cannot be interpreted simply in terms of coverage or thickness; herein we demonstrate that barrier properties are very likely to be correlated to the hydrophobic/hydrophilic character of the ITO-aryl surfaces as shown in Figure 4a,b.

3.4. Practical Adhesion Aspects of Polypyrrole Thin Layers

Practical adhesion of PPy thin layers has been qualitatively investigated by solvent washing, ultrasonication and bend stress. Simple solvent washing using a laboratory wash bottle demonstrates the efficiency of the diazonium adhesive layer in anchoring PPy to ITO. This is shown in a video available in the Supplementary Material SM Video1. This video shows that without any aryl adhesive layer, PPy is simply and easily removed upon washing with deionized water. The middle part of the video (starting at 0.48 min) shows that another PPy film grown on bare ITO is very easily delaminated using tweezers. In the last part of the video (starting at 1.07 min), prolonged washing of PPy topcoat of ITO-NH$_2$ electrode does not induce any adhesion failure. We have then compared adhesion of PPy to bare ITO and ITO-NH$_2$ using ultrasonication (video not shown): again, the aryl layer prevents adhesion failure of the PPy, whereas on bare ITO the film is removed previously reported [22]. Actually, the removal of PPy from ITO-NH$_2$ was only possible through scratching. Such robust adhesion of polypyrrole to aminophenyl-grafted glassy carbon electrode found by Patterson and Ignaszak [23] confirms our previous findings on ITO-NH$_2$ and supports the actual work which generalizes the strong adhesion of polypyrrole to flexible ITO and other electrodes.

One of the important aspects of polymer adhesion is to define the type of adhesion bonding. In a previous publication, Samanta et al. [47] have demonstrated that photochemically-prepared PPy/silver nanocomposite film was removed using adhesive tape from bare ITO, whereas no failure occurred in the case of pyrrole-functionalized ITO sheets. A failure within the polymer film is defined as "cohesive failure". Herein, ultrasonication of ITO-PPy resulted in the delamination of the conductive polymer topcoat. XPS inspection of the ITO fracture surface, after ITO-PPy adhesion failure, brought strong supporting evidence for a neat ITO surface without any attenuation of the characteristic peaks from ITO (see Survey region in Figure 8a). It is worthwhile to note that the C1s peak is weakly intense and is partly due to unavoidable adventitious contamination during the transfer from atmosphere to the vacuum chamber of the XPS apparatus. The spectrum displayed in Figure 8a is a clear sign of an "adhesive failure", ca exactly at the ITO-PPy interface. Both materials are split apart upon ultrasonication.

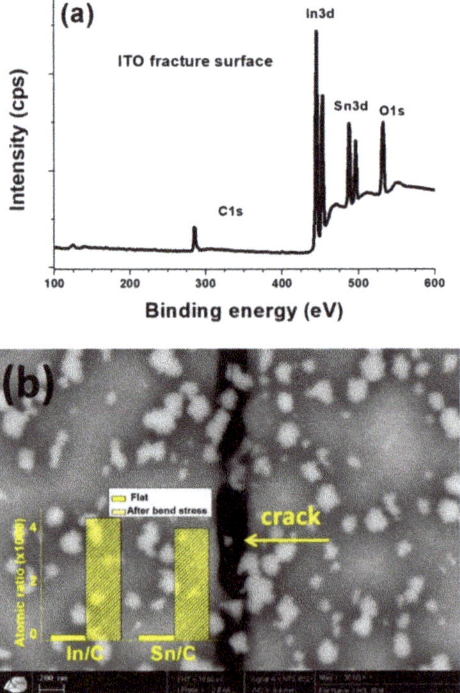

Figure 8. Adhesion testing of polypyrrole thin films on (**a**) bare ITO using XPS; and (**b**) ITO-NH$_2$ using SEM and XPS. Inset of 8b shows the bar graph of In/C atomic.

We also examined the propensity of the ITO-NH$_2$-PPy hybrid electrodes to withstand bending stress. The ITO-NH$_2$-PPy test electrode was prepared as described above and stuck on the outer wall of a tube for 24 h bending stress; the bending radius was 3.2 mm as measured using an electronic digital caliper (Fowler, model S225; Newton, MA, USA). This is a small radius and accounts for an important stress [20]. Compared to the unbent sample, the PPy film withstood the very high bending stress and remained stuck to the underlying ITO-NH$_2$ substrate. However, we found a slight increase in the In/C and Sn/C atomic ratios, which probably resulted from a possible diffusion of In and Sn in the top coat. Indeed, SEM inspection of the ITO-NH$_2$-PPy surface after bending showed a crack in the ITO layer (Figure 8b).

From these qualitative tests, clearly the conductive polymer films strongly adhere to the ITO surface due to the molecular interactions that operate at interfaces between PPy and all diazonium-modified ITO flexible electrodes. The aryl groups were shown to be bridged to the oxide surface through metal-O-C bonds [54,55]. We have selected functional groups in *para* position, that undergo either electrostatic interactions or electron-donor/acceptor bonds (Figure 9). Although not covalent in nature, these interfacial interactions are sufficiently strong to keep PPy anchored to the ITO via the aryl layer. The deposition of PPy on aryl-modified ITO is simple, fast, and provides strong adhesion that withstands severe conditions, even under high mechanical stress. This contrasts with the poor adhesion of polypyrrole to bare ITO, as previously reported in the literature [19].

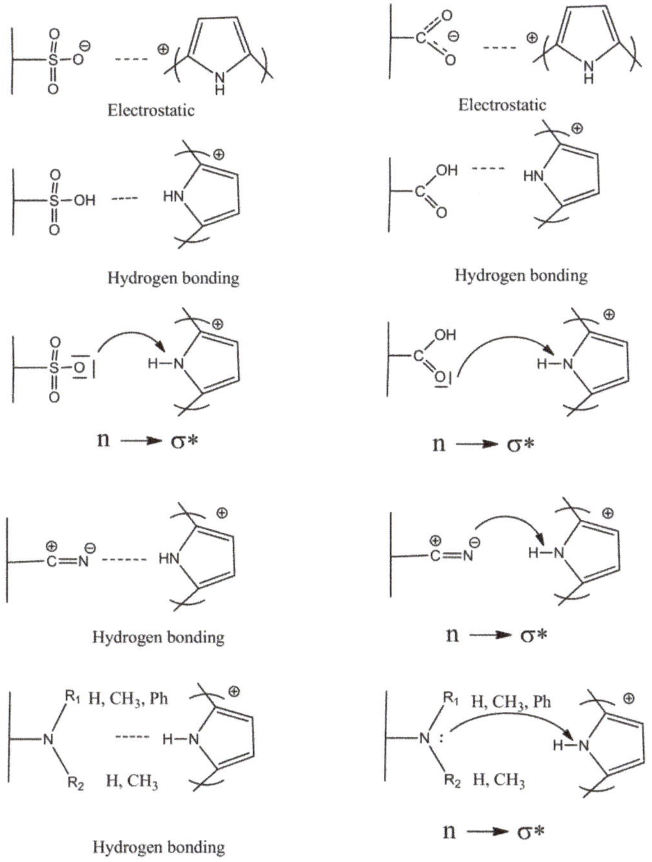

Figure 9. Possible molecular interactions at aryl-polypyrrole interfaces.

4. Conclusions

In this work, flexible ITO sheets were grafted using in situ-generated diazonium salts bearing electron acceptor (COOH, SO_3H and CN) or electron-donor (NH_2, $N(CH_3)_2$, NH-Phenyl) groups. We have carefully grafted monolayers which were thin enough to prevent passivation but which permitted to prepare adherent polypyrrole top coatings by electropolymerization. The aryl layers were characterized by contact angle measurements and by XPS. Interestingly, the blocking effects and charge transfer resistance of the aryl layers were correlated with their hydrophobic character. All aryl layers served as remarkable coupling agents for the polypyrrole top coats which withstand sonication, solvent washing and bend stress. On bare ITO, PPy could be removed by simple washing with water or using tweezers, without any trace of polymer left on the surface as judged from XPS analysis of the fracture surface. This accounts for adhesive failure of the coating. In contrast, on all aryl-modified ITO sheets no delamination took place, and the polymer could be only removed by scratching the polypyrrole film. The adherent polypyrrole exhibited a better redox activity in terms of high current output on the most hydrophobic surfaces obtained with aminated, electron-donor groups, thus indicating the possibility to tune both the adhesion strength and redox properties of the conductive polymer. Adhesion of polypyrrole to ITO via aryl layers, is likely to be due to interfacial electrostatic interactions between the electron acceptor groups $COOH/COO^-$, SO_3H/SO_3H and $CN/C^+ = N^-$ and the polypyrrole positively charged backbone. As far as the nitrogen-containing electron-donor groups, such as NH_2, $N(CH_3)_2$ and NH-Phenyl, are concerned, the adhesion is attributed to n-σ^* electron-donor-acceptor interactions (the pyrrole N-H bond being the σ^* acceptor).

This work sheds a new light on the use of diazonium salts to make adhesive primer layers for electroactive polymers and other organic coatings. As silanes in the 20th century, so diazonium salts could be regarded as magic molecular glues for tuning interfacial properties of next generation advanced materials of relevance to sensors, printed soft electronics and flexible thermoelectrics to name but these timely remarkable applications.

Supplementary Materials: The following are available online at http://www.mdpi.com/2571-9637/1/1/5/s1.

Author Contributions: Conceptualization of the research work by M.L., M.M.C., D.G.-S. and K.D.; Methodology by M.L. and M.M.C.; Validation by M.M.C., K.D., D.G.-S., J.-J.A.; Formal Analysis by M.L., R.P., M.M.C.; Writing of Original Draft was done by M.L. and M.M.C.; Writing: Review and Editing, M.L., M.M.C., J.-J.A. and M.A.O.; Supervision, M.M.C. and D.G.-S.; Funding Acquisition, J.-J.A., M.A.O. and D.G.-S.

Funding: One of us (Momath Lo) gratefully thanks the Cooperation and Cultural Action Service of the French Embassy in Senegal for a PhD grant.

Acknowledgments: All authors thank S. Chehimi (EFREI, Villejuif, France) for the video making presented as Supplementary Material.

Conflicts of Interest: The authors declare no conflict of interest.

References

1. MacDiarmid, A.G. Polyaniline and polypyrrole: Where are we headed? *Synth. Met.* **1997**, *84*, 27–34. [CrossRef]

2. George, P.M.; Lyckman, A.W.; LaVan, D.A.; Hegde, A.; Leung, Y.; Avasare, R.; Testa, C.; Alexander, P.M.; Langer, R.; Sur, M. Fabrication and biocompatibility of polypyrrole implants suitable for neural prosthetics. *Biomaterials* **2005**, *26*, 3511–3519. [CrossRef] [PubMed]

3. Bharti, M.; Singh, A.; Samanta, S.; Debnath, A.K.; Aswal, D.K.; Muthe, K.P.; Gadkari, S.C. Flexo-green Polypyrrole—Silver nanocomposite films for thermoelectric power generation. *Energy Convers. Manag.* **2017**, *144*, 143–152. [CrossRef]

4. Sun, X.; Zhang, H.; Zhou, L.; Huang, X.; Yu, C. Polypyrrole-coated zinc ferrite hollow spheres with improved cycling stability for lithium-ion batteries. *Small* **2016**, *12*, 3732–3737. [CrossRef] [PubMed]

5. Omastová, M.; Trchová, M.; Kovárová, J.; Stejskal, J. Synthesis and structural study of polypyrroles prepared in the presence of surfactants. *Synth. Met.* **2003**, *138*, 447–455. [CrossRef]

6.	Sadki, S.; Schottland, P.; Brodie, N.; Sabouraud, G. The mechanisms of pyrrole electropolymerization. *Chem. Soc. Rev.* **2000**, *29*, 283–293.

7.	Cruz-Silva, R.; Amaro, E.; Escamilla, A.; Nicho, M.E.; Sepulveda-Guzman, S.; Arizmendi, L.; Romero-Garcia, J.; Castillon-Barraza, F.F.; Farias, M.H. Biocatalytic synthesis of polypyrrole powder colloids and films using horseradish peroxidase. *J. Colloid Interface Sci.* **2008**, *328*, 263–269. [CrossRef] [PubMed]

8.	Saad, A.; Cabet, E.; Lilienbaum, A.; Hamadi, S.; Abderrabba, M.; Chehimi, M.M. Polypyrrole/Ag/mesoporous silica nanocomposite particles: Design by photopolymerization in aqueous medium and antibacterial activity. *J. Taiwan Inst. Chem. Eng.* **2017**, *80*, 1022–1030. [CrossRef]

9.	Hamouma, O.; Oukil, D.; Omastová, M.; Chehimi, M.M. Flexible paper@carbonnanotube@polypyrrole composites: The combined pivotal roles of diazonium chemistry and sonochemicalpolymerization. *Colloids Surf. Physicochem. Eng. Asp.* **2018**, *538*, 350–360. [CrossRef]

10.	Omastová, M.; Mičušík, M. Polypyrrole coating of inorganic and organic materials by chemical oxidative polymerization. *Chem. Pap.* **2012**, *66*, 392–414. [CrossRef]

11.	Baibarac, M.; Gómez-Romero, P. Nanocomposites based on conducting polymers and carbon nanotubes: From fancy materials to functional applications. *J. Nanosci. Nanotechnol.* **2006**, *6*, 289–302. [CrossRef] [PubMed]

12.	Lee, S.; Sung, H.; Han, S.; Paik, W. Polypyrrolefilm formation by solution-surface electropolymerization: Influence of solvents and doped anions. *J. Phys. Chem.* **1994**, *98*, 1250–1252. [CrossRef]

13.	Ferreira, C.A.; Aeiyach, S.; Delamar, M.; Lacaze, P.C. Electropolymerization of pyrrole on iron electrodes: Influence of solvent and electrolyte on the nature of the deposits. *J. Electroanal. Chem. Interfacial Electrochem.* **1990**, *284*, 351–369. [CrossRef]

14.	Kaplin, D.A.; Qutubuddin, S. Electrochemically synthesized polypyrrolefilms: Effects of polymerization potential and electrolyte type. *Polymer* **1995**, *36*, 1275–1286. [CrossRef]

15.	Raso, M.A.; González-Tejera, M.J.; Carrillo, I.; Sanchez De La Blanca, E.; García, M.V.; Redondo, M.I. Electrochemical nucleation and growth of poly-N-methylpyrrole on copper. *Thin Solid Films* **2015**, *19*, 2387–2392. [CrossRef]

16.	Sharifirad, M.; Omrani, A.; Rostami, A.A.; Khoshroo, M. Electrodeposition and characterization of polypyrrolefilms on copper. *J. Electroanal. Chem.* **2010**, *645*, 149–158. [CrossRef]

17.	Alfaro-López, H.M.; Aguilar-Hernandez, J.R.; Garcia-Borquez, A.; Hernandez-Perez, M.A.; Contreras-Puente, G.S. Electropolymerization of polypyrrole films in aqueous solution with side-coupler agent to hydrophobic groups. *Interface Control. Org. Thin Films* **2009**, *129*, 73–78.

18.	Wang, Y.; Northwood, D.O. An investigation into the nucleation and growth of an electropolymerized polypyrrole coating on a 316L stainless steel surface. *Thin Solid Films* **2008**, *516*, 7427–7432. [CrossRef]

19.	Castro-Beltran, A.; Dominguez, C.; Bahena-Uribe, D.; Sepulveda-Guzmana, S.; Cruz-Silva, R. Effect of non-electroactive additives on the early stage pyrrole electropolymerization on indium tin oxide electrodes. *Thin Solid Films* **2014**, *566*, 23–31. [CrossRef]

20.	Singh, A.; Salmi, Z.; Joshi, N.; Jha, P.; Kumar, A.; Lecoq, H.; Lau, S.; Chehimi, M.M.; Aswal, D.K.; Gupta, S.K. Photo-induced synthesis of polypyrrole-silver nanocomposite films on *N*-(3-trimethoxysilylpropyl)pyrrole-modified biaxially oriented polyethylene terephthalate flexible substrates. *RSC Adv.* **2013**, *3*, 5506–5523. [CrossRef]

21.	Bouktit, B.; Salmi, Z.; Decorse, P.; Lecocq, H.; Jouini, M.; Aswal, D.; Singh, A.; Chehimi, M.M. Polypyrrole/Ag nanocomposite films on diazonium salt modified indium tin oxide substrate. *J. Colloid Sci. Biotechnol.* **2013**, *2*, 200–210. [CrossRef]

22.	Lo, M.; Diaw, A.K.D.; Gningue-Sall, D.; Aaron, J.-J.; Oturan, M.A.; Chehimi, M.M. The role of diazonium interface chemistry in the design of high performance polypyrrole-coated flexible ITO sensing electrodes. *Electrochem. Commun.* **2017**, *77*, 14–17. [CrossRef]

23.	Patterson, N.; Ignaszak, A. Modification of glassy carbon with polypyrrole through an aminophenyl linker to create supercapacitive materials using bipolar electrochemistry. *Electrochem. Commun.* **2018**, *93*, 10–14. [CrossRef]

24.	Mevellec, V.; Rousse, S.; Tessier, L.; Chancolon, J.; Mayne-L'Hermite, M.; Deniau, G.; Viel, P.; Palacin, S. Grafting polymers on surfaces: A new powerful and versatile diazonium salt-based one-step process in aqueous media. *Chem. Mater.* **2007**, *19*, 6323–6330. [CrossRef]

25. Viel, P.; Le, X.T.; Huc, V.; Bar, J.; Benedetto, A.; Le Goff, A.; Filoramo, A.; Alamarguy, D.; Noël, S.; Baraton, L.; et al. Covalent grafting onto self-adhesive surfaces based on aryldiazonium salt seed layers. *J. Mater. Chem.* **2008**, *18*, 5913–5920. [CrossRef]

26. Alageel, O.; Abdallah, M.N.; Luo, Z.Y.; Del-Rio-Highsmith, J.; Cerruti, M.; Tamimi, F. Bonding metals to poly(methyl methacrylate) using aryldiazonium salts. *Dent. Mater.* **2015**, *31*, 105–114. [CrossRef] [PubMed]

27. Watts, J.F.; Rattana, A.; Abel, M.-L. Interfacial chemistry of adhesives on hydrated aluminium and hydrated aluminium treated with an organosilane. *Surf. Interface Anal.* **2004**, *36*, 1449–1468. [CrossRef]

28. Jesson, D.A.; Abel, M.-L.; Hay, J.N.; Smith, P.A.; Watts, J.F. Organic-inorganic hybrid nanoparticles: Surface characteristics and interactions with a polyester resin. *Langmuir* **2006**, *22*, 5144–5151. [CrossRef] [PubMed]

29. Tashiro, H.; Nakaya, M.; Hotta, A. Enhancement of the gas barrier property of polymers by DLC coating with organosilane interlayer. *Diam. Relat. Mater.* **2013**, *35*, 7–13. [CrossRef]

30. Bélanger, D.; Pinson, J. Electrografting: A powerful method for surface modification. *Chem. Soc. Rev.* **2011**, *40*, 3995–4048. [CrossRef] [PubMed]

31. Wang, J.; Firestone, M.A.; Auciello, O.; Carlisle, J.A. Surface functionalization of ultrananocrystalline diamond films by electrochemical reduction of aryldiazonium salts. *Langmuir* **2004**, *20*, 11450–11456. [CrossRef] [PubMed]

32. Mirkhalaf, F.; Graves, J.E. Nanostructured electrocatalysts immobilised on electrode surfaces and organic film templates. *Chem. Pap.* **2012**, *66*, 472–483. [CrossRef]

33. Bakas, I.; Yilmaz, G.; Ait-Touchente, Z.; Lamouri, A.; Lang, P.; Battaglini, N.; Carbonnier, B.; Chehimi, M.M.; Yagci, Y. Diazonium salts for surface-confined visible light radical photopolymerization. *J. Polym. Sci. Part A Polym. Chem.* **2016**, *54*, 3506–3515. [CrossRef]

34. Stockhausen, V.; Nguyen, V.Q.; Martin, P.; Lacroix, J.-C. Botom-up electrochemical fabrication of conjugated ultrathin layers with tailored switchable properties. *ACS Appl. Mater. Interfaces* **2017**, *9*, 610–617. [CrossRef] [PubMed]

35. Wang, Y.; Fantin, M.; Park, S.; Gottlieb, E.; Fu, L.; Matyjaszewski, K. Electrochemically mediated reversible addition−fragmentation chain-transfer polymerization. *Macromolecules* **2017**, *50*, 7872–7879. [CrossRef] [PubMed]

36. Gui, A.L.; Luais, E.; Peterson, J.R.; Godding, J.J. Zwitterionic phenyl layers: Finally, stable, anti-biofouling coatings that do not passivate electrodes. *ACS Appl. Mater. Interfaces* **2013**, *5*, 4827–4835. [CrossRef] [PubMed]

37. Cao, C.; Zhang, Y.; Jiang, C.; Qi, M.; Liu, G. Advances on aryldiazonium salt chemistry based interfacial fabrication for sensing applications. *ACS Appl. Mater. Interfaces* **2017**, *9*, 5031–5049. [CrossRef] [PubMed]

38. Guselnikova, O.; Postnikov, P.; Elashnikova, R.; Trusova, M.; Kalachyova, Y.; Libansky, M.; Barek, J.; Kolska, Z.; Švorčík, V.; Lyutakov, O. Surface modification of Au and Ag plasmonic thin films via diazonium chemistry: Evaluation of structure and properties. *Colloids Surf. A Physicochem. Eng. Asp.* **2017**, *516*, 274–285. [CrossRef]

39. Jlassi, K.; Chandran, S.; Poothanari, M.A.; Benna-Zayani, M.; Thomas, S.; Chehimi, M.M. Clay/polyaniline hybrid through diazonium chemistry: Conductive nanofiller with unusual effects on interfacial properties of epoxy nanocomposites. *Langmuir* **2016**, *32*, 3514–3524. [CrossRef] [PubMed]

40. Sandomierski, M.; Strzemiecka, B.; Chehimi, M.M.; Voelkel, A. Reactive diazonium-modified silica fillers for high-performance polymers. *Langmuir* **2016**, *32*, 11646–11654. [CrossRef] [PubMed]

41. Jacques, A.; Chehimi, M.M.; Poleunis, C.; Delcorte, A.; Delhalle, J.; Mekhalif, Z. Grafting of 4-pyrrolyphenyldiazonium in situ generated on NiTi, an adhesion promoter for pyrrole electropolymerisation. *Electrochim. Acta* **2016**, *211*, 879–890. [CrossRef]

42. Kullapere, M.; Mirkhalaf, F.; Tammeveski, K. Electrochemical behaviour of glassy carbon electrodes modified with aryl groups. *Electrochim. Acta* **2010**, *56*, 166–173. [CrossRef]

43. Jacques, A.; Devillers, S.; Delhalle, J.; Mekhalif, Z. Electrografting of in situ generated pyrrole derivative diazonium salt for the surface modification of nickel. *Electrochim. Acta* **2013**, *109*, 781–789. [CrossRef]

44. Shul, G.; Weissmann, M.; Bélanger, D. Electrochemical characterization of glassy carbon electrode modified with 1,10-phenanthroline groups by two pathways: Reduction of the corresponding diazonium ions and reduction of phenanthroline. *Electrochim. Acta* **2015**, *162*, 146–155. [CrossRef]

45. Seck, S.M.; Charvet, S.; Fall, M.; Baudrin, E.; Geneste, F.; Lejeune, M.; Benlahsen, M. Functionalization of amorphous nitrogenated carbon thin film electrodes for improved detection of cadmium vs. copper cations. *J. Electroanal. Chem.* **2015**, *738*, 154–161. [CrossRef]

46. Jiang, C.; Moraes Silva, S.; Fan, S.; Wu, Y.; Tanzirul Alam, M.; Liu, G.; Gooding, J.J. Aryldiazonium salt derived mixed organic layers: From surface chemistry to their applications. *J. Electroanal. Chem.* **2017**, *785*, 265–278. [CrossRef]

47. Samanta, S.; Bakas, I.; Singh, A.; Aswal, D.K.; Chehimi, M.M. In situ diazonium-modified flexible ITO-coated PEN substrates for the deposition of adherent silver–polypyrrole nanocomposite films. *Langmuir* **2014**, *30*, 9397–9406. [CrossRef] [PubMed]

48. Baranton, S.; Belanger, D. Electrochemical derivatization of carbon surface by reduction of in situ generated diazonium cations. *J. Phys. Chem. B* **2005**, *109*, 24401–24410. [CrossRef] [PubMed]

49. Downard, A.J.; Prince, M.J. Barrier properties of organic monolayers on glassy carbon electrodes. *Langmuir* **2001**, *17*, 5581–5586. [CrossRef]

50. Wu, J.-S.; Gu, D.-W.; Huang, D.; Shen, L.-J. Chemical in situ polymerization of polypyrrole nanoparticles on the hydrophilic/hydrophobic surface of SiO_2 substrates. *Synth. React. Inorg. Met. Org. Nano Met. Chem.* **2012**, *42*, 1064–1070. [CrossRef]

51. Wang, P.C.; Huang, Z.; MacDiarmid, A.G. Critical dependency of the conductivity of polypyrrole and polyaniline films on the hydrophobicity/hydrophilicity of the substrate surface. *Synth. Met.* **1999**, *101*, 852–853. [CrossRef]

52. Perruchot, C.; Chehimi, M.M.; Delamar, M.; Cabet-Deliry, E.; Miksa, B.; Slomkowski, S.; Khan, M.A.; Armes, S.P. Chemical deposition and characterization of thin polypyrrole films on glass plates: Role of organosilane treatment. *Colloid Polym. Sci.* **2000**, *278*, 1139–1154. [CrossRef]

53. Azioune, A.; Chehimi, M.M.; Miksa, B.; Basinska, T.; Slomkowski, S. Hydrophobic protein−polypyrrole interactions: The role of van der Waals and Lewis acid−base forces as determined by contact angle measurements. *Langmuir* **2002**, *18*, 1150–1156. [CrossRef]

54. AïtAtmane, Y.; Sicard, L.; Lamouri, A.; Pinson, J.; Sicard, M.; Masson, C.; Nowak, S.; Decorse, P.; Piquemal, J.-Y.; Galtayries, A.; et al. Functionalization of aluminum nanoparticles using a combination of aryl diazonium salt chemistry and iniferter method. *J. Phys. Chem. C* **2013**, *117*, 26000–26006. [CrossRef]

55. Jacques, A.; Devillers, S.; Arrotin, B.; Delhalle, J.; Mekhalif, Z. Polyelectrolyte multilayers deposition on Nitinol modified by in situ generated diazonium in gentle conditions. *J. Electrochem. Soc.* **2014**, *161*, G55–G62. [CrossRef]

Article

On the Contrasting Effect Exerted by a Thin Layer of CdS against the Passivation of Silver Electrodes Coated with Thiols

Emanuele Salvietti [1,*], Walter Giurlani [1], Maria Luisa Foresti [1], Maurizio Passaponti [1], Lorenzo Fabbri [1], Patrick Marcantelli [1], Stefano Caporali [2,3], Stefano Martinuzzi [2,3], Nicola Calisi [2,3], Maddalena Pedio [4] and Massimo Innocenti [1,3,*]

[1] Dipartimento di Chimica, Università degli Studi di Firenze, Via della Lastruccia 3, 50019 Sesto Fiorentino (FI), Italy; walter.giurlani@unifi.it (W.G.); foresti@unifi.it (M.L.F.); maurizio.passaponti@unifi.it (M.P.); lorenzo.fabbri2@stud.unifi.it (L.F.); patrick.marcantelli@unifi.it (P.M.)

[2] Dipartimento di Ingegneria Industriale, Università degli Studi di Firenze, via S. Marta, 50139 Florence, Italy; stefano.caporali@unifi.it (S.C.); stefano.martinuzzi@unifi.it (S.M.); nicola.calisi@unifi.it (N.C.)

[3] Consorzio Interuniversitario Nazionale per la Scienza e Tecnologia dei Materiali, via Giusti 9, 50121 Florence, Italy

[4] Istituto Officina dei Materiali, Consiglio Nazionale delle Ricerche, TASC Laboratory, 34012 Trieste, Italy; pedio@iom.cnr.it

* Correspondence: emanuele.salvietti@unifi.it (E.S.); m.innocenti@unifi.it (M.I.); Tel.: +39-055-457-3102 (M.I.)

Received: 14 June 2018; Accepted: 27 July 2018; Published: 31 July 2018

Abstract: The passivation of metal electrodes covered by self-assembled monolayers of long-chain thiols is well known. The disappearance of the voltammetric peak of redox species in solution is a classical test for the formation of full layers of thiols. Similar studies on semiconductors are still very limited. We used silver surfaces covered by an ultrathin layer of CdS as substrate for self-assembling of n-hexadecanethiol ($C_{16}SH$), and we compared the experimental results with those obtained by using the bare silver surface as substrate. The strong insulating effect of $C_{16}SH$ deposited on Ag(III) is shown by the inhibition of the voltammetric peak of $Ru(NH_3)_6^{3+/2+}$. On the contrary, the voltammogram obtained on CdS-covered Ag(III) is very similar to that obtained on the bare Ag(III) electrode, thus suggesting that the presence of CdS exerts a contrasting effect on the passivation of the silver electrode. A crucial point of our work is to demonstrate the effective formation of $C_{16}SH$ monolayers on Ag(III) covered by CdS. The formation of full layers of $C_{16}SH$ was strongly suggested by the inhibition of the stripping peak of Cd from the CdS deposit covered by $C_{16}SH$. The presence of $C_{16}SH$ was confirmed by electrochemical quartz crystal microbalance (EQCM) measurements as well as by Auger electron spectroscopy (AES) analysis.

Keywords: ECALE; CdS; silver single crystals; alkanthiols; SAMs; EQCM; AES

---

1. Introduction

The assembly of alkanethiols on metals has been the focus of numerous studies in recent years [1,2]. Most of the work was carried out on gold substrates, although compact alkanethiols films form spontaneously upon immersion of a variety of other metals like silver and copper. Similar studies on semiconductors are still very limited. To date, semiconductive substrates used for alkanethiols SAMs formation are InP [3–6], GaAs [7–9], and Ge [10].

Systematic investigations performed on n-alkanethiols SAMs on gold, with n ranging between 1 and 21, showed marked differences between long- and short-chain thiols monolayers [11]. In particular, long-chain thiols were found to form a densely packed, crystalline-like assembly with fully extended alkyl chains tilted from the surface normal by 20–30°. These long-chain thiols provide substantial barriers to electron transfer and are strongly resistant to ion penetration. On the contrary, the barrier becomes weaker while decreasing the chain length. At the same time, the structures become less ordered.

Self-assembled monolayers of hydroxythiols have been used as insulating barriers between the electrode surface and redox species in solution [12–15]. The closest approach of the redox species to the electrode surface is limited by the thickness of the insulating SAM layer [16,17]. This barrier decreases the electron transfer rate by increasing the separation between the electrode surface and redox molecules. Electron transfer in these cases proceeds via electron tunneling through the insulator, resulting in relatively slow kinetics. Electron tunneling through the full thickness of the self-assembled layer is strongly suggested by the dependence of the electron transfer on the thickness of the monolayer film [12]. From temperature-dependent current–voltage measurements carried out on n-alkanemonothiol SAMs (with $n = 8$, 12 and 16), electron tunneling was also shown to be the dominant transport mechanism [18]. More generally, the passivation of metal electrodes covered by self-assembled monolayers of long-chain thiols is well known. The disappearance of the voltammetric peak of redox species such as $Ru(NH_3)_6^{3+/2+}$ or $Fe(CN)_6^{3-/4-}$ in solution is a standard test for the formation of full layers of long-chain organic SAMs [11–14,19]. On the contrary, self-assembled monolayers of alkanethiols on semiconducting surfaces seem to exhibit different properties. In this respect, studies performed on electron tunneling at InP electrodes covered by alkanethiols showed that the distance dependence is a factor of two softer than that of gold [4–6]. A softer distance dependence for the electronic coupling for hole transfer from an alkanethiol covered n-InP electrodes to a redox species in solution was also found. This observation can be explained by consideration of the change in tunneling energy between the two types of electrodes. Earlier studies performed on organic SAMs on gold explained the faradaic contributions by the electron transfer at defects sites and electron tunneling at "collapsed" sites in the monolayer [20,21]. The hypothesis of possible contributions arising from defects or pinholes of the alkanethiols films on InP was taken into account to explain the softer distance dependence of the electron transfer. However, these contributions are likely to be very small [4].

The deposition of alkanethiols on semiconductors could open new perspectives in the vast field of self-assembling phenomena. Our group has had great experience in the growth of II–VI [22–24] and III–V compound semiconductors [25] on silver single crystals by the electrochemical atomic layer epitaxy (ECALE) method [23,26–29]. This is a method based on surface-limited phenomena such as underpotential depositions [30]. In the ECALE method, the UPD of the metallic element is alternated with that of the non-metallic element to form a single monolayer of the compound per cycle. The number of cycles determines how thick the deposit will be. The characteristics of composition, morphology and structure of the compounds grown by ECALE, as well as the bandgap values determined by photoelectrochemical measurements, indicate the high quality of these compounds [23,24]. Hence, the ECALE technique holds the promise of being able to provide low-cost, structurally well-ordered solids whose composition can be controlled at the nanoscopic level along the direction perpendicular to the substrate. Therefore, the compound semiconductors grown by ECALE seem to exhibit all of the needed characteristics for their use as a substrate for the self-assembling of alkanethiols. In this paper we report on the self-assembly of 1-hexadecanethiol ($C_{16}SH$) on Ag(III) covered by an ultrathin film of CdS deposited by ECALE, and we compare the experimental results with those obtained by using the bare silver surface as substrate. Hexaaminerutenium ion was used as a redox probe to test the electron transfer.

We chose CdS to exploit the high affinity of Cd towards the sulfur of thiol and due to the simplicity of the deposition of this compound in a highly ordered form through ECALE. In fact, CdS grows on Ag(III) along the Cd-terminated basal plane of the wurtzite [23], which is its most common crystalline form in nature.

A long chain alkanethiol such as 1-hexadecanethiol ($C_{16}SH$) was chosen to ensure a strong inhibition of the redox processes on the Ag(III) surface covered by the SAM.

The self-assembling processes of thiols on a metal Me, such as Au, Ag and Cu, form stable surface structures [1]. Complete electrodesorption of the *n*-alkanethiol SAMs occurs according to the electroreduction reaction that yields a small peak preceding the hydrogen evolution reaction [31]:

$$C_nS^-Me + e^- \rightarrow C_nS^- + Me \tag{1}$$

2. Materials and Methods

2.1. Materials

Merck analytical reagent grade $3CdSO_4 \cdot 8H_2O$ and absolute ethanol, Aldrich analytical reagent grade Na_2S, KCl, NaOH, $Ru(NH_3)_6Cl_3$ and purity grade Fluka 95% hexadecanethiol were used without further purification. Merck analytical reagent grade $HClO_4$ and NH_3 were used to prepare the pH 9.2 ammonia buffer. The working electrodes were silver single crystal discs grown in a graphite crucible, oriented by X-rays and cut according to the Bridgman technique [32,33]. An automated deposition apparatus consisting of Pyrex solution reservoirs, solenoid valves, a distribution valve and a flow-cell was used under the control of a computer. The electrolytic cell was a Teflon cylinder with about a 5 mm inner diameter and a 30 mm outer diameter, whose inner volume, 0.5 mL, was delimited by the working electrode on one side and the counter electrode on the other side. The inlet and the outlet for the solutions were placed on the side walls of the cylinder. The counter electrode was gold foil, and the reference electrode was an Ag/AgCl/sat. KCl placed on the outlet tubing. Both the distribution valve and the cell were designed and realized in the workshop of the Department of Chemistry in Florence [22].

2.2. ECALE Cycles for CdS Deposition

The procedure for CdS deposition is thoroughly described in reference [20]. Briefly, CdS growth was obtained by depositing sulfur at −0.65 V from a Na_2S solution, washing the cell, injecting the cadmium solution while keeping the electrode at the same potential, waiting 60 s to deposit Cd at underpotential, washing the cell, and repeating this cycle as many times as desired. In this paper we performed experiments on CdS deposited with 10 and 30 ECALE cycle. According to the thickness found for CdS deposits obtained with 100 deposition cycles (20–25 nm) [24], the corresponding thicknesses should range between 2 and 7.5 nm.

2.3. Self-Assembly of Alkanethiols on Ag(III) and on CdS-Covered Ag(III)

For the attainment of full layers formation, both substrates were immersed in a 0.3 mm solution of alkanethiol in pure ethanol for at least 12 h. To avoid formation of thiols multilayers the substrates were later kept for 1 h in pure ethanol. The effect of shorter treatments will be shown in the results section. Alternatively, more dilute thiol solutions (0.03 mm) with longer times of modifications (more than 18 h) were used.

2.4. EQCM Measurement

EQCM measurements were carried out using the basic instrument supplied by Seiko EG&EG (QCA917). The working electrode for the EQCM measurements was a 9 MHz AT-cut quartz crystal with silver electrode furnished by Ditta Nuova Mistral (Latina, Italy). The diameter of the quartz crystal was 14.0 mm, and the silver electrode diameter was 7.4 mm. The area of the working electrode in contact with solution was limited to 0.43 cm^2 by an O-ring. The silver electrode on the crystals consisted of 300 nm Ag sputter deposited on an adhesion layer of 50 nm Ti. A suitable flow-cell entirely made of Teflon was designed and realized in the workshop of the Department of Chemistry in Florence. The counter electrode was gold foil, and the reference electrode was an Ag/AgCl sat KCl placed on the outlet tubing of the cell.

2.5. Auger Electron Spectroscopy (AES) Analysis

AES measurements were performed introducing the ex situ prepared samples, in an ultra-high vacuum chamber equipped by a cylindrical mirror analyzer (CMA) with a coaxial electron gun. The spectra were taken in the counting mode. The spectra in the range of 50–600 eV measure the Auger features, namely C KLL, S MLL, Cd MNN and Ag MNN, and have been measured by 1000 and 3000 eV primary electron beams.

3. Results

3.1. SAM of $C_{16}SH$ on Ag(III)

In Figure 1a a cyclic voltammogram (CV) of 0.1 mm $C_{16}SH$ on Ag(III) electrodes using 0.1 M NaOH in 95% ethanol and 5% water as recorded from -0.8 V to -1.6 V, after keeping the electrode in the solution for 15 min a second voltammogram was recorded. The second CV overlaps completely with the first, indicating that this time is insufficient to produce significant variation on the amount of adsorbed $C_{16}SH$. However, even this reduced adsorption time is sufficient to strongly reduce the capacitive current as shown by the flattening of the cyclic voltammogram in the supporting electrolyte (inset of Figure 1a). Moreover, a significant inhibition of the $Ru(NH_3)_6^{3+}$ reduction is also shown. Figure 1b shows the cyclic voltammograms obtained from 1mm $Ru(NH_3)_6^{3+}$ in 0.1 M KCl on the bare Ag(III) (curve a), and on Ag(III) covered by $C_{16}SH$ deposited from a 0.03 mm solution for 10 min (curve b), 2 h (curve c) and 26 h (curve d). Similar dependence on the adsorption time was observed for the cyclic voltammograms of $Fe(CN)_6^{3-/4-}$ on a gold electrode coated by thiols [34]. This behavior is consistent with the kinetic model of thiol adsorption proposed by Nuzzo and coworkers [35].

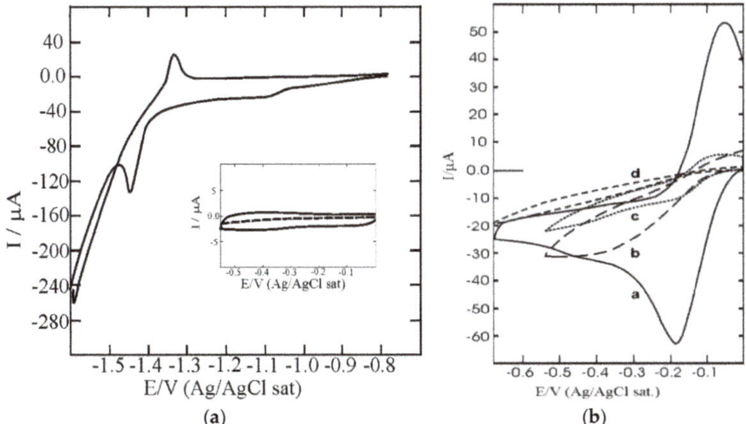

Figure 1. (**a**) Cyclic voltammograms of 0.1 mm $C_{16}SH$ on Ag(III) electrodes using 0.1 M NaOH in 95% ethanol and 5% water as recorded from -0.8 V to -1.6 V/Ag/AgCl sat; inset: cyclic voltammogram of a bare Ag(III) (solid curve) and on Ag(III) modified for 15 min in $C_{16}SH$ solution (dashed curve); (**b**) cyclic voltammograms obtained from 1 mm $Ru(NH_3)_6^{3+}$ in 0.1 M KCl on the bare Ag(III) (curve a), and on Ag(III) covered by $C_{16}SH$ deposited from a 0.03 mm solution for 10 min (curve b), 2 h (curve c) and 26 h (curve d). The scan rate was 50 mV/s.

It must be noted that the curves in Figure 1b were obtained from different experiments, and not in succession. This means that the auto-assembling process was not disturbed at all. Analogous experiments carried out in succession showed that the same degree of inhibition as that shown by curve d is reached after three consecutive 10 min treatments of the electrode.

Unfortunately, the reduction peak of $C_{16}SH$ after SAM formation takes place at potentials similar to those of Figure 1a. The partial overlap with hydrogen evolution hinders the accurate estimation of the charge involved, and hence of the amount of $C_{16}SH$. Alternatively, the amount of the deposited $C_{16}SH$ can be obtained with electrochemical quartz crystal microbalance (EQCM) measurements by measuring the mass decrease involved in thiol dissolution. Systematic EQCM measurements were performed at increasingly longer modification times. The associated mass variation increases up to reach a limiting value for times greater than 12 h. Repeated washings with ethanol are also necessary to avoid formation of multilayers. Curve a in Figure 2 shows the frequency variation obtained while recording a cyclic voltammogram from −0.1 to −1.8 V. Keeping in mind that a frequency increase corresponds to a mass decrease, it is easy to verify that the SAM layer dissolution starts at −1.25 V. However, in spite of the very low scan rate (2 mV/s), and the very negative final potential, the dissolution process is not yet completed in the forward scan but continues during the backward scan. Curve b in Figure 2 is obtained on a silver electrode not covered by SAM. Both curves were recorded in a 0.1 M NaOH in 95% ethanol and 5% water. By comparison with curve a, the entire frequency variation of the modified electrode seems to be ascribable to thiol dissolution. The mass variation, Δm, is given by the Sauerbrey equation:

$$\Delta f = -\frac{C_f \Delta m}{A} \tag{2}$$

where Δf is the frequency variation, A is the electrode area, and C_f is a coefficient that depends on the quartz properties and on the fundamental resonance frequency. In our case, $C_f = 0.183$ Hz* cm^2/ng, then, the frequency variation involved in curve a of Figure 2 (about 200 Hz for a mass coverage of 3.9 $Å^2$/molecule) gives a mass variation higher than that deduced from the molecular coverage, 18.5 $Å^2$/molecule, found for alkanethiols on Ag(III) [1] and must be considered roughly approximate. The EQCM measurements were performed using commercial polycrystalline silver electrodes (there is no Ag monocrystalline (111) electrode for EQCM that are commercially available) with a very high roughness factor, with the direct consequence of a larger real area respect to the geometrical one. Therefore, the experimental mass variation involved in surface phenomena such as self-assembling is scarcely indicative of what happens on a single crystal. Therefore, rather than to calculate the exact amount of the deposited $C_{16}SH$, EQCM measurements were performed to establish the time necessary for the complete formation of the SAM, and to constitute a reference for the analogous measurements carried out in the presence of the CdS film, the measured amount of SAM in the two cases are consistent meaning that the system is reproducible even if not approximal to an ideal flat surface.

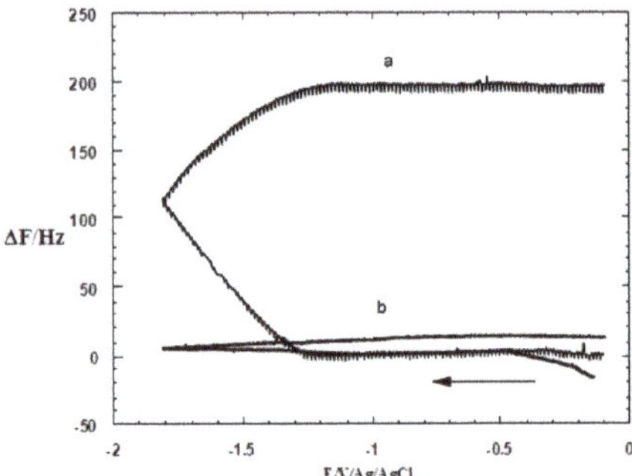

Figure 2. Frequency change during a single potential scan from −0.1 to 1.8 V recorded on the silver EQCM electrode modified with $C_{16}SH$ for times greater than 12 h (curve a) and on a silver electrode not covered by SAM (curve b). Both curves were recorded in a 0.1 M NaOH in 95% ethanol and 5% water. The scan rate was 2 mV/s.

The slowness of desorption process is confirmed by successive experiments: the silver electrode was modified for 16 h. Then, it was repeatedly treated at E = −1.8 V up to a total time of 8 min. The effect of each potential treatment was verified on a typical surface phenomenon such as the underpotential deposition of sulfur. Figure 3a shows S UPD on a bare Ag(III) electrode, whereas curves a, b and c of Figure 3b are obtained after each potential treatment. The S UPD peak does not increase with further potential treatments. The attainment of a limiting behavior, together with the quasi-complete disappearance of the thiol reduction peak in NaOH 0.1 M, suggests the quasi-complete desorption of the thiol SAM layer. This experiment leads to two considerations. First, it confirms the incomplete dissolution during the forward scan of Figure 2; in fact, at a scan rate of 2 mV/s, it takes about 5 min to cover the potential range −1.25 V ÷ −1.8 V. Secondly, the partial inhibition of S UPD indicates that traces of thiol are still present, even though the EQCM data suggest the complete dissolution.

It is interesting to note that the traces of thiol that are sufficient to inhibit surface phenomena don't significantly affect redox processes occurring in solution. To this purpose, we performed another experiment: the electrode was modified for 2 h, and subsequently treated at E = −1.6 V for 3 min and then for further 2 min (total time 5 min). Both the shorter time of modification and less negative applied potential were chosen to better scale the experimental results. Then, the UPD of S (Figure 3c) and the reduction curve of $Ru(NH_3)_6^{3+}$ (Figure 3d) were alternatively recorded. The dashed curves are recorded on the bare Ag(III) electrode; curves a′ and a″ on the electrode modified for 2 h; curves b′ and b″ after the first treatment at −1.6 V, and curves c′ and c″ after the second treatment at −1.6 V. From Figure 3d it is evident that curve c″ is not far from the dashed curve, thus indicating that most of the thiol has been removed. At the same time, curve c′ of Figure 3c is much lower than the dashed one, thus again indicating that traces of thiols are sufficient to inhibit surface phenomena such as underpotential depositions, providing to an over estimation of the thiol still present. This result is in good agreement with the observation that *n*-alkanethiols with *n* > 6 could remain physisorbed on the terraces [36]. However, our results seem to contradict the possibility of completely removing the physisorbed micelles during electroreduction, at least for $C_{16}SH$.

The important findings of this experiment are that the criterion for SAM layer formation cannot be based on surface limited phenomena (UPD), because of the interactions with other species on the surface but must be based on the inhibition of the redox processes (Ru^{3+}/Ru^{2+}) occurring in solution, which seems be related only to the area actually available for the reaction. More precisely, curve d in Figure 1b corresponds to the limiting shape of the inhibited process. Note that curve c in Figure 1b, albeit not far from curve d, shows the typical inflection that is characteristic of a SAM layer not yet completed [37].

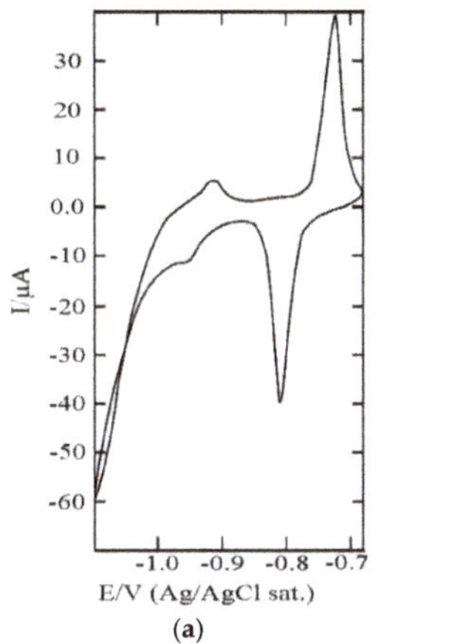

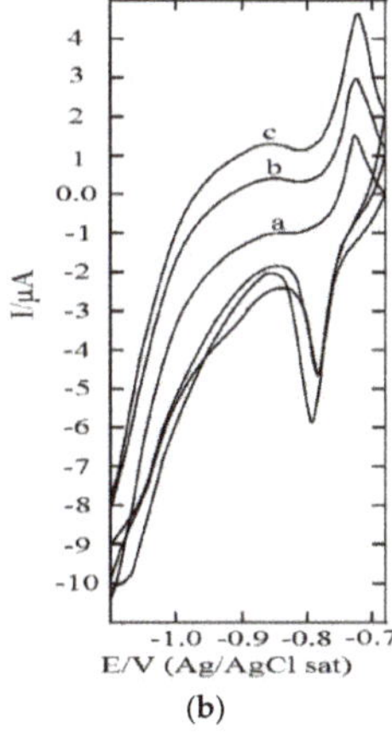

Figure 3. *Cont.*

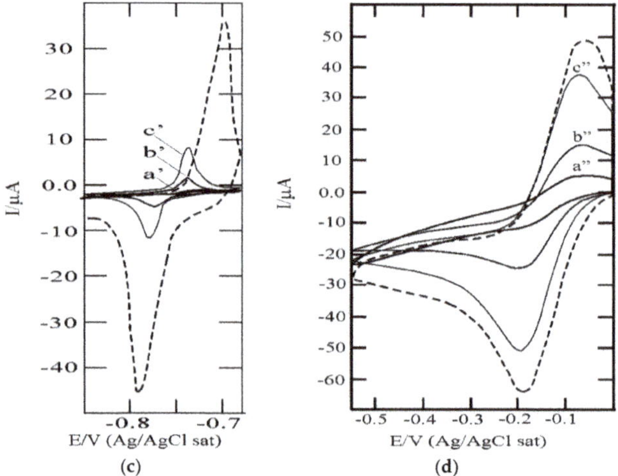

Figure 3. (**a**) Oxidative UPD of S on Ag(III) from 1 mm Na_2S, as recorded from −1.1 to −0.68 V; (**b**) oxidative UPD of S on Ag(III) modified with $C_{16}SH$ for 16 h and successively treated at at E = −1.8 for 1 min (curve a), 3 min (curve b) and 4 min (curve c) up to a total time of 8′; (**c**) oxidative UPD of S on Ag(III) modified with $C_{16}SH$ for 2 h (curve a′) and successively treated at E = −1.6 V for 3 min (curve b′) and further 2 min (curve c′) up to a total time of 5 min; as comparison, the dashed curve is the curve on the bare Ag(III); (**d**) cyclic voltammograms obtained from 1 mm $Ru(NH_3)_6{}^{3+}$ in 0.1 M KCl on Ag(III) modified with $C_{16}SH$ for 2 h (curve a″) and successively treated at E = −1.6 V for 3 min (curve b″) and further 2 min (curve c″) up to a total time of 5 min; as comparison, the dashed curve is the curve on the bare Ag(III). All curves were recorded in pH 9.2 ammonia buffer solutions. The scan rate was 50 mV/s.

3.2. SAM of $C_{16}SH$ on Ag(III) Covered by an Ultrathin Film of CdS

The first experiments were performed on Ag(III) covered by 10 ECALE cycles of CdS. As stated in the experimental section, the thickness estimated for this deposit is about 2 ÷ 2.5 nm. Curve a in Figure 4 shows the cyclic voltammogram of $Ru(NH_3)_6{}^{3+}$ as obtained after having modified this substrate with $C_{16}SH$ for about 26 h. This curve is very close to that obtained on the bare Ag(III) (curve b). The apparent lowering is actually due to the unavoidable decrease of the capacitive contribution due to the SAM layer formation. For a comparison, curve c in the figure is the cyclic voltammogram obtained on a Ag(III) covered by $C_{16}SH$, which is the same as curve d in Figure 1b.

3.3. $C_{16}SH$ SAM Formation on CdS

Various evidences for SAM formation were obtained. EQCM measurements were performed on a Ag(III) substrate covered by 10 ECALE cycles of CdS and modified with $C_{16}SH$ in the same experimental conditions as those of Figure 2. Figure 5 shows the comparison between the frequency curves obtained in the presence of CdS (curve a) and in the absence (curve b). The almost similar frequency variation suggests that the same amount of $C_{16}SH$ is deposited on the two different substrates. Incidentally, it is interesting to note that the only difference is in a more favored dissolution process of $C_{16}SH$ in the presence of CdS. The reduction process begins to occur at −0.8 V and it is over at −1.7 V. Then, the dissolution process is completed during the forward scan. This observation is in good agreement with the hypothesis of a contrasting behavior, exercised by the presence of CdS, on surface reactivity limited by $C_{16}SH$. Other evidences for the SAM formation were given by AES measurements. Table 1 reports the film thickness of CdS and $C_{16}SH$ as estimated by the analysis of the attenuated AES signal amplitudes [38,39] of substrates using the electron mean free path database of NIST [40].

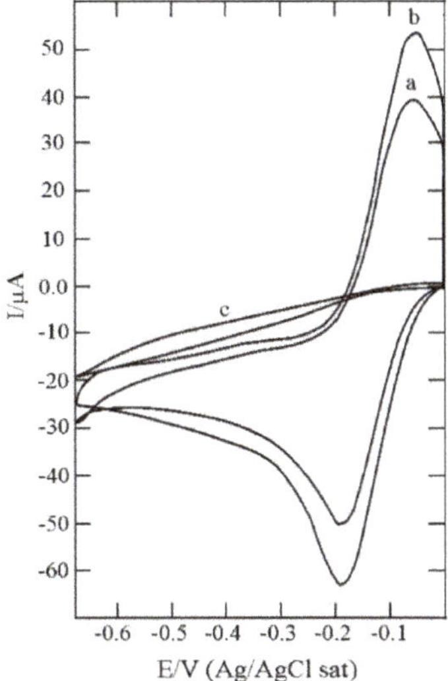

Figure 4. Cyclic voltammograms obtained from 1 mm $Ru(NH_3)_6^{3+}$ in 0.1 M KCl on Ag(III) covered by 10 ECALE cycles and modified with $C_{16}SH$ for 26 h (curve a), on bare Ag(III) (curve b), and on Ag(III) modified with $C_{16}SH$ in the absence of CdS (curve c).The scan rate was 50 mV/s.

Table 1. Film thicknesses of CdS and $C_{16}SH$ as estimated by the analysis of the attenuated AES signal amplitudes using the electron mean free path database of NIST.

Sample	Total Thickness (Å)	CdS Thickness (Å)	$C_{16}SH$ Thickness (Å)
30 CdS/Ag(III)	85 + 7	75 + 10	-
$C_{16}SH$/Ag(III)	25 + 5	-	25 + 5
$C_{16}SH$ + 10CdS/Ag(III)	40 + 6	20 + 3	20 + 3
$C_{16}SH$ + 30CdS/Ag(III)	88 + 8	68 + 7	20 + 3

The thickness values estimated for the CdS films are in good agreement with the preceding measurements [24]. It must be noted that the $C_{16}SH$ SAM layer thickness on CdS is lower than that on Ag(III), thus suggesting a higher tilt angle. The presence of traces of oxygen (1%) localized on the surface and of a small amount of carbon (3%) probably arising from atmospheric contamination were found. The latter value is difficult to be determined since the C signal is superimposed to the silver peak.

The presence of the intact thiol molecule on Ag(III) and pre-covered CdS Ag(III) samples was verified by X-ray absorption measurements on the C K-edge and S $L_{2,3}$ edges and will be discussed in a forthcoming paper.

Both EQCM and AES measurements confirm the presence of the $C_{16}SH$ SAM layer. However, none of them are able to detect details of the maximum coverage and the structure of SAM layer. This latter measurement is provided by the inhibition of the stripping peak of Cd from the CdS deposit covered by $C_{16}SH$. Curve a in Figure 6a is the stripping peak of Cd from CdS deposited on Ag(III) with 10 ECALE cycles. When such a substrate was modified with $C_{16}SH$ for about 18 h, the dissolution of Cd is completely hindered (curve b). For a comparison, curve c refers to only a partial formation of SAM, as obtained when the substrate was modified for insufficiently long times. In this case, the stripping peak is only partially inhibited. The stripping of Cd occurs at potentials progressively more positive as the number of ECALE cycles is increased. Curve a in Figure 6b is the stripping curve of Cd from CdS deposited with 30 ECALE cycles and curve b shows again the blocking effect of the thiol SAM Layer.

It must be noted that the described behavior indicates that the thiol SAM physically blocks the underlying CdS deposits. From this point of view, the possibility that curve a of Figure 4 was simply due to a total or partial SAM removal and that, therefore, the effect of CdS was only apparent, has to be disregarded. In fact, any partial Cd dissolution from the underlying CdS deposit should distort the reduction curve of $Ru(NH_3)_6^{3+}$. As an example, Figure 7 shows the cyclic voltammogram of $Ru(NH_3)_6^{3+}$ on a silver electrode covered by 30 ECALE cycles of CdS not protected by the thiol. The slightly positive current at the initial potential E = −0.08 V indicates that a small amount of Cd is dissolved, what was reasonably expected on the basis of curve a of Figure 6b. This amount of dissolved Cd is again reduced during the negative scan, giving rise to the small bump at about −0.33 V.

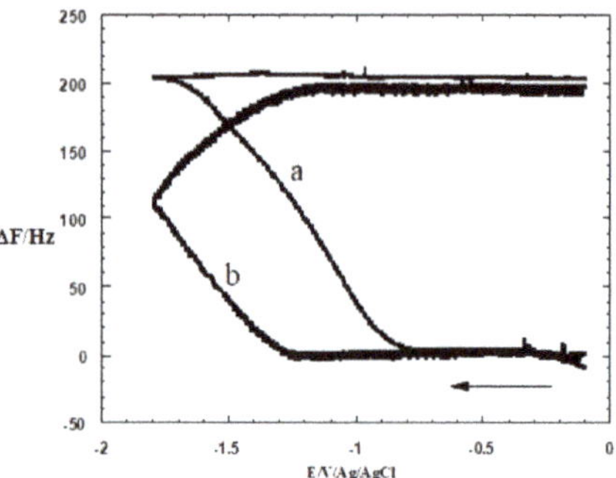

Figure 5. Frequency change during a single potential scan from −0.1 to 1.8 V recorded on the silver EQCM electrode modified with $C_{16}SH$ for times greater than 12 h in presence of CdS (curve a) and in the absence (curve b). The scan rate was 2 mV/s.

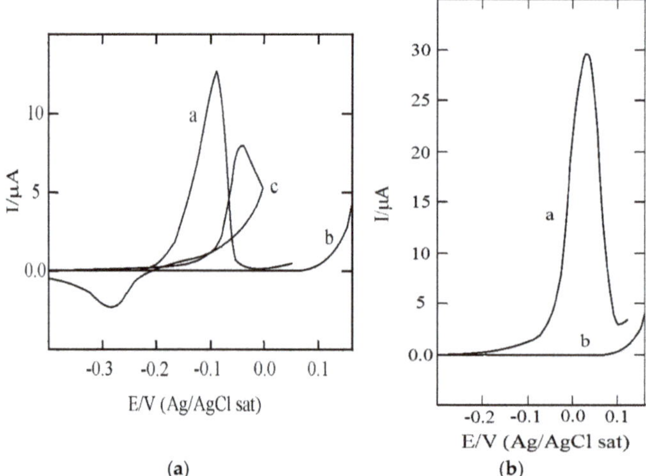

Figure 6. (a) Linear sweep voltammograms for the oxidative strippings of Cd from CdS deposited with 10 ECALE cycles on Ag(III) not covered by thiol (curve a); modified for about 18 h (curve b) and modified for short times (curve c); (b) linear sweep voltammograms for the oxidative strippings of Cd from CdS deposited with 30 ECALE cycles on Ag(III) not covered by thiol (curve a); modified for about 18 h (curve b). The scan rate was 10 mV/s.

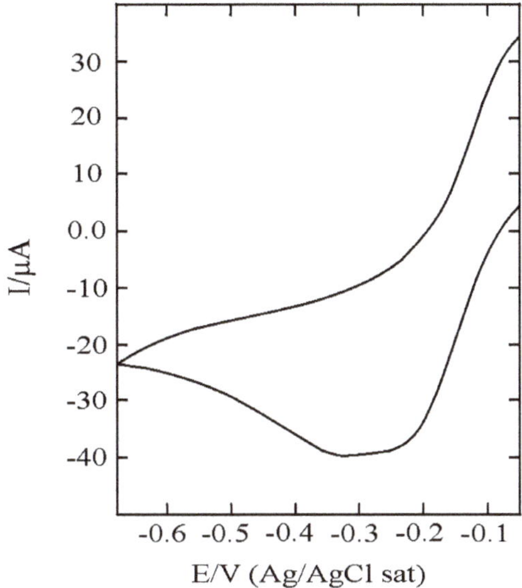

Figure 7. Cyclic voltammograms obtained from 1 mm $Ru(NH_3)_6{}^{3+}$ in 0.1 M KCl on Ag(III) covered by 30 ECALE cycles in the absence of thiol. The scan rate was 50 mV/s.

3.4. Estimate of the SAM Layer Quality

An exhaustive discussion on the criteria used for defect level assessment is given by Miller et al. in reference [12]. Here, electron tunneling through the full thickness of self-assembled organic monolayers of ω–hydroxy thiols is demonstrated. Briefly, the possibility of either small pinholes exposing the electrode surface or collapse sites [21] causing the monolayer to be significantly thinner than the bulk monolayer SAM layer are taken into account. Because of the strong dependence of the electron transfer kinetic on the thickness of the insulating SAM layer, even a small coverage of such defects could be entirely responsible for the current measured. In order to assess the effect of such defects, Miller et al. used the theoretical model of Amatore et al. for the redox kinetics at partially blocked electrodes [37]. According to this model, the presence of pinholes or collapse sites separated by distances greater than the characteristic diffusion length of the experiment (which, in our experiments, is in the order of tens of microns) should give rise to sigmoidal voltammetric waves characteristic of an array of microelectrodes. The behavior as an array of microelectrodes was also obtained after having deliberately perforated monolayers-coated electrodes [41]. As a matter of fact, in our experiments, sigmoidal waves were only observed with insufficient modification time (see for instance curve c of Figure 1b). On the other hand, the presence of defects separated by distances smaller than the characteristic diffusion length should give rise to voltammetric curves indistinguishable in shape from those obtained at electrodes with continuous insulating layer.

Further evidences were obtained by modifying the Ag(III) covered by 30 ECALE cycles of CdS for increasing time, in a way similar to the experiment of Figure 3. The thicker CdS film is necessary to limit Cd dissolution during the recording of $Ru(NH_3)_6{}^{3+}$ cyclic voltammogram. Figure 8a. shows the curves of $Ru(NH_3)_6{}^{3+}$ obtained after having modified the electrode for 20 min (curve a), 30 min (curve b), 90 min (curve c) and 4 h (curve d). Surprisingly, the redox process shows an inhibition effect that increases with the adsorption time. This result is in contrast with curve a of Figure 4 obtained after having modified the substrate for 17 h. However, a further modification of 17 h yielded the dashed curve in Figure 8a. This latter curve, although not coincident with curve a of Figure 4, indicates a reversed trend. This behavior is consistent with a higher level of order ensured by the additional modification time. Then, it is reasonable to assume that the occurring of charge transfer is connected to the order of the SAM rather than to the adsorption time.

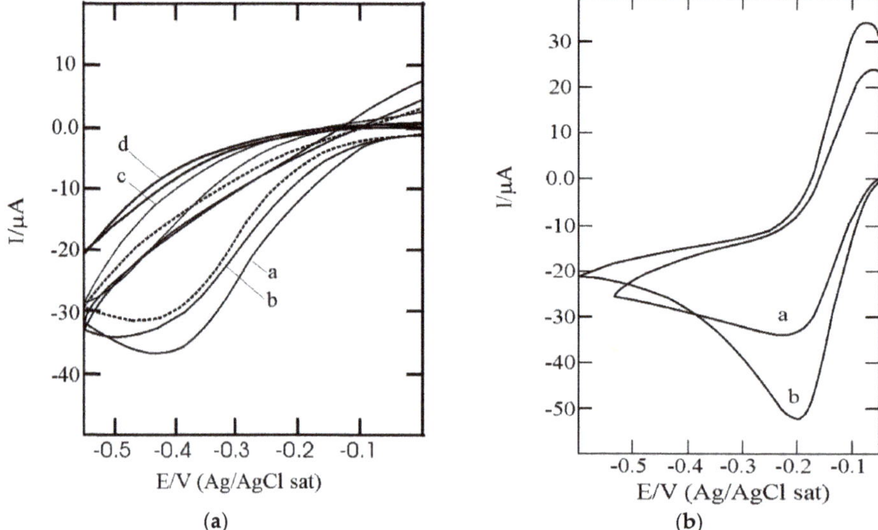

(a)   (b)

Figure 8. (a) Cyclic voltammograms obtained from 1 mm $Ru(NH_3)_6^{3+}$ in 0.1 M KCl on Ag(III) covered by 30 ECALE cycles and modified with $C_{16}SH$ for 20 min (curve a), 30 min (curve b), 90 min (curve c) and 4 h (curve d); the dashed curve was obtained after having further modified the substrate for 17 h; (b) cyclic voltammograms obtained from 1 mm $Ru(NH_3)_6^{3+}$ in 0.1 M KCl on Ag(III) covered by 30 ECALE cycles and modified with $C_{16}SH$ for 3 h (curve a), and more than 12 h (curve b).The scan rate was 50 mV/s.

As a matter of fact, curve a in Figure 8b shows the $Ru(NH_3)_6^{3+}$ reduction peak obtained on Ag(III) covered by 30 ECALE cycles and modified for a time (3 h), which, according to Nuzzo and co-workers [35], is surely insufficient to produce an ordered monolayer. The $Ru(NH_3)_6^{3+}$ reduction is much less inhibited than in the experiments of Figure 8a. At the same time it is much more inhibited that in the presence of a SAM formed on a similar substrate with an adsorption time longer than 12 h (curve b). To avoid Cd dissolution, the initial potential was shifted to -0.05 V. Then, it is reasonable to assume that the necessary condition for charge transfer is that the monolayer is ordered.

4. Conclusions

The behavior of self-assembled monolayers of 1-hexanedecanthiol ($C_{16}SH$) formed on Ag(III) covered by ultrathin films of CdS was compared with that of analogous SAMs formed on the bare Ag(III). The strong insulating effect of $C_{16}SH$ deposited on Ag(III) is shown by the inhibition of the voltammetric peak of $Ru(NH_3)_6^{3+/2+}$. On the contrary, the voltammogram obtained on Ag(III) covered by an ultrathin film of CdS is very similar to that obtained on the bare Ag(III) electrode. The crucial point of our work is to demonstrate the effective formation of $C_{16}SH$ monolayers on Ag(III) covered by CdS. A number of different experiments were performed to confirm the presence of $C_{16}SH$. In particular, EQCM measurements revealed that an identical mass decrease is involved in thiol dissolution both in the presence and in the absence of CdS. Then, AES measurements confirmed the presence of $C_{16}SH$ SAM layer both on Ag(III) and on CdS. The thickness of the SAM layer formed on CdS, 20 Å, is lower than that of the SAM layer formed on Ag(III), 25 Å, thus indicating a higher tilt angle.

Neither EQCM nor AES are able to detect the formation of a complete SAM layer of thiol. However, the evidence for a well-formed SAM layer is given by the physical blocking of the stripping peak of Cd from the CdS deposit covered by $C_{16}SH$. For this reason, only partial inhibition of the Cd stripping peak is observed with SAM layer of $C_{16}SH$ obtained with not sufficiently long adsorption times. Moreover, the presence of the intact thiol molecule on Ag(III) and pre-covered CdS Ag(III) samples was verified by X-ray absorption measurements on the C K-edge and S $L_{2,3}$ edges and will be discussed in a forthcoming paper.

Our results suggest that self-assembly on semiconductors could become interesting whenever a charge transfer is required. As an example, they could be used to anchor molecules that change their properties following an electrochemical reduction or oxidation. The hypothesis that the observed behavior was simply due to a total or partial SAM removal and that, therefore, the effect of CdS was only apparent, has to be disregarded, since in this case the concomitant CdS dissolution should be observed.

While the structure of thiols on Ag(III) is known to give well packed and ordered structures [1,2,16,17,31], none of the measurements presented here allow us to determine the exact structure or degree of crystallinity of the SAM layers obtained in the presence of CdS. Our results on the occurrence of electron transfer are purely phenomenological, since they are limited to the system studied, particularly Ag(III) covered by n-CdS using a reduction process such as that of $Ru(NH_3)_6{}^{3+}$. Any interpretation would require the extension of the measurements to other substrates (for example gold) and to other semiconductors (for example other II–VI or III–V compound semiconductors both *n*-type and *p*-type). Finally, other electrochemical probes, such as $Fe(CN)_6{}^{3-}$ or $Fe(CN)_6{}^{4-}$, should be used to explore possible discrepancies towards reduction or oxidation processes.

Author Contributions: Funding acquisition, S.C.; Investigation, W.G., M.P. (Maurizio Passaponti), L.F. and P.M.; Methodology, M.L.F., S.M. and N.C.; Supervision, M.P. (Maddalena Pedio) and M.I.; Writing-original draft, E.S.

Acknowledgments: The authors are grateful to Ferdinando Capolupo for the preparation of the silver single crystal electrodes. The technical services of TASC laboratory are kindly acknowledged. The financial support of the Italian CNR, of the Murst and of the Consorzio Interuniversitario per la Scienza e Tecnologia (INSTM) is gratefully acknowledged. We are also grateful to Regione Toscana—Programma Operativo Regionale POR FESR Toscana 2014–2020, BANDO N. 2: Progetti di ricerca e sviluppo delle MPMI, Project GADGET "Gioielli in Argento da Galvanica, Ecologica e Tecnologica".

Conflicts of Interest: The authors declare no conflicts of interest.

References

1. Schreiber, R. Structure and growth of self-assembled monolayers. *Prog. Surf. Sci.* **2000**, *65*, 151–256. [CrossRef]
2. Ulman, A. Formation and structure of self-assembled monolayers. *Chem. Rev.* **1996**, *96*, 1533–1554. [CrossRef] [PubMed]
3. Gu, Y.; Lin, Z.; Butera, R.A.; Smentkowski, V.S.; Waldeck, D.H. Preparation of Self-Assembled Monolayers on InP. *Langmuir* **1995**, *11*, 1849–1851. [CrossRef]
4. Gu, Y.; Waldeck, D.H. Studies of electron tunneling at semiconductor electrodes. *J. Phys. Chem.* **1996**, *100*, 9573–9576. [CrossRef]
5. Gu, Y.; Kumar, K.; Read, I.; Zimmt, M.B. Studies into the character of electronic coupling in ectron transfer. *J. Photochem. Photobiol. A* **1997**, *105*, 189–196. [CrossRef]
6. Gu, Y.; Waldeck, D.H. Electron Tunneling at the Semiconductor−Insulator−Electrolyte Interface. Photocurrent Studies of the *n*-InP−Alkanethiol−Ferrocyanide System. *J. Phys. Chem. B* **1998**, *102*, 9015–9028. [CrossRef]
7. Sheen, C.W.; Shi, J.X.; Martensson, J.; Parikh, A.N.; Allara, D.L. A New Class of Organized Self-Assembled Monolayers: Alkane Thiols on GaAs(100). *J. Am. Chem. Soc.* **1992**, *114*, 1514–1515. [CrossRef]
8. Baum, T.; Ye, S.; Uosaki, K. Formation of Self-Assembled Monolayers of Alkanethiols on GaAs Surface with in Situ Surface Activation by Ammonium Hydroxide. *Langmuir* **1999**, *15*, 8577–8579. [CrossRef]

9. Ye, S.; Li, G.; Noda, H.; Uosaki, K.; Osawa, M. Characterization of self-assembled monolayers of alkanethiol on GaAs surface by contact angle and angle-resolved XPS measurements. *Surf. Sci.* **2003**, *529*, 163–170. [CrossRef]

10. Kosuri, M.R.; Cone, R.; Li, Q.; Han, S.M.; Bunker, B.C.; Mayer, T.M. Adsorption kinetics of 1-alkanethiols on hydrogenated Ge(111). *Langmuir* **2004**, *20*, 835–840. [CrossRef] [PubMed]

11. Porter, M.D.; Bright, T.B.; Allara, D.L.; Chidsey, C.E.D. Spontaneously organized molecular assemblies. 4. Structural characterization of *n*-alkyl thiol monolayers on gold by optical ellipsometry, infrared spectroscopy, and electrochemistry. *J. Am. Chem. Soc.* **1987**, *109*, 3559–3568. [CrossRef]

12. Miller, C.; Cuendet, P.; Graetzel, M. Adsorbed .omega.-hydroxy thiol monolayers on gold electrodes: Evidence for electron tunneling to redox species in solution. *J. Phys. Chem.* **1991**, *95*, 877–886. [CrossRef]

13. Miller, C.; Graetzel, M. Electrochemistry at .omega.-hydroxythiol coated electrodes. 2. Measurement of the density of electronic states distributions for several outer-sphere redox couples. *J. Phys. Chem.* **1991**, *95*, 5225–5233. [CrossRef]

14. Becka, A.M.; Miller, C.J. Electrochemistry at .omega.-hydroxy thiol coated electrodes. 3. Voltage independence of the electron tunneling barrier and measurements of redox kinetics at large overpotentials. *J. Phys. Chem.* **1992**, *96*, 2657–2668. [CrossRef]

15. Terrettaz, S.; Becka, A.M.; Traub, M.J.; Fettinger, J.C.; Miller, C.J. .omega.-Hydroxythiol Monolayers at Au Electrodes. 5. Insulated Electrode Voltammetric Studies of Cyano/Bipyridyl Iron Complexes. *J. Phys. Chem.* **1995**, *99*, 11216–11224. [CrossRef]

16. Hatchett, D.W.; Uibel, R.H.; Stevenson, K.J.; Harris, J.M.; White, H.S. Electrochemical measurement of the free energy of adsorption of *n*-alkanethiolates at Ag(111). *J. Am. Chem. Soc.* **1998**, *120*, 1062–1069. [CrossRef]

17. Azzaroni, O.; Vela, M.E.; Andreasen, G.; Carro, P.; Salvarezza, R.C. Electrodesorption potentials of self-assembled alkanethiolate monolayers on Ag(111) and Au(111). An electrochemical, scanning tunneling microscopy and density functional theory study. *J. Phys. Chem. B* **2002**, *106*, 12267–12273. [CrossRef]

18. Lee, T.; Wang, W.; Reed, M.A. Mechanism of Electron Conduction in Self-Assembled Alkanethiol Monolayer Devices. *Ann. N. Y. Acad. Sci.* **2003**, *1006*, 21–35. [CrossRef] [PubMed]

19. Krysiński, P.; Brzostowska-Smolska, M. Three-probe voltammetric characterisation of octadecanethiol self-assembled monolayer integrity on gold electrodes. *J. Electroanal. Chem.* **1997**, *424*, 61–67. [CrossRef]

20. Finklea, H.O.; Robinson, L.R.; Blackburn, A.; Richter, B.; Allara, D.; Bright, T. Formation of an Organized Monolayer by Solution Adsorption of Octadecyltrichlorosilane on Gold: Electrochemical Properties and Structural Characterization. *Langmuir* **1986**, *2*, 239–244. [CrossRef]

21. Finklea, H.O.; Avery, S.; Lynch, M.; Furtsch, T. Blocking oriented monolayers of alkyl mercaptans on gold electrodes. *Langmuir* **1987**, *3*, 409–413. [CrossRef]

22. Innocenti, M.; Pezzatini, G.; Forni, F.; Foresti, M.L. CdS and ZnS Deposition on Ag(111) by Electrochemical Atomic Layer Epitaxy. *J. Electrochem. Soc.* **2001**, *148*, C357–C362. [CrossRef]

23. Cecconi, T.; Atrei, A.; Bardi, U.; Forni, F.; Innocenti, M.; Loglio, F.; Foresti, M.L.; Rovida, G. X-ray photoelectron diffraction (XPD) study of the atomic structure of the ultrathin CdS phase deposited on Ag(111) by electrochemical atomic layer epitaxy (ECALE). *J. Electron Spectros. Relat. Phenom.* **2001**, *114–116*, 563–568. [CrossRef]

24. Innocenti, M.; Cattarin, S.; Cavallini, M.; Loglio, F.; Foresti, M.L. Characterisation of thin films of CdS deposited on Ag(111) by ECALE. A morphological and photoelectrochemical investigation. *J. Electroanal. Chem.* **2002**, *532*, 219–225. [CrossRef]

25. Innocenti, M.; Forni, F.; Pezzatini, G.; Raiteri, R.; Loglio, F.; Foresti, M.L. Electrochemical behavior of As on silver single crystals and experimental conditions for InAs growth by ECALE. *J. Electroanal. Chem.* **2001**, *514*, 75–82. [CrossRef]

26. Foresti, M.L.; Milani, S.; Loglio, F.; Innocenti, M.; Pezzatini, G.; Cattarin, S. Ternary CdS$_x$Se$_{1-x}$ deposited on Ag(111) by ECALE: Synthesis and characterization. *Langmuir* **2005**, *21*, 6900–6907. [CrossRef] [PubMed]

27. Loglio, F.; Innocenti, M.; D'Acapito, F.; Felici, R.; Pezzatini, G.; Salvietti, E.; Foresti, M.L. Cadmium selenide electrodeposited by ECALE: Electrochemical characterization and preliminary results by EXAFS. *J. Electroanal. Chem.* **2005**, *575*, 161–167. [CrossRef]

28. Cavallini, M.; Facchini, M.; Albonetti, C.; Biscarini, F.; Innocenti, M.; Loglio, F.; Salvietti, E.; Pezzatini, G.; Foresti, M.L. Two-dimensional self-organization of CdS ultra thin films by confined electrochemical atomic layer epitaxy growth. *J. Phys. Chem. C* **2007**, *111*, 1061–1064. [CrossRef]

29. Wang, L.; Bevilacqua, M.; Chen, Y.X.; Filippi, J.; Innocenti, M.; Lavacchi, A.; Marchionni, A.; Miller, H.; Vizza, F. Enhanced electro-oxidation of alcohols at electrochemically treated polycrystalline palladium surface. *J. Power Sources* **2013**, *242*, 872–876. [CrossRef]

30. Gregory, B.W.; Stickney, J.L. Electrochemical atomic layer epitaxy (ECALE). *J. Electroanal. Chem. Interfacial Electrochem.* **1991**, *300*, 543–561. [CrossRef]

31. Azzaroni, O.; Schilardi, P.L.; Salvarezza, R.C. Metal electrodeposition on self-assembled monolayers: A versatile tool for pattern transfer on metal thin films. *Electrochim. Acta* **2003**, *48*, 3107–3114. [CrossRef]

32. Foresti, M.L.; Capolupo, F.; Innocenti, M.; Loglio, F. Visual Detection of Crystallographic Orientations of Face-Centered Cubic Single Crystals. *Cryst. Growth Des.* **2002**, *2*, 73–77. [CrossRef]

33. White, R.E.; Bockris, J.O.; Conway, B.E. *Modern Aspects of Electrochemistry*; Springer: New York, NY, USA, 1995; ISBN 978-1-4899-1726-3.

34. Diao, P.; Jiang, D.; Cui, X.; Gu, D.; Tong, R.; Zhong, B. Studies of structural disorder of self-assembled thiol monolayers on gold by cyclic voltammetry and ac impedance. *J. Electroanal. Chem.* **1999**, *464*, 61–67. [CrossRef]

35. Bain, C.D.; Troughton, E.B.; Tao, Y.T.; Evall, J.; Whitesides, G.M.; Nuzzo, R.G. Formation of monolayer films by the spontaneous assembly of organic thiols from solution onto gold. *J. Am. Chem. Soc.* **1989**, *111*, 321–335. [CrossRef]

36. Vericat, C.; Andreasen, G.; Vela, M.E.; Martin, H.; Salvarezza, R.C. Following transformation in self-assembled alkanethiol monolayers on Au(111) by in situ scanning tunneling microscopy. *J. Chem. Phys.* **2001**, *115*, 6672–6678. [CrossRef]

37. Amatore, C.; Savéant, J.M.; Tessier, D. Charge transfer at partially blocked surfaces: A model for the case of microscopic active and inactive sites. *J. Electroanal. Chem. Interfacial Electrochem.* **1983**, *147*, 39–51. [CrossRef]

38. Seah, P.M.; Dench, A.W. Quantitative electron spectroscopy of surfaces: A standard data base for electron inelastic mean free paths in solids. *Surf. Interface Anal.* **1979**, *1*, 2–11. [CrossRef]

39. Woodruff, D.P.; Delchar, T.A. *Modern Techniques of Surface Science*; Cambridge Solid State Science Series; Cambridge Univercity Press: Cambridge, UK, 1986.

40. Powell, C.J.; Jablonski, A. *NIST Electron Inelastic-Mean-Free-Path Database*; U.S. Department of Commerce: Washington, DC, USA, 2010.

41. Finklea, H.O.; Snider, D.A.; Fedyk, J.; Sabatani, E.; Gafni, Y.; Rubinstein, I. Characterization of Octadecanethiol-Coated Gold Electrodes as Microarray Electrodes by Cyclic Voltammetry and ac Impedance Spectroscopy. *Langmuir* **1993**, *9*, 3660–3667. [CrossRef]

Article

X-ray Absorption under Operating Conditions for Solid-Oxide Fuel Cells Electrocatalysts: The Case of LSCF/YSZ

Francesco Giannici [1],*, Giuliano Gregori [2], Alessandro Longo [3,4], Alessandro Chiara [1], Joachim Maier [2] and Antonino Martorana [1]

[1]   Dipartimento di Fisica e Chimica, Università di Palermo, Viale delle Scienze Ed. 17, 90128 Palermo, Italy; alessandro.chiara31@gmail.com (A.C.); antonino.martorana@unipa.it (A.M.)

[2]   Max Planck Institute for Solid State Research, 70569 Stuttgart, Germany; giulianogregori@hotmail.com (G.G.); s.weiglein@fkf.mpg.de (J.M.)

[3]   Netherlands Organization for Scientific Research (NWO), Dutch-Belgian Beamline, ESRF—The European Synchrotron, CS40220, 38043, 71 Avenue des Martyrs, 38000 Grenoble, France; alessandro.longo@cnr.it

[4]   Istituto per lo Studio dei Materiali Nanostrutturati (ISMN)-CNR, UOS Palermo, Via Ugo La Malfa, 153, 90146 Palermo, Italy

*   Correspondence: francesco.giannici@unipa.it

Received: 5 December 2018; Accepted: 2 January 2019; Published: 8 January 2019

Abstract: We describe a novel electrochemical cell for X-ray absorption spectroscopy (XAS) experiments during electrical polarization suitable for high-temperature materials such as those used in solid oxide fuel cells. A half-cell LSCF/YSZ was then investigated under cathodic and anodic conditions (850 °C and applied electrical bias ranging from +1 V to −1 V in air). The in situ XAS measurements allowed us to follow the LSCF degradation into simple oxides. The rapid deterioration of LSCF is ascribed to the formation of excess of oxygen vacancies leading to the collapse of the mixed perovskite structure.

Keywords: SOFC; cathode; XAFS; in situ

---

1. Introduction

Solid-oxide fuel cells (SOFC) represent a forefront theme in materials research, since they allow the production of electrical energy from hydrogen or hydrocarbon fuels with very high thermodynamic efficiency. SOFC electrodes act as catalysts (for oxidation of the fuel at the anode, and for reduction of oxygen at the cathode), and as mixed electron/ion conductors. Although a great deal of performance testing is routinely carried out on a wide variety of electrocatalysts to optimize their composition and microstructure, advanced structural characterization studies have been relatively uncommon so far. In this paper, we studied, under operating conditions (at high temperature and under an external electrical potential), the atomic and electronic structure of $La_{0.6}Sr_{0.4}Co_{0.8}Fe_{0.2}O_3$ (LSCF), which is a technologically relevant electrode material for high-temperature and intermediate-temperature SOFC. This material has been the subject of a large number of studies in the last several decades.

Kawada et al. [1] reported on the chemical stability of LSCF, ultimately concluding that it is stable at high temperatures and under different atmospheres. However, cation segregation on one surface was observed when an oxygen partial pressure gradient was applied to LSCF, which resulted in Co-rich and Sr-rich precipitates on the surface exposed to higher $P(O_2)$.

In general, few examples of characterization in operating conditions have been reported in the literature to date, regarding both in-house and large-scale facilities techniques. Among the few studies in the literature on this topic, Baumann et al. [2] observed a strong activation for the oxygen reduction reaction in an LSCF/YSZ model system, associated with a steep decrease of the resistivity of LSCF, due to an applied electrical potential (either cathodic or anodic) of the order of several volts. This effect was found to last for several hours after the polarization, which suggests a more profound transformation rather than a simple change of oxidation state. X-ray photoelectron spectroscopy (XPS) performed on films subject to different treatments were found to be enriched with Sr and Co on the surface.

An ever-increasing interest is developing at synchrotrons around the world to improve in situ/operando experimental setups in order to control the chemical/physical environment and the electrical loading applied to the sample under the beam. This allows for accurately monitoring the chemical changes in technologically relevant materials under service conditions. This approach was used, for instance, to study several SOFC cathode-electrolyte-air triple boundary surfaces using photoelectron microscopy under polarization [3]. More recently, Nenning et al. [4] examined the electrochemistry and XPS spectra of lanthanum strontium ferrite (LSF) and lanthanum strontium cobaltite (LSC) thin film electrodes over YSZ, during polarization (around 0.5 V either cathodic or anodic) and in different atmospheres (H_2 or O_2). Given the surface-sensitive conditions of such an XPS investigation, a pronounced difference was found in the behavior of cobalt compared to all other elements, which was attributed by the authors to the higher reducibility of the Co ions. In fact, in an earlier soft X-ray absorption spectroscopy study of LSCF under different oxygen partial pressures at 800 °C, it was already found that the Co absorption edge is modified by the oxygen non-stoichiometry, while the Fe edge is not [5].

Among synchrotron techniques, X-ray absorption spectroscopy allows one to obtain information about the atomic and electronic structure of the samples, so that it is particularly suited for SOFC electrode materials, often constituted by variable valence elements that can modify their oxidation state and the local structure in dependence of reaction environment and applied potential [6–9]. Regarding this concern, the development of reliable experimental protocols [10] is essential to obtain significant information on SOFC materials [11–14]. Among the issues addressed by the above cited authors, Lai et al. investigated catalyst poisoning, arguing that deactivation of an LSCF cathode by CO_2 is enhanced under a cathodic bias. Skinner et al. observed that cathodes constituted by Ruddlesden-Popper phases did not undergo significant redox phenomena under polarization. Irvine et al. studied the processes occurring at the electrode-electrolyte interface, and, in particular, assessed the reversible exsolution phenomena occurring during the operation.

We have recently developed an electrochemical cell for the X-ray structural characterization of SOFC electrocatalysts under operating conditions: high temperature (up to 800 °C), controlled atmosphere, and electrical loading. To this aim, we modified a thermochemical cell for in-situ X-ray spectroscopy with an electrically insulating quartz sample holder, and current collectors in order to impose a voltage bias on the sample.

In this paper, we present our first results with this setup on a LSCF/YSZ cathode/electrolyte half-cell. By applying an external voltage bias, ranging from −1 V to +1 V, and employing X-ray absorption spectroscopy at a high temperature, we were able to investigate the different role of iron and cobalt concerning the oxidation state and structural rearrangement, as a function of time, temperature, and electrical polarization in operating conditions.

2. Materials and Methods

A dense $La_{0.6}Sr_{0.4}Co_{0.8}Fe_{0.2}O_3$ (LSCF) film of 200 nm thickness was deposited by pulsed laser deposition on yttria-stabilized zirconia (100) oriented crystal (YSZ, 9.5 mol% Y_2O_3, 5 mm × 5 mm × 0.5 mm, CrysTec GmbH, Berlin, Germany) from a commercial LSCF target (HITEC, Karlsruhe, Germany). The deposition was carried out with a laser pulse energy of 1.6 J/cm^2 and a pulse frequency of 5 Hz, which keeps the substrate temperature at 770 °C under an oxygen partial pressure of 0.4 mbar. After deposition, the films were annealed at 640 °C for 30 min in oxygen.

Platinum current collectors were deposited by sputtering on opposite sides of the bilayer, covering half of the surface, and enabling the other half to be studied in the spectroscopy experiment (see Figure 1). The sample was mounted on a fused quartz sample holder inside the "Microtomo" furnace developed at the European Synchrotron Radiation Facility (Grenoble, France) for in situ X-ray absorption spectroscopy. The voltage across the electrode-electrolyte bilayer is applied through an external I-V source, through two platinum wires ending with platelets, which are brought in contact with the current collectors on the sample. For this experiment, a constant electrical bias was applied to the half-cell, which ranged from −1 V (cathodic) to +1 V (anodic). In this convention, "cathodic" refers to a negative polarization of the LSCF side electrode with respect to the counter-electrode on the YSZ side (oxygen ions flowing from LSCF to YSZ).

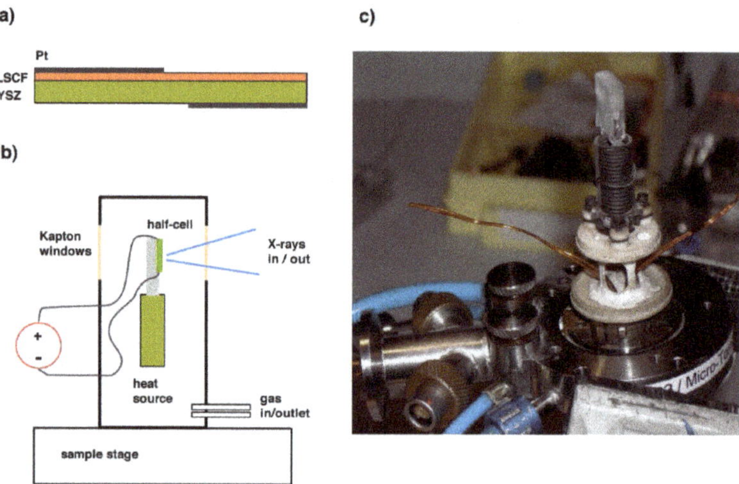

Figure 1. (**a**) LSCF/YSZ half-cell. (**b**) Schematic of the X-ray absorption spectroscopy setup. (**c**) Inner view of the cell.

A steady air flux flowed inside the cell during the measurement. In this cell, the sample holder can be mounted at either 45 degrees, or parallel to the incoming beam, which allows X-ray absorption spectroscopy in a transmission or a fluorescence mode. In particular, the X-ray absorption near edge structure (XANES) data presented in this paper were acquired at the Fe K-edge (7112 eV) and at the Co K-edge (7709 eV) in a fluorescence mode using a multi-element solid state Ge detector.

3. Results

During the heating of LSCF from RT to 800 °C, the Co K-edge shows a gradual variation, which is evident from Figures 2 and 3. In particular, from Figure 3, where the intensity ratio between the features at 7729 eV and 7734 eV is plotted, we get an impression how continuous the evolution of the XANES spectrum is with the temperature, without evident discontinuities due to the known rhombohedral to a cubic structural transition below 500 °C [15]. It is likely that the progressive changes in the edge are linked to the structural rearrangement, but also to the expected increase of oxygen vacancies concentration with the temperature.

The applied electrical bias as a function of time is plotted in Figure 4. The position of the absorption K-edge of cobalt (i.e., its oxidation state) is plotted in the same way, and it follows almost instantaneously the bias changes (Figure 5). On the other hand, it can be observed in the plot that, within the first cycle of cathodic/anodic polarization, there is a continuous transformation of the edge that takes about 30 min, which suggests a diffusive process and/or a chemical transformation different from a simple change of oxidation state. One can appreciate that, after an initial polarization at +1 V leading to a slight shift towards a larger Co K-edge value and an abrupt change to −1 V yielding to a reduction of Co (lower values of the Co K-edge), the return to +1 V does not bring the

Co K-edge to the initial energy values. Although the energy difference is rather small, this might be indicative of irreversible processes occurring in the LSCF layer after only 60 min. This is supported by the subsequent voltage step. In this case, during the application of −1 V, the edge energy does not reach the initial values, which suggests that the LSCF film already underwent some permanent structural modifications.

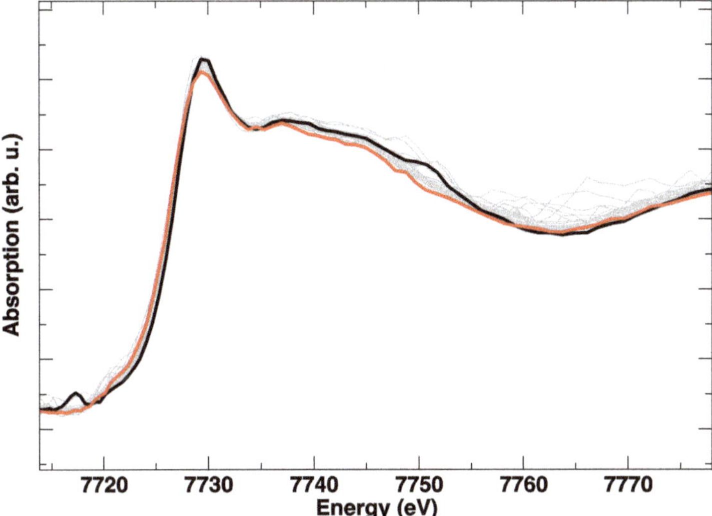

Figure 2. Co K-edge XANES spectra during heating, from room temperature (black line) to 800 °C (red line). Intermediate spectra are plotted in grey.

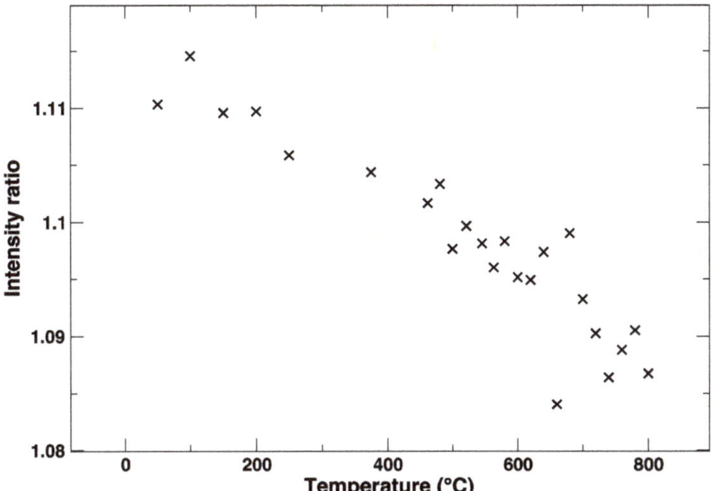

Figure 3. Intensity ratio between the Co K-edge XANES maximum at 7729 eV and the local minimum at 7734 eV.

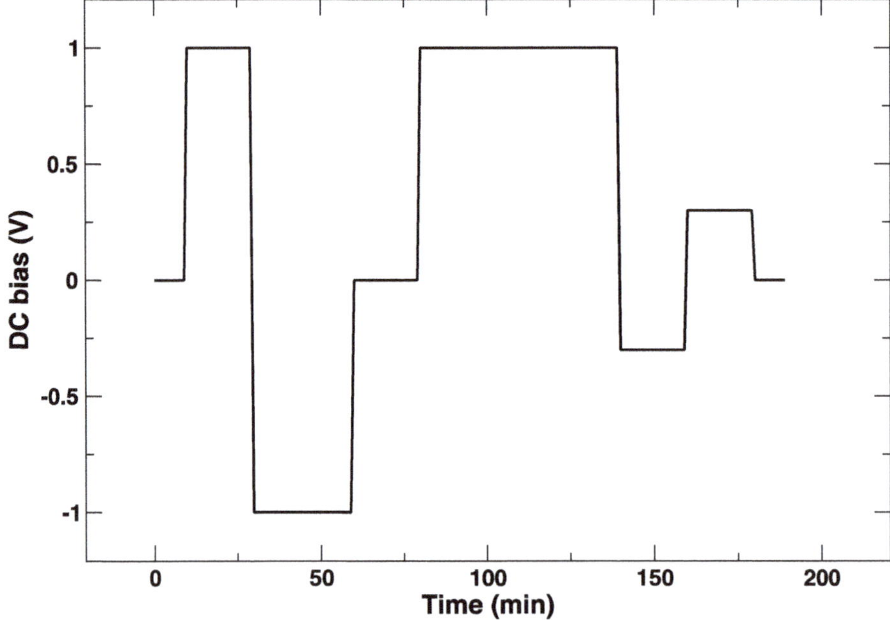

Figure 4. DC bias applied during the in situ Co K-edge XANES experiment at 850 °C.

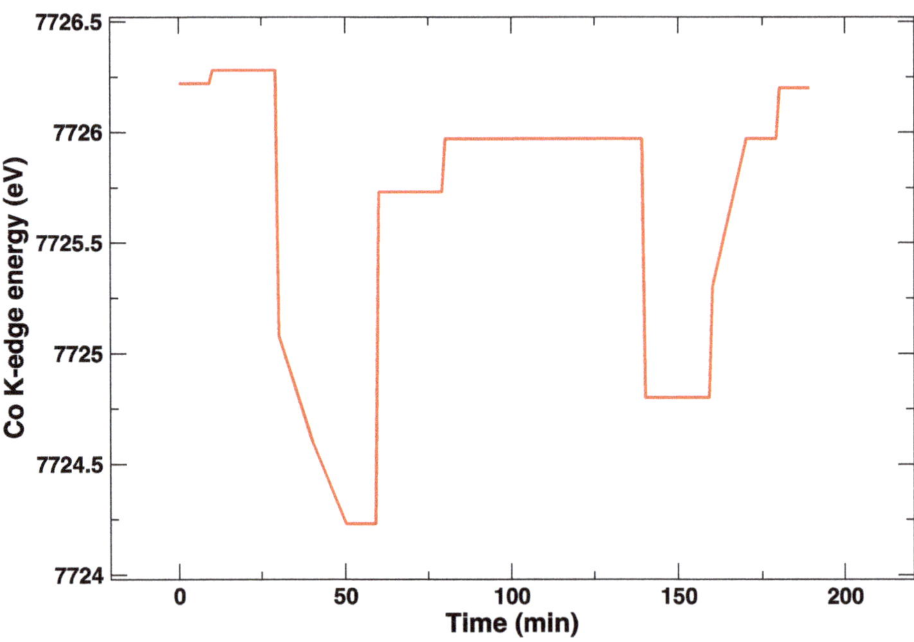

Figure 5. Energy of the Co K-edge, measured as the maximum of the first derivative of the XANES spectrum, as a function of time at 850 °C.

These edge transformations show a clear irreversibility, which is evident when comparing the red spectrum in Figure 6 and corresponds to the initial anodic bias and the final (green) spectrum, taken in anodic polarization, but after the cathodic step. The postmortem Co K-edge resembles that of the spinel Co_3O_4 [15]: it is completely different from the starting LSCF and this modification is maintained irrespective of the applied potential. The room temperature spectra, taken before and after polarization cycles, in Figure 7, confirm with higher resolution the final transformation of LSCF.

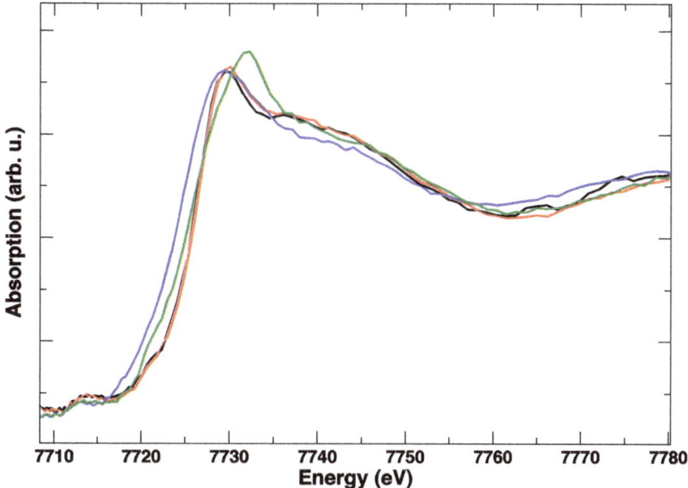

Figure 6. Representative Co K-edge XANES spectra at 850 °C during two cycles of the electrical polarization experiment. No applied bias (black), +1 V (red), −1 V (blue), +1 V (green).

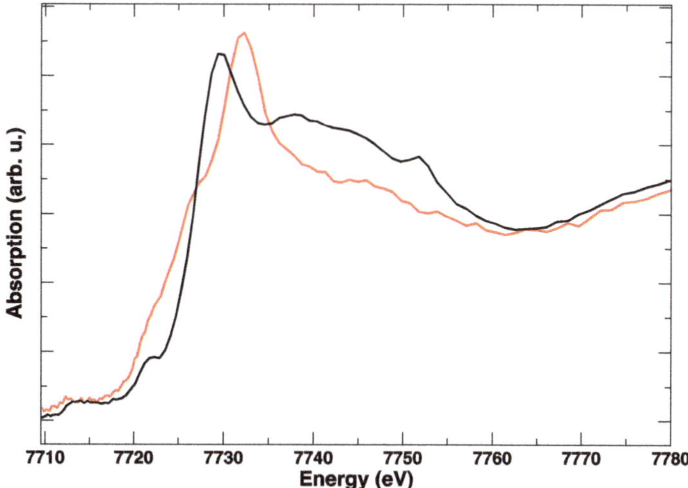

Figure 7. Room temperature Co K-edge XANES spectra of LSCF before treatment (black line) and after the in situ electrical polarization experiment (red line).

A subsequent polarization cycle (Figure 8) with simultaneous acquisition of the Fe K-edge spectra highlights that: (1) the Fe edge is almost unchanged in response to the electric bias, (2) the spectrum is different from LSCF, and resembles Fe_2O_3 nanoparticles [16,17].

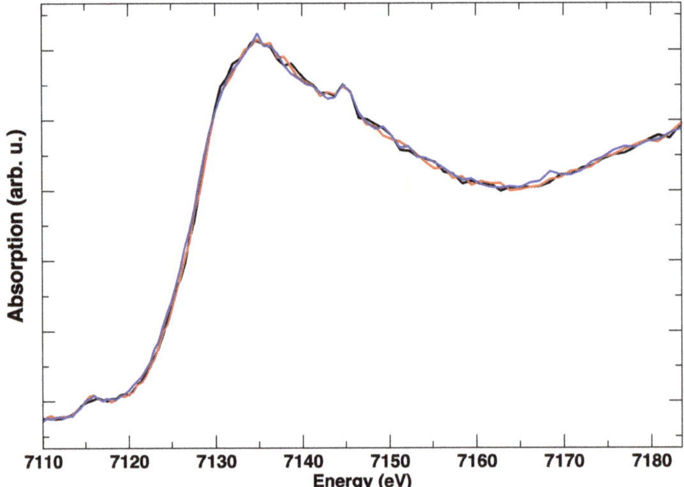

Figure 8. Fe K-edge XANES spectra measured at 850 °C with an applied electrical bias: +1 V (black), −1 V (red), 0 V (blue), over the course of 40 minutes.

The evidence points toward the complete and irreversible transformation of the material. The *post mortem* XANES spectrum on the iron K-edge shows a deep modification with respect to the fresh sample (Figure 9), with the treated sample resembling Fe_2O_3 [16,17]. It appears that the polarization cycling has produced a total breakdown of the LSCF structure, with de-mixing of the starting material in iron oxides, cobalt oxides, and other compounds. We propose that such a rapid structural breakdown is induced both by the high cathodic voltage (−1 V) and by the abrupt cycling between cathodic and anodic (+1 V) conditions. It has been noted in the past that the stability of the LSCF phase is tested and optimized for SOFC operation (that is, cathodic polarization), and not for the inverse operation that is characteristic of an electrolyzer. In fact, the current density curves show a pronounced asymmetry between cathodic and anodic polarization for LSCF cathodes compared to other materials such as lanthanum strontium manganites [18].

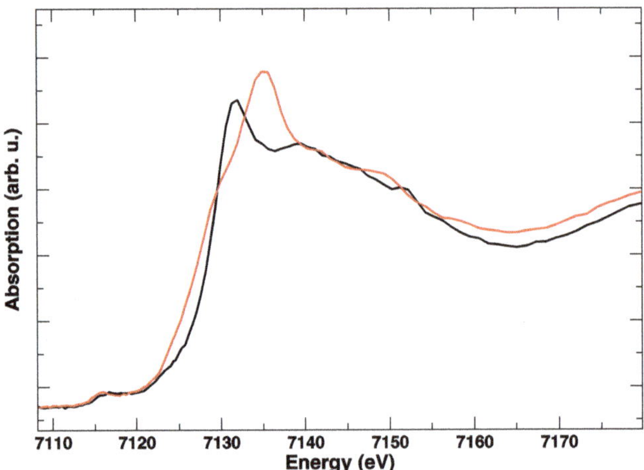

Figure 9. Room temperature Fe K-edge XANES spectra of LSCF before treatment (black line) and after the in situ electrical polarization experiment (red line).

4. Discussion

This paper presents one of the few structural studies of SOFC electrodes during operation. In particular, we investigated a LSCF/YSZ half-cell under electrical polarization. On the basis of the *post mortem* XANES spectra of iron and cobalt, we can state that, as a consequence of the high temperature polarization cycling, the LSCF structure was disrupted, producing the de-mixing of the spinel Co_3O_4, with a Co oxidation state intermediate between +2 and +3, and of hematite Fe_2O_3 with Fe oxidation state +3. This means that both cationic species are reduced compared to the LSCF mixed oxide [19]. It has been reported before that an applied cathodic bias increases the concentration of oxygen vacancies [20]. It appears then that the increased concentration of oxygen vacancies above a certain level induces the collapse of the perovskite mixed oxide, which is also favored by the high temperature. This eventually leads to a rearrangement of the mixed oxide toward the simple oxides retaining an oxidation state lower than that of LSCF.

Author Contributions: Supervision, J.M. and A.M.; Writing—original draft, F.G., G.G., A.L. and A.C.

Funding: We acknowledge funding through MIUR projects Futuro in Ricerca "INnovative Ceramic and hybrid materials for proton-conducting fuel cells at Intermediate Temperature: design, characterization, and device assembly (INCYPIT)", PON R&C 2007-1013 "Tecnologie ad alta Efficienza per la Sostenibilità Energetica ed ambientale On-board (TESEO)", PRIN2010 "Celle a combustibile ad ossido solido operanti a temperatura intermedia alimentate con biocombustibili (BIOITSOFC)".

Acknowledgments: We acknowledge the ESRF for provision of beamtime. We thank H. Vitoux (ESRF) for assistance in the cell design and during the experiment, B. Stuhlhofer (MPI) for thin film deposition, R. Merkle (MPI) for helpful discussion, and the Glass workshop of MPI for the sample holder fabrication.

Conflicts of Interest: The authors declare no conflict of interest.

References

1. Kawada, T.; Oh, M.Y.; Watanabe, H.; Kimura, Y.; Fujimaki, Y.; Masumitsu, T.; Watanabe, S.; Hashimoto, S.; Amezawa, K. Compositional and Mechanical Stabilities of a $(La,Sr)(Co,Fe)O_{3-\delta}$ Cathode under SOFC Operation. *ECS Trans.* **2012**, *45*, 307–312. [CrossRef]

2. Baumann, F.S.; Fleig, J.; Konuma, M.; Starke, U.; Habermeier, H.-U.; Maier, J. Strong Performance Improvement of $La_{0.6}Sr_{0.4}Co_{0.8}Fe_{0.2}O_{3-\delta}$ SOFC Cathodes by Electrochemical Activation. *J. Electrochem. Soc.* **2005**, *152*, A2074–A2079. [CrossRef]

3. Backhaus-Ricoult, M.; Adib, K.; St. Clair, T.; Luerssen, B.; Gregoratti, L.; Barinov, A. In-situ study of operating SOFC LSM/YSZ cathodes under polarization by photoelectron microscopy. *Solid State Ion.* **2008**, *179*, 891–895. [CrossRef]

4. Nenning, A.; Opitz, A.K.; Rameshan, C.; Rameshan, R.; Blume, R.; Havecker, M.; Knop-Gericke, A.; Rupprechter, G.; Klötzer, B.; Fleig, J. Ambient Pressure XPS Study of Mixed Conducting Perovskite-Type SOFC Cathode and Anode Materials under Well-Defined Electrochemical Polarization. *J. Phys. Chem. C* **2016**, *120*, 1461–1471. [CrossRef] [PubMed]

5. Orikasa, Y.; Ina, T.; Fukutsuka, T.; Amezawa, K.; Kawada, T.; Uchimoto, Y. X-ray Absorption Spectroscopic Studies on Electronic Structure in $La_{0.6}Sr_{0.4}Co_{0.8}Fe_{0.2}O_{3-\delta}$ Perovskite-Type Oxides. *ECS Trans.* **2008**, *13*, 201–205.

6. Giannici, F.; Longo, A.; Balerna, A.; Martorana, A. Dopant−Host Oxide Interaction and Proton Mobility in $Gd:BaCeO_3$. *Chem. Mater.* **2009**, *21*, 597–603. [CrossRef]

7. Lupetin, P.; Giannici, F.; Gregori, G.; Martorana, A.; Maier, J. Effects of grain boundary decoration on the electrical conduction of nanocrystalline CeO_2. *J. Electrochem. Soc.* **2012**, *159*, B417–B425. [CrossRef]

8. Giannici, F.; Gregori, G.; Aliotta, C.; Longo, A.; Maier, J.; Martorana, A. Structure and oxide ion conductivity: Local order, defect interactions and grain boundary effects in acceptor-doped ceria. *Chem. Mater.* **2014**, *26*, 5994–6006. [CrossRef]

9. Giannici, F.; Canu, G.; Gambino, M.; Longo, A.; Salomé, M.; Viviani, M.; Martorana, A. Electrode−Electrolyte Compatibility in Solid-Oxide Fuel Cells: Investigation of the LSM−LNC Interface with X-ray Microspectroscopy. *Chem. Mater.* **2015**, *27*, 2763–2766. [CrossRef]

10. Hagen, A.; Traulsen, M.L.; Kiebach, W.R.; Johansen, B.S. Spectroelectrochemical cell for in situ studies of solid oxide fuel cells. *J. Synchrotron Rad.* **2012**, *19*, 400–407. [CrossRef] [PubMed]

11. Woolley, R.J.; Ryan, M.P.; Skinner, S.J. In Situ Measurements on Solid Oxide Fuel Cell Cathodes— Simultaneous X-ray Absorption and AC Impedance Spectroscopy on Symmetrical Cells. *Fuel Cells* **2013**, *13*, 1080–1087. [CrossRef]

12. Longo, A.; Liotta, L.F.; Banerjee, D.; La Parola, V.; Puleo, F.; Cavallari, C.; Sahle, C.J.; Moretti Sala, M.; Martorana, A. The Effect of Ni Doping on the Performance and Electronic Structure of LSCF Cathodes Used for IT-SOFCs. *J. Phys. Chem. C* **2018**, *122*, 1003–1013. [CrossRef]

13. Lai, S.Y.; Ding, D.; Liu, M.; Liu, M.; Alamgir, F.M. Operando and In Situ X-ray Spectroscopies of Degradation in $La_{0.6}Sr_{0.4}Co_{0.2}Fe_{0.8}O_{3-\delta}$ Thin Film Cathodes in Fuel Cells. *ChemSusChem* **2014**, *7*, 3078–3087. [CrossRef] [PubMed]

14. Irvine, J.T.S.; Neagu, D.; Verbraeken, M.C.; Christodoulos, C.; Graves, C.; Mogensen, M.B. Evolution of the electrochemical interface in high-temperature fuel cells and electrolysers. *Nat. Energy* **2016**, *1*, 15014. [CrossRef]

15. Wang, S.; Katsuki, M.; Dokiya, M.; Hashimoto, T. High temperature properties of $La_{0.6}Sr_{0.4}Co_{0.8}Fe_{0.2}O_{3-\delta}$ phase structure and electrical conductivity. *Solid State Ion.* **2003**, *159*, 71–78. [CrossRef]

16. Fröba, M.; Köhn, R.; Bouffaud, G.; Richard, O.; van Tendeloo, G. Fe_2O_3 Nanoparticles within Mesoporous MCM-48 Silica: In Situ Formation and Characterization. *Chem. Mater.* **1999**, *11*, 2858–2865. [CrossRef]

17. Kuzmin, A.; Chaboy, J. EXAFS and XANES analysis of oxides at the nanoscale. *IUCrJ* **2014**, *1*, 571–589. [CrossRef] [PubMed]

18. Kim-Lohsoontorn, P.; Bae, J. Electrochemical performance of solid oxide electrolysis cell electrodes under high-temperature coelectrolysis of steam and carbon dioxide. *J. Power Sources* **2011**, *196*, 7161–7168. [CrossRef]

19. Soldati, A.L.; Baqué, L.; Napolitano, F.; Serquis, A. Cobalt–iron red–ox behavior in nanostructured $La_{0.4}Sr_{0.6}Co_{0.8}Fe_{0.2}O_{3-\delta}$ cathodes. *J. Solid State Chem.* **2013**, *198*, 253–261. [CrossRef]

20. Baumann, F.S.; Fleig, J.; Habermeier, H.-U.; Maier, J. Impedance spectroscopic study on well-defined $(La,Sr)(Co,Fe)O_{3-\delta}$ model electrodes. *Solid State Ion.* **2006**, *177*, 1071–1081. [CrossRef]

Article

Switchable Interfaces: Redox Monolayers on Si(100) by Electrochemical Trapping of Alcohol Nucleophiles

Long Zhang [1,2], Ruth Belinda Domínguez Espíndola [1], Benjamin B. Noble [3], Vinicius R. Gonçales [4], Gordon G. Wallace [2], Nadim Darwish [1,*], Michelle L. Coote [3,*] and Simone Ciampi [1,*]

[1] School of Molecular and Life Sciences, Curtin Institute of Functional Molecules and Interfaces, Curtin University, Bentley, Western Australia 6102, Australia; long.zhang@unsw.edu.au (L.Z.); rdominguez@uaem.mx (R.B.D.E.)
[2] ARC Centre of Excellence for Electromaterials Science, Intelligent Polymer Research Institute, University of Wollongong, Wollongong, New South Wales 2500, Australia; gwallace@uow.edu.au
[3] ARC Centre of Excellence for Electromaterials Science, Research School of Chemistry, Australian National University, Canberra, Australian Capital Territory 2601, Australia; benjamin.noble@anu.edu.au
[4] School of Chemistry, The University of New South Wales, Sydney, New South Wales 2052, Australia; v.goncales@unsw.edu.au
* Correspondence: nadim.darwish@curtin.edu.au (N.D.); michelle.coote@anu.edu.au (M.L.C.); simone.ciampi@curtin.edu.au (S.C.); Tel.: +61-8-9266-9009 (S.C.)

Received: 29 June 2018; Accepted: 17 July 2018; Published: 20 July 2018

Abstract: Organic electrosynthesis is going through its renaissance but its scope in surface science as a tool to introduce specific molecular signatures at an electrode/electrolyte interface is under explored. Here, we have investigated an electrochemical approach to generate in situ surface-tethered and highly-reactive carbocations. We have covalently attached an alkoxyamine derivative on an Si(100) electrode and used an anodic bias stimulus to trigger its fragmentation into a diffusive nitroxide (TEMPO) and a surface-confined carbocation. As a proof-of-principle we have used this reactive intermediate to trap a nucleophile dissolved in the electrolyte. The nucleophile was ferrocenemethanol and its presence and surface concentration after its reaction with the carbocation were assessed by cyclic voltammetry. The work expands the repertoire of available electrosynthetic methods and could in principle lay the foundation for a new form of electrochemical lithography.

Keywords: electrosynthesis; switchable surfaces; alkoxyamine surfaces; redox monolayers

---

1. Introduction

Synthetic organic electrochemistry traces its origin back to the work of Faraday and Kolbe and its green credentials are currently prompting a renaissance [1–3]. Chemical reactions that are coupled to the flow of electricity allow, for instance, chemists to generate unstable intermediates in situ, to control very precisely and accurately the supply of reactants and to monitor reaction processes in real-time [4]. While the large majority of the work on synthetic electrochemistry has focused on bulk synthesis [5,6], there is also a strong motivation to expand electrochemical synthetic methods toward the chemical modification of interfaces [7–10]. Addressing the molecular details of a surface [11], especially those of semiconductors, has been central to the development of fields such as molecular electronics [12], sensing [13], energy conversion [14] and cell biology [15]. Semiconductor electrodes and in particular silicon electrodes, have the advantage of being readily available in a crystalline form, they have unique photo-electrochemical properties and can form covalently-bound monolayers [16]. Silicon like all non-oxide semiconductors is thermodynamically unstable and especially under anodic polarization in aqueous environments tends to grow an electrically-insulating silica layer [17,18]. A common laboratory approach to attach covalently organic molecules on silicon substrates is a three-step wet chemical process. Removal of the native silica layer with fluoride-containing solutions is followed by chemical passivation [19] of the hydrogen-terminated silicon surface by means of hydrosilylation of terminal alkenes or alkynes [20–22] and finally chemical derivatization of the aliphatic monolayer [16]. In this work we seek to expand the chemical repertoires of this last step and explore the synthetic scope of reactive carbocations in the context of surface chemistry.

There are several methods to generate carbocations in solution [23] but only few examples are available for electrochemical generation of carbocations at electrodes. We have recently reported on the putative generation of carbocations at metallic electrodes after the fragmentation of anodic intermediates of alkoxyamines. Alkoxyamines are heat-labile molecules, widely used as an in situ source of nitroxides in polymer and materials sciences. We have shown that the one-electron oxidation of an alkoxyamine leads to an anodic intermediate that rapidly fragments releasing a nitroxide species at room temperature [24].

In the current work we have developed a surface model system to explore the feasibility of using surface-tethered carbocations to trap solution nucleophiles. The carbocation is electro-generated in situ from an alkoxyamine molecule that is exposed at the distal end of an organic monolayer grown on a Si(100) electrode. By way of applying a positive bias to the Si(100) electrode, the alkoxyamine is anodically cleaved to release a 2,2,6,6-tetramethylpiperidine-1-oxyl (TEMPO) molecule and generate a surface-confined carbocation species. The latter is attacked by electron donors via nucleophilic substitutions. As proof-of-principle we have prepared redox-active monolayers by generating the surface carbocations in the presence of ferrocenemethanol (**2**) molecules.

2. Materials and Methods

2.1. Chemicals

All chemicals, unless specified otherwise, were of analytical grade and used as received. Milli-Q™ water (>18 MΩ cm, Millipore Corporation, Burlington, MA, USA) was used to prepare solutions and to clean all glassware. Anhydrous solvents used in chemical reactions were purified under nitrogen by a solvent drying system from LC Technology Solutions Inc. Dichloromethane (DCM), methanol (MeOH) and 2-propanol were redistilled prior to use. Hydrogen peroxide (30 wt % in water), ammonium fluoride (Puranal™, 40 wt % in water) and sulfuric acid (Puranal™, 95–97%) were used to clean the wafers and were of semiconductor grade. 1,8-nonadiyne (Sigma-Aldrich, 98%) was redistilled from sodium borohydride (Sigma-Aldrich, 99+%) under reduced pressure (80 °C, 10−12 Torr) and stored under a high purity argon atmosphere prior to use. Tetrabutylammonium hexafluorophosphate salt (Bu$_4$NPF$_6$, Sigma-Aldrich, ≥98%) was used as a supporting electrolyte. 4-Vinylbenzyl chloride (90%), ammonium sulphite, ferrocene (98%), ferrocenemethanol (**2**, 97%) and 2,2,6,6-tetramethylpiperidine-1-oxyl (TEMPO hereafter, 98%) were purchased from Sigma-Aldrich. The surface reactive (azide-tagged) alkoxyamine 1-(1-(4-azidomethyl)phenyl)ethoxy-2,2,6,6-tetramethylpiperidine (**1**) was synthesized according to minor modification of previously reported procedure (see Scheme 1) [25]. Prime grade, single-side polished silicon wafers of 100-orientation (<100> ± 0.5°), p-type (boron-doped) of 100 mm diameter, 500–550 μm thickness and of a nominal resistivity of 0.001–0.003 Ω cm were obtained from Siltronix, S.A.S. (Archamps, France).

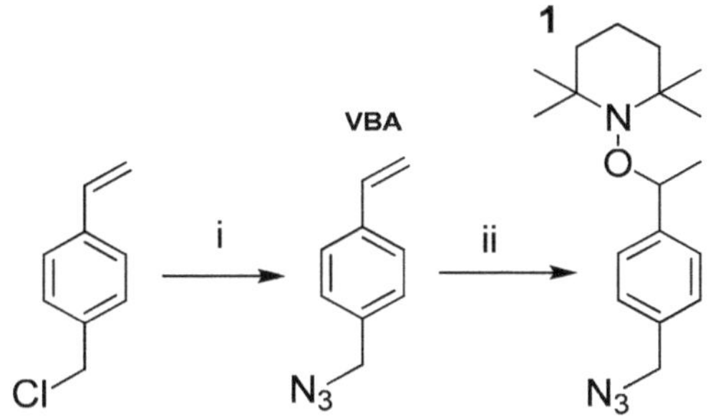

Scheme 1. Synthesis of the azide-tagged alkoxyamine **1**. i. NaN$_3$, DMF, 82%. ii. TEMPO, Mn(OAc)$_3$·2H$_2$O, NaBH$_4$, Toluene/EtOH, 50%.

2.2. Synthetic Methods

Thin-layer chromatography (TLC) was performed on silica gel Merck aluminium sheets (60 F$_{254}$). Merck 60 Å silica gel (220–400 mesh particle size) was used for column chromatography. Nuclear magnetic resonance (NMR) spectra were recorded on a Bruker Avance 400 spectrometer in deuterated dimethyl sulfoxide (d-DMSO) using the residual solvent signal as internal reference. High-resolution mass spectral data (HRMS, mass accuracy 2–4 ppm) of alkoxyamine **1** were obtained using a Waters Xevo QTof MS via ESI experiments and infusing the sample at 8 μL/min.

Synthesis of 4-Vinylbenzyl azide (**VBA**). Sodium azide (1.30 g, 20 mmol) was added in one portion to a stirred solution of 4-vinylbenzyl chloride (1.53 g, 10 mmol) in *N,N*-dimethylformamide (DMF, 20 mL). The reaction mixture was stirred at room temperature for 12 h under argon then poured into a large excess water (100 mL). The mixture was extracted with DCM (3 × 50 mL). The combined organic layers were then washed with brine (2 × 100 mL), dried over MgSO$_4$, filtered and dried under vacuum to afford the crude title compound as a brown oily residue. The crude material was purified by silica gel column chromatography (hexane) to give **VBA** as a light yellow oil (1.31 g, 82%). ^{1}H NMR (400 MHz, d-DMSO): δ 7.50 p.p.m. (d, Ar-H, *J* = 8.12 Hz, 2H), 7.35 (d, Ar-H, *J* = 8.04 Hz, 2H), 6.74 (dd, Ar-CH=CH$_2$, *J* = 17.66 Hz, 1H), 5.85 (dd, Ar-CH=CH$_2$, *J* = 17.68 Hz, 0.96 Hz, 1H), 5.28 (dd, Ar-CH=CH$_2$, *J* = 11.84 Hz, 0.92 Hz, 1H), 4.43 (s, Ar-CH$_2$-N$_3$, 2H); ^{13}C NMR (100 MHz, d-DMSO): δ 136.96, 136.11, 135.13, 128.69, 126.39, 114.64, 53.32.

Synthesis of 1-(1-(4-Azidomethyl) phenyl)ethoxy-2,2,6,6-tetramethylpiperidine (**1**). Alkoxyamine **1** was synthesized from **VBA** via the following procedure. To an ice-cold solution of TEMPO (0.31 g, 2 mmol) in toluene/ethanol (60 mL, 1:1, v/v) **VBA** (3.20 g, 20 mmol) and Mn(OAc)$_3$·2H$_2$O (5.36 g, 20 mmol) were added in one portion while stirring in air. Stirring was continued for one min and then a 15-fold molar excess (with respect to TEMPO) of NaBH$_4$ was added in portions over 15 min. After stirring overnight under nitrogen atmosphere, the residue was isolated by filtration, the filtrate was suspended in water and the aqueous solution was then extracted three times with DCM. The combined organic layers were evaporated under vacuum and the crude material was purified by silica gel column chromatography (ethyl acetate/hexane, 1:40, v/v) to yield alkoxyamine **1** as a colourless oil liquid (0.32 g, 50%).

^{1}H NMR (400 MHz, d-DMSO): δ 7.36–7.29 p.p.m. (m, Ar-H, 4H), 4.75 (q, NO-CH-Ar, *J* = 13.50 Hz, 1H), 4.42 (s, N$_3$-CH$_2$-Ar, 2H), 1.54–1.38 (m, 6H), 1.37–1.19 (m, 6H), 1.12 (s, 3H), 0.97 (s, 3H), 0.57 (s, 3H); ^{13}C NMR (100 MHz, d-DMSO): δ 145.08, 134.08, 128.21, 126.76, 82.09, 59.22, 58.98, 53.40, 33.99, 33.76, 23.00, 20.04, 16.66; HRMS (**1**, *m/z*): [M + H]$^+$ calcd for C$_{18}$H$_{29}$N$_4$O 317.2336, found 317.2335.

2.3. Surface Modification

2.3.1. Light-Assisted Hydrosilylation of 1,8-Nonadiyne on Si(100) (**S-1**)

The assembly of the acetylenylated Si(100) surface by covalent attachment of 1,8-nonadiyne on hydrogen-terminated silicon is based on a photochemical hydrosilylation method [26]. In brief, silicon wafers were mechanically cut into pieces (approximately 10 × 10 mm in size), rinsed several times with small portions of DCM, MeOH and Milli-Q™ water. The samples were then immersed in hot Piranha solution (100 °C, a 3:1 (v/v) mixture of concentrated sulfuric acid to 30% hydrogen peroxide, Caution: piranha solution reacts violently with organic substances) for 20 min. The samples were then rinsed with water and immediately etched with a deoxygenated 40% aqueous ammonium fluoride solution for 5 min under a stream of argon. A small amount (ca. 5 mg) of ammonium sulphite was added to the etching bath for building an anaerobic environment and hence avoiding the in situ oxidation of the hydrogen-terminated silicon. The freshly etched samples were washed sequentially with Milli-Q™ water and DCM and blown dry with argon before dropping a small deoxygenated sample of 1,8-nonadiyne (approximate 50 µL) on the hydrogen-terminated wafer and covering it with a quartz slide to minimize evaporation. The wafer was then rapidly transferred to an air-tight UV reaction chamber and kept under positive argon pressure. A collimated LED source (λ = 365 nm, nominal power output >190 mW, Thorlabs part M365L2 coupled to a SM1P25-A collimator adapter) was fixed over the sample at a distance of about 10 cm. After illumination for a 2 h period, the resulting acetylene-functionalized sample (**S-1**, Scheme 2) was removed from the reaction chamber, rinsed several times with DCM and rested for 12 h in a sealed vial at +4 °C under DCM before being further reacted with alkoxyamine **1**.

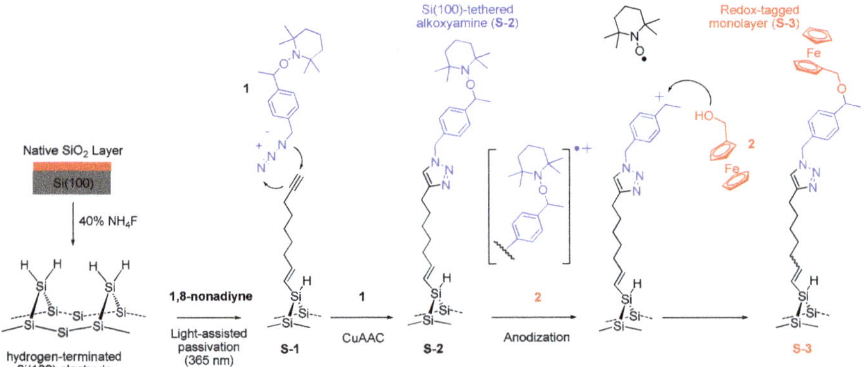

Scheme 2. Light-assisted (365 nm) hydrosilylation of 1,8-nonadiyne to passivate an hydrogen-terminated Si(100) surface (**S-1**) and covalent attachment of alkoxyamine **1** via CuAAC "click" reactions to yield an alkoxyamine monolayer (**S-2**). Anodization of **S-2** in the presence of the alcohol nucleophile **2** leads to release of TEMPO in the electrolyte with formation of a redox-active monolayer (**S-3**) by reaction of **2** with the putative surface-tethered carbocation intermediate.

2.3.2. Click Immobilization of Alkoxyamine 1 (**S-2**)

Surface **S-1** was reacted with molecule **1** to yield the alkoxyamine monolayers (**S-2**) via a copper(I)-catalysed "click" alkyne-azide cycloaddition (CuAAC) reaction. In brief, to a reaction vial containing the alkyne-functionalized silicon surface (**S-1**) was added (i) the azide (alkoxyamine **1**, 0.5 × 10^{-3} M, 2-propanol/water, 1:1, v/v), (ii) copper(II) sulphate pentahydrate (1.0 × 10^{-4} M) and (iii) sodium ascorbate (5 mg/mL). The reaction was carried out without excluding air from the reaction environment, at room temperature and under ambient light. The samples were removed from the reaction vessel after a reaction time of 2 h and were rinsed thoroughly with copious amounts of 2-propanol, water, 2-propanol and DCM and blown dry with argon before being analysed or further reacted (Scheme 2).

2.4. Surface Characterization

2.4.1. X-ray Photoelectron Spectroscopy

X-ray photoelectron spectroscopy (XPS) characterization was performed on an ESCALab 250 Xi (Thermo Scientific, Waltham, MA, USA) spectrometer with a monochromated Al Kα source to characterize the formation of an alkoxyamine **1** monolayer on silicon. The pressure in the analysis chamber during measurement was $<10^{-8}$ mbar. The pass energy and step size for narrow scans were 20 eV and 0.1 eV respectively, with a take-off angle normal to the sample surface. Spectral analysis was performed using Avantage 4.73 software and curve fitting was carried out with a mixture of Gaussian–Lorentzian functions after background subtraction. Peaks were calibrated to C–C at 284.8 eV.

2.4.2. Electrochemical Measurements

Electrochemical experiments were performed in a single-compartment, three-electrode PTFE cell with the modified silicon surface as the working electrode, a platinum mesh (ca. 200 mm^2) as the counter electrode and silver/silver chloride "leakless" electrode in 3 M sodium chloride as the reference electrode. A rectilinear cross-section gasket defined the geometric area of the working electrode to 28 mm^2. Electrical contact between the silicon substrate and a copper plate was ensured by rapidly rubbing a gallium indium eutectic onto the back of the silicon electrode. Cyclic voltammetry (CV) measurements were performed using a CHI 910B electrochemical analyser. All potentials are reported versus the reference electrode. Tetrabutylammonium hexafluorophosphate (Bu$_4$NPF$_6$, Sigma-Aldrich, $\geq$98%) was recrystallized twice from 2-propanol and used at the concentration of 1.0×10^{-1} M as the supporting electrolyte in DCM solutions. The reaction between the anodically-cleaved **S-2** samples and ferrocenemethanol (**2**), as well as control experiments were **2** is replaced by ferrocene, were performed under argon and at room temperature. **S-3** samples (Scheme 2) were thoroughly washed with DCM prior to analysis. Cyclic voltammetry of **S-3** samples was performed under ambient conditions in aqueous 1.0 M HClO$_4$ electrolytes.

3. Results and Discussion

As a proof-of-principle, we have devised "trapping" experiments on an electrode surface modified by alkoxyamine **1** (**S-2**, Scheme 2). The anodization of the **S-2** monolayers is expected to generate benzylic cations tethered on the electrode surface while the nitroxide fragment is released into the electrolyte. The covalent attachment of alkoxyamine **1** to the silicon surface was first verified by the XPS (Figure 1). Peaks at 99.3 and 100.0 eV for **S-2** samples were respectively ascribed to the Si 2p$_{3/2}$ and Si 2p$_{1/2}$ emissions (Figure 1b). The presence of a band at 101.1-104.3 eV is due to SiO$_x$ formed during the hydrosilylation and/or CuAAC processes. Two signals in the carbon envelope that are centred at 284.6 and 286.3 eV indicates the formation of C-C and C-O/-N, respectively [27], as expected for the putative structure of **S-2** samples (Figure 1c). As shown in Figure 1d, refinement of the N 1s emission gives two fitted peaks with binding energies of 400.5 and 401.7 eV in an approximate 3:1 ratio. The position of the low binding energy line in the **S-2** samples is in agreement with the 400.6 eV previously assigned to nitroxide nitrogen atoms in thin films [22]. Electrons from the triazole heterocycle also contributed to the 400.5 and 401.7 emissions with an expected 2:1 ratio.

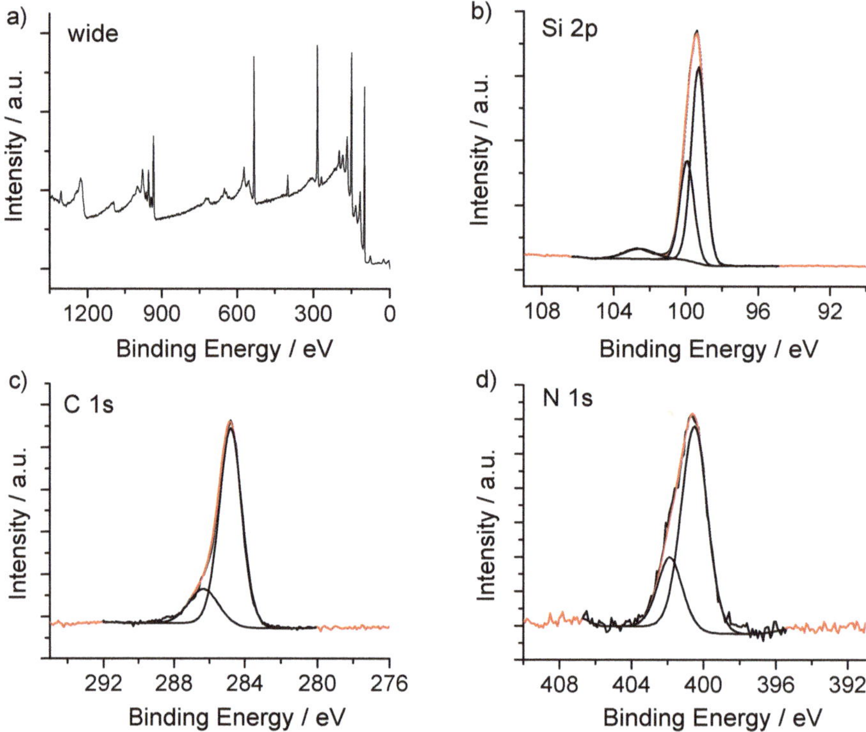

Figure 1. XPS survey spectra (**a**) and high-resolution Si 2p (**b**), C 1s (**c**) and N 1s (**d**) scans of **S-2** samples assembled on high-doped p-type Si(100) surface by CuAAC reactions between alkoxyamine **1** and alkyne-terminated monolayers (**S-1**).

As shown Figure 2, the anodization of the alkoxyamine monolayer (**S-2**) via a potentiostatic experiment (electrode held at 1.0 V versus reference for 30 s) in 1.0×10^{-1} M DCM/Bu$_4$NPF$_6$ in the presence of an excess of **2** is effective in trapping the alcohol at the distal end of the monolayer. The newly-formed redox-active surface (**S-3**) is washed with copious DCM and then analysed by cyclic voltammetry in an electrolyte that does not contain a redox-active species. Figure 2a shows the well-defined pair of redox waves (solid red lines) which are ascribed to the surface-confined redox tag (**S-3**). The surface coverage of ferrocenyl units, derived from integration of the anodic wave, ranges over almost on order of magnitude from 5.5×10^{-12} mol cm^{-2}, as the low end, up to the value of 3.1×10^{-11} mol cm^{-2} for our highest current yield. Our best result is still only about 10% of close-packed packed layer of redox ferrocenes; defining the exact nature of any competitive side-reactions is beyond the scope of this work. Two control experiments were carried out as follows. We used ferrocene, which lacks the hydroxyl function, to rule out non-specific adsorption of the redox probe as well as trying to remove the anodic step. That is, the working electrode modified bearing the alkoxyamine **1** is rested at open circuit potential (OCP) conditions in the presence of the nucleophile (**2**). In both controls (Figure 2b) we found no evidence of redox peaks when either the nucleophile is removed from the system (+, Figure 2b) or when the **S-2** surface is not anodized to trigger the switch masked alkoxyamine/reactive carbocation ($\bigcirc$, Figure 2b).

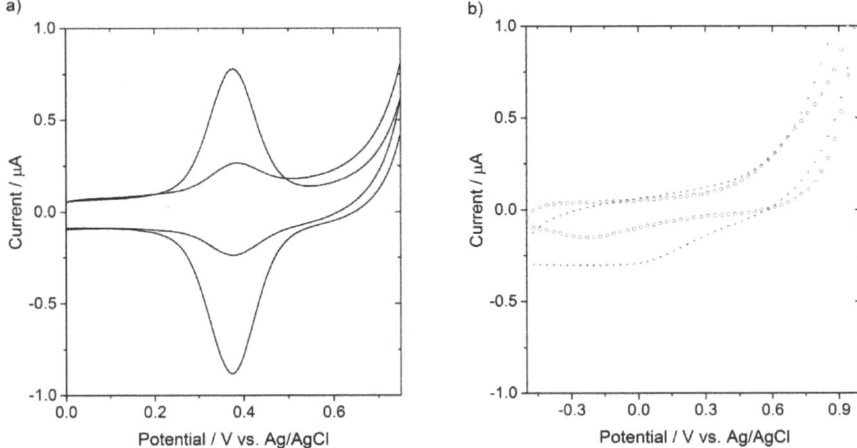

Figure 2. Cyclic voltammograms (CVs) of redox nucleophiles reacted with the anodized alkoxyamines monolayer (conversion of **S-2** into **S-3**). CVs were acquired at a voltage sweep rate of 100 mV s^{-1}. (a) CVs analysis of the Si(100) electrode in 1.0 M HClO$_4$ after the anodization of **S-2** monolayers in the presence of ferrocene methanol (**2**) with the surface coverage of ferrocene units ranging from 5.5 × 10^{-12} mol cm^{-2}, as the low end, up to the value of 3.1 × 10^{-11} mol cm^{-2}. (b) Control experiments for prolonged standing of **S-2** electrodes at open circuit potential in the presence of **2** in the electrolyte (○) or CVs for the experiments as in (**a**) but with ferrocene replacing **2** (+).

4. Conclusions

We have demonstrated a method to achieve a switchable interface—from unreactive to electrophilic—and used it to attach alcohol species as monolayers on Si(100) electrodes. Unlike conventional esterification approaches on surfaces (e.g., DCC/DMAP), here we show an electrochemical alternative that involves a surface carbocation. This reactive intermediate is generated in situ under an external potential stimulus that switches a stable alkoxyamine into a reactive surface trap for nucleophiles. This proof-of-principle expands to the realm of surfaces some of the recent advances made in synthetic organic electrochemistry. This knowledge may aid the development of cationic polymerization on surfaces and this surface chemistry platform can in principle be coupled to photo-electrochemical reactions on semiconductors and photoconductors [26] in order to control electron transfer, hence chemical reactivity, in two dimensions [28,29].

Author Contributions: Conceptualization, S.C.; Data curation, L.Z. and S.C.; Formal analysis, L.Z., B.B.N., M.L.C. and S.C.; Funding acquisition, G.G.W., N.D., M.L.C. and S.C.; Investigation, L.Z., R.B.D.E., V.R.G. and S.C.; Methodology, S.C.; Project administration, S.C.; Resources, G.G.W., M.L.C. and S.C.; Supervision, G.G.W. and S.C.; Validation, R.B.D.E.; Visualization, L.Z.; Writing—original draft, L.Z.; Writing—review & editing, V.R.G., N.D., M.L.C. and S.C.

Funding: This work was supported by grants from the Australian Research Council (DE160100732 (S.C.), DE160101101 (N.D.), CE 140100012 (L.Z., B.B.N., M.L.C. and G.G.W.)).

Acknowledgments: We acknowledge support from Curtin University's Microscopy & Microanalysis Facility, whose instrumentation has been partially funded by the University, State and Commonwealth. L.Z. would like to thank AINSE Ltd. for providing financial assistance. M.L.C. gratefully acknowledges an ARC Georgina Sweet ARC Laureate Fellowship.

Conflicts of Interest: The authors declare no conflict of interest.

References

1. Horn, E.J.; Rosen, B.R.; Baran, P.S. Synthetic organic electrochemistry: An enabling and innately sustainable method. *ACS Cent. Sci.* **2016**, *2*, 302–308. [CrossRef] [PubMed]
2. Yan, M.; Kawamata, Y.; Baran, P.S. Synthetic organic electrochemical methods since 2000: On the verge of a renaissance. *Chem. Rev.* **2017**, *117*, 13230–13319. [CrossRef] [PubMed]
3. Yan, M.; Kawamata, Y.; Baran, P.S. Synthetic organic electrochemistry: Calling all engineers. *Angew. Chem. Int. Ed.* **2018**, *57*, 4149–4155. [CrossRef] [PubMed]
4. Frontana-Uribe, B.A.; Little, R.D.; Ibanez, J.G.; Palma, A.; Vasquez-Medrano, R. Organic electrosynthesis: A promising green methodology in organic chemistry. *Green Chem.* **2010**, *12*, 2099–2119. [CrossRef]
5. Gütz, C.; Selt, M.; Bänziger, M.; Bucher, C.; Römelt, C.; Hecken, N.; Gallou, F.; Galvão, T.R.; Waldvogel, S.R. A novel cathode material for cathodic dehalogenation of 1, 1-dibromo cyclopropane derivatives. *Chem. Eur. J.* **2015**, *21*, 13878–13882. [CrossRef] [PubMed]
6. Gütz, C.; Bänziger, M.; Bucher, C.; Galvão, T.S.R.; Waldvogel, S.R. Development and scale-up of the electrochemical dehalogenation for the synthesis of a key intermediate for NS5A inhibitors. *Org. Process Res. Dev.* **2015**, *19*, 1428–1433. [CrossRef]
7. Andrieux, C.P.; Gonzalez, F.; Savéant, J.-M. Derivatization of carbon surfaces by anodic oxidation of arylacetates. Electrochemical manipulation of the grafted films. *J. Am. Chem. Soc.* **1997**, *119*, 4292–4300. [CrossRef]
8. Robins, E.; Stewart, M.; Buriak, J. Anodic and cathodic electrografting of alkynes on porous silicon. *Chem. Commun.* **1999**, 2479–2480. [CrossRef]
9. Hurley, P.T.; Ribbe, A.E.; Buriak, J.M. Nanopatterning of alkynes on hydrogen-terminated silicon surfaces by scanning probe-induced cathodic electrografting. *J. Am. Chem. Soc.* **2003**, *125*, 11334–11339. [CrossRef] [PubMed]
10. Belanger, D.; Pinson, J. Electrografting: A powerful method for surface modification. *Chem. Soc. Rev.* **2011**, *40*, 3995–4048. [CrossRef] [PubMed]
11. Gooding, J.J.; Ciampi, S. The molecular level modification of surfaces: From self-assembled monolayers to complex molecular assemblies. *Chem. Soc. Rev.* **2011**, *40*, 2704–2718. [CrossRef] [PubMed]
12. Aragonès, A.C.; Darwish, N.; Ciampi, S.; Sanz, F.; Gooding, J.J.; Díez-Pérez, I. Single-molecule electrical contacts on silicon electrodes under ambient conditions. *Nat. Commun.* **2017**, *8*, 15056–15063. [CrossRef] [PubMed]
13. Scheres, L.; ter Maat, J.; Giesbers, M.; Zuilhof, H. Microcontact printing onto oxide-free silicon via highly reactive acid fluoride-functionalized monolayers. *Small* **2010**, *6*, 642–650. [CrossRef] [PubMed]
14. Bard, A.J.; Bocarsly, A.B.; Fan, F.R.F.; Walton, E.G.; Wrighton, M.S. The concept of Fermi level pinning at semiconductor/liquid junctions. Consequences for energy conversion efficiency and selection of useful solution redox couples in solar devices. *J. Am. Chem. Soc.* **1980**, *102*, 3671–3677. [CrossRef]
15. Ciampi, S.; James, M.; Le Saux, G.; Gaus, K.; Justin Gooding, J. Electrochemical "switching" of Si(100) modular assemblies. *J. Am. Chem. Soc.* **2012**, *134*, 844–847. [CrossRef] [PubMed]
16. Ciampi, S.; Harper, J.B.; Gooding, J.J. Wet chemical routes to the assembly of organic monolayers on silicon surfaces via the formation of Si−C bonds: Surface preparation, passivation and functionalization. *Chem. Soc. Rev.* **2010**, *39*, 2158–2183. [CrossRef] [PubMed]
17. Ciampi, S.; Eggers, P.K.; Le Saux, G.; James, M.; Harper, J.B.; Gooding, J.J. Silicon (100) Electrodes resistant to oxidation in aqueous solutions: An unexpected benefit of surface acetylene moieties. *Langmuir* **2009**, *25*, 2530–2539. [CrossRef] [PubMed]
18. Noufi, R.; Frank, A.J.; Nozik, A.J. Stabilization of n-type silicon photoelectrodes to surface oxidation in aqueous electrolyte solution and mediation of oxidation reaction by surface-attached organic conducting polymer. *J. Am. Chem. Soc.* **1981**, *103*, 1849–1850. [CrossRef]
19. Nemanick, E.J.; Hurley, P.T.; Webb, L.J.; Knapp, D.W.; Michalak, D.J.; Brunschwig, B.S.; Lewis, N.S. Chemical and electrical passivation of single-crystal silicon(100) surfaces through a two-step chlorination/alkylation process. *J. Phys. Chem. B* **2006**, *110*, 14770–14778. [CrossRef] [PubMed]
20. Linford, M.R.; Chidsey, C.E.D. Alkyl monolayers covalently bonded to silicon surfaces. *J. Am. Chem. Soc.* **1993**, *115*, 12631–12632. [CrossRef]

21. Sun, Q.-Y.; de Smet, L.C.P.M.; van Lagen, B.; Wright, A.; Zuilhof, H.; Sudhöelter, E.J.R. Covalently attached monolayers on hydrogen-terminated Si(100): Extremely mild attachment by visible light. *Angew. Chem. Int. Ed.* **2004**, *43*, 1352–1355. [CrossRef] [PubMed]

22. Zhang, L.; Vogel, Y.B.; Noble, B.B.; Gonçales, V.R.; Darwish, N.; Brun, A.L.; Gooding, J.J.; Wallace, G.G.; Coote, M.L.; Ciampi, S. TEMPO monolayers on Si(100) electrodes: Electrostatic effects by the electrolyte and semiconductor space-charge on the electroactivity of a persistent radical. *J. Am. Chem. Soc.* **2016**, *138*, 9611–9619. [CrossRef] [PubMed]

23. Yoshida, J.-I.; Kataoka, K.; Horcajada, R.; Nagaki, A. Modern strategies in electroorganic synthesis. *Chem. Rev.* **2008**, *108*, 2265–2299. [CrossRef] [PubMed]

24. Zhang, L.; Laborda, E.; Darwish, N.; Noble, B.B.; Tyrell, J.H.; Pluczyk, S.; Le Brun, A.P.; Wallace, G.G.; Gonzalez, J.; Coote, M.L. Electrochemical and electrostatic cleavage of alkoxyamines. *J. Am. Chem. Soc.* **2018**, *140*, 766–774. [CrossRef] [PubMed]

25. Danko, M.; Szabo, E.; Hrdlovic, P. Synthesis and spectral characteristics of fluorescent dyes based on coumarin fluorophore and hindered amine stabilizer in solution and polymer matrices. *Dyes Pigments* **2011**, *90*, 129–138. [CrossRef]

26. Vogel, Y.B.; Zhang, L.; Darwish, N.; Gonçales, V.R.; Le Brun, A.; Gooding, J.J.; Molina, A.; Wallace, G.G.; Coote, M.L.; Gonzalez, J. Reproducible flaws unveil electrostatic aspects of semiconductor electrochemistry. *Nat. Commun.* **2017**, *8*, 2066. [CrossRef] [PubMed]

27. Ciampi, S.; Böcking, T.; Kilian, K.A.; James, M.; Harper, J.B.; Gooding, J.J. Functionalization of acetylene-terminated monolayers on Si(100) surfaces: A click chemistry approach. *Langmuir* **2007**, *23*, 9320–9329. [CrossRef] [PubMed]

28. Choudhury, M.H.; Ciampi, S.; Yang, Y.; Tavallaie, R.; Zhu, Y.; Zarei, L.; Gonçales, V.R.; Gooding, J.J. Connecting electrodes with light: One wire, many electrodes. *Chem. Sci.* **2015**, *6*, 6769–6776. [CrossRef] [PubMed]

29. Vogel, Y.B.; Gonçales, V.R.; Gooding, J.J.; Ciampi, S. Electrochemical microscopy based on spatial light modulators: A projection system to spatially address electrochemical reactions at semiconductors. *J. Electrochem. Soc.* **2018**, *165*, H3085–H3092. [CrossRef]

Review

Time-Resolved X-ray Absorption Spectroscopy in (Photo)Electrochemistry

Martina Fracchia [1], Paolo Ghigna [1], Alberto Vertova [2], Sandra Rondinini [2] and Alessandro Minguzzi [2,*

[1] Dipartimento di Chimica, Università degli Studi di Pavia, Viale Taramelli 13, 27100 Pavia, Italy; martina.fracchia02@universitadipavia.it (M.F.); paolo.ghigna@unipv.it (P.G.)

[2] Laboratory of Applied Electrochemistry, Dipartimento di Chimica, Università degli Studi di Milano, Via Golgi 19, 20133 Milano, Italy; alberto.vertova@unimi.it (A.V.); sandra.rondinini@unimi.it (S.R.)

* Correspondence: alessandro.minguzzi@unimi.it; Tel.: +39-025-0314-224

Received: 7 November 2018; Accepted: 1 December 2018; Published: 5 December 2018

Abstract: This minireview aims at providing a complete survey concerning the use of X-ray absorption spectroscopy (XAS) for time-resolved studies of electrochemical and photoelectrochemical phenomena. We will see that time resolution can range from the femto-picosecond to the second (or more) scale and that this joins the valuable throughput typical of XAS, which allows for determining the oxidation state of the investigated element, together with its local structure. We will analyze four different techniques that use different approaches to exploit the in real time capabilities of XAS. These are quick-XAS, energy dispersive XAS, pump & probe XAS and fixed-energy X-ray absorption voltammetry. In the conclusions, we will analyze possible future perspectives for these techniques.

Keywords: X-ray absorption spectroscopy; energy dispersive; quick-XAS; FEXRAV; free electron laser; electrochemistry; photoelectrochemistry; photochemistry; pump & probe

---

1. Introduction

The research of efficient materials in chemistry and physics requires new criteria for their rational design and thus the definition of new paradigms for the elucidation of structure-performance guidelines.

In catalysis, the investigation on reaction mechanisms and on the concomitant changes in the catalyst nature becomes therefore crucial. Thus, the quest for analytical tools capable of directly detecting any information on the reactant-intermediate-product sequence and on the catalyst role in the course of the reaction is imperative.

In this respect, operando spectroscopies are obviously central, since they can assume the role of an independent, uncoupled source of information with respect to the primary ones (such as products yield, selectivity, or potential/current characteristics in the case of electrochemistry).

In the recent past, consequently, a flourishing use of operando spectroscopies in the study of electrochemical and photoelectrochemical phenomena has emerged, particularly within the framework of energy and environmental applications [1,2].

Among the different spectroscopic techniques, X-ray absorption (XAS) is particularly attractive being able to provides a particularly complete picture of the system of interest. In fact, and despite the need of large-scale synchrotron radiation facilities, it provides fundamental information on the local electronic and atomic structure of a selected element. Indeed, the latter characteristic is one of the most intriguing for XAS: tuning the energy of the incident X-ray beam, it is possible to select the element under consideration. This is vital for operando experiments, since the spectroscopic response is independent on any element in the beam different from that under investigation.

Another enabling feature of XAS is the high energy of X-rays: especially in the so-called "hard" X-ray regime (roughly, with energies higher than 8–10 keV), the absorption by all light elements (e.g., C, H, O, N) is negligible. Thus, most of the electrolytes adopted in electrochemistry as well as selected cell materials (especially if polymer-based) are transparent, and therefore operando studies are quite undemanding.

The principles of X-ray absorption spectroscopy can be briefly described as follows. When an X-ray photon with proper energy is shined over a sample, it can excite a core electron. Core electrons are in localized states, the energy of which is almost independent of the chemical environment. This is the reason for which XAS is element selective. At energy corresponding to the binding energy of the core level, the absorption coefficient has an abrupt change, called absorption edge. The excited electron can be promoted either to unoccupied bound levels or to continuum unbound levels, as pictorially described in Figure 1.

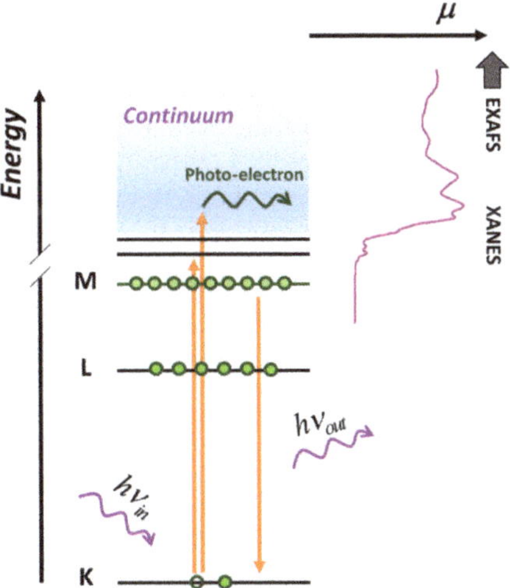

Figure 1. Scheme of the excitation and de-excitation processes involved in X-ray absorption spectroscopy (XAS). An X-ray photon ($h\nu_{in}$) can promote a core electron either to bound or continuum empty states. Transitions to bound states result into spectral features in the X-ray Absorption Near Edge Structure (XANES) region, while transitions to continuum states give rise to the Extended X-ray Absorption Fine Structure (EXAFS) oscillations. Radiative de-excitation processes result into X-rays fluorescence lines ($h\nu_{out}$). Core levels in the figure are named according to the Sommerfeld notation.

In this case, it can be back-diffused (scattered) by surrounding atoms. For a given photon energy, the interaction between the emitted electron and back-scattered waves can be destructive or constructive depending on the interatomic distance. This effect leads to an oscillation of the absorption coefficient for X-ray energy, extending to several hundreds of eV above the absorption edge. This is the so-called Extended X-ray Absorption Fine Structure (EXAFS), and allows for determining the local structure (number, nature and distance of surrounding atoms) around the photoabsorber. At energies close to the edge, the absorption coefficient has very strong modulations. This is the X-ray Absorption Near Edge Structure (XANES), that is due to electronic transitions to bound unoccupied states, and/or by multiple scattering events. XANES gives important information on the coordination environment (type of coordination and its distortions). Finally, the exact energy at which the absorption edge falls finely depends on the electron density on the photoabsorber, and thus reflects the oxidation state. Therefore, XAS gives an independent determination of the oxidation state of atoms in compounds.

Time resolved spectroscopies are of fundamental importance in studying the kinetics and dynamics of chemical reactions. With the introduction of third generation synchrotrons and, more recently, of free-electron lasers, time-revolved XAS became accessible thanks to the extremely high photon flux that can be impinged on the sample.

Time-resolved XAS include different techniques that allow spanning between picoseconds/femtoseconds and seconds/hours timescales, thus permitting the study of a wide range of phenomena, from the formation of photogenerated charges to charge transfers, from reaction mechanisms to slowly activity-losing systems.

In this review, we aim at summarizing the most important contributions achieved so far in the use of time-resolved XAS in electrochemistry and photoelectrochemistry.

For a more comprehensive, extended and didactic source of information, the Reader is invited to consider a recent book chapter [3] and earlier reviews [4–6].

Here we recall the fact that XAS is a bulk technique and averages on all atoms in the beam. However, a proper sample choice allows being sensitive to surface atoms: the most adopted strategies are (i) the study of small nanoparticles, where the surface/bulk atom ratio is particularly high and (ii) the use of "bulk active" materials. A classic example of the last case are hydrous oxides where, up to a certain load, all atoms (external surface and bulk) are addressable by charged and uncharged species from/to the electrolyte, thus being in the condition to act as active sites [7].

For what concerns the different strategies that can be adopted for time resolved experiment, we need to consider all variables that are into play: X-ray energy, X-ray absorption coefficient, applied potential (or scan rate, if in coupling with electrochemical methods), presence/absence of illumination and, obviously, time.

The most obvious approach is to record full XAS spectra at different times. This leads us to discuss: quick-XAS, energy dispersive XAS and pump & probe XAS.

2. Quick-XAS

Usually, an XAS spectrum is recorded by stepping the monochromator and counting each point for a relatively long time, typically in the order of seconds or several seconds. In the quick-XAS approach, the monochromator is continuously moved over an energy interval and the absorption coefficient, μ, is continuously measured. In this way, a typical XANES spectrum can be recorded in few seconds or even in less than a second [8–11]. One of the very first examples of fast acquisition using a quick-XAS approach is reported in Figure 2.

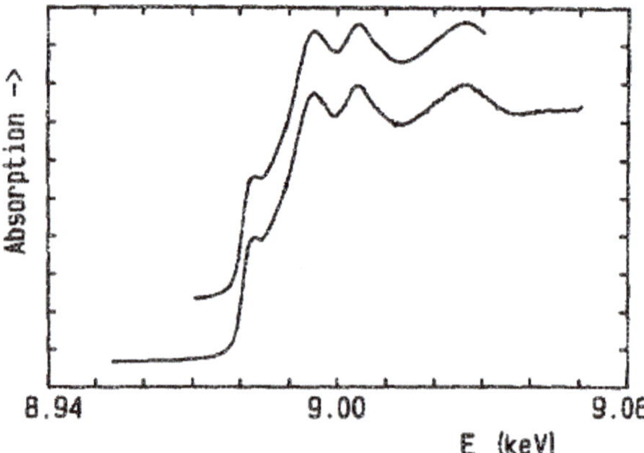

Figure 2. Cu-K edge spectra of a metal foil at room temperature measured in 72 s using a "conventional" step-by-step scan (upper spectrum) and in 6 s using the quick-XAS approach (lower spectrum. Reprinted from [9], with the permission of AIP Publishing.

Quick XAS has been largely employed in electrochemistry for dynamic studies in energy storage systems. For example, Yu et al. monitored the electrochemical delithiation of $LiFePO_4$, a well-known cathode material in lithium batteries, by exploiting the capability of XANES to easily distinguish between the $LiFePO_4$ and the $FePO_4$ phases [12]. They found that the delithiation proceeds with a "two-phase reaction", in agreement with other works in the literature. In addition, assuming that the kinetic is limited by lithium diffusion, the authors could estimate an apparent Li^+ diffusion coefficient of 1.9×10^{-13} $cm^2 \cdot s^{-1}$.

Ishiguro et al. monitored surface events occurring on a Pt_3Co/C cathode catalyst in proton exchange membrane fuel cell (PEMFC), with a time resolution of 500 ms [13]. With quick XAS (acquired both at the Pt L_{III}- and the Co K-edges) they could achieve a complete understanding of the structural and electronic changes occurring in this material during the fuel-cell operating processes. In addition, they compared the reaction mechanism of Pt_3Co/C with that of Pt/C catalyst. They found that while the two catalysts have similar reactions mechanisms, large increases in the rate constants are observed when Co is added to Pt.

Quick XAS has been employed in the study of electrocatalysts for various reactions, like Hydrogen Evolution Reaction (HER), Oxygen Evolution Reaction (OER) or Oxygen Reduction Reaction (ORR). The aim is to assess the effective chemical nature of an electrocatalyst under working conditions. An example is the work by Gorlin Y. et al., where a bifunctional manganese oxide (MnO_x) was studied as catalyst for both OER and ORR reactions [14]. At a potential value of 0.7 V vs. RHE, where the ORR takes place, a disordered $Mn(II,III)O_4$ spinel is found. When the potential is increased at 1.8 V, relevant to the OER, 80% of the film is oxidized to Mn(III,IV) oxide, while the remaining part consists of $Mn(II,III)O_4$, thus suggesting that a Mn(III,IV) oxide is the actual responsible for the OER. As quick XAS cannot reach the *ms* timescale, the authors could not provide any insight into the dynamics of the process. However, they were able to prove that the catalyst reached a stationary condition at all the potentials under investigation.

3. Energy Dispersive XAS

While the quick-XAS approach opens the way to time-resolved XAS, the sub-second timescale is not accessible, thus hindering the study of fast chemical reaction. In the energy dispersive approach, an XAS spectrum can be obtained in a timescale of milliseconds, which is quite satisfactory for a large part of fast chemical phenomena. In a dispersive XAS beamline, a bent monochromator is used to focus a polychromatic (pink) beam on the sample. The beam then diverges after the sample and the different wavelengths are spatially separated. A position sensitive detector placed after the sample is therefore used to acquire the spectrum. Examples of the use of energy dispersive XAS (EDXAS) in electrochemistry date back to late 1980s and early 1990s [15,16]. However, only after the introduction of fast and linear detectors [17] it was possible to reach the sub-second timescale. Since then, the technique has evolved into a powerful tool to investigate fast chemical phenomena [18–20]. From what concerns electrochemistry, EDXAS was employed to monitor the dynamics of oxidation and reduction on electrodes for different applications. In a seminal experiment in 1989, EDXAS was coupled to cyclic voltammetry to investigate the XANES at the Ni K-edge in α-$Ni(OH)_2$ and β-$Ni(OH)_2$ in KOH [16]. A continuous shift to higher values of the edge energy position was observed in concomitance with the oxidation from Ni(II) to Ni(III) and, at the same time, a change in the spectral profile related to a distortion of the octahedral coordination of the six oxygen atoms surrounding Ni was detected.

Rose et al. investigated the kinetics of the electrochemical formation of palladium β-hydride and β-deuteride in carbon-supported Pd catalyst nanoparticles [21]. The rate determining step of the formation of β-hydride was found to be the diffusion of H through the bulk of the Pd particles; in case of β-deuteride, the authors observed that the reaction was limited by the interfacial reduction of D^+. In addition, they were able to determine in both cases the stoichiometry of the two phases.

It has to be noted that EDXAS can also be employed to study time-resolved processes occurring in solution, particularly in the immediate proximity of the electrode. As an example, O' Malley et al.

considered the electrodesorption of copper on a platinum electrode in order to study the diffusion of Cu^{2+} ions as a function of time and distance from the electrode surface [22].

Great effort has been devoted in the recent years to investigate the electrochemical behavior of Pt-based electrodes. Allen et al. investigated both the oxidation and reduction of dispersed Pt catalyst, studying in real time the modifications of the Pt oxidation state and of the number of O and Pt nearest neighbors [15]. Moreover, they found evidences for a different mechanism of oxidation/reduction in case of clusters with respect to bulk electrodes. In another paper, the same authors describe the use of EDXAS to record full XAS spectra while the electrode potential follows a cyclic voltammetric profile [23]. This approach allowed for determining structural parameters (such as the average coordination number of Pt and O atoms around Pt) as a function of the applied potential. This, in turn, allowed for distinguishing between oxygen adsorbing in the double-layer region from the Pt oxidation that is typically observed in CV at higher potentials.

The electrochemical oxidation of the surfaces of Pt nanoparticles was further investigated by Imai et al. with a sub-second time resolution [24]. As shown in Figure 3, after applying a potential of 1.4 V vs. RHE in acidic media, they exploited the high sensitivity of the EXAFS region towards the local Pt-Pt and Pt-O distances, to monitor their change in time.

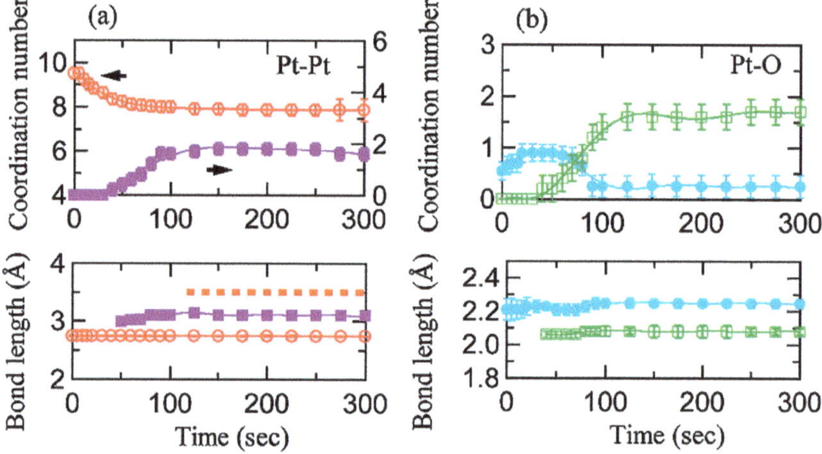

Figure 3. Example of outcome from an operando energy dispersive-XAS experiments on Pt nanoparticles. Bond lengths and coordination numbers as a function of time for (**a**) Pt-Pt and (**b**) Pt-O bonds during potential step at 1.4 V. In (**a**), it is possible to observe Pt-Pt bonds with 2.7 Å (metallic platinum) plus two types of longer Pt-Pt bonds (Pt-Pt bonds in platinum oxides, 3.1 and 3.5 Å), the latter appearing at about 40 and 100–120 s, respectively. In (**b**), longer Pt-O bonds (due adsorbed oxygen species) turn into shorter Pt-O (Pt-O bonds in oxides). Reprinted with permission [24]. Copyright 2009 American Chemical Society.

The authors found that longer Pt-O bonds are initially formed, indicating the formation of Pt-OHH and/or Pt-OH species; afterwards, these species turn to shorter Pt-O bonds, and the formation of α-PtO$_2$ was detected. Finally, after 100 s, a precursor of β-PtO$_2$ is formed.

EDAXS was also used to study the oxidation of Pt nanoparticles in presence of competing halide ions (Cl^- and Br^-) [25]. The kinetic parameters obtained by the time resolved experiment confirmed the two step model proposed by Zolfaghari, Conway and Jerkiewicz [26]. As evident from Figure 4, the competing effect of the halides was found to be significant mainly on the first step. In addition, the partial discrepancy, particularly at short times, between the integrated quantity of charge and the occupancy of Pt 5*d* states obtained by XAS suggests a non-negligible role of anion redox activity.

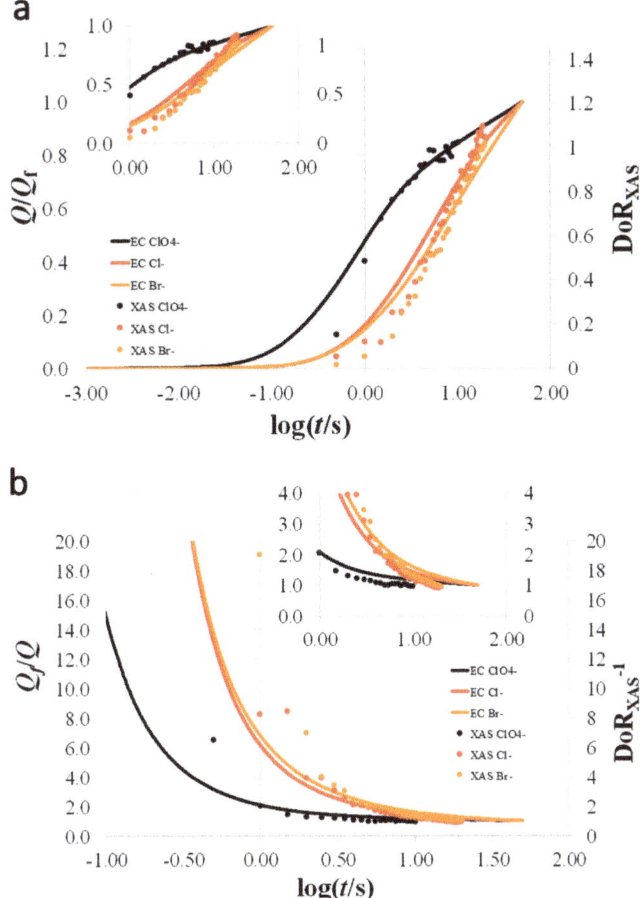

Figure 4. IR corrected results for operando ED-XAS in the case of Pt nanoparticles oxidation (and the relevant following reduction) in aqueous in 0.1 M $HClO_4$ (black), 0.1 M $HClO_4$ + 10 mM KCl (red) and 0.1 M $HClO_4$ + 10 mM KBr (orange). (**a**) Q/Qf (full lines) and Degree of reaction (DoR, by XAS) (dots) and (**b**) $(Q/Q_f)^{-1}$ (full lines) and $DoRXAS^{-1}$ (dots) as a function of $\log(t)$ for a 0.5 e 1.4 V (reversible hydrogen electrode) step. Reprinted from [25]. Copyright 2018, with permission from Elsevier.

The oxidation and reduction of highly hydrated IrO_x, one of the most active electrocatalysts for OER, was recently investigated by EDXAS [27]. The complementary information between XAS and chronoamperometry evidenced parasitic and side reactions. In addition, the role and time evolution of hydration degree and persistence of iridium in different charge states was discussed.

Finally, a very recent paper discussed the stability of the copper(II) lactate complex in alkaline solution, used as a precursor for the electrodeposition of Cu_2O, finding that the complex is stable in a wide range of applied potentials [28]. The authors were also able to determine the structure of the complex in solution by fitting the XANES data.

4. Pump & Probe XAS

This method can be applied in photochemistry or in photoelectrochemistry. It consists in exciting the system with photons of desired energy ("pump", typically in the visible or in the UV range) before probing the system with X-ray photons. The X-ray beam from a synchrotron radiation source has a characteristic time pattern that reflects the pattern of the electron bunches in the storage ring. The sub-µs timescale is then achievable with this approach. The probe is delayed with respect to the pump in dependence on the timescale of the phenomena under investigation. This is evident in the study of photochemistry, where pump & probe XAS was initially used instead of time-resolved X-ray diffraction to have structural information on disordered or dissolved species [29]. The technique was recently used to probe the charge carrier dynamics in semiconductors used as photocatalysts, such as TiO_2 and WO_3, as discussed in the following.

In the case of TiO_2, thanks to picosecond XAS, Rittmann and co-workers performed an experiment at the Ti K-edge and the Ru L_{III}-edge to study the lifetime of photogenerated trapped electrons for photoexcited bare and N719-dye-sensitized anatase and amorphous TiO_2 nanoparticles [30]. The outcomes point to the existence of different types of defects (either at the surface or in the bulk) in dependence on the type of particles and on the presence/absence of the dye. In case of dye-sensitized anatase or amorphous TiO_2 they could observe the formation of Ti^{3+} defects in trapping sites that are pentacoordinated, i.e. defective, meaning that the trapping center is localized at the outer surface of the crystallite. The trap lifetime is in the order of nanoseconds. This is a typical result possible only with XAS, which, being element-specific, allows for obtaining a direct information on the electronic and local structure of Ti.

Pump & probe XAS was also used in photochemistry and photoelectrochemistry to investigate different metal complexes, especially in the fields of energy storage and conversion. Many literature works concern the usage of Cobalt-based complexes as water oxidation or water reduction catalysts for artificial photosynthesis. In a work by Song et al. two cubane-like catalysts for water oxidation were investigated through in situ XANES under photocatalytic conditions: it was observed that the $\{Co(II)_4O_4\}$ core undergoes a rapid oxidation from Co(II) to Co(III) or higher valent states, and then Co(II) is slowly restored [31]. Moonshiram and co-workers investigated through transient XAS the electronic and structural dynamics of $[Ru(bpy)_3]^{2+}/[LCo(III)Cl_2]^+$ hybrid systems used as catalysts for water reduction in the photosplitting of water [32]. When the chromophore is excited, the cobalt (III) octahedral complex is reduced to Co(II). In addition, they repeated the experiment in the presence of sodium ascorbate/ascorbic acid electron donor, finding the experimental proof of the formation of a Co(I) square planar species, followed by the formation of the octahedral Co(III) starting complex with two aquo-ligands.

Canton et al. employed ultrafast optical and X-ray techniques (coupled to DFT calculations) to identify the light-induced electron transfer and the associated structural and spin changes occurring in two photoexcited heterobimetallic ruthenium-cobalt complexes [33].

The quest for higher temporal resolution has been boosted in the very recent year by the development of free-electron lasers (FEL) [34–36]. In FEL facilities, ultrashort pulses of X-rays paved the way for spectroscopic experiment in the femtosecond timescale [37–39]. This allowed for investigating the local electronic and structural modifications occurring in a transition metal compound immediately after the photoexcitation. For example, Bressler et al. used femtosecond XANES to solve a long-standing issue about the population mechanism of quintet states in iron(II)-based complexes, which resulted to be a $^1MLCT \rightarrow {}^3MLCT \rightarrow {}^5T$ cascade from the initially excited state [29].

In the field of semiconducting oxides, Santomauro and co-workers extended the investigation of Ritmann et al. by performing the XAS experiment at an XFEL facility to probe the dynamics of the trapping of the photogenerated electrons in TiO_2 [40]. They concluded that the electrons are localized in correspondence of Ti atoms in a time lower than 300 fs, leading to the formation of Ti^{3+} centers. In addition, they confirmed that electrons are localized at Ti penta-coordinate sites.

The dynamics of the photoexcitation was also investigated at an XFEL in the case of WO_3 nanoparticles. Uemura et al. observed that immediately after the excitation the tungsten is initially reduced from W(VI) to a mean oxidation state of 5.3 [38]. Subsequently, the tungsten oxide undergoes a change in its local structure within the following 200 ps. Some of the results of this paper are reported in Figure 5.

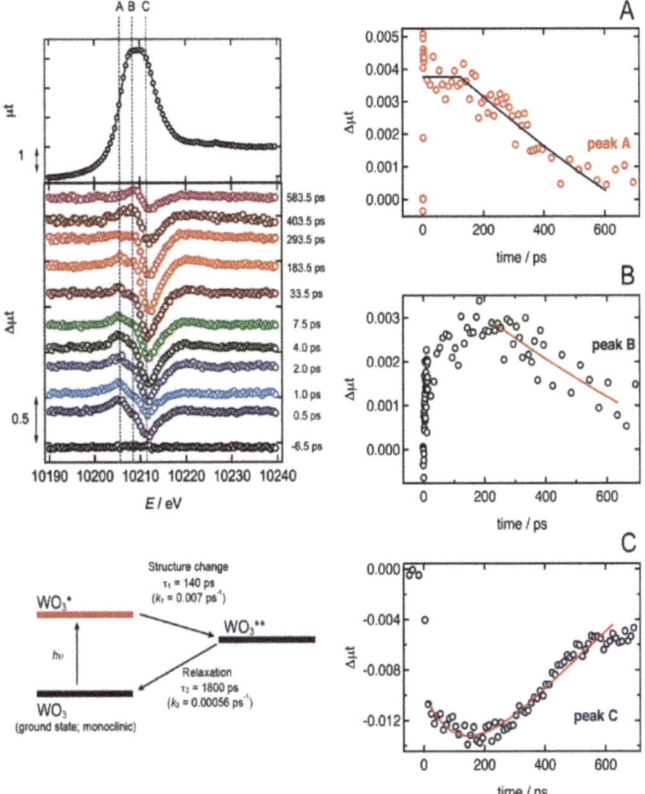

Figure 5. Top, left: W L_{III} XANES spectra of WO3 in the ground state together with differential XANES spectra of WO_3, each being the subtraction of the XANES spectrum of an excited state (at the indicated time) and the spectrum of a ground state. **Right**: Absorption intensities of W L_{III} XANES for peaks A, B, and C (as shown in the figure at the left) as a function of time. **Bottom, left**: A proposed scheme for the photoexcitation process of WO_3. Reprinted from [38]. Copyright 2016, with permission from Wiley.

All the listed cases refer to ex-situ or in-situ (on solvent dispersed particles) experiments, where, in most cases, the semiconductors or the complexes are photoexcited and the dynamics of the photogenerated carriers is followed. As a matter of fact, pump & probe XAS, either at synchrotron or at FEL facilities, is still in its very early days, to the point that studies under operando condition are very rare, and are mainly obtained by adding a sacrificial donor or acceptor and not by applying an external potential. In fact, to the authors' best knowledge, only one contribution was published, limited to synchrotron radiation pump & probe XAS [41].

In this work, the authors investigated the α-Fe_2O_3/IrO_x architecture as a model photoelectrode, and they could detect a charge transfer between hematite and IrO_x occurring in the nanosecond timescale, resulting in reduction of Ir or in an increased density of empty Ir $5d$ states depending on the applied potential. Figure 6 represents schematically the adopted setup (left) together with the results obtained when an anodic photocurrent is observed. In this case, difference spectra indicate a hole transfer from the semiconductor to IrO_x when no delay between pump & probe is applied. When pump & probe are delayed for 600 ns, the Ir $5d$ state results to be depleted, indicating a more intense hole transfer.

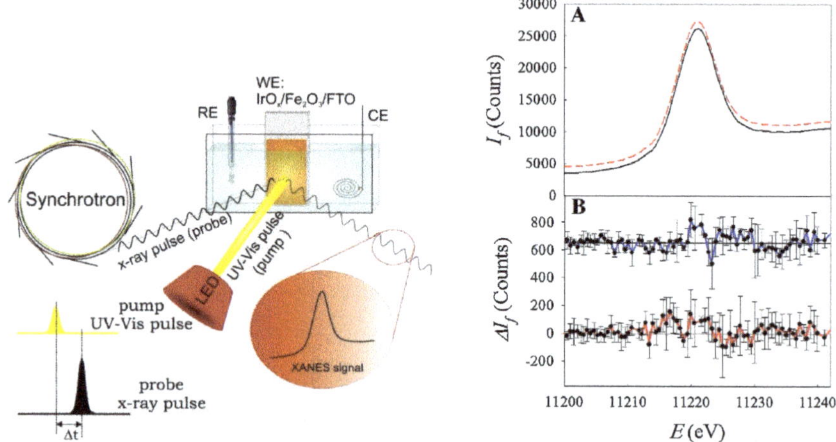

Figure 6. **Left**: schematic setup for an operando pump & probe (P & P) experiment on a photoelectrochemical system using synchrotron light. The UV-Vis pump is synchronized with the probe pulses, the latter generated by single bunches from the synchrotron. **Right**: (**A**) Ir-L$_{III}$ XANES spectrum in the presence (red dotted line) and in the absence (black full line) of 400 nm light acquired in presence of 600 ns delay between the pump and the probe; spectra are shifted along the y axis for the sake of better clarity. (**B**) difference spectra (light on-light off, with error bars) in the presence (red line) and in the absence (blue line) of delay; the two spectra are shifted for clarity, and the zero is defined by black horizontal lines. Reprinted from [41]. Copyright 2016, with permission from Elsevier.

5. FEXRAV

Finally, an alternative approach is to fix the X-ray energy and record the corresponding absorption coefficient as a function of time. In this case, in the framework of hyphenated techniques, it would be convenient to collect joint information from electrochemical methods. This is the principle behind the fixed energy X-ray absorption voltammetry (FEXRAV). This technique is an application of the single energy X-ray absorption detection first proposed by Filipponi et al. [42]. It consists in recording the X-ray absorption coefficient at a fixed energy while the applied potential in an electrochemical cell is scanned according to a cyclic voltammetry triangular-wave program [43]. If the energy is properly selected to give the maximum contrast between the spectral features of different oxidation states, any shift from the original oxidation state determines a variation of the absorption coefficient, and can then be detected. FEXRAV allows for quickly mapping the oxidation states of the element under consideration within the selected potential window, and this can be preliminary to deeper X-ray absorption spectroscopy (XAS) characterizations, like XANES or EXAFS. Moreover, the time-length of the experiment is much shorter than a series of XAS spectra and opens the door to kinetic analysis. Initially introduced for the study of IrO$_x$ materials, FEXRAV has been applied to operando studies of the electrochemistry of a variety of materials, from Ag nanoparticles [44], to copper oxide materials [45], to iron oxide and oxo-hydroxide materials [46], to palladium based electrocatalysts [47]. Figure 7 reports an example of a FEXRAV experiment in the case of Ag nanoparticles studied as electrocatalysts for the electrodehalogenation of trichloromethane in aqueous media. Here the X-ray energy is set at a value that guarantees the increase of μ when Ag is oxidized. The potential is swept from 0 to -1.1 V and there's a visible decrease of μ due to the presence (adsorption) of trichloromethane or of its reduction intermediates. The release of Cl$^-$ causes Ag to oxidize at about 0.2 V. This is witnessed by XAS, by means of an increase of μ. The latter decreases when the potential is reversed, in correspondence to the reduction peak at about 0.1 V.

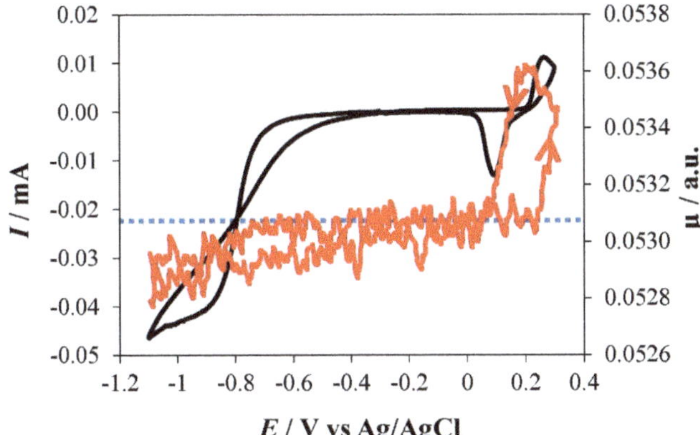

Figure 7. Example of a fixed energy X-ray absorption voltammetry experiment on Ag nanoparticles in aqueous of 10 mM trichloromethane (chloroform): CV (black) and FEXRAV (red) at 1 mV·s^{-1}. Reprinted from [44]. Copyright 2016, with permission from Elsevier.

6. Conclusions and Perspectives

This minireview aims at providing a comprehensive and updated overview of the state of the art concerning use and capabilities of time resolved XAS in electrochemistry and photoelectrochemistry.

It is evident that the number of published papers on these topics is not very large, especially if compared to other types of spectroscopies. This is due to two main factors: (i) the dependence on large facilities, which limits the number of users that can approach and that have access to these techniques; (ii) the general unawareness to electrochemists of the mere existence of XAS, notwithstanding its very high potentialities. Of course, these two factors are strongly related.

Indeed, as we hope to have successfully demonstrated in this work, the potentialities of time resolved XAS are very high. In fact, time resolved XAS can help to clarify reaction mechanisms, metastable (transient) structures and to identify the nature of intermediate reacting states, and we hope that the range of applications will extend progressively in the close future.

Fortunately, we are witnessing a progressive increase of available synchrotron radiation sources (for a general outlook at the existing facilities, the reader is referred to the website https://lightsources. org/lightsources-of-the-world/). In parallel, it is worth noting that innovative benchtop laboratory X-ray spectrophotometers are commercially available. Even if the photon flux of these instruments is still not comparable with that available at a synchrotron beamline, we believe that technical advances will allow for rapid improvements which, together with an increasing knowledge on the correct design of samples and cells [48], will likely make XAS available also in academics and companies worldwide.

Funding: A.M. is thankful to Università degli Studi di Milano through the "Piano di Sostegno alla Ricerca 2015/2017".

Conflicts of Interest: The authors declare no conflict of interest.

References

1. Simpson, B.H.; Rodríguez-López, J. Emerging techniques for the in situ analysis of reaction intermediates on photo-electrochemical interfaces. *Anal. Methods* **2015**, *7*, 7029–7041. [CrossRef]
2. Caudillo-Flores, U.; Muñoz-Batista, M.J.; Kubacka, A.; Fernández-García, M. Operando Spectroscopy in Photocatalysis. *Chem. Photo Chem.* **2018**, 777–785.
3. Minguzzi, A.; Ghigna, P. X-ray absorption spectroscopy in electrochemistry from fundamentals to fixed energy X-ray absorption voltammetry. In *Electroanalytical Chemistry: A Series of Advances, Volume 27*; Bard, A.J., Zoski, C.G., Eds.; CRC Press–Taylor and Francis Group: Boca Raton, FL, USA, 2017; pp. 119–181, ISBN 9781351981675.

4. Russell, A.E.; Rose, A. X-ray absorption spectroscopy of low temperature fuel cell catalysts. *Chem. Rev.* **2004**, *104*, 4613–4635. [CrossRef]

5. Abruña, H.D.; Bommarito, G.M.; Yee, H.S. X-ray Standing Waves and Surface EXAFS Studies of Electrochemical Interfaces. *Acc. Chem. Res.* **1995**, *28*, 273–279. [CrossRef]

6. Sharpe, L.R.; Heineman, W.R.; Elder, R.C. EXAFS Spectroelectrochemistry. *Chem. Rev.* **1990**, *90*, 705–722. [CrossRef]

7. Doyle, R.L.; Godwin, I.J.; Brandon, M.P.; Lyons, M.E.G. Redox and electrochemical water splitting catalytic properties of hydrated metal oxide modified electrodes. *Phys. Chem. Chem. Phys.* **2013**, *15*, 13737–13783. [CrossRef] [PubMed]

8. Frahm, R. Quick scanning exafs: First experiments. *Nucl. Inst. Methods Phys. Res. A* **1988**, *270*, 578–581. [CrossRef]

9. Frahm, R. New method for time dependent X-ray absorption studies. *Rev. Sci. Instrum.* **1989**, *60*, 2515–2518. [CrossRef]

10. Stötzel, J.; Lützenkirchen-Hecht, D.; Frahm, R. A new flexible monochromator setup for quick scanning X-ray absorption spectroscopy. *Rev. Sci. Instrum.* **2010**, *81*, 073109. [CrossRef]

11. Müller, O.; Nachtegaal, M.; Just, J.; Lützenkirchen-Hecht, D.; Frahm, R. Quick-EXAFS setup at the SuperXAS beamline for in situ X-ray absorption spectroscopy with 10ms time resolution. *J. Synchrotron Radiat.* **2016**, *23*, 260–266. [CrossRef]

12. Yu, X.; Wang, Q.; Zhou, Y.; Li, H.; Yang, X.Q.; Nam, K.W.; Ehrlich, S.N.; Khalid, S.; Meng, Y.S. High rate delithiation behaviour of LiFePO4studied by quick X-ray absorption spectroscopy. *Chem. Commun.* **2012**, *48*, 11537–11539. [CrossRef]

13. Ishiguro, N.; Saida, T.; Uruga, T.; Nagamatsu, S.I.; Sekizawa, O.; Nitta, K.; Yamamoto, T.; Ohkoshi, S.I.; Iwasawa, Y.; Yokoyama, T.; et al. Operando time-resolved X-ray absorption fine structure study for surface events on a Pt3Co/C cathode catalyst in a polymer electrolyte fuel cell during voltage-operating processes. *ACS Catal.* **2012**, *2*, 1319–1330. [CrossRef]

14. Gorlin, Y.; Lassalle-Kaiser, B.; Benck, J.D.; Gul, S.; Webb, S.M.; Yachandra, V.K.; Yano, J.; Jaramillo, T.F. In situ X-ray absorption spectroscopy investigation of a bifunctional manganese oxide catalyst with high activity for electrochemical water oxidation and oxygen reduction. *J. Am. Chem. Soc.* **2013**, *135*, 8525–8534. [CrossRef] [PubMed]

15. Allen, P.G.; Conradson, S.D.; Wilson, M.S.; Gottesfeld, S.; Raistrick, I.D.; Valerio, J.; Lovato, M. Direct observation of surface oxide formation and reduction on platinum clusters by time-resolved X-ray absorption spectroscopy. *J. Electroanal. Chem.* **1995**, *384*, 99–103. [CrossRef]

16. McBreen, J.; O'Grady, W.E.; Tourillon, G.; Dartyge, E.; Fontaine, A.; Pandya, K.I. In situ time-resolved X-ray absorption near edge structure study of the nickel oxide electrode. *J. Phys. Chem.* **1989**, *93*, 6308–6311. [CrossRef]

17. Dent, A.; Evans, J.; Newton, M.; Corker, J.; Russell, A.; Abdul Rahman, M.B.; Fiddy, S.; Mathew, R.; Farrow, R.; Salvini, G.; et al. High-quality energy-dispersive XAFS on the 1 s timescale applied to electrochemical and catalyst systems. *J. Synchrotron Radiat.* **1999**, *6*, 381–383. [CrossRef] [PubMed]

18. Pascarelli, S.; Mathon, O.; Muñoz, M.; Mairs, T.; Susini, J. Energy-dispersive absorption spectroscopy for hard-X-ray micro-XAS applications. *J. Synchrotron Radiat.* **2006**, *13*, 351–358. [CrossRef] [PubMed]

19. Labiche, J.-C.; Mathon, O.; Pascarelli, S.; Newton, M.A.; Ferre, G.G.; Curfs, C.; Vaughan, G.B.M.; Homs, A.; Carreiras, D.F. The fast readout low noise camera as a versatile X-ray detector for time resolved dispersive extended X-ray absorption fine structure and diffraction studies of dynamic problems in materials science, chemistry, and catalysis. *Rev. Sci. Instrum.* **2007**, *78*, 91301. [CrossRef] [PubMed]

20. Pascarelli, S.; Mathon, O. Advances in high brilliance energy dispersive X-ray absorption spectroscopy. *Phys. Chem. Chem. Phys.* **2010**, *12*, 5535–5546. [CrossRef]

21. Rose, A.; South, O.; Harvey, I.; Diaz-Moreno, S.; Owen, J.R.; Russell, A.E. In situ time resolved studies of hydride and deuteride formation in Pd/C electrodes via energy dispersive X-ray absorption spectroscopy. *Phys. Chem. Chem. Phys.* **2005**, *7*, 366–372. [CrossRef]

22. O'Malley, R.; Vollmer, A.; Lee, J.; Harvey, I.; Headspith, J.; Diaz-Moreno, S.; Rayment, T. Time-resolved studies of diffusion via energy dispersive X-ray absorption spectroscopy. *Electrochem. commun.* **2003**, *5*, 1–5. [CrossRef]

23. Allen, P.G.; Conradson, S.D.; Wilson, M.S.; Gottesfeld, S.; Raistrick, I.D.; Valerio, J.; Lovato, M. In situ structural characterization of a platinum electrocatalyst by dispersive X-ray absorption spectroscopy. *Electrochim. Acta* **1994**, *39*, 2415–2418. [CrossRef]

24. Imai, H.; Izumi, K.; Matsumoto, M.; Kubo, Y.; Kato, K.; Imai, Y. In Situ and Real-Time Monitoring of Oxide Growth in a Few Monolayers at Surfaces of Platinum Nanoparticles in Aqueous Media In Situ and Real-Time Monitoring of Oxide Growth in a Few Monolayers at Surfaces of Platinum Nanoparticles in Aqueous. *J. Am. Chem. Soc.* **2009**, *131*, 6293–6300. [CrossRef] [PubMed]

25. Minguzzi, A.; Montagna, L.; Falqui, A.; Vertova, A.; Rondinini, S.; Ghigna, P. Dynamics of oxide growth on Pt nanoparticles electrodes in the presence of competing halides by operando energy dispersive X-ray absorption spectroscopy. *Electrochim. Acta* **2018**, *270*, 378–386. [CrossRef]

26. Zolfaghari, A.; Conway, B.E.; Jerkiewicz, G. Elucidation of the effects of competitive adsorption of Cl-and Br-ions on the initial stages of Pt surface oxidation by means of electrochemical nanogravimetry. *Electrochim. Acta* **2002**, *47*, 1173–1187. [CrossRef]

27. Rondinini, S.; Minguzzi, A.; Achilli, E.; Locatelli, C.; Agostini, G.; Spinolo, G.; Vertova, A.; Ghigna, P. The dynamics of pseudocapacitive phenomena studied by Energy Dispersive XAS on hydrous iridium oxide electrodes in alkaline media. *Electrochim. Acta* **2016**, *212*, 247–253. [CrossRef]

28. Achilli, E.; Vertova, A.; Visibile, A.; Locatelli, C.; Minguzzi, A.; Rondinini, S.; Ghigna, P. Structure and Stability of a Copper(II) Lactate Complex in Alkaline Solution: A Case Study by Energy-Dispersive X-ray Absorption Spectroscopy. *Inorg. Chem.* **2017**, *56*. [CrossRef] [PubMed]

29. Bressler, C.; Milne, C.; Pham, V.-T.; Elnahhas, A.; van der Veen, R.M.; Gawelda, W.; Johnson, S.; Beaud, P.; Grolimund, D.; Kaiser, M.; Borca, C.N.; et al. Femtosecond XANES study of the light-induced spin crossover dynamics in an iron(II) complex. *Science* **2009**, *323*, 489–492. [CrossRef]

30. Rittmann-Frank, M.H.; Milne, C.J.; Rittmann, J.; Reinhard, M.; Penfold, T.J.; Chergui, M. Mapping of the photoinduced electron traps in TiO2 by picosecond X-ray absorption spectroscopy. *Angew. Chem. Int. Ed.* **2014**, *53*, 5858–5862. [CrossRef] [PubMed]

31. Song, F.; Moré, R.; Schilling, M.; Smolentsev, G.; Azzaroli, N.; Fox, T.; Luber, S.; Patzke, G.R. {Co4O4} and {CoxNi4-xO4} Cubane Water Oxidation Catalysts as Surface Cut-Outs of Cobalt Oxides. *J. Am. Chem. Soc.* **2017**, *139*, 14198–14208. [CrossRef]

32. Moonshiram, D.; Gimbert-Suriñach, C.; Guda, A.; Picon, A.; Lehmann, C.S.; Zhang, X.; Doumy, G.; March, A.M.; Benet-Buchholz, J.; Soldatov, A.; et al. Tracking the Structural and Electronic Configurations of a Cobalt Proton Reduction Catalyst in Water. *J. Am. Chem. Soc.* **2016**, *138*, 10586–10596. [CrossRef] [PubMed]

33. Canton, S.E.; Zhang, X.; Liu, Y.; Zhang, J.; Pápai, M.; Corani, A.; Smeigh, A.L.; Smolentsev, G.; Attenkofer, K.; Jennings, G.; et al. Watching the dynamics of electrons and atoms at work in solar energy conversion. *Faraday Discuss.* **2015**, *185*, 51–68. [CrossRef] [PubMed]

34. Chergui, M. Emerging photon technologies for chemical dynamics. *Faraday Discuss.* **2014**, *171*, 11–40. [CrossRef] [PubMed]

35. Patterson, B.D.; Abela, R. Novel opportunities for time-resolved absorption spectroscopy at the X-ray free electron laser. *Phys. Chem. Chem. Phys.* **2010**, *12*, 5647–5652. [CrossRef] [PubMed]

36. Lemke, H.T.; Bressler, C.; Chen, L.X.; Fritz, D.M.; Gaffney, K.J.; Galler, A.; Gawelda, W.; Haldrup, K.; Hartsock, R.W.; Ihee, H.; et al. Femtosecond X-ray absorption spectroscopy at a hard X-ray free electron laser: Application to spin crossover dynamics. *J. Phys. Chem. A* **2013**, *117*, 735–740. [CrossRef] [PubMed]

37. Oloff, L.P.; Chainani, A.; Matsunami, M.; Takahashi, K.; Togashi, T.; Osawa, H.; Hanff, K.; Quer, A.; Matsushita, R.; Shiraishi, R.; et al. Time-resolved HAXPES using a microfocused XFEL beam: From vacuum space-charge effects to intrinsic charge-carrier recombination dynamics. *Sci. Rep.* **2016**, *6*, 1–10. [CrossRef] [PubMed]

38. Uemura, Y.; Kido, D.; Wakisaka, Y.; Uehara, H.; Ohba, T.; Niwa, Y.; Nozawa, S.; Sato, T.; Ichiyanagi, K.; Fukaya, R.; et al. Dynamics of Photoelectrons and Structural Changes of Tungsten Trioxide Observed by Femtosecond Transient XAFS. *Angew. Chemie Int. Ed.* **2016**, *55*, 1364–1367. [CrossRef] [PubMed]

39. Picón, A.; Lehmann, C.S.; Bostedt, C.; Rudenko, A.; Marinelli, A.; Osipov, T.; Rolles, D.; Berrah, N.; Bomme, C.; Bucher, M.; et al. Hetero-site-specific X-ray pump-probe spectroscopy for femtosecond intramolecular dynamics. *Nat. Commun.* **2016**, *7*, 5–10. [CrossRef]

40. Santomauro, F.G.; Lübcke, A.; Rittmann, J.; Baldini, E.; Ferrer, A.; Silatani, M.; Zimmermann, P.; Grübel, S.; Johnson, J.A.; Mariager, S.O.; et al. Femtosecond X-ray absorption study of electron localization in photoexcited anatase TiO2. *Sci. Rep.* **2015**, *5*, 14834. [CrossRef]

41. Baran, T.; Fracchia, M.; Vertova, A.; Achilli, E.; Naldoni, A.; Malara, F.; Rossi, G.; Rondinini, S.; Ghigna, P.; Minguzzi, A.; et al. Operando and Time-Resolved X-ray Absorption Spectroscopy for the Study of Photoelectrode Architectures. *Electrochim. Acta* **2016**, *207*, 16–21. [CrossRef]

42. Filipponi, A.; Borowski, M.; Loeffen, P.W.; De Panfilis, S.; Di Cicco, A.; Sperandini, F.; Minicucci, M.; Giorgetti, M. Single-energy X-ray absorption detection: A combined electronic and structural local probe for phase transitions in condensed matter. *J. Phys. Condens. Matter* **1998**, *10*, 235–253. [CrossRef]

43. Minguzzi, A.; Lugaresi, O.; Locatelli, C.; Rondinini, S.; D'Acapito, F.; Achilli, E.; Ghigna, P. Fixed energy X-ray absorption voltammetry. *Anal. Chem.* **2013**, *85*, 7009–7013. [CrossRef] [PubMed]

44. Rondinini, S.; Lugaresi, O.; Achilli, E.; Locatelli, C.; Vertova, A.; Ghigna, P.; Minguzzi, A.; Comninellis, C. Fixed Energy X-ray Absorption Voltammetry and Extended X-ray Absorption Fine Structure of Ag Nanoparticle Electrodes. *J. Electroanal. Chem.* **2016**, *766*, 71–77. [CrossRef]

45. Baran, T.; Wojtyła, S.; Lenardi, C.; Vertova, A.; Ghigna, P.; Achilli, E.; Fracchia, M.; Rondinini, S.; Minguzzi, A. An Efficient CuxO Photocathode for Hydrogen Production at Neutral pH: New Insights from Combined Spectroscopy and Electrochemistry. *ACS Appl. Mater. Interfaces* **2016**, *8*, 21250–21260. [CrossRef] [PubMed]

46. Fracchia, M.; Visibile, A.; Ahlberg, E.; Vertova, A.; Minguzzi, A.; Ghigna, P.; Rondinini, S. α- and γ-FeOOH: Stability, Reversibility, and Nature of the Active Phase under Hydrogen Evolution. *ACS Appl. Energy Mater.* **2018**, *1*, 1716–1725. [CrossRef]

47. Montegrossi, G.; Giaccherini, A.; Berretti, E.; Di Benedetto, F.; Innocenti, M.; d'Acapito, F.; Lavacchi, A. Computational Speciation Models: A Tool for the Interpretation of Spectroelectrochemistry for Catalytic Layers under Operative Conditions. *J. Electrochem. Soc.* **2017**, *164*, E3690–E3695. [CrossRef]

48. Achilli, E.; Minguzzi, A.; Visibile, A.; Locatelli, C.; Vertova, A.; Naldoni, A.; Rondinini, S.; Auricchio, F.; Marconi, S.; Fracchia, M.; et al. 3D-printed photo-spectroelectrochemical devices for in situ and in operando X-ray absorption spectroscopy investigation. *J. Synchrotron Radiat.* **2016**, *23*, 622–628. [CrossRef] [PubMed]

MDPI

St. Alban-Anlage 66

4052 Basel

Switzerland

Tel. +41 61 683 77 34

Fax +41 61 302 89 18

www.mdpi.com

Surfaces Editorial Office

E-mail: surfaces@mdpi.com

www.mdpi.com/journal/surfaces

9 783039 216420